SECOND EDITION

EARTH STRUCTURE

AN INTRODUCTION TO
STRUCTURAL GEOLOGY AND TECTONICS

Ben A. van der Pluijm

UNIVERSITY OF MICHIGAN

Stephen Marshak

UNIVERSITY OF ILLINOIS

With contributions by

RICHARD W. ALLMENDINGER

MARK T. BRANDON

B. CLARK BURCHFIEL

FREDERICK A. COOK

DAVID A. FOSTER

DAVID R. GRAY

JAMES P. HIBBARD

PAUL F. HOFFMAN

TERESA E. JORDAN

ELIZABETH L. MILLER

BORIS A. NATAL'IN

KEVIN T. PICKERING

LEIGH H. ROYDEN

STEFAN M. SCHMID

A. M. CÊLAL ŞENGÖR

ALAN G. SMITH

M. SCOTT WILKERSON

SECOND EDITION

EARTH STRUCTURE

AN INTRODUCTION TO
STRUCTURAL GEOLOGY AND TECTONICS

W • W • NORTON & COMPANY
NEW YORK • LONDON

W. W. Norton & Company has been independent since its founding in 1923, when William Warder Norton and Mary D. Herter Norton first published lectures delivered at the People's Institute, the adult education division of New York City's Cooper Union. The Nortons soon expanded their program beyond the Institute, publishing books by celebrated academics from America and abroad. By mid-century, the two major pillars of Norton's publishing program—trade books and college texts—were firmly established. In the 1950s, the Norton family transferred control of the company to its employees, and today—with a staff of four hundred and a comparable number of trade, college, and professional titles published each year—W. W. Norton & Company stands as the largest and oldest publishing house owned wholly by its employees.

The text of this book is composed in Times, with the display set in Conduit ITC.
Composition by Shepherd Incorporated
Manufacturing by Courier, Westford

Editor: Leo A. W. Wiegman
Project editor: Thomas Foley
Director of manufacturing: Roy Tedoff
Copy editor: Philippa Solomon
Photography editors: Nathan Odell and Erin O'Brien
Layout artists: Shepherd Incorporated
Editorial assistants: Erin O'Brien and Rob Bellinger
Book Designer: Rubina Yeh

Library of Congress Cataloging-in-Publication Data

Van der Pluijm, Ben A., 1955-
    Earth structure : an introduction to structural geology and tectonics / Ben A. van der
Pluijm, Stephen Marshak ; with contributions by Richard W. Allmendinger . . . [et al.]--
2nd ed.
        p. cm.
    Includes bibliographical references and index.

    ISBN 0-393-92467-X

    1. Geology, Structural. 2. Plate tectonics.   I. Marshak, Stephen, 1955- II. Title.

QE601.V363 2003
551.8--dc22   2003063957

W. W. Norton & Company, Inc., 500 Fifth Avenue, New York, N.Y. 10110
www.wwnorton.com

W. W. Norton & Company Ltd., Castle House, 75/76 Wells Street, London W1T 3QT

2 3 4 5 6 7 8 9 0

Brief Contents

Preface

This book is concerned with the deformation of rock in the Earth's lithosphere, as viewed from the atomic scale, through the grain scale, the hand specimen scale, the outcrop scale, the mountain range scale, and the tectonic plate scale. A deformational feature observed on one scale typically reflects processes occurring on other scales. For example, we can't understand continental deformation without understanding mountains, we can't understand mountains without understanding folding and faulting, and we can't understand folding and faulting without understanding ductile and brittle deformation mechanisms at the atomic scale. This book attempts to integrate topics pertaining to all scales of rock deformation, emphasizing the linkages between structural geology and tectonics.

Every month, perhaps a thousand pages of new ideas and observations relevant to structural geology and tectonics are published in the major scholarly journals. The amount of material on structural geology and tectonics that has appeared over the past 150 years is staggering. We have purposely decided to write this book with a novice to the field in mind. We, as instructors, face a massive challenge when trying to distill an introductory course out of this ever-changing and ever-growing mountain of information. We want students to be comfortable with certain basic concepts (say, fault terminology or stress theory), and at the same time, we want them to experience the excitement of discovery and to build their own "big picture" of how the Earth works. And all this must be done in a few short months! Rather than loading the text with excessive detail and peppering it with extensive referencing, we opted instead to present a distillation that offers a perspective on most aspects of the field. The reason for this approach is to highlight the "guts" of structural geology and tectonics, thereby providing a foundation for future study and a platform for further discussion. When reading the text, the reader should maintain a critical and questioning attitude toward the concepts discussed, which will not only stimulate the mind but also aid in absorbing the material. Concepts are remembered better when their interrelationships are recognized, rather than being presented as just a series of definitions. In some cases we may have advanced a controversial position and perhaps future readers will be the ones to prove some of our viewpoints either right or wrong.

Structural geology and tectonics are a lot of fun once one has waded through the initial terminology morass. Our personal approach to teaching structural geology and tectonics is reflected in the fairly informal writing style of this text. Whenever possible, we use familiar analogies such as rubber bands, syrup, and pool balls. Similarly, we have kept illustrations simple in the early chapters so that the point of the figure is obvious. Terms and definitions related to topics that we do not introduce in the main body of the text are included in tables as a reference. The subject index will direct you to the appropriate location in the text for any specific term.

There's no single right way to teach structural geology and tectonics. Moreover, we increasingly see that structural geology and tectonics is one of the first classes for students who plan to major in geological sciences. We decided to write this book because we found both that existing books did not suit the changing needs of the courses that we teach ourselves and that many other instructors shared our views. Some books try to be a lab manual and a lecture text at the same time, while others are slanted too much toward the research interests of the particular writer(s). Some books are organized in such a way that a reading assignment on a single topic must include splices from all over the book, and others provide more detail than can possibly be covered in a single semester course so that students are, frankly, overwhelmed. We have deleted topics that are generally taught in laboratory sections because these topics cannot be treated adequately within the framework of a lecture textbook. We also do not burden the narrative with references, but rather provide introductory reading lists at the end of each chapter.

In order to provide instructors with optimal freedom to develop their own course outlines, we've made sure that most chapters are self-contained modules that can be presented in various sequences. Ben, for example, starts his course with a description of rocks, via primary structures, faults and fractures, folds, and fabrics, before introducing stress, strain, rheology, and deformation mechanisms. Steve, by contrast, teaches stress, strain, and rheology immediately after primary structures and presents brittle deformation theory before discussing faults and fractures. We both concentrate on tectonics at the end of our courses, but tectonic implications are typically interwoven with the discussion of different classes of structures earlier in the course. In the end, instructors work hard to make their lectures comprehensive yet comprehensible, accurate yet enjoyable. We have tried to do the same with this book.

CHANGES IN THE SECOND EDITION

All chapters were revised for the Second Edition, but the general organization remains the same. New sections have been added, while some old ones have been removed or combined. The new edition also includes a chapter on "Geophysical Imaging" and four new essays in Chapters 21 and 22 on the European Alps, the Altaids, the Appalachians, and the Cascadia wedge. The remaining essays were updated and revised. The new and revised art offers an even more informative illustration of concepts and topics and will give instructors the opportunity for modern classroom use (see ancillaries).

ANCILLARIES

Earth Structure is supported by a Norton Resource Library offering teachers hundreds of digital copies of the figures from the new edition. The Norton Resource Library images may be used in classrooms as overhead transparencies, computer presentations, and student worksheets incorporated in exams, or course websites. Instructors may either download figures by chapter from the Norton Resource Library, after obtaining a password from Norton, or request the images on a CD-ROM. Both password and CD-ROM requests are located at the Norton Resource Library web address: www.wwnorton.com/college/nrl/welcome.htm.

THANKS!

This book could not have been written without the help of the students in our classes, who, through their successes and mistakes, have shown us which explanations work and which do not. We are indebted to the following colleagues for their expert contributions to this new edition: Rick Allmendinger, Mark Brandon, Clark Burchfiel, A. M. Cêlal Şengör, Fred Cook, David Foster, David Gray, Jim Hibbard, Paul Hoffman, Teresa Jordan, Elizabeth Miller, Boris Natal'in, Kevin Pickering, Leigh Royden, Stefan Schmid, Alan Smith, and Scott Wilkerson.

We are also grateful to our many colleagues who have provided generous dollops of advice and from whom we have borrowed data and interpretations. Colleagues who have commented on and/or contributed to one or more chapters include (in alphabetical order): Mark Fisher, Jerry Magloughlin, Klaus Mezger, Carl Richter, Mike Sandiford, and John Stamatakos. Formal reviews of chapters in the First Edition were given by David Anastasio, Stanley Cebull, Bill Dunne, Terry Engelder, Karl Karlstrom, Win Means, Jim Talbot, Adolph Yonkee, and Vincent Cronic. The Second Edition was revised based on our own experiences with the First Edition, a better appreciation of some of the topics, and the informal feedback from many users of the First Edition, which received formal reviews from Roy Schlische and Bill Dunne. The editorial and production staff for W. W. Norton, particularly copy editor Philippa Solomon and editor Leo Wiegman, as well as Erin O'Brien, Thom Foley, Rubina Yeh, and Jack Repcheck, have been most helpful and accommodating. Stan Maddock and Dale Austin produced the artwork, most of which has been redrafted and updated from the First Edition. We also thank our graduate advisors (Paul Williams, Henk Zwart, and Terry Engelder, respectively) for helping us enter this business and for guiding our first uncertain steps. We thank all of our graduate students for many lively and interesting discussions. And finally, but foremost, we thank our wives, Lies and Kathy, and our children. Wouter and Robbie, and David and Emma, respectively, for not grumbling too much about the absences in body and spirit that writing this book has required. To them we thankfully dedicate this book and hope that one day they may even read it.

Ben van der Pluijm, Ann Arbor, Michigan
Stephen Marshak, Urbana, Illinois
September 12, 2003

PART A

FUNDAMENTALS

CHAPTER ONE

Overview

1.1 INTRODUCTION

Did you ever take a cross-country drive? Hour after hour of tedious driving, as the highway climbed hills and dropped into valleys? The monotonous gray rocks exposed in road cuts largely went unnoticed, right? You passed pretty scenery, but it was static and seemed to tell no story simply because you did not have a basis in your mind with which to interpret your natural surroundings. It was much the same for scholars of generations past, before the establishment of modern science. The Earth was a closed book, hiding its secrets in a language that no one could translate. Certainly, ancient observers marveled at the enormity of mountains and oceans, but with the knowledge they had at hand they could do little more than dream of supernatural processes to explain the origin of these features. Gods and monsters contorted the Earth and spit flaming rock; and giant turtles and catfish shook the ground. Then, in fifteenth-century Europe, an intellectual renaissance spawned an age of discovery, during which the Earth was systematically charted, and the pioneers of science cast aside dogmatic views of our universe that had closed peoples' minds for the previous millennia and began to systematically observe their surroundings and carry out experiments to create new knowledge. The scientific method was born.

In geology, the stirrings of discovery are evident in the ink sketches of the great artist and inventor Leonardo da Vinci (1452–1519), who carefully drew the true shapes of rock bodies in sketches to understand the natural shape of the Earth (Figure 1.1). In the seventeenth century came the first description of rock deformation. Nicholas Steno (1631–1686) examined outcrops where the bedding of rock was not horizontal, and speculated that strata that do not presently lie in horizontal layers must have in some way been *dislocated* (the term he used for deformed). Perhaps Steno's establishment of the **principle of original horizontality** can be viewed as the birth of structural geology. By the beginning of the eighteenth century, the structural complexity of rocks in mountain ranges like the Alps was widely recognized (Figure 1.2), and it became clear that such features demanded explanation.

The pace of discovery quickened during the latter half of the eighteenth century and through the nineteenth century. In his "Theory of the Earth with Proofs and Illustrations," James Hutton (1726–1797) proposed the concept of **uniformitarianism** and provided an explanation for the nature of **unconformities.** Since the publication of this book in 1785 there has been a group of scientists who recognize themselves as geologists. These new geologists defined the geometry of structures in mountain ranges, learned how to make geologic maps, discovered the processes involved in the formation of rocks, and speculated on the origins of specific structures and on mountain ranges in general.

Ideas about the origin of mountains have evolved gradually. At first, mountain ranges were thought to be a consequence of a *vertical push* from below, perhaps

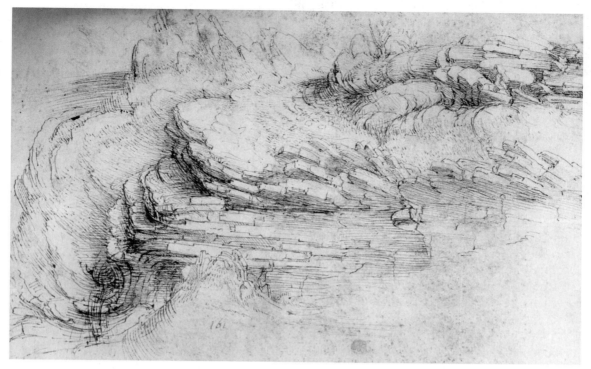

FIGURE 1.1 Sketch by Leonardo da Vinci showing details of folded strata in the mountains of Italy (ca. 1500 AD). In recent years it was discovered that in addition to his careful observations of the natural world, da Vinci also completed insightful friction experiments.

FIGURE 1.2 Aerial view of the European Alps (France).

FIGURE 1.3 Model of mountain building and associated deformation as represented by G. P. Scrope (1825). The uplift is caused by intrusion of an igneous core, and the folds are generated by down-slope movement.

associated with intrusion of molten rock along preexisting zones of weakness, and folds and faults in strata were attributed to gravity sliding down the flanks of these uplifts (Figure 1.3). Subsequently, the significance of *horizontal forces* was emphasized, and geologists speculated that mountain ranges and their component structures reflected the contraction of the Earth that resulted from progressive cooling. In this model, the shrinking of the Earth led to wrinkling of the surface. One of the more notable discoveries (about 1850) was the recognition by James Hall (1811–1898) that Paleozoic strata in the Appalachian Mountains of North America were much thicker than correlative strata in the interior of the continent. This discovery led to the development of the **geosyncline theory,** a model in which deep sedimentary basins, called geosynclines, evolved into mountain ranges. Contraction theory and geosynclinal theory, or various combinations of the two, were widely accepted until the 1960s, when the views of Alfred Wegener (1880–1930), Arthur Holmes (1898–1965), and Harry Hess (1906–1969) led to the formulation of a very different model. Building on the work of Alfred Wegener's **continental drift theory** and Arthur Holmes's mantle convection model, Harry Hess proposed the revolutionary idea of a mobile seafloor (**seafloor spreading hypothesis**) that lead to the formulation of **plate tectonic theory.** In this theory, the Earth consists of several, rigid plates that change in space and time. The interaction between these plates offers a unifying explanation for the occurrence of mountain ranges, ocean basins, earthquakes, volcanoes, and other previously disparate geologic phenomena.

As the foundations of geology grew, diverse features of rocks and mountains gained names, and the once amorphous, nondescript masses of rock exposed on our planet became history books that preserve the Earth's biography. Perhaps your concept of the planet has evolved rapidly as well, because of the courses in geology and other sciences that you have taken thus far. Now, as you drive across the countryside, you scare the daylights out of your passengers as you twist to see and discuss roadside outcrops. The rocks are no longer gray masses to you, but they contain recognizable patterns and shapes and fabrics. The purpose of this book is to increase your ability to interpret these features, and particularly to use them as clues to understanding the processes that have shaped and continue to change the outer layers of the Earth.

1.2 CLASSIFICATION OF GEOLOGIC STRUCTURES

When you finished your introductory geology course, you probably had a general concept of what a geologic structure is. The term probably brings to mind images of folds and faults. Perhaps you had the opportunity to take a field trip where you saw some of these structures in the wild. These features are formed in response to pushes and pulls associated with the forces that arise from the movement of tectonic plates or as a consequence of differential buoyancy between parts of the lithosphere. But what about bedding in a sedimentary rock and flow banding in a rhyolite flow; are these structures? And what about slump folds in a debris flow; are they structures? Well . . . yes, but the link between their formation and plate motion is less obvious. So, maybe we need to have a more general concept of a geologic structure.

The most fundamental definition of a **geologic structure** is a geometric feature in rock whose shape, form, and distribution can be described. From this definition it is obvious that there are several ways in which geologic structures can be subdivided into groups. In other words, by necessity there are several different, yet equally valid classification schemes that can be used in organizing the description of geologic structures. Different schemes are relevant for different purposes, so we will briefly look at various classification schemes for geologic structures that will return in subsequent chapters. At first, these various classification schemes may seem very confusing. Thus, we recommend that you start by recognizing the basic geometric classes as the foundation of your understanding. As you learn about these classes, refer back to the lists below, and see how a particular geometric class fits into one or more of the classification schemes.

I. Classification based on *geometry,* that is, on the shape and form of a particular structure
- *Planar (or subplanar) surface*
- *Curviplanar surface*
- *Linear feature*

This subdivision represents perhaps the most basic classification scheme. In this scheme we include the following classes of structures: joint, vein, fault, fold, shear zone, foliation, and lineation.

II. Classification based on geologic *significance*
- *Primary:* formed as a consequence of the formation process of the rock itself
- *Local gravity-driven:* formed due to slip down an inclined surface; slumping at any scale driven by local excess gravitational potential
- *Local density-inversion driven:* formed due to local lateral variations in rock density, causing a local buoyancy force
- *Fluid-pressure driven:* formed by injection of unconsolidated material due to sudden release of pressure
- *Tectonic:* formed due to lithospheric plate interactions, due to regional interaction between the asthenosphere and the lithosphere, due to crustal-scale or lithosphere-scale gravitational potential energy and the tendency of crust to achieve isostatic compensation

The first four items in this scheme can be grouped as *primary and nontectonic structures,* meaning that they are not directly related to the forces associated with moving plates. We purposely say "can" because in many circumstances these categories of structures do form in association with tectonic activity. For example, gravity sliding may be triggered by tectonically generated seismicity, and salt domes may be localized by movement of tectonic normal faults. These first four categories will be discussed in Chapter 3. The fifth category of structures is very large and forms the primary focus of this book.

III. Classification based on *timing* of formation
- *Syn-formational:* formed at the same time as the material that will ultimately form the rock
- *Penecontemporaneous:* formed before full lithification, but after initial deposition

- *Post-formational:* formed after the rock has fully formed, as a consequence of phenomena not related to the immediate environment of rock formation

IV. Classification based on the *process* of formation, that is, the deformation mechanism
- *Fracturing:* related to development or coalescence of cracks in rock
- *Frictional sliding:* related to the slip of one body of rock past another, or of grains past one another, both of which are resisted by friction
- *Plasticity:* resulting from deformation by the internal flow of crystals without loss of cohesion, or by non-frictional sliding of crystals past one another
- *Diffusion:* resulting from material transport either solid-state or assisted by a fluid (dissolution)

V. Classification based on the mesoscopic *cohesiveness* during deformation
- *Brittle:* formed by loss of cohesion across a mesoscopic discrete surface
- *Ductile:* formed without loss of cohesion across a mesoscopic discrete surface
- *Brittle/ductile:* involving both brittle and ductile aspects

Note that the scale of observation (in this case, mesoscopic) is critical in the distinction between brittle and ductile deformation, because ductile deformation can involve microscopic-scale fracturing and frictional sliding.

VI. Classification based on the *strain* significance, in which a reference frame, usually the Earth's surface, is defined
- *Contractional:* resulting in shortening of a region
- *Extensional:* resulting in extension of a region
- *Strike-slip:* resulting from movement without either shortening or extension

Note that shortening in one direction can be, but does not have to be, accompanied by extension in a different direction, and vice versa. Also, regional deformation usually results in the vertical displacement of the Earth's surface, a component of deformation that is commonly overlooked.

VII. Classification based on the *distribution of deformation* in a volume of rock
- *Continuous:* occurs through the rock body at all scales
- *Penetrative:* occurs throughout the rock body, at the scale of observation; up close, there may be spaces between the structures

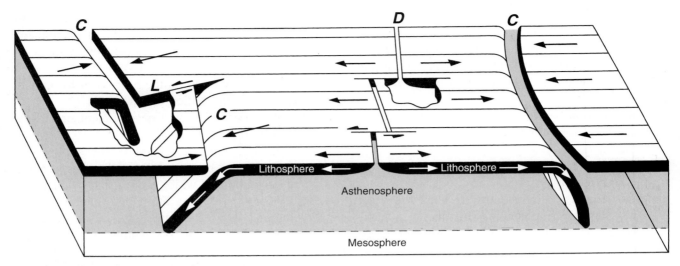

FIGURE 1.4 The principal features of plate tectonics. Three types of plate boundaries arise from the relative movement (arrows) of lithospheric plates: *C*—convergent boundary, *D*—divergent boundary, and *L*—lateral slip (or transform) boundary.

- *Localized:* continuous or penetrative structure occurs only within a definable region
- *Discrete:* structure occurs as an isolated feature

We can conveniently consider the basic geometric classes of structures to be a manifestation of the mesoscopic cohesiveness of deformation. Joints, veins, and certain types of faults are manifestations of primarily brittle deformation, whereas cleavage, foliation, and folding are largely manifestations of ductile deformation processes. Thus, in this book we subdivide our discussions of specific structures into two parts: "Brittle Structures" (Part B) and "Ductile Structures" (Part C). As a first approximation, brittle deformation is more common in the upper part of the crust, where temperatures and pressures are relatively low, and ductile deformation is more common in the deeper part of the crust, because it is favored under conditions of greater pressure and temperature. Also, ductile deformation is commonly a manifestation of plastic deformation and diffusion, whereas brittle deformation is a consequence of fracturing and frictional sliding. However, it is important to emphasize right from the start that different processes can act in the same places in the Earth. The processes that occur at any given time may reflect geologic variables such as **strain rate,** which is the rate of displacement in the rock body (Chapter 5). For example, a sudden increase in strain rate may cause rock that is deforming in a ductile manner (by folding) to suddenly behave in a brittle manner (by fracturing). You can see this remarkable effect by, respectively, slow and quick pulling of a piece of SillyPutty®.

Ultimately, most crustal structures are a consequence of plate tectonic activity, which is the slow (on the order of centimeters per year) but steady motion of segments of the outer, stiff layer of the Earth, called the **lithosphere,** over the weaker **asthenosphere.** The forces that this motion generates, especially those from interactions at plate boundaries, produce the structures we study in the field and in the laboratory. The three types of plate motions are **convergence, divergence, and lateral slip** (Figure 1.4). Without the activity that arises from these plate motions, such as deformation, volcanism, and earthquakes, the Earth would be as dead as the Moon. In other words, plate tectonics provides the global framework to examine the significance of structures that occur on local and regional scales.

1.3 STRESS, STRAIN, AND DEFORMATION

We have already used the words stress, strain, and deformation without definition, because these are common English words and most people have an intuitive grasp of what they mean. Stress presumably has something to do with pushing and pulling, and strain and deformation have something to do with bending, breaking, stretching, or squashing. But in standard English, stress and strain are often used interchangeably; for example, advertisements for aspirin talk about "the stress and strain of everyday life." In structural geology, however, these terms have more exact

Before

After

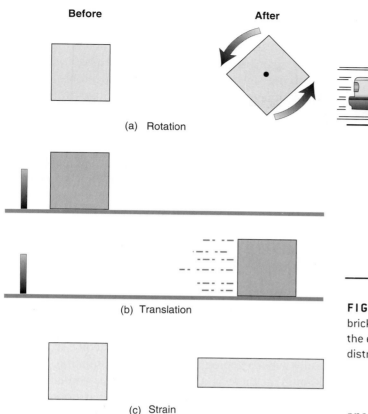

(a) Rotation

(b) Translation

(c) Strain

FIGURE 1.5 The three components of deformation: (a) rotation, (b) translation, and (c) strain.

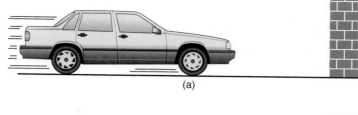

(a)

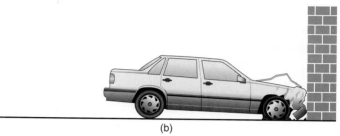

(b)

FIGURE 1.6 Strain in the real world. (a) A car approaches a brick wall in a crash test; (b) the same car after impact. Note the extreme distortion of the front (i.e., inhomogeneous strain distribution).

meanings, so right from the start we want to clarify their usage (and avoid headaches).

The **stress** (σ) acting on a plane is the force per unit area of the plane ($\sigma = F$/area). We will see in Chapter 3 that when referring to the stress at a point in a body, a more complicated definition is needed. **Deformation** refers to changes in shape, position, or orientation of a body resulting from the application of a differential stress (i.e., a state in which the magnitude of stress is not the same in all directions). More specifically, deformation consists of three components (Figure 1.5): (1) a **rotation,** which is the pivoting of a body around a fixed axis, (2) a **translation,** which is a change in the position of a body, and (3) a **strain,** which is a distortion or change in shape of a body (Chapter 3). To visualize a strain, consider the test crash of a car that is rapidly approaching a brick wall (Figure 1.6a). In Figure 1.6b, the car and the wall have attempted to occupy the same space at the same time, with variable success. Since the structural integrity of the car is less than that of the wall, the push between car and wall squashed the car, thereby resulting in a strain. In *homogeneous strain,* the strain exhibited at

one point in the body is the same as the strain at all other points in the body. Cars are designed so that strain is *heterogeneous,* meaning that the strain is not equal throughout the body, and the passengers are protected from some of the impact.

What about translation and rotation? These components of deformation are a bit harder to recognize, but they do occur. For example, a rigid body of rock that has moved along a fault plane clearly has been translated relative to the opposing side of the fault (Figure 1.7a), and a fault block in which strata are inclined relative to horizontal strata on the opposing wall of the fault has clearly been rotated (Figure 1.7b). Such rotations occur at all scales, as emphasized by work in paleomagnetism, which demonstrates that continental blocks have been rotated around a vertical axis as a consequence of shear along major strike-slip faults and plate boundaries.

In order to describe deformation, it is necessary to define a *reference frame.* The reference frame used in structural geology is loosely called the undeformed state. We can't know whether a rock body has been moved or distorted unless we know where it originally was and what its original shape was. Ideally, if we know both the original and final positions of an array of points in a body of rock, we can describe a deformation with mathematical precision by defining a *coordinate transformation.* For example, in Figure 1.8a, four points (labeled *m, n, o,* and *p*) define a

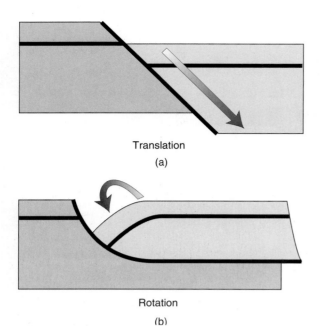

FIGURE 1.7 The translational and rotational components of deformation shown schematically along a fault. (a) A translated fault block; (b) a rotated fault block in the hanging wall.

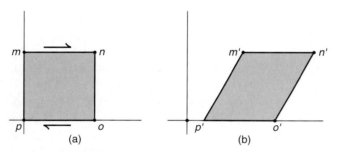

FIGURE 1.8 Deformation represented as a coordinate transformation. Points *m*, *n*, *o*, and *p* move to new positions *m'*, *n'*, *o'*, and *p'*.

square in a Cartesian coordinate system. If the square is sheared by stresses acting on the top and bottom surfaces, as indicated by the arrows, and moved from its original location, it changes into a parallelogram that is displaced from the origin (Figure 1.8b). The deformation can be described by saying that points *m*, *n*, *o*, and *p* moved to points *m'*, *n'*, *o'*, and *p'*, respectively. In other words, coordinates of all four corners of the square have been transformed. If you are mathematically adept, you will probably realize that this transformation can be described by a mathematical function, but we won't get into that now . . . wait until Chapter 3.

In many real circumstances, we don't have an external reference frame, so we can only partially describe a deformation. For example, at an isolated outcrop we

may be able to describe strain—say, because of the presence of deformed fossils—but we have no absolute record of translations or rotations. Then we may talk about *relative displacement* and *relative rotation*. A flat-lying bed of Paleozoic limestone in the Midcontinent region of the United States was at one time below sea level and, because of plate motion, it was formed at a different latitude than today, but we can't immediately characterize these movements.[1] If, however, we see a fault offset a limestone bed by 2 meters, we say that one side of the fault has moved 2 m relative to the other side.

1.4 STRUCTURAL ANALYSIS AND SCALES OF OBSERVATION

At this point, we know what a structure is and we know what a geologist means by deformation. We also know that there is a group of people who call themselves **structural geologists.** But what do structural geologists do? One way to gain insight into the subject of structural geology is to think about the type of work that structural geologists carry out. Not surprisingly, structural geologists do structural analysis, which involves many activities (outlined in Table 1.1). Throughout the book you see that we use tables like this to summarize concepts and terms. Many terms not specifically mentioned in the text can be found in these tables, which serve as convenient reference points throughout the text.

Looking at Table 1.1 you will note that in many of the definitions we have to refer to the **scale of observation.** For the results of a structural analysis to be interpretable, the scale of our analysis must be taken into account. For example, a bed of sandstone in a single outcrop in a mountain may appear to be undeformed. But the outcrop may display only a small part of a huge fold that cannot be seen unless you map at the scale of the whole mountain. Structural geologists commonly refer to these relative scales of observation by a series of subjective prefixes. **Micro** refers to features

---

[1]Paleomagnetic and paleontologic methods are primarily used for this in the Paleozoic. In the Mesozoic and Tertiary, ocean-floor magnetic anomalies are available as well, but in the Precambrian only the paleomagnetic approach remains.

TABLE 1.1	CATEGORIES OF STRUCTURAL ANALYSIS
Descriptive analysis	The characterization of the shape and appearance of geologic structures. It includes development of a precise vocabulary (jargon) that permits one geologist to create an image of a structure that any other geologist can understand, and development of methods for uniquely describing the orientation of a structure in three-dimensional space.
Kinematic analysis	The determination of the movement paths that rocks or parts of rocks have taken during transformation from the undeformed to the deformed state. This subject includes, for example, use of features in rocks to define the direction of movement on a fault.
Strain analysis	The development of mathematical tools for quantifying the strain in a rock. This activity includes the search for features in rock that can be measured to define strain.
Dynamic analysis	The development of an understanding of stress and its relation to deformation. This activity includes the use of tools for measuring the present-day state of stress in the Earth, and the application of techniques for interpreting the state of stress responsible for microstructures in rocks.
Mechanism analysis	The study of processes on the atomic scale to grain scale that allow structures to develop. This activity includes study of both fracture and flow of rock.
Tectonic analysis	The study of the relationship between structures and global tectonic processes. This activity includes the study and interpretation of regional-scale or megascopic structural features, and the study of relationships among structural geology, stratigraphy, and petrology.

that are visible optically at the scale of thin sections, or that may only be evident with the electron microscope; the latter is sometimes referred to as submicroscopic. **Meso** refers to features that are visible in a rock outcrop, but cannot necessarily be traced from outcrop to outcrop. **Macro** refers to features that can be traced over a region encompassing several outcrops to whole mountain ranges. In some circumstances, geologists use the prefix **mega** to refer to continental-scale deformational, such as the movements of tectonic plates over time. Of course there are no sharp boundaries between these scales, and their usage will vary with context, but a complete structural analysis tries to integrate results from several scales of observation.

Each scale of observation has its own set of tools. For example, optical and electron microscopes are used for observations on the microscale, and satellite imaging may be used for observations on the macroscale. The mesoscopic recognition and description of rocks and their structures are of fundamental importance to field analysis, which requires a set of eyes,[2] a hammer, a compass, and a hand lens. Field work is, in fact, pretty much a low-tech, low-budget

FIGURE 1.9 Field area in Antarctica.

affair unless you are working in the High Himalayas, Antarctica (Figure 1.9), or some similarly remote setting that requires extensive logistics (like expeditions, planes, and helicopters). For structural field work we record observations on lithologies and rock structures in notebooks or on portable devices and we measure the orientation of geometric elements with a compass. The compass to a structural geologist is like the stethoscope to a doctor: it is the professional's tool (and should be clearly visible at all times).

---

[2]Aided by corrective lenses in the case of the authors.

TABLE 1.2	TERMINOLOGY RELATED TO GEOMETRY AND REPRESENTATION OF GEOLOGIC STRUCTURES
Apparent dip	Dip of a plane in an imaginary vertical plane that is not perpendicular to the strike. The apparent dip is less than or equal to the *true dip*.
Attitude	Orientation of a geometric element in space
Cross section	Plane perpendicular to the Earth's surface
(True) dip	The slope of a surface; formally, the angle of a plane with the horizontal measured in an imaginary vertical plane that is perpendicular to the strike (Figure 1.10a)
Dip direction	Azimuth of the horizontal line that is perpendicular to the strike (Figure 1.10a)
Foliation	General term for a surface that occurs repeatedly in a body of rock (e.g., bedding, cleavage)
Lineation	General term for a penetrative linear element, such as the intersection between bedding and cleavage or alignment of elongate grains
Pitch	Angle between a linear element that lies in a given plane and the strike of that plane (also *rake*) (Figure 1.10b)
Plunge	Angle of linear element with earth's surface in imaginary vertical plane
Plunge direction	Azimuth of the plunge direction
Position	The geographic location of a geometric element (e.g., an outcrop)
Profile plane	Plane perpendicular to a given geometric element; for example, the plane perpendicular to the hinge line of a fold
Rake	Angle between a linear element that lies in a given plane and the strike of that plane (also *pitch*)
Strike	Azimuth of the horizontal line in a dipping plane or the intersection between a given plane and the horizontal surface (also *trend*) (Figure 1.10a)
Trace	The line of intersection between two nonparallel surfaces
Trend	Azimuth of any feature in map view; sometimes used as synonym for *strike*

Basic geometric principles, the ways of describing geometric features, and concepts related to constructions such as structure contours and spherical projections are explained in structural geology laboratory manuals. So, in this text we will limit ourselves to the short descriptions of terms and associated concepts in Table 1.2, some of which are illustrated in Figure 1.10.

1.5 SOME GUIDELINES FOR STRUCTURAL INTERPRETATION

In closing this introductory chapter, we make a few general comments on structural analysis. Good scientific work requires that one separates *observations* from *interpretations,* which equally holds for geologic mapping. Yet, a geologic map, a cross section, or a block diagram without interpretation misses the unique insights of the investigator. If you spend the time collecting and digesting data, you are best suited to make the interpretations (or, educated guesses). The following suggestions may help with map interpretation, but note that they hardly do justice to the intricate process of interpretation; our aim with these guidelines is mostly to point you in the right direction.

The assumptions on which interpretations are based do not hold universally; in fact, after some field experience you may disagree with one or more of the points listed. In our experience, however, the guidelines in Table 1.3 enable a reasonable, first-order interpretation of the geometry of an area.

Each individual guideline in Table 1.3 is valid under a given set of circumstances, but remember that, except for the laws, they remain mere assumptions; no

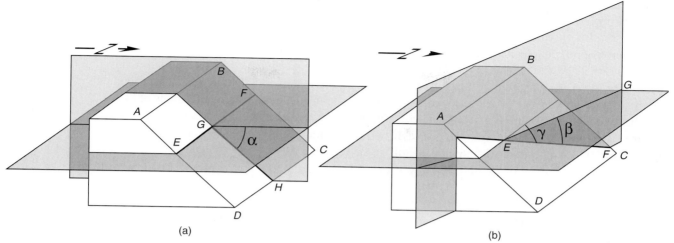

FIGURE 1.10 (a) Attitude of a plane: dip, dip direction, and strike. The strike of dipping plane *ABCD* is the intersection with an imaginary horizontal plane (line *EF*). The dip of plane *ABCD* is given by the angle α, while line *GH* represents the dip direction. Note that there are two possible directions of dip for a given strike. Thus, when using dip and strike, the general direction of dip must also be included (e.g., 090°/45° N). Alternatively, the attitude of plane *ABCD*, using dip direction and dip, is uniquely described by 000°/45°. (b) Attitude of a line: plunge, plunge direction, and pitch. Line *EF* lies in dipping plane *ABCD*. The plunge of line *EF* is the angle β, which is measured from the horizontal (line *EG*) in an imaginary vertical plane that contains both line *EF* and line *EG*; line *EG* is called the plunge direction. The attitude of line *EF*, using plunge and direction of plunge, is given by 30°/315°. The pitch (angle γ) is the angle that line *EF* makes with the horizontal (strike) of a plane (here: plane *ABCD*) containing the line. Note that when the pitch of a line is recorded, the attitude of the reference plane must be given, as well as the side from which the pitch angle is measured (here: 000°/45°, 40° W).

TABLE 1.3	SOME GUIDELINES FOR THE INTERPRETATION OF DEFORMED AREAS

- Strata are deposited horizontally. This is the Law of Original Horizontality, which makes bedding an internal reference frame.

- Strata follow one another in chronological, but not necessarily continuous, order.[3] This is known as the Law of Superposition.

- Separated but aligned outcrops of the same lithologic sequence imply stratigraphic continuity.

- Strata occur in laterally continuous and parallel layers in a region.

- Sharp discontinuities in lithologic patterns are faults, unconformities, or intrusive contacts.

- Deformed areas can be subdivided into a number of regions that contain consistent structural attitudes (structural domains). For example, an area with folded strata can be subdivided into regions with relatively constant dip direction (or even dip), such as the limbs and hinge areas of large-scale folds.

- The simplest but internally consistent interpretation is most correct.[4] This is also known as the least-astonishment principle.

more, no less. Whenever possible your assumptions should be tested by adding more observations, and when the assumptions continue to hold, only then may your interpretation be valid. This approach follows a proven scientific method, called the **testable working hypothesis,** which eventually leads to a **model.** If the model is very successful it may become a **law,** but this is rare in geology. Regardless, there is always room for alternative interpretations and models.

Increasingly, subsurface data from drilling and geophysical methods are available to structural geologists, and they should be used to test and constrain your interpretation. Drilling is restricted to the upper 10 km of the crust, but provides samples of deeply buried

---

[3]See the description of *facing* in Chapters 2 and 10. If younging directions are unknown, *transposition* may present complications (Chapter 12).
[4]But no simpler than that (paraphrasing Albert Einstein).

layers that can be compared with exposed rock units. This is a powerful test for the cross sections and block diagrams you construct. Using two-dimensional and three-dimensional deep seismic reflection imaging we get an indirect view of the deeper parts of the Earth (see Chapter 15). Seismic reflection profiles are obtained by recording the travel times of sound waves that bounce off layers in the Earth. The technique requires careful data processing such as **stacking** and **migration,** which improve the signal-to-noise ratio and localize reflectors. Correlation of these reflectors with features that are exposed at the surface or obtained from drilling gives important information on the nature of the deep structure.

1.6   CLOSING REMARKS

In this opening chapter, we quickly traveled through the history of structural geology; the types of geologic structures; the meaning of stress, strain, and deformation; and the nature of structural analysis. Of course we only scraped the surface of these topics. Our goal was to give you a first idea of what the field of structural geology and tectonics entails. Except for the historical considerations, all these topics will return in detail throughout the text. The chapters of the book are grouped into "Fundamentals" (Part A), describing the theory and background that are needed for the interpretation of natural structures, "Brittle Structures" and "Ductile Structures" (Parts B and C), reflecting a distinction that is based on the distribution of strain in deformed bodies, and

"Tectonics" and "Regional Geology" (Parts D and E), discussing some of the fundamentals of plate tectonics and plate boundaries, and perspectives on the geology of selected regions around the world. Ultimately, we will have examined structures on all scales, ranging from single atoms to the outcrop, to the mountain belt, and to the whole Earth. The relationships of these structures provide us with a relatively recent but remarkably complete picture of the tectonic evolution and inner workings of our dear planet.

ADDITIONAL READING

Many books explore field and geometric analysis, and computer and geophysical applications in structural geology. The following selection is based on our own usage, which is by no means complete.

Groshong, R. H., 1999. *3-D structural geology: A practical guide to surface and subsurface map interpretation.* Springer Verlag.

Lisle, R. J., 1996. *Geological structures and maps: A practical guide* (2nd edition). Oxford: Butterworth-Heinemann.

Marshak, S., and Mitra, G., 1988. *Basic methods of structural geology.* Englewood Cliffs: Prentice Hall.

McClay, K., 1987. *The mapping of geological structures. Geological society of London handbook.* Berkshire: Open University Press.

Rowland, S. M., and Duebendorfer, E. M., 1994. *Structural analysis and synthesis* (2nd edition). Boston: Blackwell.

CHAPTER TWO

Primary and Nontectonic Structures

2.1 INTRODUCTION

In the first chapter, "Overview," we introduced a general definition for a **geologic structure** as any definable shape or fabric in a rock body. Most of the scope of structural geology is focused on **tectonic structures,** meaning those structures that form in response to forces generated by plate interactions (such as convergence, collision, rifting, strike-slip movement, subduction, buoyancy—concepts that will be discussed later). But before we get to tectonic structures, we have to examine structures that form during or shortly after the deposition of rocks, and we want to mention structures that are not an immediate consequence of plate interactions. Collectively, these structures may be called **nontectonic structures,** because it is inferred that they form in the absence of tectonic stresses, but strictly speaking this usage is misleading. Most geologic structures are indirectly, if not directly, a consequence of tectonic activity. Examples are the creation of slopes down which sediments slide and the occurrence of volcanic activity leading to the flow of basalt, both of which are manifestations of movements in the Earth. All of these phenomena are ultimately a manifestation of movement in the Earth. Under the broad masthead of nontectonic structures we discuss depositional, penecontemporaneous, intrusive, and gravity-slide structures for both sedimentary and igneous rocks. Impact structures, which result from the collision of extraterrestrial objects with Earth's surface, are briefly discussed at the end of the chapter.

2.2 SEDIMENTARY STRUCTURES

When you look at an outcrop of sedimentary rock, the most obvious fabric to catch your eye is the primary layering or stratification (Figure 2.1), which is called **bedding.** Some of the more commonly used terms that are related to stratification are listed in Table 2.1.

FIGURE 2.1 Differential erosion of bedding surfaces in the Wasatch Formation, Bryce Canyon (Utah).

TABLE 2.1	SOME TERMINOLOGY OF STRATIFICATION
Bedding	Primary layering in a sedimentary rock, formed during deposition, manifested by changes in texture, color, and/or composition; may be emphasized in outcrop by the presence of parting
Compaction	Squeezing unlithified sediment in response to pressure exerted by the weight of overlying layers
Overturned beds	Beds that have been rotated past vertical in an Earth–surface frame of reference; as a consequence, facing is down
Parting	The tendency of sedimentary layers to split or fracture along planes parallel to bedding; parting may be due to weak bonds between beds of different composition, or may be due to a preference for bed-parallel orientation of clay
Strata	A sequence composed of layers of sedimentary rock
Stratigraphic facing	The direction to younger strata, or, in other words, the direction to the depositional top of beds
Younging direction	Same as stratigraphic facing

What defines bedding in an outcrop? In the Painted Desert of Arizona, beds are defined by spectacular variations in colors, and outcrops display garish stripes of maroon, red, green, and white. In the Grand Canyon and the Rocky Mountains, beds are emphasized by contrasts in resistance to erosion; sandstone and limestone beds form vertical cliff faces while shale layers form shallow slopes. On the cliffs that form the east edge of the Catskill Mountains in New York State, there are abrupt contrasts in grain size between adjacent beds, with a coarse conglomerate juxtaposed against siltstone or shale. Strictly speaking, a bed is the smallest subdivision of a sedimentary unit.[1] It has a definable top or bottom and can be distinguished from adjacent beds by differences in grain size, composition, color, sorting, and/or by a physical parting surface.

[1]Stratigraphers divide sequences of strata into the following units: supergroup, group, formation, member, bed. Several beds make a member, several members make a formation, and so on. Criteria for defining units are somewhat subjective. Basically, a unit is defined as a sequence of strata that can be identified and mapped at the surface or in the subsurface over a substantial region. The basis for recognizing a unit can be its age, its component sequence of lithologies, and the character of its bedding.

All of the features defining bedding, with the exception of parting, are a consequence of changes in the source of the sediment (or provenance) or the depositional environment.

In some outcrops, bedding is enhanced by the occurrence of **bedding-parallel parting.** Parting forms when beds are unroofed (i.e., overlying strata are eroded away) and uplifted to shallower depths in the crust. As a consequence, the load pushing down on the strata decreases and the strata expand slightly. During this expansion, fractures form along weak bedding planes and define the parting. This fracturing reflects the weaker bonds between contrasting lithologies of adjacent beds, or the occurrence of a preferred orientation of sedimentary grains (e.g., mica). If a sedimentary rock has a tendency to have closely spaced partings, it is said to display **fissility.** Shale, which typically has a weak bedding-parallel fabric due to the alignment of constituent clay or mica flakes, is commonly fissile.

There are three reasons why platy grains like mica have a preferred orientation in a sedimentary rock. First, the alignment of grains can reflect settling of asymmetric bodies in Earth's gravity field. Platy grains tend to lie down flat. To understand why, throw a handful of coins or cards into the air: after the coins fall they lie flat against the floor. You will rarely see a coin stand on its edge. Second, the alignment of grains can reflect flow of the fluid in which the grains were deposited. In a moving fluid, grains are reoriented so that they are hydrodynamically stable, meaning that the traction caused by the moving fluid is minimized (as is the case if the broad face of the grain parallels the flow direction). Typically, grains end up being imbricated, meaning that they overlap one another like roof shingles. **Imbrication** is a useful primary sedimentary feature that can be used to define **paleocurrent direction,** which is the direction of the current when the sediment was deposited. Third, the alignment of grains can form as a consequence of compaction subsequent to deposition. As younger sediment is piled on top, water is progressively squeezed out of the older sediment below and the grains mechanically rotate into an orientation with their flat surfaces roughly perpendicular to the applied load.

Bedding in outcrops is often highlighted by differential weathering and erosion (Figure 2.1). For example, chemical weathering (like dissolution of carbonate) of a sequence of strata containing alternating limestones and quartz sandstones will result in an outcrop face on which the quartz-rich layers stand out in relief. Fresh limestone and dolomite are almost identical in color, but weathering of a sequence of alternating lime-

TABLE 2.2	TERMS TO DESCRIBE TYPES OF BEDDING
Massive beds	Beds that are relatively thick (typically several m) and show no internal layering. Massive bedding develops in sedimentary environments where large quantities of sediment are deposited very rapidly or in environments where bioturbation (churning of the sediments by worms and other organisms) occurred.
Medium beds	Beds that are 10–30 cm thick
Rhythmic beds	A sequence of beds in which the contrast between adjacent beds is repeated periodically for a substantial thickness of strata
Thick beds	Beds that are 30–100 cm thick. Very thick beds are tens of m thick.
Thin beds	Beds that are less than 3 cm thick
Thinly laminated beds	Beds that are less than 0.3 cm thick

stones and dolostones will result in a color-banded outcrop, because the dolostones tend to weather to a buff-tan color and the limestones become grayish. Erosion of a sequence of alternating sandstones and shales may result in a stair-step outcrop face, because the relatively resistant sandstone beds become vertical cliffs and the relatively weak shale beds create slopes. Geologists have developed a jargon for describing specific types of bedding, as defined in Table 2.2.

2.2.1 The Use of Bedding in Structural Analysis

Recognition of bedding is critical in structural analysis. Bedding provides a reference frame for describing deformation of sedimentary rocks, because when sediments are initially deposited, they form horizontal or nearly horizontal layers, a concept referred to as the **Law of Original Horizontality.** Thus, if we look at an outcrop and see tilting or folding, what we are noticing are deviations in bedding attitude from original horizontality.

In complexly deformed and metamorphosed sedimentary rocks, geologists have to search long and hard to find subtle preserved manifestations of original bedding (e.g., variations in grain size, color, composition). Only by finding bedding can a geologist unravel the folding history of a region. When found, bedding is labeled S_0 (pronounced ess-zero), where the S is an abbreviation for surface. Later we will discuss surfaces in rocks that are formed by deformation, like rock cleavage, which are labeled S_1, S_2, and so on (Chapter 11).

The study of certain depositional structures within beds and on bedding surfaces is useful in tectonic analysis because they may provide important information on **depositional environment** (the setting in which the sediment was originally deposited), on **stratigraphic facing** or **younging direction** (the direction in which strata in a sequence are progressively younger), and on **current direction** (the direction in which fluid was flowing during deposition). Facing indicators allow you to determine whether a bed is right-side-up (facing up) or overturned (facing down), with respect to the Earth's surface. Recognition of facing is powerful both for stratigraphic studies and for structural studies. For example, the structural interpretation of a series of parallel beds in two adjacent outcrops depends on the facing—if the facing is the same in both outcrops, then the strata are probably **homoclinal,** meaning that they have a uniform dip. However, if the facing is opposite, then the two outcrops are likely on different limbs of a fold whose hinge area is not exposed. We'll return to the use of facing in structural analysis in Section 10.2.

2.2.2 Graded Beds and Cross Beds

Patterns within beds may contain information about stratigraphic facing and current directions that are often critical for tectonic interpretations. **Graded beds** display progressive fining of clast/grain size from the base to the top (Figure 2.2), and are a consequence of deposition from turbidity flows. A **turbidity flow** is a cloud of sediment that moves down a slope under water because the density of the sediment-water mixture is greater than that of clean water, and denser liquids sink through less dense liquids. Turbidite flows are triggered by major storms or earthquakes (because of their association with seismicity, the occurrence of turbidites may indicate that the sediment source region was tectonically active), and move down gentle slopes at considerable speed. Typically, a flow is confined to a submarine channel or canyon; when a broadening of the channel or a decrease in slope slows the speed of a turbidity current, the sediment cloud settles. During settling, the largest grains fall first, and the finest grains last. Each turbidity flow produces a separate graded sequence or a **turbidite,** which is often capped by pelagic sediment, meaning deep-marine sediments like

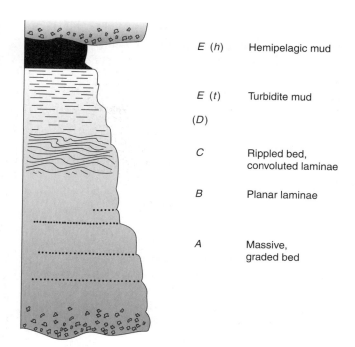

E (h)		Hemipelagic mud
E (t)		Turbidite mud
(D)		
C		Rippled bed, convoluted laminae
B		Planar laminae
A		Massive, graded bed

FIGURE 2.2 Graded bedding in a turbidite (or Bouma) sequence.

clay and plankton shells. Turbidites display an internal order, called a **Bouma sequence**[2] (Figure 2.2), which reflects changing hydrodynamic conditions as the turbidity current slows down.

In pre–plate tectonics geological literature (i.e., pre-Beatles), thick sequences of turbidites were referred to as **flysch,** a term originating from Alpine geology (see Part E). Flysch was considered to be an orogenic deposit, meaning a sequence of strata that was deposited just prior to and during the formation of a mountain range. Exactly why such strata were deposited, however, was not understood. Modern geologists now realize that the classical flysch sequences are actually turbidites laid down in a deep trench marking an active plate boundary (like a subduction zone). Turbidite flows are common in ocean trenches, which represent seismically active, convergent plate margins. During and after deposition, the trench turbidites are scraped up and deformed by continued convergence between plates, and may eventually be caught in a continental collision zone (see Chapter 16).

Cross beds are surfaces within a bed that are oblique to the overall bounding surfaces of the bed (Figure 2.3). Cross beds, which are defined by subtle partings or concentrations of grains, form when sediment moves from the windward or upstream side of a dune, ripple, or delta to a face on the leeward or downstream side, where the current velocity is lower and the

---

[2]Named after the sedimentologist Arnold Bouma.

sediment settles out. Thin beds parallel to the upper bounding surface are called **topset beds,** the inclined layers deposited parallel to the slip face are called **foreset beds,** and the thin beds parallel to the lower bounding surface are called **bottomset beds** (Figure 2.3a). The foreset beds, which typically are curved (concave up) and merge with the topset and bottomset beds, are the cross beds. If the topset beds and the upper part of the foreset beds are removed by local erosion, the bottomset beds of the next higher layer of sediment are juxtaposed against the foreset beds of the layer below. Thus, cross beds tend to be truncated at the upper bedding surface, whereas they are asymptotic to the lower bedding surface (Figure 2.3b). This geometry provides a clear stratigraphic facing indicator. The current direction in a cross-bedded layer is taken to be approxi-

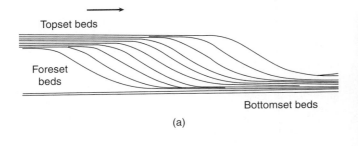

Topset beds

Foreset beds

Bottomset beds

(a)

FIGURE 2.3 (a) Terminology of cross bedding, and (b) cross beds in the Coconino Sandstone, Oak Creek Canyon (Arizona).

(b)

TABLE 2.3	COMMON SURFACE MARKINGS
Animal tracks	Patterns formed when critters like trilobites, worms, and lizards tromp over and indent the surface (the characteristic trails of these organisms are a type of *trace fossil*).
Clast imbrication	The shingle-like overlapping arrangement of tabular clasts on the surface of a bed in response to a current. Imbrication develops because tabular clasts tend to become oriented so that the pressure exerted on them by the moving fluid is minimized.
Flute casts	Asymmetric troughs formed by vortices (mini tornadoes) within the fluid that dig into the unconsolidated substrate. The troughs are deeper at the upstream end, where the vortex was stronger. They get shallower and wider at the downstream end, because the vortex dies out. Flute casts can be used as facing indicators.
Mudcracks	Desiccation of mud causes the mud to crack into an array of polygons and intervening mudcracks. Each polygon curls upwards along its margins, so that the mudcracks taper downwards and the polygons resemble shallow bowls. Mudcracks can be used as facing indicators, because an individual crack tends to taper downwards (Figure 2.4).
Raindrop impressions	Circular indentations on the bed-surface of mudstone, formed by raindrops striking the surface while it was still soft.
Ripple marks	Ridges and valleys on the surface of a bed formed as a consequence of fluid flow. If the current flows back and forth, as along a beach, the ripples are *symmetric,* but if they form in a uniformly flowing current, they are *asymmetric* (Figure 2.5). The crests of symmetric ripples tend to be pointed, whereas the troughs tend to be smooth curves. Thus, symmetric ripples are good facing indicators. Asymmetric ripples are not good facing indicators, but do provide current directions.
Traction lineation	Subtle lines on the surface of a bed formed either by trails of sediment that collect in the lee of larger grains, or by alignment of inequant grains in the direction of the current to diminish hydraulic drag.
Worm burrows	The traces of worms or other burrowing organisms that live in unconsolidated sediment. They stand out because of slight textural and color contrasts with the unburrowed rock.

mately perpendicular to the intersection between the truncated foresets and overlying bed.

2.2.3 Surface Markings

Local environmental phenomena, such as rain, desiccation (i.e., drying), current traction, and the movement of organisms, affect the surface of a bed of sediment. If the sediment is unlithified, these phenomena leave an imprint that is known as a **surface marking.** Some of the more common surface markings are listed in Table 2.3.

FIGURE 2.4 Mudcracks, separating a mud layer into platelets. The circular indentations are raindrop imprints.

FIGURE 2.5 Asymmetric ripple marks. The arrow indicates approximate current direction during deposition.

2.2.4 Disrupted Bedding

Load casts, which are also called ball-and-pillow structures, are bulbous protrusions extending downward from a sand layer into an underlying mud or very fine sand layer (Figure 2.6). They form prior to lithification where a denser sand lies on top of less dense mud and a disturbance by a storm or an earthquake causes blobs of sand to sink into the underlying mud. Load casts are useful stratigraphic facing indicators when they retain some connection to the host layer.

Where sand and mud layers are progressively buried, it is typical for the mud layers to compact and consolidate before the sand layers do. As a consequence, the water in the sand layer is under pressure. If an earthquake, storm, or slump suddenly cracks the permeability barrier surrounding the sand, water and sand are released and forced into the mud along cracks. When this happens near the Earth's surface, little mounds of sand, called **sand volcanoes,** erupt at the ground surface. The resulting wall-like intrusions of sand (or in some localities, even conglomerate) are called **clastic dikes** (Figure 2.7). At depth, partially consolidated beds of sand and mud break into pieces, resulting in a chaotic layering that is known, simply, as **disrupted bedding** (Figure 2.8).

Studies of disrupted bedding, sedimentary dikes, and sand volcanoes in lake and marsh deposits provide an important basis for determining the recurrence interval of large earthquakes. In these studies, investigators dig a trench across the deposit and then look for disrupted intervals within the sequence. Radiocarbon dating of organic matter in the disrupted layers defines the absolute age of disruption events and allows us to estimate the recurrence of earthquakes.

2.2.5 Conformable and Unconformable Contacts

Earlier we defined a contact as any surface between two geologic units. There are three basic types of contacts: (1) **depositional contacts,** where a sediment layer is deposited over preexisting rock; (2) **fault contacts,** where two units are juxtaposed by a fracture on which sliding has occurred; and (3) **intrusive contacts,** where one rock body cuts across another rock body. In this section, we consider in more detail the nature and interpretation of depositional contacts.

Relatively continuous sedimentation in a region leads to the deposition of a sequence of roughly parallel

FIGURE 2.6 Load cast or ball-and-pillow structure; wine stakes for scale (Eifel, Germany).

FIGURE 2.7 Clastic dike in Proterozoic sandstones. Note that the dike sharply cuts across bedding and that very coarse clasts are preserved in its center (Sudbury, Ontario).

FIGURE 2.8 Disrupted bedding in turbidite; hammer for scale (Cantabria, Spain).

sedimentary units in which the contacts between adjacent beds do not represent substantial gaps in time. Gaps in this context can be identified from gaps in the fossil succession. The boundary between adjacent beds or units in such a sequence is called a **conformable contact.** For example, we say, "In eastern New York, the Becraft Limestone was deposited conformably over the New Scotland Formation." The New Scotland Formation is an argillaceous limestone representing marine deposition below wave base, whereas the Becraft Limestone is a pure, coarse-grained limestone representing deposition in a shallow-marine beach environment. Bedding in the two units is parallel, and the contact between these two units is gradational.

If there is an interruption in sedimentation, such that there is a measurable gap in time between the base of the sedimentary unit and what lies beneath it, then we say that the contact is **unconformable.** For example, we say, "In eastern New York, the Upper Silurian Rondout Formation is deposited unconformably on the Middle Ordovician Austin Glen Formation," because Upper Ordovician and Lower Silurian strata are absent. Unconformable contacts are generally referred to as **unconformities,** and the gap in time represented by the unconformity (that is, the difference in age between the base of the strata above the unconformity and the top of the unit below the unconformity) is called a **hiatus.** In order to convey a meaningful description of a specific unconformity, geologists distinguish among four types of unconformities that are schematically shown in Figure 2.9 and defined in Table 2.4.

Unconformities represent gaps in the rock record that can range in duration from thousands of years to billions of years. Examples of great unconformities, representing millions or billions of years, occur in the Canadian shield, where Pleistocene till buries

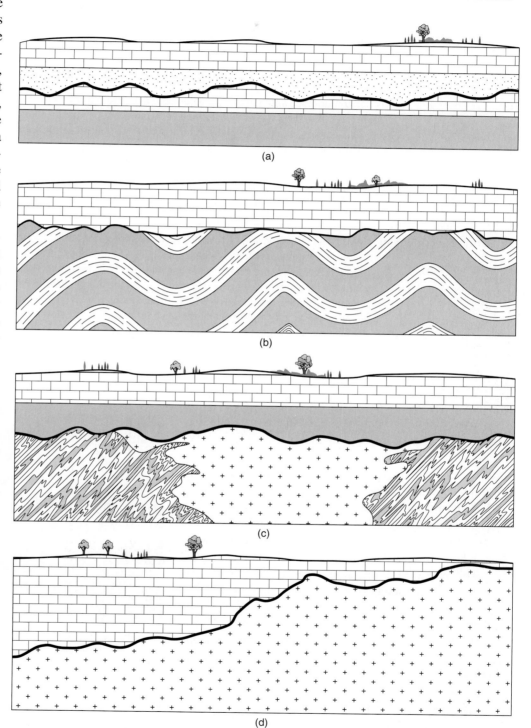

FIGURE 2.9 The principal types of unconformities: (a) disconformity, (b) angular unconformity, (c) nonconformity, (d) buttress unconformity.

TABLE 2.4	TYPES OF UNCONFORMITIES
Disconformity	At a disconformity, beds of the rock sequence above and below the unconformity are parallel to one another, but there is a measurable age difference between the two sequences. The disconformity surface represents a period of nondeposition and/or erosion (Figure 2.9a).
Angular unconformity	At an angular unconformity, strata below the unconformity have a different attitude than strata above the unconformity. Beds below the unconformity are truncated at the unconformity, while beds above the unconformity roughly parallel the unconformity surface. Therefore, if the unconformity is tilted, the overlying strata are tilted by the same amount. Because of the angular discordance at angular unconformities, they are quite easy to recognize in the field. Their occurrence means that the sub-unconformity strata were deformed (tilted or folded) and then were truncated by erosion prior to deposition of the rocks above the unconformity. Therefore, angular unconformities are indicative of a period of active tectonism. If the beds below the unconformity are folded, then the angle of discordance between the super- and sub-unconformity strata will change with location, and there may be outcrops at which the two sequences are coincidentally parallel (Figure 2.9b).
Nonconformity	Nonconformity is used for unconformities at which strata were deposited on a basement of older crystalline rocks. The crystalline rocks may be either plutonic or metamorphic. For example, the unconformity between Cambrian strata and Precambrian basement in the Grand Canyon is a nonconformity (Figure 2.9c).
Buttress unconformity	A buttress unconformity (also called *onlap unconformity*) occurs where beds of the younger sequence were deposited in a region of significant predepositional topography. Imagine a shallow sea in which there are islands composed of older bedrock. When sedimentation occurs in this sea, the new horizontal layers of strata terminate at the margins of the island. Eventually, as the sea rises, the islands are buried by sediment. But along the margins of the island, the sedimentary layers appear to be truncated by the unconformity. Rocks below the unconformity may or may not parallel the unconformity, depending on the pre-unconformity structure. Note that a buttress unconformity differs from an angular unconformity in that the younger layers are truncated at the unconformity surface (Figure 2.9d).

Proterozoic and Archean gneisses. In Figure 2.10 the classic unconformity between Paleozoic sedimentary rocks and Precambrian gneisses is shown and many introductory geology books show this contact in the Grand Canyon.

It is a special experience to put your finger on a major unconformity and to think about how much of Earth's history is missing at the contact. Imagine how James Hutton felt when, in the late eighteenth century, he stood at Siccar Point along the coast of Scotland (Figure 2.11), and stared at the Caledonian unconformity between shallowly dipping Devonian Red Sandstone and vertically dipping Silurian strata and, as the present-day waves lapped on and off the outcrop and deposited new sand, suddenly realized what the contact meant. His discovery is one of the most fundamental in field geology.

How do you recognize an unconformity (Figure 2.12) in the field today? Well, if it is an angular unconformity or a buttress unconformity, there is an angular discordance between bedding above and below

FIGURE 2.10 Unconformable contact between mid-Proterozoic Grenville gneiss (dark gray) and Cambrian sandstone and Pleistocene soils (southern Ontario, Canada).

the unconformity. A nonconformity is obvious, because crystalline rocks occur below the contact. Disconformities, however, can be more of a challenge to recognize. If strata in the sequence are fossiliferous, and you can recognize the fossil species and know their age, then you can recognize a gap in the fossil succession. Commonly, an unconformity may be marked by a surface of erosion, as indicated by scour features, or by a **paleosol,** which is a soil horizon that formed from weathering prior to deposition of the overlying sequence. Some unconformities are marked by the occurrence of a **basal conglomerate,** which contains clasts of the rocks under the unconformity. Recognition of a basal conglomerate is also helpful in determining whether the contact between strata and a plutonic rock is intrusive or whether it represents a nonconformity.

2.2.6 Compaction and Diagenetic Structures

When a clastic sediment initially settles, it is a mixture primarily of grains and water. The proportion of solid to fluid varies depending on the type of sediment. Gooey mud, which consists of clay and water, contains more water than well-packed sand. Progressive burial of sediment squeezes the water out, and the sediment compacts. **Compaction** results in a decrease in porosity (>50% in shale and >20% in sand) that results in an increase in the density of the sediment.

Lateral variation in the amount of compaction within a given layer, or contrasts in the amount of compaction in a vertical section, is a phenomenon called **differential compaction.** Differential compaction within a layer can lead to lateral variation in thickness that is called **pinch-and-swell structure.** Pinch-and-swell structure can also form as a consequence of tectonic stretching, so again, you must be careful when you see the structure to determine whether it is a depositional structure or a tectonic structure.

FIGURE 2.11 Angular unconformity in the Caledonides at Siccar Point (Scotland). The hammerhead rests on the unconformity, which is tilted due to later deformation.

The compaction of mud leads to development of a preferred orientation of clay in the resulting mudstone. Clay occurs in tiny flakes shaped like playing cards. In a wet sediment the flakes are not all parallel to one another, as in a standing house of cards, but after compaction the flakes are essentially parallel to one another, as in a collapsed house of cards. The preferred orientation of clay flakes, as we have seen, leads to bedding plane fissility that produces a **shale.** For example, in the Gulf Coast sequence of the southern United States this progression is preserved in drill cores that were obtained to study the relationship between oil maturation and clay mineralogy. In contrast, the compaction of sand composed of equant grains causes the grains to pack together more tightly, but produces little rock fabric.

Deeper compaction can cause **pressure solution,** a process by which soluble grains preferentially dissolve along the faces at which stress is the greatest. In pure limestones or sandstones, this process causes grains to suture together, meaning that grain surfaces interlock with

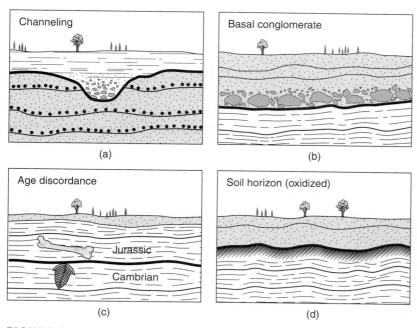

FIGURE 2.12 Some features used to identify unconformities: (a) scour channels in sediments, (b) basal conglomerate, (c) age discordance from fossil evidence, and (d) soil horizon or paleosol.

one another like jigsaw puzzle pieces. In conglomerates, the squeezing together of pebbles results in the formation of indentations on the pebble surfaces creating **pitted pebbles** (Figure 2.13). In limestones and sandstones that contain some clay, the clay enhances the pressure solution process. Specifically, pressure solution occurs faster where the initial clay concentration is higher. As a result, distinct seams of clay residue develop in the rock. These seams are called **stylolites** (Figure 2.14). In rocks with little clay (<10%), stylolites are jagged and tooth-like in cross section, like the sutures in your skull. The teeth are caused by the distribution of grains of different solubility along the stylolite. In rocks with more clay, the stylolites are wavy and the teeth are less pronounced, because the clay seams become thicker than tooth amplitude. Some of the dissolved ions that are removed at pressure-solved surfaces precipitate locally in the rock in veins or as cement in pore spaces, whereas some get transported out of the rock by moving groundwater. The proportion of reprecipitated to transported ions is highly variable, but as much as 40% of the rock can be dissolved and removed during formation of stylolites.

Some sedimentary rocks exhibit color banding that cuts across bedding. This color banding, which is called **Liesegang banding,** is the result of diffusion of impurities, or of reactions leading to alternating bands of oxidized and reduced iron. Because it can be mistaken for bedding or cross bedding, it is mentioned in the context of primary sedimentary structures. To avoid mistaken identity, search the outcrop to determine whether sets of bands cross each other (possible for Liesegang bands, but impossible for bedding), and whether the bands are disrupted at fractures or true bedding planes, because these are places where the diffusion rate changes.

2.2.7 Penecontemporaneous Structures

If sediment layers have an initial dip, meaning a gentle slope caused by deposition on a preexisting slope or tilting prior to full lithification in a tectonically active region, gravity can pull the layers down the slope. The ease with which sediments move down a slope is increased by fluid pressure in the layers, which effectively keeps the layers apart. Movement is resisted by weak electrostatic adhesion between grains, but this resistance can be overcome by the energy of an earthquake or a storm, and the sediment will move down the slope. If the sediment completely mixes with water and becomes a turbid suspension flowing into deeper water, then all of the preexisting primary structure is lost and the grains are resedimented as a new graded bed (turbidite; see Figure 2.2) farther down the slope. If the flowing mixture of sediment and water is dominantly sediment, it churns into a slurry containing chunks and clasts that are suspended in a matrix. Such slurries are called **debris flows,** and where preserved in a stratigraphic sequence, they become matrix-supported, poorly sorted conglomerates containing a range of clast sizes and shapes.

If the beds were lithified sufficiently prior to movement, so that they maintain cohesion, then the movement is called **slumping.** During slumping, the sedimentary layers tend to be folded and pulled apart and are thrust over one another. The folds and faults formed during this slumping are called **penecontemporaneous structures,** because they formed almost (Greek prefix *pene*) at the same time as the original deposition of the layers. Penecontemporaneous folds and faults are characteristically chaotic. The folds display little symmetry, and folds in one layer are of a different size and orien-

FIGURE 2.13 Pitted pebble; coin for scale.

FIGURE 2.14 Suture-like stylolites in limestone. Pocket knife for scale.

tation than the structures in adjacent layers. Penecontemporaneous faults are not associated with pronounced zones of brittle fracturing (we turn to the characteristics of brittle behavior in Chapter 6). One key to the recognition of slump structures in a sedimentary sequence is that the deformed interval is *intraformational*, meaning that it is bounded both above and below by relatively undeformed strata (Figure 2.15). Commonly, intervals of penecomtemporaneous structures occur in a sequence that also includes debris flows and turbidites, all indicative of an unstable depositional environment. While slump structures can be mistaken for local folding adjacent to a tectonic detachment fault, the opposite, tectonic folds mistakenly interpreted as slump structures, may also occur. Not a simple matter to distinguish between the two!

We tend to think of debris flows and landslides as being relatively small structures, capable of disrupting a hillslope and perhaps moving a cottage or two, but generally not much more. However, the geologic record shows that catastrophic land-slides of enormous dimension have occurred on occasion. In northern Wyoming, for example, a giant Eocene slide in association with volcanic eruption displaced dozens of mountain-sized blocks and hundreds of smaller blocks. One such large block, Heart Mountain, moved intact for several tens of kilometers, apparently riding on a cushion of compressed air above a nearly planar subhorizontal (detachment) fault (Figure 2.16).

FIGURE 2.15 Penecontemporaneous folds in the Maranosa Arenaci (Italian Apennines).

FIGURE 2.16 Heart Mountain detachment; with Paleozoic carbonates on Eocene deposits (Wyoming).

2.3 SALT STRUCTURES

Salt is a sedimentary rock that forms by the precipitation of evaporite minerals (typically halite [NaCl] and gypsum or anhydrite calcium sulfates) from saline water. Salt deposits accumulate in any **sedimentary basin,** meaning a low region that is the site of deposition, where saline water, such as seawater, evaporates sufficiently for salt to precipitate. Particularly thick salt deposits lie at the base of **passive-margin basins,** so named because they occur along tectonically inactive edges of continents. To see how these basins form, imagine a supercontinent that is being pulled apart, like Pangea in the early Mesozoic. This process, called **rifting,** involves brittle and ductile faulting, the net result of which is to thin the continental lithosphere until it breaks and an oceanic ridge is formed. During the early stages of rifting, the rift basin is dry or contains freshwater lakes. Eventually, the floor of the rift drops below sea level and a shallow sea forms. If evaporation rates are high, various salts (typically, halite and gypsum/anhydrite) precipitate out of the seawater and are deposited on the floor of the rift. When the rift evolves into an open ocean, the continental margins become passive margins that gradually subside. With continued subsidence, the layer of evaporite (salt) is buried by clastic sediments and carbonates typical of continental-shelf environments. We'll discuss this tectonic environment in more detail later (Extension Tectonics, Chapter 16), but for now we leave you a picture of a thick pile of sediment with a layer of salt near its base. This is the starting condition for the formation of **salt intrusions.**

Salt differs from other sedimentary rocks in that it is much weaker and, as a consequence, is able to flow like a viscous fluid under conditions in which other sedimentary rocks behave in a brittle fashion.[3] In some cases, deformation of salt is due to tectonic faulting or folding, but because salt is so weak, it may deform solely in response to gravity, and thereby cause deformation of surrounding sedimentary rock. If gravity is the only reason for salt movement, the deformation resulting from its movement is called **halokinesis** (combining the Greek words for salt and movement, respectively) and the resulting body of salt is called a **salt structure.**

2.3.1 Why Halokinesis Occurs

Halokinesis begins in response to three factors: (1) the development of a density inversion, (2) differential loading, and (3) the existence of a slope at the base of a salt layer. All three of these factors occur in a passive-margin basin setting. Salt is a nonporous and essentially incompressible material. So when it gets buried deeply in a sedimentary pile, it does not become denser. In fact, salt actually gets less dense with depth, because at greater depths it becomes warmer and expands. Other sedimentary rocks (like sandstone and shale), in contrast, form from sediments that originally had high porosity and thus become denser with depth because the pressure caused by overburden makes them compact. This contrast in behavior, in which the density of other sedimentary rocks exceeds the density of salt at depths greater than about 6 km, results in a **density inversion,** meaning a situation where denser rock lies over less dense rock. Salt density is about $2200 \ kg/m^3$, whereas the density of the sedimentary rocks is about $2500 \ kg/m^3$. A density inversion is an unstable condition because the salt has positive buoyancy. **Positive buoyancy** means that forces in a gravity field cause lower density materials to try to rise above higher density materials, thereby decreasing the overall gravitational potential energy of the system. **Negative buoyancy,** the reverse, is a force that causes a denser material to sink through a less dense material (see Section 2.2.4). A familiar example of positive buoyancy forces is the push that your hand feels when you try to hold an air-filled balloon under water. Holding a brick under water illustrates negative buoyancy. When the positive buoyancy force exceeds the strength of the salt and is sufficient to upwarp strata that lie over the salt structure, then it will contribute to the formation of the salt structure.

Differential loading of a salt layer takes place when the downward force on the salt layer caused by the weight of overlying strata varies laterally. This may occur where there are primary variations in the thickness or composition of the overlying strata, primary variations in the original surface topography of the salt layer, or changes in the thickness of the overlying strata due to faulting. Regardless of its cause, differential loading creates a situation in which some parts of the salt layer are subjected to a greater vertical load than other parts, and the salt is squeezed from areas of higher pressure to areas of lower pressure. For example, imagine a layer of salt whose upper surface initially bulges upward to form a small "dome." The weight of a column of rock and water from sea level down to a horizontal surface in the salt layer on either

---

[3]In fact, dry salt moves crystal plastically, whereas movement of damp salt involves pressure solution. We discuss these deformation mechanisms in Chapter 9.

side of the dome is greater than the weight of the column on the top of the dome, because salt is less dense than other sedimentary rocks. As a consequence, salt is squeezed into the dome, making it grow upwards. A salt layer that has provided salt for the production of a salt structure, and thus has itself been changed by halokinesis, is called the **source layer.**

The combination of differential loading and buoyancy force drives salt upward through the overlying strata until it reaches a level of **neutral buoyancy,** meaning the depth at which it is no longer buoyant. At this level, salt has the same density as surrounding strata. The density of mildly compacted clastic strata equals that of salt at depths of around 500–1500 m below the surface of the basin, depending on composition. At the level of neutral buoyancy, salt may begin to flow laterally, much like a thick pile of maple syrup flows laterally over your pancakes. This process, which is also driven by gravity (above the level of neutral buoyancy, the salt is subjected to a negative buoyancy force), is known as **gravity spreading.** Where salt is extruded at the land surface, it becomes a salt glacier (Figure 2.17). At the seafloor, salt also spreads like a salt glacier, except that during movement it continues to be buried by new sediment.

2.3.2 Geometry of Salt Structures and Associated Processes

In response to positive buoyancy force and to differential loading, salt will flow upward from the source bed, which thins as a consequence. If the source bed thins to the point of disappearing and the strata above the source bed and below the source bed become juxtaposed, we say that the contact between these two beds is a **primary weld.** In general, a weld is any contact between strata that were once separated by salt. At any given time, a region may contain salt structures at many stages of this development. Geologists working with salt structures have assigned a rich vocabulary to these structures based on their geometry; some are described in Table 2.5 and shown in Figure 2.18. The name assigned to a specific structure depends on its shape today, but in the context of geologic time, this shape may be only temporary.

Because salt both rises up into preexisting strata and rises during the time of deposition of overlying strata, geologists distinguish between two types of salt structure growth. If the salt rises after the overlying strata have already been deposited, then the rising salt will warp and eventually break through the overlying

FIGURE 2.17 A salt glacier that originates from a salt dome (western Iran).

TABLE 2.5	TERMINOLOGY OF SALT STRUCTURES
Detached bulb	Stems or walls connecting salt stocks to the source bed may pinch out, so that salt that was originally separated by the stem or wall becomes juxtaposed. Flow of the salt into bulbs results in folds that wrap around themselves and thus have circular profiles.
Detached canopy	Stems or walls connecting salt canopies to the source bed may pinch out, so that salt that was originally separated by the stem or wall becomes juxtaposed.
Salt anticline	An upward salt bulge relative to the source layer that is elongate in plan view. Strata overlying these structures conformably warp around the structure.
Salt canopy	In regions where the source bed was quite thick so that many diapirs form, salt walls or salt stocks may spread out and merge at a higher stratigraphic level. Salt canopies may flow dominantly in one direction in response to gravity, and if so are sometimes called *salt-tongue canopies*.
Salt diapir	A salt structure that pierces bedding in overlying strata.
Salt dome	An upward salt bulge relative to the source layer that is roughly symmetric in plan view. Strata overlying these structures conformably warp around the structure.
Salt glacier	Salt that flows out over the land when the salt diapir pierces ground surface (Figure 2.17).
Salt pillow	Same as *salt dome*.
Salt roller	An asymmetric salt dome that resembles an ocean wave.
Salt stock	Equant salt diapir in plan view. Mature salt diapirs are generally narrower at depth and broader at the top; the lower part is called the *stem* and the upper part is called the *bulb*.
Salt wall	Elongate salt diapir in plan view.

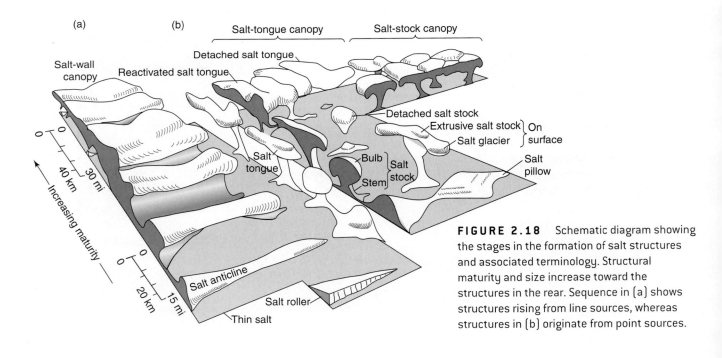

FIGURE 2.18 Schematic diagram showing the stages in the formation of salt structures and associated terminology. Structural maturity and size increase toward the structures in the rear. Sequence in (a) shows structures rising from line sources, whereas structures in (b) originate from point sources.

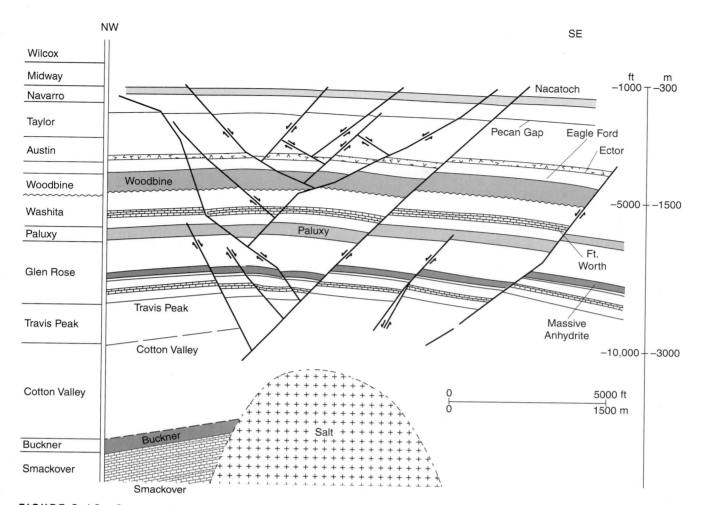

FIGURE 2.19 Cross section illustrating a normal fault array over the top of a salt dome in Texas.

strata. This process is called **upbuilding.** If, however, the rise of the salt relative to the source layer occurs coevally with further deposition, the distance between the source layer and the surface of the basin also increases. This process is called **downbuilding.** As salt moves, it deforms adjacent strata and creates complex folds and local faults. When salt diapirs approach the surface, the overlying strata are arched up and therefore are locally stretched, resulting in the development of normal faults in a complex array over the crest of the salt structure (Figure 2.19).

The structural geometry of passive-margin basins is complicated because sedimentation continues during salt movement. Sedimentary layers thicken and thin as a consequence of highs and lows in elevations caused by the salt, and the resulting differential compaction causes further salt movement. The sedimentation pattern may change in a locality when a salt structure drains out and flows into a structure at another locality. Thus, in regions where halokinesis occurs, it is common to find places where an arch evolves into a basin, or vice versa, a process we call **inversion.**

2.3.3 Gravity-Driven Faulting and Folding

The formation of salt structures is a dynamic process that is intimately linked to faulting in the overlying strata. Salt is so weak that it makes a good glide horizon on which detachment and movement of overlying strata occurs. In fact, on many passive margins, a thick package of sedimentary rocks tends to detach and slump seaward, gliding on a detachment fault in the layer of salt at its base. This movement resembles the slumping of sediment of a hillslope, though the scale of displacement it quite different. As slumping occurs, the landward portion of the basin is stretched and is therefore broken by a series of normal faults whose dip tends to decrease with depth. This change in dip with depth makes the faults concave up, which are called **listric faults.** As movement occurs on a listric fault, the strata above the fault arch into **rollover folds** (Figure 2.20). Many listric normal faults intersect the ground surface in southern Texas, because this region is part of the passive-margin basin along the Gulf

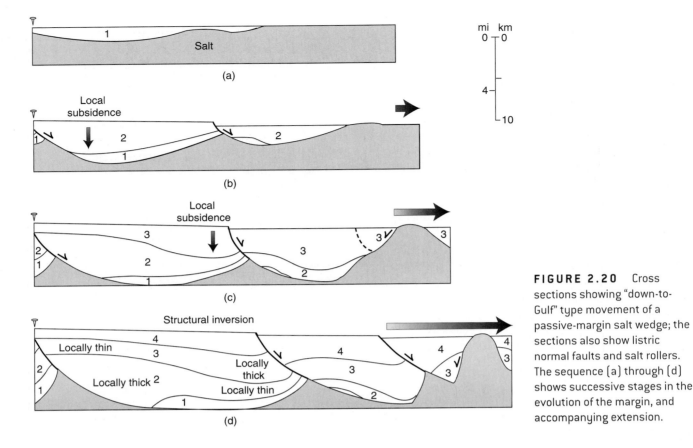

FIGURE 2.20 Cross sections showing "down-to-Gulf" type movement of a passive-margin salt wedge; the sections also show listric normal faults and salt rollers. The sequence (a) through (d) shows successive stages in the evolution of the margin, and accompanying extension.

Coast. Because the faults dip south and they transport rock toward the Gulf of Mexico, they are sometimes called **down-to-Gulf faults.** Slip on these faults thins the stratigraphic section above the salt layer and thus results in differential loading. As a result, the salt can rise beneath the fault, evolving from a salt dome to a salt roller, to a diapir that eventually cuts the overlying fault. At the toe of the passive margin wedge, a series of thrust faults develop to accommodate displacement of the seaward-moving section, just as thrusts develop at the toe of a hillside slump.

2.3.4 Practical Importance of Salt Structures

Why spend so much time dealing with salt structures and passive margins? Simply because these regions are of great economic and societal importance. Passive-margin basins are major oil reservoirs, and much of the oil in these reservoirs is trapped adjacent to salt bodies. Oil rises in the upturned layers along the margins of the salt body and is trapped against the margin of the impermeable salt. In recent years, salt bodies in passive-margin basins are being used as giant storage tanks for gas or oil, and are being considered as poten-

tial sites for the storage of nuclear waste. This is one of many examples where structural geology is central to reaching important societal decisions.

2.4   IGNEOUS STRUCTURES

You may recall from your introductory geology course that there are two principal classes of igneous rocks, and that these classes are distinguished from one another based on the environment in which the melt cools. **Extrusive rocks** are formed either from lava that flowed over the surface of the Earth and cooled under air or water, or from ash that exploded out of a volcanic vent. **Intrusive rocks** cooled beneath the surface of the Earth. During the process of intruding, flowing, settling and/or cooling, igneous rocks can develop primary structures. In this context, we use the term "primary structure" to refer to a fabric that is a consequence of igneous processes.

Where do **magmas,** the melt phase of igneous rocks, come from? If you have not had a course in igneous petrology, we'll quickly outline the nature of magmatic activity. Magma forms where conditions of heat and pressure cause existing rock (either in the crust or in the mantle) to melt. Commonly, only certain minerals within the solid rock melt (the ones that melt at a lower temperature), in which case we say that the rock has undergone **partial melting.** A magma formed by partial melting has a composition that differs from that of the rock from which it was extracted. For example, a 1–6% partial melt of ultramafic rock (peridotite) in the mantle yields a mafic magma which, when solidified, forms the gabbro and basalt that characterizes oceanic crust. Melting of an intermediate-composition crustal rock (diorite) yields a silicic magma which, when solidified, forms granite or rhyolite. Once formed, magma is less dense than the surrounding rock, and buoyancy forces cause it to rise. The density decrease is a consequence of the expansion that accompanies heating and melting, the formation of gas bubbles within the magma, and the difference in composition between magma and surrounding rock.

Magma moves by oozing up through a network of cracks and creeping along grain surfaces. The difference between the pressure within the magma and the pressure in the surrounding rock is so substantial that, as magma enters the brittle crust, it can force open new cracks. Magma continues to rise until it reaches a level of neutral buoyancy, defined as the depth where pressure in the magma equals lithostatic pressure in the surrounding rock, meaning that the buoyancy force is zero. At the level of neutral buoyancy, the magma may form a **sheet intrusion,** or may pool in a large **magma chamber** that solidifies into a bloblike intrusion called a **pluton.** We describe common types of igneous intrusions in Table 2.6. If the magma pressure is sufficiently high, the magma rises all the way to the surface of the Earth, like water in an artesian well, and is extruded at a volcano.

2.4.1 Structures Associated with Sheet Intrusions

One important aspect of sheet intrusions that is of interest to structural geologists is their relationship to stress. In Chapter 3, we will introduce the concept of stress in detail, but for now, we point out that stress acting on a plane is defined as the force per unit area of the plane. Intuitively, therefore, you can picture that the Earth's crust is held together least tightly in the direction of the smallest stress, which we call the direction of least principal stress. Sheet intrusions, in general, form perpendicular to the direction of the least principal stress, assuming no preexisting planes of weakness (such as faults). For example, in regions

TABLE 2.6	TERMINOLOGY OF IGNEOUS INTRUSIONS
Batholith	A huge bloblike intrusion; usually a composite of many plutons.
Dike	A sheet intrusion that cross cuts stratification in a stratified sequence, or is roughly vertical in an unstratified sequence.
Hypabyssal	An intrusion formed in the upper few km of the Earth's crust; hypabyssal intrusions cool relatively quickly, and thus are generally fine grained.
Laccolith	A hypabyssal intrusion that is concordant with strata at its base, but bows up overlying strata into a dome or arch.
Pluton	A moderate-sized bloblike intrusion (several km in diameter). Sometimes the term is used in a general sense to refer to any intrusion, regardless of shape or size.
Sill	A sheet intrusion that parallels preexisting stratification in a stratified sequence, or is roughly subhorizontal in an unstratified sequence.
Stock	A small, bloblike intrusion (a few km in diameter).

where the greatest stress is caused by the weight of the overlying rocks, and is therefore vertical, the least principal stress is horizontal, so vertical dikes form. **Dike swarms,** which are arrays of subparallel dikes occurring over broad regions of the crust, probably represent intrusion at depth in association with horizontal extension, which causes the horizontal stress to be tensile.

Not all dikes occur in parallel arrays. In the immediate vicinity of volcanoes, the ballooning and/or collapse of a magma chamber locally modifies the stress field and causes a complex pattern of fractures. As a result, the pattern of dikes around a volcano (Figure 2.21) includes **ring dikes,** which have a circular trace in map view, and **radial dikes,** which run outward from the center of the volcano like spokes of a wheel. At a distance from the volcano, where the local effects of the volcano on the stress field are less, radial dikes may change trend to become perpendicular to the regional least principal stress.

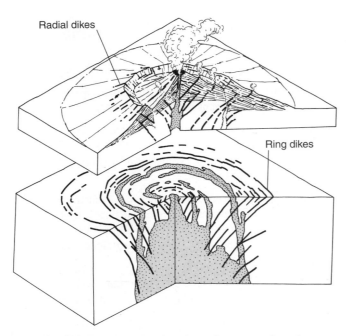

FIGURE 2.21 Types of sheet intrusions around a volcano.

Sills, or subhorizontal sheetlike intrusions, form where local stress conditions cause the least principal stress to be vertical, and/or where there are particularly weak horizontal partings in a stratified sequence. Sill intrusion can result in the development of faults. If the thickness of the intrusion changes along strike, there is differential movement of strata above the intrusion. **Laccoliths** resemble sills in that they are concordant with strata at their base, but unlike sills, they bow up the overlying strata to create a dome. For example, the laccoliths of the Henry Mountains in Utah are several kilometers in diameter.

Sheet intrusions occur in all sizes, from thin seams measured in centimeters, to the Great Dike of Zimbabwe, which is nearly 500 km long and several kilometers wide. Considering the dimensions of large intrusions (tens to hundreds of kilometers long), it is important to keep in mind that a large volume of magma can flow past a given point. The occurrence of flow may be recorded as primary igneous structures in the rock. For example, examination of dike-related structures may show the presence of drag folds, scour marks, imbricated phenocrysts, and flow foliation, particularly along the walls of an intrusion.

2.4.2 Structures Associated with Plutons

The nature of primary structures found in plutonic rocks depends on the depth in the Earth at which the intrusion solidified, because these structures reflect the temperature contrast between the intrusion and the country rock.[4] Remember that the Earth gets warmer with depth: at the surface, the average temperature is ~10°C, whereas at the center it may be as much as 4000°C. The change in temperature with depth is called the **geothermal gradient.** In the shallow crust, the geothermal gradient is in the range of 20°C/km to 40°C/km. At greater depths, however, the gradient must be less, because temperatures at the continental **Moho** (at about 40 km depth) are in the range of about 700°C (~15°C/km), and temperatures at the base of the lithosphere (at about 150 km) are in the range of about 1280°C (<10°C/km). The origin of the increased geothermal gradient near the surface is the concentration of radioactive elements in the minerals of more silicic rocks. Granitic magma begins to solidify at temperatures between 550°C to 800°C. Therefore, the temperature contrast between magma and country rock decreases with depth. In the case of shallow-level plutons, which intrude at depths of less than about 5 km, the contrast between magma and country rock is several hundred degrees. Contacts at the margins of such shallow intrusions are sharp, so that you can easily place your finger on the contact. Angular blocks of country rock float in the magma near the contact, and the country rock adjacent to the contact may be altered by fluids expelled by the magma or may be baked (a rock type called hornfels). At greater depth, the temperature contrast between magma and country rock decreases, and contacts are more gradational, until the country rock itself is likely to be undergoing partial melting. Minerals that melt at lower temperatures (like quartz and feldspar) turn to liquid, while refractory minerals (that is, minerals that melt at higher temperatures, such as amphibole and pyroxene) are still solid, though quite soft. Movement of the melt causes the soft solid layers to be contorted into irregular folds. When this mass eventually cools, the resulting rock, which is composed of a marble-cake-like mixture of light and dark contorted bands, is called a **migmatite** (Figure 2.22).

Many plutons exhibit an **intrusion foliation** that is particularly well developed near the margin of the pluton and is subparallel to pluton–host rock contact. This foliation is defined by alignment of **inequant** crystals

---

[4]To a geologist, "country rock" is not only a type of music, but also a casual term for the rock that was in a locality before the intrusion.

FIGURE 2.22 A migmatite from the North Cascades (Washington State, USA) showing complex folding and disruption.

and by elongation of chunks of **country rock** or early phases of pluton that were incorporated in the magma (called **xenoliths,** which means "foreign rocks"). Such fabric is a consequence of shear of the magma against the walls of the magma chamber and of the flattening of partially solidified magma along the chamber walls in response to pressure exerted as new magma pushes into the interior of the chamber. Similarly, intrusion foliation can be developed along the margins of dikes.

Because intrusion foliation forms during the formation of the rock and is not a consequence of tectonic movements, it is a nontectonic structure. However, it may be difficult to distinguish from schistosity resulting from tectonic forces. Plutons tend to act as mechanically strong blocks, so that regional deformation is deflected and concentrated along the margins of the pluton. Interpretation of a particular foliation therefore depends on regional analysis and the study of deformation microstructures (Chapter 9). For example, if the fabric remains parallel to the boundary of the intrusion, even when the boundary changes and individual grains show no evidence for solid-state deformation, the foliation is likely a primary igneous structure. The distinction between tectonic and primary structures in plutons has proven to be quite difficult and more often than not is ambiguous.

2.4.3 Structures Associated with Extrusion

As basaltic lava flows along the surface of the Earth, the surface of the flow may wrinkle into primary folds that resemble coils of rope, or may break into a jumble of jagged blocks that resembles a breccia. Lava flows with the rope-coil surface are called **Pahoehoe flows,** and lavas with the broken-block surface are called **Aa flows.** The wrinkles in a Pahoehoe lava should not be mistaken for tectonic folds, and the jumble of blocks caused by **autobrecciation** (that is, breaking up during flow) of Aa lavas should not be mistaken for a tectonic breccia related to faulting.

If basaltic lava is extruded beneath seawater, the surface of the flow cools quickly, and a glassy skin coats the surface of the flow. Eventually, the pressure in the glass-encased flow becomes so great that the skin punctures, and a squirt of lava pushes through the hole and then quickly freezes. The process repeats frequently, resulting in a flow composed of blobs (centimeters to meters in diameter) of lava. Each blob, which is called a pillow, is coated by a rind of fine-grained to glassy material. As the pillows build out into a large pile, creating a **pillow basalt,** successive pillows flow over earlier pillows and, while still soft, conform to the shape of the earlier flow surface (Figure 2.23). As a result, pillows commonly have a rounded top and a pointed bottom (the "apex") in cross section, and this shape can be used as a stratigraphic facing criterion.

In 1902, Mt. Pelee on the Caribbean island of Martinique erupted. It was a special kind of eruption, for instead of lava flows, a spine of rhyolite rose day by day from the peak of the volcano. This spine, as it turned out, was like the cork of a champagne bottle slowly being worked out. When the cork finally pops out of the champagne bottle, a froth of gas and liquid flows down the side of the bottle. Likewise, when, on the morning of May 2, the plug exploded off the top of Mt. Pelee, a froth of hot (>800°C) volcanic gas and ash floated on a cushion of air and rushed down the side of the mountain at speeds of up to 100 km/h. This ash flow engulfed the town of St. Pierre, and in an instant, almost 30,000 people were dead. When the ash stopped moving, it settled into a hot layer that welded together. Such a layer of welded tuff is called an **ignimbrite,** often displaying a foliation. Volcanic ash is composed of tiny glass shards with jagged spinelike forms that are a consequence of very rapid cooling. When the ash settles, the glass shards are still hot and soft, so the compaction pressure exerted by the weight of overlying ash causes the shards to flatten, thereby creating a primary foliation in ignimbrite that is comparable to bedding.

Rhyolitic lavas commonly display subtle color banding, called **flow foliation,** that has been attributed to flow of the lava before complete solidification. The banding forms because lavas are not perfectly homogeneous materials. Since the temperature is not perfectly uniform, there may be zones in which crystals have formed, while adjacent regions are still molten. Shear resulting from movement of the lava smears out these initial inhomogeneities into subparallel bands. To visualize this, think of a bowl of pancake batter into which you have dripped spoonfuls of chocolate batter. If you slowly stir the mixture, the blobs of chocolate smear out into sheets. Chocolate blobs that were initially nearby would smear into parallel sheets with an intervening band of pancake batter. In the flow, movement of the lava smears out blobs of contrasting texture into layers, which, when the rock finally freezes, have a slightly different texture than adjacent bands and thus are visible markers in outcrop. Commonly, continued movement causes previously formed layers to fold, so flow-banded outcrops typically display complex primary folds.

FIGURE 2.23   Pillow basalt from the Point Sal Ophiolite, California (USA). The asymmetric shape of the pillows and location of the "points" (apex) indicate that the stratigraphic top of the flow is up.

2.4.4 Cooling Fractures

As shallow intrusions and extrusive flows cool, they contract. Because of their fine grain size, these bodies are susceptible to forming natural cracks, or **joints,** in response to the thermal stress associated with cooling. When such joints are typically arranged in roughly hexagonal arrays that isolate columns of rock, the pattern is called **columnar jointing** (Figure 2.24). Popular tourist and movie director destinations like Devil's Tower in Wyoming, Giant's Causeway in Ireland, or the Massif Central in southern France offer spectacular examples. The long axes of columns are perpendicular to isotherms (surfaces of constant temperature) and thus they are typically perpendicular to the boundaries of the shallow intrusion or flow. If you look closely at unweathered columnar joints, the surfaces of individual joints are ribbed. We will learn later that this feature is a consequence of the way in which fractures propagate through rock (Chapter 6).

2.5 IMPACT STRUCTURES

Glancing at the Moon through a telescope, the most obvious landforms that you see are craters. Like early Earth, the Moon has been struck countless times by meteors, and each impact has left a scar which, because the Moon is tectonically inactive and has no atmosphere or water, has remained largely unchanged through succeeding eons. The Earth has been pummeled at least as frequently as the Moon, but many objects disintegrate and burn in the atmosphere before reaching the surface, and the scars of many that did strike the surface have been erased by erosion and particularly by tectonics. The vast majority of impacts on the Earth-Moon system occurred prior to about 3.9 Ga, when the solar system contained a multitude of fragments that were not yet incorporated into planets. Considering that 70% of today's Earth's surface is underlain by oceanic lithosphere, most of which is less than 200 million years old, and that all but a relatively small portion of the continental crust has either been covered by younger strata or has been involved in plate tectonics, it is not surprising that impact structures are so rare on Earth.

But even though **impact structures** are rare, and in some cases difficult to recognize, they do exist. For our discussion we distinguish three categories, based on the most obvious characteristic of the impact: (1) relatively recent surficial impacts that are defined by a visible crater, (2) impacts whose record at the present Earth surface is the disruption of sedimentary strata, and (3) impacts whose record is a distinctive map-view circular structure in basement.

(a)

(b)

FIGURE 2.24 Columnar jointing in the Massif Central (France); (a) side view, (b) top view.

Today there are about 150 impacts recognized on our planet but only a few dozen obvious impact craters can be seen. One of the largest and perhaps most famous is the approximately 50,000-year-old Barringer Meteor Crater in Arizona, which is 1.2 km in diameter and 180 m deep, and is surrounded by a raised rim that is about 50 m high (Figure 2.25a). The size of the impacting object is estimated only to be 30–50 m! As shown in the cross section (Figure 2.25b), the impact created a breccia that is about 200 m thick beneath the floor of the crater. The raised rims of the crater are not only composed of shattered rock ejected from the crater, but are also sites where bedrock has been upturned. Ancient impact sites that are no longer associated with a surficial crater dot the Midcontinent region of the United States. These sites are defined by relatively small (less than a few kilometers across) semicircular disruption zones, in which the generally flat-lying Paleozoic strata of the region are fractured, faulted, and tilted. They were originally called **crypto-volcanic structures** (from the Greek *crypto,* meaning "hidden"), because it was assumed that they were the result of underlying explosive volcanism. Typically, steeply dipping normal faults, whose map traces are roughly circular, define the outer limit of these structures. These faults are cross cut by other steep faults

(a)

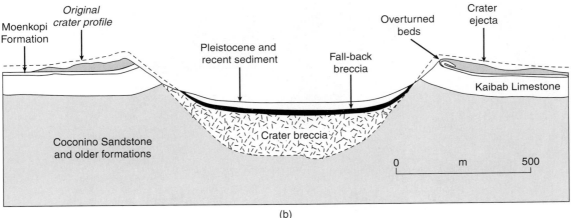

(b)

FIGURE 2.25 (a) Barringer Meteor Crater of Arizona (USA) and (b) geologic cross section showing distribution of impact structures.

that radiate from the center of the structure like spokes of a wagon wheel. This fracture geometry is similar to that around volcanoes (Figure 2.21), which appeared to support the volcanic interpretation. Near the center of the structure, bedding is steeply dipping, and faulting juxtaposes units of many different ages. Locally, the strata are broken into huge blocks that jumbled together to create an **impact breccia.** Throughout this region, rocks are broken into distinctive **shatter cones,** which are conelike arrays of fractures similar to those found next to a blast hole in rock (Figure 2.26). The apex of the cone points in the direction from which the impacting object came. In impact structures, shatter cones point up, confirming that they were caused by impact from above, as would be the case if the structure was due to an incoming meteor.

Why do impact structures have the geometry that they do? To see why, think of what happens when you drop a pebble into water. Initially, the pebble pushes down the surface of the water and creates a depression, but an instant later, the water rushes in to fill the depression, and the place that had been the center of the depression rises into a dome. In the case of meteor impact against rock, the same process takes place. The initial impact gouges out a huge crater and elastically compresses the rock around the crater. But an instant later, the rock rebounds. At the margins of the affected zone it pulls away from the walls, creating normal faults, and in the center of the zone, it flows upward, creating the steeply tilted beds. Because the rock in the near surface behaves in a brittle manner when this

occurs, this movement is accompanied by faulting and brecciation.

The incredibly high pressures that develop during an impact create distinctive changes in the rocks of the impact site. The shock wave that passes through the rock momentarily subjects rocks to very high pressures, a condition that causes **shock metamorphism.** Shock metamorphism of quartz yields unusual high-density polymorphs like stishovite, and characteristic deformation microstructures. In addition, the kinetic energy of impact is suddenly transformed into heat, with the result that rocks of the impact site are momentarily heated to temperatures as high as 1700°C. At such temperatures, the whole rock melts, only to freeze quickly into glass of the same composition as the original rock. In some cases, melt mixes with impact breccia, and injects into cracks between larger breccia fragments, forming a glasslike rock called **pseudotachylyte.**

Impact structures affecting now exposed basement crystalline rocks characteristically cause distinctive circular patterns of erosion in the basement that stand out in satellite imagery or through geophysical methods. One of the best known basement impact structures is the Sudbury complex in southern Ontario, Canada. Not only are the characteristic features of impact, like shatter cones and pseudotachylyte, readily visible in the field, but the Sudbury impact, occurring about 1.85 Ga, was large enough to affect the whole crust and cause an impact melt that produced valuable economic deposits that have been mined for many years.

FIGURE 2.26 Shatter cones that were formed by the Sudbury impact that occurred around 1.85 Ga (Ontario, Canada). The apex of a shatter cone points in the direction from which the impacting object came. Pocket knife for scale.

2.6 CLOSING REMARKS

In this chapter we explored the description of various types of sedimentary, igneous, and impact structures whose formation is not an immediate consequence of plate tectonic forces, of isostatic consequences, or of thickening or thinning of the crust. A discussion of nontectonic structures is a good way to start a structural geology course, because it gets you thinking about geometries and shapes. However, these structures are the topic of entire classes in sedimentology and igneous petrology, so you will realize that we cannot do justice to the richness of these topics. Before delving into tectonic structures, we will first introduce the fundamentals of stress and deformation. Although the description of tectonic structures does not require an understanding of these concepts, they are needed when examining how and why tectonic structures form. So, stress, deformation, and rheology are the topics of the next few chapters.

ADDITIONAL READING

Collinson, J. D., and Thompson, D. B., 1989. *Sedimentary structures* (2nd edition). Unwin Hyman: London.

Jackson, M. P. A., and Talbot, C. J., 1994. Advances in salt tectonics. In Hancock, P. L., ed., *Continental deformation.* Pergamon Press, pp. 159–179.

Melosh, H. J., 1996. *Impact cratering: A geologic process.* Oxford University Press: New York.

Paterson, S. R., Vernon, R. H., and Tobisch, O. T., 1989. A review of criteria for the identification of magmatic and tectonic foliations in granitoids. *Journal of Structural Geology,* 11, 349–363.

Selley, R. C., 1988. *Applied sedimentology.* Academic Press: New York.

Shrock, R. R., 1948. *Sequence in layered rocks.* McGraw-Hill: New York.

Worrall, D. M., and Snelson, S., 1989. Evolution of the northern Gulf of Mexico, with emphasis on Cenozoic growth faulting and the role of salt. In *The geology of North America—an overview,* v. A, pp. 97–138. Geological Society of America.

3.1 INTRODUCTION

We frequently use the words force and stress in casual conversation. Stress from yet another deadline, a test, or maybe an argument with a roommate or spouse. Appropriate force is applied to reach our goal, and so on. In science, however, these terms have very specific meanings. For example, the force of gravity keeps us on the Earth's surface and the force of impact destroys our car. Like us, rocks experience the pull of gravity, and forces arising from plate interactions result in a range of geologic structures, from microfabrics to mountain ranges (Figure 3.1). In this chapter we begin with the fundamentals of force and stress, followed by a look at the components of stress that eventually produce tectonic structures. In later chapters we will use these concepts to examine the relationship between geologic structures and stress.

To understand tectonic processes we must be familiar with the fundamental principles of mechanics. **Mechanics** is concerned with the action of forces on bodies and their effect; you can say that mechanics is the science of motion. **Newtonian**[1] (or *classical*) **mechanics** studies the action of forces on rigid bodies. The equations of Newtonian mechanics adequately describe a range of *movements* in the natural world, from the entertaining interaction between colliding balls at a game of pool (Figure 3.2a) to the galactic dance of the planets in our solar system. When reaching the subatomic level, Newtonion mechanics starts to break down and we enter the complex realm of quantum mechanics. In tectonic structures we commonly deal with interactions that involve not only movement, but also **distortion;** material displacements occur both between and within bodies. Imagine playing pool with balls made up of jelly rather than solids (Figure 3.2b). The theory associated with this type of behavior is the focus of **continuum mechanics.** In continuum mechan-

---

[1]After Isaac Newton (1642–1727).

FIGURE 3.1 Aerial view of the Karakoram range of the Himalaya.

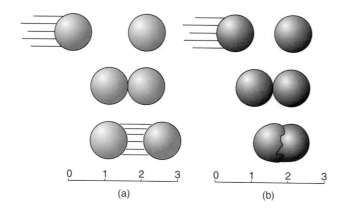

(a)   (b)

FIGURE 3.2 The interaction of nondeformable bodies is described by Newtonian (or classical) mechanics (a) and that between deformable bodies by continuum mechanics (b). Imagine the difference between playing pool with regular balls and balls made of jelly.

ics, a material is treated as a continuous medium (hence the name), that is, there are no discontinuities that appreciably affect its behavior. This may seem inappropriate for rocks at first, because we know that they consist of grains whose boundaries are material discontinuities by definition. Yet, on the scale of a rock body containing thousands or more grains we may consider the system statistically homogeneous. Indeed, the predictions from continuum mechanics theory give us adequate first-order descriptions of displacements in many natural rocks. The primary reason to use the simplifications of continuum mechanics is that it provides us with a mathematical description of deformation in relatively straightforward terms. When the behavior of rocks is dominated by discrete discontinuities, like fractures, continuum mechanics theory no longer holds. Then we need to resort to more complex modeling methods that fall outside the scope of this book.

By the time you have reached the end of this chapter, a good number of terms and concepts will have appeared.

TABLE 3.1	**TERMINOLOGY AND SYMBOLS OF FORCE AND STRESS**
Force	Mass times acceleration ($F = m \cdot a$; Newton's second law); symbol F
Stress	Force per unit area (F/A); symbol σ
Anisotropic stress	At least one principal stress has a magnitude unequal to the other principal stresses (describes an ellipsoid)
Deviatoric stress	Component of the stress that remains after the mean stress is removed; this component of the stress contains the six shear stresses; symbol σ_{dev}
Differential stress	The difference between two principal stresses (e.g., $\sigma_1 - \sigma_3$), which by definition is ≥ 0; symbol σ_d
Homogeneous stress	Stress at each point in a body has the same magnitude and orientation
Hydrostatic stress/pressure	Isotropic component of the stress; strictly, the pressure at the base of a water column
Inhomogeneous stress	Stress at each point in a body has different magnitude and/or orientation
Isotropic stress	All three principal stresses have equal magnitude (describes a sphere)
• Lithostatic stress/pressure	Isotropic pressure at depth in the Earth arising from the overlying rock column (density × gravity × depth, $\rho \cdot g \cdot h$); symbol P_l
Mean stress	$(\sigma_1 + \sigma_2 + \sigma_3)/3$; symbol σ_{mean}
Normal stress	Stress component oriented perpendicular to a given plane; symbol σ_n
Principal plane	Plane of zero shear stress; three principal planes exist
Principal stress	The normal stress on a plane with zero shear stress; three principal stresses exist, with the convention $\sigma_1 \geq \sigma_2 \geq \sigma_3$
Shear stress	Stress parallel to a given plane; symbol σ_s (sometimes the symbol τ is used)
Stress ellipsoid	Geometric representation of stress; the axes of the stress ellipsoid are the principal stresses
Stress field	The orientation and magnitudes of stresses in a body
Stress tensor	Mathematical description of stress (stress is a second-order tensor)
Stress trajectory	Principal stress directions in a body

For convenience and future reference, therefore, some of the more common terms are described in Table 3.1.

3.2 UNITS AND FUNDAMENTAL QUANTITIES

When measuring something you must select a unit for the quantity that is to be measured. The physical properties of a material can be expressed in terms of four fundamental quantities: *mass, length, time,* and *charge.* For our purposes we can ignore the quantity charge, which describes the electromagnetic interaction of particles. It plays a role, however, when we try to understand the behavior of materials at the atomic scale. The units of mass, length, and time are the kilogram (kg), the meter (m), and the second (s), respectively. This notation follows the Système International (French), better known as SI units. Throughout the text we will use SI units, but other conventions remain popular in geology (such as "kilobar," which is a measure of pressure). Where appropriate we will add these units in parentheses. In Table 3.2 the SI units of stress and some common conversions are given.

The symbol for mass is [m], for length [l], and for time [t]. Velocity [v], which combines the fundamental quantities of length and time, has the units of length divided by time. In conventional symbols this is written as

$$[v] : [lt^{-1}]$$

in which the colon means "has the quantity of." Such **dimensional analysis** is a check on the relevance of an equation. We begin by using it in the case of a force.

| TABLE 3.2 | UNITS OF STRESS AND THEIR CONVERSIONS |

	bar	dynes/cm^2	atmosphere	kg/cm^2	pascal (Pa)	pounds/in^2 (psi)
bar		10^6	0.987	1.0197	10^5	14.503
dynes/cm^2	10^{-6}		0.987×10^{-6}	1.919×19^{-6}	0.1	14.503×10^{-6}
atmosphere	1.013	1.013×10^6		1.033	1.013×10^5	14.695
kg/cm^2	0.981	0.981×10^6	0.968		0.981×10^5	14.223
pascal (Pa)	10^{-5}	10	0.987×10^{-5}	1.0197×10^{-5}		14.503×10^{-5}
pounds/in^2 (psi)	6.895×10^{-2}	6.895×10^4	6.81×10^{-2}	7.03×10^{-2}	6.895×10^3	

To use this table start in the left-hand column and read along the row to the column for which a conversion is required. For example, 1 bar = 10^5 Pa or 1 Pa = 14.5×10^{-5} psi.

3.3   FORCE

Kicking or throwing a ball shows that a force changes the velocity of an object. Newton's first law of motion, also called the **Law of Inertia,** says that in the absence of a force a body moves either at constant velocity or is at rest. Stated more formally: a free body moves without acceleration. Change in velocity is called acceleration [a], which is defined as velocity divided by time:

$$[a] : [vt^{-1}] : [lt^{-2}]$$

The unit of acceleration, therefore, is m/s^2.
Force [F], according to Newton's **Second Law of Motion,** is mass multiplied by acceleration:

$$[F] : [ma] : [mlt^{-2}]$$

The unit of force is kg · m/s^2, called a *newton* (N) in SI units. You can feel the effect of mass when you throw a tennis ball and a basketball and notice that a different force is required to move each of them.

Force, like velocity, is a vector quantity, meaning that it has both magnitude and direction. So it can be graphically represented by a line with an arrow on one side. Manipulation of forces conforms to the rules of vector algebra. For example, a force at an angle to a given plane can be geometrically resolved into two components; say, one parallel and one perpendicular to that plane.

Natural processes can be described with four basic forces: (1) the gravity force, (2) the electromagnetic force, (3) the nuclear or strong force, and (4) the weak force. Gravity is a special force that acts over large dis-

tances and is always attractive; for example, the ocean tides reflect the gravitational interaction between the Moon and the Earth. The other three forces act only over short ranges (atomic scale) and can be attractive or repulsive. The electromagnetic force describes the interaction between charged particles, such as the electrons around the atomic nucleus; the strong force holds the nucleus of an atom together; and the weak force is associated with radioactivity. It is quite possible that only one fundamental force exists in nature, but, despite the first efforts of Albert Einstein[2] and much progress since then, it has not so far been possible to formulate a **Grand Unified Theory** to encompass all four forces. The force of gravity has proved to be a particular problem.

Forces that result from action of a field at every point within the body are called **body forces.** Bungee jumping gives you a very vivid sensation of body forces though the action of gravity. The magnitude of body forces is proportional to the mass of the body. Forces that act on a specific surface area in a body are called **surface forces.** They reflect the pull or push of the atoms on one side of a surface against the atoms on the other side. Examples are a cuestick's force on a pool ball, the force of expanding gases on an engine piston, and the force of the jaws of a vice. The magnitude of surface forces is proportional to the area of the surface.

Forces that act on a body may change the velocity of (that is, accelerate) the body, and/or may result in a shape change of the body, meaning acceleration of one part of the body with respect to another part. Although force is an important concept, it does not distinguish

---

[2]German-born theoretical physicist (1879–1955).

the effect of an equal force on bodies of equal mass but with different shapes. Imagine the effect of the same force applied to a sharp object and a dull object. For example, a human is comfortably supported by a water bed, but when you place a nail between the person and the water bed, the effect is quite dramatic. Using a more geologic experience, consider hitting a rock with a pointed or a flat hammer using the same force. The rock cracks more easily with the pointed hammer than with the flat-headed hammer; in fact, we apply this principle when we use a chisel rather than a sledge hammer to collect rock samples. These examples of the intensity of force lead us into the topic of stress.

3.4 STRESS

Stress, represented by the symbol σ (sigma), is defined as the force per unit area [A], or $\sigma = F/A$. You can, therefore, consider stress as the *intensity of force,* or a measure of how concentrated a force is. A given force acting on a small area (the pointed hammer mentioned previously) will have a greater intensity than that same force acting on a larger area (a flat-headed hammer), because the stress associated with the smaller area is greater than that with the larger area. Those of you remembering turntables and vinyl records (ask your parents) are familiar with this effect. The weight of the arm holding the needle is only a few grams, but the stress of the needle on the vinly record is orders of magnitude greater because the contact area between needle and record is very small. The high stresses at the area of contact eventually gave rise to scratches and ticks in the records, so it is little wonder that we have embraced digital technologies.

You will see that stress is a complex topic, because its properties depend on the reference system. Stress that acts on a plane is a vector quantity, called **traction,** whereas stress acting on a body is described by a higher order entity, called a **stress tensor.** In the next few pages we will gradually develop the pertinent concepts and components of stress.

Because stress is force per unit area it is expressed in terms of the following fundamental quantities:

$$[\sigma] : [mlt^{-2} \cdot l^{-2}] \text{ or } [ml^{-1} \cdot t^{-2}]$$

The corresponding unit of stress is kg/m · s² (or N/m²), which is called a pascal (Pa).[3] Instead of this SI unit,

however, many geologists continue to use the unit bar, which is approximately 1 atmosphere. These units are related as follows:

$$1 \text{ bar} = 10^5 \text{ Pa} \approx 1 \text{ atmosphere}$$

In geology you will generally encounter their larger equivalents, the kilobar (kbar) and the megapascal (MPa):

$$1 \text{ kbar} = 1000 \text{ bar} = 10^8 \text{ Pa} = 100 \text{ MPa}$$

The unit gigapascal (1 GPa = 1000 MPa = 10 kbar) is used to describe the very high pressures that occur deep in the Earth. For example, the pressure at the core-mantle boundary, located at a depth of approximately 2900 km, is ~135 GPa, and at the center of the Earth (at a depth of 6370 km) the pressure exceeds 350 GPa. Later we will see how these values can be calculated (Section 3.9).

3.5 TWO-DIMENSIONAL STRESS: NORMAL STRESS AND SHEAR STRESS

Stress acting on a plane is a vector quantity, meaning that it has both magnitude and direction; it is sometimes called *traction.* Stress on an arbitrarily oriented plane, however, is not necessarily perpendicular to that plane, but, like a vector, it can be resolved into components normal to the plane and parallel to the plane (Figure 3.3). The vector component normal to the plane is called the **normal stress,** for which we use the symbol σ_n (some-

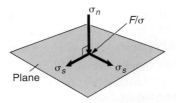

σ_n is normal stress
σ_s is shear stress
F is force; σ is stress

FIGURE 3.3 The stress on a two-dimensional plane is defined by a stress acting perpendicular to the plane (the normal stress) and a stress acting along the plane (the shear stress). The normal stress and shear stress are perpendicular to one another.

---

[3]After Blaise Pascal (1623–1662).

FIGURE 3.4 The relationship between force (F) and stress (σ) on a plane. Section through a cube showing face $ABCD$ with ribs of length AB on which a force F is applied. This force is resolved into orientations parallel (F_s) and perpendicular (F_n) to a plane that makes an angle θ with the top and bottom surface $(EF$ is the trace of this plane). The magnitudes of vectors F_s and F_n are a function of the angle θ: $F_n = F \cdot \cos \theta$, $F_s = F \cdot \sin \theta$. The magnitude of the normal (σ_n) and shear stress (σ_s) is a function of the angle θ *and* the area: $\sigma_n = \sigma \cos^2 \theta$, $\sigma_s = \sigma \frac{1}{2} (\sin 2\theta)$. (a) Force F on plane; (b) stress σ on plane; (c) normalized values of F_n and σ_n on plane with angle θ; (d) normalized values of F_s and σ_s on a plane with angle θ.

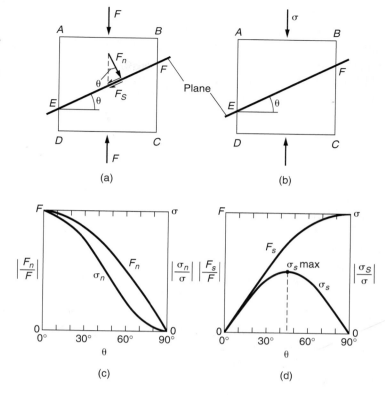

times just the symbol σ is used); the vector component along the plane is the **shear stress** and has the symbol σ_s (sometimes the symbol τ (tau) is used).

In contrast to the resolution of forces, the resolution of stress into its components is not straightforward, because the area changes as a function of the orientation of the plane with respect to the stress vector. Let us first examine the resolution of stress on a plane in some detail, because, as we will see, this has important implications.

In Figure 3.4, stress σ has a magnitude F/AB and makes an angle θ with the top and bottom of our square. The forces perpendicular (F_n) and parallel (F_s) to the plane EF are

$F_n = F \cos \theta = \sigma AB \cos \theta = \sigma EF \cos^2 \theta$
$(AB = EF \cos \theta)$ Eq. 3.1
$F_s = F \sin \theta = \sigma AB \sin \theta =$
$\sigma EF \sin \theta \cos \theta = \sigma EF \frac{1}{2}(\sin 2\theta)$ Eq. 3.2

Thus the corresponding stresses are

$\sigma_n = F_n/EF = \sigma \cos^2 \theta$ Eq. 3.3
$\sigma_s = F_s/EF = \sigma \frac{1}{2}(\sin 2\theta)$ Eq. 3.4

You notice that the equation for the normal stress and the normal force are different, as are the equations for F_s and σ_s. We graphically illustrate this difference between forces and stresses on an arbitrary plane by

plotting their normalized values as a function of the angle θ in Figure 3.4c and d, respectively. In particular, the relationship between F_s and σ_s is instructive for gaining an appreciation of the area dependence of stress. Both the shear force and the shear stress initially increase with increasing angle θ; at 45° the shear stress reaches a maximum and then decreases, while F_s continues to increase.

Thus, the stress vector acting on a plane can be resolved into vector components perpendicular and parallel to that plane, but their magnitudes vary as a function of the orientation of the plane. Let us further examine the properties of stress by determining the stress state for a three-dimensional body.

3.6 THREE-DIMENSIONAL STRESS: PRINCIPAL PLANES AND PRINCIPAL STRESSES

Previously, we discussed stress acting on a single plane (the two-dimensional case), recognizing two vector components, the normal stress and the shear stress (Figure 3.3). However, to describe stress on a randomly oriented plane in space we need to consider the three-dimensional case. We limit unnecessary complications by setting the condition that the body containing the plane is at rest. So a force applied to the body is balanced by an opposing force of equal magnitude but opposite sign; this condition is known as Newton's **Third Law of Motion.** Using another Newtonian sports analogy, kick a ball that rests against a wall and notice how the ball (the wall, in fact) pushes back with equal enthusiasm.

3.6.1 Stress at a Point

We shrink our three-dimensional body containing the plane of interest down to the size of a point for our analysis of the stress state of an object. Why this seemingly obscure transformation? Recall that two nonparallel planes have a line in common and that three or more nonparallel planes have a point in common. In other words, a point defines the intersection of an infinite number of planes with different orientations. The stress state at a point, therefore, can describe the stresses acting on all planes in a body.

In Figure 3.5a the normal stresses (σ) acting on four planes (*a–d*) that intersect in a single point are drawn. For clarity, we limit our illustrations to planes that are all perpendicular to the surface of the page, allowing the use of slice through the body. You will see later that this geometry easily expands into the full three-dimensional case. Because of Newton's Third Law of Motion, the stress on each plane must be balanced by one of opposite sign ($\sigma = -\sigma$). Because stress varies as a function of orientation, the magnitude of the normal stress on each plane (represented by the vector length) is different. If we draw an envelope around the end points of these stress vectors (heavy line in Figure 3.5a), we obtain an ellipse. Recall from geometry that an ellipse is defined by at least three nonperpendicular axes, which are shown in Figure 3.5a. This means that the magnitude of the stress for all possible planes is represented by a point on this **stress ellipse.** Now, the same can be done in three dimensions, but this is hard to illustrate on a piece of flat paper. Doing the same analysis in three dimensions, we obtain an envelope that is the three-dimensional equivalent of an ellipse, called an ellipsoid (Figure 3.5b). This **stress ellipsoid** fully describes the stress state at a point and enables us to determine the stress for any given plane. Like all ellipsoids, the stress ellipsoid is defined by three mutually perpendicular axes, which are called the **principal stresses.** These principal stresses have two properties: (1) they are orthogonal to each other, and (2) they are perpendicular to three planes that do not contain shear stresses; these planes are called the **prin-**

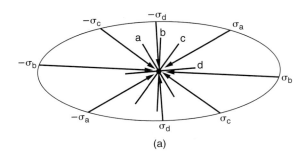

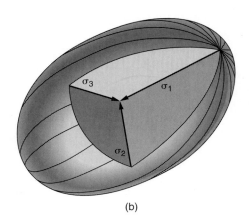

FIGURE 3.5 (a) A point represents the intersection of an infinite number of planes, and the stresses on these planes describe an ellipse in the two-dimensional case. In three dimensions this stress envelope is an ellipsoid (b), defined by three mutually perpendicular principal stress axes ($\sigma_1 \geq \sigma_2 \geq \sigma_3$). These three axes are normal to the principal planes of stress.

cipal planes of stress. So, we can describe the stress state of a body simply by specifying the orientation and magnitude of three principal stresses.

3.6.2 The Components of Stress

The orientation and magnitude of the stress state of a body is defined in terms of its components projected in a Cartesian reference frame, which contains three mutually perpendicular coordinate axes, *x*, *y*, and *z*. To see this, instead of a point representing an infinite number of planes on which our stress acts, we draw our point as an infinitely small cube whose ribs are perpendicular to each of the coordinate axes, *x*, *y*, and *z*. We resolve the stress acting on each face of a cube into three components (Figure 3.6). For a face normal to the *x*-axis the components are σ_{xx}, which is the component *normal* to that face, and σ_{xy} and σ_{xz}, which are the two components *parallel* to that face. These last two stresses are shear stress components, acting along one of the other coordinate axes *y* and *z*, respectively.

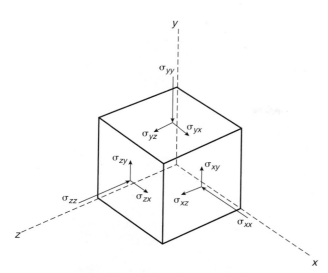

FIGURE 3.6 Resolution of stress into components perpendicular (three normal stresses, σ_n) and components parallel (six shear stresses, σ_s) to the three faces of an infinitesimally small cube, relative to the reference system x, y, and z.

Applying this same procedure for the faces normal to y and z, we obtain a total of nine stress components (Figure 3.6):

	In the direction of		
	x:	y:	z:
stress on the face normal to x:	σ_{xx}	σ_{xy}	σ_{xz}
stress on the face normal to y:	σ_{yx}	σ_{yy}	σ_{yz}
stress on the face normal to z:	σ_{zx}	σ_{zy}	σ_{zz}

The columns, from left to right, represent the components in the x, y, and z directions of the coordinate system, respectively. σ_{xx}, σ_{yy}, and σ_{zz} are normal stress components and the other six are shear stress components. Because we specified that the body itself is at rest, three of the six shear stress components must be equivalent (σ_{xy} and σ_{yx}, σ_{yz} and σ_{zy}, and σ_{xz} and σ_{zx}). If these components were unequal, the body would move, which violates our at-rest condition. So, rather than nine components, we are left with *six independent stress components* to describe the stress acting on any arbitrary infinitesimal body:

	In the direction of		
	x:	y:	z:
stress on the face normal to x:	σ_{xx}	σ_{xy}	σ_{xz}
stress on the face normal to y:	σ_{xy}	σ_{yy}	σ_{yz}
stress on the face normal to z:	σ_{xz}	σ_{yz}	σ_{zz}

The only ingredient left in our description is a *sign convention*. In physics and engineering, tensile stress is considered positive, and compressive stress negative. In geology, however, it is customary to make compression positive and tension negative, because compression is more common in the Earth. We will, therefore, use the geologic sign convention throughout the text; however, don't confuse this with the engineering sign convention used in some other textbooks.[4]

For any given state of stress there is at least one set of three mutually perpendicular planes on which the shear stresses are zero. In other words, you can rotate our infinitesimal cube such that the shear stresses on each of its three faces are zero. In this orientation, these three faces are the **principal planes of stress** (the same ones that we described in our stress ellipsoid; Section 3.6.1) and they intersect in three mutually perpendicular axes that are the **principal axes of stress** (which are the same as the axes of the stress ellipsoid in Section 3.6.1). The stresses acting along them are called the **principal stresses** for a given point or homogeneous domain within a body.

3.6.3 Stress States

If the three principal stresses are equal in magnitude, we call the stress **isotropic.** This stress state is represented by a sphere rather than an ellipsoid, because all three radii are equal. If the principal stresses are unequal in magnitude, the stress is called **anisotropic.** By convention, the maximum principal stress is given the symbol σ_1, the intermediate and minimum principal stresses acting along the other two axes are given the symbols σ_2 and σ_3, respectively. Thus, by (geologic) convention:

$$\sigma_1 \geq \sigma_2 \geq \sigma_3$$

By changing the relative values of the three principal stresses we define several common **stress states:**

General triaxial stress:	$\sigma_1 > \sigma_2 > \sigma_3 \neq 0$
Biaxial (plane) stress:	one axis $= 0$
	(e.g., $\sigma_1 > 0 > \sigma_3$)
Uniaxial compression:	$\sigma_1 > 0$; $\sigma_2 = \sigma_3 = 0$
Uniaxial tension:	$\sigma_1 = \sigma_2 = 0$; $\sigma_3 < 0$
Hydrostatic stress (pressure):	$\sigma_1 = \sigma_2 = \sigma_3$

So, we learned that the stress state of a body is defined by nine components. Mathematically this

---

[4] A further source of possible confusion is that elastic constants of materials are given with the engineering convention, so their sign needs to be reversed in our use.

ellipsoid is described by a 3 × 3 matrix (called a **second-rank tensor**). Although it may seem easier at first to use a geometric representation of stress, as we just did, for the analysis of stress in bodies it is better to apply mathematical operations. We will return to this later in the chapter (Section 3.10).

3.7 DERIVING SOME STRESS RELATIONSHIPS

Now that we can express the stress state of a body by its principal stresses, we can derive several useful relationships. Let's carry out a simple classroom experiment in which we compress a block of clay between two planks (Figure 3.7). As the block of clay develops a fracture, we want to determine what the normal and the shear stresses on the fracture plane are. To answer this question our approach is similar to our previous one (Equations 3.1 to 3.4), but now we express the normal and shear stresses in terms of the principal stress axes.

The principal stresses acting on our block of clay are σ_1 (maximum stress), σ_2 (intermediate stress), and σ_3 (minimum stress). Since we carry out our experiment under atmospheric conditions, the values of σ_2 and σ_3 will be equal, and we may simplify our analysis by neglecting σ_2 and considering only the σ_1-σ_3 plane, as shown in Figure 3.7. The fracture plane makes an angle θ (theta) with σ_3. This plane makes the trace AB in Figure 3.7b, which we assign unit length (that is, 1) for convenience. We can resolve AB along AC (parallel to σ_1) and along BC (parallel to σ_3). Then, by trigonometry, we see that the area represented by $AC = \sin \theta$, and the area represented by $BC = \cos \theta$. Note that if we assign dimension L to AB then $AC = L \cdot \sin \theta$ and $BC = L \cdot \cos \theta$.

Next we consider the forces acting on each of the surface elements represented by AB, BC, and AC. Since force equals stress times the area over which it acts, we obtain

force on side $BC = \sigma_1 \cdot \cos \theta$
force on side $AC = \sigma_3 \cdot \sin \theta$

The force on side AB consists of a normal force (i.e., $\sigma_n \cdot 1$) and a shear force (i.e., $\sigma_s \cdot 1$); recall that force is stress times area.

For equilibrium, the forces acting in the direction of AB must balance, and so must the forces acting per-

(a)

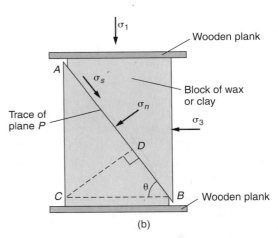

(b)

FIGURE 3.7 Determining the normal and shear stresses on a plane in a stressed body as a function of the principal stresses. (a) An illustration from the late nineteenth-century fracture experiments of Daubrée using wax. (b) For a classroom experiment, a block of clay is squeezed between two planks of wood. AB is the trace of fracture plane P in our body that makes an angle θ with σ_3. The two-dimensional case shown is sufficient to describe the experiment, because σ_2 equals σ_3 (atmospheric pressure).

pendicular to AB (which is parallel to CD). Hence, resolving along CD:

force $\perp AB$ = force $\perp BC$ resolved on CD +
force $\perp AC$ resolved on CD

or

$$1 \cdot \sigma_n = \sigma_1 \cos \theta \cdot \cos \theta + \sigma_3 \sin \theta \cdot \sin \theta \quad \text{Eq. 3.5}$$
$$\sigma_n = \sigma_1 \cos^2 \theta + \sigma_3 \sin^2 \theta \quad \text{Eq. 3.6}$$

Substituting these trigonometric relationships in Equation 3.6, we obtain

$$\cos^2 \theta = \tfrac{1}{2}(1 + \cos 2\theta)$$
$$\sin^2 \theta = \tfrac{1}{2}(1 - \cos 2\theta)$$

Simplifying, gives

$$\sigma_n = \tfrac{1}{2}(\sigma_1 + \sigma_3) + \tfrac{1}{2}(\sigma_1 - \sigma_3) \cos 2\theta \quad \text{Eq. 3.7}$$

and,

force parallel AB = force $\perp BC$ resolved on AB +
force $\perp AC$ resolved on AB

or

$$1 \cdot \sigma_s = \sigma_1 \cos \theta \cdot \sin \theta - \sigma_3 \sin \theta \cdot \cos \theta \quad \text{Eq. 3.8}$$

Note that the force perpendicular to AC resolved on AB acts in a direction opposite to the force perpendicular to BC resolved on AB, hence a negative sign is needed in Eq. 3.8, which further simplifies to

$$\sigma_s = (\sigma_1 - \sigma_3) \sin \theta \cdot \cos \theta \quad \text{Eq. 3.9}$$

Substituting this trigonometric relationship in Eq. 3.9, we get

$$\sin \theta \cdot \cos \theta = \tfrac{1}{2} \sin 2\theta$$

which gives

$$\sigma_s = \tfrac{1}{2}(\sigma_1 - \sigma_3) \sin 2\theta \quad \text{Eq. 3.10}$$

From Equations 3.7 and 3.10 we can determine that the planes of maximum normal stress are at an angle θ of $0°$ with σ_3, because $\cos 2\theta$ reaches its maximum value ($\cos 0° = 1$). Secondly, planes of maximum shear stress lie at an angle θ of $45°$ with σ_3 because $\sin 2\theta$ reaches its maximum value ($\sin 90° = 1$) (see also Figure 3.2c and d). Whereas faulting resulted in a shearing motion along the fault plane, we find that the fault plane in our experiment is not parallel to the plane of maximum shear stress ($\theta > 45°$). This perhaps suprising result

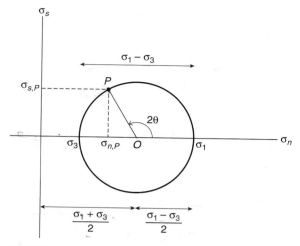

FIGURE 3.8 The Mohr diagram for stress. Point P represents the plane in our clay experiment of Figure 3.7.

reflects a fundamental property of solids that we analyze in Chapter 6.

3.8 MOHR DIAGRAM FOR STRESS

The equations we derived for σ_n and σ_s do not offer an obvious sense of their values as a function of orientation of a plane in our block of clay. Of course, a programmable calculator or simple computer program will do the job, but a convenient graphical method, known as the **Mohr diagram**[5] (Figure 3.8), was introduced over a century ago to solve Equations 3.7 and 3.10. A Mohr diagram is an "XY"-type (Cartesian) plot of σ_s versus σ_n that graphically solves the equations for normal stress and shear stress acting on a plane within a stressed body. In our experiences, many people find the Mohr construction difficult to comprehend. So we'll first examine the proof and underlying principles of this approach to try to take the magic out of the method.

If we rearrange Equations 3.7 and 3.10 and square them, we get

$$[\sigma_n - \tfrac{1}{2}(\sigma_1 + \sigma_3)]^2 = [\tfrac{1}{2}(\sigma_1 - \sigma_3)]^2 \cos^2 2\theta \quad \text{Eq. 3.11}$$
$$\sigma_s^2 = [\tfrac{1}{2}(\sigma_1 - \sigma_3)]^2 \sin^2 \theta \quad \text{Eq. 3.12}$$

[5]Named after Otto Mohr (1835–1918).

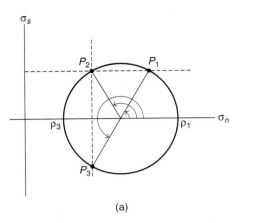

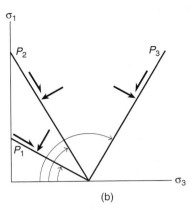

FIGURE 3.9 For each value of the shear stress and the normal stress there are two corresponding planes, as shown in the Mohr diagram (a). The corresponding planes in $\sigma_1 - \sigma_3$ space are shown in (b).

Adding Equations 3.11 and 3.12 gives

$$[\sigma_n - \tfrac{1}{2}(\sigma_1 + \sigma_3)]^2 + \sigma_s^2 = [\tfrac{1}{2}(\sigma_1 - \sigma_3)]^2 \cdot (\cos^2 2\theta + \sin^2 2\theta) \qquad \text{Eq. 3.13}$$

Using the trigonometric relationship

$$(\cos^2 2\theta + \sin^2 2\theta) = 1$$

in Equation 3.13 gives

$$[\sigma_n - \tfrac{1}{2}(\sigma_1 + \sigma_3)]^2 + \sigma_s^2 = [\tfrac{1}{2}(\sigma_1 - \sigma_3)]^2 \qquad \text{Eq. 3.14}$$

Note that Equation 3.14 has the form $(x - a)^2 + y^2 = r^2$, which is the general equation for a circle with radius r and centered on the x-axis at distance a from the origin. Thus the Mohr circle has a radius $\tfrac{1}{2}(\sigma_1 - \sigma_3)$ that is centered on the σ_n axis at a distance $\tfrac{1}{2}(\sigma_1 + \sigma_3)$ from the origin. The construction is shown in Figure 3.8. You also see from this figure that the Mohr circle's radius, $\tfrac{1}{2}(\sigma_1 - \sigma_3)$, is the maximum shear stress, $\sigma_{s,\,max}$. The stress difference $(\sigma_1 - \sigma_3)$, called the **differential stress,** is indicated by the symbol σ_d.

3.8.1 Constructing the Mohr Diagram

To construct a Mohr diagram we draw two mutually perpendicular axes; σ_n is the abscissa (x-axis) and σ_s is the ordinate (y-axis). In our clay deformation experiment, the maximum principal stress (σ_1) and the minimum principal stress (σ_3) act on plane P that makes an angle θ with the σ_3 direction (Figure 3.7); in the Mohr construction we then plot σ_1 and σ_3 on the σ_n-axis

(Figure 3.8). These principal stress values are plotted on the σ_n axes because they are normal stresses, but with the special condition that the planes on which they act, the principal planes, have zero shear stress ($\sigma_s = 0$). We then construct a circle through points σ_1 and σ_3, with O, the midpoint, at $\tfrac{1}{2}(\sigma_1 + \sigma_3)$ as center, and a radius of $\tfrac{1}{2}(\sigma_1 - \sigma_3)$. Next, we draw a line OP such that angle $PO\sigma_1$ is equal to 2θ. This step often gives rise to confusion and errors. First, remember that we plot *twice* the angle θ, which is the angle between the plane and σ_3, because of the equations we are solving. Second, remember that we measure 2θ from the σ_1-side on the σ_n-axis.[6] When this is done, the Mohr diagram is complete and we can read off the value of $\sigma_{n,P}$ along the σ_n-axis, and the value of $\sigma_{s,P}$ along the σ_s-axis for our plane P, as shown in Figure 3.8. We see that

$$\sigma_{n,P} = \tfrac{1}{2}(\sigma_1 + \sigma_3) + \tfrac{1}{2}(\sigma_1 - \sigma_3) \cos 2\theta$$
$$\text{and}$$
$$\sigma_{s,P} = \tfrac{1}{2}(\sigma_1 - \sigma_3) \sin 2\theta$$

A couple of additional observations can be made from the Mohr diagram (Figure 3.9). There are two planes, oriented at angle θ and its complement $(90 - \theta)$, with equal shear stresses but different normal stresses. Also, there are two planes with equal normal stress, but with shear stresses of opposite sign (that is, they act in different directions on these planes).

In general, for each orientation of a plane, defined by its angle θ, there is a corresponding point on the circle. The coordinates of that point represent the normal and shear stresses that act on that plane. For example, when $\theta = 0°$ (that is, for a plane parallel to σ_3), P coincides with σ_1, which gives $\sigma_n = \sigma_1$ and $\sigma_s = 0$. In other words, for any value of σ_1 and σ_3 ($\sigma_3 = \sigma_2$ in our compression experiment), we can determine σ_n and σ_s graphically for planes that lie at an angle θ with σ_3. If we decide to change our earlier experi-

---

[6]Alternative conventions for this construction are also in use, but be careful that they are not mixed.

ment by gluing the planks to the clay block and then moving the planks apart (a tension experiment), we must use a negative sign for the least principal stress (in this case, $\sigma_1 = \sigma_2$ and σ_3 is negative). So the center O of the Mohr circle can lie on either side of the origin, but is always on the σ-axis.

The Mohr diagram also nicely illustrates the attitude of planes along which the shear stress is greatest for a given state of stress. The point on the circle for which σ_s is maximum corresponds to a value of $2\theta = 90°$. For the same point, the magnitude of σ_s is equal to the radius of the circle, that is, $\frac{1}{2}(\sigma_1 - \sigma_3)$. Thus the $(\sigma_1 - \sigma_3)$, the differential stress, is twice the magnitude of the maximum shear stress:

$$\sigma_d = 2\sigma_{s,\ max} \qquad \text{Eq. 3.15}$$

When there are changes in the principal stress magnitudes without a change in the differential stress, the Mohr circle moves along the σ_n-axis without changing the magnitude of σ_s. In our experiment, this would be achieved by increasing the air pressure in the classroom or carrying out the experiment under water;[7] this "surrounding" pressure is called the **confining pressure** (P_c) of the experiment. In Chapter 6 we return to the Mohr stress diagram and the role of the confining pressure for fracturing of rocks, but let's get comfortable with the construction with a simple assignment. Figure 3.10a shows six planes in a stressed body at different angles with σ_3. Using the graph in Figure 3.10b, draw the Mohr circle and estimate the normal and shear stresses for these six planes. You can check your estimates by using Equations 3.7 and 3.10.

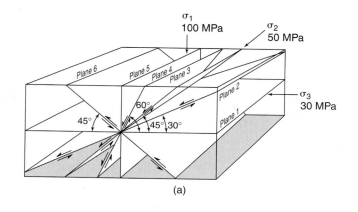

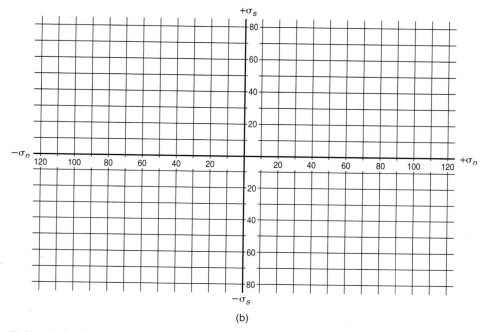

FIGURE 3.10 Adventures with the Mohr circle. To estimate the normal and shear stresses on the the six planes in (a) apply the Mohr construction in (b). The principal stresses and angles θ are given in (a). You can check your estimates from the construction in $\sigma_n - \sigma_s$ space by using Equations 3.7 and 3.10.

3.8.2 Some Common Stress States

Now that you are familiar with the Mohr construction, let's look at its representation of the various stress states that were mentioned earlier. The three-dimensional Mohr diagrams in Figure 3.11 may at first appear a lot more complex than those in our earlier examples, because they represent three-dimensional stress states rather than two-dimensional conditions. Three-dimensional Mohr constructions simply combine three two-dimensional Mohr circles for $(\sigma_1 - \sigma_2)$, $(\sigma_1 - \sigma_3)$, and $(\sigma_2 - \sigma_3)$, and each of these three Mohr circles adheres to the procedures outlined earlier.

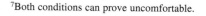

[7]Both conditions can prove uncomfortable.

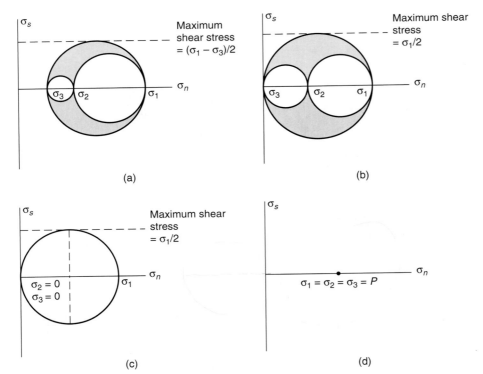

FIGURE 3.11 Mohr diagrams of some representative stress states: (a) triaxial stress, (b) biaxial (plane) stress, (c) uniaxial compression, and (d) isotropic stress or hydrostatic pressure, P (compression is shown).

Figure 3.11a shows the case for general triaxial stress, in which all three principal stresses have nonzero values ($\sigma_1 > \sigma_2 > \sigma_3 \neq 0$). Biaxial (plane) stress, in which one of the principal stresses is zero (e.g., $\sigma_3 = 0$) is shown in Figure 3.11b. Uniaxial compression ($\sigma_2 = \sigma_3 = 0$; $\sigma_1 > 0$) is shown in Figure 3.11c, whereas uniaxial tension ($\sigma_1 = \sigma_2 = 0$; $\sigma_3 < 0$) would place the Mohr circle on the other side of the σ_n-axis. Finally, isotropic stress, often called hydrostatic pressure, is represented by a single point on the σ_n-axis of the Mohr diagram (positive for compression, negative for tension), because all three principal stresses are equal in magnitude ($\sigma_1 = \sigma_2 = \sigma_3$; Figure 3.11d).

3.9 MEAN STRESS AND DEVIATORIC STRESS

In Chapters 4 and 5 we will explore how stresses result in deformation and how stress and strain are related. Because of a body's response to stress, we subdivide the stress into two components, the mean stress and the deviatoric stress (Figure 3.12). The **mean stress** is defined as ($\sigma_1 + \sigma_2 + \sigma_3$)/3, using the symbol σ_m. The

difference between mean stress and total stress is the **deviatoric stress** (σ_{dev}), so

$$\sigma = \sigma_{mean} + \sigma_{dev}$$

The mean stress is often called the hydrostatic component of stress or the **hydrostatic pressure,** because a fluid is stressed equally in all directions. Because the magnitude of the hydrostatic stress is equal in all directions it is an isotropic stress component. When we consider rocks at depth in the Earth we generally refer to **lithostatic pressure,**[8] P_l, rather than the hydrostatic pressure. The lithostatic stress component is best explained by a simple but powerful calculation. Consider a rock at a depth of 3 km in the middle of a continent. The lithostatic pressure at this point is a function of the weight of the overlying rock column because other (tectonic) stresses are unimportant. The local pressure is a function of rock density, depth, and gravity:

$$P_l = \rho \cdot g \cdot h \qquad \text{Eq. 3.16}$$

If ρ (density) equals a representative crustal value of 2700 kg/m^3, g (gravity) is 9.8 m/s^2, and h (depth) is 3000 m, we get

$$P_l = 2700 \cdot 9.8 \cdot 3000 = 79.4 \cdot 10^6 \text{ Pa} \approx 80 \text{ MPa}$$
(or 800 bars)

In other words, for every kilometer in the Earth's crust the lithostatic pressure increases by approximately 27 MPa. With depth the density of rocks increases, so you cannot continue to use the value of 2700 kg/m^3. For crustal depths greater than approximately 15 km the average density of the crust is 2900 kg/m^3. Deeper into Earth the density increases further, reaching as much as 13,000 kg/m^3 in the solid inner core.

---

[8]Also called overburden pressure.

Because the lithostatic pressure is of equal magnitude in all directions, it follows that $\sigma_1 = \sigma_2 = \sigma_3$. The actual state of stress on a body at depth in the Earth is often more complex than only that from the overlying rock column. Anisotropic stresses that arise from tectonic processes, such as the collision of continental plates or the drag of the plate on the underlying material, contribute to the stress state at depth. The differential stresses of these anisotropic stress components,

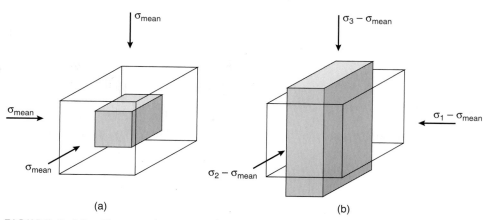

(a)   (b)

FIGURE 3.12   The mean (hydrostatic) and deviatoric components of the stress. (a) Mean stress causes volume change and (b) deviatoric stress causes shape change.

however, are many orders of magnitude less than the lithostatic stress. In the crust, differential stresses may reach a few hundred megapascals, but in the mantle, where lithostatic pressure is high, they are only on the order of tens of megapascals or less (see Section 3.13). Yet, such low differential stresses are responsible for the slow motion of "solid" mantle that is a critical element of our planet's plate dynamics.

Let's return to Figure 3.12 and the preceding comments. Why divide a body's stress state into an isotropic (lithostatic/hydrostatic) and an anisotropic (deviatoric) component? For our explanation we return to look at the deformation of a stressed body. Because isotropic stress acts equally in all directions, it results in a *volume change* of the body (Figure 3.12a). Isotropic stress is responsible for the consequences of increasing water pressure at depth on a human body or air pressure changes during take-off and landing of a plane (remember those painfully popping ears?). Place an air-filled balloon under water and you will see that isotropic stress maintains the spherical shape of the balloon, but reduces the volume. Deviatoric stress, on the other hand, changes the *shape* of a body (Figure 3.12b). As we will see in Chapter 4, distortion of a body can often be measured in structural geology, but volume change is considerably more difficult to determine. As in determining distortions, knowledge about the original volume of a body is the obvious way to determine any volume change. Reliable volume markers, however, are rare in rocks and we resort to indirect approaches such as chemical contrasts between deformed and undeformed samples. The division between the isotropic and anisotropic components of stress provides the connection between the volumetric and distortional components of deformation, respectively.

3.10   THE STRESS TENSOR

The stress ellipsoid is a convenient way to visualize the state of stress, but it is cumbersome for calculations. For example, it is difficult to determine the stresses acting on a randomly chosen plane in a three-dimensional body, or the corresponding stresses when we change the reference system (e.g., by a rotation). In contrast, the **stress tensor**, which mathematically describes the stress state in terms of three orthogonal stress axes, makes such determinations relatively easy. So let us take a look at the stress tensor in a little more detail.

A vector is a physical quantity that has magnitude and direction; it is visualized as an arrow with length and orientation at a point in space. A vector is represented by three coordinates in a Cartesian reference frame that we describe by a matrix consisting of three components. Figure 3.4 showed that stress at a point is a physical quantity that is defined by nine components, which is called a *second-rank tensor*. This is represented by an ellipsoid with orientation, size, and shape at a point in space. The rank of a tensor reflects the number of matrix components and is determined by raising the number 3 to the power of a tensor's rank; for the stress tensor this means, $3^2 = 9$ components. It follows that a vector is a first-rank tensor ($3^1 = 3$ components) and a scalar is a zero-rank tensor ($3^0 = 1$ component). Geologic examples of zero-rank tensors are pressure, temperature, and time; whereas force, velocity, and acceleration are examples of first-rank tensors.

Consider the transformation of a point P in three-dimensional space defined by coordinates $P(x, y, z)$ to point $P'(x', y', z')$. The transformed condition is identified by adding the prime symbol ('). We can describe the transformation of the three coordinates of P as a function of P' by

$$\begin{bmatrix} x' = ax + by + cz \\ y' = dx + ey + fz \\ z' = gx + hy + iz \end{bmatrix}$$

The tensor that describes the transformation from P to P' is the matrix

$$\begin{bmatrix} a & b & c \\ d & e & f \\ g & h & i \end{bmatrix}$$

In matrix notation, the nine components of a stress tensor are

$$\begin{bmatrix} \sigma_{11} & \sigma_{12} & \sigma_{13} \\ \sigma_{21} & \sigma_{22} & \sigma_{23} \\ \sigma_{31} & \sigma_{32} & \sigma_{33} \end{bmatrix} = [\sigma_{ij}]$$

with σ_{11} oriented parallel to the 1-axis and acting on a plane perpendicular to the 1-axis, σ_{12} oriented parallel to the 1-axis and acting on a plane perpendicular to the 2-axis, and so on. The systematics of these nine components make for an unnecessarily long notation, so in shorthand we write

$$[\sigma_{ij}]$$

where i refers to the row (component parallel to the i-axis) and j refers to the column (component acting on the plane perpendicular to the j-axis).

You will notice the similarity between our approach to the stress tensor and our earlier approach to the description of stress at a point, consisting of one normal stress ($i = j$) and two shear stresses ($i \neq j$) for each of three orthogonal planes. The stress tensor is simply the mathematical representation of this condition. Now we use this notation for decomposing the total stress into the *mean stress* and *deviatoric stress*

$$[\sigma_{ij}] = \begin{bmatrix} \sigma_m & 0 & 0 \\ 0 & \sigma_m & 0 \\ 0 & 0 & \sigma_m \end{bmatrix} + \begin{bmatrix} \sigma_{11} - \sigma_m & \sigma_{12} & \sigma_{13} \\ \sigma_{21} & \sigma_{22} - \sigma_m & \sigma_{23} \\ \sigma_{31} & \sigma_{32} & \sigma_{33} - \sigma_m \end{bmatrix}$$

where $\sigma_m = (\sigma_{11} + \sigma_{22} + \sigma_{33})/3$

Decomposing the stress state in this manner demonstrates the property that shear stresses ($i \neq j$) are restricted to the deviatoric component of the stress, whereas the mean stress contains only normal stresses. Because $\sigma_{ij} = \sigma_{ji}$, both the mean stress and the deviatoric stress are **symmetric tensors.**

Once you have determined the stress tensor, it is relatively easy to change the reference system. In this context, you are reminded that the values of the nine stress components are a function of the reference frame. Thus, when changing the reference frame, say by a rotation, the components of the stress tensor are changed. These transformations are greatly simplified by using mathematics for stress analysis, but we'd need another few pages explaining vectors and matrix transformations before we could show some examples. If you would like to see a more in-depth treatment of this topic, several useful references are given in the reading list.

3.11 A BRIEF SUMMARY OF STRESS

Let's summarize where we are in our understanding of stress. You have seen that there are two ways to talk about stress. First, you can refer to stress on a plane (or *traction*), which can be represented by a vector (a quantity with magnitude and direction) that can be subdivided into a component normal to the plane (σ_n, the *normal stress*) and a component parallel to the plane (σ_s, the *shear stress*). If the shear stress is zero, then the stress vector is perpendicular to the plane, but this is a special case; in general, the stress vector is not perpendicular to the plane on which it acts. It is therefore meaningless to talk about stress without specifying the plane on which it is acting. For example, it is wrong to say "the stress at 1 km depth in the Earth is 00°/070°," but it is reasonable to say "the stress vector acting on a vertical, north-south striking joint surface is oriented 00°/070°." In this example there must be a shear stress acting on the fracture; check this for yourself. If the magnitude of this shear stress exceeds the frictional resistance to sliding along the fracture, then there might be movement.

The stress state at a point cannot be described by a single vector. Why? Because a point represents the intersection of an infinite number of planes, and without knowing which plane you are talking about, you cannot define the stress vector. If you want to describe the stress state at a point you must have a tool that will allow you to calculate the stress vector associated with any of the infinite number of planes. We introduced three tools: (1) the stress ellipsoid, (2) the three principal stress axes, and (3) the stress tensor. The stress ellipsoid is the envelope containing the tails or tips (for compression and tension, respectively) of the stress vectors associated with the infinite number of planes passing through the point, with each of the specified vectors and its opposite associated with one plane. On all but three of these planes the vectors have shear stress components. As a rule, there will be three mutually perpendicular planes on which the shear component is zero; the stress vector acting on each of these planes is perpendicular to the plane. These three planes are called the *principal planes of stress,* and the associated stress vectors are the *principal axes of stress,* or *principal stresses* ($\sigma_1 \geq \sigma_2 \geq \sigma_3$). Like any ellipsoid, the stress ellipsoid has three axes, and the principal stresses lie parallel to these axes. Given the three principal stresses, you have uniquely defined the stress ellipsoid; given the stress ellipsoid, you can calculate the stress acting on any random plane that passes through the center of the ellipsoid (which is the point for which we defined the stress state). So, the stress ellipsoid and the principal stresses give a complete description of the stress at a point. Structural geologists find these tools convenient to work with because they are easy to visualize. Thus we often represent the stress state at a point by picturing the stress ellipsoid, or we talk about the values of the principal stresses at a location. For example, we would say that "the orientation of the maximum principal stress at the New York–Pennsylvania border trends about 070°."

For calculations, these tools are a bit awkward and a more general description of stress at a point is needed; this tool is the *stress tensor.* The stress tensor consists of the components of three stress vectors, each associated with a face of an imaginary cube centered in a specified Cartesian frame of reference. Each face of the cube contains two of the Cartesian axes. If it so happens that the stress vectors acting on the faces of the cube have no shear components, then by definition they are the principal stresses, and the axes in your Cartesian reference frame are parallel to the principal stresses. But if you keep the stress state constant and rotate the reference frame, then the three stress vectors will have shear components. The components of the three stress vectors projected onto the axes of your reference frame (giving one normal stress and two shear stresses) are written as components in a 3×3 matrix (a second-rank tensor). If the axes of the reference frame happen to be parallel to the principal stresses, then the diagonal terms of the matrix are the principal stresses and the off-diagonal terms are zero (that is, the shear stresses are zero). If the axes have any other orientation, then the diagonal terms are not the principal stresses and some, or all, of the off-diagonal terms are

not equal to zero. When using the three principal stresses or the stress ellipsoid, you are merely specifying a special case of the stress tensor at a point.

3.12 STRESS TRAJECTORIES AND STRESS FIELDS

By connecting the orientation of principal stress vectors at several points in a body, you obtain trajectories that show the variation in orientation of that vector within the body, which are called **stress trajectories.** Generally, stress trajectories for the maximum and minimum principal stresses are drawn, and a change in trend means a change in orientation of these principal stresses. Collectively, principal stress trajectories represent the orientation of the **stress field** in a body. In some cases the magnitude of a particular stress vector is represented by varying the spacing between the trajectories. An example of the stress field in a block that is pushed on one side is shown in Figure 3.13. If the stress at each point in the field is the same in magnitude and orientation, the stress field is *homogeneous;* otherwise it is *heterogeneous,* as in Figure 3.13. Homogeneity and heterogeneity of the stress field should not be confused with isotropic and anisotropic stress. Isotropic means that the principal stresses are equal (describing a sphere), whereas homogeneous stress implies that the orientation and shape of the stress ellipsoids are equal throughout the body. In a homogeneous stress field, all principal stresses have the same orientation and magnitude. The orientation of stress trajectories under natural conditions typically varies, arising from the presence of discontinuities in

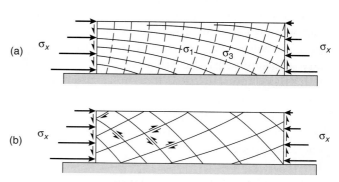

FIGURE 3.13 (a) Theoretical stress trajectories of σ_1 (full lines) and σ_3 (dashed lines) in a block that is pushed from the left resisted by frictional forces at its base. Using the predicted angle between maximum principal stress (σ_1) and fault surface of around 30° (Coulomb failure criterion; Chapter 6) we can predict the orientation of faults, as shown in (b).

TABLE 3.3	SOME STRESS MEASUREMENT METHODS
Borehole breakouts	The shape of a borehole changes after drilling in response to stresses in the host rock. Specifically, the hole becomes elliptical with the long axis of the ellipse parallel to minimum horizontal principal stress $(\sigma_{s, hor})$.
Hydrofracture	If water is pumped under sufficient pressure into a well that is sealed off, the host rock will fracture. These fractures will be parallel to the maximum principal stress (σ_1), because the water pressure necessary to open the fractures is equal to the minimum principal stress.
Strain release	A strain gauge, consisting of tiny electrical resistors in a thin plastic sheet, is glued to the bottom of a borehole. The hole is drilled deeper with a hollow drill bit (called *overcoring*), thereby separating the core to which the strain gauge is connected from the wall of the hole. The inner core expands (by elastic relaxation), which is measured by the strain gauge. The direction of maximum elongation is parallel to the direction of maximum compressive stress and its magnitude is proportional to stress according to Hooke's Law (see Chapter 5).
Fault-plane solutions	When an earthquake occurs, records of the first motion on seismographs around the world enable us to divide the world into two sectors of compression and two sectors of tension. These zones are separated by the orientation of two perpendicular planes. One of these planes is the fault plane on which the earthquake occurred, and from the distribution of compressive and tensile sectors, the sense of slip on the fault can be determined. Seismologists assume that the bisector of the two planes in the tensile sector represents the minimum principal stress (σ_3) and the bisector in the compressive field is taken to be parallel to the maximum compressive stress (σ_1).

rocks, the complex interplay of more than one stress field (like gravity), or contrasts in rheology (Chapter 5).

3.13 METHODS OF STRESS MEASUREMENT

Up to this point our discussion of stress has been pretty theoretical, except perhaps for our classroom experiment with clay and the example of kicking a ball around. Before you forget that stress is a physical quantity rather than an abstract concept (as in psychology), we will close this chapter with a few notes on stress measurements and an application. Because the methods of present-day stress measurements are explained in most engineering texts on rock mechanics, the more common methods are briefly described in Table 3.3. At the end of this section we'll offer a few general comments on geologic stress and give a sense of stress magnitudes.

3.13.1 Present-Day Stress

As it turns out, it is quite difficult to obtain a reliable measure of present-day stress in the Earth. The deter-

mination of the absolute magnitude of the stress is particularly challenging. Generally, stress determinations give the stress differences (the differential stress) and the orientation of the principal stresses, using earthquake focal mechanisms, well-bore enlargements (or "breakouts"), and other in situ stress measurements (Table 3.3), and the analysis of faults and fractures. Earthquake focal mechanisms define a set of two possible fault planes and slip vectors, which are assumed to parallel the maximum resolved shear stress on these planes. Several focal mechanisms and slip vectors on faults of different orientation are used to determine the (best-fit) principal stress axes. The magnitude of stress is based on the energy release of earthquakes, but this relationship is incomplete. The analysis of the orientation of exposed faults and their observed slip uses a similar inversion approach. The elliptical distortion of vertical wells that were drilled for petroleum and gas exploration is a direct gauge of the local stress field and is widely available; the long axis of the distortion is parallel to the horizontal, minimum principal stress. Other in situ stress measurements, such as hydraulic fracturing, in which a hole is capped and pressurized by a fluid until fracturing releases the fluid pressure, reflect the local stress field. Whether this local field reflects the regional (or, remote) tectonic stress or merely the conditions surrounding the particular geologic feature, including the role of pore pressure, is a topic of ongoing debate.

We can get an intuitive sense of differential stress magnitudes in nature from a simple consideration of mountainous regions. We have all looked in awe at steep walls of rock, especially when they are scaled by climbers. In the western Himalayas, vertical cliffs rise up to 2000 m above the valleys. Using Equation 3.16, we can calculate the vertical stress at the base of such cliffs is >50 MPa,[9] while the minimum horizontal stress (that is, atmospheric pressure) is only about 0.1 MPa. Of course mountain ranges of 6–9 km high are not vertical cliffs, so we require a modification of Equation 3.16 to get the differential stress from the load of mountains. Using a triangular load with height h on an elastic medium, we get (without showing the derivation)

$$\sigma_d \approx 0.5 \, \rho \cdot g \cdot h \qquad \text{Eq. 3.17}$$

Given that mountain ranges are up to 9 km in height, this implies differential stresses that exceed 100 MPa.

Present-day stress determinations, like borehole measurements, give differential stress magnitudes that likewise range from tens to hundreds of megapascals. Realize, though, that these methods only record stress magnitudes in the outermost part (upper crust) of Earth (Figure 3.14). The magnitude of differential stresses deep in Earth can only be understood when we know something about Earth's thermal structure and mechanisms of rock deformation, where differential stresses are one to two orders of magnitude less (see Section 3.13.3).

3.13.2 Paleostress

If we wish to determine the ancient stress field from rocks, most of the approaches listed above are not suitable. For the analysis of **paleostress** we are essentially limited to the analysis of fault and fracture data, and to microstructural approaches such as grain-size determinations and grain deformation analysis. Fault-slip analysis requires some understanding of fault mechanics (Chapter 6), but, in short, it uses fault orientation and the sense of slip on that fault with the assumption that the slip direction parallels the maximum shear stress in that plane. Numerical analysis of sufficiently large fault data sets can provide the **reduced stress tensor** that contains the orientations of the three principal stresses and

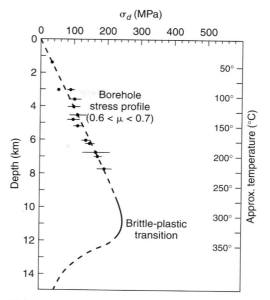

FIGURE 3.14 In situ borehole measurements of differential stress (σ_d) with depth, indicating a friction coefficient (μ) in the range of 0.6–0.7 for the upper crust.

the differential stress ratio, $(\sigma_2 - \sigma_3)/(\sigma_1 - \sigma_2)$, which ranges from 0 (when $\sigma_2 = \sigma_3$) to 1 (when $\sigma_1 = \sigma_2$).

Microstructural methods require an understanding of crystal plastic processes and the development of microstructures that are discussed in Chapter 9. The grain size of plastically deformed rocks appears to be nonlinearly related to the differential stress magnitude, based on laboratory experiments at elevated pressure and temperature conditions. Similarly, the development of deformation microstructures of individual grains, like crystal twins, is a function of the differential stress. Collectively, these approaches broadly constrain the differential stress magnitudes at many depths in Earth, complementing upper crustal data from fault studies and present-day stress determinations. Whereas these methods remain unexplained at this point, some of the information that is obtained from them is incorporated in the final section on stress in Earth.

3.13.3 Stress in Earth

From large data sets of present-day stress measurements we find that the results are generally in good agreement about the orientation of the principal stresses and that they compare reasonably well in magnitude. An application of these approaches and the information that they provide about regional stress patterns and plate dynamics is shown in Figure 3.15. This global synthesis of stress data, part of the World Stress Map

---

[9] 2700 kg/m^3 × 9.8 m/s^2 × 2000 m = 53 MPa.

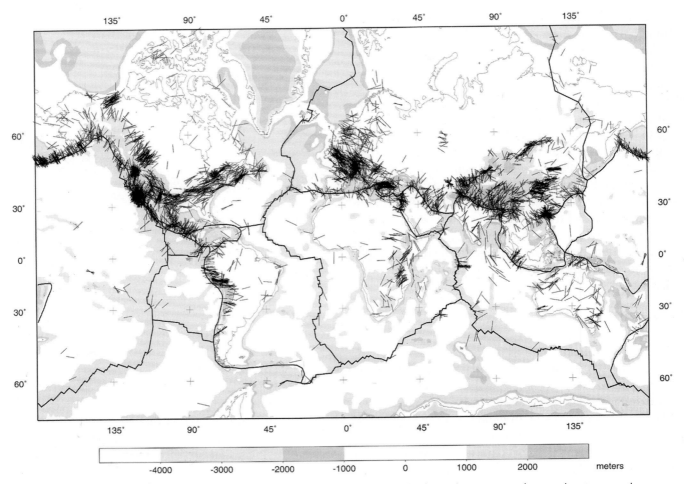

FIGURE 3.15 [a] World Stress Map showing orientations of the maximum horizontal stress superimposed on topography.

project, reflects an international effort that catalogs present-day stress patterns around the world. The majority of stress determinations are from earthquake focal mechanism solutions.

The global stress summary map (Figure 3.15b) shows regionally systematic stress fields in the upper crust, despite the geologic complexity found at Earth's surface. The orientation and magnitudes of horizontal principal stresses are uniform over areas hundreds to thousands of kilometers in extent. These data also show that the upper crust is generally under compression, meaning that maximum compressive stresses are horizontal, resulting in either reverse or strike-slip faulting. For example, the maximum stress in the eastern half of North America is oriented approximately NE–SW with differential stresses on the order of a hundred megapascals. Areas where horizontal tensile stresses dominate are regions of active extension, such as the East African Rift zone, the Basin and Range of western North America, and high plateaus in Tibet and western South America. Using this compilation we can divide the global stress field into *stress provinces,* which generally correspond to active geologic provinces. From this, a pattern emerges that is remarkably consistent with the broad predictions from the main driving forces of plate tectonics (such as the pull of the downgoing slab in subduction zones and "push" at ocean ridges) and with the effects of plate interactions (such as continent–continent collision). In Chapter 14 we'll revisit the driving forces of plate tectonics. When studying these global patterns you must realize that they only reflect the present-day stress field. Many of the world's geologic provinces reflect ancient tectonic activity, with configurations and processes that are no longer active today, and the present-day global stress pattern is unrelated to this past activity. For example, the orientation of today's compressive stresses in eastern North America reflect the opening of the Atlantic Ocean. They are at a high angle to late Paleozoic compressive stresses, which resulted from compressional Appalachian-Caledonian activity at the margin.

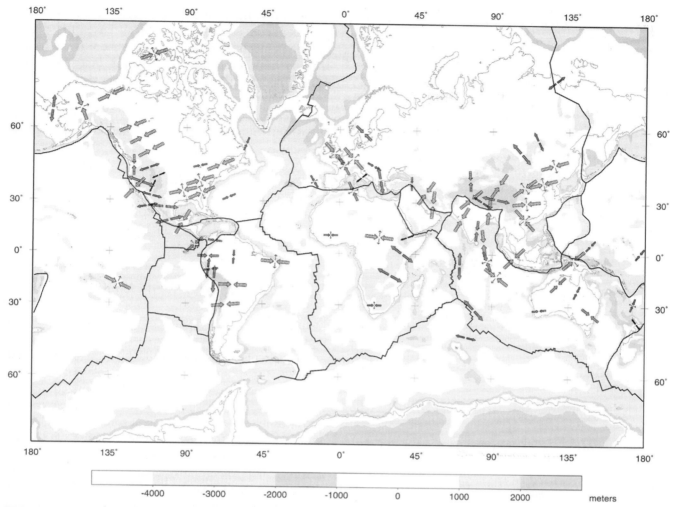

FIGURE 3.15 (Continued.) (b) The generalized pattern based on (a) shows stress trajectories for individual plates; an inward pointing arrow set reflects reverse faulting, an outward pointing arrow set reflects normal faulting, double sets indicate strike-slip faulting.

What happens at depth? While lithostatic pressure increases with depth (see Equation 3.17), differential stress cannot increase without bounds, because the rocks that comprise the Earth do not have infinite strength. **Strength** is the ability of a material to support differential stress; in other words, it is the maximum stress before rocks fail by fracturing or flow. Combining present-day stress and paleostress data with experimental data on flow properties of rocks and minerals (Chapter 9) gives generalized strength profiles for Earth. These **strength curves** represent the differential stress magnitude with depth, given assumptions on the composition and temperature of rocks. At this point we include representative strength curves without much explanation just to give you an idea of stress magnitudes with depth.

You remember from your introductory geology class that the outermost rheologic layer of the Earth is called the **lithosphere,** comprising the crust and part of the upper mantle, which overlies the **asthenosphere.** Taking a quartzo-feldspathic crust and an olivine-rich upper mantle, a low geothermal gradient (about 10°C/km), and a crustal thickness of about 40 km, produces the lithospheric strength curve in Figure 3.16a. You will notice that a sharp decrease in strength occurs around 25 km, which reflects the change from brittle to plastic flow (called the *brittle–plastic transition*) in quartzo-feldspathic rocks. The properties of an olivine-rich mantle are quite different from the crust, and this in turn produces a sharp increase in strength and, therefore, a return to brittle behavior at the crust-mantle boundary

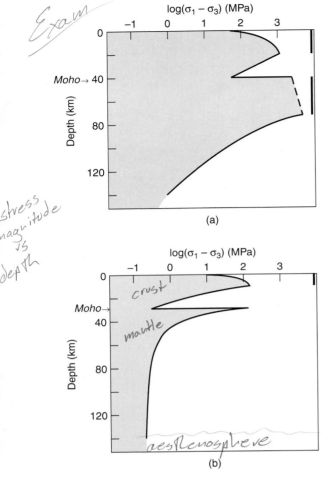

(handwritten annotations: Exam; stress magnitude vs depth; crust; mantle; asthenosphere)

FIGURE 3.16 Strength curves showing the variation in differential stress magnitude with depth in the Earth for (a) a region characterized by a low geothermal gradient (e.g., Precambrian shield areas) and (b) a region with a high geothermal gradient (e.g., areas of continental extension). Differential stresses are largely based on experimental data for brittle failure and ductile flow, which change as a function of composition and temperature. In these diagrams the only compositional change occurs at the crust–mantle boundary (the Moho); in the case of additional compositional stratification, more drops and rises will be present in the strength curve. The bar at the right side of each diagram indicates where seismic activity may occur.

(the Moho). As with the crustal profile, strength decreases as plastic behavior replaces the brittle regime of mantle rocks with depth. This brittle versus plastic behavior of rock is strongly dependent on temperature, as demonstrated by creating a strength profile at a higher geothermal gradient (Figure 3.16b), which promotes plastic flow and reduces the strength up to one order of magnitude. With a geothermal gradient

of 20°C/km, the brittle–plastic transition now occurs at a depth of about 10 km. In all cases the deeper mantle is mechanically weak, because it is characterized by high temperatures, meaning that the mantle only supports differential stresses on the order of a few megapascals. When considering strength profiles, remember the distinction between differential stress and lithostatic pressure. The lithostatic pressure always becomes greater with depth in the Earth, and is orders of magnitude greater than the differential stress. For example, the lithostatic stress at a depth of 100 km in the Earth is several thousand megapascals (use Equation 3.16), but the differential stress is only on the order of 1–10 MPa!

3.14 CLOSING REMARKS

Dynamic analysis, the study of stresses in a body, is a topic whose relevance goes well beyond structural geology. Societal challenges like building collapse and mass wasting come to mind as examples of phenomena whose disastrous effects can be minimized by adequate knowledge of stress states. In this chapter you have learned the fundamentals of force and stress, and obtained an intuitive sense of the meaning of stresses on a body, the stress ellipsoid, and stress conditions in Earth. A more quantitative analysis of the material is left for advanced texts (see reading list). Throughout this book we mainly focus on the general relationship (or lack thereof) between the geometry of geologic structures and their origins. Although many aspects of dynamic analysis remained unmentioned in this chapter, you now have the basic tools needed for the next step in our journey: the analysis of deformation and strain.

ADDITIONAL READING

Anderson, D. L., 1989. *Theory of the Earth.* Blackwell Scientific: Oxford.

Angelier, J., 1994. Fault slip analysis and Paleostress reconstruction. In Hancock, P. L., ed., *Continental deformation.* Pergamon, pp. 53–100.

Engelder, T., 1993. *Stress regimes in the lithosphere.* Princeton University Press.

Jaeger, J. C., and Cook, N. G. W., 1976. *Fundamentals of rock mechanics.* Chapman and Hall: London.

Means, W. D., 1976. *Stress and strain—basic concepts of continuum mechanics for geologists.* Springer-Verlag: New York.

Nye, J. F., 1985. *Physical properties of crystals, their representation by tensors and matrices* (2nd edition). Oxford University Press: Oxford.

Turcotte, D. L., and Schubert, G., 1982. *Geodynamics—applications of continuum physics to geological problems.* J. Wiley & Sons: New York.

Zoback, M. L., 1992. First and second order patterns of stress in the lithosphere: the World Stress Map project. *J. Geophysical Research,* 97, 11703–11728.s

4.1 INTRODUCTION

The geologic history of most crustal rocks involves significant changes in the shape of original features like sedimentary bedding, igneous structures, rock inclusions, and grains. The formation of folds or faults springs to mind as an example of this deformation, which we examine in detail later in the book. A small-scale example of deformation is illustrated in the slab of rock in Figure 4.1, which contains Cambrian trilobites that were distorted from the original shape of these fossils. Deformed fossils, folds, and other features document the permanent shape changes that occur in natural rocks. The study and quantification of these distortions, which occur in response to forces acting on bodies, is the subject of "Deformation and Strain."

Recall the force of gravity, for example. Pouring syrup on pancakes is easy because of Earth's gravity, but in a space station it is quite difficult to keep the syrup in its preferred place. That you are able to read this text sitting down is another convenient effect of Earth's gravity; in a space station you would be floating around (possibly covered by syrup). Let us consider a more controlled experiment to analyze the response of materials to an applied force. We can change the shape of a block of clay or plasticine by the action of, say, your hands or a vise. When forces affect the spatial geometry of a body (syrup, you, plasticine, or rocks) we enter the realm of deformation. Most simply stated: *deformation of a body occurs in response to forces*. We will see later that deformation affects stress (force acting on an area), so there is no simple stress–deformation relationship. The response of a body to forces may have many faces. In some cases, the body is merely displaced or rotated, such as when you get up from the chair and move around the room. In other cases, the body becomes distorted, as in the

FIGURE 4.1   Deformed trilobites (*Angelina sedgwicki*) in a Cambrian slate from Wales. Knowledge about their original symmetry enables us to quantify the strain.

clay block experiment or with the flow of syrup. In this chapter we will examine these responses both qualitatively and quantitatively.

4.2   DEFORMATION AND STRAIN

Deformation and strain are closely related terms that are sometimes used as synonyms, but they are not the same. **Deformation** describes the collective displacements of points in a body; in other words, it describes the complete transformation from the initial to the final geometry of a body. This change can include a **translation** (movement from one place to the other), a **rotation** (spin around an axis), and a **distortion** (change in shape). **Strain** describes the changes of points in a body relative to each other; so, it describes the distortion of a body. This distinction between deformation and strain may not be immediately obvious from these abstract descriptions, so we use an example. In Figure 4.2 we change the shape and position of a square, say, a slice of the clay cube we used in Chapter 3. We arbitrarily choose a reference frame with axes that parallel the margins of the printed page. The displacement of points within the body, represented by the four corner points of the square, are indicated by vectors. These vectors describe the displacement field of the body from the ini-

tial to the final shape. The displacement field can be subdivided into three components:

1. A distortion (Figure 4.2b)
2. A rotation (Figure 4.2c)
3. A translation (Figure 4.2d)

Each component in turn can be described by a vector field (shown for point *A* only) and their sum gives the total **displacement field.** Importantly, a change in the order of addition of these vector components affects the final result. Deformation, therefore, is not a vector entity, but a second-order tensor (similar to stress). We will return to this later.

When the rotation and distortion components are zero, we only have a translation. This translation is formally called **rigid-body translation** (RBT), because the body undergoes no shape change while it moves. For convenience, we will simply refer to this component as **translation,** and the deformation is called translational. When the translation and distortion components are zero, we have only rotation of the body. By analogy to translation, we call this component **rigid-body rotation** (RBR), or simply **spin,** and the corresponding deformation is called rotational. Recalling the pool table example of the previous chapter (Figure 3.2), the deformation of a pool ball that is hit is fully described by a translational and a spin component. When translation and spin

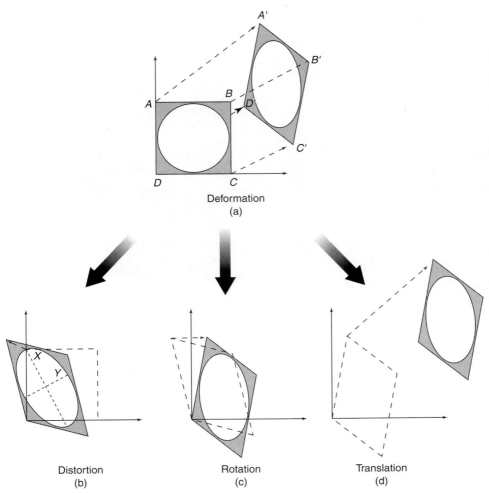

FIGURE 4.2 The components of deformation. The deformation of a square (a) is subdivided into three independent components: (b) a distortion; (c) a rotation; and (d) a translation. The displacement of each material point in the square, represented by the four corners of the initial square, describes the displacement field. The corresponding strain ellipse is also shown. The distortion occurs along the axes *X* and *Y*, which are the principal strain axes.

Deformation
(a)

Distortion
(b)

Rotation
(c)

Translation
(d)

are both zero, the body undergoes distortion; this component is described by strain. So, strain is a component of deformation and therefore not a synonym. In essence, we have defined deformation and strain relative to a **frame of reference.** Deformation describes the complete displacement field of points in a body relative to an *external reference frame,* such as the edges of the paper on which Figure 4.2 is drawn. Strain, on the other hand, describes the displacement field of points relative to each other. This requires a reference frame within the body, an *internal reference frame,* like the edges of the square. Place yourself in the square and you would be unaware of any translation, just as when you are flying in an airplane or riding a train.[1] Looking out of

the window, however, makes you aware of the displacement by offering an external reference frame.

One final element is missing in our description of deformation. In Figure 4.2 we have constrained the shape change of the square by maintaining a constant area. You recall that shape change results from the deviatoric component of the stress, meaning where the principal stresses are unequal in magnitude (see Figure 3.12). The hydrostatic component of the total stress, however, contributes to deformation by changing the area (or volume, in three dimensions) of an object. Area or volume change is called **dilation**[2] and is positive or negative, as the volume increases or decreases, respectively. Because dilation results in changes of line lengths it is similar to strain, except that the relative lengths of the lines remain the same. Thus, it is useful to distinguish strain from volume change. In summary, deformation is described by:

1. Rigid-body translation (or translation)
2. Rigid-body rotation (or spin)
3. Strain
4. Volume change (or dilation)

It is relatively difficult in practice to determine the translational, spin, and dilational components of deformation. Only in cases where we are certain about the original position of a body can translation and spin be determined, and only when we know the original volume of a body can dilation be quantified. On the

---

[1]We assume constant velocity; your stomach would notice acceleration.

[2]Or dilatation in the United Kingdom.

other hand, we often do know the original shape of a body, so the quantification of strain is a common activity in structural geology.

4.3 HOMOGENEOUS STRAIN AND THE STRAIN ELLIPSOID

Strain describes the distortion of a body in response to an applied force. Strain is **homogeneous** when any two portions of the body that were similar in form and orientation before are similar in form and orientation after strain. This can be illustrated by drawing a square and a circle on the edge of a deck of cards; homogeneous strain changes a square into a parallelogram and a circle into an ellipse (Figure 4.3b). We define homogeneous strain by its geometric consequences:

1. Originally straight lines remain straight.
2. Originally parallel lines remain parallel.
3. Circles become ellipses; in three dimensions, spheres become ellipsoids.

When one or more of these three restrictions does not apply, we call the strain **heterogeneous** (Figure 4.3c). Because conditions (1) and (2) are maintained during the deformation components of translation and rotation, deformation is homogeneous by definition if the strain is homogeneous. Conversely, heterogeneous strain implies heterogeneous deformation. Homogeneous and heterogeneous deformation should not be confused with rotational and nonrotational deformation; the latter reflect the presence of a spin component.

Because heterogeneous strain is more complex to describe than homogeneous strain, we try to analyze heterogeneously strained bodies or regions by separating them into homogeneous portions. In other words, homogeneity of deformation is a matter of scale. Consider a heterogeneous deformation feature like a fold, which can be approximated by three essentially homogeneous sections: the two limbs and the hinge (see Chapter 10 for fold terminology). The heterogeneously deformed large square of Figure 4.3c consists of nine smaller squares for which the strain conditions are approximately homogeneous. Given the scale dependence of homogeneity and not to complicate our explanations unnecessarily, we will limit our discussion in this chapter to homogeneous strain.

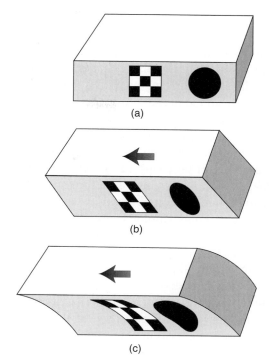

(a)

(b)

(c)

FIGURE 4.3 Homogeneous and heterogeneous strain. A square and a circle drawn on a stack of cards (a) transform into a parallelogram and an ellipse when each card slides the same amount, which represents homogeneous strain (b). Heterogeneous strain (c) is produced by variable slip on the cards, for example by increasing the slip on individual cards from bottom to top.

In a homogeneously strained, two-dimensional body there will be at least two **material lines** that do not rotate relative to each other, meaning that their angle remains the same before and after strain. What is a material line? A material line connects features, such as an array of grains, that are recognizable throughout a body's strain history. The behavior of four material lines is illustrated in Figure 4.4 for the two-dimensional case,

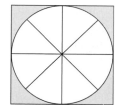 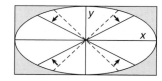

FIGURE 4.4 Homogeneous strain describes the transformation of a square to a rectangle or a circle to an ellipse. Two material lines that remain perpendicular before and after strain are the principal axes of the strain ellipse (solid lines). The dashed lines are material lines that do not remain perpendicular after strain; they rotate toward the long axis of the strain ellipse.

in which a circle changes into an ellipse. In homogeneous strain, two orientations of material lines remain perpendicular before and after strain. These two material lines form the axes of an ellipse that is called the **strain ellipse.** Note that the lengths of these two material lines change from the initial to the final stage; otherwise we would not strain our initial circle. Analogously, in three dimensions we have three material lines that remain perpendicular after strain and they define the axes of an ellipsoid, the **strain ellipsoid.** The lines that are perpendicular before and after strain are called the **principal strain axes.** Their lengths define the strain magnitude and we will use the symbols X, Y, and Z to specify them, with the convention that $X \geq Y \geq Z$. In a more intuitive explanation, you may consider the strain ellipsoid as the modified shape of an initial sphere embedded in a body after the application of a homogeneous strain. We describe strain in two-dimensional space by the two axes of the strain ellipse and an angle describing the rotation of this ellipse. In three-dimensional space, therefore, we use the three axes of the strain ellipsoid and three rotation angles. This means that the strain ellipsoid is defined by six independent components, which is reminiscent of the stress ellipsoid (see Section 3.6.2). Indeed, the strain ellipsoid is a visual representation of a second-rank tensor, but keep in mind that the stress and strain

ellipsoids are not the same. We dedicate an entire chapter to the relationship between the stress and strain ellipsoids (Chapter 5, "Rheology"), where you will explore its complexity.

4.4 STRAIN PATH

The measure of strain that compares the initial and final configuration is called the **finite strain,** identified by subscript f, which is independent of the details of the steps toward the final configuration. When these intermediate strain steps are determined they are called **incremental strains,** identified by subscript i. The summation of all incremental strains (that is, their product), therefore, is the finite strain. We will see that there are many ways to measure finite strain in a rock, but measurement of strain increments is more difficult. Yet, incremental strain may be more crucial for unraveling the deformation history of a rock or region than finite strain. Let us explore this with a simple example (Figure 4.5).

Finite strains for the distortion of a square in Figures 4.5a and 4.5b are the same, because the initial and final configurations are identical. The steps or strain increments by which these final shapes were

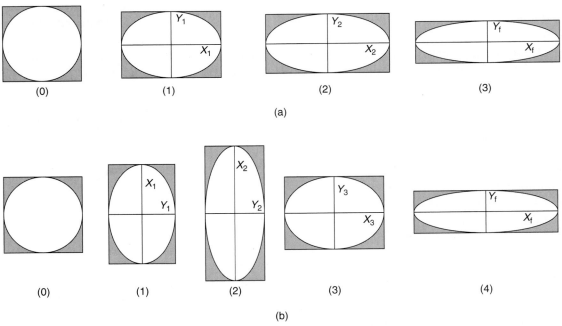

FIGURE 4.5 The finite strains, X_f and Y_f, in (a) and (b) are the same, but the strain path by which each was reached is different. This illustrates the importance of understanding the incremental strain history (here, X_i and Y_i) of rocks and regions and inherent limitation of finite strain analysis.

reached, however, are very different. We say that the **strain path** of the two examples is different, but the finite strains are the same. The path presented in Figure 4.5a has incremental strains that reflect a strain ellipse that becomes increasingly elliptical; in other words, the ratio of the long over the short axis (X/Y) becomes greater. The path in Figure 4.5b, on the other hand, shows that the orientation of the X_i and Y_i axes was perpendicular to those of the finite strain ellipse during part of the history. Consider this in a geologic context: the two paths would represent very different strain histories of a region, yet their finite strains would be identical. Obviously, an important piece of information is lost without knowledge of the strain path. It is therefore critical for structural analysis to distinguish finite strain from incremental strain. But because incremental strains are harder to determine, structural geologists imply finite strain when they loosely discuss the "strain" of a rock or a region.

4.5 COAXIAL AND NON-COAXIAL STRAIN ACCUMULATION

Earlier (Figure 4.4) we saw that strain involves the rotation of material lines. Recall that a material line is made up of a series of points in a body; for example, a row of calcium atoms in a calcite crystal or an array of grains in a quartzite. There is no mechanical contrast between the material line and the body as a whole, so that material lines behave as **passive markers.** All material lines in the body, except those that remain perpendicular before and after a strain increment (that is, the principal strain axes), rotate relative to each other. In the general case for strain, the principal incremental strain axes are not necessarily the same throughout the strain history. In other words, the principal incremental strain axes rotate relative to the finite strain axes, a scenario that is called **non-coaxial strain accumulation.** The case in which the same material lines remain the principal strain axes at each increment is called **coaxial strain accumulation.** These important concepts are not obvious, so we turn to a classroom experiment for further exploration.

First we examine non-coaxial strain accumulation. Take a deck of playing cards (or a thick phone book) and draw a circle on the face perpendicular to the cards. By sliding the cards past one another by roughly equal amounts, the initial circle changes into an ellipse (Figure 4.6a). Draw the ellipse axes (i.e., incremental strain axes) X_1 and Y_1 on the face. Continuing to slide the cards produces a more elliptical shape. Again mark the ellipse axes X_2 and Y_2 of this second step on the cards, but use another color. Continue this action a third time so that in the end you have three steps (increments) and three X-Y pairs. Note that the last ellipse represents the finite strain. Now, as you return

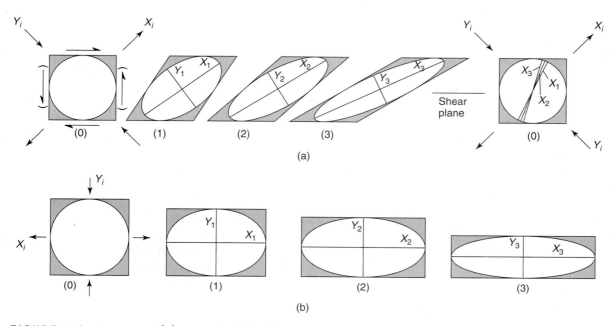

FIGURE 4.6 Non-coaxial (a) and coaxial (b) strain. The incremental strain axes are different material lines during non-coaxial strain. In coaxial strain the incremental strain axes are parallel to the same material lines. Note that the magnitude of the strain axes changes with each step.

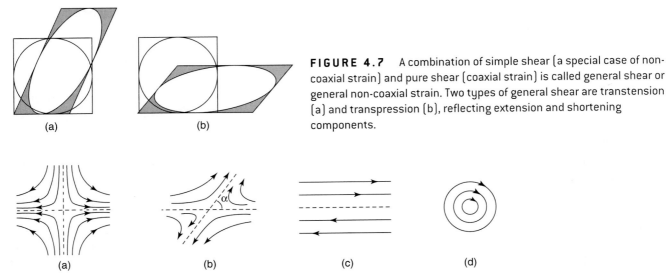

FIGURE 4.7 A combination of simple shear (a special case of non-coaxial strain) and pure shear (coaxial strain) is called general shear or general non-coaxial strain. Two types of general shear are transtension (a) and transpression (b), reflecting extension and shortening components.

FIGURE 4.8 Particle paths or flow lines during progressive strain accumulation. These flow lines represent pure shear (a), general shear (b), simple shear (c), and rigid-body rotation (d). The cosine of the angle α is the kinematic vorticity number, W_k, for these strain histories; $W_k = 0$, $0 < W_k < 1$, $W_k = 1$, and $W_k = \infty$, respectively.

the cards to their starting configuration, by restoring the original circle, you will notice that the pairs of strain axes of the three increments do not coincide. For each step a different set of material lines maintained perpendicularity, and thus the incremental strain axes do not coincide with the finite strain axes. You also see that with each step the long axis of the finite strain ellipse rotated more toward the shear plane over which the cards slide. You can imagine that a very large amount of sliding will orient the long axis of the finite strain ellipse nearly parallel to (meaning a few degrees off) the shear plane.

In the case of coaxial strain accumulation we return to our earlier experiment with clay (Figure 4.6b). Take a slice of clay with a circle drawn on its front surface and press down on the top and bottom. When you draw the incremental strain axes at various steps, you will notice that they coincide with one another, while the ellipticity (the *X/Y* ratio) increases. So, with coaxial strain accumulation there is no rotation of the incremental strain axes with respect to the finite strain axes.[3]

The component describing the rotation of material lines with respect to the principal strain axes is called the **internal vorticity,** which is a measure of the degree of non-coaxiality. If there is zero internal vorticity, the strain history is coaxial (as in Figure 4.6b), which is sometimes called **pure shear.** The non-coaxial strain

history in Figure 4.6a describes the case in which the distance perpendicular to the shear plane (or the thickness of our stack of cards) remains constant; this is also known as **simple shear.** In reality, a combination of simple shear and pure shear occurs, which we call **general shear** (or **general non-coaxial strain accumulation;** Figure 4.7). Internal vorticity is quantified by the **kinematic vorticity number,** W_k, which relates the angular velocity and the stretching rate of material lines. Avoiding the math, a convenient graphical way to understand this parameter is shown in Figure 4.8. When tracking the movement of individual points within a deforming body relative to a reference line, we obtain a displacement field (or **flow lines**) that enables us to quantify the internal vorticity. The angular relationship between the asymptote and the reference line defines W_k:

$$W_k = \cos \alpha \qquad \text{Eq. 4.1}$$

For pure shear $W_k = 0$ (Figure 4.8a), for general shear $0 < W_k < 1$ (Figure 4.8b), and for simple shear $W_k = 1$ (Figure 4.8c). Rigid-body rotation or spin can also be described by the kinematic vorticity number (in this case, $W_k = \infty$; Figure 4.8d), but remember that this rotational component of deformation is distinct from the internal vorticity of strain. Using Figure 4.6 as an example, the deformation history shown in Figure 4.6a represents non-coaxial, nonrotational deformation. The orientation of the shear plane does not rotate between each step, but the incremental strain axes do

---

[3]In kinematic theory we use the *infinitesimally* small incremental strain axes or the *instantaneous* strain axes.

TABLE 4.1	TYPES OF STRAIN
Coaxial strain	Strain in which the incremental strain axes remain parallel to the finite strain axes during progressive strain
Heterogeneous strain	Strain in which any two portions of a body similar in form and orientation before strain undergo relative change in form and orientation (also: *inhomogeneous strain*)
Homogeneous strain	Strain in which any two portions of a body similar in form and orientation before strain remain similar in form and orientation after strain
Incremental strain	Strain state of one step in a progressive strain history
Instantaneous strain	Incremental strain of vanishingly small magnitude (a mathematical descriptor); also called *infinitesimal incremental strain*
Finite strain	Strain that compares the initial and final strain configurations; sometimes called *total strain*
Non-coaxial strain	Strain in which the incremental strain axes rotate relative to the finite strain axes during progressive strain

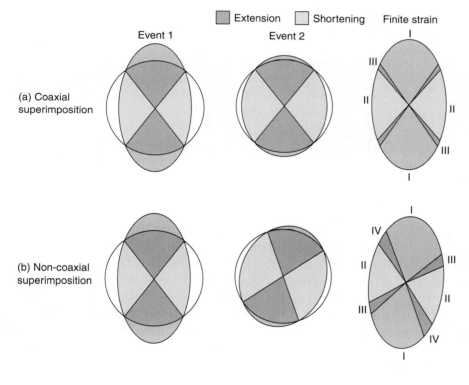

FIGURE 4.9 Superimposed strain. The strain ellipse contains regions in which lines are extended and regions where lines are shortened, which are separated by lines of zero finite elongation. (a) Coaxial superimposition of a strain event 2 produces regions in which lines continue to be extended (I), and regions where lines continue to be shortened (II), separated by regions where shortening during event 1 is followed by extension during event 2 (III). (b) Non-coaxial superimposition additionally produces a region where extension is followed by shortening during event 2 (IV). Practically this means that structures reflecting different strain states may be formed, especially in non-coaxial strain regimes such as shear zones.

rotate. The strain history in Figure 4.6b represents coaxial, nonrotational deformation, because the incremental axes remain parallel.

Already, several types of strain have been introduced, so we summarize them in Table 4.1 before continuing.

4.6 SUPERIMPOSED STRAIN

The strain path describes the superimposition of a series of strain increments. For each of these incre-

ments the body is divided into regions containing material lines that extend and shorten; these regions are separated by planes containing lines with zero length change. Considering only the two-dimensional case (Figure 4.9) we recognize regions of extension and regions of shortening separated along two lines of zero length change in the strain ellipse. When we coaxially superimpose a second strain increment on the first ellipse, we obtain three regions (Figure 4.9a): (I) a region of continued extension, (II) a region of continued shortening, and (III) a region of initial shortening

FIGURE 4.10 A small fold in turbidites from the Newfoundland Appalachians (Canada). An axial plane cleavage is visible in the mica-rich layer. Lens cover for scale.

that is now in extension. The geometry is a little more complex when the incremental strain history is non-coaxial. Superimposing an increment non-coaxially on the first strain state results in four regions (Figure 4.9b): (I) a region of continued extension, (II) a region of continued shortening, (III) a region of initial shortening that is now in extension, and (IV) a region of extension that is now in shortening. Clearly, superimposition of significant strain increments can produce complex deformation patterns in rocks. For example, extensional structures formed during one part of the deformation history may become shortened during a later part of the history, resulting in outcrop patterns that, at first glance, may seem contradictory.

4.7 STRAIN QUANTITIES

Having examined the necessary fundamentals of strain, we can now turn our attention to practical applications in structural analysis using the quantification of strain.

Take the small fold with axial plane cleavage shown in Figure 4.10. How much strain does this deformation feature represent? How do we go about determining this? In the next sections we will examine strain quantification using three measures: length change or **longitudinal strain,** volume change or **volumetric strain,** and angular change or **angular strain.** You'll find that all of these approaches are pertinent to the analysis of our little fold.

4.7.1 Longitudinal Strain

Longitudinal strain is defined as a change in length divided by the original length. Longitudinal strain is expressed by the **elongation, e,** which is defined as

$$\mathbf{e} = (l - l_o)/l_o \qquad \text{or} \qquad \mathbf{e} = \delta l/l_o \qquad \text{Eq. 4.2}$$

where l is the final length, l_o is the original length, and δl is the length change (Figure 4.11a). Because we divide values with the same units, longitudinal strain is a dimensionless quantity. A longitudinal strain of

0.3 for a stretched rod or a continental region is independent of the original dimensions of the object. This definition of elongation implies that negative values of **e** reflect shortening whereas positive values of **e** represent extension. We label the maximum, intermediate, and minimum elongations, e_1, e_2, and e_3, respectively, with $e_1 \geq e_2 \geq e_3$. Remember the sign convention we just described! In practice, geologists commonly give the elongation in percent, using the absolute value, $|e| \times 100\%$, and the terms shortening and extension instead of a negative or positive sign; for example, 30% extension or 40% shortening.

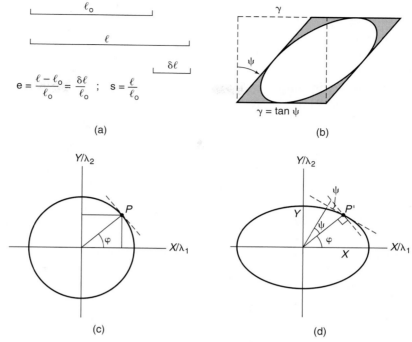

FIGURE 4.11   Strain quantities. The elongation, **e**, and stretch, **s**, in (a); the angular shear, ψ, and the shear strain, γ, in (b). In (c) the relationships between quadratic elongation (λ), stretch (**s**), and angular shear (ψ) are shown for line OP that transforms into OP' (d).

4.7.2   Volumetric Strain

A relationship similar to that for length changes holds for three-dimensional (volume) change. For *volumetric strain*, Δ, the relationship is

$$\Delta = (V - V_0)/V_0 \quad \text{or} \quad \Delta = \delta V/V_0 \qquad \text{Eq. 4.3}$$

where V is the final volume, V_0 is the original volume, and δV is the volume change. Like longitudinal strain, volumetric strain is a ratio of values with the same units, so it also is a dimensionless quantity. Positive values for Δ represent volume gain, whereas negative values represent volume loss.

4.7.3   Angular Strain

Longitudinal and volumetric strain are relatively straightforward and easily defined strain parameters. Angular strains are slightly more difficult to handle as they measure the change in angle between two lines that were initially perpendicular. The change in angle is called the **angular shear**, ψ, but more commonly we use the tangent of this angle, called the **shear strain**, γ (Figure 4.11b):

$$\gamma = \tan \psi \qquad \text{Eq. 4.4}$$

Like the longitudinal and volumetric strains, the shear strain is a dimensionless parameter.

4.7.4   Other Strain Quantities

In calculations such as those associated with the Mohr circle for strain (see Section 4.8), we make use of a quantity called the **quadratic elongation,** λ, which is defined as

$$\lambda = (l/l_0)^2 = (1 + e)^2 \qquad \text{Eq. 4.5}$$

where l is the final length, l_0 is the original length, and **e** is the elongation.

The root of the quadratic elongation is called the **stretch, s,** which is a convenient strain parameter that directly relates to the dimensions of the strain ellipsoid:

$$s = \lambda^{1/2} = l/l_0 = 1 + e \qquad \text{Eq. 4.6}$$

The quadratic elongation, λ, and especially the stretch, **s,** are convenient measures because they describe the lengths of the principal axes (X, Y, and Z) of the strain ellipsoid:

$$X = s_1, \, Y = s_2, \, Z = s_3 \qquad \text{Eq. 4.7}$$

with $X \geq Y \geq Z$, and

$$X^2 = \lambda_1, \quad Y^2 = \lambda_2, \quad Z^2 = \lambda_3 \qquad \text{Eq. 4.8}$$

This relationship between the quadratic elongation, stretches, and the strain ellipse is illustrated in two dimensions in Figure 4.11c. A circle with unit radius ($r = 1$) becomes distorted into an ellipse that is defined by the length of axes $\sqrt{\lambda_1}$ (i.e., $= X$) and $\sqrt{\lambda_2}$ (i.e., $=Y$). As a consequence of this distortion, a line OP at an initial angle of φ with the X-axis becomes elongated (OP') with an angle φ' to the λ_1/X-axis.[4] From Figure 4.11c you can determine that the relationship between φ and φ' is described by

$$\tan \varphi' = Y/X \cdot \tan \varphi = (\lambda_2/\lambda_1)^{1/2} \cdot \tan \varphi \qquad \text{Eq. 4.9}$$

or, rearranging this equation

$$\tan \varphi = X/Y \cdot \tan \varphi' = (\lambda_1/\lambda_2)^{1/2} \cdot \tan \varphi' \qquad \text{Eq. 4.10}$$

In Section 4.2 we introduced the concept of volume change as a fourth component of deformation in which length changes occur proportionally. It is instructive to see how this affects Equations 4.9 and 4.10. In the two-dimensional case, the area of an ellipse is $\pi(X \cdot Y)$, thus an ellipse derived from a unit circle with area π ($r = 1$, so $\pi r^2 = \pi$) implies that $X \cdot Y = 1$. Thus, if we assume no volume change in deformed objects, we simplify Equations 4.9 and 4.10 to

$$\tan \varphi' = X^{-2} \cdot \tan \varphi \qquad \text{Eq. 4.11}$$

or

$$\tan \varphi = X^2 \cdot \tan \varphi' \qquad \text{Eq. 4.12}$$

because $Y = 1/X$.

Go ahead and apply these relationships to the geologic situation posed in Figure 4.12 to get some hands-on experience.[5] You can further explore your understanding of the above manipulations by showing that Equations 4.11 and 4.12 also apply to the three-dimensional case when we assume that $Y = 1$, as well as that the volume is constant. The condition $Y = 1$ means that zero elongation occurs in the direction perpendicular to the sectional ellipse containing axes X

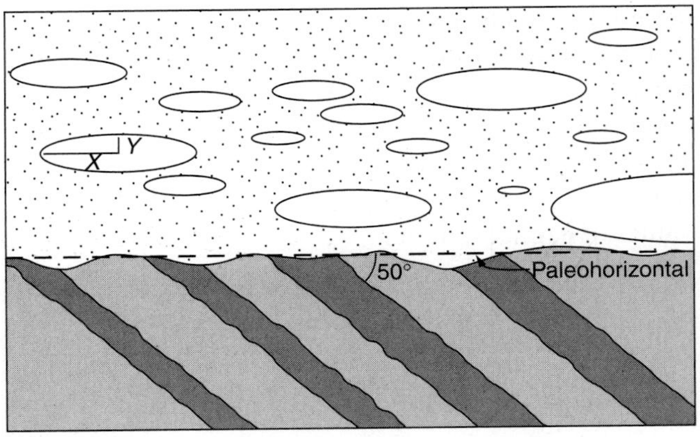

FIGURE 4.12 A sequence of tilted sandstone beds is unconformably overlain by a unit containing ellipsoidal inclusions (e.g., clasts in a conglomerate). The strain ratio of the inclusions in sectional view is $X/Y = 4$, and the dip of the underlying beds is $50°$. What was the angle of dip for the beds in sectional view if the inclusions were orginally spherical?

and Z, which is a commonly made assumption in strain analysis.

Earlier we distinguished between information contained in finite strain analysis from incremental strain analysis. The measure of strain that is most suitable for incremental strain histories is the **natural strain.** This measure of strain is no more or less "natural" than any of the other measures, but derives its name from the natural logarithm.[6] Natural strain does not compare the initial and final strain states, but is the summation of individual strain increments. Recall that the elongation is defined as $\delta l/l_0$ (Equation 4.2). This also holds for incremental strains, in which l_0 represents the length at the beginning of each increment. For a vanishingly small increment (or **infinitesimal strain**), the elongation is defined as

$$\mathbf{e}_i = \delta l/l_0 \qquad \text{Eq. 4.13}$$

The natural strain, ε (epsilon), is the summation of these increments:

$$\varepsilon = \sum_{l=l_0}^{l=l} \delta l/l_0 = \int_{l_0}^{l} \delta l/l_0 \qquad \text{Eq. 4.14}$$

Integrating Equation 4.14 gives

$$\varepsilon = \ln l/l_0 = \ln \mathbf{s} \qquad \text{Eq. 4.15}$$

---

[4]It is customary to use the ' (prime) to mark the deformed state.

[5]The angle of the dipping beds before deformation of the overlying conglomerate is about 78°.

---

[6]The base of the natural logarithm (ln) is 2.72; the base 10 (log) is also used.

Using Equation 4.6, you can express the natural strain in terms of the elongation, **e,** and the quadratic elongation, λ:

$$\varepsilon = \ln(1 + \mathbf{e}) \qquad \text{Eq. 4.16}$$
$$\varepsilon = \tfrac{1}{2}\ln\lambda \qquad \text{Eq. 4.17}$$

4.8 THE MOHR CIRCLE FOR STRAIN

We learned that the strain state is described geometrically by an ellipsoid, so strain is a *second-rank tensor.* We can use the same mathematics for strain that we have used for stress in Chapter 3, but do remember that the stress and strain ellipsoids are not the same. Because of the mathematical similarities, a Mohr circle construction for strain can be used to represent the relationship between longitudinal and angular strain in a manner similar to that for σ_n and σ_s in the Mohr diagram for stress. Usually the quadratic elongation, λ, and the shear strain, γ, are used, for which we need to rewrite some relationships and introduce a few convenient substitutions.

Considering Figure 4.11c and applying some trigonometric relationships, we get

$$\lambda = \lambda_1 \cos^2\varphi + \lambda_3 \sin^2\varphi \qquad \text{Eq. 4.18}$$
$$\lambda = \tfrac{1}{2}(\lambda_1 + \lambda_3) + \tfrac{1}{2}(\lambda_1 - \lambda_3)\cos 2\varphi \qquad \text{Eq. 4.19}$$

and

$$\gamma = [(\lambda_1/\lambda_3) - \lambda_3/\lambda_1 - 2)]^{1/2}\cos\varphi\sin\varphi \qquad \text{Eq. 4.20}$$
$$\gamma = -\tfrac{1}{2}(\lambda_1 - \lambda_3)\sin 2\varphi \qquad \text{Eq. 4.21}$$

But this expresses strain in terms of the undeformed state. We observe a body after strain has occurred, so it is more logical to express strain in terms of the deformed state. We therefore need to express the equations in terms of the angle φ' that we measure rather than the original angle φ, which is generally unknown. To this end we introduce the parameters $\lambda' = 1/\lambda$ and $\gamma' = \gamma/\lambda$ and use the equations for double angles. We then get

$$\lambda' = \tfrac{1}{2}(\lambda_1' + \lambda_3') - \tfrac{1}{2}(\lambda_3' - \lambda_1')\cos 2\varphi' \quad \text{Eq. 4.22}$$
$$\gamma' = \tfrac{1}{2}(\lambda_3' - \lambda_1')\sin 2\varphi' \qquad \text{Eq. 4.23}$$

If you compare these equations with Equations 3.7 and 3.10 for the normal stress (σ_n) and the shear stress (σ_s)

and follow their manipulation in Section 3.8, you will find that Equations 4.22 and 4.23 describe a circle with a radius $\tfrac{1}{2}(\lambda_3' - \lambda_1')$, whose center is located at $\tfrac{1}{2}(\lambda_1' + \lambda_3')$ in a reference frame with γ' on the vertical axis and λ' on the horizontal axis. This is the Mohr circle construction for strain.

At first glance these manipulations appear unnecessarily confusing and they tend to discourage the application of the construction. So, let's look at an example (Figure 4.13). Assume that a unit square is shortened by 50% and extended by 100% (Figure 4.13a). Thus, $\mathbf{e}_1 = 1$ and $\mathbf{e}_3 = -0.5$, respectively; consequently, $\lambda_1 = 4$ and $\lambda_3 = 0.25$. Note that the area remains constant because $\lambda_1^{1/2} \cdot \lambda_3^{1/2} = 1$. Using the parameter λ', we get $\lambda_1' = 0.25$ and $\lambda_3' = 4$. Plotting these values on the Mohr diagram results in a circle with radius $r = \tfrac{1}{2}(\lambda_3' - \lambda_1') = 1.9$, whose center is at $\tfrac{1}{2}(\lambda_1' + \lambda_3') = 2.1$ on the λ' axis. It is now quite simple to obtain a measure of the longitudinal strain and the angular strain for any line oriented at an angle φ' to the strain axes. For example, for a line in the $\lambda_1\lambda_3$-plane (i.e., *XZ*-plane) of the strain ellipsoid at an angle of 25° to the maximum strain axis, we plot the angle $2\varphi'$ (50°) from the λ_1' end of the circle and draw line *OP'* (Figure 4.13b). The corresponding strain values are

$$\lambda' = 0.9 \quad \text{and} \quad \gamma = 1.4$$

thus

$$\lambda = 1.1 \quad \text{and} \quad \gamma = 1.5$$

This means that if line *OP* represented the long axis of a fossil (e.g., a belemnite), it will have extended and also rotated from this original configuration. Using Equation 4.10 (applied in Figure 4.13), we can also calculate the original angle φ that our belemnite made with our reference frame:

$$\varphi = \arctan[(\lambda_1/\lambda_3)^{1/2}\cdot\tan\varphi'] = 62°$$

This latter calculation highlights the easily misunderstood relationship between the angular shear and the rotation angle of a particular element in a deforming body. The rotation of line *OP* to *OP'* in the deformed state occurred over an angle of 37° (62° − 25°). However, this angle is not equal to the angular shear, ψ, of that element, which is derived from Equation 4.4, and gives $\psi = 56°$. We plot these various angles in $\lambda_1\lambda_3$-space (i.e., *XZ*-space) in Figure 4.13c.

You may have noticed that we consider coaxial strain in our example of the Mohr circle for strain

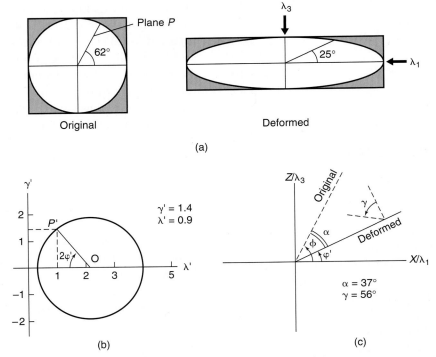

FIGURE 4.13 The Mohr circle for strain. For a deformed object (a), the reciprocal values of the principal strains are plotted in $\gamma'\lambda'$-space, where $\lambda' = 1/\lambda$ and $\gamma' = \gamma/\lambda$ (b). The corresponding rotation of line OP in XZ-space (or $\lambda_1\lambda_3$-space) is shown in (c).

TABLE 4.2	STRAIN STATES
General strain[7] (triaxial strain)	A state in which all three strain axes have different lengths, as defined by the relationship $X > Y > Z$ (Figure 4.14a). This strain state does not imply anything about volume change.
Axially symmetric elongation	An axial strain that includes *axially symmetric extension,* where $X > Y = Z$ (Figure 4.14b), and *axially symmetric shortening,* where $X = Y > Z$ (Figure 4.14c). Axially symmetric extension results in a prolate strain ellipsoid, with extension occurring only in the X direction accompanied by equal amounts of shortening in the Y and Z directions ($Y/Z = 1$). This geometry is sometimes referred to as a cigar-shaped ellipsoid. Axially symmetric shortening requires equal amounts of extension ($X/Y = 1$) in the plane perpendicular to the shortening direction, Z. The strain ellipsoid assumes an oblate or hamburger shape.[8]
Plane strain	A state where one of the strain axes (commonly Y) is of the same length before and after strain: $X > Y (= 1) > Z$ (Figure 4.14d). Thus plane strain is a special type of triaxial strain, but it can be conveniently described by a two-dimensional strain ellipse with axes X and Z, because no change occurs in the third dimension (Y). In many studies this particular strain state is assumed.
Simple elongation	A state where all material points move parallel to a straight line, defined by $X > Y = Z = 1$ or $X = Y (= 1) > Z$. In these two cases, a sphere becomes a prolate ellipsoid in extension and an oblate ellipsoid in shortening (Figure 4.14e), respectively. Because two strain axes remain of equal length before and after deformation, simple elongation must involve a change in volume ($\Delta \neq 0$), a volume decrease in the case of simple shortening and a volume increase in the case of simple extension.

[7]Not to be confused with *general shear* (p. 68).

[8]Both smoking and fatty foods are bad for your health.

construction, in which the incremental strain axes are parallel to the finite strain axes. The construction for non-coaxial strain adds a component of rotation to the deformation (Section 4.5).[9] In Mohr space, this rotational component moves the center of the Mohr circle off the λ' (reciprocal longitudinal strain) axis. In fact, the rotational component of strain can be quantified from the off-axis position of the Mohr circle, but this application takes us well beyond our introduction to the Mohr circle for strain. We direct you to the reading list for more advanced treatments.

4.9 STRAIN STATES

Having defined the various strain parameters as well as their mathematical descriptions, we close by listing several characteristic strain states in Table 4.2, based on various relationships between the principal strain axes X, Y, and Z of the strain ellipsoid. Except for simple elongation, the strain states may represent constant volume conditions (that is, $\Delta = 0$), which means that $X \cdot Y \cdot Z = 1$. These strain states are illustrated in Figure 4.14.

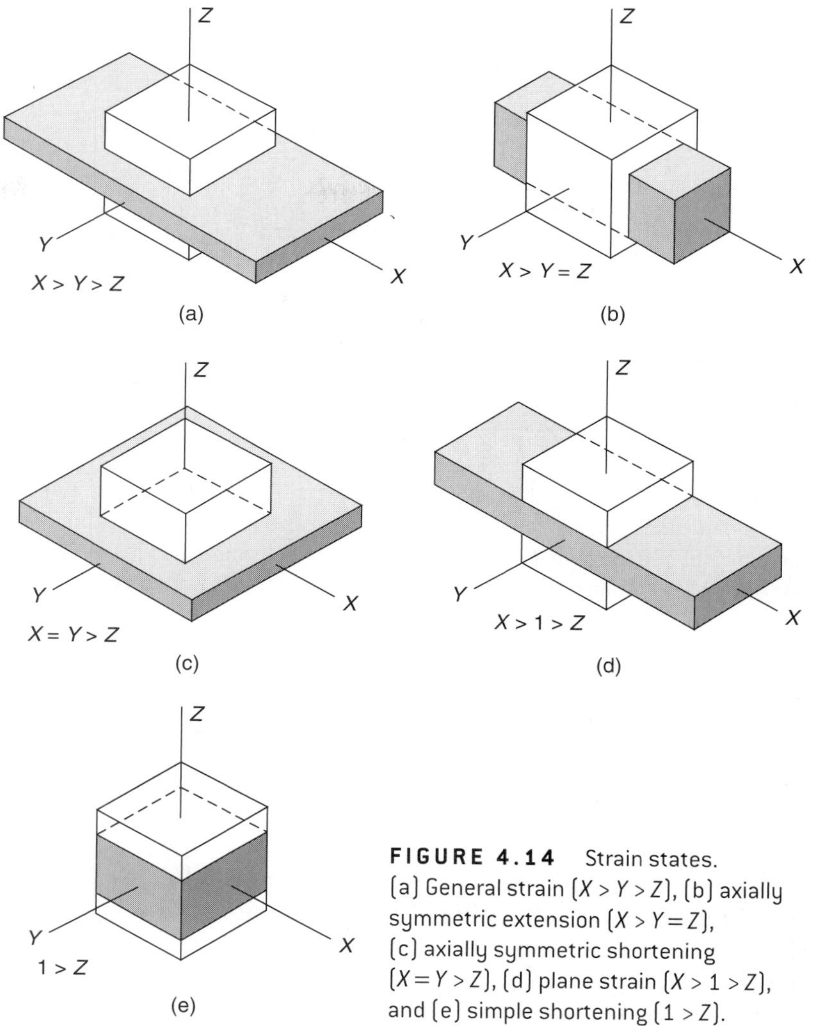

FIGURE 4.14 Strain states. (a) General strain ($X > Y > Z$), (b) axially symmetric extension ($X > Y = Z$), (c) axially symmetric shortening ($X = Y > Z$), (d) plane strain ($X > 1 > Z$), and (e) simple shortening ($1 > Z$).

4.10 REPRESENTATION OF STRAIN

A common goal in strain analysis is to compare results obtained in one place with those obtained elsewhere in an area or in an outcrop, or even to compare data from several different regions. When we determine strain at a locality, we assume that the volume of rock analyzed is homogeneous. Strain, however, is typically heterogeneous on the scale of a single structure, such as a fold or a shear zone, and it always is heterogeneous on the scale of mountains and orogens. Nonetheless,

given a sufficiently large spatial distribution of data points we can draw important conclusions on the state of strain in a structure or a region. Whereas a listing of numerical values and orientation for X, Y, and Z, the principal strain axes, allows direct comparisons, a graphical approach to the presentation of strain data may be more informative. We'll explore orientation and magnitude data for strain in the next sections.

4.10.1 Orientation

A visually appealing way to illustrate the orientation of the strain ellipsoid is to project the orientation of one or more strain axes at their respective locations in an area. This "area" may of course represent many scales, ranging from a thin section to an entire orogenic belt. You are not limited to plotting the orientation of the strain axes, but can also include information on

---

[9]This rotation is different from the line rotation in our example; all lines, except material lines that parallel the principal strain axes, rotate in coaxial strain.

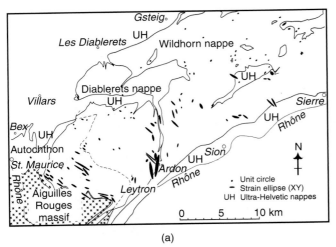

(a)

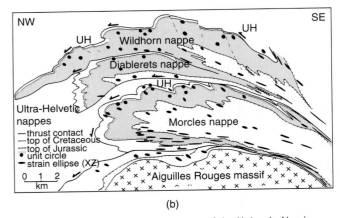

(b)

FIGURE 4.15 In (a) shapes of the *XY* sectional ellipse of the finite strain ellipsoid are shown on a map of the Helvetic Alps in Switzerland. In (b) *XZ* sectional ellipses are shown in a profile constructed from this map (note that the profile plane is not perpendicular to the map surface). The map and profile show a series of NW-directed, low-angle reverse faults or thrusts in which the long axis of the strain ellipsoid lies in the direction of thrust transport and the strain ratio general increases with depth.

the magnitude of strain, for example by using the strain ratio *X/Z* or absolute strain magnitudes as a scaling tool.

A limitation of this two-dimensional mode of representation is that we only show sections through the strain ellipsoid, something we call **sectional strain ellipses.** These sectional ellipses do not necessarily coincide with a plane containing one or more of the principal strain axes, the principal planes of strain, meaning that the axes of the sectional ellipse do not parallel the principal strain axes. In such cases one is left with two reasonable options: (1) draw the shape of the sectional ellipse on the surface, or (2) show the orientation of the strain axes using plunge and direction of plunge. Luckily we can often find a projection plane that is close to a principal plane of strain, so that small deviations are relatively unimportant.

Figure 4.15 shows an example from an area in the Swiss Alps of Europe that illustrates the informative approach of superimposing strain data on a map and on a section. We discuss the details of the European Alps in Chapter 23, but you can see that the degree of strain generally increases with depth in this stack of Helvetic thrust sheets (i.e., the ratio of *X/Z* increases with depth). The orientation of the strain ellipsoid also varies through the stack, but the long axis, *X,* generally points in the direction of northwesterly thrust transport and parallels the boundaries of high strain regions (shear zones; see Chapter 12) that separate individual thrust sheets.

4.10.2 Shape and Intensity

The inherently three-dimensional strain data can be conveniently represented in a two-dimensional plot, called a **Flinn diagram,**[10] by using ratios of the principal strain axes. In fact, we'll see later (Section 4.11) that strain analysis often produces strain ratios rather than absolute magnitudes of the strain axes. In the Flinn diagram for strain (Figure 4.16a) we plot the ratio of the maximum stretch over the intermediate stretch on the vertical (*a*) axis and the ratio of the intermediate stretch over the minimum stretch on the horizonal (*b*) axis:

$$a = X/Y = (1 + \mathbf{e}_1)/(1 + \mathbf{e}_2) \qquad \text{Eq. 4.24}$$
$$b = Y/Z = (1 + \mathbf{e}_2)/(1 + \mathbf{e}_3) \qquad \text{Eq. 4.25}$$

The shape of the strain ellipsoid is represented by the parameter, **k:**

$$\mathbf{k} = (a - 1)/(b - 1) \qquad \text{Eq. 4.26}$$

The value of **k** describes the slope of a line that passes through the origin (angle β in Figure 4.16a). A strain sphere lies at the origin of this plot (coordinates 1,1), representing *a* = *b* = 1 or *X* = *Y* = *Z* = 1. Ellipsoid shapes are increasingly **oblate** for values of **k** approaching 0 and increasingly **prolate** for values of **k** approaching ∞. If **k** = 0, the strain is uniaxially oblate (*a* = *X/Y* = 1), and if **k** = ∞, the

---

[10]Named after Derek Flinn.

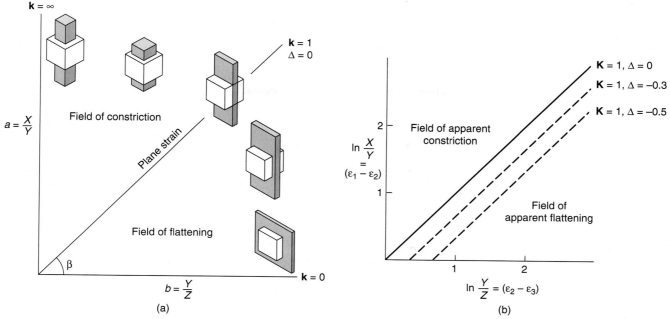

FIGURE 4.16 The Flinn diagram plots the strain ratios X/Y versus Y/Z (a). In the Ramsay diagram (b) the logarithm of these ratios is used; in the case of the natural logarithm (ln) it plots $(\varepsilon_1 - \varepsilon_2)$ versus $(\varepsilon_2 - \varepsilon_3)$. The parameters **k** and **K** describe the shape of the strain ellipsoid. Note that volume change produces a parallel shift of the line **k** = 1 or **K** = 1.

strain is uniaxially prolate ($b = Y/Z = 1$). The value **k** = 1 represents the special case for which a equals b, which is called plane strain ($X \geq Y = 1 \geq Z$). The line represented by **k** = 1 states separates the field of *constriction* ($\infty > $ **k** $ > 1$) from the field of *flattening* ($1 > $ **k** $ > 0$) in the Flinn diagram.[11]

A useful modification of the Flinn diagram, called the **Ramsay diagram**,[12] uses the natural logarithm of the values a and b (Figure 4.16b):

$$\ln a = \ln (X/Y) = \ln [(1 + e_1)/(1 + e_2)] \qquad \text{Eq. 4.27}$$
$$\ln b = \ln (Y/Z) = \ln [(1 + e_2)/(1 + e_3)] \qquad \text{Eq. 4.28}$$

Using

$$\ln x/y = \ln x - \ln y, \text{ and}$$
$$\varepsilon = \ln (1 + e) \qquad \text{Eq. 4.16}$$

where ε is natural strain, we convert Equations 4.27 and 4.28 to

$$\ln a = \varepsilon_1 - \varepsilon_2 \qquad \text{Eq. 4.29}$$
$$\ln b = \varepsilon_2 - \varepsilon_3 \qquad \text{Eq. 4.30}$$

[11]Later we will modify this by adding "apparent" to flattening and constriction, which reflects the role of volume change.
[12]Named after John Ramsay.

The parameter **k** of the Flinn diagram becomes **K** in the Ramsay diagram:

$$\mathbf{K} = \ln a/\ln b = (\varepsilon_1 - \varepsilon_2)/(\varepsilon_2 - \varepsilon_3) \qquad \text{Eq. 4.31}$$

Both logarithmic plots with base e (natural logarithm, ln) and base 10 (log) are used in the Ramsay diagram. The Ramsay diagram is similar to the Flinn diagram in that the line **K** = 1 separates the fields of constriction ($\infty > $ **K** $ > 1$) and flattening ($1 > $ **K** $ > 0$) and the unit sphere lies at the origin ($\ln a = \ln b = 0$). Note that the origin in the Ramsay diagram has coordinates (0, 0). There are a few advantages to the Ramsay diagram. First, small strains that plot near the origin and large strains that plot away from the origin are more evenly distributed. Second, the Ramsay diagram allows a graphical evaluation of the incremental strain history, because equal increments of progressive strain (the strain path) plot along straight lines, whereas unequal increments follow curved trajectories. In the Flinn diagram both equal and unequal strain increments plot along curved trajectories.

The foregoing description assumes that volume change (Δ) does not occur during the strain history. To consider the effect of volume change, recall that if $\Delta = 0$ (i.e., there is no volume change) then $X \cdot Y \cdot Z = 1$.

Thus, using Equation 4.3 $\Delta = (V - V_o)/V_o$, and substituting $V = X \cdot Y \cdot Z$ and $V_o = 1$, we get

$$\Delta + 1 = X \cdot Y \cdot Z = (1 + e_1) \cdot (1 + e_2) \cdot (1 + e_3) \quad \text{Eq. 4.32}$$

Expressed in terms of natural strains, this becomes

$$\ln(\Delta + 1) = \varepsilon_1 + \varepsilon_2 + \varepsilon_3 \quad \text{Eq. 4.33}$$

Further rearrangement of this expression in a form that uses the axes of the Ramsay diagram gives

$$(\varepsilon_1 - \varepsilon_2) = (\varepsilon_2 - \varepsilon_3) - 3\varepsilon_2 + \ln(\Delta + 1) \quad \text{Eq. 4.34}$$

Prolate and oblate ellipsoids are separated by plane strain conditions ($\varepsilon_2 = 0$)

$$(\varepsilon_1 - \varepsilon_2) = (\varepsilon_2 - \varepsilon_3) + \ln(\Delta + 1) \quad \text{Eq. 4.35}$$

which represents a straight line ($y = mx + b$) with unit slope ($m = 1$, or angle β is 45°). If $\Delta > 0$ the line intersects the $(\varepsilon_1 - \varepsilon_2)$ axis (volume gain), and if $\Delta < 0$ it intersects the $(\varepsilon_2 - \varepsilon_3)$ axis (volume loss). In all cases, the slope of the line remains at 45° and $\varepsilon_2 = 0$. In Figure 4.16b the case for various percentages of volume loss is illustrated. We use the terms **apparent flattening** and **apparent constriction** for the fields in the diagram that are separated by the line representing plane strain conditions ($K = 1$). The term "apparent" is used because volume change must be known to determine the actual strain state of a body. For example, the location of a strain ellipsoid in the "flattening field" may represent true flattening, but it may also represent plane strain or even true constriction conditions. These simply depend on the degree of volume loss. Like the natural strain parameter, the Ramsay plot offers a distinction between strain increments with constant ratios of volume change and those with varying ratios. In the former case the strain path is straight, whereas in the latter case the path is curved.

The parameters **k** and **K** describe the shape of the ellipsoid, but the position of a data point in the Flinn and Ramsay diagrams not only describes the shape of the strain ellipsoid, but also reflects the degree (or intensity) of strain. This second element is not so readily appreciated. The farther a point in the Flinn/Ramsay diagram is located from the origin, the more the strain ellipsoid deviates from a sphere. But the same deviation from a sphere (i.e., the same degree of strain), occurs for different shapes of the ellipsoid, that is, for different values of **k** (or **K**). Similarly, the same shape of the strain ellipsoid may occur for different degrees of strain. The parameter that describes the degree or **intensity** of strain is

$$i = \{[(X/Y) - 1]^2 + [(Y/Z) - 1]^2\}^{1/2} \quad \text{Eq. 4.36}$$

or in the case of natural strains

$$I = (\varepsilon_1 - \varepsilon_2)^2 + (\varepsilon_2 - \varepsilon_3)^2 \quad \text{Eq. 4.37}$$

So, the ability of the Flinn diagram to represent strain states and volume change and the added convenience of the Ramsay diagram to evaluate incremental strain histories present sufficient flexibility to graphically present most strain data. Listing the corresponding shape (**k** or **K**) and intensity (**i** or **I**) parameters allows for numerical comparisons between strain analyses in the same structure and/or those made over a large region.[13]

Other graphical strain representations have been proposed, but in practice they offer little improvement over the Flinn or Ramsay diagrams. In closing, we explore a new method that combines both orientation and magnitude data. The three principal axes of the strain ellipsoid are plotted in lower hemisphere projection, whereas their relative magnitudes are represented by intensity contours. The number of contours separating the axes is proportional to the difference in magnitude. Figure 4.17 shows a number of these plots for a constant orientation of the principal strain axes, but with ellipsoids of different shape (defined by **k**) and degree (defined as X/Z). As the shape changes from prolate (ellipsoids 1–8) to oblate (ellipsoids 13–20), the distribution changes from a point maximum to a girdle. As the degree of strain increases, the number of contours increases (from left column to right column). In Figure 4.17b the corresponding strain states are plotted in a Flinn-type diagram.

4.11 FINITE STRAIN MEASUREMENT

We are now ready to tackle the practical aspects of strain analysis. The determination of strain is a common task for structural geologists unravelling the geologic history of an area or examining the development

---

[13]There are no simple mathematical relationships between **k** and **K**, and **i** and **I**.

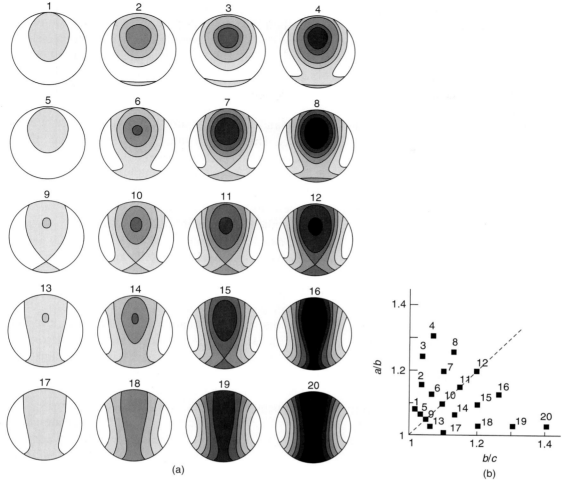

FIGURE 4.17 Magnitude-orientation diagrams (a), showing oblate, plane strain, and prolate ellipsoids of varying intensity. The maximum, intermediate, and minimum axes in all diagrams are 60°/000°, 30°/180°, and 00°/090°, respectively. The corresponding position of each ellipsoid in a Flinn-type diagram is shown in (b).

of individual structures. Over the years, many methods have been proposed, some being more useful and rigorous than others. In this section, therefore, you will not find an exhaustive treatment of all methods of strain analysis; rather we limit our description to the principles, and advantages and disadvantages, of the more widely used methods. Advanced structural geology textbooks and laboratory manuals discuss related strain methods in greater detail and often present step-by-step instructions for each method. There is no doubt that the best way to learn about each method (and appreciate the assumptions) is to apply them to data sets.

The availability of personal computers and peripherals (digitizer, scanner, image analyzer) has added a new and powerful dimension to strain analysis. Increasingly sophisticated programs are available that carry out many of the more time-consuming tasks. It

has become easy to obtain lots of strain data without knowing much about the methods. This "black box" approach is dangerous, because you may not realize the implicit assumptions and consider the limitations of a method. Understanding the basic tenets of strain determinations is necessary to interpret the data, so in the next few sections we focus on the fundamentals of the art of strain analysis.

4.11.1 What Are We Really Measuring in Strain Analysis?

Strain analysis attempts to quantify the magnitude and/or the orientation of the strain ellipsoid(s) in rocks and regions. In its most complete form, each strain analysis gives the lengths and orientations of the three principal strain axes. More commonly, however, we obtain strain ratios (X/Y, Y/Z, and X/Z), because we do not know the absolute dimensions of the original state. Powerful as strain quantification may appear, we need to ask two important questions. The first question arises because the strain analyzed may only represent a part of the total strain history that the rock or region has undergone. So the question is: *how complete is our measurement?* The second question comes about because the results of a particular strain analysis only pertain to the objects that were used for the analysis. A difference in behavior between the objects and other parts of the rock, say conglomerate clasts and their matrix, may result in a strain value for the whole rock that differs from that for the conglomerate clasts. So our second question is: *how representative is our analysis?* We explore this second consideration by two simple experiments that are shown in Figure 4.18.

We place a marble within a cube of clay, and as we load the block it deforms into a rectangular box. After sectioning, however, you learn that the marble remains undistorted! So, using the marble as the basis for strain analysis you conclude there is zero percent strain (or $X/Y = 1$), yet looking at the clay you clearly see that strain has accumulated. In fact, you determine a strain ratio, X/Y, of 1.8 for the clay. Now we carry out a second experiment,

in which we replace the space occupied by the marble with a fluid. The result is quite different from that of the previous run. In the second case both materials show finite strain, but the strain measured from the bubble, which now has the shape of an ellipsoid, is higher than that measured from the shape of the deformed clay block. These obviously contrasting results do not represent a paradox, because all answers are correct in their own way. There is zero strain for the marble, and finite strain for the fluid bubble is greater than that for the clay block. These different results simply reflect the response of materials with different strengths. The clay is weaker than the marble, but the fluid bubble is weaker than the clay. We therefore identify strain markers of two types: passive and active markers.

Passive strain markers are elements in the body that have no mechanical contrast; they deform in a manner indistinguishable to that of the whole body. For example, a circle drawn on our clay cube would constitute a passive marker. Such markings are rare in nature, but inclusions of the same composition as the matrix are close to this condition; for example, quartz grains in a quartzite or oolites in a carbonate. In the case of passive markers, we say that our body behaves as a homogeneous system for strain.

Active strain markers have mechanical contrast with their matrix and may behave quite differently. The marble and fluid inclusions in the above experiments are both examples of active markers. Conglomerate clasts in a shale matrix or garnets in a mica schists are natural examples of active strain markers, which represent a heterogeneous system for strain.

The active role of markers need not be a crippling problem as long as the strain results are considered with the limitations of each method in mind. Finally, strain analysis can lead to misinterpretation because of an unjustified confidence in hard numbers, which has been poetically described as "garbage

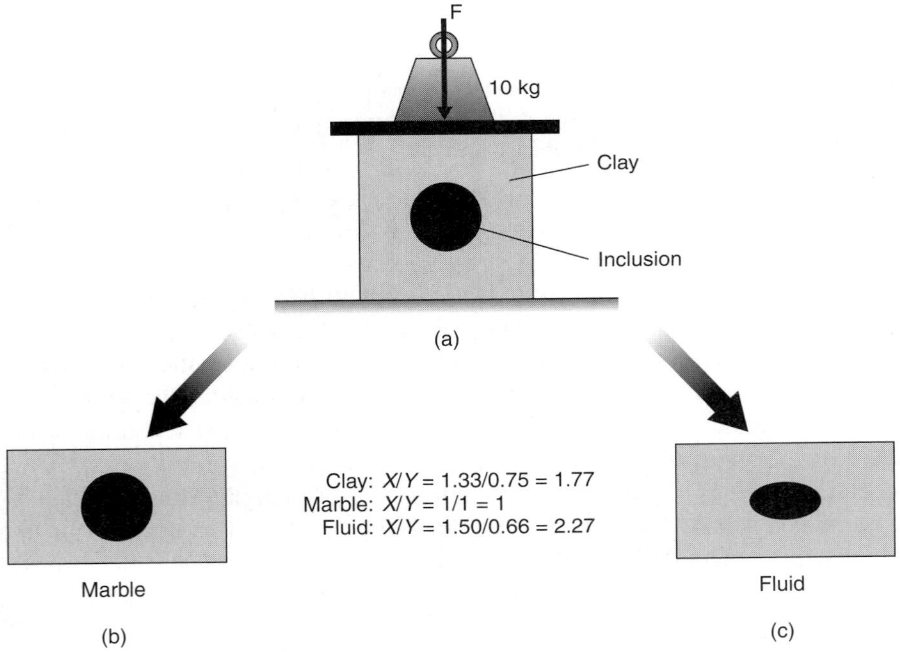

Clay: $X/Y = 1.33/0.75 = 1.77$
Marble: $X/Y = 1/1 = 1$
Fluid: $X/Y = 1.50/0.66 = 2.27$

FIGURE 4.18 In two simple experiments we deform a cube of clay with a marble inclusion and a fluid inclusion. If we require that the elongations in the clay are the same for both runs, the resulting elongations for the marble and fluid bubble are quite different. This illustrates the different response to stress of materials with mechanical contrast or heterogeneous systems. A practical example is the determination of strain using (strong) conglomerate clasts in a (weak) clay matrix.

in, garbage out." Try to remember our simple clay experiments when you get to the interpretation of your strain data.

4.11.2 Initially Spherical Objects

Recall that homogeneous strain is defined as the change in shape from a sphere to an ellipsoid. Thus, geologic features with spherical shapes are perfect for strain analysis and some of the first analyses were indeed carried out using spherical objects. The classic example involves ooids in (oolitic) limestones (Figure 4.19). Ooids are particles that have grown radially by accretion around a nucleus; commonly they are calcareous in composition. These spheres become increasingly ellipsoidal with increasing strain (Figure 4.19b and c). Some other examples of objects that are approximately spherical are vesicles and amygdules in basalt flows and reduction spots (areas where chemical change has occurred due to the presence of impurities) in shales. But keep in mind that depositional conditions may affect these strain markers; the flow of lava may stretch amygdules and compaction may affect the shape of reduction spots.

Once we are convinced that our markers are initially spherical, the strain analysis of these objects is relatively straightforward. If we preserve the deformed shapes in three dimensions, we can obtain a direct representation of the shape of the strain ellipsoid by measuring the shape of the objects in three mutually perpendicular sections. In fact, these sections do not necessarily have to be perpendicular, but this simplifies the procedure. In each section we measure the long

and the short axis of the sectional ellipse for several objects, which gives us ratios. The term **sectional strain ellipse** was used before to emphasize the fact that you do not measure the principal axes of the objects, but only their lengths in section. We combine the ratios from each of the three sections to determine the shape of the strain ellipsoid. By choosing the sections such that they coincide with the principal planes of strain (that is, the planes containing two of the principal strain axes) we simplify our procedure. In this case, we directly obtain the strain ratios (X/Y, Y/Z, and X/Z) in our three sections. Otherwise we need to use trigonometric relationships to determine the ellipsoid.

An example is given in Figure 4.20. Say, you measure a ratio of $X/Y = 2$ in one section and in a second section, parallel to the YZ-plane of the strain ellipsoid, you measure a ratio of $Y/Z = 1.2$. In three dimensions this gives a ratio of $X/Y/Z$ of 2.4/1.2/1. You can check this in a third section, containing X and Z, which should give a ratio $X/Z = 2.4$. Assuming that the volume remained constant ($\Delta = 0$), you can fully specify the strain state of your sample. If

$$X/Y/Z = 2.4/1.2/1 \quad \text{and} \quad \Delta + 1 = X \cdot Y \cdot Z = 1$$

then

$$X = 2Y \quad \text{and} \quad Z = Y/1.2$$

thus

$$2Y \cdot Y \cdot (Y/1.2) = 1.7\ Y^3 = 1$$

so

$$Y = 0.8 \quad \text{and} \quad X/Y/Z = 1.7/0.8/0.7$$

Now determine what happens to the strain ratio $X/Y/Z$ if you allow for approximately 50% volume loss ($\Delta = -0.5$), representing a volume change that has been suggested in many studies.[14] You see that, irrespective of the amount of volume change, the strain ratios remain the same, but that the magnitudes of the individual axes become smaller with volume loss (and vice versa, i.e., greater with volume gain). Consequently, the position of the strain states in Flinn and Ramsay diagrams remains the same with or without volume change, and thus values of **k** (or **K**) and **i**

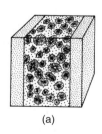

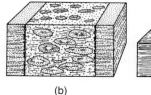
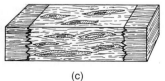

(a)

(b)   (c)

FIGURE 4.19 Ooids (a) that are deformed after (b) 25% ($X/Z = 1.8$) and (c) 50% ($X/Z = 4.0$) shortening.

---

[14]$X/Y/Z = 1.3/0.7/0.5$

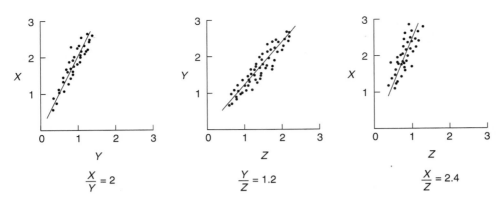

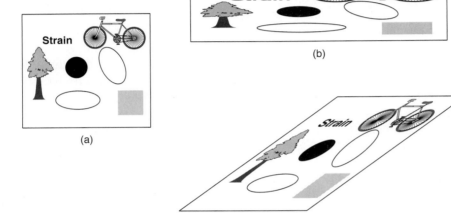

FIGURE 4.20 Strain from initially spherical objects. The long and short axes of elliptical objects are measured in three orthogonal sections. For convenience we assume that these sectional ellipses are parallel to the principal planes. The slope of regression lines through these points (which should intersect the origin) is the strain ratio in that section. In this example: $X/Y = 2$, $Y/Z = 1.2$, and $X/Z = 2.4$, which gives $X/Y/Z = 2.4/1.2/1$.

FIGURE 4.21 Changes in markers of various shapes under (a) homogeneous strain, (b) coaxial, constant volume strain, and (c) non-coaxial, constant volume strain. Note especially the change from circle to ellipse and from an initial ellipse to an ellipse of different ratio, and the relative extension and shortening of the markers.

(or **I**) also remain the same. So we have no indication of volume change from our analysis of the shape of markers unless we know their original volume. To solve the critical problem of volume change we may use geochemical approaches, to which we will return in another chapter.

4.11.3 Initially Nonspherical Objects

Strained particles of all shapes have several properties in common with the finite strain ellipsoid. Material lines (meaning, arrays of particles) that are not parallel to any one of the principal strain axes will tend to rotate toward the maximum strain axis (X) and away from the minimum strain axis (Z). Material lines that are parallel to the Z-axis will undergo the greatest amount of shortening, whereas material lines that are parallel to the X-axis will have the greatest amount of extension. For an intuitive understanding of strain from nonspherical objects these two properties are most important. Figure 4.21 illustrates what happens to markers with variable shapes. Although they look different, they all change in a predictable manner; but actually quantifying strain from these changes is another matter. Strain quantification is greatly simplified if the initial shape of our marker is an ellipsoid (or ellipse in the two-dimensional case of Figure 4.21), because the superimposition of ellipsoids, by definition, produces another ellipsoid. Mathematically, is it true that adding two tensors of the same rank produces a third tensor of that rank? Let's look at this first for low-rank tensors: adding two zero-rank tensors (scalars) gives another scalar; for example, $3 + 5 = 8$. Adding two first-rank tensors (vectors) produces another vector. So this indeed is a property of tensors, and therefore adding two second-rank tensors will produce another second-rank tensor.[15] In other words, superimposing the finite strain ellipsoid on an ellipsoidal body gives an ellipsoid that contains the proper-

---

[15]Mathematically it is the product of two tensors.

FIGURE 4.22 Deformed clasts in a late Paleozoic conglomerate (Narragansett, RI, USA).

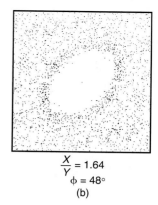

$$\frac{X}{Y} = 1.64$$
$$\phi = 48°$$

(a)   (b)

FIGURE 4.23 Strain from initially spherical or nonspherical objects using the center-to-center method. The distance between object centers is measured as a function of the angle with a reference line (a). A graphical representation of this method using 250 deformed ooids in a natural sample section is shown in (b). The empty area defines an ellipse with the ratio X/Y; the angle of the long axis with the reference line is also given.

ties of both the initial ellipsoid (initial marker shape) and the superimposed ellipsoid (finite strain ellipsoid). We already showed this graphically when discussing superimposed strain (Figure 4.9). Although most objects in nature are not perfect spheres, their shapes may be reasonably approximated by ellipsoids; for example, ellipsoidal clasts of a conglomerate (Figure 4.22). You will realize that we cannot simply determine the strain by measuring shapes in the deformed state of the ellipsoids, without knowing the initial shape of the objects. Two techniques are commonly used to resolve this problem: the center-to-center method and the R_f/Φ method.

4.11.3.1 CENTER-TO-CENTER METHOD
The basic principle of the **center-to-center method** of strain analysis is that the distances between the centers of objects are systematically related to the orientation of the finite strain ellipsoid.[16] For example, in the XY-plane of the strain ellipsoid, grain centers that lie along the direction of the shortening axis (Y) tend to become closer during deformation than grain centers aligned with the extension axis (X). Measuring the distances of centers as a function of an arbitrary reference orientation produces maximum and minimum values that correspond to the orientation and the strain ratio of X and Y (Figure 4.23). Computers have greatly eased the application of this principle by using a digitizer to determine grain centers and a method to graphically analyze the results. Dividing the center-to-center distance between two objects by the sum of their mean radii is a normalization procedure that better constrains the shape of the ellip-

soid.[17] Figure 4.23b shows the result of a **normalized center-to-center analysis** using a computer program.

4.11.3.2 R_f/Φ METHOD
The R_f/Φ **method** utilizes the systematic shape changes that occur in deformed ellipsoidal objects. In a given section we measure the long and short axes of sectional ellipses and the orientation of the long axis with respect to a reference line (Figure 4.24a). For convenience we again assume that the section is parallel to the XY-plane, but this is not a prerequisite of the method. Plotting the ratio of these axes (R_f) versus the orientation (Φ) yields a cloud of data points, reflecting the addition of the initial ellipsoid (the original shape of the body) and the finite strain ellipsoid. The maximum and minimum values of R_f are related to the initial ratio (R_o) of the objects and the strain ratio (R_s; in this example X/Y), as follows:

$$R_{f,max} = R_s \cdot R_o \qquad \text{Eq. 4.38}$$
$$R_{f,min} = R_o/R_s \qquad \text{Eq. 4.39}$$

which gives the strain ratio, R_s, by dividing Equation 4.38 by Equation 4.39:

$$R_s = (R_{f,max}/R_{f,min})^{1/2} \qquad \text{Eq. 4.40}$$

The maximum initial ratio is determined by multiplying Equations 4.38 and 4.39:

$$R_o = (R_{f,max} \cdot R_{f,min})^{1/2} \qquad \text{Eq. 4.41}$$

---

[16]Requiring an anticlustered initial distribution of centers.

[17]Known as the Fry and norm(alized)-Fry methods, respectively, after Norm Fry.

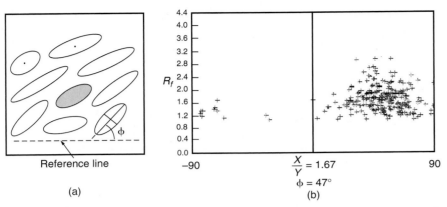

(a)
Reference line

R_f

$\frac{X}{Y}$ = 1.67

ϕ = 47°

(b)

FIGURE 4.24 Strain from initially spherical or nonspherical objects using the R_f/Φ method. The ratios of elliptical objects are plotted as a function of their orientation relative to a reference line (a). Application of this method to 250 ooids (same sample as in Figure 4.23) shows the cloud of data points that characterizes this method. From the associated equations we determine the strain ratio and the angle of the long axis with the reference line. The values obtained from using the center-to-center (Figure 4.23) and R_f/Φ (Figure 4.24) methods are indistinguishable within the resolution of these methods.

These relationships hold when the original ratio, R_o, is greater than the strain ratio, R_s ($R_o > R_s$). On the other hand, if $R_o < R_s$, then

$$R_{f,max} = R_s \cdot R_o \text{ (same as above), but} \qquad \text{Eq. 4.42}$$
$$R_{f,min} = R_s/R_o \qquad \text{Eq. 4.43}$$

and the strain ratio and the maximum original ratio become

$$R_s = (R_{f,max} \cdot R_{f,min})^{1/2} \qquad \text{Eq. 4.44}$$
$$R_o = (R_{f,max}/R_{f,min})^{1/2} \qquad \text{Eq. 4.45}$$

The orientation of the X-axis relative to the reference line in our particular example is determined by the position of $R_{f,max}$ and the Y-axis is perpendicular to the X-axis in the section surface. Evaluation of the scatter of points in R_f/Φ plots is greatly eased by reference curves that are used as overlays on the data. An example of the application of this method is shown in Figure 4.24b.

4.11.4 Objects with Known Angular Relationships or Lengths

The transformation of a sphere to an ellipsoid involves changes in line lengths as well as angular changes (Figure 4.4). Thus, methods that allow us to determine either or both of these parameters are suitable for strain determination. We will explore the principles behind some methods in the following sections.

4.11.4.1 ANGULAR CHANGES Recall the relationship between the original angle ϕ of a line, the angle after deformation (ϕ'), and the strain ratio (X/Y):[18]

$$X/Y = \tan \phi/\tan \phi' \qquad \text{Eq. 4.46}$$

The original angular relationships are well known in geologic objects that have natural symmetry. One suitable group of objects is fossils, such as trilobites (Figure 4.1) and brachiopods, because they contain easily recognizable elements that are originally perpendicular. When we have several deformed objects available that preserve these geometric relationships, we can reliably obtain both the orientation of the strain axes and the strain ratios from their distortion. One approach is shown in Figure 4.25, which requires several deformed fossils in different orientations. Note that the method is independent of the size of the individual fossils (big ones mixed with small ones) and the presence of different species (as long as the symmetry element remains present), because only the angular changes in these objects are used in the analysis. In the plane of view, which often is the bedding plane, we measure the angular shear, ψ, for each deformed object. We plot the value of ψ versus a reference orientation (angle α), making sure that we carefully keep track of positive and negative angles.[19] The resulting

---

[18]We only consider the two-dimensional case here, so X/Z (see Equation 4.9) becomes X/Y.

[19]This is known as the Breddin method, after Hans Breddin.

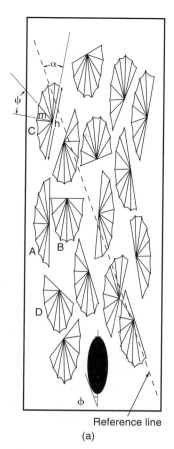

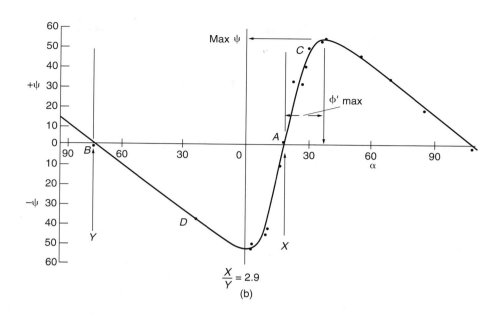

$$\frac{X}{Y} = 2.9$$

(b)

FIGURE 4.25 Strain is determined from a collection of deformed bilaterally symmetric brachiopods (a), by constructing a graph that correlates the angular shear with the orientation (b). From this (Breddin) graph the orientation of the principal axes and the strain ratio can be determined.

relationship is a curve that intersects the α-axis at two points (Figure 4.25b). These intersections represent conditions of zero angular shear, which coincide with the orientations of the principal strain axes in this surface, because the principal strain axes are defined as the material lines that are perpendicular before and after strain (Section 4.3). The strain ratio is determined using Equation 4.9 at conditions for maximum angular shear, which occurs at an angle $\varphi = 45°$. Substituting $\tan \varphi = 1$ in Equation 4.46, gives

$$Y/X = \tan \varphi'_{max} \qquad \text{Eq. 4.47}$$

where φ'_{max} is the angle between the position of the principal axis and maximum angular shear on the α-axis (Figure 4.25b). In cases where not enough distorted fossils are available to derive a reliable curve, the strain ratio can be determined by comparison with a set of curves that are predetermined for various strain ratios. Figure 4.25 also illustrates that a population of fossils with different orientations in the plane of view allow direct determination of the orientation of the

principal strain axes. By definition, fossils that do not show any angular shear must have symmetry elements that coincide with the orientation of the principal strain axes (fossils A and B in Figure 4.25).

4.11.4.2 LENGTH CHANGES Change in line lengths is possibly the easiest strain method to understand, but we postponed its discussion until now because its practical application is limited to special circumstances. A classic example of the circumstances in which longitudinal strain analysis is appropriate is shown in Figure 4.26, which illustrates two sections through a rock containing deformed belemnites. Recalling how strain induces a change from a circle to an ellipse, you will remember that length changes occur in a systematic manner (Figure 4.4). Given measurements of several length changes we can determine a best-fit strain ellipsoid, or ellipse in two dimensions. For this type of analysis it is most convenient to use ratios of the longitudinal strain quantities, the stretch ($\mathbf{s} = l/l_o$) or its square, the quadratic elongation ($\lambda = [l/l_o]^2$), because

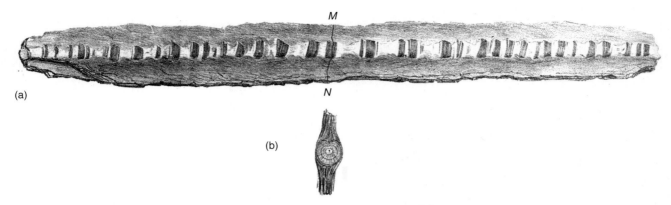

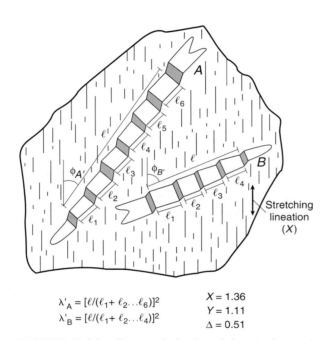

FIGURE 4.26 Stretched belemnite from the Swiss Alps. In longitudinal view (a), calcite filling is seen between the fossil segments, whereas in cross-sectional view (b) the marker remains circular. MN is the location of the sectional view.

$$\lambda'_A = [\ell/(\ell_1 + \ell_2 \ldots \ell_6)]^2$$
$$\lambda'_B = [\ell/(\ell_1 + \ell_2 \ldots \ell_4)]^2$$

$X = 1.36$
$Y = 1.11$
$\Delta = 0.51$

FIGURE 4.27 Changes in line length from broken and displaced segments of two once continuous fossils (A and B; belemnites) assuming that the extension direction (and thus φ') is known. Simultaneously solving equation $\lambda' = \lambda_1' \cos^2 \varphi' + \lambda_2' \sin^2 \varphi'$ (the reciprocal of Equation 4.18) for each element gives the principal strains X and Y. An advantage of this method is that volume change (Δ) can be directly determined from the relationship $X \cdot Y = I + \Delta$.

these quantities correspond directly to the lengths of the ellipsoid axes for orientations parallel to the principal strain axes. Individual stretches are readily measured if the original length of the object is known. To determine the strain ratios we need at least two different stretches and some indication of the orientation of one of the principal strain axes within that plane (Figure 4.27); without an indica-

tion of the orientation, we need at least three longitudinal strain measures. Optimally we apply this analysis in a perpendicular surface to obtain the three-dimensional strain, but this is rare in natural rocks. Strain determinations from length (and angular) changes typically give two-dimensional strain, but given appropriate assumptions (like plane strain or constant volume) the results are nevertheless quite useful. For example, the belemnites in Figure 4.27 give a X/Y strain ratio of 1.23, but also indicate volume gain within this plane. Assuming constant volume strain for the rock as a whole implies that $Z = 1/(X \cdot Y) = 0.66$.

4.11.5 Rock Textures and Other Strain Gauges

So far we have mainly considered strain analysis from objects that change shape. But what about obtaining strain from rigid objects, that is objects that react to strain without changing shape? Imagine that our clay cube in Figure 4.18 contains a few hundred steel needles that are randomly oriented. After deformation these needles will have changed orientation, but they will not have changed shape. We learned that reorientation is a function of the amount of strain, and that all material lines not parallel to the principal strain axes change angle in an orderly fashion. We should be able, therefore, given a few assumptions, to quantify the strain by measuring the preferred orientation of grains. Such measurements can be obtained in a number of ways, including methods that measure grain shape and grain crystallography. Two representative methods are X-ray goniometry (Figure 4.28) and magnetic anisotropy. Without getting into the details, goniometry uses the diffraction of X-rays in a sample to measure the crystallographically preferred orientation of a popula-

FIGURE 4.28 X-ray pole-figure device, consisting of an X-ray diffractometer (in this case a single-crystal device) with a special attachment to rotate a sample that is located within the holder (middle) through the X-ray beam. The X-ray source is to the right.

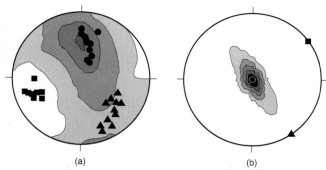

FIGURE 4.29 X-ray goniometry data showing the c-axis preferred orientation of (a) chlorite and (b) mica in two slates; the contours are multiples of random distribution. Superimposed on these X-ray textures are the magnetic susceptibility fabrics from each sample, using a square/triangle/circle for the maximum/intermediate/minimum susceptibility axis; the minimum susceptibility axis coincides with the orientation of the c-axis.

tion of grains. Magnetic anisotropy measures the response of a sample to an external field, which reflects the shape or crystal structure of minerals in the sample. Examples of the results from these two methods on chlorite-dominated and mica-dominated slate samples are shown in Figure 4.29 in lower hemisphere projections. The preferred orientation of the minerals is determined relative to a uniformly distributed grain population. In one analysis no mechanical contrast between markers and matrix is assumed, meaning that the markers respond to strain in the same way as the rock as a whole (such objects are called passive markers).[20] From theoretical considerations that we will not develop here, corresponding strains are defined by

$$e_i = \rho_i^{-1/3} - 1 \qquad\qquad \text{Eq. 4.48}$$

in which e is the "elongation" and ρ is obtained by the normalization of principal pole density to the average pole density for each of three perpendicular directions ($i = 1$–3), which determines the strain ellipsoid. Note our use of quotation marks around "elongation" to indicate that only under specific conditions does e agree with **e** (the elongation of Equation 4.2). Regardless, the method gives a reliable quantitative measure of the intensity of the grain

fabric. Magnetic anisotropy data is used analogously to determine a strain ellipsoid, but this method is less robust.

A few other strain methods are worth mentioning but they require some knowledge that will not be presented until later chapters in this book. So, unless you have already studied that material, they are mentioned here for completion only. The change in orientation of a foliation may be used as a strain gauge in shear zones and provides an estimate of the shear strain, given knowledge of the width of the zone, degree of coaxiality, and origin of the foliation (see Chapter 12). Fibrous overgrowths on rigid grains or fibrous vein-fillings may track the strain history, and are used as incremental strain gauges (Chapter 7). Finally, crystal plastic processes like twinning and recrystallization (Chapter 9) can provide a measure of finite strain.

4.11.6 What Do We Learn from Strain Analysis?

We close this chapter on deformation and strain by exploring the significance of strain analysis. After decades of work, structural geologists have produced a vast amount of finite and incremental strain data, on structures ranging from the microscale (thin sections) to the macroscale (mountain belts and continents). The aim of most of these studies is to measure strain magnitude across a region, an outcrop, or a hand specimen, in order to quantify the structural history or the development of a particular feature. For example, a sharp increase in strain

---

[20]Note that there is a strong mechanical contrast between steel needles and clay in the hypothetical experiment. In natural rocks this mechanical contrast is much less, and at low strains there is little difference between a passive-marker model (March model) and models involving active markers (e.g., Jeffery model).

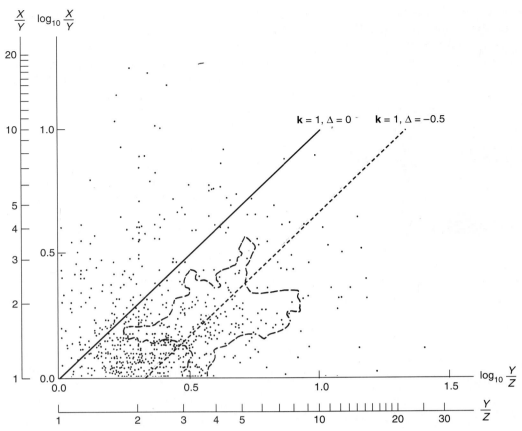

FIGURE 4.30 Compilation of finite strain values from natural markers in a Ramsay diagram that plots $\log_{10} X/Y$ versus $\log_{10} Y/Z$; the corresponding X/Y and Y/Z ratios are also indicated. The field inside the dashed line contains an additional 1000 analyses, primarily from slates, that have been omitted for clarification. Plane strain conditions at constant volume ($\mathbf{k} = 1, \Delta = 0$) separate the field of apparent constriction from the field of apparent flattening, which contains most of the strain values. Plane strain conditions accompanied by 50% volume loss are indicated by the line $\mathbf{k} = 1, \Delta = -0.5$.

magnitude in an area may define a region of high strain, like a ductile shear zone (Chapter 12). Strain analysis provides general constraints on the strain required to form specific geologic structures, such as folds (Chapter 10), and foliations and lineations (Chapter 11), which can then be incorporated into the larger regional history. For example, strain determinations in samples with slaty cleavage (Chapter 11) indicate that shortening strain is greater than 50%. We could describe the results of numerous other studies, but instead we will draw a few general conclusions from the large data set that is shown in Figure 4.30.

A compilation of hundreds of finite strain values that were obtained using a variety of techniques are plotted in a Ramsay diagram with the axes $\log_{10} X/Y$ versus $\log_{10} Y/Z$ (Figure 4.30). This large data set gives a good idea of the magnitudes of strain that are typical in natural rocks. Axial ratios range from 1 to 20 and

stretches are in the range $1 < X < 3$, and $0.13 < Z < 1$. You also notice that most of the analyses lie in the field of apparent flattening (compare with Figure 4.16), which may be interpreted as true flattening strain or the involvement of volume loss.[21] If we assume that plane strain conditions ($\mathbf{k} = 1$) equally divide the number of measurements, then a volume loss on the order of 50% ($\Delta = -0.5$) is required. One wonders, of course, where that volume of rock material went, and the role of mass transfer continues to be a topic of lively debate in the literature. The problem is further aggravated when you consider that strain values reflect the strain of the marker that was used in the determination and

---

[21]As a third alternative, this may be explained by the superimposition of non-coaxial strain increments, such as vertical compaction followed by horizontal shortening.

not necessarily that of the host rock. For many practical examples this means that the values may actually underestimate the finite strain. Finally, most of these data represent regional strains and exclude strains measured in shear zones. The axial ratio increases exponentially with increasing shear strain, and reported values for γ of as much as 40 in shear zones (meaning that φ is approaching 90°; Equation 4.3) represent extremely large axial strain ratios that lie well outside the plot—a wide range in magnitudes, indeed.

4.12 CLOSING REMARKS

The analysis of strain is a popular approach by structural geologists to understand and quantify the geologic history of rocks and regions. One should be cautious, however, with the interpretation of these data, and with numerical values in particular. Whereas numbers tend to provide a sense of reliability, adequate interpretation requires a solid understanding of the underlying assumptions as well as the inherent errors in these numbers. Always remember that the numbers generated are no better than the input parameters (assumptions, measurements). One aspect that was ignored thus far is the duration over which a certain amount of strain accumulates, which is called the **strain rate.** In basic terms, (longitudinal) strain rate is elongation divided by time (e/t). In the next chapter ("Rheology"), this important concept is explored. The rheology of rocks describes the relationship between stress and strain, and offers an opportunity to integrate much of what we learned thus far.

ADDITIONAL READING

Elliott, D., 1972. Deformation paths in structural geology. *Geological Society of America Bulletin,* 83, 2621–2638.

Erslev, E. A., 1988. Normalized center-to-center strain analysis of packed aggregates. *Journal of Structural Geology,* 10, 201–209.

Fry, N., 1979. Random point distributions and strain measurement in rocks. *Tectonophysics,* 60, 89–104.

Groshong, R. H., Jr., 1972. Strain calculated from twining in calcite. *Geological Society of America Bulletin,* 83, 2025–2038.

Lisle, R. J., 1984. *Geological strain analysis. A manual for the R_f/Φ method.* Pergamon Press: Oxford.

Lister, G. S., and Williams, P. F., 1983. The partitioning of deformation in flowing rock masses. *Tectonophysics,* 92, 1–33.

Means, W. D., 1976. *Stress and strain. Basic concepts of continuum mechanics for geologists.* Springer-Verlag: New York.

Means, W. D., 1990. Kinematics, stress, deformation and material behavior. *Journal of Structural Geology,* 12, 953–971.

Means, W. D., 1992. *How to do anything with Mohr circles (except fry an egg): A short course about tensors for structural geologists.* Geological Society of America Short-Course Notes (2 volumes).

Oertel, G., 1983. The relationship of strain and preferred orientation of phyllosilicate grains in rocks— a review. *Tectonophysics,* 100, 413–447.

Pfiffner, O. A., and Ramsay, J. G., 1982. Constraints on geologic strain rates: Arguments from finite strain states of naturally deformed rocks. *Journal of Geophysical Research,* 87, 311–321.

Ramsay, J. G., and Wood, D. S., 1973. The geometric effects of volume change during deformation processes. *Tectonophysics,* 16, 263–277.

Ramsay, J. G., and Huber, M. I., 1983. *The techniques of modern structural geology. Volume 1: Strain analysis.* Academic Press: London.

Simpson, C., 1988. Analysis of two-dimensional finite strain. In Marshak, S., and Mitra G., eds., *Basic methods of structural geology.* Edited by: S. Marshak and G. Mitra. Prentice Hall: Englewood Cliffs, pp. 333–359.

Rheology

5.1 INTRODUCTION

Earlier we defined strain as the shape change that a body undergoes in the presence of a stress field. But what do we really know about the corresponding stress? And is stress independent of strain? In this chapter we turn to the final and perhaps most challenging aspect of fundamental concepts: the relationship between stress and strain. Whereas it is evident that there is no strain without stress, the relationship between stress and strain is not easy to define on a physical basis. In other words, realizing that stress and strain in rocks are related is quite a different matter from physically determining their actual relationship(s). In materials science and geology we use the term **rheology** (from the Greek *rheos,* meaning "stream" or "flow") to describe the ability of stressed materials to deform or to flow, using fundamental parameters such as strain rate (strain per unit of time; Section 5.1.1), elasticity (Section 5.3.1), and viscosity (Section 5.3.2). These and other concepts will be discussed in this chapter, where we look especially at their significance for understanding rock deformation, rather than focusing on the associated mathematics.

Recalling that stress and strain are second-order tensors, their proportionality is therefore a fourth-order tensor (that is, there are $3^4 = 81$ components). Up front we give a few brief, incomplete descriptions of the most important concepts that will appear throughout this chapter (Table 5.1) to help you to navigate through some of the initial material, until more complete definitions can be given.

Rheology is the study of flow of matter. Flow is an everyday phenomenon and in the previous chapter (Section 4.1) we used syrup on pancakes and human motion as day-to-day examples of deformation. Rocks don't seem to change much by comparison, but remember that geologic processes take place over hundreds of thousands to millions of years. For example, yearly horizontal displacement along the San Andreas Fault (a strike-slip fault zone in California) is on the order of a few centimeters, so considerable deformation has accumulated over the last 700,000 years. Likewise, horizontal displacements on the order of tens to hundreds of kilometers have occurred in the Paleozoic Appalachian fold-and-thrust belt of eastern North America over time period of a few million years (m.y.). Geologically speaking, time is available

TABLE 5.1	BRIEF DESCRIPTIONS OF FUNDAMENTAL CONCEPTS AND TERMS RELATED TO RHEOLOGY
Elasticity	Recoverable (non-permanent), instantaneous strain
Fracturing	Deformation mechanism by which a rock body or mineral loses coherency by simultaneously breaking many atomic bonds
Nonlinear viscosity	Permanent strain accumulation where the stress is exponentially related to the strain rate
Plasticity	Deformation mechanism that involves progressive breaking of atomic bonds without the material losing coherency
Strain rate	Rate of strain accumulation (typically, elongation, e, over time, t); shear strain rate, $\dot{\gamma}$ (gamma dot), is twice the longitudinal strain rate
Viscosity	Non-recoverable (permanent) strain that accumulates with time; the strain rate–stress relationship is linear

in large supply, and given sufficient amounts of it, rocks are able to flow, not unlike syrup. Glacier ice offers an example of flow in a solid material that shows relatively large displacements on human time scales (Figure 5.1). The flow of window glass, on the other hand, is an urban legend that you can refute with the information presented in this chapter.[1] When you look through the windows of an old house you may find that the glass distorts your view. The reason is, as the story goes, that the glass has sagged under its own weight with time (driven by gravity), giving rise to a wavy image. One also finds that the top part of the glass is often thinner than the bottom part. Using the viscous properties of glass (Table 5.5), however, you will see that this is likely due to old manufacturing processes rather than solid flow at surface temperatures. But before examining the mechanical behavior of materials, we need first to introduce the concept of strain rate.

5.1.1 Strain Rate

The time interval it takes to accumulate a certain amount of strain is described by the **strain rate,** symbol $\dot{e}$, which is defined as elongation (e) per time (t):[2]

$$\dot{e} = \mathbf{e}/t = \delta l/(l_o t) \qquad \text{Eq. 5.1}$$

You recall that elongation, length change divided by original length, $\delta l/l_o$ (Equation 4.2), is a dimensionless quantity; thus the dimension of strain rate is $[t]^{-1}$; the

unit is second^{-1}. This may appear to be a strange unit at first glance, so let's use an example. If 30% finite longitudinal strain ($|\mathbf{e}| = 0.3$) is achieved in an experiment that lasts one hour (3600 s), the corresponding strain rate is $0.3/3600 = 8.3 \times 10^{-5}$/s. Now let's see what happens to the strain rate when we change the time duration of our experiment, while maintaining the same amount of finite strain.

Time interval for 30% strain	$\dot{e}$
1 day (86.4 × 10³ s)	3.5 × 10⁻⁶/s
1 year (3.15 × 10⁷ s)	9.5 × 10⁻⁹/s
1 m.y. (3.15 × 10¹³ s)	9.5 × 10⁻¹⁵/s

Thus, the value of the strain rate changes as a function of the time period over which finite strain accumulates. Note that the percentage of strain did not differ for any of the time intervals. So what is the strain rate for a fault that moves 50 km in 1 m.y.? It is not possible to answer this question unless the displacement is expressed relative to another dimension of the body, that is, as a strain. We try again: What is the strain rate of an 800-km long fault moving 50 km in 1 m.y.? We get a strain rate of $(50/800)/(3.15 \times 10^{13}) = 2 \times 10^{-15}$/s. In many cases, commonly involving faults, geologists prefer to use the **shear strain rates** ($\dot{\gamma}$). The relationship between shear strain rate and (longitudinal) strain rate is

$$\dot{\gamma} = 2\dot{e} \qquad \text{Eq. 5.2}$$

A variety of approaches are used to determine characteristic strain rates for geologic processes. A widely used estimate is based on the Quaternary displacement along the San Andreas Fault of California, which gives a strain rate on the order of 10^{-14}/s. Other observations

---

[1]For an exposition of the legend, see the first edition of this book, p. 79.
[2]Here we consider only *longitudinal* strain rate.

FIGURE 5.1 Malaspina Glacier in Alaska showing moraines (dark bands) that are folded during differential flow of the ice.

(such as isostatic uplift, earthquakes,[3] and orogenic activity) support similar estimates and typical geologic strain rates therefore lie in the range of 10^{-12}/s to 10^{-15}/s. Now consider a small tectonic plate with a long dimension of 500 km at a divergent plate boundary. Using a geologic strain rate of 10^{-14}/s, we obtain the yearly spreading rate by multiplying this dimension of the plate by 3.15×10^{-7}/year, giving 16 cm/year, which is the order of magnitude of present-day plate velocities. On a more personal note, your 1.5-cm long fingernail grows 1 cm per year, meaning a growth rate of 0.67/year (or 2×10^{-8}/s). Your nail growth is therefore much, much faster than geologic rates, even

though plates "grow" on the order of centimeters as well. We can offer many more geologic examples, but at this point we hope to leave you acquainted with the general concept of strain rate and typical values of 10^{-12}/s to 10^{-15}/s for geologic processes. Note that exceptions to this geologic range are rapid events like meteorite impacts and explosive volcanism, which are on the order of 10^{-2}/s to 10^{-4}/s.

5.2 GENERAL BEHAVIOR: THE CREEP CURVE

Compression tests on rock samples illustrate that the behavior of rocks to which a load is applied is not simple. Figure 5.2a shows what is called a **creep curve,** which plots strain as a function of time. In this experi-

---

[3]Remember that earthquakes, typically lasting only a few seconds, reflect discrete displacements (on the order of 10–100 cm); during periods in between there is no seismic activity and thus little or no slip.

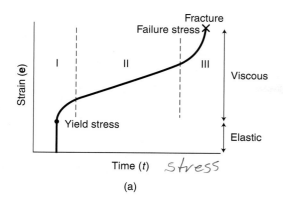

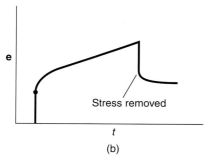

FIGURE 5.2 Generalized strain–time or creep curve, which shows primary (I), secondary (II), and tertiary (III) creep. Under continued stress the material will fail (a); if we remove the stress, the material relaxes, but permanent strain remains (b).

ment the differential stress is held constant. Three creep regimes are observed: (1) **primary** or **transient creep,** during which strain rate decreases with time following very rapid initial accumulation; (2) **secondary** or **steady-state creep,** during which strain accumulation is approximately linear with time; and (3) **tertiary** or **accelerated creep,** during which strain rate increases with time; eventually, continued loading will lead to failure. Restating these three regimes in terms of strain rate, we have regimes of (1) decreasing strain rate, (2) constant strain rate, and (3) increasing strain rate. The strain rate in each regime is the slope along the creep curve.

Rather than continuing our creep experiment until the material fractures, we decide to remove the stress sometime during the interval of steady-state creep. The corresponding creep curve for this second experiment is shown in Figure 5.2b. We see a rapid drop in strain when the stress is removed, after which the material relaxes a little more with time. Eventually there is no more change with time but, importantly, permanent strain remains. In order to examine this behavior of natural rocks we turn to simple analogies and rheologic models.

5.3 RHEOLOGIC RELATIONSHIPS

In describing the various rheologic relationships, we first divide the behavior of materials into two types, **elastic behavior** and **viscous behavior** (Figure 5.3). In some cases, the flow of natural rocks may be approximated by combinations of these **linear rheologies,** in which the ratio of stress over strain or stress over strain rate is a constant. The latter holds true for part of the mantle, but correspondence between stress and strain rate for many rocks is better represented by considering **nonlinear rheologies,** which we discuss after linear rheologies. For each rheologic model that is illustrated in Figure 5.3 we show a physical analog, a creep (strain–time) curve and a stress–strain or stress–strain rate relationship, which will assist you with the descriptions below. Such equations that describe the linear and nonlinear relationships between stress, strain, and strain rate are called **constitutive equations.**

5.3.1 Elastic Behavior

What is **elastic behavior** and is it relevant for deformed rocks? Let's first look at the relevance. In the field of seismology, the study of earthquakes, elastic properties are very important. As you know, seismic waves from an earthquake pass through the Earth to seismic monitoring stations around the world. As they travel, these seismic waves briefly deform the rocks, but after they have passed, the rocks return to their undeformed state. To imagine how rocks are able to do so we turn to a common analog: a rubber band. When you pull a rubber band, it extends; when you remove this stress, the band returns to its original shape. The greater the stress, the farther you extend the band. Beyond a certain point, called the failure stress, the rubber band breaks and brings a painful end to the experiment. This ability of rubber to extend lies in its atomic structure. The bond lengths between atoms and the angles between bonds in a crystal structure represent a state of lowest potential energy for a crystal. These bonds are able to elongate and change their relative angles to some extent, without introducing permanent changes in the crystal structure. Rubber bands extend particularly well because rubber can accommodate large changes in the angular relationships between bonds; however, this causes a considerable increase in the potential energy, which is recovered when we let go of the band, or when it snaps. So, once the stress is released, the atomic structure returns to its energetically most stable configuration, that is, the lowest potential energy. Like the elasticity of a rubber band, the ability of rocks to deform elastically also resides in nonpermanent distortions of

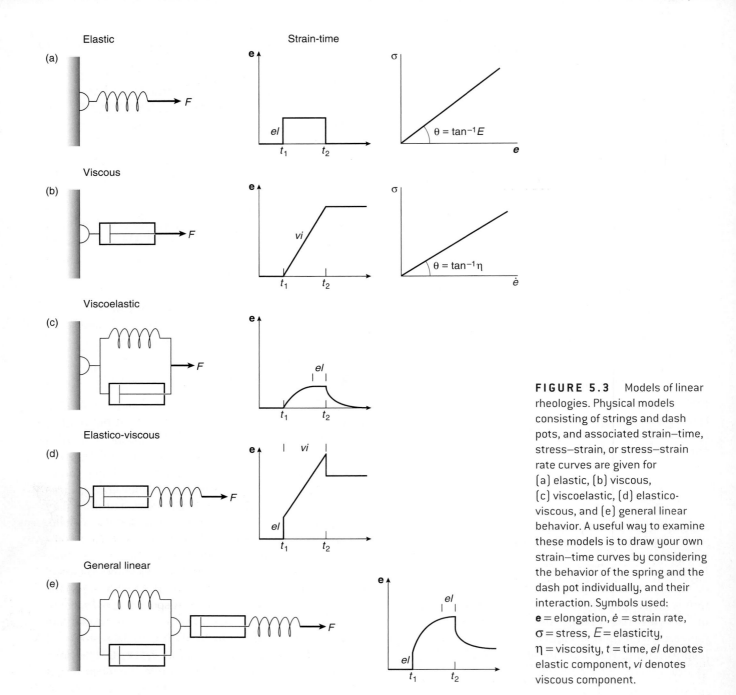

FIGURE 5.3 Models of linear rheologies. Physical models consisting of strings and dash pots, and associated strain–time, stress–strain, or stress–strain rate curves are given for (a) elastic, (b) viscous, (c) viscoelastic, (d) elastico-viscous, and (e) general linear behavior. A useful way to examine these models is to draw your own strain–time curves by considering the behavior of the spring and the dash pot individually, and their interaction. Symbols used: e = elongation, $\dot{e}$ = strain rate, σ = stress, E = elasticity, η = viscosity, t = time, el denotes elastic component, vi denotes viscous component.

the crystal lattice, but unlike rubber, the magnitude of this behavior is relatively small in rocks.

Expressing elastic behavior in terms of stress and strain, we get

$$\sigma = E \cdot \mathbf{e} \qquad \text{Eq. 5.3}$$

where E is a constant of proportionality called **Young's modulus** that describes the slope of the line in the σ–$\mathbf{e}$ diagram (tangent of angle θ; Figure 5.3a). The unit of this elastic constant is Pascal (Pa = kg/m · s^{-2}), which

is the same as that of stress (recall that strain is a dimensionless quantity). Typical values of E for crustal rocks are on the order of -10^{11} Pa.[4] Linear Equation 5.3 is also known as **Hooke's Law,**[5] which describes elastic behavior. We use a spring as the physical model for this behavior (Figure 5.3a).

---

[4]We require a negative sign here to produce a negative elongation (shortening) from applying a compressive (positive) stress.
[5]After the English physicist Robert Hooke (1635–1703).

TABLE 5.2	SOME REPRESENTATIVE BULK MODULI (K) AND SHEAR MODULI (OR RIGIDITY, G) IN 10^{11} Pa AT ATMOSPHERIC PRESSURE AND ROOM TEMPERATURE		
Crystal		K	G
Iron (Fe)		1.7	0.8
Copper (Cu)		1.33	0.5
Silicon (Si)		0.98	0.7
Halite (NaCl)		0.14	0.26
Calcite (CaCO$_3$)		0.69	0.37
Quartz (SiO$_2$)		0.3	0.47
Olivine (Mg$_2$SiO$_4$)		1.29	0.81
Ice (H$_2$O)		0.073	0.025
From Poirier (1985).			

TABLE 5.3	SOME REPRESENTATIVE POISSON'S RATIOS (AT 200 MPa CONFINING PRESSURE)
Basalt	0.25
Gabbro	0.33
Gneiss	0.27
Granite	0.25
Limestone	0.32
Peridotite	0.27
Quartzite	0.10
Sandstone	0.26
Schist	0.31
Shale	0.26
Slate	0.30
Glass	0.24
Sponge	$\ll 0.10$
From Hatcher (1995) and other sources.	

We can also write elastic behavior in terms of the shear stress, σ_s:

$$\sigma_s = G \cdot \gamma \qquad \text{Eq. 5.4}$$

where G is another constant of proportionality, called the **shear modulus** or the **rigidity,** and γ is the shear strain.

The corresponding constant of proportionality in volume change (dilation) is called the **bulk modulus, K:**

$$\sigma = K \cdot [(V - V_o)/V_o] \qquad \text{Eq. 5.5}$$

Perhaps more intuitive than the bulk modulus is its inverse, $1/K$, which is the **compressibility** of a material. Representative values for bulk and shear moduli are listed in Table 5.2.

It is quite common to use an alternative to the bulk modulus that expresses the relationship between volume change and stress, called **Poisson's ratio,**[6] represented by the symbol ν. This elastic constant is defined as the ratio of the elongation perpendicular to the compressive stress and the elongation parallel to the compressive stress:

$$\nu = e_{perpendicular}/e_{parallel} \qquad \text{Eq. 5.6}$$

---

[6]Named after the French mathematician Simeon-Denis Poisson (1781–1840).

Poisson's ratio describes the ability of a material to shorten parallel to the compression direction without corresponding thickening in a perpendicular direction. Therefore the ratio ranges from 0 to 0.5, for fully compressible to fully incompressible materials, respectively. Incompressible materials maintain constant volume irrespective of the stress. A sponge has a very low Poisson's ratio, while a metal cylinder has a relatively high value. A low Poisson's ratio also implies that a lot of potential energy is stored when a material is compressed; indeed, if we remove the stress from a sponge it will jump right back to its original shape. Values for Poisson's ratio in natural rocks typically lie in the range 0.25–0.35 (Table 5.3).

A central characteristic of elastic behavior is its *reversibility:* once you remove the stress, the material returns to its original shape. Reversibility implies that the energy introduced remains available for returning the system to its original state. This energy, which is a form of potential energy, is called the **internal strain energy.** Because the material is undistorted after the stress is removed, we therefore say that strain is **recoverable.** Thus, elastic behavior is characterized by **recoverable strain.** A second characteristic of elastic behavior is the instantaneous response to stress: finite

TABLE 5.4	ELASTIC CONSTANTS
Bulk modulus (K)	Ratio of pressure and volume change
Compressibility ($1/K$)	The inverse of the bulk modulus
Elasticity (E)	Young's modulus
Poisson's ratio (v)	A measure of compressibility of a material. It is defined as the ratio between **e** normal to compressive stress and **e** parallel to compressive stress.
Rigidity (G)	Shear modulus
Shear modulus (G)	Ratio of the shear stress and the shear strain
Young's modulus (E)	Ratio of compressive stress and longitudinal strain

strain is achieved immediately. Releasing the stress results in an instantaneous return to a state of no strain (Figure 5.3a). Both these elastic properties, recoverable and instantaneous strain, are visible in our rubber band or sponge experiments. However, elastic behavior is complicated by the granular structure of natural rocks, where grain boundaries give rise to perturbations from perfect elastic behavior in nongranular solids like glass. A summary of the elastic constants is given in Table 5.4.

Now we return to our original question about the importance of elastic behavior in rocks. With regard to finite strain accumulation, elastic behavior is relatively unimportant in naturally deformed rocks. Typically, elastic strains are less than a few percent of the total strain. So the answer to our question on the importance of elasticity depends on your point of view; a seismologist will say that elastic behavior is important for rocks, but a structural geologist will say that it is not very important.

5.3.2 Viscous Behavior

The flow of water in a river is an example of **viscous behavior** in which, with time, the water travels farther downstream. With this viscous behavior, strain accumulates as a function of time, that is, strain rate. We describe this relationship between stress and strain rate as

$$\sigma = \eta \cdot \dot{e} \qquad \text{Eq. 5.7}$$

where η is a constant of proportionality called the **viscosity** (tan θ, Figure 5.3b) and $\dot{e}$ is the strain rate. This ideal type of viscous behavior is commonly referred to as *Newtonian*[7] or **linear viscous behavior,** but do not confuse the use of "linear" in linear viscous behavior with that in linear stress–strain relationships in the previous section on elasticity. The term *linear* is used here to emphasize a distinction from nonlinear viscous (or non-Newtonian) behavior that we discuss later (Section 5.3.6).

To obtain the dimensional expression for viscosity, remember that strain rate has the dimension of $[t^{-1}]$ and stress has the dimension $[ml^{-1}t^{-2}]$. Therefore η has the dimension $[ml^{-1}t^{-1}]$. In other words, the SI unit of viscosity is the unit of stress multiplied by time, which is Pa · s (kg/m · s). In the literature we often find that the unit Poise[8] is used, where 1 Poise = 0.1 Pa · s.

The example of flowing water brings out a central characteristic of viscous behavior. Viscous flow is irreversible and produces **permanent** or **non-recoverable strain.** The physical model for this type of behavior is the dash pot (Figure 5.3b), which is a leaky piston that moves inside a fluid-filled cylinder.[9] The resistance encountered by the moving piston reflects the viscosity of the fluid. In the classroom you can model viscous behavior by using a syringe with one end open to the air. To give you a sense of the enormous range of viscosities in nature, the viscosities of some common materials are listed in Table 5.5.

How does the viscosity of water, which is on the order of 10^{-3} Pa · s (Table 5.5), compare with that of rocks? Calculations that treat the mantle as a viscous medium produce viscosities on the order of 10^{20}–10^{22} Pa · s. Obviously the mantle is much more viscous than water (>20 orders of magnitude!). You can demonstrate this graphically when calculating the slope of the lines for water and mantle material in the stress–strain rate diagram; they are 0.06° and nearly 90°, respectively. The much higher viscosity of rocks implies that motion is transferred over much larger distances. Stir water, syrup, and jelly in a jar to get a sense of this implication of viscosity. Obviously there is an enormous difference between materials that flow in our daily experience, such as water and syrup, and the "solids" that make up the Earth. Nevertheless, we can approximate the behavior of the Earth as a viscous

---

[7]Named after the British physicist Isaac Newton (1642–1727).

[8]Named after the French physician Jean-Louis Poiseuille (1799–1869).

[9]Just like us you have probably never heard of a dash pot until now, but an old V8-engine that uses equal amounts of oil and gas will also do.

TABLE 5.5	REPRESENTATIVE VISCOSITIES (IN Pa · s)
Air	10^{-5}
Water	10^{-3}
Olive oil	10^{-1}
Honey	4
Glycerin	83
Lava	$10-10^4$
Asphalt	10^5
Pitch	10^9
Ice	10^{12}
Glass	10^{14}
Rock salt	10^{17}
Sandstone slab	10^{18}
Asthenosphere (upper mantle)	$10^{20}-10^{21}$
Lower mantle	$10^{21}-10^{22}$

From several sources, including Turcotte and Schubert (1982).

medium over the large amount of time available to geologic processes (we will return to this with modified viscous behavior in Section 5.3.4). Considering an average mantle viscosity of 10^{21} Pa · s and a geologic strain rate of 10^{-14}/s, Equation 5.7 tells us that the differential (or flow) stresses at mantle conditions are on the order of tens of megapascals. Using a viscosity of 10^{14} Pa · s for glass, flow at atmospheric conditions produces a strain rate that is much too slow to produce the sagging effect that is ascribed to old windows (see Section 5.1).

5.3.3 Viscoelastic Behavior

Consider the situation in which the deformation process is reversible, but in which strain accumulation as well as strain recovery are delayed; this behavior is called **viscoelastic behavior.**[10] A simple analog is a water-soaked sponge that is loaded on the top. The load on the soaked sponge is distributed between the water (viscous behavior) and the sponge material (elastic behavior). The water will flow out of the sponge in response to the load and eventually the sponge will support the load elastically. For a physical model we place a spring (elastic behavior) and a dash pot (viscous behavior) in parallel (Figure 5.3c). When stress is applied, both the spring and the dash pot move simultaneously. However, the dash pot retards the extension of the spring. When the stress is released, the spring will try to return to its original configuration, but again this movement is delayed by the dash pot.

The constitutive equation for viscoelastic behavior reflects this addition of elastic and viscous components:

$$\sigma = E \cdot \mathbf{e} + \eta \cdot \dot{e} \qquad \text{Eq. 5.8}$$

5.3.4 Elastico-Viscous Behavior

Particularly instructive for understanding earth materials is **elastico-viscous**[11] **behavior,** where a material behaves elastically at the first application of stress, but then behaves in a viscous manner. When the stress is removed the elastic portion of the strain is recovered, but the viscous component remains. We can model this behavior by placing a spring and a dash pot in series (Figure 5.3d). The spring deforms instantaneously when a stress is applied, after which the stress is transmitted to the dash pot. The dash pot will move at a constant rate for as long as the stress remains. When the stress is removed, the spring returns to its original state, but the dash pot remains where it stopped earlier. The constitutive equation for this behavior, which is not derived here, is

$$\dot{e} = \dot{\sigma}/E + \sigma/\eta \qquad \text{Eq. 5.9}$$

where $\dot{\sigma}$ is the stress per time unit (i.e., stress rate).

When the spring is extended, it stores energy that slowly relaxes as the dash pot moves, until the spring has returned to its original state. The time taken for the stress to reach $1/e$ times its original value is known as the **Maxwell relaxation time,** where e is the base of natural logarithm ($e = 2.718$). Stress relaxation in this situation decays exponentially. The Maxwell relaxation time, t_M, is obtained by dividing the viscosity by the shear modulus (or rigidity):

$$t_M = \eta/G \qquad \text{Eq. 5.10}$$

---

[10]Also known as firmo-viscous or Kelvinian behavior, after the Irish-born physicist William Kelvin (1824–1907).

[11]Also called Maxwellian behavior after the Scottish physicist James C. Maxwell (1831–1879).

In essence the Maxwell relaxation time reflects the dominance of viscosity over elasticity. If t_M is high then elasticity is relatively unimportant, and vice versa. Because viscosity is temperature dependent, t_M can be expressed as a function of temperature. Figure 5.4 graphs this relationship between temperature and time for appropriate rock properties and shows that mantle rocks typically behave in a viscous manner (as a fluid). The diagram also suggests that crustal rocks normally fail by fracture (elastic field), but lower crustal rocks deform by creep as well. This discrepancy reflects the detailed properties of crustal materials and their non-linear viscosities, as discussed later.

Maxwell proposed this model to describe materials that initially show elastic behavior, but given sufficient time display viscous behavior, which matches the behavior of Earth rather well. Recall that seismic waves are elastic phenomena (acting over short time intervals) and that the mantle is capable of flowing in a viscous manner over geologic time (acting over long time intervals). Taking a mantle viscosity of 10^{21} Pa · s and a rigidity of 10^{11} Pa, and assuming an olivine-dominated mantle (Table 5.2), we get a Maxwell relaxation time for the mantle of 10^{10} s, or on the order of 1000 years. This time agrees well with the uplift that we see following the retreat of continental glaciers after the last Ice Age, which resulted in continued uplift of regions like Scandinavia over thousands of years after the ice was removed.

5.3.5 General Linear Behavior

So far we have examined two fundamental and two combined models and, with some further fine-tuning, we can arrive at a physical model that fairly closely approaches reality while still using linear rheologies. Such **general linear behavior** is modeled by placing the elastico-viscous and viscoelastic models in series (Figure 5.3e). Elastic strain accumulates at the first application of stress (the elastic segment of the elastico-viscous model). Subsequent behavior displays the interaction between the elastico-viscous and viscoelastic models. When the stress is removed, the elastic strain is first recovered, followed by the viscoelastic component. However, some amount of strain (permanent strain) will remain, even after long time intervals (the viscous component of the elastico-viscous model). The creep (**e**–t) curve for this general linear behavior is shown in Figure 5.3e and closely mimics the creep curve that is observed in experiments on natural rocks (compare with Figure 5.2b). We will not present the lengthy equation describing general linear behavior here, but you realize that it represents some combination of viscoelastic and elastico-viscous behavior.

5.3.6 Nonlinear Behavior

The fundamental characteristic that is common to all previous rheologic models is a linear relationship between strain rate and stress (Figure 5.5a): $\dot{e} \propto \sigma$. Experiments on geologic materials (like silicates) at elevated temperature show, however, that the relationship between strain rate and stress is often *nonlinear* (Figure 5.5b): $\dot{e} \propto \sigma^n$; where the exponent n is greater than 1. In other words, the proportionality of strain rate and stress is not constant. Rather, strain rate changes as a function of stress, and vice versa. This behavior is

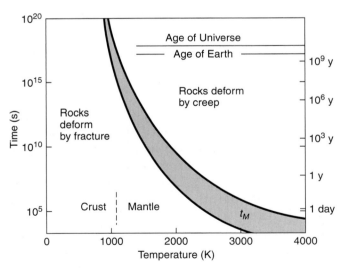

FIGURE 5.4 The Maxwell relaxation time, t_M, is plotted as a function of time and temperature. The curve is based on experimentally derived properties for rocks and their variabilities. The diagram illustrates that hot mantle rocks deform as a viscous medium (fluid), whereas cooler crustal rocks tend to deform by failure. This first-order relationship fits many observations reasonably well, but is incomplete for crustal deformation.

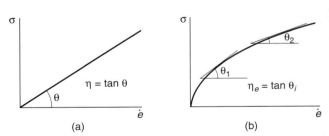

FIGURE 5.5 Linear (a) and nonlinear rheologies (b) in a stress–strain rate plot. The viscosity is defined by the slope of the linear viscous line in (a) and the effective viscosity by the slope of the tangent to the curve in (b).

also displayed by wet paint, which therefore serves as a suitable analog. In order to understand the physical basis for nonlinear behavior in rocks we need to understand the processes that occur at the atomic scale during the deformation of minerals. This requires

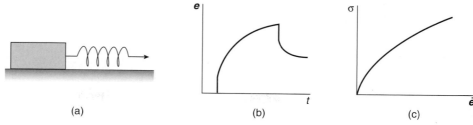

FIGURE 5.6   Nonlinear rheologies: elastic-plastic behavior. (a) A physical model consisting of a block and a spring; the associated strain–time curve (b) and stress–strain rate curve (c).

the introduction of many new concepts that we will not discuss here. We will return to this aspect in detail in Chapter 9, where we examine the role of crystal defects and their mobility.

A physical model representative of rocks showing nonlinear behavior is shown in Figure 5.6, and is known as **elastic-plastic behavior.** In this configuration a block and a spring are placed in series. The spring extends when a stress is applied, but only elastic (recoverable) strain accumulates until a critical stress is reached (the **yield stress**), above which the block moves and permanent strain occurs. The yield stress has to overcome the resistance of the block to moving (friction), but, once it moves, the stress remains constant while the strain accumulates. In fact, you experience elastic-plastic behavior when towing your car with a nylon rope that allows some stretch. Removing the elastic component (e.g., a sliding block with a nonelastic rope) gives **ideal plastic behavior,**[12] but this has less relevance to rocks.

A consequence of nonlinear rheologies is that we can no longer talk about (Newtonian) viscosity, because, as the slope of the stress–strain rate curve varies, the viscosity also varies. Nevertheless, as it is convenient for modeling purposes to use viscosity at individual points along the curve, we define the **effective viscosity** (η_e) as

$$\eta_e = \sigma/\dot{e} \qquad \text{Eq. 5.11}$$

This relationship is the same as that for viscous or Newtonian behavior (Equation 5.7), but in the case of effective viscosity you have to remember that η_e changes as the stress and/or the strain rate changes. In Figure 5.6 you see that the effective viscosity (the tangent of the slope) decreases with increasing stress and strain rate, which means that flow proceeds faster under these conditions. Thus, effective viscosity is

not a material property like Newtonian viscosity, but a convenient description of behavior under prescribed conditions of stress or strain rate. For this reason, η_e is also called **stress-dependent** or **strain rate–dependent viscosity.**

The constitutive equation describing the relationship between strain rate and stress for nonlinear behavior is

$$\dot{e} = A\,\sigma^n \exp(-E^*/RT) \qquad \text{Eq. 5.12}$$

This relationship introduces several new parameters, some of which we used earlier without explanation (as in Figure 5.4). E^*, the activation energy, is an empirically derived value that is typically in the range of 100–500 kJ/mol, and n, the stress exponent, lies in the range $1 > n > 5$ for most natural rocks, with $n = 3$ being a representative value. The crystal processes that enable creep are temperature dependent and require a minimum energy before they are activated (see Chapter 9); these parameters are included in the exponential part of the function[13] as the activation energy (E^*) and temperature (T in degrees Kelvin); A is a constant, and R is the gas constant. Table 5.6 lists experimentally derived values for A, n, and E^* for many common rock types.

From Table 5.6 we can draw conclusions about the relative strength of the rock types, that is, their **flow stresses.** If we assume that the strain rate remains constant at 10^{-14}/s, we can solve the constitutive equation for various temperatures. Let us first rewrite Equation 5.12 as a function of stress:

$$\sigma = (\dot{e}/A)^{1/n} \exp(E^*/RT) \qquad \text{Eq. 5.13}$$

If we substitute the corresponding values for A, n, and E^* at constant T, we see that the differential stress value for rock salt is much less than that for any of the

---

[12]Or Saint-Venant behavior, after the French nineteenth-century physicist A. J. C. Barre de Saint-Venant.

[13]exp(a) means e^a, with e = 2.72.

TABLE 5.6	EXPERIMENTALLY DERIVED CREEP PARAMETERS FOR COMMON MINERALS AND ROCK TYPES		
Rock type	A $(MPa^{-n}s^{-1})$	n	E^* $(kJ \cdot mol^{-1})$
Albite rock	2.6×10^{-6}	3.9	234
Anorthosite	3.2×10^{-4}	3.2	238
Clinopyroxene	15.7	2.6	335
Diabase	2.0×10^{-4}	3.4	260
Granite	1.8×10^{-9}	3.2	123
Granite (wet)	2.0×10^{-4}	1.9	137
Granulite (felsic)	8.0×10^{-3}	3.1	243
Granulite (mafic)	1.4×10^{-4}	4.2	445
Marble (< 20 MPa)	2.0×10^{-9}	4.2	427
Orthopyroxene	3.2×10^{-1}	2.4	293
Peridotite (dry)	2.5×10^{4}	3.5	532
Peridotite (wet)	2.0×10^{3}	4.0	471
Plagioclase (An75)	3.3×10^{-4}	3.2	238
Quartz	1.0×10^{-3}	2.0	167
Quartz diorite	1.3×10^{-3}	2.4	219
Quartzite	6.7×10^{-6}	2.4	156
Quartzite (wet)	3.2×10^{-4}	2.3	154
Rock salt	6.29	5.3	102

From Ranalli (1995) and other sources.

other rock types. This fits the observation that rock salt flows readily, as we learn from, for example, the formation of salt diapirs (Chapter 2). Limestones and marbles are also relatively weak and therefore regional deformation is often localized in these rocks. Quartz-bearing rocks, such as quartzites and granites, in turn are weaker than plagioclase-bearing rocks. Olivine-bearing rocks are among the strongest of rock types, meaning that they require large differential stresses to flow. But if this is true, how can the mantle flow at differential stresses of tens of megapascals? The answer is obtained by solving Equation 5.13 for various temperatures. In Figure 5.7 this is done using a "cold" geothermal gradient of 10 K/km[14] for some of the rock types in Table 5.6. The graph supports our first-order conclusion on the relative strength of various rock types, but it also shows that with increasing depth, the strength of all rock types decreases significantly. The latter is an important observation for understanding the nature of deformation processes in the deep Earth.

5.4 ADVENTURES WITH NATURAL ROCKS

Our discussion of rheology so far has been mostly abstract. We treated rocks as elastic springs and fluids, but we have not really looked at the behavior of natural rocks under different environmental conditions. The

---

[14]A representative "hot" geothermal gradient is 30 K/km.

results from decades of experiments on rocks will help us to get a better appreciation of the flow of rock. The reason for doing experiments on natural rocks is twofold: (1) we observe the actual behavior of rocks rather than that of syrup or elastic bands, and (2) we can vary several parameters in our experiments, such as pressure, temperature, and time, to examine their role in rock deformation. A vast amount of experimental data is available to us and many of the principles have, therefore, been known for several decades. Here we will limit our discussion by looking only at experiments that highlight particular parameters. By combining the various responses, we can begin to understand the rheology of natural rocks.

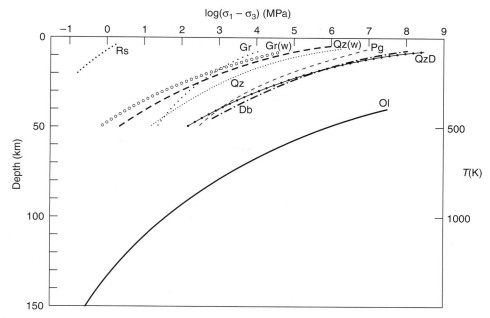

FIGURE 5.7 Variation of creep strength with depth for several rock types using a geothermal gradient of 10 K/km and a strain rate of 10^{-14}/s. Rs = rock salt, Gr = granite, Gr(w) = wet granite, Qz = quartzite, Qz(w) = wet quartzite, Pg = plagioclase-rich rock, QzD = quartz diorite, Db = diabase, Ol = Olivine-rich rock.

An alternative approach to examining the flow of rocks is to study material behavior in scaled experiments. Scaling brings fundamental quantities such as length [l], mass [m], and time [t] to the human scale. For example, we can use clay as a model material to study faulting, or wax to examine time-dependent behavior. Each analog that is used in scaled experiments has advantages and disadvantages, and the experimentalist has to make trade-offs between experimental conditions and geologic relevance. We will not use scaled experiments in the subsequent section of this chapter.

5.4.1 The Deformation Apparatus

A deformation experiment on a rock or a mineral can be carried out readily by placing a small sample in a vise, but when you try this experiment you have to be careful to avoid being bombarded with randomly flying chips as the material fails. If you ever cracked a hard nut in a nutcracker, you know what we mean. In rock deformation experiments we attempt to control the experiment a little better for the sake of the experimentalist, as well as to improve the analysis and interpretation of the results. A typical deformation apparatus is schematically shown in Figure 5.8. In this

rig, a cylindrical rock specimen is placed in a pressure chamber, which is surrounded by a pressurized fluid that provides the **confining pressure,** P_c, on the specimen through an impermeable jacket. This experimental setup is known as a **triaxial testing apparatus,** named for the triaxial state of the applied stress, in which all three principal stresses are unequal to zero. For practical reasons, two of the principal stresses are equal. In addition to the fluid that provides the confining pressure, a second fluid may be present in the specimen to provide **pore-fluid pressure,** P_f. The difference between confining and pore pressure, $P_c - P_f$, is called the **effective pressure** (P_e). Adjusting the piston at the end of the test cylinder results in either a maximum or minimum stress along the cylinder axis, depending on the magnitudes of fluid pressure and axial stress. The remaining two principal stresses are equal to the effective pressure. By varying any or all of the axial stress, the confining pressure, or the pore-fluid pressure, we obtain a range of stress conditions to carry out our deformation experiments. In addition, we can heat the sample during the experiment to examine the effect of temperature. This configuration allows a limited range of finite strains, so a torsion rig with rotating plates is increasingly used for experiments at high shear strains.

A triaxial apparatus enables us to vary stress, strain, and strain rate in rock specimens under carefully controlled parameters of confining pressure, temperature, pore-fluid pressure, and time (that is, duration of the

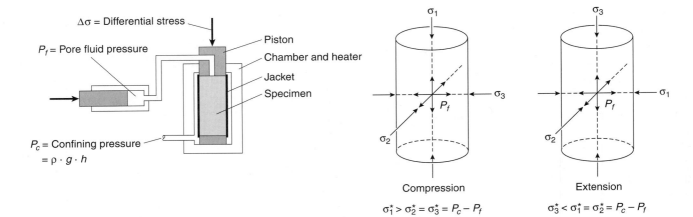

FIGURE 5.8 Schematic diagram of a triaxial compression apparatus and states of stress in cylindrical specimens in compression and extension tests. The values of P_c, P_f, and σ can be varied during the experiments.

Compression

$$\sigma_1^* > \sigma_2^* = \sigma_3^* = P_c - P_f$$
$$\sigma_1^* = \Delta\sigma + P_c - P_f$$

Extension

$$\sigma_3^* < \sigma_1^* = \sigma_2^* = P_c - P_f$$
$$\sigma_3^* = P_c - P_f - \Delta\sigma$$

$\sigma_1^*, \sigma_2^*, \sigma_3^* =$ Maximum, intermediate, minimum effective principal stresses

$P_c - P_f =$ Effective pressure

experiment). What happens when we vary these parameters and what does this tell us about the behavior of natural rocks? Before we dive into these experiments it is useful to briefly review how these environmental properties relate to the Earth. Both confining pressure and temperature increase with depth in the Earth (Figure 5.9). The confining pressure is obtained from the simple relationship

$$P_c = \rho \cdot g \cdot h \qquad \text{(Eq. 4.15)}$$

where ρ is the density, g is gravity, and h is depth. This is the pressure from the weight of the overlying rock column, which we call the **lithostatic pressure.** The temperature structure of the Earth is slightly more complex than the constant gradient of 10 K/km used earlier in Figure 5.7. At first, temperature increases at an approximately constant rate (10°C/km–30°C/km),[15] but then the thermal gradient is considerably less (Figure 5.9). Additional complexity is introduced by the heat generated from compression at high pressures, which is reflected in the **adiabatic gradient** (dashed line in Figure 5.9). But if we limit our considerations to the crust and uppermost mantle, a linear **geothermal gradient** in the range of 10°C/km–30°C/km is a reasonable approximation. In Section 5.1 we learned that most geologic processes occur at strain rates on the order of 10^{-14}/s, with the exception of meteoric impacts, seismic events, and explosive volcanism. In contrast to geologic strain rates, experimental work is typically limited by the patience and the life expectancy of the experimentalist. Some of the slowest experiments are

carried out at strain rates of 10^{-8}/s (i.e., 30% shortening in a year), which is still four to seven orders of magnitude greater than geologic rates. Having said all this, now let's look at the effects of varying environmental conditions, such as confining pressure, temperature, strain rate, and pore-fluid pressure, during deformation experiments.

5.4.2 Confining Pressure

You recall from Chapter 3 that confining pressure acts equally in all directions, so it imposes an isotropic stress on the specimen. When we change the **confining pressure** during our experiments we observe a very important characteristic: increasing confining pressure results in greater strain accumulation before failure (Figure 5.8). In other words, increasing confining pressure increases the viscous component and therefore the rock's ability to flow. What is the explanation for this? If you have already read Chapter 6, the Mohr circle for

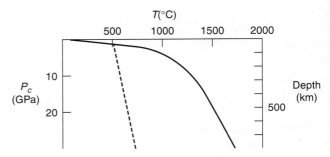

FIGURE 5.9 Change of temperature (T) and pressure (P_c) with depth. The dashed line is the adiabatic gradient, which is the increase of temperature with depth resulting from increasing pressure and the compressibility of silicates.

---

[15]Because we are dealing with rates, K and °C are interchangeable.

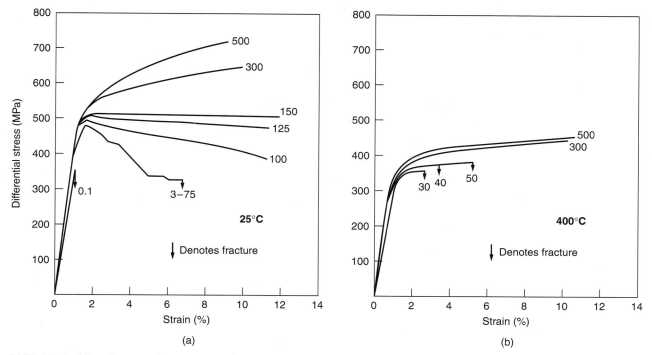

FIGURE 5.10 Compression stress–strain curves of Solnhofen limestone at various confining pressures (indicated in MPa) at (a) 25°C and (b) 400°C.

stress and failure criteria give an explanation, but we will assume that this material is still to come. So, we'll take another approach. Moving your arm as part of a workout exercise is quite easy, but executing the same motion under water is a lot harder. Water "pushes" back more than air does, and in doing so it resists the motion of your arms. Similarly, higher confining pressures resist the opening of rock fractures, so any shape change that occurs is therefore viscous (ignoring the small elastic component).

The effect of confining pressure is particularly evident at elevated temperatures, where fracturing is increasingly suppressed (Figure 5.10b). When we compare common rock types, the role of confining pressure varies considerably (Figure 5.11); for example, the effect is much more pronounced in sandstone and shale than in quartzite and slate. Thus, it appears that larger strains can be achieved with increasing depth in the Earth, where we find higher lithostatic pressures.

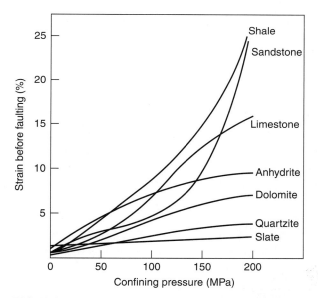

FIGURE 5.11 The effect of changing the confining pressure on various rock types. For these common rocks, the amount of strain before failure (ductility) differs significantly.

5.4.3 Temperature

A change in **temperature** conditions produces a marked change in behavior (Figure 5.12). Using the same limestone as in the confining pressure experiments (Figure 5.10), we find that the material fails rapidly at low temperatures. Moreover, under these conditions most of the strain prior to failure is recoverable (elas-

tic). When we increase the temperature, the elastic portion of the strain decreases, while the ductility increases, which is most noticeable at elevated confining pressures (Figure 5.12b). You experience this temperature-dependence of flow also if you pour syrup on pancakes in a tent in the Arctic or in the Sahara: the ability of syrup to flow increases with temperature. Furthermore,

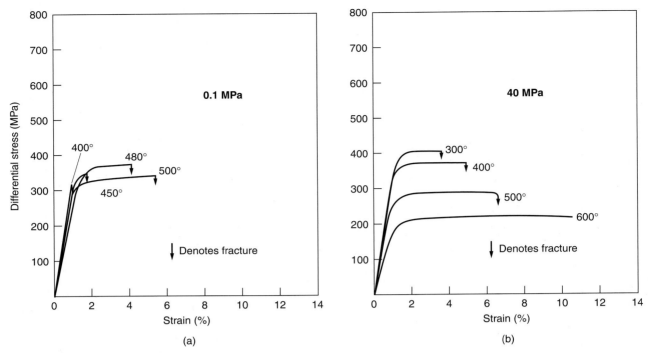

FIGURE 5.12 Compression stress–strain curves of Solnhofen limestone at various temperatures (indicated in °C) at (a) 0.1 MPa confining pressure and at (b) 40 MPa confining pressure.

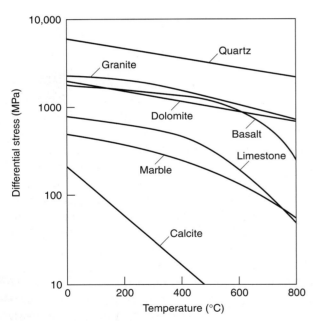

FIGURE 5.13 The effect of temperature changes on the compressive strength of some rocks and minerals.

the maximum stress that a rock can support until it flows (called the **yield strength** of a material) decreases with increasing temperature. The behavior of various rock types and minerals under conditions of increasing temperature is shown in Figure 5.13, from which we see that calcite-bearing rocks are much more affected than, say, quartz-bearing rocks. Collectively these experiments demonstrate that rocks have lower strength and are more ductile with increasing depth in the Earth, where we find higher temperatures.

5.4.4 Strain Rate

It is impossible to carry out rock deformation experiments at geologic rates (Section 5.4.1), so it is particularly important for the interpretation of experimental results to understand the role of **strain rate.** The effect is again best seen in experiments at elevated temperatures, such as those on marble (Figure 5.14). Decreasing the strain rate results in decreased rock strength and increased ductility. We again turn to an analogy for our understanding. If you slowly press on a small ball of Silly Putty®,[16] it spreads under the applied stress (ductile flow). If, on the other hand, you deform the same ball by a blow from a hammer, the material will shatter into many pieces (brittle failure). Although the environmental conditions are the same, the response is dramatically different because the strain rate differs.[17]

---

[16]A silicone-based material that offers hours of entertainment if you are a small child or a professional structural geologist.

[17]Assuming that the stress from a slow push and a rapid blow are equal.

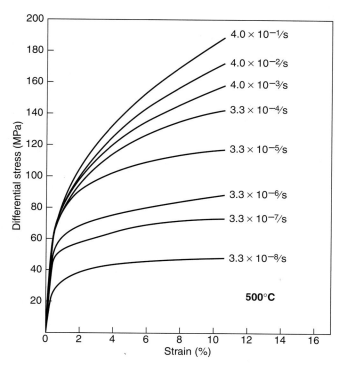

FIGURE 5.14 Stress versus strain curves for extension experiments in weakly foliated Yule marble for various constant strain rates at 500°C.

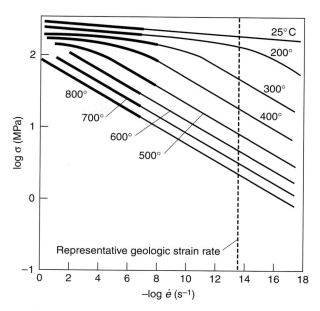

FIGURE 5.15 Log stress versus −log strain rate curves for various temperatures based on extension experiments in Yule marble. The heavy lines mark the range of experimental data; the thinner part of the curves is extrapolation to slower strain rates. A representative geologic strain rate is indicated by the dashed line.

Because rocks show similar effects from strain rate variation, the Silly Putty® experiment highlights a great uncertainty in experimental rock deformation. Extrapolating experimental results for strain rates over many orders of magnitude has significant consequences (Figure 5.15). Consider a strain rate of 10^{-14}/s and a temperature of 400°C, where ductile flow occurs at a differential stress of 20 MPa. At the same temperature, but at an experimental strain rate of 10^{-6}/s, the flow stresses are nearly an order of magnitude higher (160 MPa). Comparing this with the results of temperature experiments, you will notice that temperature change produces effects similar to strain rate variation in rock experiments (higher $t \propto$ lower $\dot{e}$); $\dot{e}$ has therefore been used as a substitute for geologic strain rates. In spite of the uncertainties, the volume of experimental work and our understanding of the mechanisms of ductile flow (Chapter 9) allow us to make reasonable extrapolations to rock deformation at geologic strain rates.

5.4.5 Pore-Fluid Pressure

Natural rocks commonly contain a fluid phase that may originate from the depositional history or may be secondary in origin (for example, fluids released from prograde metamorphic reactions). In particular, low-grade sedimentary rocks such as sandstones and shales, contain a significant fluid component that will affect their behavior under stress. To examine this parameter, the deformation rig shown in Figure 5.8 contains an impermeable jacket around the sample. Experiments show that increasing the **pore-fluid pressure** produces a drop in the sample's strength and reduces the ductility (Figure 5.16a). In other words, rocks are weaker when the pore-fluid pressure is high. We return to this in Chapter 6, but let us briefly explore this effect here. Pore-fluid pressure acts equally in all directions and thus counteracts the confining pressure, resulting in an effective pressure ($P_e = P_c - P_f$) that is less than the confining pressure. Thus, we can hypothesize that increasing the pore-fluid pressure has the same effect as decreasing the confining pressure of the experiment. We put this to the test by comparing the result of two experiments on limestone in Figure 5.16a and 5.16b that vary pore-fluid pressure and confining pressure, respectively. Clearly, there is remarkable agreement between the two experiments, supporting our hypothesis.

The role of fluid content is a little more complex than is immediately apparent from these experiments,

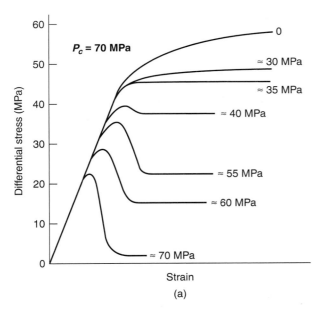

$P_c = 70$ MPa

Strain

(a)

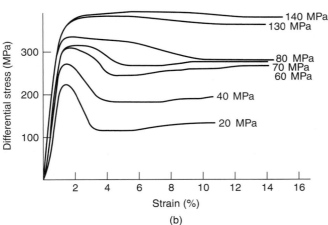

(b)

FIGURE 5.16 Comparing the effect on the behavior of limestone of (a) varying pore-fluid pressure and (b) varying confining pressure.

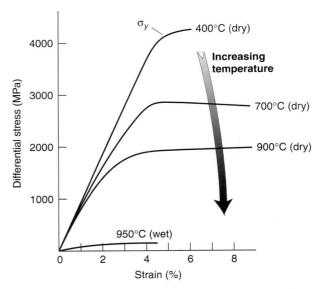

FIGURE 5.17 The effect of water content on the behavior of natural quartz. Dry and wet refer to low and high water content, respectively. The curves also show the effect of temperature on single crystal deformation, which is similar to that for rocks (Figure 5.12).

because of fluid-chemical effects. While ductility decreases with increasing pore-fluid pressure, the corresponding decreased strength of the material will actually promote flow. The same material with low fluid content ("dry" conditions) would resist deformation, but at high fluid content ("wet" conditions) flow occurs readily. This is nicely illustrated by looking at the deformation of quartz with varying water content (Figure 5.17). The behavior of quartz is similar to that in the previous rock experiment: the strength of "wet" quartz is only about one tenth that of "dry" quartz at the same temperature. The reason for this weakening lies in the substitution of OH groups for O in the silicate crystal lattice, which strains and weakens the

Si-O atomic bonds (see Chapter 6).[18] In practice, fluid content explains why many minerals and rocks deform relatively easily even under moderate stress conditions.

5.4.6 Work Hardening–Work Softening

Laboratory experiments on rocks bring an interesting property to light that we first noticed in the general creep curve of Figure 5.2a, namely, that the relationship between strain and time varies in a single experiment. The strain rate may decrease, increase, or remain constant under constant stress. When we carry out experiments at a constant strain rate we often find that the stresses necessary to continue the deformation may increase or decrease, phenomena that engineers call **work hardening** (greater stress needed) and **work softening** (lower stress needed), respectively. In a way you can think of this as the rock becoming stronger or weaker with increasing strain; therefore, we also call this effect **strain hardening** and **strain softening,** respectively. A practical application of work hardening is the repeated rolling of metal that gives it greater strength, especially when it is heated. While you may not realize this when seeing the repair bill after an

---

[18]This is called **hydrolytic weakening** (Chapter 6).

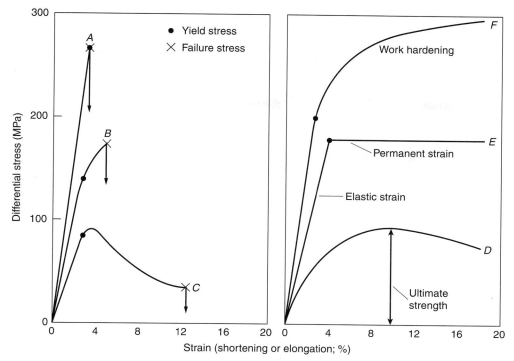

FIGURE 5.18 Representative stress–strain curves of brittle (*A* and *B*), brittle-ductile (*C*), and ductile behavior (*D–F*). *A* shows elastic behavior followed immediately by failure, which represents brittle behavior. In *B*, a small viscous component (permanent strain) is present before brittle failure. In *C*, a considerable amount of permanent strain accumulates before the material fails, which represents transitional behavior between brittle and ductile. *D* displays no elastic component and work softening. *E* represents ideal elastic-plastic behavior, in which permanent strain accumulates at constant stress above the yield stress. *F* shows the typical behavior seen in many of the experiments, which displays a component of elastic strain followed by permanent strain that requires increasingly higher stresses to accumulate (work hardening). The yield stress marks the stress at the change from elastic (recoverable or nonpermanent strain) to viscous (nonrecoverable or permanent strain) behavior; failure stress is the stress at fracturing.

unfortunate encounter with a moving tree, car makers use this to strengthen some metal parts. This effect was already known in ancient times, when Japanese samurai sword makers were producing some of the hardest blades available using repeated heating and hammering of the metal. Work softening is the opposite effect, in which the stress required to continue the experiment is less; in constant stress experiments this results in a strain rate increase. Both processes may occur in the same materials as shown in, for example, Figure 5.10a. At low confining pressure a stress drop is observed after about 2% strain, representing work softening. At high confining pressures, the limestone displays work hardening, because increasingly higher stresses are necessary to continue the experiment. Schematically this is summarized in Figure 5.18, where curve *D* displays work softening and curve *F* displays work hardening.

Work hardening and work softening are understood from the atomic-scale processes that enable rocks to flow. However, unless you already have some background in crystal plasticity a quick explanation here would be insufficient. We therefore defer explanation of the physical basis of these processes until the time we discuss crystal defects and their movement (Chapter 8).

5.4.7 Significance of Experiments to Natural Conditions

We close this section on experiments with a table that summarizes the results of varying the confining pressure, temperature, fluid pressure, and strain rate (Table 5.7) in experiments on rocks and by examining the significance for geologic conditions.

From Table 5.7 we see that increasing the confining pressure (P_c) and the fluid pressure (P_f) have

TABLE 5.7	EFFECT OF ENVIRONMENTAL PARAMETERS ON RHEOLOGIC BEHAVIOR	
	Effect	Explanation
High P_c	Suppresses fracturing; increases ductility; increases strength; increases work hardening	Prohibits fracturing and frictional sliding; higher stress necessary for fracturing exceeds that for ductile flow
High T	Decreases elastic component; suppresses fracturing; increases ductility; reduces strength; decreases work hardening	Promotes crystal plastic processes
Low $\dot{e}$	Decreases elastic component; increases ductility; reduces strength; decreases work hardening	Promotes crystal plastic processes
High P_f	Decreases elastic component; promotes fracturing; reduces ductility; reduces strength *or* promotes flow	Decreases P_c ($P_e = P_c - P_f$) and weakens Si-O atomic bonds

opposing effects, while increasing temperature (T) and lowering strain rate ($\dot{e}$) have the same effect. Confining pressure and temperature, which both increase with depth in the Earth, increasingly resists failure, while promoting larger strain accumulation; that is, they increase the ability of rocks to flow. High pore-fluid content is more complicated as it favors fracturing if P_f is high or promotes flow in the case of intracrystalline fluids. From these observations we would predict that brittle behavior (fracturing) is largely restricted to the upper crust, while ductile behavior (flow) dominates at greater depth. A natural test supporting this hypothesis is the realization that faulting and earthquakes generally occur at shallow crustal levels (<15 km depth),[19] while large-scale ductile flow dominates the deeper crust and mantle (e.g., mantle convection).

5.5   CONFUSED BY THE TERMINOLOGY?

By now perhaps a baffling array of terms and concepts have passed before your eyes. If you think that you are the only one who is confused, you should look at the scientific literature on this topic. Let us, therefore, try to bring some additional order to the terminology. Table 5.8 lists brief descriptions of terms that are commonly used in the context of rheology, and contrasts the mechanical behavior of rock deformation with operative deformation mechanisms. A schematic diagram of some representative stress–strain curves summarizes the most important elements (Figure 5.18).

Perhaps the two most commonly used terms in the context of rheology are brittle and ductile. In fact, we use them to subdivide rock structures in Parts B and C of this text, so it is important to understand their meaning. **Brittle behavior** describes deformation that is localized on the mesoscopic scale and involves the formation of fractures. For example, a fracture in a tea cup is brittle behavior. In natural rocks, brittle fracturing occurs at finite strains of 5% or less. In contrast, **ductile behavior** describes the ability of rocks to accumulate significant permanent strain with deformation that is distributed on the mesoscopic scale. The shape change that we achieved by pressing our clay cube is ductile behavior. In nature, a faulted rock and a folded rock are examples of brittle and ductile behavior, respectively. Importantly, these two modes of behavior do not define the mechanism by which deformation occurs. This distinction between behavior and mechanism is important, and can be explained with a simple example.

Consider a cube that is filled with small undeformable spheres (e.g., marbles) and a second cube that is filled with spheres consisting of clay (Figure 5.19). If we deform these cubes into rectangular blocks, the mechanism by which this shape change is achieved is

---

[19]Excluding deep earthquakes in subduction zones, which represent special conditions.

TABLE 5.8	TERMINOLOGY RELATED TO RHEOLOGY, WITH EMPHASIS ON BEHAVIOR AND MECHANISMS
Brittle-ductile transition	Depth in the Earth below which brittle behavior is replaced by ductile processes (see under "behavior" below)
Brittle-plastic transition	Depth in the Earth where the dominant deformation mechanism changes from fracturing to crystal plastic processes (see under "mechanisms" below)
Competency	Relative term comparing the resistance of rocks to flow
Failure stress	Stress at which failure occurs
Fracturing	Deformation mechanism by which a rock body or mineral loses coherency
Crystal plasticity	Deformation mechanism that involves breaking of atomic bonds without the material losing coherency
Strength	Stress that a material can support before failure
Ultimate strength	Maximum stress that a material undergoing work softening can support before failure
Work hardening	Condition in which stress necessary to continue deformation experiment increases
Work softening	Condition in which stress necessary to continue deformation experiment decreases
Yield stress	Stress at which permanent strain occurs
Material behavior	
Brittle behavior	Response of a solid material to stress during which the rock loses continuity (cohesion). Brittle behavior reflects the occurrence of brittle deformation mechanisms. It occurs only when stresses exceed a critical value, and thus only occurs after the body has already undergone some elastic and/or plastic behavior. The stress necessary to induce brittle behavior is affected strongly by pressure (stress-sensitive behavior); brittle behavior generally does not occur at high temperatures.
Ductile behavior	A general term for the response of a solid material to stress such that the rock appears to flow mesoscopically like a viscous fluid. In a material that has deformed ductilely, strain is distributed, i.e., strain develops without the formation of mesoscopic discontinuities in the material. Ductile behavior can involve brittle (cataclastic flow) or plastic deformation mechanisms.
Elastic behavior	Response of a solid material to stress such that the material develops an instantaneous, recoverable strain that is linearly proportional to the applied stress. Elastic behavior reflects the occurrence of elastic deformation mechanisms. Rocks can undergo less than a few percent elastic strain before they fail by brittle or plastic mechanisms, and conditions of failure are dependent on pressure and temperature during deformation.
Plastic behavior	Response of a solid material to stress such that when stresses exceed the yield strength of the material, it develops a strain without loss of continuity (i.e., without formation of fractures). Plastic behavior reflects the occurrence of plastic deformation mechanisms, is affected strongly by temperature, and requires time to accumulate (strain rate–sensitive behavior).
Viscous behavior	Response of a liquid material to a stress. As soon as the differential stress becomes greater than zero, a viscous material begins to flow, and the flow rate is proportional to the magnitude of the stress. Viscous deformation takes time to develop.
Deformation mechanisms	
Brittle deformation mechanisms	Mechanisms by which brittle deformation occurs, namely fracture growth and frictional sliding. Fracture growth includes both joint formation and shear rupture formation, and sliding implies faulting. If fracture formation and frictional sliding occur at a grain scale, the resulting deformation is called cataclasis; if cataclasis results in the rock "flowing" like a viscous fluid, then the process is called cataclastic flow.
Elastic deformation mechanisms	Mechanisms by which elastic behavior occurs, namely the bending and stretching, without breaking, of chemical bonds holding atoms or molecules together.
Plastic deformation mechanisms	Mechanisms by which plastic deformation occurs, namely dislocation glide, dislocation creep (glide and climb; including recovery, recrystallization), diffusive mass transfer (grain-boundary diffusion or Coble creep, and diffusion through the grain or Herring-Nabarro creep), grain–boundary sliding/superplasticity.

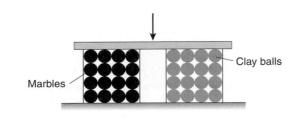

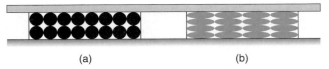

FIGURE 5.19 Deformation experiment with two cubes containing marbles (a), and balls of clay (b), showing ductile strain accumulation by different deformation mechanisms. The finite strain is equal in both cases and the mode of deformation on the scale of the block is distributed (ductile behavior). However, the mechanism by which the deformation occurs is quite different: in (a) frictional sliding of undeformed marbles occurs, while in (b) individual clay balls distort into ellipsoids.

very different. In Figure 5.19a, the rigid spheres slide past one another to accommodate the shape change without distortion of the individual marbles. In Figure 5.19b, the shape change is achieved by changes in the shape of individual clay balls to ellipsoids. In both cases the deformation is not localized, but distributed throughout the block; at the sphere boundaries in Figure 5.19a, and within the spheres in Figure 5.19b. Both experiments are therefore expressions of ductile behavior, although the mechanisms by which deformation occurred are quite different.[20]

Commonly you will encounter the term **brittle-ductile transition** in the literature, but it is not always clear what is meant. For example, seismologists use this term to describe the depth below which non-subduction zone earthquakes no longer occur. If they mean that deformation occurs by a mechanism other than faulting, this usage is incorrect because ductility is not a mechanism. Ductile behavior can occur by faulting if thousands of small cracks take up the strain. Alternatively, ductile behavior may represent deformation in which crystallographic processes are important. We have seen that ductility simply reflects the ability of a material to accumulate significant permanent strain. Clearly, we have to separate behavior (brittle vs. ductile) from mechanism (e.g., fracturing, frictional sliding, crystal plasticity, diffusion, all of which are

discussed in later chapters), and in many instances terms such as brittle-ductile transition lead to unnecessary confusion. Instead, a useful solution is to contrast the mechanisms, for example **brittle-plastic transition** or localized versus nonlocalized deformation. Returning to our experiment in Figure 5.19, the marble-filled cube (a) deforms by frictional sliding whereas the clay balls (b) deform plastically. In a rheologic context this is normal stress-dependent and strain rate–dependent behavior, respectively.

In some instances we appear to be faced with a **brittle-ductile paradox.** It is quite common in the field to find folded beds that appear to be closely associated with faults (such as fault-propagation and fault-bend folds, see Chapters 11 and 18). If we assume that faulting and folding occurred simultaneously, then we are left with a situation where both brittle faulting and ductile folding took place at essentially the same level in the Earth. How can we explain this situation and why does it only *appear* to be contradictory? There is no reason to expect that P_c, T, and P_f are sufficiently different over the relatively small volume of rock to account for the simultaneous occurrence of these two behavioral modes of deformation. Strain rate and strain rate gradients, on the other hand, may vary considerably in any given body of rock. Recall the Silly Putty® experiment in which fracturing occurred at high strain rates and ductile flow at lower rates. Similarly, faulting in nature may occur at regions of high strain rates, but some distance away the strain rate may be sufficiently low to give rise to ductile folding. So strain rate gradients are one explanation for the simultaneous occurrence of brittle and ductile structures, which resolves the apparent brittle-ductile paradox.

Competency and strength are two related terms that describe the relationship of rocks to stress. *Strength* is the stress that a material can support before failure.[21] **Competency** is a relative term that compares the resistance of rocks to flow (for example, Figures 5.7 and 5.13). Experiments and general field observations have given us a qualitative *competency scale* for rocks. The competency of sedimentary rocks increases in the order:

rock salt → shale → limestone → greywacke → sandstone → dolomite

---

[20]These mechanisms are frictional sliding and crystal plasticity, respectively.

[21]In the case of work softening, we use the term *ultimate strength* (Figure 5.18).

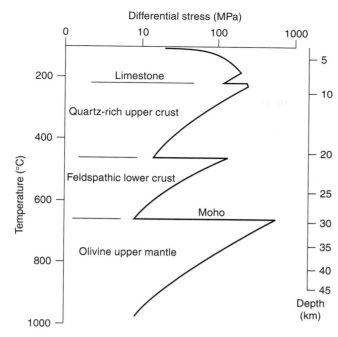

FIGURE 5.20 Rheologic stratification of the lithosphere based on the mechanical properties of characteristic minerals. Computed lithospheric strength (i.e., the differential stress) changes not only as a function of composition, but also as a function of depth (i.e., temperature).

For metamorphic/igneous rocks the order of increasing competency is:

schist → marble → quartzite → gneiss → granite → basalt

Note that competency is not the same as the amount of strain that can accumulate in a body. Therefore ductility contrast between materials should not be used as a synonym for competency contrast.

5.6 CLOSING REMARKS

With this chapter on rheology we conclude Part A on the fundamentals of rock deformation, containing the necessary background to examine the significance of natural deformation structures on all scales, from mountain belts to thin sections. In fact, we can predict broad rheologic properties of the whole earth, given assumptions on mineralogy and temperature. Figure 5.20 shows a composite strength curve for the upper part of the mantle and crust, based on experimental data such as those discussed in this chapter. These "Christmas tree" strength profiles emphasize the role of Earth's compositional stratification and provide reasonable, first-order predictions of rheology for use in numerical models of whole earth dynamics.

Much of the material that we presented in this chapter on rheology takes a basic approach. We examined mainly monomineralic rocks, such as marbles, as opposed to polymineralic rocks, such as granites. Intuitively you will realize that polymineralic rocks require an understanding of the behavior of each of the various constituents, but it is even more complicated than that. Consider the behavior of glass. Glass itself breaks quite easily; however, when glass needles are embedded in resin (epoxy), the **composite material** is unexpectedly strong and resistant to breaking. Glass fiber, consisting of glass and resin, combines the strength of glass with the flexibility of resin. Thus, the behavior of composite materials, and by inference polymineralic rocks, is not simply a matter of knowing the behavior of its constituents, but reflects a complex interplay of properties.

We have chosen to subdivide our discussion of natural deformation in subsequent chapters into brittle and ductile structures. In spite of the sometimes confusing usage of these terms, they do allow a convenient distinction for describing the behavior of natural rocks in the field. Figure 5.21 schematically highlights the mesoscopic aspect of this subdivision, without inferring the mechanisms by which the features form; the subsequent chapters will take care of the latter. This figure also illustrates the broad separation of three types of common geologic structures, faults, folds, and shear zones.

Aside from all the detailed information that deformation structures provide about Earth's history (and that is discussed at length in the subsequent chapters), you should not forget to simply enjoy the sheer beauty and enormity of deformation structures such as faults and folds. This is what attracted many of us to this field of study in the first place.

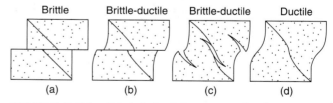

FIGURE 5.21 Brittle (a) to brittle-ductile (b, c) to ductile (d) deformation, reflecting the general subdivision between faults and folds that is used in the subsequent parts of the text.

ADDITIONAL READING

Carmichael, R. S., ed., 1989. *Practical handbook of physical properties of rocks and minerals.* CRC Press: Boca Raton.

Jaeger, J. C., and Cook, N. G. W., 1976. *Fundamentals of rock mechanics.* Chapman and Hall: London.

Kirby, S. H., and Kronenberg, A. K., 1987. Rheology of the lithosphere: selected topics. *Reviews of Geophysics,* 25, 1219–1244.

Poirier, J.-P., 1985. *Creep of crystals. High-temperature deformation processes in metals, ceramics and minerals.* Cambridge University Press: Cambridge.

Ranalli, G., 1995. *Rheology of the Earth* (2nd edition). Allen and Unwin: Boston, 366 pp.

Rutter, E., 1986. On the nomenclature of mode of failure transitions in rocks. *Tectonophysics,* 122, 381–387.

BRITTLE STRUCTURES

Brittle Deformation

6.1 INTRODUCTION

Drop a glass on a tile floor and watch as it breaks into dozens of pieces; you have just witnessed an example of brittle deformation! Because you've probably broken a glass or two (or a plate or a vase) and have seen cracked buildings, sidewalks, or roads, you already have an intuitive feel for what brittle deformation is all about. In the upper crust of the Earth, roughly 10 km in depth, rocks primarily undergo brittle deformation, creating a myriad of geologic structures. To understand why these structures exist, how they form in rocks of the crust, and what they tell us about the conditions of deformation, we must first learn why and how brittle deformation takes place in materials in general. Our purpose in this chapter is to introduce the basic terminology used to describe brittle deformation, to explain the processes by which brittle deformation takes place, and to describe the physical conditions that lead to brittle deformation. This chapter, therefore, provides a basis for the discussion of brittle structures that we present in Chapters 7 and 8.

6.2 VOCABULARY OF BRITTLE DEFORMATION

Research in the last few decades has changed the way geologists think about brittle deformation and, as a consequence, the vocabulary of brittle deformation has evolved. To avoid misunderstanding, therefore, we begin the chapter on brittle deformation by defining our terminology. Table 6.1 summarizes definitions of terms that we use in discussing brittle deformation, and includes additional terms that you may come across when reading articles on this topic. Remember that some of the brittle deformation vocabulary is

TABLE 6.1	TERMINOLOGY OF BRITTLE DEFORMATION
Brittle deformation	The permanent change that occurs in a solid material due to the growth of fractures and/or due to sliding on fractures. Brittle deformation only occurs when stresses exceed a critical value, and thus only after a rock has already undergone some elastic and/or plastic behavior.
Brittle fault zone	A band of finite width in which slip is distributed among many smaller discrete brittle faults, and/or in which the fault surface is bordered by pervasively fractured rock.
Brittle fault	A single surface on which movement occurs specifically by brittle deformation mechanisms.
Cataclasis	A deformation process that involves distributed fracturing, crushing, and frictional sliding of grains or of rock fragments.
Crack	*Verb:* to break or snap apart. *Noun:* a fracture whose displacement does not involve shear displacement (i.e., a joint or microjoint).
Fault	*Broad sense:* a surface or zone across which there has been measurable sliding parallel to the surface. *Narrow sense:* a brittle fault. The narrow definition emphasizes the distinctions between faults, fault zones, and shear zones.
Fracture zone	A band in which there are many parallel or subparallel fractures. If the fractures are wavy, they may anastomose with one another. *Note:* The term has a somewhat different meaning in the context of ocean-floor tectonics.
Fracture	A general term for a surface in a material across which there has been loss of continuity and, therefore, strength. Fractures range in size from grain-scale to continent-scale.
Healed microcrack	A microcrack that has cemented back together. Under a microscope, it is defined by a plane containing many fluid inclusions. (Fluid inclusions are tiny bubbles of gas or fluid embedded in a solid).
Joint	A natural fracture which forms by tensile loading, i.e., the walls of the fracture move apart very slightly as the joint develops. *Note:* A minority of geologists argue that joints can form due to shear loading.
Microfracture	A very small fracture of any type. Microfractures range in size from the dimensions of a single grain to the dimensions of a thin section.
Microjoint	A microscopic joint; microjoints range in size from the dimensions of a single grain to the dimensions of a hand-specimen. Synonymous with *microcrack*.
Shear fracture	A macroscopic fracture that grows in association with a component of shear parallel to the fracture. Shear fracturing involves coalescence of microcracks.
Shear joint	A surface that originated as a joint but later became a surface of sliding. *Note:* A minority of geologists consider a shear joint to be a joint that initially formed in response to shear loading.
Shear rupture	A shear fracture.
Shear zone	A region of finite width in which ductile shear strain is significantly greater than in the surrounding rock. Movement in shear zones is a consequence of ductile deformation mechanisms (cataclasis, crystal plasticity, diffusion).
Vein	A fracture filled with minerals precipitated from a water solution.

FIGURE 6.1 A geologist measuring fractures in an outcrop, near Tuross Point on the southeastern coast of Australia. The more intensely fractured rock is a fine-grained mafic intrusive and the less fractured rock is a coarse-grained felsic intrusive.

controversial, and you will find that not all geologists agree on the definitions that we provide.

Brittle deformation is simply the permanent change that occurs in a solid material due to the growth of fractures and/or due to sliding on fractures once they have formed. By this definition, a **fracture** is any surface of discontinuity, meaning a surface across which the material is no longer bonded (Figure 6.1). If a fracture fills with minerals precipitated out of a hydrous solution, it is a **vein,** and if it fills with (igneous or sedimentary) rock originating from elsewhere, it is a **dike.** A **joint** is a natural fracture in rock across which there is no measurable shear displacement. Because of the lack of shear involved in joint formation, joints can also be called **cracks** or **tensile fractures.** **Shear fractures,** in contrast, are mesoscopic fractures across which there has been displacement. Sometimes geologists use the term "shear fracture" instead of "fault" when they wish to imply that the amount of shear displacement on the fracture is relatively small, and that the shear displacement accompanied the formation of the fracture in once intact rock. In Chapter 7 we show several examples of fractures and veins in natural rocks.

In a broad sense, a **fault** is a surface or zone on which there has been measurable displacement. In a narrower sense, geologists restrict use of the term fault to a fracture surface on which there has been sliding. When using this narrow definition of fault, we apply the term **fault zone** to refer either to a band of finite width across which the displacement is partitioned among many smaller faults, or to the zone of rock bordering the fault that has fractured during faulting. Chapter 8 shows many examples of natural faults. We apply the term **shear zone** to a band of finite width in which the ductile shear strain is significantly greater than in the surrounding rock. Movement in shear zones can be the consequence of **cataclasis** (distributed fracturing, crushing, and frictional sliding of grains of rock or rock fragments), crystal-plastic deformation mechanisms (dislocation glide, dislocation climb), and diffusion. We'll mention cataclastic shear zones in this chapter, but other types of shear zones are discussed in Chapter 12, after we have introduced ductile deformation (Chapter 9).

Regardless of type, a fracture does not extend infinitely in all directions (Figure 6.2). Some fractures intersect the surface of a body of rock, whereas others terminate within the body. The line representing the intersection of the fracture with the surface of a rock body is the **fracture trace,** and the line that separates the region of the rock which has fractured from the

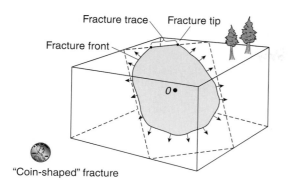

Fracture trace   Fracture tip

Fracture front

O•

"Coin-shaped" fracture

(a)

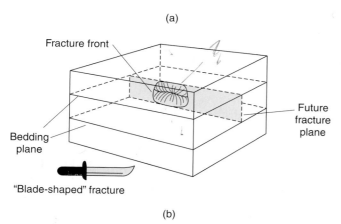

Fracture front

Bedding
plane

Future
fracture
plane

"Blade-shaped" fracture

(b)

FIGURE 6.2   (a) A block diagram illustrating that a fracture surface terminates within the limits of a rock body. The top surface of the block is the ground surface; erosion exposes the fracture trace. Note that the trace of the fracture on the ground surface is a line of finite length (with a fracture tip at each end). The arrows indicate that this particular fracture grew radially outward from an origin, labeled O, the dot in the center of the fracture plane. (b) A blade fracture which has propagated in a sedimentary layer and terminates at the bedding plane.

region that has not fractured is the **fracture front.** The point at which the fracture trace terminates on the surface of the rock is the **fracture tip.** In three dimensions, some fractures have irregular surfaces whereas others have geometries that roughly resemble coins or blades.

6.3   WHAT IS BRITTLE DEFORMATION?

To understand brittle deformation we need to look at the atomic structure of materials. Solids are composed of atoms or ions that are connected to one another by chemical bonds that can be thought of as tiny springs. Each chemical bond has an equilibrium length, and the angle between any two chemical bonds connected to the same atom has an equilibrium value (Figure 6.3a). During elastic strain, the bonds holding the atoms together within the solid stretch, shorten, and/or bend, but they do not break (Figure 6.3b)! When the stress is removed, the bonds return to their equilibrium conditions and the elastic strain disappears. In other words, elastic strain is fully *recoverable*. Recall from Chapter 5 that this elastic property of solids explains the propagation of earthquake waves though Earth.

Rocks cannot accumulate large elastic strains; you certainly cannot stretch a rock to twice its original length and expect it to spring back to its original shape!

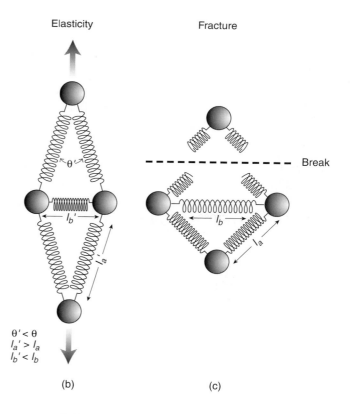

Elasticity   Fracture

Break

FIGURE 6.3   A sketch illustrating what is meant by stretching and breaking of atomic bonds. The chemical bonds are represented by springs and the atoms by spheres. (a) Four atoms arranged in a lattice at equilibrium. (b) As a consequence of stretching of the lattice, some bonds stretch and some shorten, and the angle between pairs of bonds changes. (c) If the bonds are stretched too far, they break, and elastic strain is released.

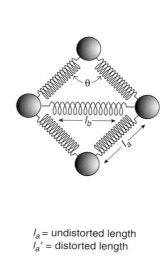

l_a = undistorted length
l_a' = distorted length

(a)

$\theta' < \theta$
$l_a' > l_a$
$l_b' < l_b$

(b)   (c)

| TABLE 6.2 | CATEGORIES OF BRITTLE DEFORMATION PROCESSES |

Cataclastic flow	This type of brittle deformation refers to macroscopic ductile flow as a result of grain-scale fracturing and frictional sliding distributed over a band of finite width.
Frictional sliding	This process refers to the occurrence of sliding on a preexisting fracture surface, without the significant involvement of plastic deformation mechanisms.
Shear rupture	This type of brittle deformation results in the initiation of a macroscopic shear fracture at an acute angle to the maximum principal stress when a rock is subjected to a triaxial compressive stress. Shear rupturing involves growth and linkage of microcracks.
Tensile cracking	This type of brittle deformation involves propagation of cracks into previously unfractured material when a rock is subjected to a tensile stress. If the stress field is homogeneous, tensile cracks propagate in their own plane and are perpendicular to the least principal stress (σ_3).

At most, a rock can develop a few percent strain by elastic distortion. If the stress applied to a rock is greater than the stress that the rock can accommodate elastically, then one of two changes can occur: the rock deforms in a ductile manner (strains without breaking), or the rock deforms in a brittle manner (that is, it breaks).

What actually happens during brittle deformation? Basically, if the stress becomes large enough to stretch or bend chemical bonds so much that the atoms are too far apart to attract one another, then the bonds break, resulting in either formation of a fracture (Figure 6.3c) or slip on a preexisting fracture. In contrast to elastic strain, brittle deformation is *nonrecoverable,* meaning that the distortion remains when the stress is removed. Again we will use earthquakes as an example; in this case, elastic strain is exceeded at the focus, resulting in failure and displacement. The pattern of breakage during brittle deformation depends on stress conditions and on material properties of the rock, so brittle deformation does not involve just one process. For purposes of our discussion, we divide brittle deformation processes into four categories that are listed in Table 6.2 and illustrated in Figure 6.4.

6.4 TENSILE CRACKING

6.4.1 Stress Concentration and Griffith Cracks

In Figure 6.5 we illustrate a crack in rock on the atomic scale. One way to create such a crack would be for all the chemical bonds across the crack surface to break at once. In this case, the tensile stress necessary for this to occur is equal to the strength of each chemical bond multiplied by all the bonds that had once crossed the area of the crack. If you know the strength of a single chemical bond, then you can calculate the stress necessary to break all the bonds simply by multiplying the bond strength by the number of bonds. Using realistic values for the elasticity (Young's modulus, E) and small strain (<10%), Equation 5.3 in Chapter 5 results in a theoretical strength of rock that is thousands of megapascals. Measurement of rock strength in the Earth's crust shows that tensile cracking occurs at crack-normal tensile stresses of less than about 10 MPa, when the confining pressure is low,[1] a value that is hundreds of times less than the theoretical strength of rock. Keeping the concept of theoretical strength in mind, we therefore face a paradox: How can natural rocks fracture at stresses that are so much lower than their theoretical strength?

The first step toward resolving the **strength paradox** came when engineers studying the theory of elasticity realized that the **remote stress** (stress due to a load applied at a distance from a region of interest) gets concentrated at the sides of flaws (e.g., holes) inside a material. For example, in the case of a circular hole in a vertical elastic sheet subjected to tensile stress at its ends (Figure 6.6a), the **local stress** (i.e., stress at the point of interest) tangent to the sides of the hole is three times the remote stress magnitude (σ_r). The magnitude of the local tangential stress at the top and bottom of the hole equals the magnitude of the remote stress, but is opposite in sign (i.e., it is compressive). If the hole has the shape of an ellipse instead

---

[1]The failure strength of rocks under tension is much less than the failure strength of rocks under compression; failure strength under compression depends on the confining pressure.

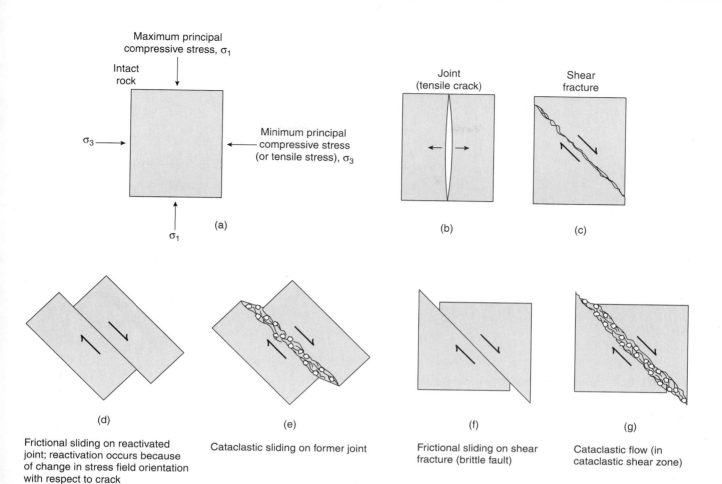

FIGURE 6.4 Types of brittle deformation. (a) Orientation of the remote principal stress directions with respect to an intact rock body. (b) A tensile crack, forming parallel to σ_1 and perpendicular to σ_3 (which may be tensile). (c) A shear fracture, forming at an angle of about 30° to the σ_1 direction. (d) A tensile crack that has been reoriented with respect to the remote stresses and becomes a fault by undergoing frictional sliding. (e) A tensile crack which has been reactivated as a cataclastic shear zone. (f) A shear fracture that has evolved into a fault. (g) A shear fracture that has evolved into a cataclastic shear zone.

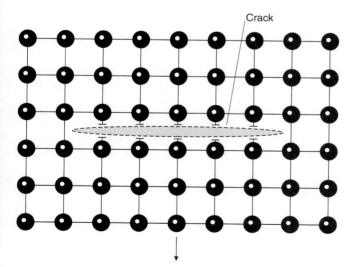

FIGURE 6.5 A cross-sectional sketch of a crystal lattice (balls are atoms and sticks are bonds) in which there is a crack. The crack is a plane of finite extent across which all atomic bonds are broken.

of a circle (Figure 6.6b), the amount of stress concentration, C, is equal to $2b/a + 1$, where a and b are the short and long axes of the ellipse, respectively. Thus, values for stress concentration at the ends of an elliptical hole depend on the axial ratio of the hole: the larger the axial ratio, the greater the stress concentration. For example, at the ends of an elliptical hole with an axial ratio of 8:1, stress is concentrated by a factor of 17, and in a 1 μm × 0.02 μm crack the stress is magnified by a factor of ~100!

With this understanding in mind, A. W. Griffith, in the 1920s, took the next step toward resolving the strength paradox when he applied the concept of stress concentration at the ends of elliptical holes to fracture development. Griffith suggested that all materials contain preexisting microcracks or flaws at which stress concentrations naturally develop, and that because of the stress concentrations that develop at the tips of these cracks, they propagate and become larger cracks

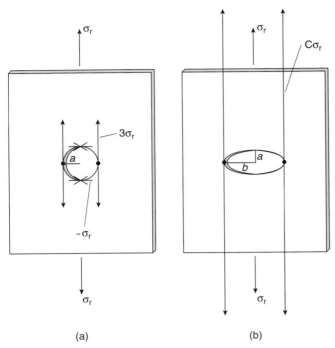

(a)   (b)

FIGURE 6.6 Stress concentration adjacent to a hole in an elastic sheet. If the sheet is subjected to a remote tensile stress at its ends (σ_r), then stress magnitudes at the sides of the holes are equal to $C\sigma_r$, where C, the stress concentration factor, is $(2b/a) + 1$. (a) For a circular hole, $C = 3$. (b) For an elliptical hole, $C > 3$.

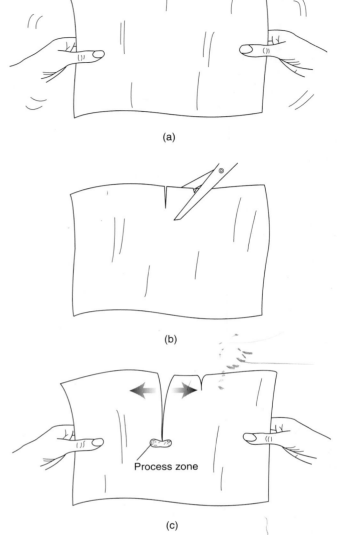

FIGURE 6.7 Illustration of a home experiment to observe the importance of preexisting cracks in creating stress concentrations. (a) An intact piece of paper is difficult to pull apart. (b) Two cuts, a large one and a small one, are made in the paper. (c) The larger preexisting cut propagates. In the shaded area, a region called the process zone, the plastic strength of the material is exceeded and deforms.

even when the host rock is subjected to relatively low remote stresses. He discovered that in a material with cracks of different axial ratios, the crack with the largest axial ratio will most likely propagate first. In other words, stress at the tips of preexisting cracks can become sufficiently large to rupture the chemical bonds holding the minerals together at the tip and cause the crack to grow, even if the remote stress is relatively small. Preexisting microcracks and flaws in a rock, which include grain-scale fractures, pores, and grain boundaries, are now called **Griffith cracks** in his honor. Thus we resolve the strength paradox by learning that rocks in the crust are relatively weak because they contain Griffith cracks.

Griffith's concept provided useful insight into the nature of cracking, but his theory did not adequately show how factors such as crack shape, crack length, and crack orientation affect the cracking process. In subsequent years, engineers developed a new approach to studying the problem. In this approach, called **linear elastic fracture mechanics,** we assume that cracks in a material have nearly infinite axial ratio (defined as long axis/short axis), meaning that all cracks are very sharp. Linear elastic fracture mechanics theory predicts that, if factors like shape and orientation are

equal, a longer crack will propagate before a shorter crack. We'll see why later in this chapter, when we discuss failure criteria.

We can examine how preexisting cracks affect the magnitude of stress necessary for tensile cracking in a simple experiment. Take a sheet of paper (Figure 6.7) and pull at both ends. You have to pull quite hard in order for the paper to tear. Now make two cuts, one that is ~0.5 cm long and one that is ~2 cm long, in the edge of the sheet near its center, and pull again. The pull that you apply gets concentrated at the tip of the preexisting cuts, and at this tip the strength of the paper is exceeded. You will find that it takes much less force to tear the paper, and that it tears apart by growth of the longer preexisting cut. The reason that sharp cracks do not propagate under extremely small stresses is that the tips of real cracks are blunted by a crack-tip **process zone,** in which the material deforms plastically (Figure 6.7c).

It is implicit in our description of crack propagation that the total area of a crack does not form instantaneously, but rather a crack initiates at a small flaw and then grows outward. If you have ever walked out on thin ice covering a pond, you are well aware of this fact. As you move away from shore, you suddenly hear a sound like the echo of a gunshot; this is the sound of a fracture forming in the ice due to the stress applied by your weight. If you have the presence of mind under such precarious circumstances to watch how the crack forms, you will notice that the crack initiates under your boot, and propagates outwards into intact ice at a finite velocity. This means that at any instant, only chemical bonds at the crack tip are breaking. In other words, not all the bonds cut by a fracture are broken at once, and thus the basis we used for calculating theoretical strength in the first place does not represent reality.

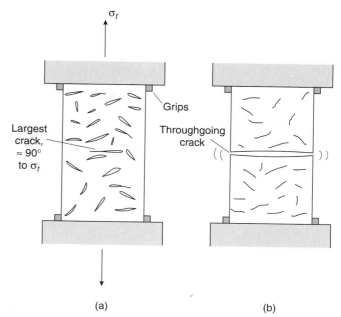

FIGURE 6.8 Development of a throughgoing crack in a block under tension. (a) When tensile stress (σ_t) is applied, Griffith cracks open up. (b) The largest, properly oriented cracks propagate to form a throughgoing crack.

6.4.2 Exploring Tensile Crack Development

Let's consider what happens during a laboratory experiment in which we stretch a rock cylinder along its axis under a relatively low confining pressure (Figure 6.8a), a process called **axial stretching.** As soon as the remote tensile stress is applied, preexisting microcracks in the sample open slightly, and the remote stress is magnified to create larger local stresses at the crack tips. Eventually, the stress at the tip of a crack exceeds the strength of the rock and the crack begins to grow. If the remote tensile stress stays the same after the crack begins to propagate, then the crack continues to grow, and may eventually reach the sample's margins. When this happens, the sample fails, meaning it separates into two pieces that are no longer connected (Figure 6.8b).

We can also induce tensile fracturing by subjecting a rock cylinder to axial compression, under conditions of low confining pressure. Under such stress conditions, mesoscopic tensile fractures develop parallel to the cylinder axis (Figure 6.9a), a process known as **longitudinal splitting.** Longitudinal splitting is similar to tensile cracking except that, in uniaxial compression, the cracks that are not parallel to the σ_1 direction are closed, whereas cracks that are parallel to the compression direction can open up. To picture

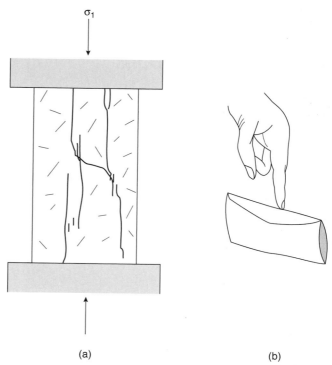

FIGURE 6.9 (a) A cross section showing a rock cylinder with mesoscopic cracks formed by the process of longitudinal splitting. (b) An "envelope" model of longitudinal splitting. If you push down on the top of an envelope (whose ends have been cut off), the sides of the envelope will move apart.

this, imagine an envelope standing on its edge. If you push down on the top edge of the envelope, the sides of the envelope pull apart, even if they were not subjected to a remote tensile stress (Figure 6.9b). In rocks, as the compressive stress increases, the tensile stress at the tips of cracks exceeds the strength of the rock, and the crack propagates parallel to the compressive stress direction.

In the compressive stress environment illustrated in Figure 6.9, the confining pressure required is very small; but tensile cracks can also be generated in a rock cylinder when the remote stress is compressive under higher confining pressure when adding fluid pressure in pores and cracks of the sample (i.e., the pore pressure; Figure 6.10). The uniform, outward push of a fluid in a microcrack can have the effect of creating a local tensile stress at crack tips, and thus can cause a crack to propagate. We call this process **hydraulic fracturing.** As soon as the crack begins to grow, the volume of the crack increases, so if no additional fluid enters the crack, the fluid pressure decreases. Crack propagation ceases when pore pressure drops below the value necessary to create a sufficiently large tensile stress at the crack tip, and does not begin again until the pore pressure builds up sufficiently. Therefore, tensile cracking driven by an increase in pore pressure typically occurs in pulses. We'll return to the role of hydraulic fracturing later in this chapter and again in Chapter 8.

6.4.3 Modes of Crack-Surface Displacement

Before leaving the subject of Griffith cracks, we need to address one more critical issue, namely, the direction in which an individual crack grows when it is loaded. So far we have limited ourselves to cracks that are perpendicular to a remote stress. But what about cracks in other orientations with respect to stress, and how do they propagate? Materials scientists identify three configurations of crack loading. These configurations result in three different modes of crack-surface displacement (Figure 6.11). Note that the "displacement" we are referring to when describing crack propagation is only the infinitesimal movement initiating propagation of the crack tip and is not measurable mesoscopic displacement as in faults.

During **Mode I displacement,** a crack opens very slightly in the direction perpendicular to the crack surface, so Mode I cracks are tensile cracks. They form parallel to the principal plane of stress that is perpendicular to the σ_3 direction, and can grow in their plane without changing orientation. During **Mode II displacement** (the sliding mode), rock on one side of the crack surface moves very slightly in the direction parallel to the fracture surface and perpendicular to the fracture front. During **Mode III displacement** (the tearing mode), rock on one side of the crack slides very slightly parallel to the crack but in a direction parallel to the fracture front.

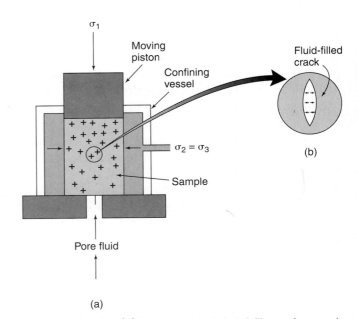

FIGURE 6.10 (a) Cross-sectional sketch illustrating a rock cylinder in a triaxial loading experiment. Fluid has access to the rock cylinder and fills the cracks. (b) A fluid-filled crack that is being pushed apart from within by pore-fluid pressure.

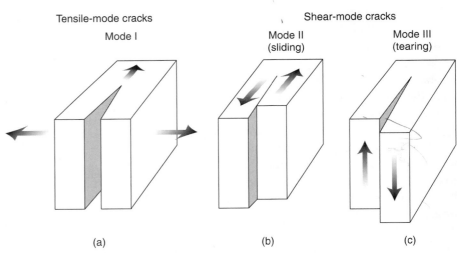

FIGURE 6.11 Block diagrams illustrating the three modes of crack surface displacement: (a) Mode I, (b) Mode II, (c) Mode III. Mode I is a tensile crack, and Mode II and Mode III are shear-mode cracks.

Although shear-mode cracks appear similar to mesoscopic faults, Mode II (and Mode III) cracks are not simply microscopic equivalents of faults. We realize the difference when we examine the propagation of shear-mode cracks. As they start growing, shear-mode cracks immediately curve into the orientation of tensile or Mode I cracks, meaning that shear-mode cracks do *not* grow in their plane. Propagating shear-mode cracks spawn new tensile cracks called **wing cracks,** a process illustrated in Figure 6.12.

6.5 PROCESSES OF BRITTLE FAULTING

A *brittle fault* is a surface on which measurable slip has developed without much contribution by plastic deformation mechanisms. Brittle faulting happens in response to the application of a differential stress, because slip occurs in response to a shear stress parallel to the fault plane. In other words, for faulting to take place, σ_1 cannot equal σ_3, and the fault surface cannot parallel a principal plane of stress. Faulting causes a change in shape of the overall rock body that contains the fault. Hence, faulting contributes to development of regional strain; however, because a brittle fault is, by definition, a discontinuity in a rock body, the occurrence of faulting does not require development of measurable ductile strain in the surrounding rock.

There are two basic ways to create a brittle fault (Figure 6.4). The first is by shear rupturing of a previously intact body of rock. The second is by shear reactivation of a previously formed weak surface (for example, a joint, a bedding surface, or a preexisting fault) in a body of rock. A preexisting weak surface may slip before the differential stress magnitude reaches the failure strength for shear rupture of intact rock. Once formed, movement on brittle faults takes place either by frictional sliding, by the growth of fault-parallel veins, or by cataclastic flow. We'll focus mostly on fault formation and friction, the central processes of brittle faulting, and only briefly mention the other processes here.

6.5.1 Slip by Growth of Fault-Parallel Veins

Not all faults that undergo displacement in the brittle field move by frictional sliding. On some faults, the opposing surfaces are separated from one another by mineral crystals (e.g., quartz or calcite) precipitated out of water solutions present along the fault during movement (Figure 6.13). Such **fault-surface veins** may be composed of mineral fibers (needle-like crystals), of blocky crystals, or of both.

The process by which fault-surface veins form is not well understood. In some cases, they may form when high fluid pressures cause a crack to develop along a weak fault surface. Immediately after cracking, one side of the fault moves slightly with respect to the other; the crack then seals by precipitation of vein material in the crack (see also Chapter 7). In other cases, vein formation may reflect gradual dissolution of steps (or asperities) on the fault surface, followed by the transfer of ions through fluid films to sites of lower stress, where mineral precipitation takes place. This second process may occur without the formation of an actual discontinuity, across which the rock loses cohesion. Whether syn-slip veining or frictional sliding takes place on a fault surface probably depends on strain rate and on the presence of water. Fault-surface veining is probably more common when water is present along the fault, and when movement occurs slowly.

6.5.2 Cataclasis and Cataclastic Flow

Cataclasis refers to movement on a fault by a combination of microcracking, frictional sliding of fragments past one another, and rotation and transport of grains. To picture the process of cataclasis, imagine what happens to corn passing between two old-fashioned mill stones. The millstones slide past one another and, in the process, transform the corn into cornmeal. Cataclasis, if affecting a relatively broad band of rock, results in mesoscopic ductile strain, in which case it is also called **cataclastic flow** (Chapter 9),

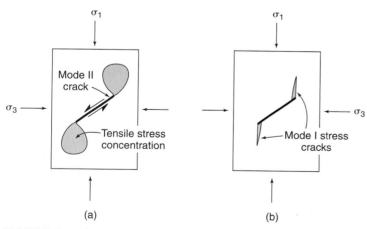

FIGURE 6.12 Propagating shear-mode crack and the formation of wing cracks. (a) A tensile stress concentration occurs at the ends of a Mode II crack that is being loaded. (b) Mode I wing cracks form in the zones of tensile-crack concentration.

FIGURE 6.13 Photograph of stepped calcite slip fibers on a fault surface; pencil indicates displacement direction.

6.6 FORMATION OF SHEAR FRACTURES

Shear fractures differ markedly from tensile cracks. A shear fracture is a surface across which a rock loses continuity when the shear stress parallel to the surface is sufficiently large. Shear fractures are initiated in laboratory rock cylinders at a typical angle of about 30° to σ_1 under conditions of confining pressure ($\sigma_1 > \sigma_2 = \sigma_3$). Because there is a component of normal stress acting on the fracture in addition to shear stress, friction resists sliding on the fracture during its formation. If the shear stress acting on the fracture continues to exceed the frictional resistance to sliding, the fracture grows and displacement accumulates. Shear fractures (or faults) are therefore not simply large shear-mode cracks, because, as we have seen, shear-mode cracks cannot grow in their own plane.

because the rock over the width of the band effectively flows. To picture cataclastic flow, think of how the cornmeal that we just produced, when poured from one container to another, behaves much like a fluid, even though the individual grains certainly are solid. Movement on a fault that involves development of a zone in which cataclastic flow occurs is often referred to as a **cataclastic shear zone.**

This conceptual difference is very important.

So how do shear fractures form? We can gain insight into the process of shear-fracture formation by generating shear ruptures during a laboratory triaxial loading experiment, using a rock cylinder under confining pressure. So to begin our search for an answer to this question, we first describe such an experiment.

In a confined-compression triaxial-loading experiment, we take a cylinder of rock, jacket it in copper or rubber, surround it with a confining fluid in a pressure chamber, and squeeze it between two hydraulic pistons. In the experiment shown in Figure 6.14, the rock itself stays dry. During the experiment, we apply a confining pressure ($\sigma_2 = \sigma_3$) to the sides of the cylinder by increasing the pressure in the surrounding fluid, and an axial load (σ_1) to the ends of the cylinder by moving the pistons together at a constant rate. By keeping the value of σ_3 constant while σ_1 gradually increases, we increase the differential stress ($\sigma_d = \sigma_1 - \sigma_3$). In this experiment we measure the magnitude of σ_d, the change in length of the cylinder (which is the axial strain, e_a), and the change in volume (Δ) of the cylinder.

A graph of σ_d versus e_a (Figure 6.14a) shows that the experiment has four stages. In Stage I, we find that as σ_d increases, e_a also increases and that the relationship between these two quantities is a concave-up curve. In Stage II of the experiment, the relationship between σ_d and e_a is a straight line with a positive slope. During Stage I and most of Stage II, the volume of the sample decreases slightly. In Stage III of the experiment, the slope of the line showing the relation between σ_d and e_a decreases. The stress at which the curve changes slope is called the **yield strength.** During the latter part of Stage II and all of Stage III, we observe a slight increase in volume, a phenomenon known as **dilatancy,** and if we had a very sensitive microphone attached to the sample during this time, we would hear lots of popping sounds that reflect the formation and growth of microcracks. Suddenly, when σ_d equals the failure stress (σ_f), a shear rupture surface develops at an angle of about 30° to the cylinder axis,

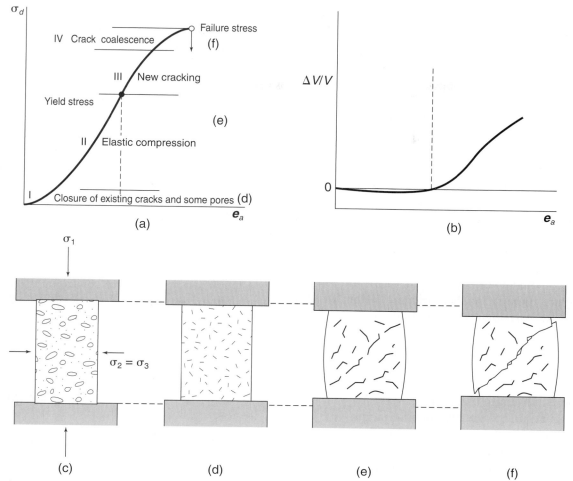

FIGURE 6.14 Fracture formation. (a) Stress–strain plot (differential stress versus axial shortening) showing the stages (I–IV) in a confined compression experiment. The labels indicate the process that accounts for the slope of the curve. (b) The changes in volume accompanying the axial shortening illustrate the phenomenon of dilatancy; left of the dashed line, the sample volume decreases, whereas to the right of the dashed line the sample volume increases. (c–f) Schematic cross sections showing the behavior of rock cylinders during the successive stages of a confined compression experiment and accompanying stress–strain plot, emphasizing the behavior of Griffith cracks (cracks shown are much larger than real dimensions). (c) Pre-deformation state, showing open Griffith cracks. (d) Compression begins and volume decreases due to crack closure. (e) Crack propagation and dilatancy (volume increase). (f) Merging of cracks along the future throughgoing shear fracture, followed by loss of cohesion of the sample (mesoscopic failure).

and there is a stress drop. A stress drop in this context means that the axial stress supported by the specimen suddenly decreases and large strain accumulates at a lower stress. To picture this stress drop, imagine that you're pushing a car that is stuck in a ditch. You have to push hard until the tires come out of the ditch, at which time you have to stop pushing so hard, or you will fall down as the car rolls away. The value of σ_d at the instant that the shear rupture forms and the stress drops is called the **failure strength for shear rupture.** Once failure has occurred, the sample is no longer

intact and frictional resistance to sliding on the fracture surface determines its further behavior.

What physically happened during this experiment? During Stage I, preexisting open microcracks underwent closure. During Stage II, the sample underwent elastic shortening parallel to the axis, and because of the Poisson effect expanded slightly in the direction perpendicular to the axis (Figure 6.14d). The Poisson effect refers to the phenomenon in which a rock that is undergoing elastic shortening in one direction extends in the direction at right angles to the shortening direction. The ratio between the amount of shortening and the amount of extension is called Poisson's ratio, ν.[2] At the start of Stage III, tensile microcracks begin to grow throughout the sample, and wing cracks grow at the tips of shear-mode cracks. The initiation and growth of these cracks causes the observed slight increase in volume, and accounts for the popping noises (Figure 6.14e). During Stage III, the tensile cracking intensifies along a narrow band that cuts across the sample at an angle of about 30° to the axial stress. Failure occurs in Stage IV when the cracks self-organize to form a throughgoing surface along which the sample loses continuity, so that the rock on one side can frictionally slide relative to the rock on the other side (Figure 6.14f). As a consequence, the cylinders move together more easily and stress abruptly drops.

The fracture development scenario described above shows that the failure strength for shear fracture is not a definition of the stress state at which a single crack propagates, but rather it is the stress state at which a multitude of small cracks coalesce to form a through-going rupture. Also note that two ruptures form in some experiments, both at ~30° to the axial stress. The angle between these **conjugate fractures** is ~60°, and the acute bisector is parallel to the maximum principal stress. With continued displacement, however, it is impossible for both fractures to remain active, because displacement on one fracture will offset the other. Thus, typically only one fracture will evolve into a throughgoing fault (see next section).

6.7 PREDICTING INITIATION OF BRITTLE DEFORMATION

We have seen that brittle structures develop when rock is subjected to stress, but so far we have been rather vague about defining the stress states in which brittle deformation occurs. Clearly, an understanding of the stress state at which brittle deformation begins is valuable, not only to geologists, who want to know when, where, and why brittle geologic structures (joints, faults, veins, and dikes) develop in the Earth, but also to engineers, who must be able to estimate the magnitude of stress that a building or bridge can sustain before it collapses. When we discuss brittle deformation, we are really talking about three phenomena: tensile crack growth, shear fracture development, and frictional sliding. In this section, we examine the stress conditions necessary for each of these phenomena to occur.

6.7.1 Tensile Cracking Criteria

A **tensile cracking criterion** is a mathematical statement that predicts the stress state in which a crack begins to propagate. All tensile cracking criteria are based on the assumption that macroscopic cracks grow from preexisting flaws (Griffith cracks) in the rock, because preexisting flaws cause stress concentrations. Griffith was one of the first researchers to propose a tensile cracking criterion. He did so by looking at how energy was utilized during cracking. Griffith envisioned that a material in which a crack forms can be modeled as a thermodynamic system consisting of an elastic plate containing a preexisting elliptical crack. If a load is applied to the ends of the sheet so that it stretches and the crack propagates, the total energy of this system can be defined by the following equation (known as the **Griffith energy balance**):[3]

$$dU_T = dU_s - dW_r + dU_E \qquad \text{Eq. 6.1}$$

where dU_T is the change in total energy of the system, dU_s is the change in surface energy due to growth of the crack (this term arises because crack formation breaks bonds, and energy stored in a broken bond is greater than the energy stored in a satisfied bond), dW_r is the work done by the load in deforming the plate, and dU_E is the change in the strain energy stored in the plate (strain energy is the energy stored by chemical bonds that have been stretched out of their equilibrium length or angle). Griffith pointed out that because the system starts with the load already in place, formation of the crack does not change the total energy of the system. Thus, the change in total energy for an increment of crack growth (dc) equals 0 (in equation form, $dU_T/dc = 0$ for an equilibrium condition). With this point in mind, Equation 6.1 can be rewritten as

$$dW_r = dU_s + dU_E \qquad \text{Eq. 6.2}$$

In words, Equation 6.2 means that the work done on a system is divided between creation of new elastic strain energy in the sheet and creation of new crack surface by propagation of the crack. Using this con-

---

[2]Since Poisson's ratio has the units length/length, it is dimensionless; a typical value of ν for rocks is 0.25.

[3]This is a plane stress criterion; for plane strain conditions, E is replaced by $E/(1 - \nu)^2$, where ν is Poisson's ratio.

cept, along with several theorems from elasticity theory, Griffith devised a tensile cracking criterion, which we present without derivation:

$$\sigma_t = (2E\gamma/\pi c)^{1/2} \qquad \text{Eq. 6.3}$$

where σ_t = critical remote tensile stress (tensile stress at which the weakest Griffith crack begins to grow), E = Young's modulus, γ = energy used to create new crack surface, and c = half-length of the preexisting crack. Reading this equation, we see that the critical remote tensile stress for a rock sample is proportional to material properties of the sample and the length of the crack. Note that as crack length increases, the value of critical remote tensile stress (σ_t) decreases.

Subsequently, researchers have utilized concepts from the engineering study of linear elastic fracture mechanics to develop tensile cracking criteria. According to this work, the following equation defines conditions at which Mode I cracks propagate:

$$K_I = \sigma_t \, Y(\pi c)^{1/2} \qquad \text{Eq. 6.4}$$

where K_I (pronounced "kay one," where the "one" represents a Mode I crack) is the **stress intensity factor,** σ_t is the remote tensile stress, Y is a dimensionless number that takes into account the geometry of the crack (e.g., whether it is penny-shaped, blade-shaped, or tunnel-shaped), and c is half of the crack's length. In this analysis, all cracks in the body are assumed to have very large ellipticity; that is, cracks are assumed to be very sharp at their tips. Note that cracks with smaller ellipticity require higher stresses to propagate.

Equation 6.4 says that the value of K_I increases when σ increases. A crack in the sample begins to grow when K_I attains a value of K_{Ic}, which is the **critical stress intensity factor** or the **fracture toughness** (a measure of tensile strength). The fracture toughness is constant for a given material. When K_I reaches K_{Ic}, the value of σ reaches σ_t, where σ_t is the **critical remote tensile stress** at the instant the crack starts to grow. We can rewrite Equation 6.4 to create an equation that more directly defines the value of σ_t at the instant the crack grows:

$$\sigma_t = K_{Ic}/[Y(\pi c)^{1/2}] \qquad \text{Eq. 6.5}$$

Note that, in this equation, the remote stress necessary for cracking depends on the fracture toughness, the crack shape, and the length of the crack. If other factors are equal, a longer crack generally propagates before a shorter crack. Similarly, if other factors are equal, crack shape determines which crack propagates

first. Because c increases as the crack grows, crack propagation typically leads to sample failure. This relationship has another interesting consequence for natural rocks. Because crack length depends on the grain size of a sample, fine-grained rocks should be stronger than coarse-grained rocks.

Equations like Equations 6.4 and 6.5 can also be written for Mode II and Mode III cracks. By comparing equations for the three different modes of cracking, you will find that, other factors being equal, a Mode I crack (i.e., a crack perpendicular to σ_3) propagates before a Mode II or Mode III crack. However, since other factors like crack shape and length come into play, Mode II or Mode III cracks sometimes propagate before Mode I cracks in a real material. Remember that the instant they propagate, Mode II and III cracks either bend and become Mode I cracks, or they develop wing cracks at their tips; shear cracks cannot propagate significantly in their own plane.

In summary, we see that the stress necessary to initiate the propagation of a crack depends on the ellipticity, the length, the shape, and the orientation of a preexisting crack. Study of crack-propagation criteria is a very active research area, and further details concerning this complex subject (such as subcritical crack growth under long-term loading or corrosive conditions) are beyond the scope of this book (see the reading list).

6.7.2 Shear-Fracture Criteria and Failure Envelopes

A **shear-fracture criterion** is an expression that describes the stress state at which a shear rupture forms and separates a sample into two pieces. Because shear-fracture initiation in a laboratory sample inevitably leads to failure of the sample, meaning that after rupture the sample can no longer support a load that exceeds the frictional resistance to sliding on the fracture surface, shear-rupture criteria are also commonly known as **shear-failure criteria.**

Charles Coulomb[4] was one of the first to propose a shear-fracture criterion. He suggested that if all the principal stresses are compressive, as is the case in a confined compression experiment, a material fails by the formation of a shear fracture, and that the shear stress parallel to the fracture surface, at the instant of failure, is related to the normal stress by the equation

$$\sigma_s = C + \mu\sigma_n \qquad \text{Eq. 6.6}$$

---

[4]An eighteenth-century French naturalist.

where σ_s is the shear stress parallel to the fracture surface at failure; C is the **cohesion** of the rock, a constant that specifies the shear stress necessary to cause failure if the normal stress across the potential fracture plane equals zero (note that this C is not the same as the c in Equations 6.3–6.5); σ_n is the normal stress across the shear fracture at the instant of failure; and μ is a constant traditionally known as the **coefficient of internal friction.** The name for μ originally came from studies of friction between grains in unconsolidated sand and of the control that such friction has on slope angles of sand piles, so the name is essentially meaningless in the context of shear failure of a solid rock; μ should be viewed simply as a constant of proportionality. Equation 6.6, also known as **Coulomb's failure criterion,** basically states that the shear stress necessary to initiate a shear fracture is proportional to the normal stress across the fracture surface.

The Coulomb criterion plots as a straight line on a Mohr diagram[5] (Figure 6.15). To see this, let's plot the results of four triaxial loading experiments in which we increase the axial load on a confined granite cylinder until it ruptures. In the first experiment, we set the confining pressure ($\sigma_2 = \sigma_3$) at a relatively low value, increase the axial load (σ_1) until the sample fails, and then plot the Mohr circle representing this **critical stress state,** meaning the stress state at the instant of failure, on the Mohr diagram. When we repeat the experiment, using a new cylinder, and starting at a higher confining pressure, we find that as σ_3 increases, the differential stress ($\sigma_1 - \sigma_3$) at the instant of failure also increases. Thus, the Mohr circle representing the second experiment has a larger diameter and lies to the right of the first circle. When we repeat the experiment two more times and plot the four circles on the diagram, we find that they are all tangent to a straight line with a slope of μ (i.e., $\tan\phi$) and a y-intercept of C, and this straight line is the Coulomb criterion. Note that we can also draw a straight line representing the criterion in the region of the Mohr diagram below the σ_n-axis.

A line drawn from the center of a Mohr circle to the point of its tangency with the Coulomb criterion defines 2θ, where θ is the angle between the σ_3 direction and the plane of shear fracture (typically about 30°). Because

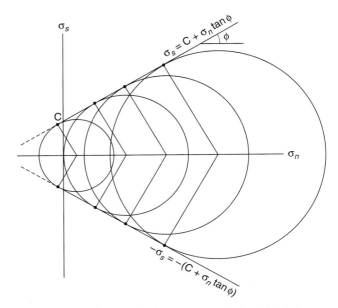

FIGURE 6.15 Mohr diagram showing a Coulomb failure envelope based on a set of experiments with increasing differential stress. The circles represent differential stress states at the instant of shear failure. The envelope is represented by two straight lines, on which the dots represent failure planes.

the Coulomb criterion is a straight line, this angle is constant for the range of confining pressures for which the criterion is valid. The reason for the 30° angle becomes evident in a graph plotting normal stress magnitude and shear stress magnitude as a function of the angle between the plane and the σ_1 direction (Figure 6.16). Notice that the minimum normal stress does not occur in the same plane as the maximum shear stress. Shear stress is at its highest on a potential failure plane oriented at 45° to σ_1, but the normal stress across this potential plane is still too large to permit shear fracturing in planes of this orientation. The shear stress is a bit lower across a plane oriented at 30° to σ_1, but is still fairly high. However, the normal stress across the 30° plane is substantially lower, favoring shear-fracture formation.

Coulomb's criterion is an empirical relation, meaning that it is based on experimental observation alone, not on theoretical principles or knowledge of atomic-scale or crystal-scale mechanisms. This failure criterion does not relate the stress state at failure to physical parameters, as does the Griffith criterion, nor does it define the state of stress in which the microcracks, which eventually coalesce to form the shear rupture, begin to propagate. The Coulomb criterion does not predict whether the fractures that form will dip to the right or to the left with respect to the axis of the rock cylinder in a triaxial loading experiment. In fact, as

---

[5]Recall that on the Mohr diagram, normal stresses (σ_n) are plotted on the x-axis and shear stresses (σ_s) are plotted on the y-axis. A Mohr circle represents the stress state by indicating the values of σ_n and σ_s acting on a plane oriented at $\theta°$ to the σ_3 direction. The circle intersects the x-axis at σ_1 and σ_3 (both of which are normal stresses, because they are principal stresses), and the angle between the x-axis and a radius from the center of the circle to a point on the circle defines the angle 2θ.

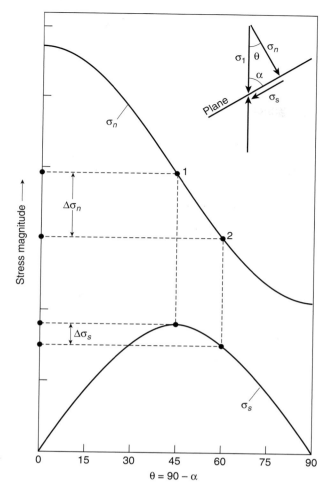

FIGURE 6.16 The change in magnitudes of the normal and shear components of stress acting on a plane as a function of the angle α between the plane and the σ_1 direction; the angle $\theta = 90 - \alpha$ is plotted for comparison with other diagrams. At point 1 ($\alpha = \theta = 45°$), shear stress is a maximum, but the normal stress across the plane is quite large. At point 2 ($\theta = 60°$, $\alpha = 30°$), the shear stress is still quite high, but the normal stress is much lower.

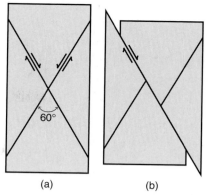

FIGURE 6.17 Cross-sectional sketch showing how only one of a pair of conjugate shear fractures (a) evolves into a fault with measurable displacement (b).

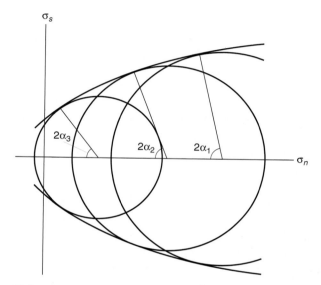

FIGURE 6.18 Mohr failure envelope. Note that the slope of the envelope steepens toward the σ_s-axis. Therefore, the value of α (the angle between fault and σ_1) is not constant (compare $2\alpha_1$, $2\alpha_2$, and $2\alpha_3$).

mentioned earlier, **conjugate shear fractures,** one with a right-lateral shear sense and one with a left-lateral shear sense, may develop (Figure 6.17). The two fractures, typically separated by an angle of ~60°, correspond to the tangency points of the circle representing the stress state at failure with the Coulomb failure envelope.

The German engineer Otto Mohr conducted further studies of shear-fracture criteria and found that Coulomb's straight-line relationship only works for a limited range of confining pressures. He noted that at lower confining pressure, the line representing the stress state at failure curved with a steeper slope, and that at higher confining pressure, the line curved with a shallower slope (Figure 6.18). Mohr concluded that

over a range of confining pressure, the failure criteria for shear rupture resembles a portion of a parabola lying on its side, and this curve represents the **Mohr-Coulomb criterion** for shear fracturing. Notice that this criterion is also empirical. Unlike Coulomb's straight-line relation, the change in slope of the Mohr-Coulomb failure envelope indicates that the angle between the shear fracture plane and σ_1 actually does depend on the stress state. At lower confining pressures, the angle is smaller, and at high confining pressures, the angle is steeper.

The Mohr-Coulomb criterion (both for positive and negative values of σ_s) defines a failure envelope on the Mohr diagram. A **failure envelope** separates the field on the diagram in which stress states are "stable" from

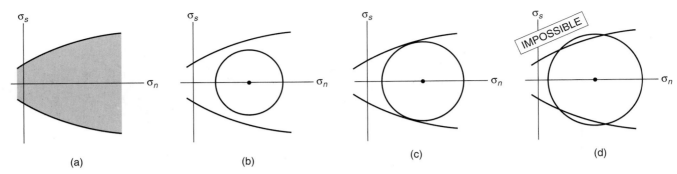

FIGURE 6.19 (a) A brittle failure envelope as depicted on a Mohr diagram. Within the envelope (shaded area), stress states are stable, but outside the envelope, stress states are unstable. (b) A stress state that is stable, because the Mohr circle, which passes through values for σ_1 and σ_3 and defines the stress state, falls entirely inside the envelope. (c) A stress state at the instant of failure. The Mohr circle touches the envelope. (d) A stress state that is impossible.

the field in which stress states are "unstable" (Figure 6.19). By this definition, a **stable stress state** is one that a sample can withstand without undergoing brittle failure. An **unstable stress state** is an impossible condition to achieve, for the sample will have failed by fracturing before such a stress state is reached (Figure 6.19). In other words, a stress state represented by a Mohr circle that lies entirely within the envelope is stable, and will not cause the sample to develop a shear rupture. A circle that is tangent to the envelope specifies the stress state at which brittle failure occurs. Stress states defined by circles that extend beyond the envelope are unstable, and are therefore impossible within the particular rock being studied.

Can we define a failure envelope representing the critical stress at failure for very high confining pressures, very low confining pressures, or for conditions where one of the principal stresses is tensile? The answer to this question is controversial. We'll look at each of these conditions separately.

At high confining pressures, samples may begin to deform plastically. Under such conditions, we are no longer really talking about brittle deformation, so the concept of a "failure" envelope no longer really applies. However, we can approximately represent the "yield" envelope, meaning the stress state at which the sample begins to yield plastically, on a Mohr diagram by a pair of lines that parallel the σ_n-axis (Figure 6.20a). This yield criterion, known as **Von Mises criterion,** indicates that plastic yielding is effectively independent of the differential stress, once the yield stress has been achieved.

If the tensile stress is large enough, the sample fails by developing a throughgoing tensile crack. The tensile stress necessary to induce tensile failure may be represented by a point, T_0, the **tensile strength,** along

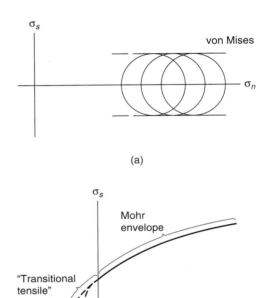

FIGURE 6.20 (a) Mohr diagram illustrating the Von Mises yield criterion. Note that the criterion is represented by two lines that parallel the σ_n-axis. (b) The extrapolation of a Mohr envelope to its intercept with the σ_n-axis, illustrating the "transitional-tensile" regime, and the tensile strength (T_0). Note that the tensile strength has a range of values, because it depends on the dimensions of preexisting flaws in the deforming sample.

the σ_n-axis to the left of the σ_s-axis (Figure 6.20b). As we have seen, however, the position of this point depends on the size of the flaws in the sample. Thus, even for the same rock type, experiments show that the tensile strength is very variable and that it is best represented by a range of points along the σ_n-axis.

There are competing views as to the nature of failure for rocks subjected to tensile stresses that are less than the tensile strength. Some geologists have suggested that failure occurs under such conditions by the formation of fractures that are a hybrid between tensile cracks and shear ruptures, and have called these fractures **transitional-tensile fractures** or **hybrid shear fractures.** The failure envelope representing the conditions for initiating transitional-tensile fractures is the steeply sloping portion of the parabolic failure envelope (Figure 6.20b). Most fracture specialists, however, claim that transitional-tensile fractures do not occur in nature, and point out that no experiments have yet clearly produced transitional-tensile fractures in the lab. We'll discuss this issue further in Chapter 8.

Taking all of the above empirical criteria into account, we can construct a **composite failure envelope** that represents the boundary between stable and unstable stress states for a wide range of confining pressures and for conditions for which one of the principal stresses is tensile (Figure 6.21). The envelope roughly resembles a cross section of a cup lying on its side. The various parts of the curve are labeled. Starting at the right side of the diagram, we have Von Mises criteria, represented by horizontal lines. (Remember that the Von Mises portion of the envelope is really a plastic yield criterion, not a brittle failure criterion.) The portion of the curve where the lines begin to slope effectively represents the brittle–plastic transition. To the left of the brittle–plastic transition, the envelope consists of two straight sloping lines, representing Coulomb's criterion for shear rupturing. For failure associated with the Coulomb criterion, remember that the angle between the shear rupture and the σ_1 direction is independent of the confining pressure. Closer to the σ_s-axis, the slope of the envelope steepens, and the envelope resembles a portion of a parabola. This parabolic part of the curve represents Mohr's criterion, and for failure in this region, the decrease in the angle between the fracture and the σ_1 direction depends on how far to the left the Mohr circle touches the failure curve. The part of the parabolic envelope with steep slopes specifies failure criteria for supposed transitional-tensile fractures formed at a very small angle to σ_1, but as we discussed, the existence of such fractures remains controversial. The point where the envelope crosses the σ_n-axis represents the failure criterion for tensile cracking, but as we have discussed, this criterion really shouldn't be specified by a point, for the tensile strength of a material depends on the dimensions of the flaws it contains. Note that for a circle tangent to the composite envelope at T_0, $2\theta = 180$ (or, $\alpha = 0$), so

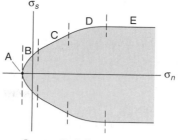

Composite failure envelope

A: Tensile failure criterion
B: Mohr (parabolic) failure criterion
C: Coulomb (straight-line) failure criterion
D: Brittle-plastic transition
E: von Mises plastic yield criterion

(a)

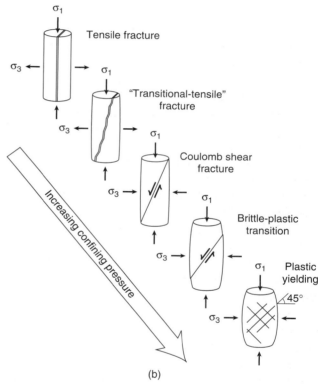

(b)

FIGURE 6.21 (a) A representative composite failure envelope on a Mohr diagram. The different parts of the envelope are labeled, and are discussed in the text. (b) Sketches of the fracture geometries that form during failure. Note that the geometry depends on the part of the failure envelope that represents failure conditions, because the slope of the envelope is not constant.

the fracture that forms is parallel to σ_1! Also, note that there is no unique value of differential stress needed to cause tensile failure, as long as the magnitude of the differential stress (the diameter of the Mohr circle) is less than about $4T_0$, for this is the circle whose curvature is the same as that of the apex of the parabola.

6.8 FRICTIONAL SLIDING

Friction is the resistance to sliding on a surface. **Frictional sliding** refers to the movement on a surface that takes place when shear stress parallel to the surface exceeds the frictional resistance to sliding. The principles of frictional sliding were formulated hundreds of years ago. We can do a simple experiment that produces, at first, counterintuitive behavior (Figure 6.22). Attach a spring to a beam of wood that is placed on a table with the flat side down. Pull the spring until the beam slides. Now place the beam on its narrow side and repeat the experiment. Surprisingly, the spring extends by the same amount before beam sliding occurs, irrespective of the area of contact. Similar experiments led to what are often called Amontons's laws[6] of friction:

- Frictional force is a function of normal force.
- Frictional force is independent of the (apparent) area of contact (in his words, "the resistance caused by rubbing only increases or diminishes in proportion to greater or lesser pressure [load] and not according to the greater or lesser extent of the surfaces").
- Frictional force is mostly independent of the material used (in his words, "the resistance caused by rubbing is more or less the same for iron, lead, copper and wood in any combination if the surfaces are coated with pork fat").

Let's explore the reason for this behavior, which has important consequences for natural faulting processes and earthquake mechanics. Note that we should distinguish between **static friction,** which is associated with first motion (associated with Coulomb failure), and **dynamic friction,** which is associated with continued motion.

Friction exists because no real surface in nature, no matter how finely polished, is perfectly smooth. The bumps and irregularities which protrude from a rough surface are called **asperities** (Figure 6.23a). When two surfaces are in contact, they touch only at the asperities, and the asperities of one surface may indent or sink into the face of the opposing surface (Figure 6.23b and c). The cumulative area of the asperities that contact the opposing face is the **real area of contact** (A_r in Figure 6.23). In essence, asperities act like an anchor holding a ship in place. In order for the ship to drift,

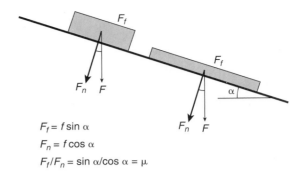

$$F_f = f \sin \alpha$$
$$F_n = f \cos \alpha$$
$$F_f/F_n = \sin \alpha/\cos \alpha = \mu$$

FIGURE 6.22 Frictional sliding of objects with same mass, but with different (apparent) contact areas. The friction coefficients and, therefore, sliding forces (F_f) are equal for both objects, regardless of (apparent) contact area.

either the anchor chain must break, or the anchor must drag along the sea floor. Similarly, in order to initiate sliding of one rock surface past another, it is necessary either for asperities to break off, or for them to plow a furrow or groove into the opposing surface.

The stress necessary to break off an asperity or to cause it to plow depends on the real area of contact, so, as the real area of contact increases, the frictional resistance to sliding (that is, the force necessary to cause sliding) increases. Again, considering our ship analogy, it takes less wind to cause a ship with a small anchor to drift than it does to cause the same-sized ship with a large anchor to drift. Thus, the frictional resistance to sliding is proportional to the normal force component across the surface, because of the relation between real area of contact and friction. An increase in the normal force (load) pushes asperities into the opposing wall more deeply, causing an increase in the real area of contact. Returning to our earlier experiment with a sliding beam, we can now understand that the object's mass, rather than the (apparent) area of contact (the side of the beam), determines the ability to slide.

6.8.1 Frictional Sliding Criteria

Because of friction, a certain critical shear stress must be achieved in a rock before frictional sliding is initiated on a preexisting fracture, and a relation defining this critical stress is the **failure criterion for frictional sliding.** Experimental work shows that failure criteria for frictional sliding, just like the Coulomb failure criterion for intact rock, plot as sloping straight lines on a Mohr diagram. A compilation of friction data from a large number of experiments, using a great variety of rocks (Figure 6.24), shows that the failure criterion for frictional sliding is basically independent of rock type:

$$\sigma_s/\sigma_n = \text{constant} \qquad\qquad \text{Eq. 6.7}$$

---

[6]Named after the seventeenth-century French physicist Guillaume Amontons; Leonardo da Vinci earlier described similar relationships in the fifteenth century.

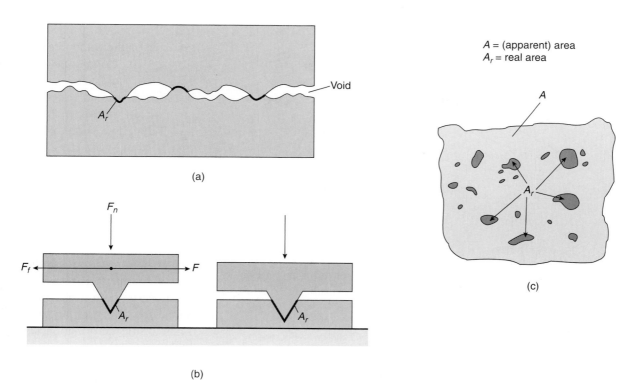

FIGURE 6.23 Concept of asperities and the real area of contact (A_r). (a) Schematic cross-sectional close-up showing the irregularity of a fracture surface and the presence of voids and asperities along the surface. (b) Idealized asperity showing the consequence of changing the load (normal force) on the real area of contact. (c) Map of a fracture surface; the shaded areas are real areas of contact.

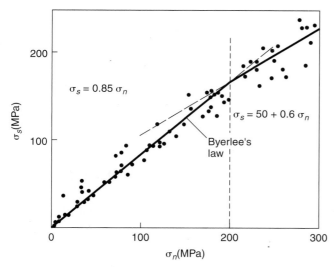

FIGURE 6.24 Graph of shear stress and normal stress values at the initiation of sliding on preexisting fractures in a variety of rock types. The best-fit line defines Byerlee's law, which is defined for two regimes.

The empirical relationship between normal and shear stress that best fits the observations, known as **Byerlee's law**,[7] depends on the value of σ_n. For $\sigma_n < 200$ MPa, the best-fitting criterion is a line described by the relationship $\sigma_s = 0.85\sigma_n$, whereas for 200 MPa $< \sigma_n <$ 2000 MPa, the best-fitting criterion is a line described by the equation $\sigma_s = 50$ MPa $+ 0.6\sigma_n$. The proportionality between normal and shear stress is, as before, called the **friction coefficient.**

6.8.2   Will New Fractures Form or Will Existing Fractures Slide?

Failure envelopes allow us to quickly determine whether it is more likely that an existing shear rupture will slip in a sample, or that a new shear rupture will form (Figure 6.25). For example, look at Figure 6.25b, which shows both the Byerlee's frictional sliding envelope and the Coulomb shear fracture envelope for Blair dolomite. Note that the slope and intercept of the two envelopes

---

[7]After the geophysicist J. Byerlee, who first proposed the equations in the 1970s.

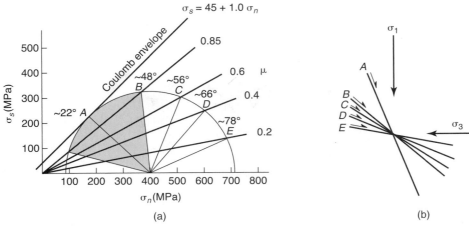

FIGURE 6.25 (a) Mohr diagram based on experiments with Blair dolomite, showing how a single stress state (Mohr circle) would contact the frictional sliding envelope before it would contact the Coulomb envelope (heavy line). Surface A in (b) is the Coulomb shear fracture that would form in an intact rock. However, sliding would occur on surfaces between intersections with the friction envelope (marked by shaded area for friction envelope μ = 0.85) before new fracture initiation. Preexisting surfaces B to E are surfaces that will slide with decreasing friction coefficients. Consider the geologic relevance of decreasing friction coefficients for stress state, failure, and fracture orientation.

are different, so that for a specific range of preexisting fracture orientations, the Mohr circle representing the stress state at failure touches the frictional envelope before it touches the fracture envelope, meaning the pre-existing fracture slides before a new fracture forms.

However, preexisting fractures do not always slide before a new fracture is initiated. Confined compression experiments indicate that if the preexisting fracture is oriented at a high angle to the σ_1 direction (generally > 75°; plane E in Figure 6.25), the normal stress component across the discontinuity is so high that friction resists sliding, and it is actually easier to initiate a new shear fracture at a smaller angle to σ_1 (plane B in Figure 6.25). Sliding then occurs on the new fracture. If a preexisting fracture is at a very small angle to σ_1 (generally < 15°; plane A in Figure 6.25), the shear stress on the surface is relatively low, so again it ends up being easier to initiate a new shear fracture than to cause sliding on a preexisting weak surface. Thus, preexisting planes whose angles to σ_1 are between 15° and 75° probably will be reactivated before new fractures form.

6.9 EFFECT OF ENVIRONMENTAL FACTORS IN FAILURE

The occurrence and character of brittle deformation at a given location in the earth depends on environmental conditions (confining pressure, temperature, and fluid

pressure) present at that location, and on the strain rate (see Chapter 5). Conditions conducive to the occurrence of brittle deformation are more common in the upper 10–15 km of the Earth's crust. However, at slow strain rates or in particularly weak rocks, ductile deformation mechanisms can also occur in this region, as evident by the development of folds at shallow depths in the crust. Below 10–15 km, plastic deformation mechanisms dominate. However, at particularly high fluid pressures or at very rapid strain rates, brittle deformation can still occur at these depths.

In this chapter, we have described brittle deformation without considering how it is affected by environmental factors. Not surprisingly, temperature, fluid pressure, strain rate, and rock anisotropy play significant roles in the stress state at failure and/or in the orientation of the fractures that form when failure occurs. Most of these factors have already been discussed in Chapter 5, so we close this chapter on brittle deformation by focusing on the effect of fluids.

6.9.1 Effect of Fluids on Tensile Crack Growth

All rocks contain pores and cracks—we've already seen how important these are in the process of brittle failure. In the upper crust of the earth below the water table, these spaces, which constitute the porosity of rock, are filled with fluid. This fluid is most commonly water, though in some places it is oil or gas.

If there is a high degree of **permeability** in the rock, meaning that water can flow relatively easily from pore to pore and/or in and out of the rock layer, then the pressure in a volume of pore water at a location in the crust is roughly hydrostatic, meaning that the pressure reflects the weight of the overlying water column (Figure 6.26). **Hydrostatic (fluid) pressure** is defined by the relationship $P_f = \rho \cdot g \cdot h$, where ρ is the density of water (1000 kg/m³), g is the gravitational constant (9.8 m/s²), and h is the depth. **Pore pressure,** which is the fluid pressure exerted by fluid within the pores of a rock, may exceed the hydrostatic pressure if permeability is restricted. For example, the fluid trapped in a

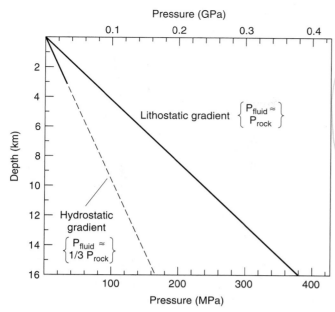

FIGURE 6.26 Graph of lithostatic versus hydrostatic pressure as a function of depth in the Earth's crust.

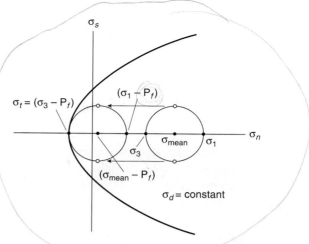

FIGURE 6.27 A Mohr diagram showing how an increase in pore pressure moves the Mohr circle toward the origin. The increase in pore pressure decreases the mean stress (σ_{mean}), but does not change the magnitude of differential stress ($\sigma_1 - \sigma_3$). In other words, the diameter of the Mohr circle remains constant, but its center moves to the left.

sandstone lens surrounded by impermeable shale cannot escape, so the pore pressure in the sandstone can approach or even equal **lithostatic pressure** (P_l), meaning that the pressure approaches the weight of the overlying column of rock (i.e., $P_f = P_l = \rho \cdot g \cdot h$, where $\rho = 2000–3000 \text{ kg/m}^3$). Note that rock, on average, is two to three times denser than water. When the fluid pressure in pore water exceeds hydrostatic pressure, we say that the fluid is **overpressured.**

How does pore pressure affect the tensile failure strength of rock? The pore pressure is an outward push that opposes inward compression from the rock, so the fluid supports part of the applied load. If pore pressure exceeds the least compressive stress (σ_3) in the rock, tensile stresses at the tips of cracks oriented perpendicularly to the σ_3 direction become sufficient for the crack to propagate. In other words, pore pressure in a rock can cause tensile cracks to propagate, even if none of the remote stresses are tensile, because pore pressure can induce a crack-tip tensile stress that exceeds the magnitude of σ_3. This process is called **hydraulic fracturing.** On a Mohr diagram, it can be represented by movement of Mohr's circle to the left (Figure 6.27). Note that rocks do not have to be overpressured in order for natural hydraulic fracturing to occur, but P_f must equal or exceed the magnitude of σ_3.

Another effect of fluids comes from the chemical reaction of the fluids with the minerals comprising a rock. Reaction with fluids may lower the tensile stress needed to cause a crack to propagate, even if the pore-fluid pressure is low. Water, for example, reacts with

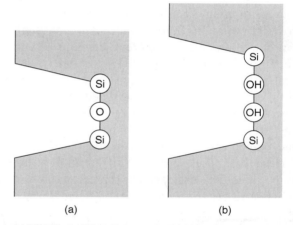

FIGURE 6.28 Sketch of the atoms at the tip of a crack that illustrates the principle of hydrolytic weakening. (a) Si atoms are bonded to an O atom at the tip of a dry crack. (b) At the tip of a wet crack, Si atoms are bonded to two OH molecules that replace the O (same charge). The Si—O—Si bonding is stronger than the Si—OH to OH—Si bonding and the Si—O—Si bond length is less than that of Si—OH—OH—Si.

quartz to bring about substitution of OH molecules for O atoms in quartz lattice at a crack tip (Figure 6.28). Since the bond between adjacent OH groups is not as strong as the bond between oxygen atoms, it breaks more easily, so it takes less remote tensile stress to cause the crack to propagate. This phenomenon is called **subcritical crack growth,** because crack propagation occurs at stresses less than the critical stress necessary to cause a crack to propagate in dry rock.

6.9.2 Effect of Dimensions on Tensile Strength

Rock tensile strength is *not* independent of scale, in that larger rock samples are inherently weaker than smaller rock samples. Why? Because larger samples are more likely to contain appropriately oriented and larger Griffith cracks that will begin to propagate when a stress is applied, thereby nucleating the throughgoing cracks that result in failure of the whole sample. Equation 6.2 emphasizes this point, because the tensile stress at failure is inversely proportional to the crack half-length. You can imagine that if a sample is so small that it consists only of a piece of perfect crystal lattice, it will be very strong indeed. In fact, the reason that turbine blades in modern jet engines are so strong is that they are grown as relatively flawless single crystals.

6.9.3 Effect of Pore Pressure on Shear Failure and Frictional Sliding

We can observe the effects of pore pressure on shear fracturing by running a confined compression experiment in which we pump fluid into the sample through a hole in one of the pistons, thereby creating a fluid pressure, P_f, in pores of the sample (Figure 6.10). The fluid creating the confining pressure acting on the sample is different from and is not connected to the fluid inside the sample. The magnitude of P_f decreases the confining pressure (σ_3) and σ_1 by the same amount. So if the pore pressure increases in the sample, the mean stress decreases but the differential stress remains the same. This effect can be represented by the Coulomb failure criterion equation; P_f decreases the magnitude of σ_n on the right side of the equation

$$\sigma_s = C + \mu(\sigma_n - P_f) \qquad \text{Eq. 6.8}$$

The term $(\sigma_n - P_f)$ is commonly labeled σ_n^*, and is called the **effective stress.**

From a Mohr diagram, we can easily see the effect of increasing the P_f in this experiment. When P_f is increased, the whole Mohr circle moves to the left but its diameter remains unchanged (Figure 6.27), and when the circle touches the failure envelope, shear failure occurs, even if the relative values of σ_1 and σ_3 are unchanged. In other words, a differential stress that is insufficient to break a dry rock, may break a wet rock, if the fluid in the wet rock is under sufficient pressure. Thus, an increase in pore pressure effectively weakens a rock. In the case of forming a shear fracture in intact rock, pore pressure plays a role by pushing open microcracks, which coalesce to form a rupture at smaller remote stresses.

Similarly, an increase in pore pressure decreases the shear stress necessary to initiate frictional sliding on a preexisting fracture, for the pore pressure effectively decreases the normal stress across the fracture surface. Thus, as we discuss further in Chapter 8, fluids play an important role in controlling the conditions under which faulting occurs.

6.9.4 Effect of Intermediate Principal Stress on Shear Rupture

Fractures form parallel to σ_2, and so the value of σ_2 does not affect values of normal stress (σ_n) and shear stress (σ_s) across potential shear rupture planes. Therefore, in this chapter we have assumed that the value of σ_2 does not have a major effect on the shear failure strength of rock, and we considered failure criteria only in terms of σ_1 and σ_3. In reality, however, σ_2 does have a relatively small effect on rock strength. Specifically, rock is stronger in confined compression when the magnitude of σ_2 is closer to the magnitude of σ_1, than when the magnitude of σ_2 is closer to the magnitude of σ_3. An increase in σ_2 has the same effect as an increase in confining pressure.

6.10 CLOSING REMARKS

Why study brittle deformation? For starters, since rocks deform in a brittle manner under the range of pressures and temperatures found at or near the Earth's surface (down to 10–15 km), fractures pervade rocks of the upper crust. In fact every rock outcrop that you will ever see contains fractures at some scale (Figure 6.29). Because they are so widespread, fractures play a major role in determining the permeability and strength of rock, and the resistance of rock to erosion. Therefore, fractures affect the velocity and direction of toxic waste transport, the location of an ore deposit, the durability of a foundation, the stability of a slope, the suitability of a reservoir, the safety of a mine shaft, the form of a landscape, and so on. Moreover, fracturing is the underlying cause of earthquakes, and contributes to the evolution of regional tectonic features. This chapter has provided an introduction to the complex and rapidly evolving subject of brittle deformation and fracture mechanics. We have

FIGURE 6.29 Aerial view of the San Andreas Fault, north of Landers, California.

tried to describe what fractures are, how they form, and under what conditions. In the next two chapters, we apply this information to developing an understanding of the major types of brittle structures: joints, veins, and faults.

ADDITIONAL READING

Atkinson, B. K., ed., 1987. *Fracture mechanics of rock.* London: Academic Press.

Bowden, F. P., and Tabor, D., 1950. *The friction and lubrication of solids.* Clarendon Press: Oxford.

Brace, W. F., and Bombolakis, E. G., 1963. A note on brittle crack growth in compression. *Journal of Geophysical Research,* 68, 3709–3713.

Brace, W. F., and Byerlee, J. D., 1966. Stick-slip as a mechanism for earthquakes. *Science,* 153, 990–992.

Engelder, T., 1993. *Stress regimes in the lithosphere.* Princeton University Press.

Hancock, P. L., 1985. Brittle microtectonics: principles and practice. *Journal of Structural Geology,* 7, 437–457.

Jaeger, J. C., and Cook, N. G. W., 1979. *Fundamentals of rock mechanics* (3rd ed.). Chapman and Hall: London.

Lockner, D. A., 1991. A multiple-crack model of brittle fracture. *Journal of Geophysical Research,* 96, 19,623–19,642.

Paterson, M. S., 1978. *Experimental rock deformation—the brittle field.* Springer-Verlag: New York.

Pollard, D. D., and Aydin, A., 1988. Progress in understanding jointing over the past century. *Geological Society of America Bulletin,* 100, 1181–1204.

Price, N. J., 1966. *Fault and joint development in brittle and semi-brittle rock.* Pergamon Press: London.

Scholz, C. H., 2002. *The mechanics of earthquakes and faulting* (2nd edition). Cambridge University Press: Cambridge.

Scholz, C. H., 1998. Earthquakes and friction laws. *Nature,* 391, 37–42.

Secor, D. T., 1965. Role of fluid pressure in jointing. *American Journal of Science,* 263, 633–646.

Joints and Veins

7.1 INTRODUCTION

Visitors from around the world trek to Arches National Park in southeastern Utah (USA) to marvel at its graceful natural arches. These arches appear to have been carved through high, but relatively thin, free-standing sandstone walls. From the air, you can see that the park contains a multitude of such walls, making its landscape resemble a sliced-up loaf of bread (Figure 7.1a). The surfaces of rock walls in Arches Park initiated as **joints,** which are natural fractures in rock across which there has been no shear displacement (see Table 7.1 for a more formal definition). Erosive processes through the ages have preferentially attacked the walls of the joints, so that today you can walk in the space between the walls. Though joints are not always as dramatic as those in Arches National Park, nearly all outcrops contain joints. At first glance, joints may seem to be simple and featureless geologic struc-

tures, but in fact they are well worth studying, not only because of their importance in controlling landscape morphology, but also because they profoundly affect rock strength, influence hydrologic properties (such as permeability), and because they can provide a detailed, though subtle history of stress and strain in a region.

Although the basic definition of the term joint is nongenetic, most contemporary geologists believe that they form during Mode I loading (see Chapter 6); that is, that they are tensile fractures that form perpendicular to the σ_3 trajectory and parallel to the principal plane of stress that contains the σ_1 and σ_2 directions. Not all geologists share this viewpoint, and some researchers use the term "joint" when referring to shear fractures as well. This second usage is discouraged, because structures that are technically faults might also be referred to as joints, so we do not use the term "joint" in reference to a shear fracture.

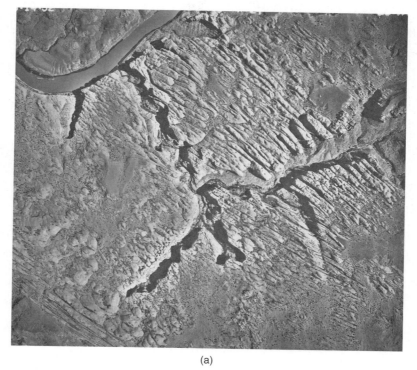
(a)

(b)

FIGURE 7.1 Examples of joints and veins on different scales. (a) Air photo of regional joints in sandstone near Arches National Park, Utah (USA). Note the Colorado River for scale. (b) Veining in limestone exposed in a road cut near Catskill (New York State, USA).

In this chapter, we begin by describing the morphology of individual joints and the geometric characteristics of groups of joints. Then, we discuss how to study joints in the field, and how to interpret them. We conclude by describing **veins,** which are fractures filled with minerals that precipitated from a fluid (Figure 7.1b). But before we begin, we offer a note of caution. The interpretation of joints and veins remains quite controversial, and it is common for field trips that focus on these structures to end in heated debate. As you read this chapter, you'll discover why.

TABLE 7.1	JOINT TERMINOLOGY
Arrest line	An arcuate ridge on a joint surface, located at a distance from the origin, where the joint front stopped or paused during propagation of the joint; also *rib marks*.
Columnar joints	Joints that break rock into generally hexagonal columns; they form during cooling and contraction in hypabyssal intrusions or lava flows.
Conjugate system	Two sets of joints oriented such that the dihedral angle between the sets is approximately 60°.
Continuous joints	Throughgoing joints that can be traced across an outcrop, and perhaps across the countryside.
Cross joints	Discontinuous joints that cut across the rock between two systematic joints, and are oriented at a high angle to the systematic joints.
Cross-strike joints	Joints that cut across the general trend of fold hinges in a region of folded rocks (i.e., the joints cut across regional bedding strike).
Dessication cracks	Joints formed in a layer of mud when it dries and shrinks; dessication cracks (or *mud cracks*) break the layer into roughly hexagonal plates.
Discontinuous joints	Short joints that terminate within an outcrop, generally at the intersection with another joint.
En echelon	An arrangement of parallel planes in a zone of fairly constant width; the planes are inclined to the borders of the zone and terminate at the borders of the zone. In an *en echelon* array, the component planes are of roughly equal length.
Exfoliation	See sheeting joints.
Hackle zone	The main part of a plumose structure, where the fracture surface is relatively rough due to microscopic irregularities in the joint surface formed when the crack surfaces get deflected in the neighborhood of grain-scale inclusions in the rock, or due to off-plane cracking (formation of small cracks adjacent to the main joint surface) as the fracture propagates.
Hooking	The curving of one joint near its intersection with an earlier formed joint.
Inclusion	A general term for any solid inhomogeneity (e.g., fossil, pebble, burrow, xenolith, amygdule, coarse grain, etc.) in a rock; inclusion may cause local stress concentrations.
Joint	A natural, unfilled, planar or curviplanar fracture which forms by tensile loading (i.e., the walls of a joint move apart very slightly as the joint develops). Joint formation does not involve shear displacement.

7.2 SURFACE MORPHOLOGY OF JOINTS

If you look at an exposed joint surface, you'll discover that the surface is not perfectly smooth. Rather, joint surfaces display a subtle roughness that often resembles the imprint of a feather. This pattern is called a plumose structure (Figure 7.2).

7.2.1 Plumose Structure

Plumose structures form at a range of scales, depending on the grain size of the host rock. In very fine-grained coal, for example, components of plumose structure tend to be much smaller than in relatively coarser siltstone. Some of the best examples of plumose structure form in fine-grained rocks like shale, siltstone, and basalt, but you might not see obvious plumose structure on joints in very coarse-grained rocks like granite. Let's look at plumose structure a little more closely (Figure 7.3a). A plumose structure spreads outward from the **joint origin,** which, as the name suggests, represents the point at which the joint started to grow. Joint origins typically look like small dimples in the fracture plane (Figure 7.3b). Several distinct morphologies surround the joint origin. In the **mirror zone,** which lies closest to the origin, the joint surface is very smooth. Further from the origin, the mirror zone merges with the **mist zone,** in which the joint surface slightly roughens. Mirror and mist zones, while they are well developed in joints formed in

TABLE 7.1	JOINT TERMINOLOGY (CONTINUED)
Joint array	Any group of joints (systematic or nonsystematic).
Joint density	The surface area of joints per unit volume of rock (also referred to as *joint intensity*).
Joint origin	The point on the joint (usually a flaw or inclusion) at which the fracture began to propagate; it is commonly marked by a dimple.
Joint set	A group of systematic joints.
Joint stress shadow	The region around a joint surface where joint-normal tensile stress is insufficient to cause new joints to form.
Joint system	Two or more geometrically related sets of joints in a region.
Mirror region	Portion of a joint surface adjacent to the joint origin where the surface is very smooth; mirrors do not occur if the rock contains many small-scale heterogeneities.
Mist region	A portion of a joint surface surrounding the mirror where the fracture surface begins to roughen.
Nonsystematic joints	A joint that is not necessarily planar, and is not parallel to nearby joints.
Orthogonal system	Two sets of joints that are at right angles to one another.
Plume axis	The axis of the plume in a plumose structure.
Plumose structure	A subtle roughness on the surface of some joints (particularly those in fine-grained rocks) that macroscopically resembles the imprint of a feather.
Sheeting joints	Joints formed near the ground surface that are roughly parallel to the ground surface; sheeting joints on domelike mountains make the mountains resemble delaminating onions; also *exfoliation*.
Strike-parallel joints	Joints that parallel the general trend of fold-hinges in a region of folded strata (i.e., the joints parallel regional bedding strike).
Systematic joints	Roughly planar joints which occur as part of a set in which the joints parallel one another, and which are relatively evenly spaced from one another.
Twist hackle	One of a set of small *en echelon* joints formed along the edge of a larger joint; a twist hackle is not parallel to the larger joint, and forms when the fracture surface twists continuously into a different orientation and then breaks up into segments.

glassy rocks, are difficult to recognize in coarser rocks. Continuing outward, the mist zone merges with the **hackle zone,** in which the joint surface is even rougher. It is the hackle zone that forms most of the plumose structure. Roughness in the hackle zone defines vague lineations, or **barbs,** that curve away from a **plume axis,** which together comprise the feather-like plume. The acute angle between the barbs and the axis points back toward the joint origin, so the plume defines the local direction of joint propagation. The median line may be fairly straight and distinct, or it may be wavy and diffuse (Figure 7.4a and b). On some joint surfaces, concentric ridges known as **arrest lines** (Figure 7.4c) form on the joint surface at a distance from the origin. These ridges represent, as the name indicates, breaks in the growth of the joint.

7.2.2 Why Does Plumose Structure Form?

Mode I loading of a perfectly isotropic and homogeneous material should yield a perfectly smooth, planar fracture that is oriented perpendicular to the remote σ_3. Real joints are not perfectly smooth for two reasons. First, real rocks are not perfectly isotropic and homogeneous, meaning that the material properties of a rock change from point to point in the rock. Inhomogeneities exist because not all grains in a rock have the same composition and because not all grains are in perfect contact with one another. The presence of inhomogeneities distorts the local stress field at the tip of a growing joint, so that the principal stresses at the tip are not necessarily parallel to the remote σ_3. As a

(a)

(b)

FIGURE 7.2 Photographs of plumose structure on joint surfaces (New York State, USA). (a) Wavy plumose structure on a joint in siltstone. (b) Plumose structure in thin bedded siltstone; pencil points to the point of origin.

propagation velocity are relatively small near the joint origin, because the crack is very short, and increase with distance from the origin, eventually reaching a maximum (called terminal velocity). If the stress magnitude at the tip exceeds a critical value, the energy available for cracking rock exceeds the energy needed to create a single surface. The excess energy goes into breaking bonds off the plane of the main joint surface, resulting in the formation of microscopic cracks that splay off the main joint. If the energy becomes very large, the crack may actually split into two separate, parallel surfaces.

With these two conditions in mind, we can now explain why plumose structure has distinct morphological features. The dimple at the origin forms because the flaw[1] at which the joint nucleated either was not perpendicular to the remote σ_3, or caused a local change in the orientation of stress trajectories. The portion of the joint that formed in the immediate vicinity of the origin was, therefore, not perpendicular to the remote σ_3. As soon as the crack propagated away from the flaw, it curved into parallelism with the $\sigma_1\sigma_2$ principal plane (Figure 7.3). In the mirror zone, the joint is still short, so the stress intensity, tensile stress magnitude, and tip-propagation velocity are all relatively small. As a consequence of the low stress, only bonds in the plane exactly perpendicular to the local σ_3 can break, so the joint surface that forms is very smooth. In the mist zone, however, the joint moves faster, stress is higher, and stress at the joint tip is sufficiently large to break off-plane bonds, thereby forming microscopic off-plane cracks that make the surface rougher than in the mirror zone. In the hackle zone, the joint tip is moving at its terminal velocity and stresses at the crack tip are so large that larger off-plane cracks propagate and the crack locally bifurcates at its tip to form microscopic splays that penetrate the joint walls. The roughness of the hackle zone also reflects the formation of tiny splays

consequence, the joint-propagation path slightly twists and tilts as the joint grows.

Second, the stress field at the tip of a crack changes as the crack tip propagates. Recall from Equation 6.4 that the stress intensity at the crack tip is proportional to the length of the crack, and that the magnitude of the local tensile stress at the tip of the crack is, in turn, proportional to the stress intensity. Thus, as the crack grows, the stress intensity at the crack tip grows, up to a limiting value. Experimental work demonstrates that the velocity of crack-tip propagation is also proportional to the stress intensity. Stress magnitude and tip-

---

[1]Flaws at which joints initiate include open pores, preexisting microcracks, irregularities on a bedding plane, inclusions (like a pebble, fossil, amygdule, or concretion), or primary sedimentary structures (a sole mark or ripple).

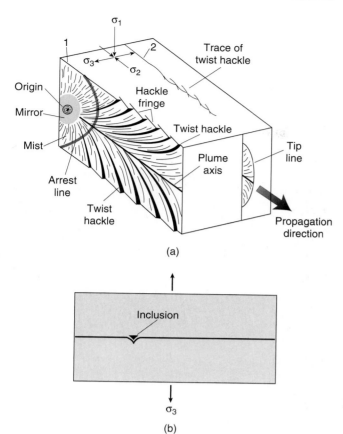

(a)

(b)

FIGURE 7.3 (a) Block diagram showing the various components of an ideal plumose structure on a joint. The face of joint 1 is exposed; joint 2 is within the rock. (b) Simple cross-sectional sketch showing the dimple of a joint origin, controlled by an inclusion.

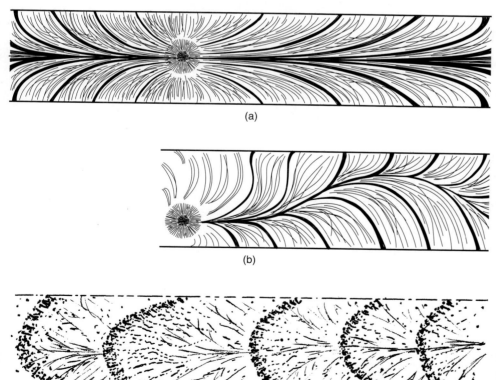

(a)

(b)

(c)

FIGURE 7.4 Types of plumose structure. (a) Straight plume. (b) Curvy plume. (c) Plume with many arrest lines, suggesting that it opened repeatedly.

and warps of the joint surface where the joint tip twists or tilts as it passes an inclusion and breaks into microscopic steps. Arrest lines on a joint surface represent places where the fracture tip pauses between successive increments of propagation. The visible ridge of the arrest line, in part, represents the contrast between the rough surface of the hackle and the relatively smooth surface of the mirror/mist zone formed as the fracture begins to propagate, and in part may be analogous to the dimple formed at a crack origin. Thus, plumose structures form because of the twisting, tilting, and splitting occurring at the tip due to variations in local stress magnitude and orientation.

7.2.3   Twist Hackle

Features such as bedding planes and preexisting fractures locally modify the orientation of principal stresses because they approximate free surfaces.[2] If a growing joint enters a region where it no longer parallels a principal plane of stress (for example, as occurs when the crack tip of a joint in a sedimentary bed approaches the bedding plane), the crack tip pivots to a new orientation. As a consequence, the joint splits into a series of small *en echelon* joints, because a joint surface cannot twist and still remain a single continuous surface. The resulting array of fractures is called **twist hackle,** and the edge of the fracture plane where twist hackle occurs is called the **hackle fringe** (Figure 7.3a). Note that if the hackle fringe intersects an outcrop face, the trace of a large planar joint within the outcrop may look like a series of small joints in an *en echelon* arrangement.

7.3   JOINT ARRAYS

7.3.1   Systematic versus Nonsystematic Joints

Systematic joints are planar joints that comprise a family in which all the joints are parallel or subparallel to one another, and maintain roughly the same average spacing over the region of observation (Figure 7.5). Systematic joints may cut through many layers of

strata, or be confined to a single layer. **Nonsystematic joints** have an irregular spatial distribution, they do not parallel neighboring joints, and they tend to be nonplanar (Figure 7.5b). Nonsystematic joints may terminate at other joints. You will often find both systematic and nonsystematic joints in the same outcrop.

(a)

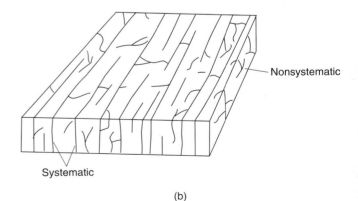

(b)

FIGURE 7.5   (a) Three sets of systematic joints controlling erosion in Cambrian sandstone (Kangaroo Island, Australia). (b) Block diagram showing occurrence of both systematic and nonsystematic joints in a body of rock.

---

[2]A "free surface" is a surface across which there is no cohesion, so it cannot transmit shear stresses. By definition, a free surface is a principal plane of stress, but if the free surface is not parallel to a principal plane of the remote stress, then the remote stress trajectories change orientation so that they are either parallel or perpendicular to the free surface.

7.3.2 Joint Sets and Joint Systems

Describing groups of joints efficiently requires a fair bit of jargon. Matters are made even worse because not all authors use joint terminology in the same way, so it's good practice to define your terminology in context. We'll describe joint patterns here and give the explanations of why various different groups of joints form later in the chapter.

A **joint set** is a group of systematic joints. Two or more joint sets that intersect at fairly constant angles comprise a **joint system,** and the angle between two joint sets in a joint system is the **dihedral angle.** If the two sets in a system are mutually perpendicular (i.e., the dihedral angle is ~90°), we call the pair an **orthogonal system** (Figure 7.6a), and if the two sets intersect with a dihedral angle significantly less than 90° (e.g., a dihedral angle of 30° to 60°), we call the pair a **conjugate system** (Figure 7.6a). Many geologists use the terms "orthogonal" or "conjugate" to imply that the pair of joint sets formed at the same time. However, as you will see later in this chapter, nonparallel joint sets typically form at different times. So, we use the terms merely to denote a geometry, not a mode or timing of origin.

As shown in Figure 7.6a, many different configurations of joint systems occur, which are distinguished from one another by the nature of the intersections between sets and by the relative lengths of the joints in the different sets. In joint systems where one set consists of relatively long joints that cut across the outcrop whereas the other set consists of relatively short joints that terminate at the long joints, the throughgoing joints are **master joints,** and the short joints that occur between the continuous joints are **cross joints** (Table 7.1).

In the flat-lying sedimentary rocks that occur in continental interior basins and platforms (e.g., the Midwest region of the United States), joint sets are

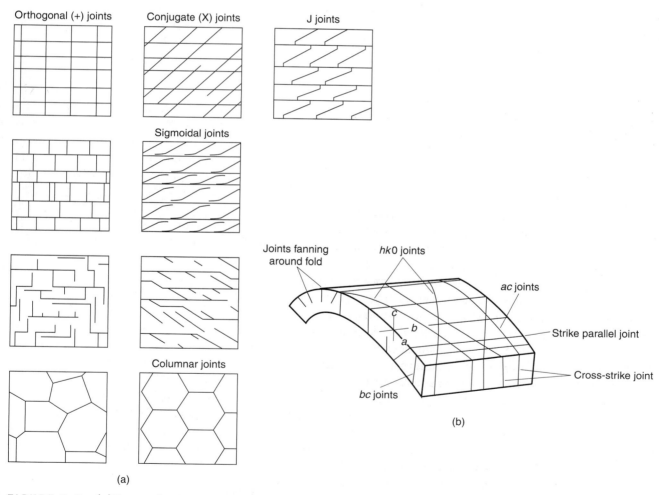

FIGURE 7.6 (a) Traces of various types of joint arrays on a bedding surface. (b) Idealized arrangement of joint arrays with respect to fold symmetry axes. The "*hk*0" label for joints that cut diagonally across the fold-hinge is based on the Miller indices from mineralogy; they refer to the intersections of the joints with the symmetry axes of the fold.

perpendicular to the ground surface (and, therefore, to bedding) and orthogonal systems are common. In gently folded sedimentary rocks, such as along the foreland margin of a mountain range (e.g., the western side of the Appalachians), strata contain both vertical joint sets that cut across the folded layers, and joints that are at a high angle to bedding and fan around the folds (Figure 7.6b). Both orthogonal and conjugate systems occur in such gently folded strata. The joint sets of an orthogonal system in folded sedimentary rocks commonly have a spatial relationship to folds of the region, so we can distinguish between **strike-parallel joints,** which parallel the general strike of bedding (roughly parallel to regional fold hinges), and **cross-strike joints,** which trend at high angles (~60° to 90°) to the regional bedding strike (Figure 7.6b).[3] Conjugate systems in gently folded rocks consist of two cross-strike sets with their acute bisector at a high angle to the fold hinge. Because both sets of joint systems need not form at the same time, a conjugate geometry of a system of joints does not require that they are conjugate shear ruptures.

In the internal portions of mountain belts, where rocks have been intensely deformed and metamorphosed, outcrops may contain so many joints that joint systems may be difficult to recognize or simply do not exist. In such regions, joints formed prior to deformation and metamorphism have been partly erased. New joints then form at different times during deformation, during subsequent uplift, or even in response to recent stress fields. Rocks in such regions are so heterogeneous that the stress field varies locally, and thus joints occur in a wide range of orientations. Nevertheless, in some cases, younger joints, meaning those formed during uplift or due to recent stress fields, may stand out as distinct sets.

Intrusive and metamorphic rocks without a strong schistosity (such as granite, migmatitic gneiss) commonly contain a set of joints that roughly parallels ground surface topography, and whose spacing decreases progressively toward the surface. Such joints are called **sheeting joints** or **exfoliation joints** (Figure 7.7). If the ground surface is not horizontal, as is the case on the sloping side of a mountain, sheeting joints curve and follow the face of the mountain, thus giving the mountain the appearance of a partially peeled onion. Rock sheets detach off the mountain along these joints, thereby creating smooth dome-shaped structures known as **exfoliation domes.** Half Dome, a challenge

FIGURE 7.7 Sheeting joints (or exfoliation) in granite of the Sierra Nevada.

that draws mountain climbers to Yosemite National Park in the Sierra Nevada Mountains of California, is an exfoliation dome, one half of which was cut away by glacial erosion.

Shallow intrusive igneous rock bodies (dikes and sills) and lava flows in many localities display **columnar jointing,** meaning that they have been broken into joint-bounded columns which, when viewed end-on, have roughly hexagonal cross sections (Figure 7.6a). In the case of sheet intrusions, the long axes of the columns tend to be perpendicular to the boundaries of the sheet (horizontal in dikes and vertical in sills); however, in some bodies the columns curve. The visual impression of columnar jointing catches people's imaginations, so these structures tend to be dubbed with unusual names like Giant's Causeway (in Ireland), Devil's Postpile (in California; see Figure 2.25), and the Spielbergian platform of Devil's Tower (in Wyoming).

7.3.3 Cross-Cutting Relations Between Joints

The way in which nonparallel joints intersect one another provides information concerning their relative ages. For example, if joint A terminates at its intersection with joint B, then joint A is younger, because a propagating fracture cannot cross a free surface, and an open preexisting joint behaves like a free surface.[4]

A younger joint's orientation also may change where it approaches an older joint that behaves like a free surface. Why? Remember that at, or near, a free

---

[3]Note that "cross-strike joints" are not necessarily the same as "cross joints" (Table 7.1).

[4]Joints that are filled, or whose faces are tightly held together by stress, can transmit at least some shear stress, and thus do not behave like perfect free surfaces.

surface, a Mode I fracture must be either parallel, or perpendicular, to the surface so as to maintain perpendicularity to σ_3. Thus, near a free surface, the local stress field differs from the remote stress field if the free surface does not parallel a principal plane of the remote stress field. If an older joint (joint B) acts as a free surface, then the younger joint (A) curves in the vicinity of joint B to become parallel to the local principal plane of stress adjacent to B, unless it already happens to parallel a principal plane of stress. The way in which the younger joint curves depends on the stress field. If the local σ_3 adjacent to the older joint is parallel to the walls of the older joint, then the younger joint curves so that it is orthogonal to the first joint at their point of intersection, a relationship called **hooking;** such a structure is called a J junction (Figure 7.6a). However, if the local σ_3 is perpendicular to the walls of the older joint, then the younger joint curves into parallelism with joints of the first set, and has a sigmoidal appearance (Figure 7.6a).

In some joint systems, two nonparallel joints appear to cross one another without any apparent interaction; in other words, they are mutually cross cutting. Such intersections are sometimes referred to as "+" intersections, if the joints are orthogonal, or "×" intersections if they are not orthogonal (Figure 7.6a). These relationships may represent situations where (1) the earlier joint did not act as a free surface, (2) the intersection of two younger joints at the same point on an older joint is simply coincidental, or (3) the cross-cutting relationship is an illusion—within the body of the outcrop, the older joint terminated, and the younger joint simply grew around it.

7.3.4 Joint Spacing in Sedimentary Rocks

When looking at jointing in a sequence of stratified sedimentary rock, you might notice that within a bed, joints are often evenly spaced. Where this occurs, we can define **joint spacing** as the average distance between adjacent members of a joint set, measured perpendicular to the surface of the joint. Informally, geologists refer to joints as being "closely spaced" or "widely spaced" in a relative sense, but to be precise, you should describe joint spacing in units of length (e.g., 5 cm).

In order to understand why joints are evenly spaced, we look at how an array of joints develops in a bed. Consider a bed of sandstone that contains five joints (Figure 7.8). Experimental work suggests that joints form in sequence; that is, first joint 1, then joint 2, then joint 3, and so on. When a new joint forms, it is at some distance greater than a minimum distance (d_m)

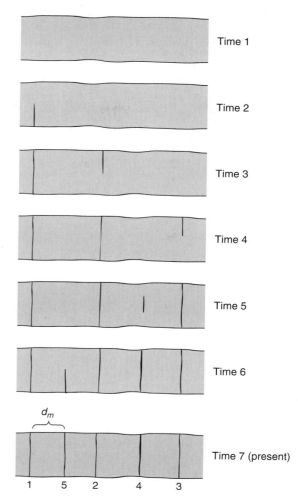

FIGURE 7.8 A model of the sequence of development of joints. Time 1 refers to the time before the first joint forms, and time 7 is the present day. This scenario suggests that joints form in a random sequence, but with regular spacing.

from a preexisting joint. Formation of a joint relieves tensile stress for a critical distance, d_m (Figure 7.9). The zone on either side of a joint in which there has been a decrease in tensile stress is called the **joint stress shadow.** Stresses sufficient to create the next joint are only achieved outside of this shadow, and are created by traction between the bed and beds above and below it, as well as by stress transmitted within the bed beyond the fracture front of the preexisting joint. The spacing between joints is determined by the width of the joint stress shadow; so, because the shadow is about the same width for all joints in the bed, the spacing ends up being fairly constant. Joint spacing depends on four parameters: bed thickness, stiffness, tensile strength, and strain. We'll examine each of these parameters in turn.

Relation between joint spacing and bed thickness. All other parameters being equal, joints are more

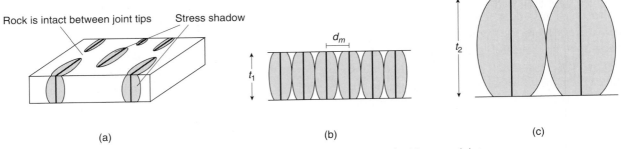

(a)   (b)   (c)

FIGURE 7.9   The concept of stress shadows around joints. The heavy vertical lines are joints; d_m refers to the average spacing between joints. (a) Block diagram illustrating stress shadow (shaded area) around each joint. Note how stress is transmitted across regions that are unfractured in the third dimension. Stresses are also exerted by tractions at bedding contacts. (b) Thin bedded sequence, containing joints with narrow stress shadows, so that the joints are closely spaced. (c) Thick bedded sequence, containing joints with wide stress shadows, so that the joints are widely spaced.

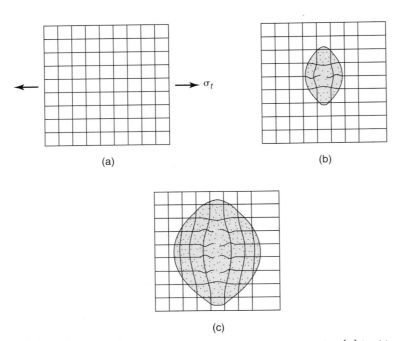

FIGURE 7.10   Illustration of why joint stress shadows exist. (a) A grid of springs. (b) Cutting one spring causes only a few springs to relax around the cut, so only a relatively small area is affected, as indicated. (c) Cutting many springs in a row causes a wider band of springs to relax; thus, a larger area is affected.

springs relax (Figure 7.10b); however, if you cut many of the springs in a row, a much wider zone of neighboring springs relaxes (Figure 7.10c). In thicker beds, joint stress shadows are wider, so joints tend to be more widely spaced.

Relation between joint spacing and lithology. Recall that the *stiffness* (i.e., the elastic value E, Young's Modulus) of a rock layer depends on lithology (i.e., Hooke's law states that $\sigma = E \cdot \mathbf{e}$; see Chapter 5). Imagine a block of rock composed of sandstone and dolomite (Figure 7.11). Dolomite is stiffer ($E \approx -600$ MPa) than sandstone ($E \approx -200$ MPa). We stretch the block under brittle conditions by a uniform amount so that all layers undergo exactly the same elongation ($\mathbf{e}$). The stress that develops in each bed is defined by Hooke's law; however, since the elongation is the same for each bed, the magnitude of σ depends on E. Thus, beds composed of rock with a larger E develop a greater stress and fracture first. In the model of Figure 7.11, the stiffer dolomite bed probably fractured a few times before the sandstone bed fractured for the first time, so more joints develop in the dolomite bed than in the sandstone bed. In sum, for a given strain, stress is larger in stiffer beds, so other factors being equal, stiffer beds have smaller joint spacings.

Relation between joint spacing and tensile strength. Predicting fracture spacing cannot be done by considering E alone, because, in some cir-

closely spaced in thinner beds, and are more widely spaced in thicker beds. The relationship is a reflection of joint-stress shadow width, because the greater the length of the joint (i.e., length of the joint trace in a plane perpendicular to bedding and joint), the wider the stress shadow (Figure 7.9b and c). To picture why this is so, imagine a net composed of springs (Figure 7.10a). If you reach into the net and cut one spring, only a few of the neighboring

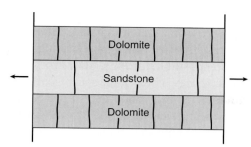

FIGURE 7.11 Cross-sectional sketch illustrating a multilayer that is composed of rocks with different values of Young's modulus. The stiffer layers (dolomite) develop more closely spaced joints.

cumstances, a rock with a smaller E may actually have a lower tensile strength than a rock with a larger E. Thus, it will crack at a lower strain than a rock with a larger E, if the rock with the larger E also has a larger tensile strength. Other factors being equal, rocks with smaller tensile strength develop more closely spaced joints.

Relation between joint spacing and the magnitude of strain. A bed that has been stretched more contains more joints than a bed that has been stretched less, as you might expect.

Overall, the spacing of well-developed joints is about equal to the bed thickness. If you ever have the chance to hike down the Grand Canyon, don't forget to look at the jointing in different units as you descend. Bedding planes tend to be weak and do not transmit shear stress efficiently, so joints typically terminate at bedding planes. Because joint spacing depends on bed thickness and lithology, joint spacing varies from bed to bed. Weak, thinly bedded shales contain such closely spaced joints that they break into tiny fragments, and, as a consequence, they tend to form slopes. In contrast, thick sandstone beds develop only widely spaced fractures, so thick sandstone beds typically protrude and hold up high cliffs.

7.4 JOINT STUDIES IN THE FIELD

Before discussing how to go about studying joints in the field, it's worth discussing *why* you might want to study joints in the field. Perhaps the most common reasons to study joints are for engineering or hydrologic applications. As we noted before, fractures affect the strength of foundations, quarrying operations, excava-

tions, groundwater and (toxic) waste flow, and slope stability. For example, if you find that a region contains a systematic joint set that is oriented north–south, you can expect groundwater to flow faster in the north–south direction than in the east–west direction, or that quarrying might be easier if the quarry walls strike north–south than if they strike east–west. But the study of jointing has applications to more academic geologic issues as well. Geologists who are interested in tectonics study joints to see if they provide information about (paleo)stress fields, and geomorphologists study joints to find out if they control the drainage patterns or the orientation of escarpments. With these goals in mind, what specifically do we look for in a joint study? In most cases the questions that we ask include the following:

1. Is jointing in the outcrop systematic or nonsystematic? In other words, can we define distinct sets of planar joints and/or regularly oriented cross joints in an outcrop, or does the outcrop contain irregular and randomly oriented joints with relatively short traces (i.e., nonsystematic joints)? If nonsystematic jointing is present, is it localized or pervasive? Formulating hypotheses on joint formation in a region needs to take into account whether the joints are systematic or not. Systematic joints likely reflect regional tectonic stress trajectories at the time of fracturing, whereas nonsystematic joints reflect local heterogeneities of the stress field. While nonsystematic joints may be important for determining rock strength and permeability, they provide no information on regional paleostress orientation.

2. What are the orientations of joint sets, if present? If there is more than one set, is there a consistent angular relationship between them, such that we can describe a joint system? Information on the orientation and distribution of joint sets and systems is critical for engineering and hydrologic analyses. For example, joint sets that run parallel to a proposed road cut would create a greater rockfall hazard than joints that are perpendicular to the cut.

3. What is the nature of cross-cutting relationships between joints of different sets, and what is the geometry of joint intersections? Do joints cross without appearing to interact, do they curve to create J intersections, or do they curve into parallelism with one another? Knowledge of cross-cutting relations allows us to determine whether one set of joints is older or younger than another set of joints, a determination that is critical to tectonic interpretations using joints.

4. What is the surface morphology of the joints? Is plumose structure visible on joint surfaces, and if so, what types of plumes (wavy or straight) are visible? Are numerous arrest lines clearly evident on the joint surface? The presence of plumose structure is taken as proof that a joint propagated as a Mode I fracture, and the geometry of the plume provides clues as to the way in which the joint propagated (e.g., in a single pulse, or in several distinct pulses). If a joint surface contains numerous origins, it probably initiated at different times along its length. Joints whose surfaces contain many arrest lines probably propagated in increments. Later we will see that starts and stops of a fracture may indicate that growth was controlled by fluctuating fluid pressure in the rock. Are there other structural features superimposed on joints (e.g., stylolitic pits or slip lineations)? If joints display surface features other than plumose structure, the features indicate post–joint formation strain. Stylolitic pitting on a joint indicates compression and resulting pressure solution (see Chapter 9) across the joint, and slip lineations suggest that the joint was reactivated as a fault later in its history.

5. What are the dimensions of joints? In other words, are the trace lengths of joints measured in centimeters or hundreds of meters? The effect that jointing has on rock strength and rock permeability over a region is significantly affected by the dimensions of the joints. For example, the presence of large throughgoing joints that parallel an escarpment contribute to the hazard of escarpment collapse more than will short nonsystematic joints. Tectonic geologists commonly focus attention on interpretation of large joints, on the assumption that these reflect regional tectonic stress conditions, which are of broad interest.

6. What is the spacing and joint density in outcrop? By joint spacing, we mean the average distance between regularly spaced joints. Information on joint spacing provides insight into the mechanical properties of rock layers and their fracture permeability. By joint density, we mean, in two dimensions, what is the trace length of joints per unit area of outcrop, or in three dimensions, what is the area of joints per unit volume of outcrop? Joint density depends both on the length of the joints and on their spacing. Information on joint density helps define the fracture-related porosity and permeability of a rock body.

7. How is the distribution of joints affected by lithology? In sedimentary rocks, do individual joints cut across a single bed, or do they cut across many beds, or even through the entire outcrop and beyond? In what way is joint spacing affected by bed composition? In the igneous rocks and their contact zones, is the joint spacing or style (both within an igneous body and in the country rock that was intruded) controlled by the proximity of the joint to contact? Information about the relationship between jointing and lithology can be related to physical characteristics such as Young's modulus (E), and can help determine variations in fracture permeability as a function of position in a stratigraphic sequence. Information on the relationship between jointing and lithology may also give insight into the cause of joint formation.

8. Are joints connected to one another or are they isolated? The connectivity of joints is critical to a determination of whether they could provide a permeable network through which fluids (for example, contaminated groundwater or petroleum) could flow.

9. How are joints related to other structures and fabrics? Are joints parallel to tectonic foliations? Are joints geometrically related to folds? Are joints reoriented by folding, or do they cut across the folds? Is there a relationship between joint orientation and measured contemporary stresses? Is the spacing or style of joints related to the proximity of faults? Information on the relation of joints to other structures provides insight into the tectonic conditions in which joints form and the timing of joint formation with respect to the formation of other structures in a region.

7.4.1 Dealing with Field Data about Joints

There are basically two ways to carry out a field study of joint orientation, spacing, and intensity. In the **inventory method,** you define a representative region and measure all joints that occur within that region. For example, you draw a circle or square on the outcrop and measure all joints that occur within its area, or you could draw a line across the outcrop and measure all joints that cross that line (Figure 7.12). The inventory method is necessary if you need to determine fracture density in a body of rock, or provide statistics on joint data. Mathematical procedures for determining joint density in three dimensions from measurements on two dimensional surfaces are available, but are beyond the scope of this book. You can use the inventory method for either systematic or nonsystematic joints.

The inventory method allows you to determine dominant joint orientations using statistical methods. But the problem with the inventory method is that a large number of nonsystematic fractures in an outcrop may obscure the existence of sets of systematic joints, especially if the systematic joints are widely spaced. For this reason, the **selection method** is a more appropriate approach. In this method, you visually scan the outcrop and subjectively decide on the dominant sets (Figure 7.12). Then, you measure a few representative joints of each set and specify

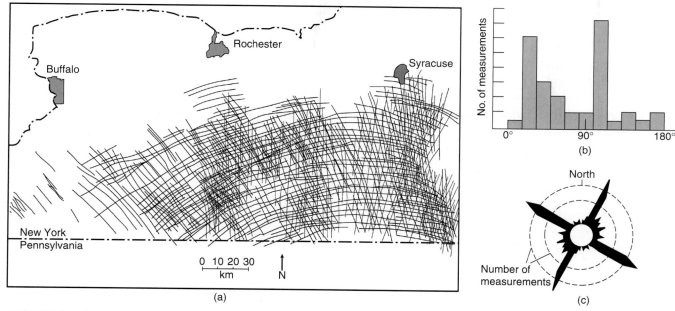

Sampling line for "inventory method"

Joint selected in "selection method"

Sampling circle for "inventory method"

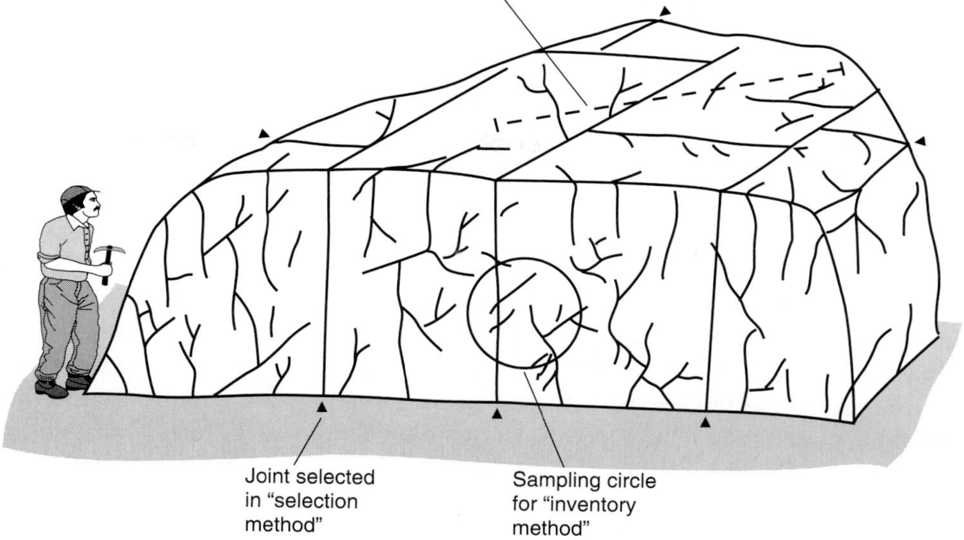

FIGURE 7.12   Joint study using the inventory and selection methods.

the spacing between joints in the set. Effectively, you are filtering your measurements in the field. While this technique won't permit determination of fracture density or provide statistics on joint orientation, it will allow you to define fracture systems in a region. The hazard with this procedure is that a careless observer may record what he or she wants to see, not what is really in the outcrop. When an observer starts scanning a new outcrop, she may subconsciously look for the same joint sets that she had seen in the last outcrop studied, and therefore may miss different, but impor-

tant, joint sets that occur in the second outcrop, just because they weren't present in the first one.

Joint data can be recorded in a number of ways. One way is to plot the strike and dip of joints on a geologic map. If the joints are vertical, a measurement of the trend of the joint may be sufficient. You can create a powerful visual impression of joint attitudes in a region, by drawing representative **joint trajectories** as trend lines on a map (Figure 7.13a). These trajectories are lines that represent the trends of joints, but they are not necessarily the map traces of individual joints.

FIGURE 7.13   Ways of representing joint arrays. (a) Joint trajectory map. (b) Frequency diagram (histogram). (c) Rose diagram. The three examples do not portray the same data sets.

Statistical diagrams that show attitudes of many different joints within a given region can help you identify dominant joint orientations in a region. What may appear to be a meaningless jumble at first sight, may resolve into significant groupings upon analysis. If joints in a particular region are not vertical, it is most appropriate to plot their attitudes on a contoured equal-area net, but if the joints are mostly vertical, a common occurrence in flat-lying sedimentary strata, their strikes can be shown on histograms. A **histogram,** in the case of joints, indicates the number of joints whose strike falls within a particular range. On a bar histogram (Figure 7.13b), the abscissa represents bearings from 0° to 180°, and the ordinate is proportional to the number of fracture-strike measurements. On a polar histogram, also called a **rose diagram,** you show the bearings directly on the diagram (Figure 7.13c). The number of joints whose strike falls within a given range is shown by a pie-slice segment whose radius is proportional to the number, or to the percentage, of joints with that orientation. Rose diagrams work better than bar histographs to give you an intuitive feel for the distribution of joint attitudes.

The orientation of joints is not the only information to record during a joint study, as you can see from the list of questions that we presented in the previous section. In modern joint studies, you should also record joint spacing, joint trace length, cross-cutting relations between joints, the relation of joints to lithology, joint surface morphology, and the relation of joints to other structures. To clarify relations among joints, it often helps to add outcrop sketches to your notes.

7.5 ORIGIN AND INTERPRETATION OF JOINTS

Why do joints form? In Chapter 6, we learned that joints develop when stress exceeds the tensile fracture strength of a rock, and Griffith cracks begin to propa-gate. But under what conditions in the Earth's brittle crust are stresses sufficient to crack rocks? In this section, we describe several possible settings to explain how stress states leading to joint formation develop in a rock body. But before you read further, a note of caution. When using these ideas as a basis for field interpretation of jointing, keep in mind that different joints in the same outcrop may have formed at different times and for different reasons. Once formed, a joint doesn't heal and disappear unless the rock gets metamorphosed or becomes pervasively deformed. Further, local variations in the stress field, which are a natural feature of inhomogeneous rock, may cause joints that formed at the same time to have different orientations at different locations. Because of these factors, joint interpretation continues to challenge geologists, and will do so for years to come.

7.5.1 Joints Related to Uplift and Unroofing

Lithostatic pressure due to the weight of overlying rock compresses rock at depth. Also, because of the Earth's geothermal gradient, rock at depth is warmer than rock closer to the surface. Subsequent regional uplift leads to erosion of the overburden and the unroofing of buried rock (Figure 7.14). This unroofing causes a change in the stress state for three reasons: cooling, the Poisson effect, and the membrane effect.

As the burial depth of rocks decreases, they *cool and contract.* The rock can shrink in a vertical direction without difficulty, because the Earth's surface is a free surface. But, because the rock is embedded in the earth, it is not free to shrink elastically in the horizontal direction as much as if it were unconfined, so horizontal tensile stress develops in the rock. Furthermore, as the overburden diminishes, rock expands (very slightly) in the vertical direction. Therefore, because of the **Poisson effect** (see Chapter 6), it contracts in the

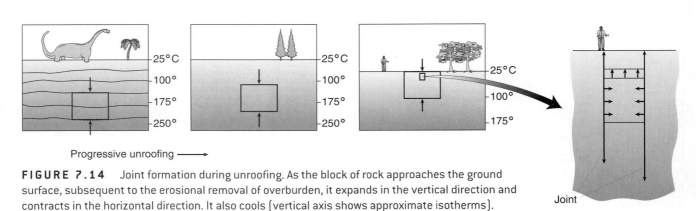

FIGURE 7.14 Joint formation during unroofing. As the block of rock approaches the ground surface, subsequent to the erosional removal of overburden, it expands in the vertical direction and contracts in the horizontal direction. It also cools (vertical axis shows approximate isotherms).

horizontal direction. Again, because the rock is embedded in the earth, it cannot shorten in the horizontal direction as much as it would if it were unconfined, so a horizontal tensional stress develops. Uplift and unroofing effectively cause rock layers to move away from the center of Earth. The layer stretches like a membrane as its radius of curvature increases, thereby creating tensile stress in the layer, called the **membrane effect.**

If the horizontal tensional stress created by any or all of these factors overcomes the compressive stresses due to burial and exceeds the rock's tensile strength, it will cause the rock to crack and to form joints. Joints formed by the processes just described tend to be vertical because they generate a horizontal σ_3. Recall that the Earth's surface is a free surface, so it must be a principal plane of stress. Therefore, the other two principal planes of stress must be vertical. Uplift and unroofing are particularly important causes of joint formation in sedimentary basins of continental interiors, which are subjected to epeirogenic movements, and in orogens that are uplifted long after collisional or convergent tectonism has ceased.

7.5.2   Formation of Sheeting Joints

Uplift and exhumation of rocks may lead to the development of sheeting joints within a few hundred meters of the Earth's surface. As we mentioned earlier, sheeting joints are commonly subparallel to topographic surfaces, and are most prominent in rocks that do not contain bedding or schistosity, particularly granitic rocks.

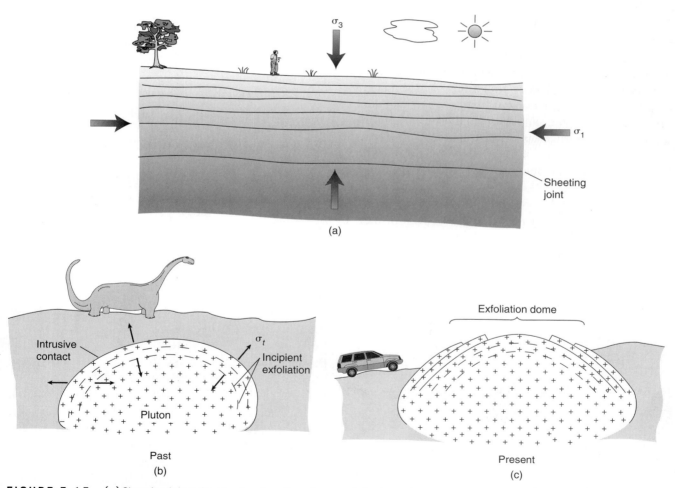

FIGURE 7.15 (a) Sheeting joints forming in a location where σ_1 is horizontal while σ_3 is vertical, near the ground surface. Note that the joints become more closely spaced closer to the ground surface. (b) Consider a situation where a pluton cools and contracts more than country rock, so σ_t (tensile stress) is oriented perpendicular to the intrusive contact. (c) Later, when the pluton is exhumed, joints form parallel to the intrusive contact and create an exfoliation dome.

The origin of sheeting joints is a bit problematic. At first glance, you might not expect joints to form parallel to the ground surface, because they are tensile fractures, and near the ground surface there is a compressive load perpendicular to the ground surface due to the weight of the overlying rock, and the lack of high fluid pressure. It appears that sheeting joints form where horizontal stress is significantly greater than the vertical load (Figure 7.15a), allowing joints to propagate parallel to the ground surface. In this regard, formation of sheeting joints resembles cracks formed by longitudinal splitting in laboratory specimens.

The stresses causing sheeting joints may, in part, be tectonic in origin, but they may also be residual stresses. A **residual stress** remains in a rock even if it is no longer loaded externally (e.g., in an unconfined block of rock sitting on a table). Residual stresses develop in a number of ways. Imagine a layer of dry sand that gets deeply buried. Because of the weight of the overburden, the sand grains squeeze together and strain elastically. If, at a later time, groundwater fills the pores between the strained grains, unstrained cement may precipitate and lock the grains together. As a consequence, the elastic strain in the grains gets locked into the resulting sandstone. When unroofing later exposes the sandstone, the grains and the cement expand by different amounts, and as a consequence stress develops in the sandstone. In the case of pluton, residual stresses develop because its thermal properties (e.g., coefficient of thermal expansion) differ from those of the surrounding wall rock, and because, during cooling, the pluton cools by a greater amount than the wall rock. The pluton and the wall rock tend to undergo different elastic strains as a result of thermal changes during cooling and later unroofing (Figure 7.15b and c). Because the pluton is welded to the surrounding country rock, the differential strain creates an elastic stress in the rock. For example, if the pluton shrinks more than the wall rock, tensile stresses develop perpendicular to the wall. At depth, compressive stress due to the overburden counters these tensile stresses, but near the surface, residual tensile stress perpendicular to the walls of the pluton may exceed the weight of the overburden and produce sheeting joints parallel to the wall of the pluton.

We earlier noted that sheeting joints tend to parallel topography. This relationship either reflects topographic control of the geometry of joints (because the vertical load is perpendicular to the ground surface), or joint control of the shape of the land surface (because rocks spall off the mountainside at the joint surface). Geologists are not sure which phenomenon is more important.

7.5.3 Natural Hydraulic Fracturing

As we saw in Chapter 3, the three principal stresses at depth in most of the continental lithosphere are compressive. Yet joints form in these regions, and these joints may be decorated with plumose structure indicating that they were driven by tensile stress. How can joints form if all three principal stresses are compressive? As we described in Chapter 6, the solution to this paradox comes from considering the effect of pore pressure on fracturing. In simple terms, the increase in pore pressure in a preexisting crack pushes outward and causes a tensile stress to develop at the crack tip that eventually exceeds the magnitude of the least principal compressive stress. If the pore pressure is sufficiently large, a tensile stress that exceeds the magnitude of σ_3 develops at the tip the crack, even if the remote principal stresses are all compressive (Figure 7.16a), and the crack propagates, a process called **hydraulic fracturing.** Oil well engineers commonly use hydraulic fracturing to create fractures and enhance permeability in the rock surrounding an oil well. They create hydraulic fractures by increasing the fluid pressure in a sealed segment of the well until the wall rock breaks. But hydraulic fracturing also occurs in nature, due to the fluid pressure of water, oil, and gas in rock, and it is this *natural* hydraulic fracturing that causes some joints to form.

If you think hard about the explanation of hydraulic fracturing that we just provided, you may wonder whether the process implies that the pore pressure in the crack becomes greater than pore pressure in the pores of the surrounding rock. It doesn't! Pore pressure in the crack can be the same as in the pores of the surrounding rock during natural hydraulic fracturing. Thus, we need to look a little more closely at the problem to understand why pore pressure can cause joint propagation.

Imagine that a cemented sandstone contains fluid-filled pores and fluid-filled cracks (Figure 7.16b). Let's focus our attention on the crack and its walls. Because the pores and the crack are connected, the fluid pressure in the pores and the crack are the same. Fluid pressure within the crack is pushing outwards, creating an opening stress, but at the same time, the fluid pressure in the pores, as well as the stress in the rock, is pushing inwards, creating a closing stress. As long as the closing stress exceeds the opening stress, the crack does not propagate. If the fluid pressure increases, the opening stress increases at the same rate as the increase in fluid pressure, but the closing stress increases at a slower rate. Eventually, the open-

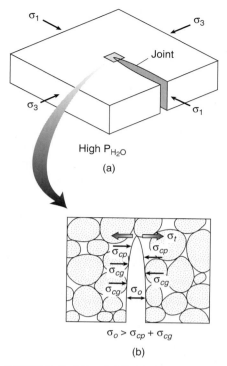

$$\sigma_o > \sigma_{cp} + \sigma_{cg}$$

(b)

FIGURE 7.16 (a) Block diagram showing the stresses in the vicinity of a crack in which there is fluid pressure that exceeds the magnitude of σ_3. As a result, there is a tensile stress, σ_t, along the crack. (b) Enlargement of the crack tip, illustrating the poroelastic effect. The opening stress (σ_o) due to fluid pressure in the crack exceeds the closing stress (σ_c), which is the sum of σ_{cp}, the closing stress where a pore is in contact with the crack, and σ_{cg}, the closing stress where a grain is in contact with the crack.

ing stress exceeds the closing stress so that the crack propagates; effectively, the outward push of the fluid in the crack creates a tensile stress at the crack tip. Why does the closing stress increase at a slower rate than the fluid pressure and the opening stress? The reason is that grains in the rock are cemented to one another, so that the grains cannot move freely in response to the increase in fluid pressure in the pores. The elasticity of the grains themselves, therefore, takes up some of the push caused by the fluid pressure. Thus, the closing stress acting on the fluid in the crack where it is in contact with a grain is less than where the fluid in the crack is in contact with a pore, but the outward push of the fluid in the crack is the same everywhere.[5] As a result, the net outward push exceeds the net inward push, and tensile stress locally develops.

---

[5]This is known as the *poroelastic effect.*

Once the crack propagates, the volume of open space between the walls of the crack increases, so the fluid pressure in the crack decreases. As a consequence, the crack stops growing until an increase in fluid pressure once again allows the stress intensity at the crack tip to drive the tip into unfractured rock. Thus, the surfaces of joints formed by natural hydraulic fracturing tend to have many arrest lines.

7.5.4 Joints Related to Regional Deformation

During a convergent or collisional orogenic event, compressive tectonic stress affects rocks over a broad region, including the continental interior. Joints form within the foreland of orogens during tectonism for a number of reasons.

Joints from natural hydrofracturing often form on the foreland margins of orogens during orogeny. The conclusion that the joints are syntectonic is based on two observations. First, the joints parallel the σ_1 direction associated with the development of tectonic structures like folds. Second, the joints locally contain mineral fill which formed at temperatures and fluid pressures found at a depth of several kilometers; thus, they are not a consequence of the recent cracking of rocks in the near surface. The origin of such joints may reflect increases in fluid pressure within confined rock layers due to the increase in overburden resulting from thrust-sheet emplacement, or from the deposition of sediment eroded from the interior of the orogen.

During an orogenic event, the maximum horizontal stress is approximately perpendicular to the trend of the orogen. As a consequence, the joints that form by syntectonic natural hydraulic fracturing are roughly perpendicular to the trend of the orogen. Because the stress state may change with time in an orogen, later-formed joints may have a different strike than earlier-formed joints, and the joints formed during a given event might not be exactly perpendicular to the fold trends where they form. Such joint patterns are typical of orogenic foreland regions, but may also occur in continental interiors.

Joints are commonly related to faulting, and these fall into three basic classes. The first class is composed of regional joints that develop in the country rock due to the stress field that is also responsible for generation and/or movement on the fault itself. Since faults are usually inclined to the remote σ_1 direction, the joints that form in the stress field that cause a fault to move will not be parallel to the fault (Figure 7.17a). The second class includes joints that develop due to the distortion of a moving fault block. For example, the hanging wall of a normal fault bock may undergo some extension, resulting in the development of joints, or the hanging block of a thrust fault may be warped as it moves over the fault, if the underlying fault surface is not planar, and thus may locally develop tensile stresses sufficient to crack the rock (Figure 7.17b). The third class includes joints that form immediately adjacent to a fault in response to tensile stresses created in the wall rock while the fault moves. Specifically, during the development of a shear rupture (i.e., a fault), an *en echelon* array of short joints forms in the rock adjacent to the rupture. These joints merge with the fault and are inclined at an angle of around 30° to 45° to the fault surface; they are called **pinnate joints** (Figure 7.17c).

The acute angle between a pinnate joint and the fault indicates the sense of shear on the fault, as we will see in Chapter 8.

When the stress acting on a region of crust is released, the crust elastically relaxes to attain a different shape. This change in shape may create tensile stresses within the region that are sufficient to create **release joints,** such as occurs in relation to folding. Folded rocks may be cut by syntectonic natural hydrofractures, a process manifested by joints oriented at a high angle to the fold-hinge, as we described previously (Figure 7.18). In addition, during the development of folds in nonmetamorphic conditions, joints often develop because of local tensile stresses associated with bending of the layers (Figure 7.18). Joints resulting from this process of outer-arc extension have a strike that is parallel to the trend of the fold-hinge, and may converge toward the core of the fold. If development of folds results in stretching of the rock layer parallel to the hinge of the fold, then cross-strike joints may develop.

Finally, joints may develop in a region of crust that has been subjected to broad regional warping, perhaps due to **flexural loading** of the crust. Like folding, joint formation reflects tensile stresses that develop when the radius of curvature of a rock layer changes.

7.5.5 Orthogonal Joint Systems

In orogenic forelands and in continental interiors, you will commonly find two systematic joint sets that are mutually perpendicular. In some cases, the joints define a **ladder pattern** (Figure 7.19a), in which the joints of one set are relatively long, whereas the joints of the other are relatively short cross joints which terminate at the long joints. In other cases, the joints define a **grid pattern** (Figure 7.19b), in which the two sets appear to be mutually cross cutting. The existence of such orthogonal systems has perplexed geologists for decades, because at first glance it seems impossible for two sets of tensile fractures to form at 90° to one another in the same regional stress field. Recent field and laboratory studies suggest a number of possible ways in which orthogonal systems develop, though their application to specific regions remains controversial.

In orogenic forelands, an orthogonal joint system typically consists of a strike-parallel and a cross-strike set, defining a ladder pattern. The two sets may have quite different origins.

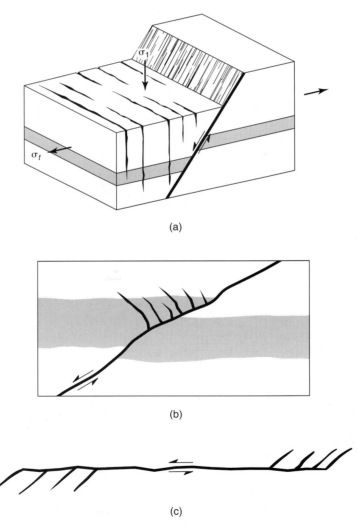

(a)

(b)

(c)

FIGURE 7.17 [a] Formation of joints in the hanging-wall block of a region in which normal faulting is taking place. [b] Formation of joints above an irregularity in a (reverse) fault surface. [c] Pinnate joints along a fault.

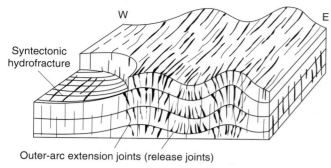

Syntectonic
hydrofracture

Outer-arc extension joints (release joints)

FIGURE 7.18 Block diagram showing outer-arc extension joints whose strike is parallel to the hinge of folds.

Cross-strike joints parallel the regional maximum horizontal stress trajectory associated with folding, and thus may have formed as syntectonic natural hydrofractures, whereas strike-parallel joints could reflect outer-arc extension of folded layers. Alternatively, the strike-parallel joints could be release joints formed when orogenic stresses relaxed.

Orthogonal joint systems may develop in regions that were subjected to a regional tensile stress that was later relaxed. During the initial stretching of the region, a set of joints develops perpendicular to the regional tensile stress. When the stress is released, the region rebounds elastically, and expands slightly in the direction perpendicular to the original stretching. A new set of joints therefore develops perpendicular to the first.

Orthogonal joint systems may also develop during uplift. Imagine that a rock layer is unloaded when the overburden erodes away. As a result, a joint set perpendicular to the regional σ_3 develops. With continued uplift and expansion, the tensile stress that develops in the layer can be relieved easily in the direction perpendicular to the existing joints; that is, they just open up. But tensile stresses cannot be relieved in the direction parallel to the existing joints, so new cross joints form, creating a ladder pattern.

Grid patterns (Figure 7.19b) suggest that the two joint sets initiated at roughly the same time, or that cracking episodes alternated between forming members of one set and then the other. If we assume that both joint sets form in the principal plane that is perpendicular to σ_3, we can interpret such occurrences as being related to the back-and-forth interchange of σ_2 and σ_3 during uplift, when σ_2 and σ_3 are similar in magnitude. To see what we mean, imagine a region where σ_1 is vertical, and σ_3 is initially north–south. When σ_3 is north–south, east–west trending joints develop. But if σ_3 switches with σ_2 and becomes east–west, then north–south trending joints develop.

7.5.6 Conjugate Joint Systems

At some localities in orogenic forelands we find that joint sets define a conjugate system in which the bisector of the dihedral angle is perpendicular to the axis of folds. The origin of such fracture systems remains one of the most controversial aspects of joint interpretation. Based on their geometry, it was traditionally assumed that conjugate joints are either shear fractures, formed at about 30° to σ_1 (representing failure when the Mohr circle touches the Coulomb failure envelope; Chapter 8), or so-called transitional-tensile fractures that are thought to form at angles less than 30° to σ_1 (representing failure when the Mohr circle touches the steep part of the failure envelope; see Chapter 8). Yet, if you examine the surfaces of the joints in these conjugate systems, you find in many cases that they display plumose structure, confirming that they formed as Mode I (extension) fractures. Further, as we noted earlier, transitional-tensile fractures have never been created in the lab, so their very existence remains suspect. The only type of crack that is known to propagate for long distances in its own plane is a Mode I crack; shear fractures form by linkage of microcracks, not by propagation of a single shear surface in its own plane. But if the members of conjugate joint systems are not shear fractures, how do they form?

Many researchers now believe that both of the two nonparallel sets in the conjugate system are cross-strike joints that initially formed perpendicular to σ_3. Thus, to explain the contrast in orientation between the two sets, they suggest that the two sets formed at different times in response to different stress fields. For example, the slightly folded Devonian strata of the Appalachian Plateau in south-central New York State contain two joint sets separated by an angle of about 60° (Figure 7.13a).

(a)	(b)

FIGURE 7.19 Two patterns of orthogonal joint systems. (a) Traces of joints defining a ladder pattern. (b) Traces of joints defining a grid pattern.

FIGURE 7.20 Large joint face in Entrada sandstone near Moab (Utah, USA). Note how the cliff face is a large joint surface. The thin bedded shale unit below the sandstone has more closely spaced joints.

The two sets are attributed to different, distinct phases of the Late Paleozoic Alleghanian orogeny; the maximum horizontal stress during the first phase of the orogeny was not parallel to that during the second phase of orogeny. Geologists who favor this model for conjugate joint systems conclude that the occurrence of slip lineations on joints of conjugate systems does not imply that the joints originated as shear fractures, but rather that they reactivated as mesoscopic faults subsequent to their formation.

7.5.7 Joint Trend as Paleostress Trajectory

Orientation data on jointing holds valuable information about the orientation of stress fields at the time of failure. Because joints propagate normal to σ_3, their planes define the trajectories of σ_3 in a region. In the case of vertical joints, the strike of the joint defines the trajectory of maximum horizontal stress (σ_H). We don't a priori know if σ_H represents σ_1 or σ_2. The maximum principal stress could be either parallel or perpendicular to the Earth's surface, depending on the depth at which the joint formed.

7.6 LIMITS ON JOINT GROWTH

Having discussed various ways in which joints initiate, we also need to address the issue of why joints stop growing. Recall from Chapter 6 that the stress intensity at the tip of a crack depends on the length of the crack. Thus, the stress intensity increases as the crack grows, and as long as the stress driving joint growth remains unchanged, the joint will keep growing. For this reason, joints that grow in large bodies of homogeneous rock can become huge surfaces, as seen in the massive beds of sandstone in southern Utah shown in Figure 7.20. But joints clearly do not propagate from one side of a continent to the other. They stop growing for one or more of the following reasons.

The joint tip may *intersect a (nearly) free surface*. Joints obviously stop growing when they reach Earth's surface. They stop growing in the subsurface where they intersect a preexisting open fracture (joint or fault) or a weak bedding plane, or where they pass downward into ductile rock. Two joints that are growing towards each other, but are not coplanar, stop growing when they enter each other's stress shadow (Figure 7.21a). In some cases, the *interaction of joint tips* causes curvature where the two joints link (Figure 7.21b). If, however, the preexisting joint is squeezed together so tightly that friction allows shear stress to be transmitted across it, or if it has been sealed by vein material, then a younger joint can cut across it.

Formation of the joint itself may cause a local *drop in fluid pressure,* because creation of the joint creates space for fluid. This increase in space temporarily causes a drop in fluid pressure, so that the stress intensity at the joint tip becomes insufficient to propagate into unfractured rock. When fluid flow into the joint increases the fluid pressure to a large enough value, the joint growth resumes. Thus, as mentioned earlier, it is

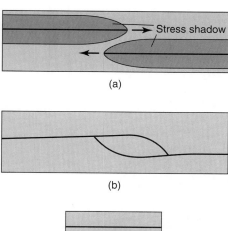

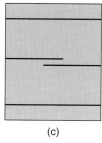

FIGURE 7.21 Joint terminations. (a) Joints terminating without curving when they approach one another. (b) Joints curving into each other and linking. (c) Map view sketch illustrating how joint spacing is fairly constant because joints that grow too close together cannot pass each other.

TABLE 7.2	VEIN TERMINOLOGY
Vein	A fracture that has filled with minerals precipitated from water solutions that passed through the fracture.
Vein array	A group of veins in a body of rock.
Planar systematic arrays	In a planar systematic array, the component veins are planar, are mutually parallel, and are regularly spaced.
Nonsystematic arrays	The veins in nonsystematic arrays tend to be nonplanar, and individual veins may vary in width.
En echelon arrays	An *en echelon* array consists of short parallel veins that lie between two parallel enveloping surfaces and are inclined at an angle to the surfaces. Typically, the veins in an *en echelon* array taper toward their terminations. The veins may be sigmoidal in cross section.
Stockwork veins	Stockwork veins comprise a cluster of irregularly shaped veins that occur in a pervasively fractured rock body. The veins are nonplanar arrays and occur in a range of orientations. In rock bodies with stockwork veins, as much as 40% or 50% of the outcrop is composed of vein material, and vein material may completely surround blocks of the host rock.

characteristic of joints being driven by high fluid pressures to grow in a start-stop manner, so their surfaces show many arrest lines.

Finally, if the joint grows into a region where energy at the crack tip can be dissipated by plastic yielding, the joint stops growing. Similarly, propagation of a joint into a rock with a *different stiffness* or tensile strength may cause it to stop growing. Also, if the joint tip enters a region where the stress intensity at the crack tip becomes too small to drive the cracking process, then the joint stops growing. The decrease in stress intensity may be due to a decrease in the tensile stress magnitude in the rock, or due to an increase in compressive stress that holds the joint together.

7.7 VEINS AND VEIN ARRAYS

In the vast desert ranges of western Arizona, there are few permanent residents, save for the snakes and scorpions, but almost every square meter of the rugged terrain has been trod upon by a dusty prospector in search of valuable deposits of gold, silver, or copper. Modern geologists mapping in the region frequently come upon traces of prospectors from years past. When you poke into many of

these excavations, you find that the focus of their efforts, the days and days of agonizing labor with pick and shovel, is nothing more than a vein of milky white quartz. What are veins? Simply speaking, a **vein** is a fracture filled with mineral crystals that precipitated from a watery solution (Figure 7.1b; Table 7.2). Quartz or calcite form the most common vein fill, but other minerals do occur in veins, including numerous ore minerals, zeolites, and chlorite. Some veins initiated as joints, whereas others initiated as faults or as cracks adjacent to faults. Veins come in all dimensions; some are narrower and shorter than a strand of hair, while others comprise massive tabular accumulations that are meters across and tens of meters long. Groups of veins are called **vein arrays** and these have a variety of forms, as described in Table 7.2.

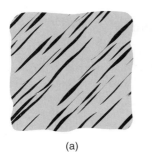 

FIGURE 7.22 Vein arrays. [a] Planar array of veins. [b] Stockwork array of veins. Vein fill is dark.

7.7.1 Formation of Vein Arrays

Planar systematic arrays (Figure 7.22a) represent mineralization of a preexisting systematic joint set or mineralization during formation of a systematic joint set. **Stockwork vein arrays** (Figure 7.22b) form where rock has been shattered, either by the existence of locally very high fluid pressure, or as a result of pervasive fracturing in association with folding and faulting.

En echelon vein arrays form in a couple of different ways. They may form by filling *en echelon* joints in the twist hackle fringe of a larger joint. As we saw earlier in this chapter, the twist hackle fringe represents the breakup of a joint into short segments when it enters a region of the rock with a different stress field. *En echelon* vein arrays also develop as a consequence of shear within a rock body that is associated with displacement across a fault zone (Figure 7.23). The fractures comprising an *en echelon* array initiate parallel to σ_1, typically at an angle of about 45° to the borders of the shear. Fractures open as displacement across the shear zone develops, and fill with vein material (Figure 7.23b). Once formed, the veins are material objects within the rock, so continued shear will rotate the veins and the angle increases. If, however, a new increment of vein growth occurs at the tip, these new increments initiate at 45° to the shear surfaces. Therefore, the veins become sigmoidal in shape (Figure 7.23c). Locally, a second generation of veins may initiate at the center of the original veins; this second set cuts obliquely across the first generation of veins. Because of the geometric relationship between *en echelon* veins and displacement, the orientation of shear-related *en echelon* veins can be used to determine shear sense (see Chapters 8 and 12).

7.7.2 Vein Fill: Blocky and Fibrous Veins

Vein fill, the mineral crystals within a vein, is either blocky (also called sparry) or fibrous. In **blocky veins,** the crystals of vein fill are roughly equant, and may

(a)

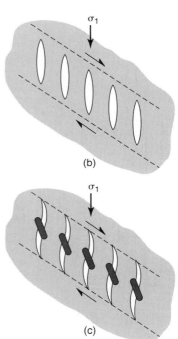

(b)

(c)

FIGURE 7.23 [a] *En echelon* veins in the Lachlan Orogen [southeastern Australia]. [b] Formation of a simple *en echelon* array. [c] Formation of sigmoidal *en echelon* veins, due to rotation of the older, central part of the veins, and the growth of new vein material at ~45° to the shear surface.

exhibit crystal faces (Figure 7.24a). The occurrence of blocky veins means that the vein was an open cavity when the mineral precipitated (this is possible only in veins formed near the surface, where rock strength is sufficient to permit a cavity to stay open or fluid pressure is great enough to hold the fracture open), that

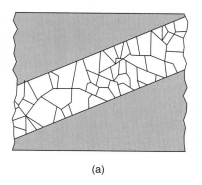
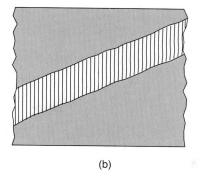

(a)   (b)

FIGURE 7.24   Vein fill types. (a) Blocky vein fill. (b) Fibrous vein fill.

previously formed vein fill later recrystallized to form blocky crystals, or that there were few nucleation sites for crystals to grow from during vein formation.

In **fibrous veins,** the crystals are long relative to their width, so that the vein has the appearance of being spanned by a bunch of hairs (Figure 7.24b). Geologists don't fully agree on the origin of fibrous veins, but some fibrous veins may form by the **crack-seal process.** The starting condition for this process is an intact rock containing pore fluid that in turn contains dissolved minerals. If the fluid pressure becomes great enough, the vein cracks and a very slight opening (only microns wide) develops. This crack immediately fills with fluid; but since the fluid pressure within the open crack is less than in the pores of the surrounding rock, the solubility of the dissolved material decreases and the mineral precipitates, thereby sealing the crack. The process repeats tens to hundreds of times, and each time the vein width grows slightly (Figure 7.25). During each increment of growth, existing grains in the vein act as nuclei on which the new vein material grows, and thus continuous crystals grow. Figure 7.25 shows microscopic evidence for this crack-seal process in the formation of a fibrous vein. Alternatively, formation of some fibrous veins may occur by a diffusion process, whereby ions migrate through fluid

FIGURE 7.25   Photomicrograph of fibrous calcite filling in tensile fractures. Width of view is ~3 cm.

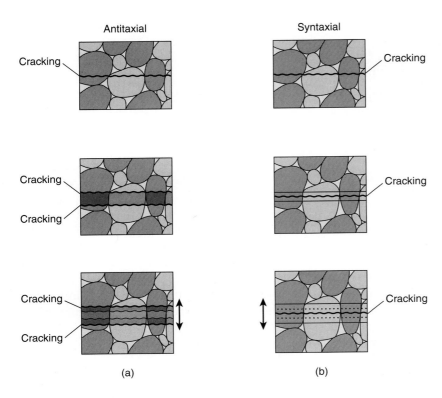

Antitaxial | Syntaxial

Cracking

Cracking

Cracking

Cracking

Cracking

Cracking

Cracking

Cracking

(a) | (b)

FIGURE 7.26 Cross-sectional sketches, at the scale of individual grains, showing the contrast in the stages of crack-seal deformation leading to antitaxial veins and syntaxial veins. (a) Formation of antitaxial veins. The increments of cracking form along the margins of the vein, and the vein composition differs from the wall rock (i.e., the fibers are not in optical continuity with the grains of the wall). During increments of cracking, tiny slices of the wall rock spall off. The slices bound the growth increments in a fiber. (b) Syntaxial vein formation. During each increment, the cracking is in the center of the vein. The composition of the fibers is the same as that of the grains in the wall rock (that is, the fibers are in optical continuity with the grains of the wall rock). Optical continuity between fiber and grain means that the crystal lattice of the grain has the same orientation as the crystal fiber of the fiber. Optically continuous fibers and grains go extinct at the same time, when viewed with a petrographic microscope.

films on grain boundaries and precipitate at the tips of fibers while the vein walls gradually move apart. During this process, an open crack never actually develops along the vein walls or in the vein.

7.7.3   Interpretation of Fibrous Veins

Fibrous veins, in particular those consisting of calcite and quartz, are interesting because they can record useful information about the progressive strain history in an outcrop. There are two end-member types of fibrous veins, syntaxial and antitaxial veins (Figure 7.26).

Syntaxial veins typically form in rocks where the vein fill is the same composition as the wall rock; for example, quartz veins in a quartz sandstone. The vein fibers nucleate on the surface of grains in the wall rock and grow inwards to meet at a median line. Each successive increment of cracking occurs at the median line, because at this locality separate fibers meet, whereas at the walls of the vein, vein fibers and grains of the wall rock form single continuous crystals. Each growth increment of a fiber is bounded by a trail of fluid inclusion.

Antitaxial veins form in rocks where the vein fill is different from the composition of the wall rock; for example, a calcite vein in a quartz sandstone. In antitaxial veins, the increments of cracking occur at the boundaries between the fibers and the vein wall, prob-

ably because that is where the bonds are weakest. Thus, antitaxial veins grow outward from the center. Increments of growth are sometimes bounded by trails of small dislodged flakes of the wall rock.

In many cases, the long axis of a fiber in a fibrous vein tracks the direction of maximum extension (stretching) at the time of growth (i.e., it parallels the long axis of the incremental strain ellipsoid). When fibers are perpendicular to the walls of the vein, the vein progressively opened in a direction roughly perpendicular to its walls (Figure 7.27a). However, vein fibers oblique to the vein walls indicate that the vein opened obliquely and that there was a component of shear displacement during vein formation (Figure 7.27b). When vein fibers are sigmoidal in shape, the extension direction rotated relative to the vein-wall orientation. Note that for identical-looking fibers, the order of the movement stages depends on whether the vein is antitaxial or syntaxial (Figure 7.27c-d). For example, if the fibers in a syntaxial vein are perpendicular to the walls in the center and oblique to the walls along the margins, it means that in the early stage of vein formation the vein had an oblique opening component, while at a later stage it did not (remember that the fibers grow toward the center). However, fibers with exactly the same shape in an antitaxial vein would indicate the opposite strain history, because in antitaxial veins the fibers grow toward the walls. The

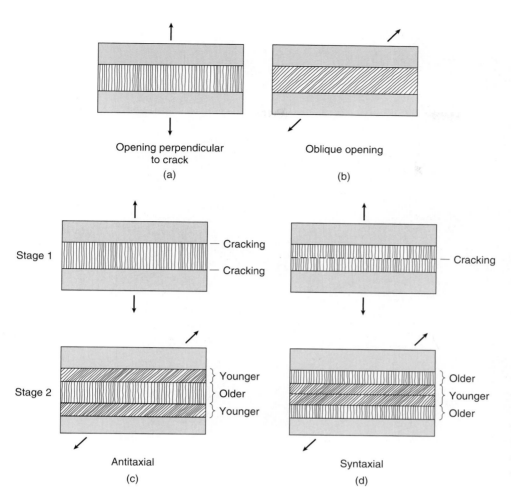

FIGURE 7.27 Cross-sectional sketches showing that the long axis of fibers in a vein tracks the direction of extension, and how a change in extension direction leads to the formation of sigmoidal fibers. (a) If the extension direction is perpendicular to the vein wall then the fibers are perpendicular to the grain. (b) If the extension direction is oblique to the vein wall, then the fibers are oblique to the vein wall. If the extension direction is effectively parallel to the vein surface (i.e., the vein is a fault surface), then the fibers are almost parallel to the vein wall, and on exposed surface would form slip lineations. (c, d) A change in extension direction leads to the formation of sigmoidal fibers. In this example, the opening is first perpendicular to the vein wall, and then is oblique to the vein. Because of the locus of vein fiber precipitation, (c) antitaxial veins and (d) syntaxial veins have different shapes.

presence of a median line helps you in recognizing this important kinematic difference, but uncertainty about the interpretation can remain.

7.8   LINEAMENTS

A geologic **lineament** is a linear feature recognized on aerial photos, satellite imagery, or topographic maps. Lineaments generally are defined only at the regional scale; that is, they are not mesoscopic or microscopic features. Structural lineaments, meaning ones that are a consequence of the localization of known geologic features, are defined by structurally controlled alignments of topographic features like ridges, depressions, or escarpments (Figure 7.28). They may also be manifested by changes in vegetation, which is, in turn, structurally controlled. Most lineaments are the geomorphologic manifestation of joint arrays, faults, folds, dikes, or contacts, but some remain a mystery and do not appear to be associated with obvious struc-

tures.[6] You should maintain a healthy skepticism when reading articles about lineaments that do not confirm imagery interpretations with ground truth. Some "lineaments" that have been described in the literature turn out to be artifacts of sunlight interaction with the ground surface, and thus do not have geologic significance. However, study of true structural lineaments often provides insight into the distribution of regional structural features, ore deposits, and seismicity.

7.9   CLOSING REMARKS

In this introduction to the rapidly evolving science of joint analysis, we hope to have conveyed not only descriptive information about the structures, but also a sense of the controversy surrounding their interpretation. One of the common questions that students ask

---

[6]Except perhaps ancient landing strips for extraterrestrials.

FIGURE 7.28 Aerial photograph of the Duncan Lake area (Northwest Territories, Canada), showing lineaments and structural control of topography. Scale is 1:14,500.

when beginning a field mapping project is, "Should I pay attention to the joints and veins?" An astute advisor might answer the question philosophically, with the words, "That depends."

If the purpose of the map is to define variations in permeability, or the location of faults, or the distribution of ore deposits, or the meaning of satellite-imagery lineaments, or the composition of fluids passing through the rock during deformation, then the joints and veins in the area should be studied. Perhaps you will find an interpretable variation in joint intensity within your map area, even if there is no systematic pattern to the jointing. Joint study is particularly important in studies of rock permeability, because the rate of fluid flow through joints may exceed the rate of flow through solid rock by orders of magnitude. Joints may make otherwise impermeable granite into a fluid reservoir, and may provide cross-formational permeability that permits oil to leak through an otherwise impermeable seal, or allow toxic waste in groundwater to leak across

an aquitard. If the purpose of your mapping is to develop an understanding of paleostress in the region, then it may only be worthwhile to study joints if you can identify systematic sets. If the purpose of your study is to develop an understanding of stratigraphy in the map area, of the history of folding in a high-grade gneiss, then joint analyses probably won't help you very much and you shouldn't spend too much time looking at them. Of course, this advice might change in the coming decade, as sophistication in our understanding of the meaning of joints continues to grow. "That depends" remains a good answer, therefore.

Joints, by definition, are rock discontinuities (fractures) on which there has been no shear. In the next chapter and the final chapter on brittle deformation, we shift our focus to fractures on which there clearly has been shear; that is, faults. As you study Chapter 8, keep in mind the features of joints that we just described, so that you can compare them with the features and meaning of faults.

ADDITIONAL READING

Ashby, M. F., and Hallam, S. D., 1986. The failure of brittle solids containing small cracks under compressive stress states. *Acta Metallurgica,* 34, 498–510.

Atkinson, B. K., 1987. *Fracture mechanics of rock.* London: Academic Press.

Bahat, D., and Engelder, T., 1984. Surface morphology on cross-fold joints of the Appalachian Plateau, New York and Pennsylvania. *Tectonophysics,* 104, 299–313.

Bai, T., and Pollard, D. D., 2000. Closely spaced fractures in layered rocks: initiation mechanism and propagation kinematics. *Journal of Structural Geology,* 22, 1409–1425.

Beach, A., 1975. The geometry of en-echelon vein arrays. *Tectonophysics,* 28, 245–263.

Engelder, T., and Geiser, P. A., 1980. On the use of regional joint sets as trajectories of paleostress fields during the development of the Appalachian Plateau, New York. *Journal of Geophysical Research,* 85, 6319–6341.

Hancock, P. L., 1985. Brittle microtectonics: principles and practice. *Journal of Structural Geology,* 7, 438–457.

Hodgson, R. A., 1961. Classification of structures on joint surfaces. *American Journal of Science,* 259, 493–502.

Kulander, B. R., Barton, C. C., and Dean, S. L., 1979. The application of fractography to core and outcrop fracture investigations: report to U.S. Department of Energy, Morgantown Energy Technology Center, METC/SP-79/3.

Narr, W., and Suppe, J., 1991. Joint spacing in sedimentary rocks. *Journal of Structural Geology,* 13, 1038–1048.

Nickelsen, R. P., and Hough, V. D., 1967. Jointing in the Appalachian Plateau of Pennsylvania. *Geological Society of America Bulletin,* 78, 609–630.

Pollard, D. D., and Aydin, A., 1988. Progress in understanding jointing over the past century. *Geological Society of America Bulletin,* 100, 1181–1204.

Price, N. J., 1966. *Fault and joint development in brittle and semi-brittle rock.* Pergamon Press: London.

Ramsay, J. G., 1980. The crack-seal mechanism of rock deformation. *Nature,* 284, 135–139.

Reches, Z., 1976. Analysis of joints in two monoclines in Israel. *Geological Society of America Bulletin,* 87, 1654–1662.

Rives, T., Rawnsley, K. D., and Petit, J.-P., 1994. Analogue simulation of natural orthogonal joint set formation in brittle varnish. *Journal of Structural Geology,* 16, 419–429.

Segall, P., 1984. Formation and growth of extensional fracture sets. *Geological Society of America Bulletin,* 94, 563–575.

Faults and Faulting

8.1 INTRODUCTION

Imagine a miner in a cramped tunnel crunching forward through a thick seam of coal. Suddenly, his pick hits hard rock. The miner chips away a bit more only to find that the seam that he's been following for the past three weeks abruptly terminates against a wall of sandstone. He says, "@%$&, my seam's cut off—there's a fault with it!" From previous experience, the miner knows that because of a fault, he will have to spend precious time digging a shaft up or down to intersect the coal seam again.

Geologists adopted the term fault, but they use the term in different ways in different contexts. In a general sense, a **fault** is any surface or zone in the Earth across which measurable **slip** (shear displacement) develops. In a more restricted sense, faults are fractures on which slip develops primarily by brittle deformation processes (Figure 8.1a). This second definition serves to distinguish a "fault" (*sensu stricto*) from a fault zone and shear zone. We use the term **fault zone** for brittle structures in which loss of cohesion and slip occurs on several faults within a band of definable width (Figure 8.1b). Displacement in fault zones can

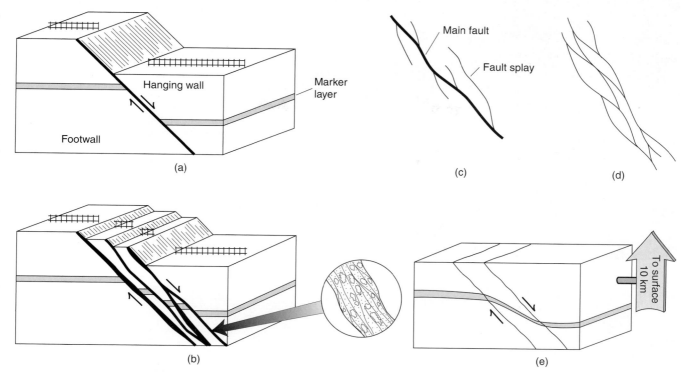

FIGURE 8.1 Sketches illustrating differences between faults, fault zones, and shear zones. (a) Fault. (b) Fault zone, with inset showing cataclastic deformation adjacent to the fault surface. (c) Sketch illustrating the relation between a principal fault and fault splays. (d) Anastomosing faults in a fault zone. (e) A shear zone, showing rock continuity across the zone. The displacements are shown to intersect the ground surface, whereas the shear zone occurs at depth in the crust.

involve formation and slip on many small, subparallel brittle faults, or slip on a principal fault from which many smaller faults diverge **(fault splays),** or slip on an anastomosing[1] array of faults (Figure 8.1c and d). **Shear zones** are ductile structures, across which a rock body does not lose mesoscopic cohesion, so that strain is distributed across a band of definable width. In ductile shear zones, rocks deform by **cataclasis,** a process involving fracturing, crushing, and frictional sliding of grains or rock fragments, or, more commonly, by **crystal plastic deformation** mechanisms (Figure 8.1e). We describe cataclastic shear zones in this chapter, but delay discussion of processes in ductile shear zones until Chapters 9 and 12.

Faults occur on all scales in the lithosphere (Figure 8.2), and geologists study them for several reasons. They control the spatial arrangement of rock units, so their presence creates puzzles that challenge even the most experienced geologic mappers. Faults affect topography and modify the landscape. Faults affect the distribution of economic resources (e.g., oil fields and ore bodies). They control the permeability of rocks and sediments, properties which, in turn, control fluid migration. Faulting creates deformation (strain ± rotation ± translation) in the lithosphere during plate interactions and intraplate movements. And, faulting may cause devastating earthquakes.

Fault analysis, therefore, plays a major role in diverse aspects of both academic and applied geology. In order to provide a basis for working with faults, this chapter introduces the terminology that is used to describe fault geometry and displacement, it discusses how to represent faults on maps and cross sections, and shows you how to recognize and interpret faults at the surface and in the subsurface. We conclude by introducing fault-system tectonics, the relation between faulting and resources, and the relation between faulting and earthquakes. Much of our discussion in this chapter focuses on the properties of mesoscopic faults; we treat the large-scale properties

---

[1]Anastomosing refers to a geometry of a group of wavy, subparallel surfaces that merge and diverge, resembling a braid of hair.

(a)

(b)

(c)

FIGURE 8.2 Photos of faults at different scales. (a) Microscopic faults, showing fractured and displaced feldspar grains. (b) Mesoscopic faults cutting thin layers in an outcrop. (c) The trace of the San Andreas (strike-slip) Fault across the countryside.

of fault systems more fully in Part D, where the basic concepts of plate tectonics are discussed.

8.2 FAULT GEOMETRY AND DISPLACEMENT

8.2.1 Basic Vocabulary

In order to discuss faults, we first need to introduce some fault vocabulary. To simplify our discussion, we treat a fault as a geometric surface in a body of rock. Rock adjacent to a fault surface is the **wall** of the fault, and the body of rock that moved as a consequence of slip on the fault is a **fault block.** If the fault is not vertical, you can distinguish between the **hanging-wall block,** which is the rock body above the fault plane, and the **footwall block,** which is the rock body below the fault plane (Figure 8.1a). This terminology is adopted from mining geology. Note that you cannot distinguish between a hanging wall and a footwall for a vertical fault.

To describe the attitude of a fault precisely, we measure the strike and dip (or dip and dip direction) of the fault (see Chapter 1). Commonly, geologists use adjectives such as steep, shallow, vertical, and so on, to convey an approximate image of fault dip (Table 8.1). Keep in mind that a fault is not necessarily a perfectly planar surface; it may curve and change attitude along strike and/or up and down dip. Where such changes occur, a single strike and dip is not sufficient to describe the attitude of the whole fault, and you should provide separate measurements for distinct segments of the fault. Faults whose dip decreases progressively with depth have been given the special name **listric faults.**

When fault movement occurs, one fault block slides relative to the other, which is described by the **net slip.** You can completely describe displacement by specifying the **net-slip vector,** which connects two formerly adjacent points that are now on opposite walls of the fault (Figure 8.3). To describe a net-slip vector, you must specify its magnitude and orientation (plunge and bearing, or rake on a plane), and the **sense of slip** (or shear sense). **Shear sense** defines the relative displacement of one wall of the fault with respect to the other wall; that is, whether one wall went up or down, and/or to the left or right of the other wall.

Like any vector, the net-slip vector can be divided into components. Generally we use the strike and dip of the fault as a reference frame for defining these components. Specifically, you measure the **dip-slip component** of net slip in the direction parallel to the

TABLE 8.1	DESCRIPTION OF FAULT DIP
Horizontal faults	Faults with a dip of about 0°; if the fault dip is between about 10° and 0°, it is called **subhorizontal.**
Listric faults	Faults that have a steep dip close to the Earth's surface and have a shallow dip at depth. Because of the progressive decrease in dip with depth, listric faults have a curved profile that is concave up.
Moderately dipping faults	Faults with dips between about 30° and 60°.
Shallowly dipping faults	Faults with dips between about 10° and 30°; these faults are also called *low-angle faults.*
Steeply dipping faults	Faults with dips between about 60° and 80°; these faults are also called *high-angle faults.*
Vertical faults	Faults that have a dip of about 90°; if the fault dip is close to 90° (e.g., is between about 80° and 90°), the fault can be called **subvertical.**

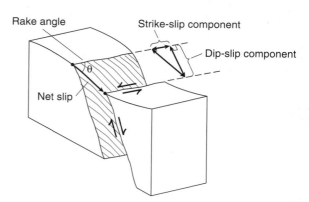

FIGURE 8.3 Block diagram sketch showing the net-slip vector with its strike-slip and dip-slip components, as well as the rake and rake angle.

dip direction, and the **strike-slip component** of net slip in the direction parallel to the strike. If the net-slip vector parallels the dip direction of the fault (within ~10°), the fault is called a **dip-slip fault;** if the vector roughly parallels the strike of the fault, the fault is called a **strike-slip fault.** If the vector is not parallel to either dip direction of the strike, we call the fault an

oblique-slip fault. As you can see in Figure 8.3, oblique-slip faults have both a strike-slip and a dip-slip component of movement.

We describe the shear sense on a dip-slip fault with reference to a horizontal line on the fault, by saying that the movement is **hanging-wall up** or **hanging-wall down** relative to the footwall. Hanging-wall down faults are called **normal faults,** and hanging-wall up faults are called **reverse faults** (Figure 8.4a and b). To define sense of slip on a strike-slip fault, imagine that you are standing on one side of the fault and are looking across the fault to the other side. If the opposite wall of the fault moves to your right, the fault is **right-lateral** (or **dextral**), and if the opposite wall of the fault moves to your left, the fault is **left-lateral** (or **sinistral;** Figure 8.4c and d). Note that this displacement does not depend on which side of the fault you are standing on. Finally, we define shear sense on an oblique-slip fault by specifying whether the dip-slip component of movement is hanging-wall up or down, and whether the strike-slip component is right-lateral or left-lateral (Figure 8.4e–h). Commonly, an additional distinction among fault types is made by adding reference to the dip angle of the fault surface; we recognize high-angle (>60° dip), intermediate-angle (30° to 60° dip), and low-angle faults (<30° dip). We provide descriptions of the basic fault types in Table 8.2, along with descriptions of other commonly used names (such as thrust and detachment).

You may be wondering where the terms "normal" and "reverse" come from. Perhaps normal faults were thought to be "normal" because the hanging-wall block appeared to have slipped down the fault plane, just like a person slips down a slide. It is a safe guess that geologists came up with the name "reverse fault" to describe faults that are the opposite of normal. Now you know!

We also distinguish among faults on the basis of whether they cause shortening or lengthening of the layers that are cut. Imagine that a fault cuts and displaces a horizontal bed marked with points *X* and *Y* (Figure 8.5a). Before movement, *X* and *Y* project to points *A* and *B* on an imaginary plane above the bedding plane. If the hanging wall moves down, then points *X* and *Y* project to *A* and *B'*. The length *AB'* is greater than the length *AB* (Figure 8.5b). In other words, movement on this fault effectively lengthens the layer. We call a fault which results in lengthening of a layer an **extensional fault.** By contrast, the faulting shown in Figure 8.5c resulted in a decrease in the distance between points *X* and *Y* (*AB* > *AB''*). We call a fault which results in shortening of a body of rock a **contractional fault.** Contractional faults result in

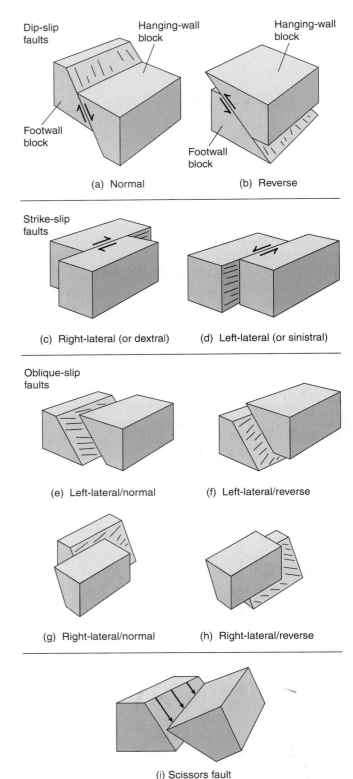

(a) Normal   (b) Reverse

(c) Right-lateral (or dextral)   (d) Left-lateral (or sinistral)

(e) Left-lateral/normal   (f) Left-lateral/reverse

(g) Right-lateral/normal   (h) Right-lateral/reverse

(i) Scissors fault

FIGURE 8.4 Block diagram sketches showing the different types of faults.

TABLE 8.2	TYPES OF FAULTS
Allochthon	The thrust sheet above a detachment is the allochthon (meaning that it is composed of *allochthonous* rock; i.e., rock that has moved substantially from its place of origin).
Autochthon	The footwall below a detachment is the autochthon; it is composed of *autochthonous* rock, or rock that is still in its place of origin.
Contractional fault	A contractional fault is one whose displacement results in shortening of the layers that the fault cuts, regardless of the orientation of the fault with respect to horizontal.
Décollement	The French word for detachment.
Detachment fault	This term is used for faults that initiate as a horizontal or subhorizontal surface along which the hanging-wall sheet of rock moved relative to the footwall. An older term "overthrust" is a regional detachment fault on which there has been a thrust sense of movement. Some detachments are listric, and on some detachments, regional normal-sense displacement occurs.
Dip-slip fault	The slip direction on a dip-slip fault is approximately parallel to the dip of the fault (i.e., has a rake between ~80° and 90°).
Extensional fault	An extensional fault is one whose displacement results in extension of the layers that the fault cuts, regardless of the orientation of the fault with respect to horizontal.
Normal fault	A normal fault is a dip-slip fault on which the hanging wall has slipped down relative to the footwall.
Oblique-slip fault	The slip direction on an oblique-slip fault has a rake that is not parallel to the strike or dip of the fault. In the field, faults with a slip direction between ~10° and ~80° are generally called oblique-slip.
Overthrust fault	This is an older term that you may find in older papers on faults, but is no longer used much today. The term is used for thrust faults of regional extent. In this context, "regional extent" means that the thrust sheet has an area measured in tens to hundreds of square km, and the amount of slip on the fault is measured in km or tens of km. Today, such faults are generally called regional detachments.
Par-autochthonous	If a fault block has only moved a small distance from its original position, the sheet is par-autochthonous (literally, relatively in place).
Reverse fault	A reverse fault is a dip-slip fault on which the hanging wall has slipped up relative to the footwall.
Scissors fault	On a scissors fault, the amount of slip changes along strike so that the hanging-wall block rotates around an axis that is perpendicular to the fault surface (Figure 8.4i).
Strike-slip fault	The slip direction on a strike-slip fault is approximately parallel to the fault strike (i.e., the line representing slip direction has a rake [pitch] in the fault plane of less than ~10°). Strike-slip faults are generally steeply dipping to vertical.
Transfer fault	A transfer fault accommodates the relative motion between blocks of rock that move because of the displacement on other faults.
Transform fault	In the preferred sense, transform faults are plate boundaries at which lithosphere is neither created nor destroyed. In a general sense, a transform fault links two other faults and accommodates the relative motion between the blocks of rock that move because of the displacement on the other two faults. However, we reserve the term *transfer fault* for this general type of displacement, independent of scale.

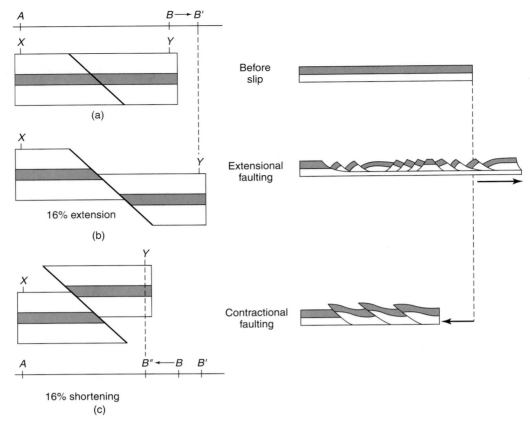

FIGURE 8.5 Extensional and contractional faulting. (a) Starting condition, (b) extension, and (c) contraction. Note the respective horizontal length changes.

duplication of section, as measured along a line that crosses the fault and is perpendicular to stratigraphic boundaries, whereas extensional faults result in loss of section. Generally, one can use the term "normal fault" as a synonym for an extensional fault, and the term "reverse fault" as synonym for contractional faults. But such usage is not always correct. Consider a normal fault that rotates during later deformation. In outcrop, this fault may have the orientation and sense of slip you would expect on a reverse fault, but, in fact, its displacement produced extensional strain parallel to layering.

8.2.2 Representation of Faults on Maps and Cross Sections

Because a fault is a type of geologic contact, meaning that it forms the boundary between two bodies of rock, faults are portrayed as a (heavy) line on geologic maps, like other contacts. We distinguish among different types of faults on maps through the use of symbols (Figure 8.6). For example, thrust faults are decorated with triangular teeth placed on the hanging-wall side of the trace. (Note that the teeth do not indicate the

direction of movement!) Normal faults, regardless of dip, are commonly portrayed by placing barbs on the hanging-wall block. We represent strike-slip faults on a map by placing arrows that indicate the sense of slip on either side of the fault (Figure 8.6c).

In cross sections, faults are also represented by a thick line (Figure 8.7). If the slip direction on the fault roughly lies in the plane of the cross section, then you indicate the sense of slip on the fault by oppositely facing half-arrows drawn on either side of the fault. If the movement on the fault is into or out of the plane of the section for a strike-slip fault, you indicate the sense of slip by drawing the head of an arrow (a circle with a dot in it) on the block moving toward you, and tail of an arrow (a circle with an X in it) on the block moving into the plane. If the movement is into or out of the page for a dip-slip fault, you place the map symbol (teeth for thrust faults and barbs for normal faults) for the fault on the hanging-wall block.

If a fault cuts across the contact between two geologic units, it must displace this contact unless the net-slip vector happens to be exactly parallel to the intersection line between the fault and the contact. The point on a map or cross section where a fault intersects

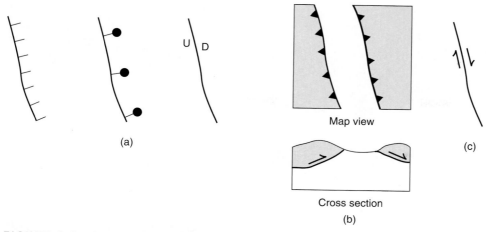

(a)

Map view

Cross section

(b)

(c)

FIGURE 8.6 Basic map symbols for (a) normal fault, (b) thrust fault, and (c) strike-slip fault.

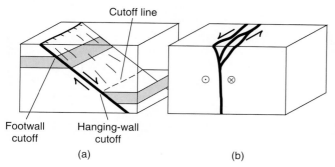

Cutoff line

Footwall
cutoff

Hanging-wall
cutoff

(a)

(b)

FIGURE 8.7 Block diagrams showing the different symbols for representing (a) dip-slip faults and (b) strike-slip faults (here, left-lateral). In (a) we also mark footwall and hanging-wall cutoffs.

a preexisting contact is called a **cutoff,** and in three dimensions (Figure 8.7), the intersection between a fault and a preexisting contact is a **cutoff line.** If the truncated contact lies in the hanging-wall block, the truncation is a **hanging-wall cutoff,** and if the truncated contact lies in the footwall, it is a **footwall cutoff.**

When combining map and cross-sectional surfaces with topography, we create a more realistic block diagram, giving us a three-dimensional representation of a region's geology. Consider an area that is characterized by a low-angle reverse faulting (a thrust). Where erosion cuts a hole through a thrust sheet, exposing rocks of the footwall, the hole is a **window** and the teeth are drawn outwards from the hole (Figure 8.8). An

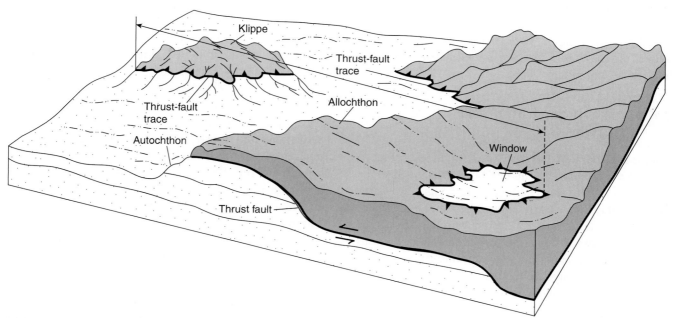

Klippe

Thrust-fault
trace

Thrust-fault
trace

Allochthon

Autochthon

Window

Thrust fault

FIGURE 8.8 Block diagram illustrating klippe, window (or fenster), allochthon (gray), and autochthon (stippled) in a thrust-faulted region. Note that the minimum fault displacement is defined by the farthest distance between thrust outcrops in klippe and window.

FIGURE 8.9 Chief Mountain in Glacier National Park (USA) is an example of a klippe. The Lewis Thrust marks the contact between resistant Precambrian rock in the hanging wall and Cretaceous shale and sandstone in the footwall of this erosional remnant.

isolated remnant of a thrust sheet surrounded by exposures of the footwall is a **klippe;** this is marked by a thrust-fault symbol with the teeth pointing inwards. An imposing example of this structural geometry is Chief Mountain in Glacier Park, Montana (Figure 8.9), where old basement rocks are emplaced on young, low-grade sediments.

8.2.3 Fault Separation and Determination of Net Slip

Imagine a **marker horizon** (a distinctive surface or layer in a body of rock, such as a bed) that has been cut and offset by slip on a fault (Figure 8.10a). We define **fault separation** as the distance between the displaced parts of the marker horizon, as measured along a specified line. Separation and net slip are not synonymous, unless the line along which we measure the separation parallels the net-slip vector. The separation for a given fault along a specified line depends on the attitude of the offset marker horizon. Therefore, separation along a specified line is not the same for two nonparallel marker horizons (Figure 8.10b). Fault separation is a

little difficult to visualize, so we will describe different types of fault separation with reference to Figure 8.10c, which shows an oblique-slip fault that cuts a steeply dipping bed. We define the types of separation illustrated in this figure in Table 8.3.

With the terms of Table 8.3 in mind, note that horizontal beds cut by a strike-slip fault have no strike separation and vertical beds cut by a dip-slip fault have no dip separation. If the fault cuts ground surface, this surface itself is a marker horizon for defining vertical separation, and linear features on the ground (e.g., fences, rows of trees, roads, railroads, river beds) serve as markers for defining horizontal separation. Note that Table 8.3 also defines **heave** and **throw,** which are old terms describing components of dip separation (Figure 8.10a).

In order to completely define the net-slip vector, you must specify its absolute magnitude, the direction of displacement (as a plunge and bearing) and the sense of slip. If you are lucky enough to recognize two points now on opposite walls of the fault that were adjacent prior to displacement, sometimes referred to as **piercing points,** then you can measure net slip

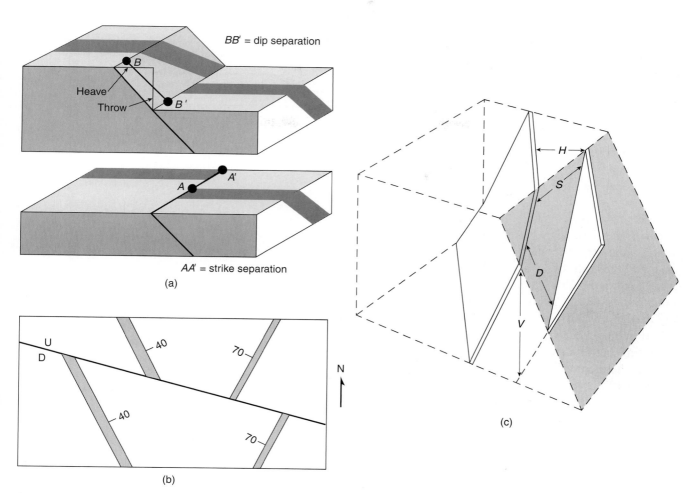

FIGURE 8.10 (a) Block diagrams showing dip separation, strike separation, heave, and throw. (b) Map view showing how separation depends on the orientation of the offset layer. The two dikes shown here dip in different directions and have, therefore, different strike separations. (c) Block diagram illustrating horizontal (*H*) and vertical (*V*) separation, as well as dip (*D*) and strike (*S*) separation.

TABLE 8.3	FAULT SEPARATION AND FAULT-SEPARATION COMPONENTS (FIGURE 8.10)
Dip separation (D)	The distance between the two bed/fault intersection points as measured along a line parallel to the dip direction.
Heave	The horizontal component of dip separation.
Horizontal separation (H)	The offset measured in the horizontal direction along a line perpendicular to the offset surface.
Stratigraphic separation	The offset measured in a line perpendicular to bedding.
Strike separation (S)	The distance between the two bed/fault intersection points as measured along the strike of the fault.
Throw	The vertical component of dip separation.
Vertical separation (V)	The distance between two points on the offset bed as measured in the vertical direction. Vertical separation is the separation measured in vertical boreholes that penetrate through a fault.

directly in the field. For example, if you observe a fence on the ground surface that has been offset by a fault, then you can define net slip, because the intersection of the fence with the ground defines a line, and the intersection of this line with the walls of the fault defines two previously adjacent points.

More commonly, however, you won't be lucky enough to observe an offset linear feature, and you must calculate the net-slip vector from other information. This can be done by measurement of (a) separation, along a specified line, of the intersection between a single marker horizon and the fault, plus information on the direction of slip; (b) separation, along two nonparallel lines, of the intersection between a single plane and the fault; or (c) separation, along a specified line, of two nonparallel marker horizons. Look at any standard structural geology methods book for an explanation of how to carry out such calculations. In the relatively rare cases where an earthquake-generating fault cuts ground surface, you can directly measure the increment of displacement accompanying a single earthquake. Generally, however, the displacement that you measure when studying ancient faults in outcrop is a **cumulative displacement** representing the sum of many incremental offsets that occurred over a long period of time.

If you do not have sufficient information to determine the net slip, you can obtain valuable information about fault displacement by searching for slip lineations and shear-sense indicators. **Slip lineations** are structures on the fault that form parallel to the net-slip vector for at least the last increment of movement on the fault and, possibly, for accumulated movement during progressive deformation. **Shear-sense indicators** are structures on the fault surface or adjacent to the fault surface that define the direction in which one block of the fault moved with respect to the other. Slip lineations alone define the plunge and bearing of the net-slip vector, and with shear-sense indicators they define the direction in which the vector points. Such information can help you interpret the tectonic significance of a fault, even if you don't know the magnitude of displacement across it. We'll discuss types of slip lineations and shear-sense indicators, and how to interpret them, later in this chapter, after we have reviewed the process that leads to their formation.

The magnitude of the net-slip vector (i.e., fault displacement) on natural faults ranges from millimeters to thousands of kilometers. For example, about 600 km of net slip occurred on the oldest part of the San Andreas fault in California. In discussion, geologists refer to faults with large net slip as **major faults** and faults with small net slip as **minor faults.** Keep in

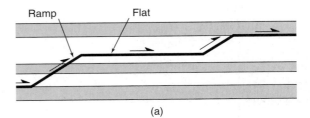

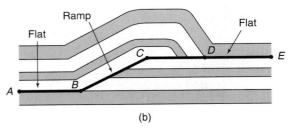

FIGURE 8.11 (a) Cross section showing the geometry of ramps and flats along a thrust fault. The fault geometry is shown prior to displacement on the fault. (b) Cross section illustrating hanging-wall and footwall flats and ramps. Segment *AB* is a hanging-wall flat on a footwall flat. Segment *BC* is a hanging-wall flat on a footwall ramp. Segment *CD* is a hanging-wall ramp on a footwall flat, and segment *DE* is a hanging-wall flat on a footwall flat.

mind that such adjectives are relative, and depend on context; a major fault on the scale of an outcrop may be a minor fault on the scale of a continent.

8.2.4   Fault Bends

As we mentioned earlier, fault surfaces are not necessarily planar. It is quite common, in fact, for the attitude of a fault to change down dip or along strike. In some cases, the change is gradual. For example, the dip of "down-to-Gulf" faults (see Chapter 2) along the coastal plain of Texas typically decreases with depth, so that the fault overall has a concave-up shape making these structures **listric faults.** Other faults have wavy traces because their attitude changes back and forth.

If the dip and/or strike of a fault abruptly changes, the location of the change is called a **fault bend.** Dip-slip faults that cut across a stratigraphic sequence in which layers have different mechanical properties typically contain numerous stratigraphically controlled bends that make the trace of the fault in cross section resemble a staircase. Some fault segments run parallel to bedding, called **flats,** and some cut across bedding, called **ramps** (Figure 8.11a). If the fault has not been folded subsequent to its formation, flats are (sub)horizontal, whereas ramps have dips of about 30° to 45°. Note that, as shown in Figure 8.11b, a segment of a

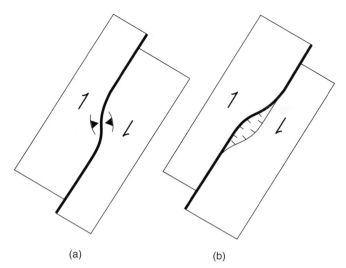

FIGURE 8.12 Map-view illustrations of (a) a restraining bend and (b) a releasing bend along a right-lateral strike-slip fault.

fault may parallel bedding in the footwall, but cut across bedding in the hanging wall. Thus, when describing stairstep faults in a stratified sequence you need to specify whether a fault segment is a ramp or flat with respect to the strata of the hanging wall, foot-wall, or both.

Fault bends (or steps) along strike-slip faults cause changes in the strike of the fault. To describe the orientation of such fault bends, imagine that you are straddling the fault and are looking along its strike; if the bend moves the fault plane to the left, you say the fault steps to the left, and if the bend moves the fault plane to the right, you say that the fault steps to the right. Note that the presence of bends along a strike-slip fault results in either contraction or extension across the step, depending on its geometry. Locations where the bend is oriented such that blocks on opposite sides of the fault are squeezed together are **restraining bends,** whereas locations where the bend is oriented such that blocks on opposite sides of the fault pull away from each other are **releasing bends** (Figure 8.12).[2] Where movement across a segment of a strike-slip fault results in some compression, we say that **transpression** is occurring across the fault, and where movement results in some extension, we say that **transtension** is occurring across the fault. Note that a step to the left on a right-lateral fault yields a restraining bend, whereas a step to the right on a right-lateral fault yields a releasing bend. Try to make up the rules for a left-lateral fault yourself.

---

[2]The terms "restraining bends" and "releasing bends" can also be applied to steps along dip-slip faults that connect two parallel fault segments.

Natural examples of these structures, such as along the San Andreas Fault of California, will be discussed in Chapter 18.

8.2.5 Fault Terminations and Fault Length

Faulting develops at all scales, from microscopic faults that offset the boundaries of a single grain, to mega-scopic faults that cut laterally across thousands of kilometers of crust. But even the biggest faults do not extend infinitely in all directions. They terminate in several ways.

Faults terminate where cut by younger structures, such as another fault, an unconformity, or an intrusion (Figure 8.13a). Application of the principle of cross-cutting relationships allows you to determine the relative age of faults with respect to the structures that cut them. Some faults link to other faults while both are active (Figure 8.13b). For example, fault splays diverge from a larger fault, and faults in an anastomosing array merge and diverge along their length. Where a fault does not terminate against another structure, it must die out, meaning that the magnitude of displacement decreases along the trace of the fault, becoming zero at its tips. In some cases, a fault splits into numerous splays near its end, thereby creating a fan of small fractures called a **horsetail** (point *B* in Figure 8.13b), or it may die in an array of pinnate fractures. Alternatively, the deformation associated with the fault dies out in a zone of ductile deformation (e.g., folding or penetrative strain; point *C* in Figure 8.13b). The boundary between the slipped and unslipped region at the end of a fault is the **tip line** of the fault (Figure 8.14a). Faults that you can map today in the field terminate at the ground surface either because the fault intersected the ground when it moved or because it has been subsequently exposed by erosion. If the fault intersected the ground surface while it was still active, it is an **emergent fault** (Figure 8.14a), but if it intersects the ground surface only because the present surface of erosion has exposed an ancient, inactive fault, it is an **exhumed fault.** Exhumed and emergent faults must die out along their strike, unless they terminate at another structure; their tip line intersects the ground surface at a point. A fault that dies out in the subsurface, and thus does not intersect the ground surface, is called a **blind fault** (Figure 8.14b).

Whereas the length of some faults is limited by their intersection with, or truncation by, other structures, for many faults the trace length changes with time as the fault evolves. To visualize this process, imagine a fault that grows outward. At a given instant of time, slip has

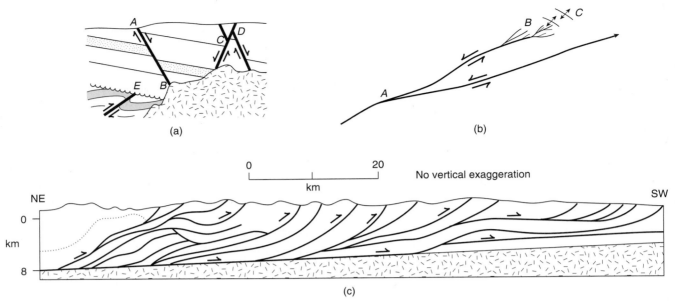

FIGURE 8.13 (a) Cross-sectional sketch showing various types of fault terminations. The fault terminates at the ground surface at point *A*; at point *B*, the fault has been cut by a pluton; at *C* and *D*, one fault cuts another; at *E*, the fault was eroded at an unconformity. (b) Termination of a fault by merging with another fault (at point *A*), or by horsetailing (at point *B*) and dying out into a zone of ductile deformation (at point *C*). (c) A series of ramps merging at depth with a basal detachment.

occurred where the fault surface already exists, but there is no slip beyond the tip of the fault (Figure 8.15a). A little later, after more fault-tip propagation, there is increased slip in the center of the fault (Figure 8.15b). As a consequence, the displacement changes along the length of the fault and the magnitude of displacement must be less than the length of the fault. Considering this relation, we might expect a general relationship between fault length and displacement: the longer the

fault trace, the greater the displacement. Indeed, recent work supports this idea, though the details remain controversial. Faults that are meters long display offsets typically on the order of centimeters or less, whereas faults with lengths on the order of tens of kilometers have typical offsets on the order of several hundreds of meters. Figure 8.15c shows a plot of fault length versus offset, based on examination of thousands of faults that occur in a variety of lithologies and range in length

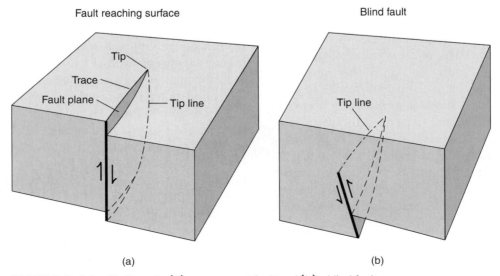

FIGURE 8.14 Tip lines for (a) an emergent fault and (b) a blind fault.

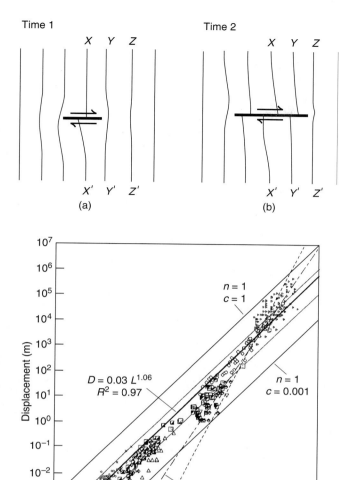

Time 1 Time 2

X Y Z   X Y Z

X' Y' Z' X' Y' Z'

(a)   (b)

$D = 0.03\, L^{1.06}$
$R^2 = 0.97$

$n = 1$
$c = 1$

$n = 1$
$c = 0.001$

$n = 1.5$

$n = 2$

Displacement (m)

Fault length (m)

(c)

FIGURE 8.15 (a, b) Map view illustrating that displacement on a fault grows as the fault length increases. At time 1, the short fault only offsets marker line *XX'* by a small amount. At time 2, the fault has grown in length, and marker line *XX'* has been offset by a greater amount. Note that the displacement decreases toward the end of the fault. (c) Log–log plot showing the apparent relationship between fault length (*L*) and fault displacement (*D*): $D = c \cdot L^n$. The exponent, *n*, is called the *fractal dimension*. Various fits are possible, but a general relationship is $D = 0.03\, L^{1.06}$, suggesting an approximate displacement–length ratio of about 0.03.

from centimeters to hundreds of kilometers. The points can be fitted to a ~45° sloping band, suggesting that, given a knowledge of fault length, we can predict displacement (or vice versa) independent of the properties of the material. Debate continues as to whether a single relationship is appropriate for both regional and mesoscopic faults, as the best-fitting straight line seemingly underestimates displacement for shorter faults and overestimates for longer faults. Nevertheless, a convenient rule of thumb emerges—that fault displacement is about 3% of fault length—based on the values in Figure 8.15. Other studies indicate similar scaling properties for fault gouge/cataclasite width; Table 8.4 lists approximate relationships for fault-zone characteristics.

8.3 CHARACTERISTICS OF FAULTS AND FAULT ZONES

8.3.1 Brittle Fault Rocks

Faulting involves either shear fracturing of a previously intact rock, in which case a multitude of cracks coalesce, or slip on a preexisting fracture, which may lead to formation of new off-plane fractures and fault splays. Thus, the process of brittle faulting tends to break up rock into fragments, producing **brittle fault rock.** We classify brittle fault rock based on whether it is **cohesive** or **noncohesive** (i.e., whether or not the fragments comprising the fault rock remain stuck together to form a coherent mass without subsequent cementation or alteration) and on the *size* of the fragments that make up the fault rock. Table 8.5 summarizes the principal terms that we use to describe brittle fault rocks.

Creation of a random array of nonsystematic mesoscopic fractures that surround angular blocks of rock creates **fault breccia** (Figure 8.16a). In general, breccias have *random fabrics,* meaning they do not contain a distinctive foliation. Continued displacement across

TABLE 8.4	FIRST-ORDER RELATIONSHIPS BETWEEN FAULT PARAMETERS		
	Length	Displacement	Fault Zone Width
Length	—	10^2	10^4
Displacement	10^{-2}	—	10^2
Fault Zone Width	10^{-4}	10^{-2}	—
Row fault property equals value times column property; for example, width = 0.01 × displacement ($W = 0.01 \times D$). *(From Scholz, 2002)*			

TABLE 8.5	CLASSIFICATION OF BRITTLE FAULT ROCK

Noncohesive Brittle Fault Rocks

Fault gouge	Rock composed of material whose grain size has been mechanically reduced by pulverization. Grains in fault gouge are less than about 1 mm in diameter. Like breccia, gouge is noncohesive. Shearing of gouge along a fault surface during progressive movement may create foliation within the gouge. Clay formed by alteration of silicate minerals in fault zones may be difficult to distinguish from true gouge.
Indurated gouge	Fault gouge that has been cemented together by minerals precipitated from circulating groundwater.
Fault breccia	Rock composed of angular fragments of rock greater than about 1 mm, and as much as several m across; fault breccia is noncohesive.
Vein-filled breccia	Fault-breccia blocks that are cemented together by vein material. Another term, *indurated breccia*, is synonymous.

Cohesive Brittle Fault Rocks

Pseudotachylyte	A glass or microcrystalline material that forms when frictional heating melts rock during slip on a fault. Pseudotachylyte commonly flows into cracks between breccia fragments or into cracks penetrating the walls of the fault. In special cases, pseudotachylyte may be several m thick (e.g., impact sites), but generally it is mm to cm in thickness.
Argille scagliose	A fault rock that forms in very fine-grained clay- or mica-rich rock (e.g., shale or slate) and is characterized by the presence of a very strong wavy anastamosing foliation. As a consequence, the rock breaks into little scales or platy flakes.
Cataclasite	A cohesive fault rock composed of broken, crushed, or rolled grains. Unlike breccia, it is a solid rock that does not disintegrate when struck with a hammer.

(a)

(b)

FIGURE 8.16 (a) Fault breccia from the Buckskin detachment (Battleship Peak, Arizona, USA). (b) Banded clay gouge from the Lewis Thrust (Alberta, Canada).

the fault zone may crush and further fragment breccia, and/or may break off microscopic asperities protruding from slip surfaces in the fault zone, thereby creating a fine-grained rock flour that we call **fault gouge** (Figure 8.16b). Gouge and (micro)breccia are noncohesive fault rocks, meaning that they easily fall apart when collected at a fault zone or hit with a hammer.

The network of fractures between fragments in breccia and gouge allows groundwater to pass through the fault zone. Minerals like quartz or calcite may precipitate out of the groundwater, thereby cementing together rock fragments in the fault zone. As a result, breccia and gouge become **indurated,** meaning that the fragments are cemented together. In coarse breccia, the cement typically also fills veins of euhedral or blocky crystals in the open spaces between rock fragments, resulting in formation of a **vein-filled breccia** (Figure 8.16b). Circulating groundwater may also have the effect of causing intense alteration and new growth of minerals in the gouge or breccia zone. Alteration rates in fault zones tend to be greater than in intact rock, because fragmentation of rock body increases the net area of reactive surfaces. As a consequence, some minerals (e.g., feldspar) transform into clay (Figure 8.16b). The layer of clay that develops in some fault zones can act as an impermeable barrier, or **seal,** to further fluid

movement. In olivine-rich rocks, such as basalt and peridotite, reaction of fault rock with water yields the mineral serpentine. As we mentioned earlier, clay and serpentine are relatively weak minerals, so their presence along a fault may allow it to slip at lower frictional stresses than it would if the original minerals were present.

Cataclasite is a cohesive brittle fault rock that differs from gouge or breccia in that the fragments interlock, allowing the fragmented rock to remain coherent even without cementation. Cataclasites generally have random fabrics (i.e., no strong foliation or lineation). Some geologists use subcategories of cataclasite, based on the proportion of matrix in the rock. In **protocataclasite,** 10–50% of the rock is matrix; in **cataclasite** (*sensu stricto*) 50–90% is matrix; and in **ultracataclasite** 90–100% is matrix.

Table 8.5 also lists two less common types of fault rock, pseudotachylyte and argille scagliose. **Pseudotachylyte** (from the prefix *pseudo-*, which means "like," and the noun tachylyte, which is a type of volcanic glass) is glass or very finely crystalline material that forms when frictional sliding generates enough heat to melt the rock adjacent to the fault (Figure 8.17). Such conditions occur during earthquakes. Because rock is not a good conductor, the heat generated by

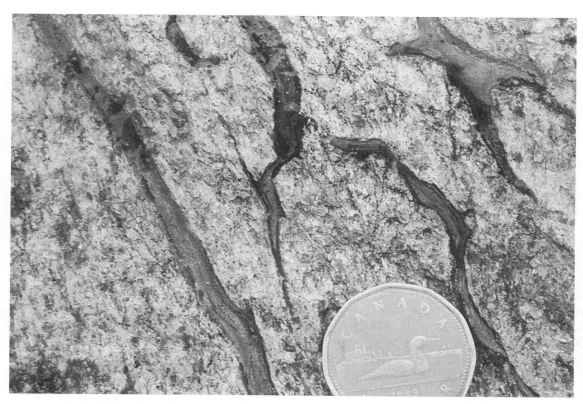

FIGURE 8.17 Pseudotachylyte (dark, wispy bands) near the Grenville Front (Ontario, Canada); looney for scale. *(Courtesy of J. Magloughlin)*

(a)

frictional sliding cannot flow away from the fault, and temperatures in the fault zone quickly become very high (>1000°C). Melt formed in such a setting squirts into cracks and pores in the fault walls where it cools so quickly that it solidifies into a glass. **Argille scagliose** refers to a strongly foliated fault rock formed by pervasive shearing of a clay-rich or very fine-grained mica-rich lithology such as shale or slate. In argille scagliose ("scaly clay" in Italian) foliation planes (microscopic shear surfaces) are anastomosing and very shiny, yielding a rock that has the overall appearance of a pile of oyster shells. Argille scagliose can develop under nonmetamorphic conditions (e.g., at the base of a sedimentary mélange) or in low-grade metamorphic conditions. Scaly fabrics also form in other fine-grained lithologies, such as coal and serpentinite.

8.3.2 Slickensides and Slip Lineations

Displacement on a fault in the brittle field involves frictional sliding and/or pressure solution slip. Each process yields distinctive structures (slickensides and slip lineations) on the fault surface, which may provide information about the direction of net slip and, in some cases, the shear sense of slip during faulting.

If slip on a fault takes place by frictional sliding, asperities on the walls of the fault break off and/or plow into the opposing surface and wear down. As a result, the two walls of the fault may become smoother and, in some cases, attain a high polish. Fault surfaces that have been polished by the process of frictional sliding are called **slickensides** (Figure 8.18). Slickensides form either on the original wall rock of the fault, or on the surface of a thin layer of gouge/cataclasite.

(b)

FIGURE 8.18 (a) Shiny slickensided surface in Paleozoic strata of the Appalachians (Maryland, USA); coin for scale. (b) Slip fibers on a fault surface, showing steps that indicate sense of shear; compass for scale.

Some asperities on one wall of the fault plow into the surface of the other wall, thereby creating **groove lineations** on the slickenside, resulting in formation of a **lineated slickenside,** also called **slickenlines.** These lineations resemble the glacial striations created when rocks entrained in the base of a moving glacier scratch across bedrock, though they are much smaller.

Groove lineations are not the only type of lineation that forms on brittle faults. On fault surfaces coated with fine-grained material, fault slip may mold gouge into microscopic linear ridges that, along with grooves, create a lineation visible on the fault surface. Also, some fault surfaces initiate with small lateral steps, whose presence gives the fault a corrugated appearance. These **corrugations** resemble grooves, but can be longer than the total displacement on the fault. The origin of corrugations is not well understood.

Some fault surfaces are coated by elongate **fibers** of vein minerals (typically quartz, calcite, or chlorite), whose long axes lie subparallel to the fault surface (Figure 8.19a). These fibers grow incrementally by the "crack-seal" deformation mechanism, or by solution mass transfer through a fluid film along the fault surface. Thus, fiber formation does not always involve brittle rupture along the fault surface. Whether frictional sliding is accompanied by fiber growth depends on the strain rate and fluid conditions during faulting. Fibers form at smaller strain rates and require the presence of water films, and typically form in imbricate sheets (Figure 8.19b). As movement continues, multiple sheets of fibers may develop on top of one another, so that a vein up to several centimeters thick eventually develops along the fault plane. In relatively thick veins (> 2 or 3 cm), the internal portion of the vein may consist of blocky growth, which forms either by recrystallization of earlier-formed fibers, or by precipitation of euhedral crystals in gaps along the fault surface.

Slip lineations trend parallel to an increment of displacement on a fault, and thus allow you to determine whether the increment resulted in strike-slip, dip-slip, or oblique-slip offset. If all movement on the fault has been in the same direction, then the slip lineation defines the orientation of the net-slip vector. You must use caution when interpreting slip lineations, however, because they may only define the last few increments of displacement on a fault surface on which slip in a range of directions occurred. Moreover, the last increment of frictional sliding sometimes erases grooves formed during earlier increments, and formation of one sheet of fiber lineations may cover and obscure preexisting sheets of fiber lineations. If more than one set of lineations are preserved on a fault surface, you can study them to distinguish among multiple nonparallel slip increments.

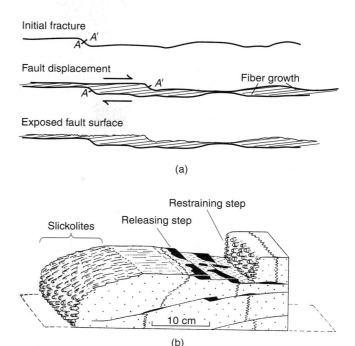

FIGURE 8.19 (a) Illustration of the growth of slip fibers along a fault, and (b) block diagram illustrating steps along a fiber-coated fault surface. Restraining steps become pitted by pressure solution, releasing steps become the locus of vein growth, and oblique restraining steps become slickolites.

Structures on a fault surface may provide constraints on the shear sense. For example, as frictional sliding occurs, an original irregularity on a slickenside becomes polished more smoothly on the upslip side, and remains rougher on the downslip side, and subtle steps may develop on the fault surface. These features create an anisotropy on a slickenside; the surface feels smoother as you slide your hand in the shear direction as opposed to sliding your hand in the direction opposite to the shear direction. However, anisotropy on a slickenside is subtle and may be ambiguous because the intersection of pinnate fractures with the fault surface may create steps that face in the opposite direction to the steps formed by differential polishing.

Sheets of mineral fibers formed during slip on a fault provide a more reliable indication of shear sense. Because the fibers composing the sheets form at a low angle to the fault surface, fiber sheets tend to overlap one another like shingles, and tilt away from the direction of shear. Features developed on mesoscopic steps along a fault that moved by pressure-solution slip may also define shear sense. Restraining steps oppose movement on the fault, and therefore become pitted by pressure solution. Pit axes on the steps are roughly parallel to the net-slip vector, and thus are subparallel to the fault surface. If the restraining-step face is not perpendicular to the fault surface, its pit axes are oblique to the step face. Surfaces containing such oblique pits look like a cross between a stylolite and a slip lineation, and thus some authors call them **slickolites** (Figure 8.19b). Releasing steps, at which the opposing walls of the fault pull away from one another, typically become coated with fibrous veins. The long axes of the fibers of these veins parallel the pit axes on releasing steps along the same fault.

8.3.3 Subsidiary Fault and Fracture Geometries

Fault zones consist of one or more major faults, along with an array of subsidiary faults, including both discrete smaller faults that occur within a larger fault zone (and may anastomose with one another) and fault splays that branch off. Such subsidiary faults may initiate when the primary rupture splits into more than one surface during its formation in intact rock, when numerous subparallel faults initiate simultaneously in a fault zone, when conjugate shear fractures develop at an angle to the principal fault, or when numerous preexisting surfaces in the fault zone reactivate during a deformation event.

In the case of emergent strike-slip faults, like the San Andreas Fault, a particularly interesting array of subsidiary faults, known as R- (or Riedel) and P-shears, develops. To picture how these develop, imagine an experiment in which you place a thin layer of clay over two wooden blocks, and then shear one block horizontally past the other (Figure 8.20a). The clay develops ductile strain before any fractures appear. When fracturing in the clay begins to develop, the first fractures are short shear fractures that are inclined at an angle to the trace of the throughgoing fault that eventually forms (Figure 8.20b). These short fractures are called *R*- or **Riedel shears.**[3] Generally, you will find two distinct sets of Riedel shears (R and R′) that together define a conjugate pair. The bisector of the acute angle between conjugate R- and R′-shears reflects the local orientation of σ_1 adjacent to the future fault. As shear continues, a third set of fractures, called P-shears, develops. P-shears link together the previously formed Riedel shears, and, eventually, a throughgoing fault zone consisting of linked R-, R′-, and P-shears develops. Whether subsidiary shears also form at depth in the earth remains debatable. In fact, some geologists suggest that they only form when a weak layer is sheared by relative displacement of

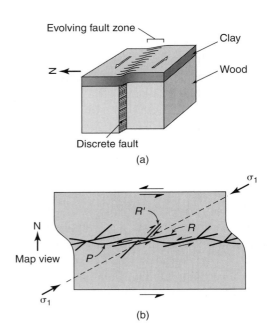

FIGURE 8.20 Growth of R-, R′-, and P-shears. (a) Schematic diagram illustrating a layer of clay that deforms when underlying blocks of wood slide past one another. (b) Map view of the top surface of the clay layer, illustrating the orientation of Riedel (R), conjugate Riedel (R′), and P-shears. Note that the acute bisector of the R- and R′-shears is parallel to the remote σ_1 direction.

two stronger blocks on either side or below, much as in our experiment.

Tensile fractures can also develop in association with faulting, which may occur in the wall rock, or in the fault zone itself. A series of parallel tensile fractures that forms within a fault zone defines an **en echelon array** or **stepped array** that tends to dilate and become veins (see Chapter 7). Typically, stepped veins in a fault zone initiate at an angle of about 45° to the zone boundary and rotate with the direction of shear. With progressive displacement of the fault-zone walls, the earlier-formed parts of the veins rotate, but new vein increments initiate at an angle of about 45° to the walls, producing a sigmoidal shape whose sense of rotation defines the sense of shear on the fault (see Figure 7.23).

8.3.4 Fault-Related Folding

Faults and folds commonly occur in the same outcrop. The spatial juxtaposition of these structures, one brittle and one ductile, may seem paradoxical at first. How can rock break at the same time that it distorts ductilely? One explanation for the juxtaposition of faults and folds in an outcrop is that these structures formed at different times under different pressure and temperature conditions. For example, imagine that the folds

---

[3]After the geologist who first discussed this relationship.

formed 500 million years ago at a depth of 15 km in the crust, but, once formed, the rock body containing the folds was unroofed and moved to shallow depths, where a later, and totally separate, deformation event created brittle faults in the body. Typically, in such examples, the faults cut across the preexisting folds and are not geometrically related to the folds. But in many instances it is clear from the spatial and geometric relation between folds and faults that the two structures formed together during the same deformation event. Folds that form in association with faults are called **fault-related folds.**

From earlier experiments we learned that the transition between brittle and ductile deformation can depend on strain rate (Chapter 5; recall the behavior of Silly Putty®). If ductile strain (e.g., folding) in a rock body develops fast enough to accommodate regional deformation, then differential stress does not get very high in the rock and faulting need not occur. But if regional deformation cannot be accommodated sufficiently fast by ductile strain, then differential stress builds until it exceeds the failure strength of the rock or the frictional strength of an existing fracture, and faulting occurs. Alternatively, strain rate may vary with position in a rock body, so that while one region in a rock body faults, another region in the rock body folds. Finally, strain rate and differential stress magnitude can vary with time at a given location during the same overall deformation event, so that episodes of faulting and folding may alternate at the location.

In this section, we introduce several types of folds that develop in association with faults, but we again return to the subject of fault-related folding when we discuss aspects of regional deformation (Chapters 17 and 18). Imagine a horizontally stratified rock that undergoes shortening due to regional compression. Initially, the ends of the body move toward one another very slowly. If, during this stage, differential stress in the body is lower than the shear failure strength of the rock, the layers will fold. Such folding may yield merely a gentle flexure of adjacent beds, or a pronounced asymmetric anticline-syncline pair (Figure 8.21). If, at a later time, folding can no longer accommodate the internal displacement of the block, the differential stress magnitude in the block increases and faulting initiates. Stress buildup leading to faulting may reflect a change in the strain rate or may reflect locking up of the folds. By locking up, we mean that the change in the geometry of the beds resulting from fold formation makes continued folding more difficult (i.e., like a stick, the layers can be bent only so far before they break). Note that the sense

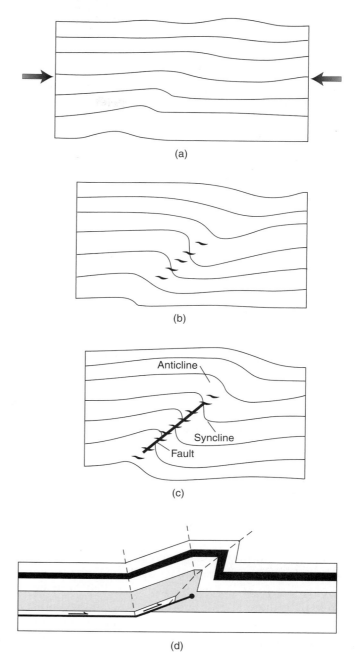

FIGURE 8.21 Progressive development of fault-related folding in a stratified sequence. (a) A small flexure develops during shortening of the layers, and a pronounced anticline-syncline pair develops. (b) *En echelon* (or stepped) gashes form in the fold. (c) A fault breaks through the fold, cutting through a gentle flexure. (d) Geometry of a fault-propagation fold.

of asymmetry of these folds[4] reflects the sense of shear on the fault. In some cases, a fault cuts through the fold along the hinge of the fold, whereas in other cases, the fault breaks through the limbs of the folds between the adjacent anticline and syncline (at the

---

[4]Sometimes called *fault-inception folds.*

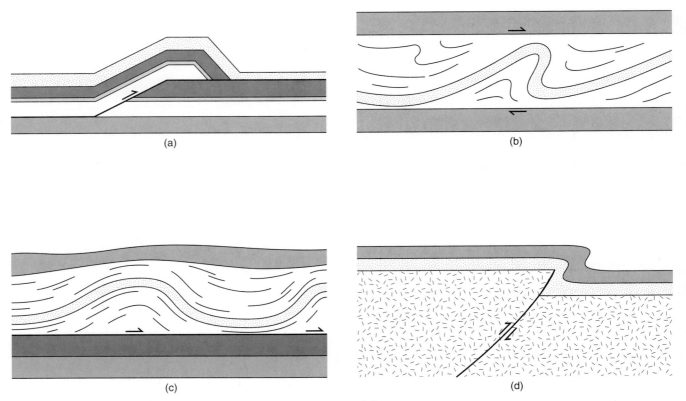

FIGURE 8.22 Fault-related folds. (a) Fault-bend fold on a thrust. (b) Folding in a fault zone. (c) Detachment fold. (d) Drape fold over faulted basement.

inflection surface of the fold); these are called **fault-propagation folds** (Figure 8.21d).

Bends in a fault cause folding of strata that move past the bend during displacement, because the moving layers must accommodate the bends without gaps. Folds that form in this manner are called **fault-bend folds.** Fault-bend folds form in association with all types of faults, but most of the literature concerning them pertains to examples developed along dip-slip faults (Figure 8.22a).

A third situation in which folding accompanies faulting occurs where a sequence of interlayered weak and strong rock layers (e.g., interbedded shale and sandstone) is caught between the opposing walls of a fault zone (Figure 8.22b). In such a circumstance, displacement of the rigid fault-zone walls with respect to one another causes the intervening rock to fold, much like a carpet that is caught between the floor and a sliding piece of furniture. Folds in fault zones formed by such a process tend to be very asymmetric. Typically, the hinges of folds within the fault zone are initially perpendicular to the overall shear direction, but with continued displacement, they may be bent, eventually curving into parallelism with the shear direction. Fault zones in which such folding occurs are effectively ductile shear zones (see Chapter 12), but we mention them here because this deformation can occur in association with brittle sliding.

The asymmetry of fault-zone folds provides a clue to the sense of shear on associated faults. However, because the hinges of fault-zone folds change during progressive shear, the geometry of a single fold may not necessarily provide the net-slip vector. You can define its direction if you measure a slip lineation on the fault surface, but if such lineations are not present, you may be able to obtain this information by measuring numerous folds with a range of hinge trends using a technique called the **Hanson slip-line method.**[5] We leave a description of this method for laboratory books on structural geology.

The sheet of rock above a detachment fault may deform independently of the rock below. The resulting folds are called **detachment folds** (Figure 8.22c). Shear on the detachment fault accommodates the contrast in strain between the folded hanging-wall and the unfolded footwall.

---

[5]Named after its originator.

| TABLE 8.6 | SHEAR-SENSE INDICATORS FOR BRITTLE FAULTS AND FAULT ZONES |

Offset markers	You can define shear sense if you are able to define the relative displacement of two piercing points on opposite walls of the fault, or can calculate the net-slip vector based on field study of the separation of marker horizons.
Fault-related folds	The sense of asymmetry of fault-related folds defines the shear sense. Typically, fault-inception folds verge in the direction of shear (see Chapter 11 for a definition of fold vergence). If the hinges of folds in a fault zone occur in a range of orientations, you may need to use the Hanson slip-line method to determine shear sense. Note that the asymmetry of rollover folds relative to shear sense is opposite to that of other fault-related folds.
Fiber-sheet imbrication	The imbrication of slip-fiber sheets on a fault provides a clear indication of shear sense. Fiber sheets tilt away from the direction of shear.
Steps on slickensides	Microscopic steps develop along slickensided surfaces. Typically the face of the step is rougher than the flat surface. However, slickenside steps may be confused with the intersection between pinnate fractures and the fault, giving an opposite shear sense.
En echelon veins	*En echelon* veins tilt toward the direction of shear. If the veins are sigmoidal, the sense of rotation defines the shear sense.
Carrot-shaped grooves	Grooves on slickensides tend to be deeper and wider at one end and taper to a point at the other, thus resembling half a carrot. The direction in which that "carrot" points defines the direction of shear.
Chatter marks	As one fault block moves past another, small wedge-shaped blocks may be plucked out of the opposing surface. The resulting indentations on the fault surface are known as chatter marks.
Pinnate fractures	The inclination of pinnate fractures with respect to the fault surface defines the shear sense.

In continental-interior platform regions (e.g., the Midcontinent of the United States), a thin and relatively weak veneer of Phanerozoic sedimentary rock was deposited over relatively rigid Precambrian crystalline basement. At various times during the Phanerozoic, steeply dipping faults in the basement reactivated, causing differential movement of basement blocks. This movement forces the overlying layer of sedimentary rocks to passively bend into a fold that drapes over the edge of the basement block. Fault-related folds that form in this way are called **drape folds** or **forced folds** (Figure 8.22d). Kinematically, they are fault-propagation folds, but they are given a separate name to emphasize their unique tectonic setting.

We close with a comment on the term **drag fold,** which is used to refer to all fault-related folds in older literature. This general application of the term is misleading, because "drag" implies that the fold formed by shear resistance on the fault that retarded movement of the hanging wall with respect to the footwall. Rather, as we've seen, fault-related folds form in a number of different ways, most of which do not involve such "drag." Thus we discourage and avoid the use of this term.

8.3.5 Shear-Sense Indicators of Brittle Faults—A Summary

So far we have explored how various features of fault zones can be used to determine the shear sense. Given the lengthy descriptions that this required, we provide a concise summary of the various brittle shear-sense indicators in Table 8.6.

8.4 RECOGNIZING AND INTERPRETING FAULTS

Displacement on a fault can alter the landscape. The nature of the disruption depends on the type of fault, on how much displacement occurred, on the rate of displacement, on how recently faulting occurred, on the climate in the area, and on whether the fault is emergent or blind. We include climate and age of faulting in our list of factors because they determine the extent to which erosion can erase the effects of displacements. For example, in a humid climate with abundant rainfall, a fault's geomorphologic manifestation may disappear

FIGURE 8.23 Fault scarp in the Basin and Range Province (Nevada, USA). Note the normal sense of offset, with person for scale.

within months or years, whereas in a desert climate, a fault's geomorphologic manifestation may last for millennia.

Seismic faulting along an emergent strike-slip fault typically creates a **surface rupture,** which is manifested by broken ground and fissures. Displacement on an emergent dip-slip fault creates a step in the ground surface, called a **fault scarp** (Figure 8.23). Both dip-slip and strike-slip faulting can offset topographic features, such as ridges or river beds, and a recent fault may offset roads, fences, and sections of buildings. You can identify the trace of the San Andreas Fault in the wine district of California, for example, by offset rows of casks in a wine cellar, and historic faults by offsets in ancient buildings in Greece. Erosion typically destroys surface ruptures and fault scarps within tens to thousands of years, depending on climate. However, the abrupt change in elevation of the ground surface caused by the faulting can be maintained even after the original scarp has eroded away. Recent movement on a blind fault, by definition, does not result in a surface rupture, but the trace of the fault may be evident from differential movement of the ground surface, which can be detected by ground-based surveying equipment or by modern GPS (global positioning system) equipment.

Displacement across regional strike-slip faults (such as the Alpine Fault of New Zealand and California's San Andreas Fault) is not necessarily exactly parallel to the fault surface, in part due to the existence of restraining and releasing bends along the fault, and in part due to the relative motion of lithospheric plates during or after faulting. Transpression across a strike-slip fault may result in the formation of **fault-parallel ridges** that may be tens to hundreds of meters high (Figure 8.24), and transtension along a large strike-slip fault results in the development of **sag ponds.** Because of these features, the traces of major strike-slip faults are so obvious that they can be seen from space. If transpression occurs along a fault for millions of years, a mountain range develops adjacent to the fault (e.g., the New Zealand Alps), and if transtension occurs for millions of years, a sizable sedimentary basin develops adjacent to the fault. As we will see in Chapter 19, displacement in a regional-scale strike-slip fault zone may result in the formation of en echelon or stepped structures near the ground surface, which may have surface manifestations in the form of local ridges or troughs oriented at an angle to the main fault trace.

Inactive emergent or exhumed faults also have a topographic manifestation, for two reasons. First, the fault zone itself may have a different resistance to erosion than the surrounding intact rock. If the fault zone consists of noncohesive breccia or gouge, it tends to be weaker than the surrounding intact rock and thus more easily erodes. As a consequence, the fault trace on the ground surface evolves into a linear trough that may control surface drainage. Alternatively, if the fault becomes indurated, it may become more resistant to erosion than the surrounding region and will stand out in relief. Second, if the fault juxtaposes two rock units with different resistance to erosion, then a topographic scarp develops along the trace of the fault because the weaker unit erodes more rapidly, and the land surface underlain by the weaker unit becomes topographically lower (Figure 8.25). Such a **fault-line scarp** differs from a fault scarp, in that it is not the plane of the fault itself. In fact, the slope direction of a fault-line scarp may even be opposite to the dip direction of the underlying fault.

Fault zones are sometimes manifested by subtle features of the landscape. For example, fault zones containing abundant fractures may control local groundwater movement and may preferentially drain the surface, an effect that can cause changes in vegetation that appear as lineaments in remote-sensing images. Similarly, faults may offset structurally controlled topographic ridges, and thus may cause an alignment of ridge terminations. Ancient and inactive faults can be identified in the field, even if they do not disrupt the landscape or if fault-related structures have been covered or removed. Features indicative of faulting include the following: juxtaposition of rock units that were not in contact when first formed, offset of marker horizons, loss or duplication of section; and the presence of slickensided surfaces, slip lineations, fault rocks, and fault-related folds. Doing some of your own fieldwork in faulted regions is simply the best way to obtain an appreciation of the many manifestations of faulting.

FIGURE 8.24 Ridges formed along a transpressional segment of the San Andreas Fault. The striped white and gray rocks are basement that is pushed up relative to the (dark) sedimentary cover, in response to shortening along this strike-slip fault.

8.4.1 Recognition of Faults from Subsurface Data

To many geologists, particularly those working in economic geology, petroleum geology, and hydrogeology, the "field" consists of the subsurface region of the crust, and "field data" includes drill-hole logs (including downhole records of lithology based on cores or cuttings, electrical-conductivity measurements, or gamma-ray measurements), geophysical measurements (e.g., regional variation in the Earth's gravity and magnetic fields), and seismic-reflection data. Analysis of drill-hole logs and geophysical data provides an important basis for identification of subsurface faults.

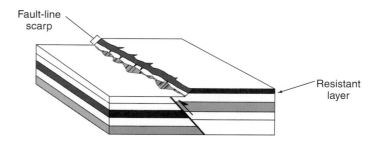

FIGURE 8.25 (a) Block diagram illustrating a fault-line scarp caused by the occurrence of a resistant stratigraphic layer (in black) that has been uplifted on one side.

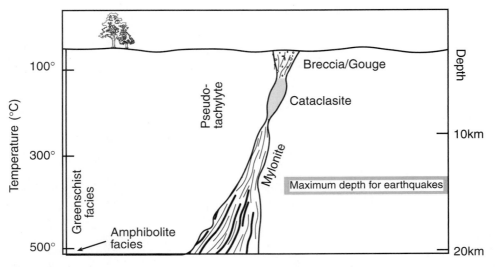

FIGURE 8.26 Change in fault character with depth for a steeply dipping fault. Note the change in fault zone width and types of structures with depth.

8.4.2 Changes in Fault Character with Depth

The characteristics of a fault depend on the magnitude of displacement on the fault, on whether or not faulting ruptures a previously intact rock or activates a preexisting surface, and on the pressure and temperature conditions (i.e., burial depth) at which faulting occurs. Figure 8.26 is a synoptic diagram of the various expressions of a major crustal fault zone. At very shallow depths in the earth (less than ~5 km), mesoscopic faults that form by reactivation of a preexisting joint or bedding surface typically result in discrete slickensided or fiber-coated surfaces. Mesoscopic shallow-level faults that break through previously intact rock tend to be bordered by thin **breccia** or **gouge zones,** and macroscopic faults, which inevitably break through a variety of rock units and across contacts, tend to be bordered by wider breccia and gouge zones, and subsidiary fault splays.

As we discussed earlier, rocks become progressively more ductile with depth in the crust, because of the increase in temperature and pressure that occurs with depth. Consequently, at depths between ~5 km and 10–15 km, faulting tends to yield a fault zone composed of **cataclasite.** Whereas cataclasite forms by brittle deformation on a grain scale, movement in the fault zone resembles viscous flow and strain is distributed across the zone (i.e., we have ductile behavior; see Chapter 9). The **brittle-plastic transition** for typical crustal rocks lies at a depth of 10–15 km in the crust. We purposely specify the transition as a range, because rocks consist of different minerals, each of which behaves plastically under different conditions, and because the depth of transition depends on the local geothermal gradient. Temperature conditions at a depth of around 10–15 km are in the range of 250°C to 350°C (i.e., lower greenschist facies of metamorphism), where plastic deformation mechanisms become the dominant contributor to strain in (quartz-rich) crustal rocks. The activity of plastic deformation mechanisms below this brittle-plastic transition yields a fine-grained and foliated fault-zone rock, called **mylonite.**

The degree of ductile deformation that accompanies faulting also depends on the strain rate and on the fluid pressure. At slower strain rates, a given rock type is weaker and tends to behave more ductilely. Thus, rocks can deform ductilely even at shallow crustal levels if strain rates are slow, whereas rocks below the brittle-plastic transition can deform brittlely if strain rates are high. A given shear-zone interval may therefore contain both mylonite and cataclasite, either as a consequence of variations in strain rate at the same depth or because progressive displacement eventually transported the interval across the brittle-plastic boundary. At very rapid strain rates (seismic slip), fault displacement in dry rock generates **pseudotachylyte.**

The width of a given fault zone typically varies as a function of rock strength; fault zones tend to be nar-

Evidence for subsurface faulting includes the following: (1) abrupt steps on structure-contour maps; (2) excess section (i.e., repetition of stratigraphy) or loss of section in a drill core; (3) zones of brecciated rock in a drill core, although weak fault rocks typically do not survive drilling intact (and thus fault zones may appear as gaps in a core); (4) seismic-reflection profiles, on which faults appear either as reflectors themselves, or as zones which offset known reflectors; and (5) linear anomalies or an abrupt change in the wavelength of gravity and/or magnetic anomalies, suggesting the occurrence of an abrupt change in depth to a particular horizon. We explore the use of geophysical methods in (regional) structural analysis in Chapter 15.

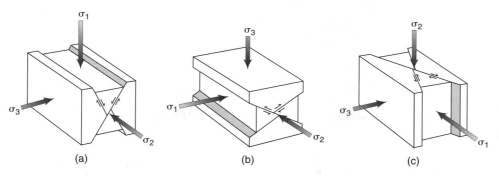

FIGURE 8.27 Anderson's theory of faulting predicts (a) (high-angle) normal faults, (b) (low-angle) reverse faults (or thrusts), and (c) (vertical) strike-slip faults.

rower in stronger rock. Thus, the width of a trans-crustal fault zone may vary with depth. Very near the surface (within a few kilometers), the fault diverges into numerous splays, because the near surface rock is weakened by jointing and by formation of alteration minerals (e.g., clay). At somewhat greater depths, rock is stronger, and the fault zone may be narrower. At still greater depths, where cataclastic flow dominates, the fault zone widens. Similarly, we expect to find widening of the mylonitic segment of the fault zone with depth, where the rocks as a whole become weaker.

8.5 RELATION OF FAULTING TO STRESS

Faulting represents a response of rock to shear stress, so it only occurs when the differential stress $(\sigma_d = \sigma_1 - \sigma_3 = 2\sigma_s)$ does not equal zero. Because the shear-stress magnitude on a plane changes as a function of the orientation of the plane with respect to the principal stresses, we should expect a relationship between the orientation of faults formed during a tectonic event and the trajectories of principal stresses during that event. Indeed, faults that initiate as Coulomb shear fractures will form at an angle of about 30° to the σ_1 direction and contain the σ_2 direction. This relationship is called **Anderson's theory of faulting.**[6] Why isn't σ_1 at 45° to the fault planes, where the shear stress is maximum? Recall the role of the normal stress, where the ratio of shear stress to normal stress on planes orientated at about 30° to σ_1 is at a maximum (see Chapter 6).

The Earth's surface is a "free surface" (the contact between ground and air/fluid) that cannot, therefore, transmit a shear stress. Therefore, regional principal stresses are parallel or perpendicular to the surface of

the Earth in the upper crust. Considering that gravitational body force is a major contributor to the stress state, and that this force acts vertically, stress trajectories in homogeneous, isotropic crust can maintain this geometry at depth. Anderson's theory of faulting states that in the Earth-surface reference frame, normal faulting occurs where σ_2 and σ_3 are horizontal and σ_1 is vertical, thrust faulting occurs where σ_1 and σ_2 are horizontal and σ_3 is vertical, and strike-slip faulting occurs where σ_1 and σ_3 are horizontal and σ_2 is vertical (Figure 8.27). Moreover, the dip of thrust faults should be ~30°, the dip of normal faults should be ~60°, and the dip of strike-slip faults should be about vertical. For example, if the σ_1 orientation at convergent margins is horizontal, Anderson's theory predicts that thrust faults should form in this environment, and indeed belts of thrust faults form in collisional mountain belts.

Anderson's theory is a powerful tool for regional analysis, but we cannot use this theory to predict all fault geometries in the Earth's crust for several reasons. First, faults do not necessarily initiate in intact rock. The frictional sliding strength of a preexisting surface is less than the shear failure strength of intact rock; thus, preexisting joint surfaces or faults may be reactivated before new faults initiate, even if the preexisting surfaces are not inclined at 30° to σ_1 and do not contain the σ_2 trajectory. Preexisting fractures that are not ideally oriented with respect to the principal stresses become oblique-slip faults. Second, a fault surface is a material feature in a rock body whose orientation may change as the rock body containing the fault undergoes progressive deformation. Thus, the fault may rotate into an orientation not predicted by Anderson's theory. Local stress trajectories may be different from regional stress trajectories because of local heterogeneities and weaknesses (e.g., contacts between contrasting lithologies, preexisting faults) in the Earth's crust. As a consequence, local fault geometry might not be geometrically related to regional stresses. Third, systematic changes in stress trajectories

---

[6]After the British geologist E. M. Anderson.

are likely to occur with depth in mountain belts (e.g., along a regional detachment), but Anderson's theory assumes that the stress field is homogeneous and that the principal stresses are either horizontal or vertical, regardless of depth.

8.5.1 Formation of Listric Faults

Let us consider changes in stress field with depth in more detail, because they explain the occurrence of **listric faults,** which are faults that have decreasing dip with depth. Deformation intensity, as manifested by strain magnitude, decreases from the interior to the margin of an orogenic belt. We can infer from this observation that the magnitude of horizontal tectonic stress similarly decreases from the interior to the margin of the belt. Now, envision a sheet of rock above a detachment that is subjected to a greater horizontal σ_1 at the hinterland than at the foreland (Figure 8.28). You might expect that such an apparent violation of equilibrium would cause the block to translate toward the foreland, but friction inhibits such movement in our model, generating shear stress at the base. Because the top of the sheet is a free surface (the contact between rock and atmosphere), principal stresses must be parallel or perpendicular to the top surface of the sheet. However, σ_1 near the bottom of the sheet cannot parallel σ_1 at the top surface, because there is shear stress at the bottom of the sheet. Calculating stress trajectories given these conditions, we find that they curve into the bottom of the sheet. If we accept

Anderson's premise that faults in the sheet initiate as Coulomb shears at about 30° to the σ_1 trajectory, then the faults must also curve; that is, their dips decrease with depth. This scenario provides a good explanation for the occurrence of listric faulting in the brittle regime, but note that the stress field is more complicated when the detachment is a ductile shear zone.

8.5.2 Fluids and Faulting

There is no doubt that fluids can play a major role in fault zones, as you will commonly find that fault rocks are altered by reaction with a fluid phase (e.g., clay forms in fault zones from the reaction of feldspar with water), and that fault zones contain abundant veins composed of minerals that precipitated from a fluid (e.g., quartz, calcite, chlorite, economic minerals). The fracturing that accompanies fault displacement creates open space within the fault zone for fluid to enter. Because of the increase in open space, fluid pressure in the fault zone temporarily drops relative to the surrounding rock. The resulting fluid-pressure gradient can actually drive groundwater into the fault zone until a new equilibrium is established. Such faulting-triggered fluid motion is known as **fault valving** or **seismic pumping.**

The presence of water in fault zones affects the stress at which faulting occurs in three ways. First, alteration minerals formed by reaction with water in the fault zone tend to have lower shear strength than minerals in the unaltered rock, and thus their presence may permit the fault to slip at a lower frictional stress than it would otherwise. Second, the presence of water in a rock may cause hydrolytic weakening of silicate minerals, and therefore allow deformation to occur at lower stresses. Third, the pore pressure of water (P_{fluid}) in the fault zone decreases the effective normal stress in a rock body, and thus decreases the magnitude of the shear stress necessary to initiate a shear rupture in intact rock or initiate frictional sliding on a preexisting surface. See Chapter 6 for more details on the mechanics of these effects.

The observation that increasing fluid pressure leads to faulting at a lower regional σ_d is illustrated by the history of earthquake activity in the Rocky Mountain Arsenal near Denver, Colorado. In the early 1960s, the U.S. military chose to dispose of large quantities (sometimes as much as 30 million liters per month) of liquid toxic waste by pumping it into the groundwater reservoir via a 4-km-deep well at the arsenal. Geophysicists noticed that when the waste was injected, dozens of small earthquakes occurred near the bottom of the well. Evidently, injection of the waste increased P_{fluid}, thereby decreasing the effective normal stress on faults such that the local stress was sufficient to cause

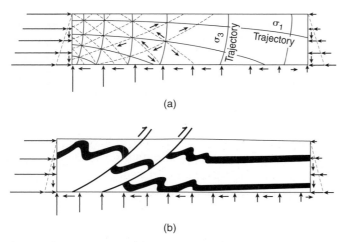

(a)

(b)

FIGURE 8.28 Curved principal stress trajectories and listric faults in a sheet of rock that is pushed from the (left) side. (a) Cross section of stress trajectories in a block bounded on the top by a free surface and on the bottom by a frictional sliding surface. (b) Predicted pattern of reverse faulting, assuming that faults form at ~30° to the σ_1 trajectory (note that only one set of reverse faults is illustrated).

preexisting faults near the well to slip until the elevated fluid pressure dissipated by fluid flow out of the well.

The concept that an increase in fluid pressure decreases effective stress across a fault helps to resolve one of the great paradoxes of structural geology, namely the movement of thrust sheets on regional-scale detachments. Look at the schematic image of a thrust sheet depicted in Figure 8.29a. The sheet is represented by a large rectangular block slipping on a detachment at its base. If you assume that the shear resistance at the base of the sheet is comparable to the frictional sliding strength of rock observed in the laboratory, then in order to move the sheet, the magnitude of the horizontal stress applied at the end of the sheet must be very large. Herein lies the paradox: the horizontal stress must be so large that it would exceed the strength of the sheet, so that the thrust sheet would deform internally (by faulting and folding) close to where the stress was applied before the whole sheet would move (Figure 8.29b). As an analogy of this paradox, picture a large Persian rug lying on a floor. If you push at one end of the rug, it simply wrinkles at that end, but it does not slip across the floor, because the shear resistance to sliding is too great. So, how do large thrust sheets form and move great distances on detachment faults?

Fluid pressure offered the first reasonable solution to this paradox. If fluid pressure (P_{fluid}) in the detachment zone approaches lithostatic pressure (i.e., the magnitude of fluid pressure approaches the weight of the overlying rock), then the effective normal stress across the fault plane approaches zero (see Chapter 6). Therefore, the shear stress necessary to induce sliding on the detachment would also become very small, so that thrust sheets can move before deforming internally (Figure 8.29c). This powerful idea, known as the **Hubbert-Rubey hypothesis,**[7] emphasizes the role of elevated fluid pressure during movement on detachments, and has wide applicability. Indeed, modern measurements confirm that in regions where detachment faults move, P_{fluid} near the detachment interval often exceeds hydrostatic conditions and may even approach lithostatic values. Note that fluid pressure cannot exceed lithostatic pressure for long timescales, as this would forcefully push the rocks upward.

Initially, geologists thought that all large thrust sheets slide down gently foreland-dipping slopes in response to gravity, effectively gliding on a cushion of fluid. But **gravity sliding** does not explain most examples of thrusting in orogenic belts, because thrust sheets typically dip away from the direction in which they move. Recall, for example, the thrust-ramp geometry that characterizes large thrust faults, leading to a regionally hinterland-dipping surface. We return to this problem in Chapter 18, where we discuss newer concepts of thrust-sheet movement and the role of the Hubbert-Rubey hypothesis in these models.

8.5.3 Stress and Faulting— A Continuing Debate

The issue of how large the shear stress (σ_s) must be in order to initiate faults or to reactivate preexisting faults remains highly controversial. The magnitude of σ_s necessary to trigger faulting depends on fluid pressure, lithology, strain rate, temperature, and the orientation of the preexisting fault. Recall that for a given range of orientations, the stress necessary to initiate frictional sliding on a preexisting fault is less than that necessary to initiate a new fault (Chapter 6). Rock mechanics experiments provide one avenue of approach into this problem. From laboratory triaxial loading experiments, we determine that, all other factors being equal, the shear stress for failure increases as confining pressure increases, and that σ_s is greatest for contractional faulting and least for extensional faulting. Limiting conditions for each of the three fault types are shown in Figure 8.30, based on the relationship:

$$\sigma_d \geq \beta \, (\rho \cdot g \cdot z) \, (1 - \lambda) \qquad \text{Eq. 8.1}$$

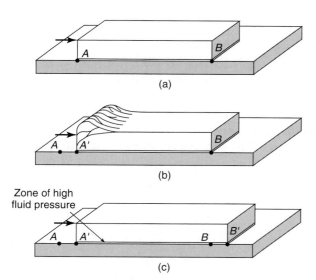

FIGURE 8.29 The thrust sheet paradox. (a) Block and sectional view of a thrust sheet on a frictional surface (heavy line). (b) Because of frictional resistance, the shear stress necessary to initiate sliding exceeds the yield strength of the frontal end of the thrust sheet, causing fracturing and folding. (c) Thrust sheet moves coherently if it rides on a "cushion" of fluid. A' and B' are displaced positions of points A and B.

[7]Named after the American geologists M. King Hubbert and William Rubey.

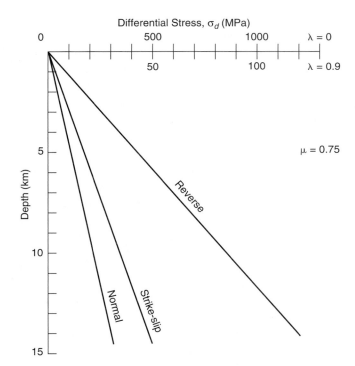

FIGURE 8.30 Graph showing variation in differential stress necessary to initiate sliding on reverse, strike-slip, and normal faults, as a function of depth. The relationship is given by Equation 8.1, assuming a friction coefficient, $\mu = 0.75$, and a fluid pressure parameter, $\lambda = 0$ (no fluid present) and $\lambda = 0.9$ (fluid pressure is 90% of lithostatic pressure).

where σ_d is differential stress (i.e., $2\sigma_s$); β is 3, 1.2, and 0.75 for reverse, strike-slip, and normal faulting, respectively; and λ is a fluid pressure parameter, defined as the ratio of pore-fluid pressure and lithostatic pressure (λ ranges from ~0.4, for hydrostatic fluid pressure, to 1 for lithostatic fluid pressure).

Geologists have questioned the validity of stress estimates based on laboratory studies, because of uncertainty about how such estimates scale up to crustal dimensions and about time-dependent changes. Thus, alternative approaches have been used to determine stress magnitudes needed to cause crustal-scale faulting, including analysis of heat generation during faulting, direct measurement of stresses near faults, and borehole data.

If movement on a fault surface in the brittle regime of the crust involves frictional sliding, some of the work done during fault movement is transformed into heat. This process, called **shear heating,** obeys the equation

$$\sigma_s \cdot u = E_e + E_s + Q \qquad \text{Eq. 8.2}$$

where σ_s is the shear stress across the fault, u is the amount of slip on the fault, E_e is the energy radiated by

earthquakes, E_s is the energy used to create new surfaces (breaking chemical bonds), and Q is the heat generated. This equation suggests that $\sigma_s \cdot u > Q$, and thus if u is known, the value of Q can provide a minimum estimate of σ_s. With this concept in mind, geologists have studied metamorphism near faults in order to calculate the amount of heat (Q) needed to cause the metamorphism and thereby provide an estimate of σ_s during faulting. Assuming that metamorphism is entirely a consequence of shear heating, these studies conclude that shear stresses across faults must be quite large. These studies, however, have been criticized, because many geologists reject the assumption that the observed metamorphism is primarily due to shear heating, and because observed heat flow adjacent to active faults is not greater than heat flow at a distance from the fault. For example, while the San Andreas Fault is a huge active fault, it is not bordered by a zone of high heat flow as predicted by the shear-heating model.[8]

In recent years, geologists have attempted to define the stress state during crustal faulting by in situ measurements of stress fields in the vicinity of faults. Recall from Chapter 3 that hydrofracture measurements in drill holes, strain-release measurements, and borehole breakouts all provide an estimate of stress orientation, and in some cases stress magnitude, in the shallow crust. In general, these measurements suggest that stresses during faulting are relatively low. For example, direct measurement of stress in the crust around the San Andreas Fault suggests that the σ_1 trajectory bends so that it is nearly perpendicular in the immediate vicinity of the fault. Such a change in stress orientation suggests that the San Andreas Fault is behaving like a surface of very low friction, and therefore could not support a large σ_s. Based on these observations, some geologists have concluded that the σ_s needed to cause movement on the San Andreas Fault, and by inference other faults, is relatively low (the **weak-fault hypothesis**).

Another way to estimate stresses during faulting comes from studies of seismicity. When an earthquake occurs, energy is released and the value of σ_s across the fault decreases. This decrease, called the **stress drop,** provides a minimum estimate of the value of σ_s that triggered the earthquake. Stress drops for earthquakes estimated in this manner range between about 0.1 MPa and 150 MPa, but are typically in the range of 1–10 MPa.

In summary, geologists do not yet agree about the stress state necessary to cause crustal faulting, because estimates derived from different approaches do not

[8]This observation is key to the strength debate on the San Andreas Fault, where low heat flow supports the suggestion of low shear stresses, but also high fluid activity.

TABLE 8.7   GEOMETRIC CLASSIFICATION OF FAULT ARRAYS

Parallel fault array	As the name suggests, a parallel fault array includes a number of fault surfaces that roughly parallel one another.
Anastomosing array	A group of wavy faults that merge and diverge along strike, thereby creating a braided pattern in map view or cross section.
En echelon array	A group of parallel fault segments that lie between two enveloping surfaces and are inclined at an angle to the enveloping surfaces.
Relay array	In map view, a relay array is a group of parallel or subparallel non-coplanar faults that are spaced at a distance from one another across strike, but whose traces overlap with one another along strike. As displacement dies out along the strike of one fault in the array, displacement increases along an adjacent fault. Thus, displacement is effectively "relayed" (transferred) from fault to fault. In a thrust belt containing a relay array of faults, regional shortening can be constant along the strike of the belt, even though the magnitude of displacement along individual faults dies out along strike.
Conjugate fault array	An array composed of two sets of faults that are inclined to one another at an angle of about 60°. Conjugate fault arrays can consist of dip-slip faults or strike-slip faults. If the faults in the array are strike-slip, then one set must be dextral and the other sinistral.
Nonsystematic fault array	In some locations, faulting occurs on preexisting fractures. If the fracture array initially had a wide range of orientations, then slip on the fractures will yield faults in a wide range of orientations. Such an array is called a nonsystematic array.

agree with one another. Most likely, there is a range of stress conditions that can cause faulting, reflecting parameters such as the geometry of the fault, the nature of the faulted material, properties of the fault rocks, and the fluid pressure in the fault zone.

8.6   FAULT SYSTEMS

Faults typically do not occur in isolation, but rather are part of a group of associated faults that develop during the same interval of deformation and in response to the same regional stress field. We classify groups of related faults either by their geometric arrangement or their tectonic significance (i.e., the type of regional deformation resulting from their movement). A group of related faults is called a *fault system* or *fault array.* Although the terms "system" and "array" can be used interchangeably, geologists commonly use array when talking about geometric classifications and system in the context of tectonic classifications. As we continue, we'll first describe a geometric classification for fault arrays, and then we'll introduce the three tectonically defined types of fault systems: normal, thrust, and strike-slip systems. In many areas, faults in normal and thrust systems merge with a detachment in sedimentary rocks at shallow depth (10–15 km), in which case

the system may be referred to as a **thin-skinned system.** If faulting involves deeper crustal rocks (i.e., basement), we call this a **thick-skinned** system. Most mountain belts show a transition between thin- and thick-skinned systems, so these are not adequate indicators of the tectonic setting.

Our description of fault systems in this chapter is meant only to be a brief introduction. We provide detailed descriptions of fault systems later in this book (Chapters 16–19), where you will also find ample illustration of associated structures and descriptions of tectonic settings.

8.6.1   Geometric Classification of Fault Arrays

Groups of faults may be classified as parallel, anastomosing, *en echelon,* relay, conjugate, or random arrays, depending on the relationships among faults in the array, and on how the faults link with one another along strike. We define these terms in Table 8.7 and illustrate them in Figure 8.31.

Typically, faults in parallel arrays of normal or thrust faults dip in the same direction over a broad region. Subsidiary faults in the array that parallel the major faults are called **synthetic faults,** whereas subsidiary faults whose dip is opposite to that of the major faults are called **antithetic faults.**

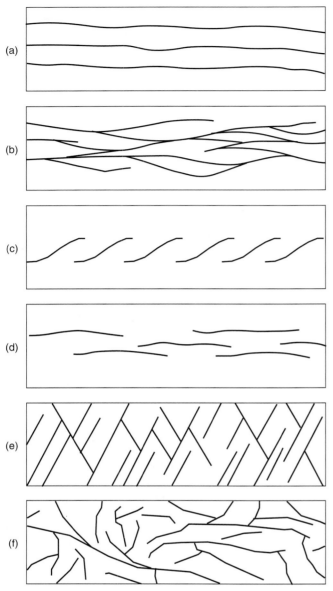

FIGURE 8.31 Map-view sketches of various types of fault arrays. (a) Parallel array, (b) anastomosing array, (c) *en echelon* or stepped array, (d) relay array, (e) conjugate array, and (f) random array. See descriptions in Table 8.7.

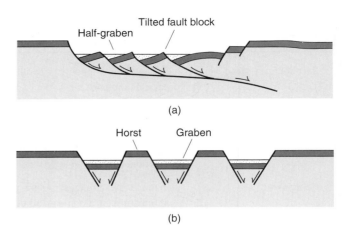

FIGURE 8.32 Normal fault systems. (a) Half-graben system. (b) Horst-and-graben system.

8.6.2 Normal Fault Systems

Regional normal fault systems form in **rifts,** which are belts in which the lithosphere is undergoing extension; along **passive margins,** which are continental margins that are not currently plate margins; and along **mid-ocean ridges.** Typically, faults in a normal fault system comprise relay or parallel arrays; they can be listric or planar, or contain distinct fault bends. Movement on both planar and listric normal fault systems generally results in rotation of hanging-wall blocks around a horizontal axis, and therefore causes tilting of overlying fault blocks and/or formation of rollover anticlines and synclines. The geometry of tilted blocks and rollover

folds developed over normal faults depends on the shape of the fault and on whether or not synthetic or antithetic faults cut the hanging-wall block.

As a consequence of the rotation accompanying displacement on a normal fault, the original top surface of the hanging-wall block tilts toward the fault to create a depression called a **half graben** (Figure 8.32a). Note that a half graben (from the German word for "trough") is bounded by a fault on only one side. Most of the basins in the Basin and Range Province of the western United States are half grabens, and the ranges consist of the exposed tips of tilted fault blocks. In places where two adjacent normal faults dip toward one another, the fault-bounded block between them drops down, creating a **graben** (Figure 8.32b). Where two adjacent normal faults dip away from one another, the relatively high footwall block between the faults is called a **horst.** Horsts and grabens commonly form because of the interaction between synthetic and antithetic faults in rift systems. We further explore normal fault systems and their tectonic settings in Chapter 16.

8.6.3 Reverse Fault Systems

Reverse fault systems are commonly arrays of thrust faults that form to accommodate large regional shortening. Not surprisingly, thrust systems are common along the margins of convergent plate boundaries and in collisional orogens. In such tectonic settings, thrusting occurs in conjunction with formation of folds, resulting in tectonic provinces called **fold-thrust belts.**

To a first approximation, fold-thrust belts resemble the wedge of snow or sand that is scraped off by a plow (Figure 8.33a). Typically, the numerous thrusts in a fold-thrust belt merge at depth with a shallowly dipping detachment. At a crustal scale, major thrust faults are

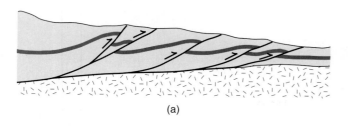

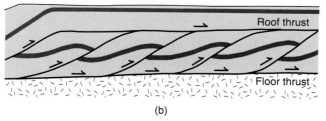

FIGURE 8.33 Reverse fault systems. (a) Imbricate fan in thrust system. (b) Duplex system with horses in between the floor and roof thrusts.

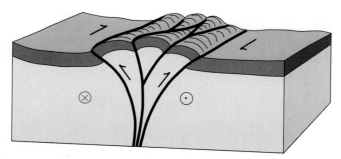

FIGURE 8.34 The formation of a (positive) flower structure from strike-slip faulting. The symbols ⊗ and ⊙ indicate motion away and toward an observer, respectively.

listic; but in detail, where thrusts cut upwards through sequences of contrasting strata, they have a stair-step profile, with **flats** following weak horizons and **ramps** cutting across beds. Ramp-flat geometries generally develop best in sequences of well-stratified sedimentary rock, such as at passive margins caught in between colliding continents, or in the sequence of sediment on the **craton** side of the orogen that is derived by erosion of the evolving orogen, called a **foreland basin.**

As is the case with normal fault systems, faults in a thrust-fault system tend to comprise relay or parallel arrays. An **imbricate fan** of thrust faults (Figure 8.33a) consists of thrusts that either intersect the ground surface or die out up dip, whereas a **duplex** (Figure 8.33b) consists of thrusts that span the interval of rock between a higher-level detachment called a **roof thrust** and a lower level detachment called a **floor thrust.** We can't say much more about thrust systems without introducing a lot of new terminology, so we'll delay further discussion of the subject until Chapter 18.

8.6.4 Strike-Slip Fault Systems

Strike-slip fault systems occur at transform boundaries, which are boundaries where two plates slide past one another without the creation or subduction of lithosphere; they can also occur within plates and as components of convergent orogens. Major continental strike-slip fault systems are complicated structures. We have already explored several associated structures of regional strike-slip faults. Typically, they splay into many separate faults in the near surface, which, in cross section, resembles the head of a flower. Because of this geometry, such arrays are called **flower structures** (Figure 8.34). We present further description of

these complexities and of the tectonic settings in which strike-slip faults occur in Chapter 19.

8.6.5 Inversion of Fault Systems

Once formed, a fault is a material discontinuity that may remain weaker than surrounding regions for long periods of geologic time. Thus, faults can be reactivated during successive pulses of deformation at different times during an area's history. If the stress field during successive pulses is different, the kinematics of movement on the fault may not be the same, and the resulting displacement from one event may be opposite to the displacement resulting from another event. For example, a normal fault formed during rifting of a continental margin may be reactivated as a thrust fault if that margin is later caught in the vice of continental collision. Likewise, the border faults of a half graben or failed rift may be reactivated later as thrust, or as strike-slip faults, if the region is later subject to compression. This reversal of displacement on a fault or fault system is called **fault inversion.** When inversion results in contraction of a previously formed basin, the process is called **basin inversion.**

8.6.6 Fault Systems and Paleostress

With the Andersonian concept of faulting in mind, the geometry of a fault system is a clue to the regional stress conditions that caused the faulting. A thrust system reflects conditions where regional σ_1 is horizontal and at a high angle to the trace of the system. A normal fault system reflects conditions where regional σ_3 is horizontal and trends at a high angle to the trend of the system. There is no single rule defining the relationship between strike-slip fault systems and stress trajectories. Oceanic transforms, for example, typically parallel the σ_3 direction, whereas continental strike-slip faults commonly trend oblique to the σ_1 direction.

Geologists study slip on variably oriented faults in a region to determine the stress field, assuming that all the faults moved in response to the same stress. Most

FIGURE 8.35 Damage from the Kobe earthquake in 1995 (Japan).

places in Earth's crust are fractured, and though there may be dominant systematic arrays of fractures in the area, there are likely to be many nonsystematic fractures as well. If a body of rock containing abundant preexisting fractures in a range of orientations is subjected to a regional homogeneous stress field, a shear stress will exist on all fractures that are not principal planes (i.e., not perpendicular to one of the principal stresses). The orientation of the resolved shear stress on each fracture is determined by the orientation and relative magnitudes of the principal stresses defining the regional stress state. Fractures whose frictional resistance is exceeded will move, and the direction of movement will be approximately parallel to the maximum resolved shear stress on the fracture surface. Fractures whose dip direction happens to be parallel to the shear stress become dip-slip faults, and fractures whose strike direction happens to be parallel to the shear stress become strike-slip faults. All other fractures that slip become oblique-slip faults.

During the past few decades, methods have been developed that permit the principal stress directions to be derived from measurements of slip trajectories on a nonsystematic array of faults, assuming that they moved in response to a regionally homogeneous stress. In principle, you need measurements of the shear sense and the trend of the net-slip vector on only four nonparallel faults to complete this **paleostress analysis,** but in prac-

tice, geologists use measurements from numerous faults and employ statistics to obtain a best-fit solution. It is beyond the scope of this book to provide the details of paleostress analysis from slip data on fault arrays, but we provide some references on the subject in Chapter 3.

8.7   FAULTING AND SOCIETY

All this terminology and theory may make us forget that the study of faulting is not just an academic avocation. Faults must be studied carefully by oil and mineral exploration geologists, because faulting controls the distribution of valuable materials. Similarly, faults and fractures play an important role in groundwater mobility and the general availability of water. Regional fault analysis is required for the localization and building of large human-made structures, such as dams and nuclear power plants. Such structures, which may have devastating impact when they fail, should clearly not be built near potentially active faults. But perhaps the most dramatic effect of fault activity on society comes in the form of earthquakes (Figure 8.35), which can be responsible for great loss of life (in some cases measured in the hundreds of thousands) and can destroy the economic stability of industrialized countries (damages may be hundreds of billions of dollars). We close this chapter by briefly looking at these more immediate consequences of faulting.

8.7.1 Faulting and Resources

Faults contribute to the development of **oil** traps by juxtaposing an impermeable seal composed of packed gouge or fault-parallel veining against a permeable reservoir rock, or by juxtaposing an impermeable unit, like shale, against a reservoir bed. Faulting may also affect oil migration by providing a highly fractured zone that serves as a fluid conduit through which oil migrates. Finally, syndepositional faulting (growth faults) affects the distribution of oil reservoir units.

Valuable **ore** minerals (e.g., gold) commonly occur in veins or are precipitated from hydrothermal fluids that were focused along fault zones, because fracturing provides enhanced permeability. Thus, fault breccias are commonly targets for mineral exploration. As we pointed out earlier, displacement on faults may control the distribution of ore-bearing horizons, and thus mining geologists map faults in a mining area in great detail.

Hydrogeologists also are cognizant of faults, because of their effect on the migration of **groundwater.** A fault zone may act as a permeable zone through which fluids migrate if it contains unfilled fractures, or the fault zone may act as a seal if it has been filled with vein material or includes impermeable gouge. In addition, faults may truncate aquifers, and/or juxtapose an aquitard against an aquifer, thereby blocking fluid migration paths.

8.7.2 Faulting and Earthquakes

Non-geologists often panic when they hear that a fault has been discovered near their home, because of the common perception that all faults eventually slip and cause earthquakes. Faulting is widespread in the crust, but fortunately, most faults are **inactive faults,** meaning that they haven't slipped in a long time, and are probably permanently stuck. Relatively few faults are **active faults,** meaning that they have slipped recently or have the potential to slip in the near future. Even when slip occurs, not all movement on active faults results in seismicity. If an increment of faulting causes an earthquake, we say that the fault is **seismic,** but if the offset occurs without generating an earthquake, we call the slip **aseismic.** Aseismic faulting is also called **fault creep** by seismologists.

Why do earthquakes occur during movement on faults? Earthquakes represent the sudden release of elastic strain energy that is stored in a rock, and can be generated when an intact rock ruptures, or when asperities on a preexisting fault snap off or suddenly plow. Rubbing two bricks while applying some pressure is a good analogy of this process. The bricks move until the indentation of asperities once again anchors them.

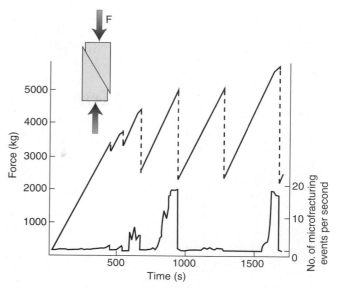

FIGURE 8.36 Laboratory frictional sliding experiment on granite, showing stick-slip behavior. The stress drops (dashed lines) correspond to slip events. Associated microfracturing activity is also indicated.

This start-stop behavior of faults is called **stick-slip behavior** (Figure 8.36). During the stick phase, stress builds up (as illustrated by the solid line), while during the slip phase the fault moves and stress at the site of faulting drops (dashed line). Typically, the stress drop is not complete, meaning that the differential stress does not decrease to zero. If we accept this model of faulting, **fault creep** occurs where a fault zone is very weak, perhaps due to the presence of weak material, hydrolytic weakening in the fault zone, or high fluid pressures in the fault zone.

Geoscientists have struggled for decades to delineate regions that have the potential to be seismic. This work involves the study of faults to determine if they are active or inactive, and whether active faults are seismic or aseismic. But fault studies alone do not provide a complete image of seismicity, because not all earthquakes occur on recognized faults. Some represent the development of new faults, some represent slip on blind faults, and some represent non-fault-related seismicity (e.g., volcanic explosions). Delineation of seismically active regions plays a major role in land-use planning. Obviously, the potential for **seismicity** must be taken into account when designing building codes for homes and when situating large public facilities like nuclear power plants, schools, hospitals, dams, and pipelines.

The primary criterion for delineating a seismically active region comes from direct measurements of seismicity. The underlying idea is that places with a potential for earthquakes in the near future probably have

FIGURE 8.37 Nearly horizontal layers of light-colored rock and gravel, and dark-colored peat have been broken and distorted as a result of earthquake activity near Banning, California, along the San Andrea Fault.

suffered earthquakes in the past. Networks of seismographs record earthquakes and provide the data needed to pinpoint the **focus** of each earthquake; that is, the region in the Earth where the seismic energy was released. Maps of earthquake **epicenters** (the point on the Earth's surface that lies directly above the focus) emphasize that most earthquakes occur along plate boundaries, but that dangerous earthquakes may also occur within plate interiors. Thus, we recognize **plate-boundary seismicity** and **intraplate seismicity.** Cross sections showing the distribution of earthquake foci show that, with the exception of convergent-margin seismicity, most earthquakes occur at depths shallower than ~15 km, which defines the lower boundary of the brittle upper crust.[9] Convergent-margin earthquakes occur along subducted slabs down to depths of about 650 km, defining the **Wadati-Benioff zone** (see Chapter 14). The deep earthquakes in a Wadati-Benioff zone occur well below the expected depth for brittle faulting. So why do these deep-focus earthquakes occur? One suggestion is that deep-focus earthquakes represent the

stress release associated with sudden mineral phase changes in the downgoing slab (e.g., olivine to spinel). Since different mineral phases occupy different volumes, a sudden phase change causes a movement in the rock body that could result in the generation of an earthquake and perhaps even in the formation of pseudotachylyte. The question is not resolved.

A reliable and detailed record of seismicity is only a few decades old, because a worldwide network of seismograph stations was not installed until after World War II. The information from these stations proved critical for the formulation of plate tectonic theory in the 1960s (Chapter 14), but governments actually funded this network to monitor underground nuclear testing in the Cold War era of the 1950s. Seismic studies cannot delineate the potential for seismicity in areas that have only infrequent earthquake activity. To identify such cryptic seismic zones, geologists rely on field data (Figure 8.37). We search for features such as cross-cutting relations with very young stratigraphic units or landforms. If the fault cuts a very young sequence of sediment or a very young volcanic flow or ash, then the fault must itself be very young. Similarly, if the fault cuts a young landform, like an alluvial fan or a glacial moraine, then the fault must be very young. Fault

---

[9]This seismically active region of the crust is also called the schizosphere, whose lower boundary is sometimes (incorrectly) called the brittle-ductile transition; we use the term brittle-plastic transition for this boundary.

scarps and triangular facets suggest that the faulting occurred so recently that erosion has not had time to erase its surface manifestation. The presence of uncemented gouge suggests that a fault was active while the rock was fairly close to the surface, a situation that implies movement on the fault occurred subsequent to uplift and exhumation of rock. The presence of pseudotachylyte may indicate that the fault was seismic. Changes in base level due to faulting at the face of a mountain range can cause a stream to cut down through alluvium that it previously deposited, creating matched terraces on opposite sides of the valley. The development of such paired terraces may indicate the occurrence of seismicity. Finally, accurate surveying of the landscape may indicate otherwise undetectable ground movements that could be a precursor to seismicity.

In special cases, it may be possible to determine the **recurrence interval** on a fault, meaning the average time between successive faulting events. This is done by studying the detailed stratigraphy of sediments deposited in marshes or ponds along the fault trace. Seismic events are recorded by layers in which sediment has been liquefied by shaking (these layers are sometimes called *seismites*), which causes disruption of bedding and sand volcanoes. By collecting organic material from the liquefied interval (e.g., wood), it may be possible to date the timing of liquefaction. Once we know the recurrence interval and the size of the earthquakes, we can estimate the seismic risk for an area.

The past few decades have seen intense study of earthquakes and related processes. As a result we have become reasonably successful in predicting *where* earthquakes will occur (if not precisely, at least the general area), but *when* they occur remains an imprecise science at best. Error margins of 50–100 years are inadequate for modern society, but being able to improve significantly on this with our current understanding of faulting seems unlikely. Geology operates on timescales much larger than our human ones. Perhaps preparation is our best bet when it comes to earthquake hazards.

8.8   CLOSING REMARKS

Chapters 6, 7, and 8 have provided an overview of brittle deformation processes and structures. At the end of Chapter 8, we also discussed briefly society's need for better understanding of fault processes. Brittle deformation is only a small part of what contributes to the development of structures and deformation of the lithosphere. We have alluded, for example, to the existence of ductile shear zones, in which displacement is not accommodated by brittle failure, and we have mentioned folding, which also deforms rocks without loss of cohesion. In the next section of this book, we turn our attention to the complementary topic of ductile deformation and resulting structures. After we have introduced these structures, we will be able to describe the relationship between faults and other geologic structures in greater detail, and describe the tectonic settings in which they form.

ADDITIONAL READING

Anderson, E. M., 1951. *Dynamics of faulting and dyke formation.* Oliver and Boyd: Edinburgh.

Bonnet, E., Bour, O., Odling, N. E., Davy, P., Main, I., Cowie, P., and Berkowitz., B., 2001. Scaling of fracture systems in geological media. *Reviews of Geophysics,* 39, 347–383.

Boyer, S. E., and Elliot, D., 1982. Thrust systems. *American Association of Petroleum Geologists Bulletin,* 66, 1196–1230.

Chester, F. M., and Logan, J. M., 1987. Composite planar fabric of gouge from the Punchbowl Fault, California. *Journal of Structural Geology,* 9, 621–634.

Hubbert, M. K., and Rubey, W. W., 1954. Role of fluid pressure in mechanics of overthrust faulting. *Bulletin of the Geological Society of America,* 70, 115–205.

Keller, E., and Pinter, N., 1996. *Active tectonics: earthquakes, uplift, and landscape.* Prentice Hall: Englewood Cliffs, 338 pp.

Kirby, S. H., 1983. Rheology of the lithosphere. *Reviews of Geophysics and Space Physics,* 21, 1458–1487.

Mandl, G., 1988. *Mechanics of tectonic faulting, models and basic concepts.* Elsevier: Amsterdam.

Petit, J. P., 1987. Criteria for the sense of movement on fault surfaces in brittle rocks. *Journal of Structural Geology,* 9, 597–608.

Scholz, C. H., 2002. *The mechanics of earthquakes and faulting* (2nd edition). Cambridge University Press: Cambridge.

Sibson, R. H., 1974. Frictional constraints on thrust, wrench and normal faults. *Nature,* 249, 542–544.

Sibson, R. H., 1977. Fault rocks and fault mechanisms. *Journal of the Geological Society of London,* 133, 190–213.

Sylvester, A. G., 1988. Strike-slip faults. *Geological Society of America Bulletin,* 100, 1666–1703.

Wise, D. U., Dunn, D. E., Engelder, J. T., Geiser, P. A., Hatcher, R. D., Kish, S. A., Odom, A. L., and Schamel., S., 1984. Fault-related rocks: suggestions for terminology. *Geology,* 12, 391–394.

DUCTILE STRUCTURES

9.1 INTRODUCTION

How can a strong layer of rock permanently bend into a tight fold (Figure 9.1)? How can a material such as ice distort, while remaining a solid? Ice is a particularly instructive example of flow in crystalline solids, because it moves on human timescales. Its behavior is directly relevant to rock deformation on geologic timescales. At first one might think that solid deformation is accomplished by bending and stretching of atomic bonds in the crystal lattice, but these movements are elastic deformations and, as described in Chapter 5, elastic deformation is recoverable (i.e., non-permanent). The movement of a glacier or the formation of a fold, however, is a permanent feature that represents ductile deformation. If we were to carefully remove a folded layer from an outcrop, or a deformed mineral from a hand specimen, they would not jump back to their original shapes. The distortions that occurred must be a result of permanent changes in the material. The principles that underlie the ability of materials like rocks to accumulate permanent strain are contained in a vast and ever-growing body of materials science literature. Structural geologists have increasingly applied concepts from materials science to geologic environments. The associated terminology, however, has not always remained consistent between these fields. In trying to keep new terms and concepts to a minimum, we've chosen to limit the coverage in this chapter; otherwise we'd lose sight of our ultimate goal: understanding the way rocks deform in the ductile regime.

In Chapter 5 we first introduced the concept of *flow*. At that point we described the topic merely in terms of stress and strain rate. We contrasted linear viscous (Newtonian) and nonlinear viscous (non-Newtonian) behavior, using analogs and simple mechanical models. In this chapter we turn to the physical processes

FIGURE 9.1 North-verging recumbent fold in mesozoic rocks of the Morcles thrust (or *Nappe*); Swiss Alps.

that allow materials to undergo appreciable, permanent distortions. To refresh your memory—strain that is distributed over the body rather than localized is what distinguishes *ductile behavior* from *brittle behavior.* But strain that appears homogeneous on one scale may represent heterogeneity on another, so again we need to include the scale of our observation. As scales of observation in structural geology range from nanometers (10^{-9} m) to kilometers, single minerals to mountain ranges, we define ductile behavior as uniform flow down to the scale of the hand specimen, that is, down to the mesoscopic scale.

We distinguish three fundamental mechanisms that produce ductile behavior in rocks and minerals: (1) **cataclastic flow,** (2) **diffusional mass transfer,** and (3) **crystal plasticity.** Which processes dominate at a given time in a rock's history is primarily a function of temperature, stress, strain rate, grain size, composition, and fluid content. Temperature, in particular, is an important parameter, but different minerals behave ductilely at different temperatures. What is considered high-temperature behavior for one mineral is low-temperature behavior for another mineral. Thus, when

talking about the relationship between temperature and deformation, we introduce a normalized parameter that is called the **homologous temperature,** T_h. The homologous temperature is a dimensionless parameter that is defined as the absolute temperature divided by the absolute melting temperature of the material:

$$T_h = T/T_m \qquad\qquad \text{Eq. 9.1}$$

where T is temperature and T_m is melting temperature of the material, both in K (kelvins). We loosely define low-temperature conditions as $0 < T_h < 0.3$, medium-temperature conditions as $0.3 < T_h < 0.7$, and high-temperature conditions as $0.7 < T_h < 1$.

After discussing the fundamental mechanisms and their associated microstructures (i.e., mineral geometries on the microscopic scale), we close this chapter by examining the interrelationship between the various rheologic parameters (such as stress and strain rate; see Chapter 5), and by introducing the powerful concept of deformation mechanism maps. Let us first turn to the three mechanisms of ductile behavior—cataclastic flow, dislocation movement, and diffusion.

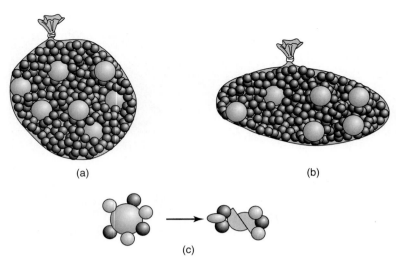

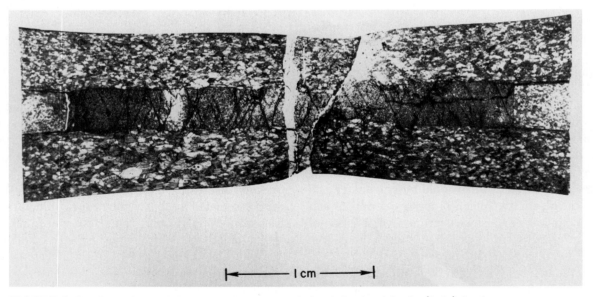

FIGURE 9.2 Bean bag experiment. Changing the shape of a bag is accomplished by the grains sliding past one another (a and b). Large grains may fracture and slide on the fracture surface (c).

9.2 CATACLASTIC FLOW

We start our examination of cataclastic flow with a simple experiment. Consider a bean bag that is originally shaped like a ball (Figure 9.2). We squash the bag so that it fits into a cube. In order for the ball-shaped bag to change shape, the beans have to slide past one another. Now imagine that the bag is strong and that you attach it to a winch that pulls the bean-filled bag through an opening that is smaller than a single bean. For the bag to pass through the small opening, all the

individual beans must fracture into smaller pieces (brittle deformation), but the bag as a whole remains coherent. Such a process, where a mesoscopic body (the bean bag) changes shape without breaking into separate pieces, but the constituents (the beans) fracture into smaller pieces and/or slide past one another, is called **cataclastic flow.** In rocks, the tiny fractures are called microcracks and the pieces move past one another by the process of frictional sliding (see Chapter 6).

During cataclastic flow a rock deforms without obvious strain localization on the scale of the hand specimen, yet the *mechanism* of deformation is (micro)fracturing and/or frictional sliding (Figure 9.3). You may now better appreciate the confusion surrounding the terms brittle and ductile (Chapter 5). Cataclasis is mesoscopic ductile behavior, yet the process by which it occurs is microscopic brittle fracturing and frictional sliding!

In rocks, microfractures may occur at grain boundaries (intergranular) or within individual grains (intragranular). In both cases the process occurs by breaking many atomic bonds at the same time. The crystal structure away from the fracture, however, remains unaffected. Frictional sliding is strongly dependent on pressure; with increasing pressure the ability of sliding to occur is reduced (see Chapter 6). Therefore, we expect to find cataclastic flow in rocks only at relatively low lithostatic pressures. This condition is met in the upper several kilometers of the crust and, indeed,

FIGURE 9.3 Extension experiment showing cataclastic flow in Luning dolomite (Italy) that is surrounded by marble that deformed by crystal plastic processes. This contrasting behavior reflects the relative strength of the materials.

we typically find cataclastic flow in shallow-crustal rocks, such as fault zones. The stress-dependence of cataclasis is one characteristic that distinguishes it from ductile mechanisms involving crystal defects, which are discussed in the next section.

9.3   CRYSTAL DEFECTS

Ductile behavior of materials at elevated temperatures is achieved by the motion of crystal defects. In simple terms, a crystal defect is an error in the crystal lattice, and there are three basic types: (1) **point defects,** (2) **line defects** or **dislocations,** and (3) **planar defects** or stacking faults. The motion of defects gives rise to permanent strain without the material losing cohesion (i.e., without fracturing). Point and line defects are most important for the deformation of rocks. Planar defects, which arise from errors in the internal layering of minerals, play only a limited role in deformation. In order to understand diffusional mass transfer and crystal plasticity, we first need to take a more detailed look at point and line defects.

9.3.1   Point Defects

There are two types of point defects: (1) **vacancies** and (2) **impurities.** Vacancies are unoccupied sites in the crystal lattice (Figure 9.4a). Impurity atoms are (a) **substitutionals,** in which an atom in a lattice site of the crystal is replaced by a different atom (Figure 9.4b), and (b) **interstitials,** in which an atom is at a nonlattice site of the crystal (Figure 9.4c). Vacancies can migrate by exchange with atoms in neighboring sites (Figures 9.4d). At first glance, the concept of migrating vacancies sounds a bit odd, but when an atom moves into a vacant site, you can equally say that the vacancy moved. The general term for this process of atom or vacancy migration is diffusion. This important process is discussed later in the chapter. When we

apply a differential stress to a crystal, this causes a gradient in the vacancy concentration. Vacancies migrate down these concentration gradients, which causes material to deform ductilely, or to flow.

9.3.2   Line Defects or Dislocations

A line defect, usually called a **dislocation,** is a linear array of lattice imperfections (Figure 9.5). More formally, a dislocation is the linear array of atoms that bounds an area in the crystal that has slipped relative to the rest of the crystal (Figure 9.6). This definition is hardly informative at this point, so we first look at the geometry of two end-member configurations, the edge dislocation and the screw dislocation, before turning to the concept of slip in crystals.

An **edge dislocation** occurs where there is an extra half-plane of atoms in the crystal lattice. As illustrated in Figure 9.7a, there are 7 vertical planes of atoms at the top half of the crystal and only 6 vertical planes of atoms at the bottom half. The termination of the *extra half-plane* (the plane that ends halfway in the crystal) is the dislocation. It extends into the crystal as the

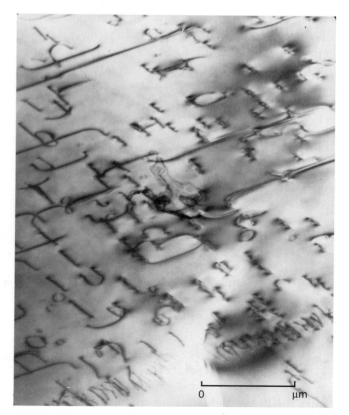

FIGURE 9.5   Transmission electron micrograph showing dislocation lines, loops, and arrays in experimentally deformed olivine.

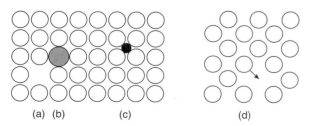

(a) (b)   (c)   (d)

FIGURE 9.4   Point defects: (a) vacancy, (b) substitutional impurity, (c) interstitial impurity, (d) vacancy migration.

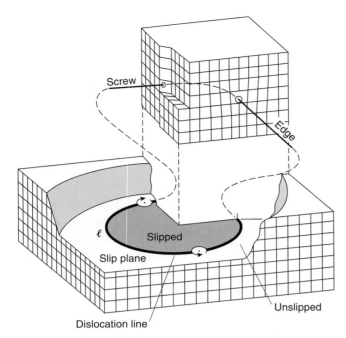

FIGURE 9.6 Geometry of a dislocation showing the edge- and screw-type dislocations and their geometrical relationship. The boundary between the unslipped and slipped portion of the crystal is the dislocation line, l.

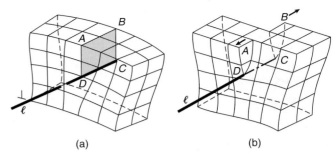

FIGURE 9.7 Types of dislocations. (a) The extra half-plane of atoms in an edge dislocation. (b) The corkscrew-like displacement of the screw dislocation. The dislocation line, l, is marked.

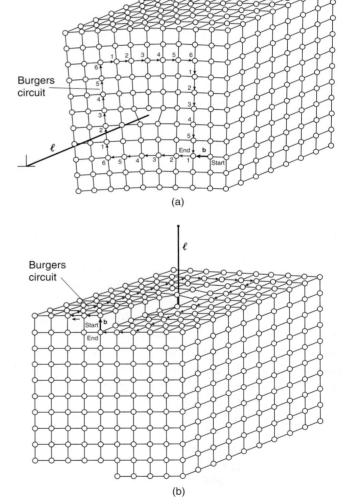

FIGURE 9.8 Determination of the Burgers vector, b, of a dislocation using a Burgers circuit. (a) The Burgers circuit around an edge dislocation (marked by l). (b) The Burgers circuit in a screw dislocation. The closure mismatch for both edge and screw dislocations is the Burgers vector, b. In the edge dislocation b⊥l, and in the screw dislocation b//l.

dislocation line, l (line *CD* in Figure 9.7a). The symbol for an edge dislocation is ⊥ or ⊤, depending on whether the location of the extra half-plane is above or below the associated glide plane of the crystal (see further). Imagine an axe that is stuck in a piece of wood. The presence of a dislocation causes a distortion of the crystal structure, just like a wedge distorts the log that is being split.

In **screw dislocations,** the atoms are arranged in a corkscrew-like fashion (Figure 9.7b); the axis of the screw marks the dislocation line (line *CD* in Figure 9.7b). A useful analogy of the geometry of a screw dislocation is a car parking deck, in which ramps carry cars up or down to individual floors. Many geologists, however, prefer a corkscrew analogy.

In a deformed crystal, an atom-by-atom circuit around the dislocation fails to close by one or more atomic distances, while a similar circuit around atoms in a perfect crystal would be complete. The arrow connecting the two ends of the incomplete circuit is called the **Burgers vector, b.** The length of the Burgers vector in most minerals is on the order of nanometers (1 nm = 1×10^{-9} m). For an edge dislocation, the Burgers circuit remains in the same plane (Figure 9.8a), while for a screw dislocation the circuit steps up or down to another plane (Figure 9.8b). Edge and screw dislocations can, therefore, be distinguished on the basis of the relationship between the Burgers vector

FIGURE 9.9 Dislocations in olivine from a Hawaiian mantle nodule. The dislocations appear by a decoration technique (described in the appendix), which allows for optical inspection. Width of view is ~200 μm.

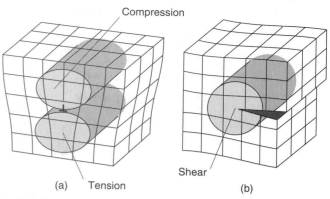

FIGURE 9.10 Geometry of the stress field (shaded region) around an edge dislocation (a) and around a screw dislocation (b).

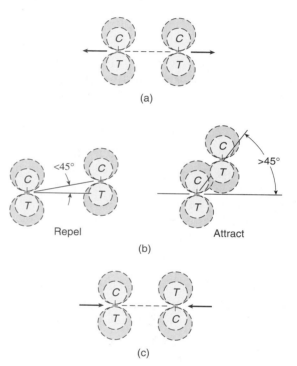

FIGURE 9.11 Interactions between neighboring edge dislocations. Regions labeled *C* and *T* are areas of compression and tension, respectively, associated with each dislocation. (a) Like dislocations on the same or nearby glide planes repel. (b) Like dislocations on widely separated glide planes may attract or repel depending on the angle between the lines joining the dislocations. (c) Unlike dislocations on the same or nearby glide planes attract.

and the dislocation line. For edge dislocations, the Burgers vector is *perpendicular* to the dislocation line, (Figure 9.8a) and for screw dislocations, the Burgers vector is *parallel* to the dislocation line (Figure 9.8b). These properties are used to determine the nature of imaged dislocations revealed by the electron microscope (see Section 9.8). Edge and screw dislocations are only end-member geometries; dislocations that consist of part edge and part screw components are called **mixed dislocations.** Besides being visible at very large magnifications in the electron microscope, crystal defect features may be indirectly seen by using a decoration technique (see appendix at the end of this chapter). Figure 9.9 shows an optical image of dislocations in the mineral olivine using a decorated sample.

Earlier we mentioned that the presence of dislocations distorts the crystal lattice, which gives rise to a local stress field around a dislocation. In an edge dislocation (Figure 9.10a) there is compressive stress on the side of the extra half-plane of atoms and tension on the opposite side. The earlier wood-splitting analogy serves to illustrate this pattern. The axe forces the wood apart, giving rise to compression, which may result in the axe becoming stuck. Just beyond the tip of the blade, however, there is tension, which is why you can split wood without the blade going all the way through. Similarly, in a screw dislocation we introduce shear stresses (Figure 9.10b). What is the effect of these local stresses? The role of compressive and tensile stresses is analogous to the behavior of magnets and charged particles. The compressive stress fields of edge dislocations repel, while the compressive and tensile fields of edge dislocations attract (Figure 9.11), just like the poles of two magnets attract or repel when

their polarities are reversed. Similarly, screw dislocations with the same sense of shear repel each other and those with opposite senses of shear attract. In a crude way you can say that dislocations are able to "see" each other by the stress fields they generate from the distortion of the crystal lattice. Later we will see that these stress fields permit dislocations to move, producing permanent distortions of the crystal, while lowering the internal strain energy. Remember that this internal strain energy is not the same as the applied stress arising from, for example, squeezing a crystal.

Edge and screw dislocations are end-member configurations, called **perfect dislocations,** because the Burgers vector has a length of one unit lattice distance (i.e., the length of one atomic bond, or multiples thereof). However, studies of minerals (e.g., calcite) have shown Burgers vectors that differ from one unit lattice distance; these are called *partial dislocations.* Partial dislocations may be formed by splitting a long Burgers vector into two or more components by the process of **dissociation.** Dissociation is energetically more favorable because it allows smaller displacements. Arrays of partial dislocations produce, for example, twinning in crystals (see Section 9.4).

9.4 CRYSTAL PLASTICITY

Dislocations are able to migrate through the crystal lattice if the activation energy for movement is achieved. The distortion of the crystal lattice around dislocations is one source of driving energy, as the system tries to achieve a lower internal strain energy. Applying a differential stress is another driving mechanism for dislocation motion. The associated distortion of solid phases is called crystal plasticity. Dislocation movement may occur by **glide** and a combination of glide and climb (**creep**), depending mainly on temperature. A third case of crystal-plastic behavior, **twinning,** occurs at low temperatures in some minerals.

9.4.1 Dislocation Glide

Deformation and temperature introduce energy into the crystal, which allows dislocations to move. However, dislocations are not free to move in any direction through the crystal. At low temperatures they are restricted to **glide planes** (or **slip planes**). The glide plane of a dislocation is the plane that contains the Burgers vector, **b,** and the dislocation line, **l.** Because a plane is defined by two nonparallel lines, each edge dislocation has one slip plane, because **b** and **l** are perpendicular. A screw dislocation on the other hand has many potential slip planes, because **b** and **l** are parallel. In crystallographic terms, a glide plane is a crystallographic plane across which bonds are relatively weak. Some crystals have only one crystallographic plane that is an easy glide plane; others may have many. Table 9.1 lists the dominant slip systems for some of the more common rock-forming minerals. Note that in many crystals more than one slip system may be active under similar conditions.

What is the actual process that allows the movement of dislocations? Nature has devised an energetically clever way for dislocations to move. Rather than simultaneously breaking all atomic bonds across a plane, such as occurs during fracturing, only bonds along the dislocation line are broken during an increment of movement. This requires much less energy than fracturing. Let us again turn to an analogy to illustrate this. The movement of dislocations is comparable to moving a large carpet across a room that contains heavy pieces of furniture. The easiest way to move the rug is to ruck up one end and propagate the ruck across the room. Energy is only needed to lift up selected furniture legs to propagate the ruck past these obstacles rather than lift all the furniture simultaneously. In nature, caterpillars and snakes move similarly by displacing one segment of their body at a time, instead of moving their entire body simultaneously. Edge dislocations move by successive breaking of bonds under the influence of a minimum stress acting on the glide plane, which is called the **critical resolved shear stress** (CRSS). If a crystal has several potential glide planes, it is likely that, for a given applied stress, the CRSS is exceeded on at least one and sometimes more than one of these glide planes. An edge dislocation moves when the unattached atoms at the bottom of the extra half-plane bond to the next atoms that are located directly below the glide plane. Thus the position of the extra half-plane moves relative to the dislocation without breaking all bonds in the extra half-plane (Figure 9.12a). A screw dislocation moves forward by shearing one atomic distance (Figure 9.12b), similar to tearing a piece of paper. While atomic bonds are broken and reattached when dislocations move toward the edge of a crystal, they leave a perfect crystal lattice behind. When a dislocation reaches the edge of the grain there are no more atoms below to attach to and the crystal becomes offset. This offset of the crystal edge produces stair-step structures on the surface of the crystal known as **slip bands,** which are sometimes visible on large crystal surfaces. Thus, the process of dislocation movement produces permanent strain without the material ever losing coherency.

9.4.2 Cross-Slip and Climb

It is not always possible for dislocations to propagate to the edge of the crystal. Point defects, such as impurity atoms that are bonded tightly to their neighbors, can resist the breaking of bonds that is required for dis-

| TABLE 9.1 | DOMINANT SLIP SYSTEMS IN COMMON ROCK-FORMING MINERALS |

Mineral	Glide plane and slip direction[a]	Comments
Calcite	$\{\bar{1}018\}<40\bar{4}1>$	e-twinning
	$\{10\bar{1}4\}<\bar{2}021>$	r-twinning
	$\{10\bar{1}4\}<\bar{2}021>$	r-glide
	$\{01\bar{1}2\}<2\bar{2}01>$ or $<\bar{2}021>$	f-glide
Dolomite	$\{\bar{1}012\}<10\bar{1}1>$	f-twinning
	$(0001)<2\bar{1}\bar{1}0>$	c-glide
	$\{01\bar{1}2\}<2\bar{2}01>$ or $<\bar{2}021>$	f-glide
Mica	$(001)<110>$	basal (c) slip
Olivine	$(001)[100]$	
	$\{110\}[001]$	
Quartz	$(0001)<11\bar{2}0>$	basal (c) slip
	$\{10\bar{1}0\}[0001]$	prism (m) slip, along c
	$\{10\bar{1}0\}<11\bar{2}0>$	prism (m) slip, along a
	$\{10\bar{1}1\}<11\bar{2}0>$	rhomb (z) slip

[a]Miller indices for equivalent glide planes from crystal symmetry are indicated by { }; specific glide planes are indicated by ( ); equivalent slip directions from crystal symmetry are indicated by < >; individual slip directions are indicated by [ ].
From: Wenk, 1985

location glide. Unfavorable stress fields of the dislocations themselves can also resist their motion, especially when many dislocations are present. Just consider trying to work your way past a car accident slowing the traffic in your lane, or even bringing it to a complete halt. Not surprisingly, obstacles that result from the presence of many immobile dislocations are called **pile-ups.** In order to overcome these obstacles, edge and screw dislocations must move out of their current glide plane, which they do by the processes of climb and cross-slip, respectively. The processes require additional energy beyond that for dislocation glide. Screw dislocations, unlike edge dislocations, are not confined to a single glide plane, because the dislocation line and Burgers vector are parallel. They can therefore leave one glide plane and move to another glide plane with relative ease, a process called **cross-slip** (Figure 9.13a). If it is so easy, why does cross-slip not occur all the time? Cross-slip requires that the dislocation abandons a favored glide plane (one with a short Burgers vector) for a less-favored one, and thus cross-slip takes place only if the CRSS on the less-favored plane is increased. Alternatively, raising the temperature lowers the CRSS that is needed for cross-slip, because atomic bonds are weakened, and cross-slip occurs more easily.

Edge dislocations cannot cross-slip because they have only one glide plane. However, they can **climb** to a different, parallel glide plane if there are vacancies to accept the lowest atoms of the extra half-plane (this is shown two-dimensionally in Figure 9.13b). Climb, therefore, involves diffusion (see further), and because the rate of vacancy production increases with rising temperature, the efficiency of dislocation climb is temperature dependent. Both cross-slip and climb are activated at temperature conditions that exceed those for dislocation glide in a mineral given the same stress conditions, and therefore they typically occur at deeper (i.e., hotter!) levels in the Earth. Although it is not possible to identify a fixed depth at which cross-slip and climb occur, because this is a function of the mineral as well as Earth's thermal structure, as a general guide, we can specify the temperature values at which these processes occur for different minerals. Glide and climb occur at temperatures greater than 300°C for quartzitic rocks and carbonates, and at

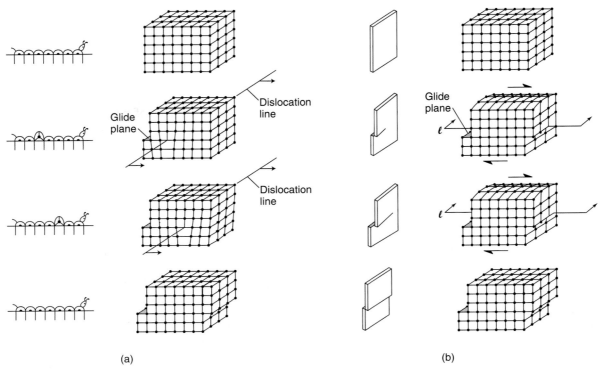

FIGURE 9.12 Dislocation glide. (a) Movement of an edge dislocation, which may be likened to the movement of a caterpillar. (b) Movement of a screw dislocation, which is analogous to tearing a sheet of paper, with the screw dislocation at the tip of the tear. After the dislocation passes through the lattice, it leaves behind a strained crystal with a perfect crystal lattice structure. The dislocation line, **l,** and the glide planes (shaded) are shown.

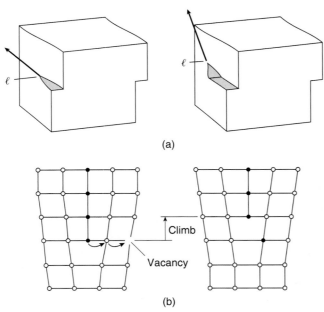

FIGURE 9.13 Cross-slip of a screw dislocation (a), and climb of an edge dislocation (b) by diffusion of atoms (arrows).

higher temperatures (>500°C) for such common minerals as dolomite, feldspar, and olivine. In the literature you find that the term **dislocation creep** is used for the combined activity of glide and climb.

While like dislocations often repel one another, unlike dislocations attract and may annihilate each other. **Dislocation annihilation** is one way of reducing the internal strain energy that arises from lattice distortions in a crystal. For example, two edge dislocations lying in the same glide plane with the extra half-plane of one dislocation inserted upwards (positive edge dislocation) and the other downward (negative edge dislocation) annihilate each other (Figure 9.14a). Similarly, convergence of screw dislocations with Burgers vectors in opposite directions also results in annihilation. Two dislocations of opposite sign but on different glide planes may still attract, but they cannot fully annihilate each other. In such cases, a point defect remains (such as a vacancy; Figure 9.14b). Because climb and cross-slip increase the probability of dislocation annihilation, the rate of dislocation annihilation is also temperature dependent.

9.4.3 Mechanical Twinning

Twins are a common feature in many minerals. You may see them with the hand lens in the minerals plagioclase and calcite. In thin section, under crossed polarizers, they are easily recognized by their extinction behavior and lathlike shape as you rotate the stage. Twins that develop during the growth of a crystal, called **growth twins,** say little or nothing about the conditions of deformation (i.e., stress and strain). In

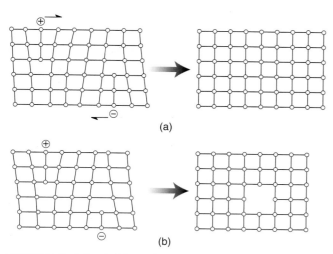

FIGURE 9.14 (a) Two edge dislocations with opposing extra half-planes that share a glide plane move in opposite direction to meet and form a perfect crystal. (b) When they move in different glide planes, a vacancy may be formed when they meet.

contrast, minerals such as calcite form twins in response to an applied stress; these are called **mechanical twins** (Figure 9.15). We'll first have a look at twinning in general and then see what information mechanical twins can provide for deformation studies.

Mechanical twinning is a type of crystal plastic process that involves the glide of **partial dislocations.** A surface imperfection, the **twin boundary,** separates two regions of a twinned crystal. The lattices in these two portions are mirror images of each other; in other words, a twin boundary is a mirror plane with a specific crystallographic orientation. As a rule, twinning planes cannot already be mirror planes in the untwinned crystal, and mechanical twinning is therefore most common in low-symmetry minerals such as trigonal calcite and dolomite, and triclinic feldspar. Recall that crystal symmetry is a geometric operation that repeats a crystal plane in another position. For reference, Table 9.2 lists the seven crystal systems and their symmetries that you may have learned in your mineralogy and/or petrology class.

Mechanical twins are produced when the resolved shear stress acting on the future twin boundary exceeds a critical value (the CRSS for twinning). During twinning, the crystal lattice rotates in the direction that produces the shortest movement (smallest linear displacement) of atoms, with a unique rotation angle. As such, mechanical twinning has similarities with dislocation glide, but differs in two aspects. First, atoms are not moved an integral atomic distance as in glide, but

FIGURE 9.15 Calcite e-twins in marble from southern Ontario (Canada). Width of view is ~4 mm.

TABLE 9.2	CRYSTAL SYSTEMS	
System	Symmetry	Crystal Axes
Triclinic	1 one-fold axis or center of symmetry	$a \neq b \neq c, \alpha \neq \beta \neq \gamma \neq 90°$
Monoclinic	1 two-fold axis or 1 symmetry plane	$a \neq b \neq c, \alpha = \gamma = 90°, \beta \neq 90°$
Orthorhombic	3 two-fold axes or 3 symmetry planes	$a \neq b \neq c, \alpha = \beta = \gamma = 90°$
Trigonal	1 three-fold axis	$a_1 = a_2 = a_3 \neq c, \beta = 90°$
Hexagonal	1 six-fold axis	$a_1 = a_2 = a_3 \neq c, \beta = 90°$
Tetragonal	1 four-fold axis	$a = b \neq c, \alpha = \beta = \gamma = 90°$
Cubic	4 three-fold axes	$a = b = c, \alpha = \beta = \gamma = 90°$

a, b, c describes the length of the crystal axes; α is the angle between b and c; β is the angle between a and c; γ is the angle between a and b.

rather only by some fraction of the atomic distance; consequently, twinning involves partial dislocations. Secondly, the twinned portion of a grain is a mirror image of the original lattice (Figure 9.16a), whereas the slipped portion of a grain has the same crystallographic orientation as the unslipped portion of the grain (Figure 9.16b). For deformation studies we are interested mostly in mechanical twinning; that is, twins produced by stress. We digress briefly to explore one application using the mineral calcite.

The fact that twinning takes place along specific crystallographic planes in a calcite crystal,[1] and that rotation occurs over a specific angle and in a specific sense, allows us to use twinning as a measure of finite strain and differential stress. The atomic structure of calcite twins is illustrated in Figure 9.17. (Note the specified rotation angle of the crystallographic c-axis, which is perpendicular to the planes containing the CO_3 groups, and that of the crystal face.) In Figure 9.18a, a deformed grain $A'B'CD$ with one twin is shown; the original grain outline is $ABCD$, whose sides are parallel to calcite crystal planes. From this figure you can see that the shear strain for the twinned grain is

$$\gamma = \tan \psi = q/T \qquad \text{Eq. 9.2}$$

For one twin, $q = p$, so

$$\gamma = \frac{2t \tan (\phi/2)}{T} \qquad \text{Eq. 9.3}$$

---

[1]We will only consider e-twins ($\{1018\}<4041>$) with a rotation angle for the c-axis of $52.5°$, and a CRSS of 10 MPa.

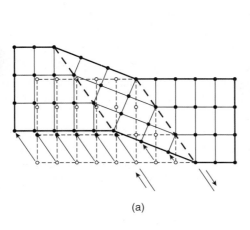

(a)

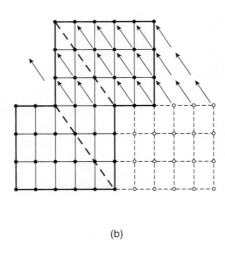

(b)

FIGURE 9.16 Schematic illustration of mechanical twinning (a). The heavy outline marks a twinned grain, in which the twin boundaries (heavy dashes) are mirror planes. The atomic displacements are of unequal length and generally do not coincide with one atomic distance. Closed circles are atoms in final structure and open circles give the original positions of displaced atoms. Twinning contrasts with dislocation glide (b), in which atoms move one or more atomic distances in the glide plane (heavy dashed line).

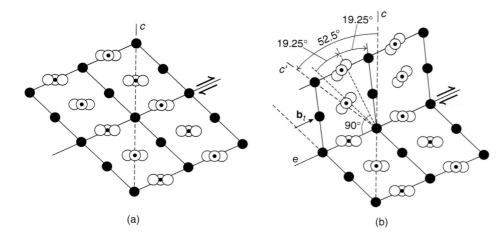

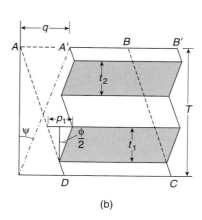

FIGURE 9.17 Calcite crystal lattice showing layers of Ca (large black dot) and CO_3 groups (C is small dot, O is large open circle); the crystallographic c-axis is marked (a). The twinned calcite lattice in (b) shows the partial dislocation (b_t) and angular rotations of the c-axis and the crystal face.

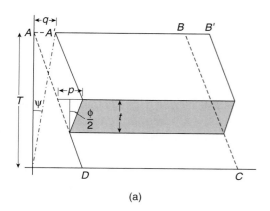

FIGURE 9.18 Calcite strain-gauge technique. An original grain ABCD (a) with a single twin of thickness, t (shaded region). In (b) a grain with multiple twins (shaded regions) is shown.

where T is the grain thickness and t is the twin thickness. For a grain containing several twins (Figure 9.18b) the shear strain is obtained by adding the strain due to each twin, or

$$\gamma = \frac{2}{T} \sum_{i=1}^{n} t_i \tan (\phi/2) \qquad \text{Eq. 9.4}$$

where n is the number of twins in the grain. Given that the angle ϕ is constant in the case of calcite ($\approx 38°$; Figure 9.18b), Equation 9.4 simplifies to

$$\gamma = \frac{0.7}{T} \pi \sum_{i=1}^{n} t_i \qquad \text{Eq. 9.5}$$

So, if we measure the total width of twins and the grain size perpendicular to the twin plane we can obtain the total shear strain for a single twinned grain. In an aggregate of grains, the shear strains will vary as a function of the crystallographic orientation of individual grains relative to the bulk strain ellipsoid, and we use this variation to determine the orientation of the principal strain axes by determining the orientations

for which the shear strains are zero[2] and maximum. This strain analysis technique is called the **calcite strain-gauge method.** Looking again at Figure 9.18 and Equation 9.5, we can now determine the maximum amount of shear strain that can be accumulated using twinning: γ_{max} occurs when the entire grain is twinned, so $t = T$; thus, $\gamma_{max} = 0.7$, or $X/Z \approx 2$. This maximum contrasts with the amount of strain that can accumulate during dislocation glide, which is unrestricted. Moreover, methods for the determination of the differential stress for an aggregate with twinned grains have been developed that use the number of activated twin planes. Thus, calcite twinning analysis can give both strain and differential stress magnitudes for naturally deformed carbonates.

The calcite strain-gauge technique has proven to be very useful in studying stress and strain fields in limestones that were subjected to small strains, the kinematics of folding, the formation of veins, the early

---

[2]Details of the method are described elsewhere, and a computer routine is normally used for the analysis.

deformation history of fold-and-thrust belts, and even deformation patterns in continental interiors. The great advantage of this method lies in the fact that twinning occurs at low homologous temperature and low differential stress, and that the orientation and magnitude of even small finite strains are recorded.

9.4.4 Strain-Producing versus Rate-Controlling Mechanisms

We saw previously that dislocations are not stationary elements of a crystal, but are able to move (glide), and that they leave behind what is called the *slipped portion* of the crystal. Perhaps surprisingly, this slipped portion has no crystallographic distortion after the dislocation has passed through this part of the crystal. This ability of dislocation to move through a crystal brings us back to the earlier definition of a dislocation that was given without much explanation at the time (Section 9.3): a dislocation is the linear array of atoms that bounds an area in the crystal that has slipped one Burgers vector more than the rest of the crystal. Having examined the various dislocation motion mechanisms, we need to recognize an important distinction between dislocation glide on the one hand, and dislocation cross-slip and climb on the other hand. Dislocation glide is the process that produces a change in the shape of grains; it is therefore the main **strain-producing mechanism** of crystal plasticity. Cross-slip and climb facilitate dislocation glide, but by themselves produce little finite strain; they allow a dislocation to leave its original glide plane, to bypass an impurity, for example. Cross-slip and climb are therefore the **rate-controlling mechanisms** of crystal plasticity, and limit the resulting strain rate. Because climb occurs at temperatures that are higher than those required for glide in a mineral, you also find the terms **low-temperature creep** for dislocation glide (and twinning) and **high-temperature creep** for dislocation glide plus climb.

9.4.5 Where Do Dislocations Come From?

Nothing in life is perfect! You have undoubtedly heard and probably experienced this yourself, and the same goes for a mineral's life. Defects, such as dislocations, are a part of all minerals, for good reasons: The small offsets that occur at the edges of crystals containing dislocations (on the order of nanometers) are used as nucleation sites during mineral growth; while for deformation, dislocations are necessary to enable the shape change during crystal plasticity. So far, we have only talked about the situation where a couple of dislocations occur at the same time, but the number of dislocations in a mineral, the **dislocation density,** N, is actually quite large. For example, "perfect" grains that have grown from a melt have a dislocation density of 10^6 cm^{-2}, and this density is several orders of magnitude larger in deformed grains. Even near perfect crystals that are grown in the laboratory still have hundreds of dislocations per square centimeter (cm^2). So what is this strange unit "cm^{-2}" for dislocation density? Dislocation density, N, describes the total length of dislocations per volume of crystal; thus $N =$ length/volume, so the unit of N is $[l]/[l^3] = [l^{-2}]$. Measuring dislocation length per unit volume is not a very convenient way to determine N, so practically we measure the number of dislocations (dimensionless scalar) that intersect an area (l^2), which gives the unit $[l^{-2}]$. Later, in Section 9.9, we will give an example of a dislocation density calculation.

In order to obtain appreciable strains from dislocation movement, we will need a great many dislocations. We have already learned that strain is produced by dislocations moving to the edge of the crystal (Figure 9.12), leaving a perfect lattice behind. So, in order for crystal plastic processes to proceed we actually need to generate dislocations. We earlier mentioned that dislocation density is greater in strained grains than in unstrained, "perfect" grains, which suggests that dislocations are generated during deformation. One mechanism for dislocation generation (or multiplication) is by **Frank-Read sources** (Figure 9.19). Consider a dislocation that is anchored at two points, A and B; this pinning may arise from impurities, climb, or interaction with other dislocations (not shown in the figure). During glide, the A–B dislocation will bow out because it is pinned at its edges (Figure 9.19b–d), and eventually this produces the kidney-shaped loop in Figure 9.19e and 9.19f. Note that the dislocation segments at a and b in Figure 9.19f are opposite in sign because their Burgers vectors are opposite. So as a and b come together they annihilate (Figure 9.19g), forming a new A–B dislocation line, while leaving the old loop present (Figure 9.19h). The process starts again for the new A–B dislocation line while the first loop continues to glide. Because there is no restriction on the number of cycles, a great many dislocation loops are generated in this manner, which occurs for both edge and screw dislocations. This and other dislocation multiplication mechanisms collectively produce the high dislocation densities that are required for grains to deform by crystal plastic processes.

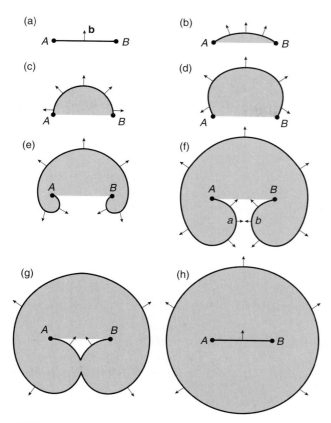

FIGURE 9.19 Dislocation multiplication in a Frank-Read source. (a) A pinned dislocation with Burgers vector, **b**, bows out during glide (b–g) to form a new dislocation (h). The slipped portion of the grain is shaded.

9.5 DIFFUSIONAL MASS TRANSFER

Flow of rocks also occurs by the transfer of material through a process called **diffusion.** We'll discuss three diffusion-related deformation mechanisms that are important for natural rocks: (1) **pressure solution,** (2) **grain-boundary diffusion,** and (3) **volume diffusion.** Diffusion occurs when an atom (or a point defect) migrates through a crystal. The process is strongly temperature dependent, because thermal energy causes atoms to vibrate, facilitating the breaking and reattachment of bonds. Increasing the temperature of a material proportionally increases the ability of individual atoms to jump to neighboring vacant sites. For example, at the melting temperature of Fe ($T_h = 1$), the **jump frequency,** Γ, of vacancies is on the order of 10^{10} per second. The jump distance, r (the distance between atoms in the crystal structure), for each jump

is 10^{-10} m (0.1 nm). We can determine the average area, R^2, for a vacancy by Einstein's equation:[3]

$$R^2 = \Gamma t r^2 \qquad \text{Eq. 9.6}$$

where t = time. If we use $t = 1$ s, then $R^2 = 0.1$ mm^2, at $T_h = 1$ for Fe metal. This area seems small, so you might think at first that the process is relatively insignificant. However, considering that geologic time is measured in millions of years, the value of R^2 becomes quite large. For example, at $t = 1$ m.y. (3.1×10^{13} s) the value of R^2 is >3000 m^2. Such areas, however, are only representative for minerals in rocks near their melting temperature, which is not the typical condition during rock deformation. At lower T_h, diffusion distances are orders of magnitude less. Another aspect of diffusion that needs to be appreciated is that R^2 does not define the linear distance between the original position of an atom and its position after time t. Diffusion is nondirectional in an isotropic stress field; it is, what we call, a **random-walk process.** So the final distance traveled is distinct from the path and area covered.

Theoretical arguments, which we will not discuss here, define a diffusion coefficient, D, for a given mineral, describing movement of a species down a concentration gradient:[4]

$$D = (\Gamma/6)\, r^2 \qquad \text{Eq. 9.7}$$

The diffusion coefficient has the dimension area/time. We can rewrite Equation 9.7 in a form that shows the temperature dependence for diffusion and a minimum energy for migration to occur:

$$D = D_o \exp(-E^*/RT) \qquad \text{Eq. 9.8}$$

where D_o is a material constant for diffusion that is empirically determined, E^* is the activation energy for migration (kJ/mol), R is the gas constant (8.31 J/mol · K), and T is absolute temperature (in K).[5] We present diffusivity in this particular form, because it is easy to compare with the constitutive equations for flow that were given in Chapter 5 and the relationships discussed later in this chapter.

---

[3]Another Albert Einstein (1879–1955) equation.
[4]Strictly speaking these equations are for vacancy movement, and define D_{vac}.
[5]We may also write this equation involving Boltzmann's constant (k), in which case E^* is given in a different form; k and R are related by the equation $R = kN_A$, where N_A is Avogadro's number (6.02×10^{23}mol^{-1}), which gives $k = 1.38 \times 10^{-23}$. Note that "exp(a)" means e^a.

Two types of solid diffusion in crystals are (1) grain-boundary diffusion or Coble creep, and (2) volume diffusion or Nabarro-Herring creep. When mass transfer involves a reactive and transporting fluid phase, the process that is geometrically similar to grain-boundary diffusion is called pressure solution. We will discuss each of these three mechanisms.

9.5.1 Volume Diffusion and Grain-Boundary Diffusion

Given sufficient time, diffusing vacancies reach the surface of the crystal where they disappear. To see how this causes deformation, consider a crystal that is being subjected to a differential stress (Figure 9.20). The vacancies migrate toward the site where stress is greatest and the atoms move to the sides where the stress is least. This results in an overall change in the distribution of mass, producing a change in shape of the crystal. But realize that this occurs without large-scale distortion of the crystal lattice. The diffusion of vacancies can occur through the entire body of a crystal or can be concentrated along a narrow region at its grain boundary (Figure 9.20a); both result in a permanent shape change as shown in Figure 9.20b. These deformation mechanisms are called **volume diffusion** (or **Nabarro-Herring creep**) and **grain-boundary diffusion** (or **Coble creep**), respectively. Thus, in the presence of a non-isotropic stress field we find that diffusion is directional.

Because both Nabarro-Herring creep and Coble creep achieve strain by the diffusion of vacancies, the strain rate for each mechanism is a function of the diffusion coefficients (volume diffusion [D_v] and grain-boundary diffusion [D_b] coefficients respectively), but also of the grain size (d):

$$\dot{e}_{Coble} \cong D_b/d^2 \qquad \text{Eq. 9.9}$$
$$\dot{e}_{Nabarro-Herring} \cong D_v/d^3 \qquad \text{Eq. 9.10}$$

These simplified relationships emphasize the critical importance of grain size in diffusional creep: a larger grain size results in a less efficient process, so a lower strain rate.

The activation energy for grain-boundary diffusion (included in D_b) is less than that for volume diffusion (included in D_v), and the grain-size dependence of volume diffusion is larger. Thus, Coble creep is a more efficient process in crustal rocks than Nabarro-Herring creep, so that the latter is restricted to high-temperature regions (e.g., temperatures in the mantle) and/or to materials with very small grain sizes.

9.5.2 Pressure Solution

Pressure solution is a mass transfer process that occurs in natural rocks at temperatures much lower than those for solid diffusion. The process is geometrically similar to grain-boundary diffusion, but involves the presence of a fluid film on grain boundaries. It is important in crustal rocks because material transfer occurs at temperatures well below those required for vacancy diffusion, thanks to a chemically active fluid film that dissolves the crystal. The dissolved ions then move along a chemical gradient that arises from differential solubility in the presence of a differential (non-isotropic) stress to regions of deposition. Recall that fluids do not support shear stresses (Chapter 3), so pressure solution only works if the fluid film is "attached" to the grain boundary by chemical bonds; thus, the fluid does not move, but dissolved atoms do. Areas of high stress, say surfaces perpendicular to the maximum principal stress, exhibit enhanced solubility and the dissolved material is transported to regions under lower stress (surfaces perpendicular to the minimum principal stress). The geometric properties of the process are very similar to our earlier description of grain-boundary diffusion and, indeed, pressure solution produces shape changes like that in Figure 9.20b, except that it occurs at the low temperatures encountered near the Earth's surface. A way to distinguish between these diffusional deformation mechanisms is to

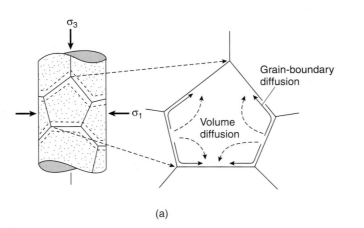

(a)   (b)

FIGURE 9.20 Diffusional flow by material transport through grains (volume diffusion or Nabarro-Herring creep) and around grains (grain-boundary diffusion or Coble creep), with a differential stress (a) that produces shape change (b).

use the terms **fluid-assisted diffusion** for pressure solution and **solid-state diffusion** for Nabarro-Herring creep and Coble creep; in colloquial terms we sometimes call them "wet diffusion" and "dry diffusion," respectively. Because fluids are abundant in shallow crustal rocks and these have low ambient temperatures, pressure solution is an important deformation mechanism in upper crustal rocks.

We infer the past activity of pressure solution diffusion in rocks from the presence of, for example, stylolites in limestones, grain overgrowths in sandstones, and cleavage and pressure shadows in some slates (Figure 9.21; see also Chapter 11). In contrast to dry diffusion, the distance over which material may be transported by fluid-assisted diffusion is not limited to individual grains, but can be substantial, particularly if the dissolved ions migrate into the pore fluid of the rock. Movement of pore fluid (i.e., groundwater flow) can flush the dissolved ions completely out of the rock, resulting in substantial volume loss. On the mesoscale, pressure solution may result in the formation of alternating layers of different composition, such as quartz-rich and mica-rich layers, in a process called **differentiation** (see Chapter 11). Alternatively, the dissolved ions may precipitate as vein fillings in cracks (Figure 9.21). The widespread occurrence of these pressure-solution structures in natural settings emphasizes the geologic importance of this deformation mechanism.

The strain rate associated with pressure solution is a function of the area and the rate of atoms that go into solution (i.e., the solubility of a material) in the fluid:

$$\dot{e}_{\text{pressure solution}} \cong D_f/d^2 \qquad \text{Eq. 9.11}$$

where D_f is the diffusion coefficient of a phase in a fluid and d is the grain size.

9.6 CONSTITUTIVE EQUATIONS OR FLOW LAWS

We defined the rate at which shape change occurs as the strain rate, $\dot{e}$ (Section 5.1). Since dislocation movement is a function of the differential stress (either arising from internal distortion or externally imposed on the system), the ambient temperature, and the activation energy for breaking bonds, the rate at which strain occurs by dislocation movement is a function of all these parameters. This relationship is described by a **constitutive equation** or **flow law,** with the general form

$$\dot{e} = A\, f(\sigma_d)\, \exp(-E^*/RT) \qquad \text{Eq. 9.12}$$

where A is a material constant, E^* is the activation energy, R is the gas constant, T is the absolute

FIGURE 9.21 Bedding-perpendicular pressure solution seams (stylolites) and veins (white structures) in argillaceous limestone (Appalachians, Pennsylvania, USA). The middle bed is pure carbonate and does not contain as many seams. Note that the stylolites cut across bedding.

temperature (in K), and $f(\sigma_d)$ represents a differential stress function; characteristic values for these parameters were given in Table 5.6. In this chapter we focus on the stress function, $f(\sigma_d)$, which is determined from experiments on natural rocks and common minerals. For dislocation glide (low-temperature creep) the function of stress is *exponential,* so the flow law is of the form

$$\dot{e} = A \exp(\sigma_d) \exp(-E^*/RT) \qquad \text{Eq. 9.13}$$

Because of the form of this relationship, dislocation glide is also called **exponential creep.**

For dislocation glide and climb (high-temperature creep), which is typical for deep crustal and mantle rocks, the stress is raised to the *power n.* This flow law takes the general form

$$\dot{e} = A \, \sigma_d^n \exp(-E^*/RT) \qquad \text{Eq. 9.14}$$

Climb-assisted glide is therefore also called **power-law creep,** and the power *n* is called the *stress exponent.*

In an earlier section we presented the diffusion coefficient for point defects (Equations 9.4–9.6). The motion of individual defects or atoms is similarly a function of differential stress and that has the form

$$\dot{e} = A \, \sigma_d \exp(-E^*/RT) \, d^{-r} \qquad \text{Eq. 9.15}$$

You will notice that the stress function of Equation 9.15 is the same as Equation 9.14, except that the stress exponent, *n,* equals 1. This means that diffusion is *linearly* related to the strain rate and, therefore, that diffusional creep is a linear viscous process (or Newtonian viscous process; Chapter 5). Note, however, that the strain rate for diffusional creep is nonlinearly related to the grain size, and that the value of *r* is in the range of 2 to 3 (Equations 9.5, 9.6, and 9.15).

We will see later that these various creep regimes produce characteristic microstructures, but let us revisit the deformation experiments of Chapter 5 and interpret their behavior in light of what we now have learned about defects and crystal plasticity.

9.7 A MICROSTRUCTURAL VIEW OF LABORATORY BEHAVIOR

While our discussion has been pretty theoretical and perhaps esoteric up to this point, it was necessary to understand that defect microstructures can explain how materials respond to stress (i.e., rheology). Deformation experiments typically show the same behavior: after an initial elastic stage, permanent (ductile) strain accumulates. The elastic component is recoverable and does not involve crystal plastic processes, but the ductile component of the curve is mostly achieved by the motion of defects. Strain accumulates at constant stress (steady-state flow), or requires increasingly higher stress (work hardening) at constant $\dot{e}$. From a microstructural perspective, **steady-state flow** implies that the generation, motion and removal of dislocations is sufficiently fast to achieve strain at a constant rate for a certain stress level. But what about a microstructural explanation for **work hardening** (Section 5.4)? Limited climb and cross-slip at low temperatures prevent dislocations from slipping past inclusions and other obstacles. Combined with a decreased frequency of dislocation annihilation, this causes dislocation density in a crystal to increase, which affects the ability of dislocations to glide because they interact with one another. Recall that the ability of dislocations to glide produces strain, so dislocation tangles restrict their motion and the rate of strain accumulation decreases (unless the differential stress increases).

With so many dislocations in a crystal, the chance of interaction is large, so let's look at this in some detail. Figure 9.22 shows a situation where one edge dislocation (D_1) moves relative to an edge dislocation (D_2) with a different slip plane (we keep dislocation D_2 stationary for the convenience of illustrating our point). As D_1 passes through D_2, the dislocation line l_2 is offset (Figure 9.22b). This offset, called a **jog,** has an important implication. Whereas the Burgers vector $\mathbf{b}_2$ for dislocation D_2 remains the same along the dislocation line, its glide plane has changed at the jog. Motion of dislocation D_2 needs a critical resolved shear stress (CRSS) that allows glide on the initial slip plane, but it also needs movement on a second slip plane for that dislocation. Because the values of the CRSS differ for crystal planes in different orientation, the ability for glide varies along the dislocation line when a jog is present, resulting in a dislocation that is held back at the jog (Figure 9.22c). This reduced ability of a dislocation to move is what causes the material to strengthen, expressed as work hardening in experiments. Diffusion of vacancies to segments of the dislocation can overcome the restriction, so work hardening is much less important in the high-temperature creep regime. Thus, the presence of impurities that pin dislocations or high dislocation densities that restrict

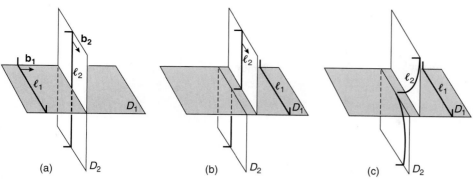

FIGURE 9.22   The formation of a jog from the interaction of two mobile edge dislocations. For simplicity, dislocation D_2 is initially kept stationary while dislocation D_1 moves; the glide planes (shaded and unshaded), Burgers vectors (**b**), and dislocation lines (**l**) for each edge dislocation are shown (a). As D_1 passes through dislocation line l_2, a small step of one Burgers vector (b_1) length is created; this small step is a jog, with a differently oriented dislocation line segment but the same b_2 (b). As a consequence, the glide plane, containing l_2 and b_2, is different along l_2. In fact, the glide plane of the jog is the same of that for D_1, but with a different Burgers vector. Assuming that the CRSS for glide differs in different directions, the ability of D_2 to move is no longer the same along l_2, and the jog pins the dislocation by anchoring a segment of l_2 (c).

dislocation motion (tangles) result in work hardening of materials, which is overcome by the activity of dislocation climb. We also observe *work softening* in some experiments (Section 5.4), but we wait until Section 9.9, where we discuss grain-size reduction, to offer an explanation.

9.8   IMAGING DISLOCATIONS

The dislocation density of an unstrained crystal is on the order of 10^6 cm^{-2} and this value is orders of magnitude higher in strained crystals. A 1 cm^3 volume of a strained quartz crystal with a dislocation density of 10^9 cm^{-2} will have a total dislocation line length of 10^9 cm or 10,000 km (the distance from Earth's equator to pole). Obviously, dislocations and other defects must be quite small to fit so many in a volume that small. We therefore need very large magnifications to see them, and this generally involves transmission electron microscopy (TEM). This technique permits imaging of microstructures at magnifications of up to 500,000×, with a resolution of better than 1 nm (1 nanometer = 10^{-9} m). Such high resolution is not usually necessary for the examination of defect microstructures and more conventional TEM work is done at magnifications of 10,000–100,000× (Figure 9.23). TEM samples require sufficient thinning of the material that it is transparent to the electron beam;

generally the thickness of the thin foil is a few hundred nanometers. Crystal defects in thin foils are revealed by diffraction contrasts that result from lattice distortions surrounding the defect, which allows us to determine both the Burgers vector of a dislocation and the crystallographic orientation of the dislocation line (Figure 9.23). Once these are established, it is possible to determine the nature of a dislocation (i.e., whether edge, screw, or mixed) from the angular relationship between **b** and **l**. Recall that for edge dislocations **b** and **l** are perpendicular, and for screw dislocations **b** and **l** are parallel.

In Figure 9.9 we observed dislocation clusters in olivine using the standard petrographic microscope and a decoration technique. Individual dislocation geometries in olivine in the transmission electron micrograph were shown in Figure 9.5, where we can distinguish arrays of parallel dislocations (lower right), straight dislocations (upper half), and dislocation loops (lower left). The terminations of the dislocations in this photomicrograph arise from the intersection of the dislocation line with the lower and upper boundaries of the thin foil; the thicker the foil the longer the dislocation would appear, until it intersects the crystal edge. Note the geometric similarity between the optical and transmission electron micrographs of Figures 9.5 and 9.23, which both contain straight dislocation lines with sharp angular bends as predicted by slip systems in olivine (Table 9.1).

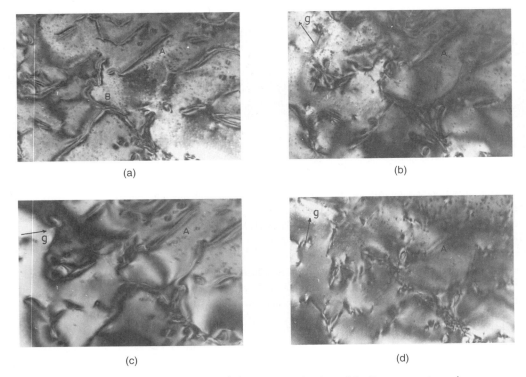

FIGURE 9.23 Dislocations in calcite (a) and determination of the Burgers vector using transmission electron microscopy (TEM). View of the same area for different diffracting lattice planes: (b) (0006), (c) (10$\bar{1}$2), (d) (10$\bar{1}$4); the orientation of the lattice plane in each image is indicated by its pole (marked by vector **g**). The presence of dislocation A in (b) and (c) rules out all possible Burgers vectors in calcite with the exception of <$\bar{2}$021>; this is confirmed by the absence of contrast from dislocation A in (d). This (time-consuming) procedure is called the *invisibility criterion*. The Burgers vector is neither exactly perpendicular nor parallel to the dislocation line, so dislocation A is a mixed dislocation. Width of view of each TEM image is ~1.7 μm.

9.9 DEFORMATION MICROSTRUCTURES

Can we recognize the (past) activity of a particular deformation mechanism and, by inference, determine the rheologic conditions during deformation? The answer to this question is mostly yes, because deformation mechanisms produce relatively characteristic microstructures that can be observed in hand specimens and thin section. However, a rock's "memory" can be incomplete; only the latest deformation mechanism may be preserved. Once we establish the operative deformation mechanism from microstructures we can proceed to make predictions about the conditions of temperature, stress, and strain rate during deformation, which is the ultimate reason to study microstructures. Throughout the book we use the term **microstructure** to describe geometric characteristics of rocks on the scale of the microscope; for example, twins are a microstructural element. We use the term **(micro)fabric,** which means different things to differ-

ent people, with an appropriate modifier (such as dimensional-preferred fabric for geometric alignments). In Chapter 12 we introduce yet another type of fabric, **crystallographic-preferred fabric,** that describes the degree of crystal lattice orientation of a mineral aggregate.

In the next several pages we look at the characteristic microstructures in deformed rocks that arise from three mechanisms: recovery, recrystallization, and superplastic creep. Mechanical twinning, a fourth mechanism, was discussed earlier (Section 9.5). To assist you with the many new concepts that will be introduced, brief descriptions of the processes, characteristic microstructures, and some related terms of crystal plastic and diffusional creep are given in Table 9.3.

9.9.1 Recovery

The presence of crystal defects such as dislocations and twins increases the **internal strain energy** of a grain, because the crystal lattice surrounding the defects is distorted. The atomic bonds are bent and

TABLE 9.3	SOME TERMS AND CONCEPTS RELATED TO CRYSTAL PLASTICITY AND DIFFUSIONAL CREEP
Annealing	Loosely used term for high-temperature grain adjustments, including static recrystallization and grain growth.
Bulge nucleation	A type of migration recrystallization in which a grain boundary bulges into a grain with higher internal strain energy, forming a recrystallized grain.
Dislocation wall	Concentration of dislocations in a planar array.
Dynamic recrystallization	Formation of relatively low-strain grains under an applied differential stress.
Foam structure	Recrystallized grain structure characterized by the presence of energetically favorable grain-boundary triple junction (at $\approx 120°$ angles).
High-angle boundary	Boundary across which the crystallographic mismatch exceeds 10°; characteristic of recrystallization.
Low-angle boundary	Tilt boundary across which the crystallographic mismatch is less than 10°; characteristic of recovery.
Migration recrystallization	Recrystallization mechanism by which grain boundaries move driven by a contrast in strain energy between neighboring grains.
Polygonized microstructure	Recovery structure showing elongate to blocky subgrains (mostly used for phyllosilicates).
Recovery	Process that forms low-angle grain boundaries by the temperature-activated rearrangement of dislocations.
Recrystallization	Mechanism that removes internal strain energy of grains remaining after recovery, producing high-angle grain boundaries that separate relatively strain-free (recrystallized) grains.
Recrystallized grains	Relatively low-strain grains that are formed by recrystallization.
Rotation recrystallization	Recrystallization mechanism by which dislocations pile up in a tilt boundary, thereby "rotating" the crystal lattice of the area that is enclosed by the tilt boundary.
Static recrystallization	Formation of strain-free grains after deformation has stopped (i.e., differential stress is removed).
Subgrain	Area of crystallographic mismatch that is less than 10° relative to the host grain.
Subgrain rotation	Rotation recrystallization mechanism by which dislocations continue to move into a low-angle tilt boundary surrounding a subgrain, thereby increasing the crystallographic mismatch and forming a high-angle grain boundary.
Superplastic creep	Grain-size-sensitive deformation mechanism by which grains are able to slide past one another without friction because of the activity of diffusion (as opposed to frictional sliding or cataclasis).
Tilt boundary	Concentration of dislocations in a planar array.
Twinning	Deformation mechanism that rotates the crystal lattice over a discreet angle such that the twin boundary becomes a crystallographic mirror plane. Such a planar defect is produced by the motion of partial dislocations.
Undulose extinction	Irregular distribution of dislocations in a grain, producing small crystallographic mismatches or lattice bending that is visible under crossed polarizers.

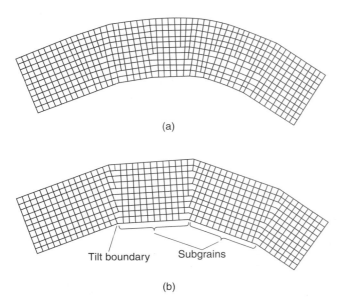

(a)

(b)

Tilt boundary   Subgrains

FIGURE 9.24 Irregularly distributed dislocations (a) are rearranged by glide and climb to form a dislocation wall (or tilt boundary) that separates subgrains (b).

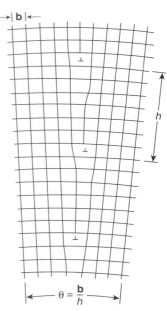

$\theta = \dfrac{\mathbf{b}}{h}$

FIGURE 9.25 A tilt boundary composed of edge dislocations at a distance h apart in a simple lattice. The crystal lattice across the boundary does not have the same orientation, but is rotated over an angle θ (in radians) = $\mathbf{b}/h$, where $\mathbf{b}$ is the Burgers vector and h is the spacing of dislocations in the tilt wall.

stretched (to give the "strain" in strain energy), so that the crystal lattice is not in its lowest energy state. Dislocation creep lowers the internal strain energy by annihilation and/or moving dislocations to the edge of crystals, so that the internal strain is minimized. This is the reason that internal strain energy is also called the **stored strain energy.** However, this does not mean that internal strain is recoverable (as in elastic strain), because permanent distortions are produced around dislocations in the crystal (recall Figure 9.12). Another way to change the internal strain energy of a grain is by localization of crystal defects. As a result of climb, cross-slip, and glide, dislocations can be arranged into a zone of dislocations, called a **dislocation wall** or **tilt boundary.** Such tilt boundaries produce a lower strain energy state in most of the grain than when dislocations are more evenly distributed across the grain (Figure 9.24). A single dislocation produces only a small crystallographic distortion that is not visible optically, but an array of dislocations in a tilt boundary makes the crystallographic mismatch optically visible (Figure 9.25). The greater the number of dislocations in the wall, that is, the closer their spacing, the greater the crystallographic mismatch across the boundary. The angular mismatch, θ, across a tilt boundary is a function of the length of the Burgers vector (**b**) of a dislocation and the spacing (h) of individual dislocations in the wall

$$2 \sin(\theta/2) = \mathbf{b}/h \qquad \qquad \text{Eq. 9.16}$$

or for small angles of θ (in radians)

$$\theta = \mathbf{b}/h \qquad \qquad \text{Eq. 9.17}$$

We call the region of a large crystal that is enclosed by a low-angle tilt boundary a **subgrain.** The convention to distinguish between low-angle (subgrains) and high-angle boundaries (recrystallized grains; discussed later) is an angular difference across the tilt boundary that is less than 10°. With this information we can estimate the number of dislocations in a tilt wall 500 μm long by 2 nm wide (using Equations 9.16 and 9.17), assuming a Burgers vector of 0.5 nm length and an angular mismatch θ of 10°. This implies that the dislocation spacing is approximately 2.9 nm and thus that there are more than 170,000 (!) dislocations in this low-angle tilt boundary, representing a dislocation density in the area of the low-angle tilt wall (1×10^{-8} cm^2) of 1.7×10^{13} cm^{-2}. This is many orders of magnitude greater than dislocation density in undeformed crystals.

In thin section, especially under cross-polarized light, undulatory extinction is one manifestation of the crystallographic mismatch that produces subgrains. It is particularly common in the minerals calcite, quartz, olivine, and pyroxene (Figure 9.26). **Recovery** is the name of the process forming low-angle grain bound-

FIGURE 9.26 Subgrain microstructure and undulose extinction in a marble mylonite from southern Ontario (Canada). Width of view is ~4 mm.

FIGURE 9.27 Recrystallization microstructure, showing relatively strain-free grains with straight grain boundaries. This image represents the most deformed stage in a marble mylonite that is also shown in Figures 9.15 and 9.26 (Ontario, Canada). Width of view is ~2 mm.

aries by the temperature-activated rearrangement of dislocation, which produces the characteristic subgrain deformation microstructure. In the case of phyllosilicates, such as muscovite, these subgrains are also called a **polygonized microstructure,** which describes the archlike geometry where each segment is oriented at a slightly different angle from the next. Experiments in which recovery dominates have shown that the stress function of the associated flow law is exponential (Equation 9.13). Materials scientists, therefore, use the term **exponential-law creep** for recovery microstructures.

9.9.2 Recrystallization

The process removing the internal strain energy that remains in grains after recovery is called **recrystallization;** it forms **high-angle grain boundaries** that separate relatively strain-free grains from each other. In rocks, a recrystallized microstructure is characterized by grains without undulatory extinction and with relatively straight grain boundaries that meet at angles of about 120° (Figure 9.27). Another example of this process and the resulting structure is found in the foam of soap. Looking closely at foam while doing the dishes or washing your hair, you will see all the geometric characteristics of a recrystallized microstructure. Because some of the same energy considerations are involved in the structure of foam,[6] we also call the microstructure of recrystallized rock a **foam structure.**

FIGURE 9.28 Microstructure of a mylonite. Note the fine-grained, quartz-rich matrix that surrounds relatively rigid feldspar clasts. Width of view is ~1 cm.

Recrystallization within an anisotropic stress field (i.e., a differential stress) is called **dynamic recrystallization.** Dynamic recrystallization results in grain-size reduction, which is well known from sheared rocks (such as mylonites; Figure 9.28). We return in more detail to mylonites in Chapter 12, but at this point we note that they have a grain size that is smaller than that of the host rock from which they formed. In fact, the term mylonite is unfortunate for these microstructures as it derives from the Greek word "mylos," meaning milling. At the time of their discovery in northern Scotland by Sir Charles Lapworth in the late 1900s (Chapter 12) it was thought that they were formed by a grinding process (which we now call cataclasis).

---

[6]In foam, however, surface energy dominates, whereas internal strain energy is more important in deformed rocks.

Since their original discovery we have learned that this is incorrect, and that dynamic recrystallization is responsible for grain-size reduction; nonetheless, the name mylonite has persisted. Based on experimental work, dynamic recrystallization can be used as a semiquantitative indicator of the temperature conditions during deformation; for example, recrystallization begins at ~300°C for calcite, ~350°C for quartz, and ~450°C for feldspar. These estimates seem to agree well with temperature estimates in deformed natural rocks.

Recrystallization occurring under isotropic stress conditions or when the differential stress is removed is called **static recrystallization;** otherwise know as **annealing.** From a microstructural perspective the only thing that distinguishes static recrystallization from dynamic recrystallization is a relatively larger recrystallized grain size. Static recrystallization reduces the internal strain energy by the formation of relatively large, strain-free grains that grow to decrease the total free energy of the rock.[7]

A closing comment about the use of the term recrystallization before we turn to the operative mechanisms. Recrystallization as used here involves changes in the strain energy of a *single phase,* whereas the term recrystallization in petrology involves multiple *phases.*

---

[7]This is sometimes called *secondary* or *exaggerated grain growth.*

In petrology the process is governed by chemical potentials rather than by strain potentials. Be sure not to confuse these very different meanings of the term recrystallization.

9.9.3 Mechanisms of Recrystallization

There are two main mechanisms for recrystallization: (1) rotation recrystallization and (2) migration recrystallization. **Rotation recrystallization** describes the progressive misorientation of a subgrain as more dislocations move into the tilt boundary, thereby increasing the crystallographic mismatch across this boundary. This produces a high-angle grain boundary without appreciable migration of the original (sub)grain boundary (Figure 9.29a). Eventually the crystallographic mismatch is sufficiently large that individual grains are recognized. Remember that progressive rotation of the subgrain occurs only by adding more dislocations in the boundary and that there is never loss of cohesion with the crystal lattice of the host grain. The convention we previously introduced to distinguish subgrains (low-angle grain boundaries) from recrystallized grains (high-angle grain boundaries) is an angle of 10°. This is admittedly an arbitrary convention, as we find a progression from low-angle to high-angle grain boundaries in rocks, but it is convenient for our purposes. Recrystallized grains are best

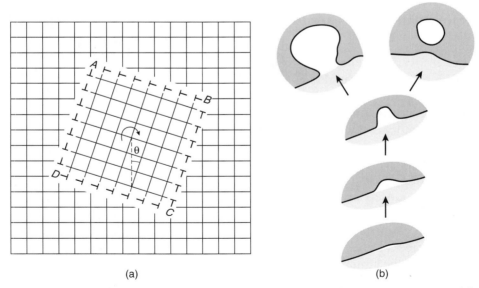

FIGURE 9.29 Recrystallization by (a) subgrain rotation and (b) bulge nucleation. In (a), a portion of a crystal is bounded by four tilt boundaries (*ABCD*); rotation by adding more dislocations of the same sign leads to a progressively greater misorientation (i.e., a recrystallized grain). In (b), growth of a dynamically recrystallized grain occurs by bulge nucleation of the grain boundary into a neighboring grain with higher internal strain energy (dark gray), leaving behind a relatively strain-free region that eventually develops into a recrystallized grain.

developed where large strain gradients exist, such as at grain boundaries. The common microstructure in which relatively deformation-free grain interiors progress to subgrains and then to recrystallized grains toward grain boundaries (Figure 9.30) is called a **core-mantle structure** or **mortar structure.** Rotation recrystallization has been observed in most common rock-forming minerals, including calcite, quartz, halite, and olivine.

Migration recrystallization is a process by which grains grow at the expense of their neighbor(s)—when grain boundaries effectively sweep through neighbors. The grain that grows has a lower dislocation density than the grain(s) consumed. Let's look at an example where the boundary of grain A migrates into grain B (Figure 9.29b). Keep in mind that a grain boundary separates two crystals whose lattices are not parallel. Migration happens when atoms in grain B near the boundary rearrange so they fit into the lattice of the crystal with lower dislocation density (grain A). As soon as this happens, these atoms become part of grain A. It is easier to rearrange atoms and bonds in grain B that are stretched and misoriented, because of its higher dislocation density. As the grain with lower dislocation density grows at

the expense of the grain with higher dislocation density, the internal strain energy of the overall system decreases. Typically the boundary of the grain with lower dislocation density bulges into the grain with higher dislocation density (Figure 9.29b). Thus, this recrystallization process has been called **bulge nucleation.** When new grains deform as they grow, this may eventually arrest their growth. In natural settings, quartz, halite, and feldspar commonly recrystallize by bulge nucleation.

The dominance of rotation recrystallization (subgrain rotation) and migration recrystallization (bulge nucleation) is largely a function of strain rate. Consider this: If you are in a hurry to get somewhere, you will try to take the fastest means of transportation. Similarly, nature uses the mechanism that produces the highest strain rate to reduce the internal strain energy of the system. Bulge nucleation is generally favored at higher strain rates and high temperatures. For both recrystallization mechanisms, the recrystallized grain size is inversely proportional to the strain rate. The smaller recrystallized grain size in mylonitic rocks, for example, is indicative of strain-rate increase. We observed **work softening,** strain-rate increase at constant stress, in some experiments (Chapter 5), which can now be understood in terms of the role of grain-size reduction during deformation.

The formation of recrystallized grains is driven by the generation and motion of dislocations, which in turn is driven by differential stress. One may, therefore, expect that a relationship exists between recrystallized grain size and differential stress magnitude. Indeed, experiments have shown that a characteristic range of grain sizes occur for a specific condition of stress and mechanism of recrystallization. This means that we can potentially estimate paleostress conditions from microstructures; that is, recrystallized grain size can be used as a **paleopiezometer** (derived from the Greek "piezo," meaning to press).[8] This is potentially a very powerful tool for understanding deformation, because paleostress is a notoriously difficult parameter to extract from rocks. Although the debate about the exact relationship is not settled, it is generally agreed that recrystallized grain size is inversely proportional to differential stress magnitude

$$\sigma_d = A d^{-i} \qquad \text{Eq. 9.18}$$

where A and i are empirically derived parameters for a mineral and d is grain size in micrometers (μm). To

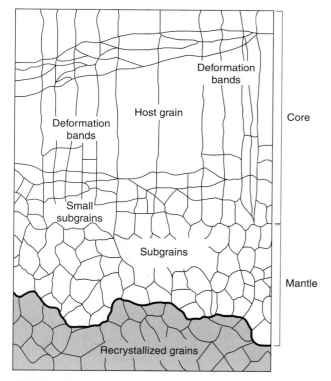

FIGURE 9.30 Core-mantle microstructure (or mortar structure). Recrystallized grains occur at the edge of the mantle by progressive misorientation of subgrains. The internal portion of the host grain (core) shows weak deformation features such as undulose extinction and deformation bands, or may even be strain-free.

---

[8]Piezometers using free dislocation density and subgrain size also exist, but these data are more difficult to obtain and the methods appear less reliable.

TABLE 9.4	EMPIRICALLY DERIVED PARAMETERS FOR RECRYSTALLIZED GRAIN SIZE–DIFFERENTIAL STRESS RELATIONSHIPS	
Mineral	A (in MPa)	i (with d in μm)
Calcite	467	1.01
Quartz	381	0.71
Quartz ("wet")	4090	1.11
Olivine	4808	0.79

Sources: Mercier et al. (1977), Ross et al. (1980), Schmid et al. (1980), Ord and Christie (1984).

give you a rough idea of these relationships, we list representative parameters for three common minerals, calcite, quartz, and olivine, in Table 9.4. These data are plotted in Figure 9.31, showing representative stress values. In considering these values, remember that considerable uncertainty surrounds paleopiezometry.

Overall, a small recrystallized grain size in a deformed rock reflects a high strain rate, a high differential stress magnitude, or a combination of both. Rock experiments show that the corresponding stress function during recrystallization has the form $f(\sigma) = \sigma^n$ (Equation 9.14), so it is also called **power-lap creep.** The value of n, the **stress exponent,** varies, but typically lies in the range of 2 to 5 for common monomineralic rocks (see Table 5.6).

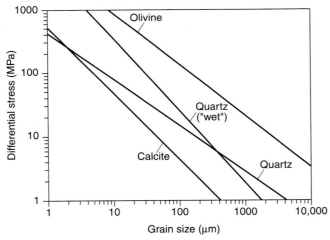

FIGURE 9.31 Empirically derived recrystallized grain size versus differential stress relationships for calcite, quartz, "wet" quartz, and olivine, using the parameters listed in Table 9.4. Note that we plot log σ_b versus log d, so that small shifts in the position of each curve reflect large changes in ambient conditions.

9.9.4 Superplastic Creep

Superplastic creep, more completely described by the somewhat cumbersome name **grain-boundary sliding superplasticity (GBSS),** will at first seem out of place after a discussion of dislocation creep mechanisms, because it returns us to the topic of diffusion. This mechanism is intentionally kept to the last, because it occurs at the highest temperature conditions. We look at the characteristics first. Superplastic creep is a grain-size-sensitive deformation mechanism in which grains change shape so they can slide past one another. This sounds like cataclastic flow, but there is an important distinction: in superplastic creep, volume and grain-boundary diffusion are sufficiently efficient to keep gaps from forming between moving grains, and therefore grains are able to slide *without friction.* Strain is produced by neighbor switching as illustrated in Figure 9.32. Superplastic creep can result in very large strains (>1000%) without appreciable internal deformation of the grains. The original definition of superplasticity is, in fact, this ability of rocks to accumulate very large strains without mesoscopic breaking. Even after large finite strains, grains are equiaxial and "fresh-looking," and show no preferred elongation or crystallographic fabric. This diffusion-assisted mechanism is mainly important in materials with relatively small grain sizes (<15 μm) that facilitate diffusion. In this context, recall Equations 9.9 and 9.10, which show the inverse exponential proportionality of diffusion to grain size.

Superplastic creep has been proposed as a natural deformation mechanism in fine-grained calcite- and quartz-rich rocks. The very high temperatures that occur in the (upper) mantle may also permit this mechanism in coarser-grained olivine-rich rocks. Superplastic creep is possible at lower differential stresses than dislocation creep, but requires rocks with relatively small grain sizes. Thus, a rock may initially deform by dynamic recrystallization until its grain size is sufficiently reduced for superplastic creep to occur. When this happens, the rock becomes much weaker; that is, the stress necessary to produce strain decreases. This weakening, known as work or strain softening, is common in ductile fault zones.

The stress function of the flow law for superplasticity approaches linearity between strain rate and stress; that is, the stress component n of Equation 9.15 approaches 1. Consequently, the strain rate is inversely proportional to grain size

$$\dot{e} \cong d^{-r}$$

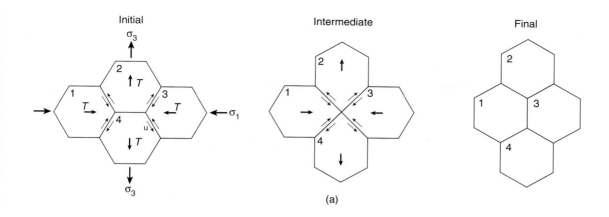

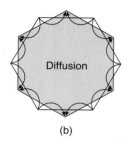

FIGURE 9.32 Grain-boundary sliding superplasticity (or superplastic creep). In (a), neighbor switching in the superplastic regime is illustrated. A group of four grains enjoy ~55% strain without appreciable deformation of each grain, except at the boundaries to accommodate grain sliding (small arrows). The required accommodation of local strain by diffusion is shown in (b), with the final grain shape shaded.

where r is 2 to 3 based on experimental work. Recalling that a linear relationship between strain rate and stress defines linear viscous rheology (Chapter 5), superplastic creep is well described by Newtonian fluid mechanics. This contrasts with dislocation creep, which typically has nonlinear rheology ($n \neq 1$). To emphasize this strong grain-size dependence of superplastic creep we also call it **grain-size-sensitive creep.**

9.10 DEFORMATION MECHANISM MAPS

Quite an array of concepts and terms have by now been introduced, so let's attempt to create, out of this information, a pattern that helps you to remember the important elements and relationships. The activity of ductile deformation mechanisms can be summarized in a diagram that shows over what ranges of stress, strain rate, temperature, and grain size each mechanism dominates for a given material; such diagrams are called **deformation mechanism maps.**[9] The variables may be stress (e.g., differential stress), temperature, and grain size, but for comparison between different materials we generally use normalized parameters. A normalized

physical quantity is the ratio between a variable and a material constant measured in the same units. In this case, stress is normalized to an elastic modulus of the material (typically the shear modulus, G), and temperature (absolute temperature, in K) is normalized to the absolute melting temperature of the material (Figure 9.33), called the homologous temperature, T_h. On

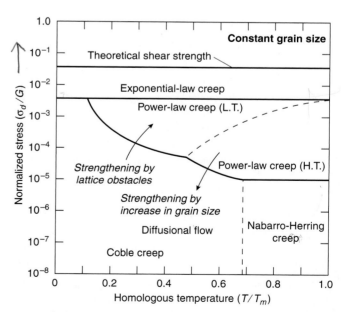

FIGURE 9.33 Schematic of a deformation mechanism map, showing normalized stress versus homologous temperature at a constant grain size.

---

[9]Also called Ashby diagrams, after the British material scientist Michael Ashby who proposed the construction in the early 1970s.

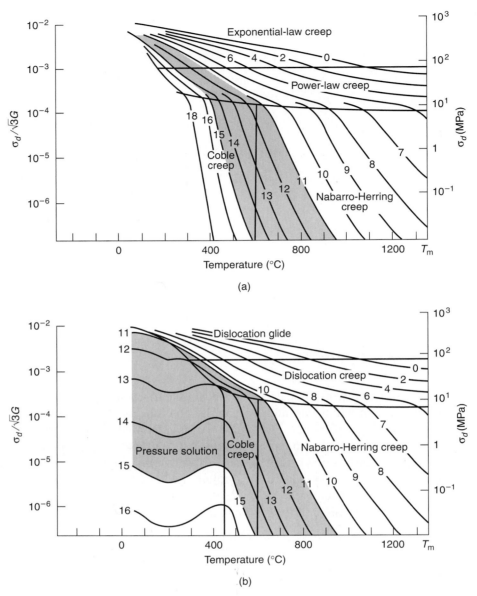

FIGURE 9.34 Deformation mechanism maps for calcite (a) without and (b) with a pressure solution field for a grain size of 100 μm. Contours of −log strain rate are shown; σ_d is differential stress; G is shear modulus; the σ_d-scale on the right is for a shear modulus, G, at 500°C. The undulation of strain rate contours in the pressure solution field arises from the competition between change in the solubility of calcite and fluid concentration with pressure as temperature increases. The range of reasonable geologic strain rates (10^{-11}–10^{-15}/s) is shaded.

the deformation mechanism map we display lines of constant strain rate, as shown in Figures 9.34 and 9.35. Only a small region of the diagram can be constrained by laboratory experiments, so we must extrapolate to most natural conditions. This is comparatively easy where an essentially linear (Newtonian) relationship exists between $\dot{e}$ and σ, such as for diffusional flow. For other regimes, such as dislocation glide (exponential-law creep), and dislocation glide and climb (power-law creep), the extrapolation of these nonlinear relation-

ships is more tenuous. Figures 9.34a and 9.35a show examples of deformation mechanism maps for two common crustal minerals, calcite and quartz, while in Figures 9.34b and 9.35b the pressure solution fields ("wet" diffusion) have been added.

The meaning of boundaries between the fields on a deformation mechanism map is not straightforward, because deformation mechanisms do not change abruptly at this boundary; rather, several mechanisms operate simultaneously. The mechanism that generates

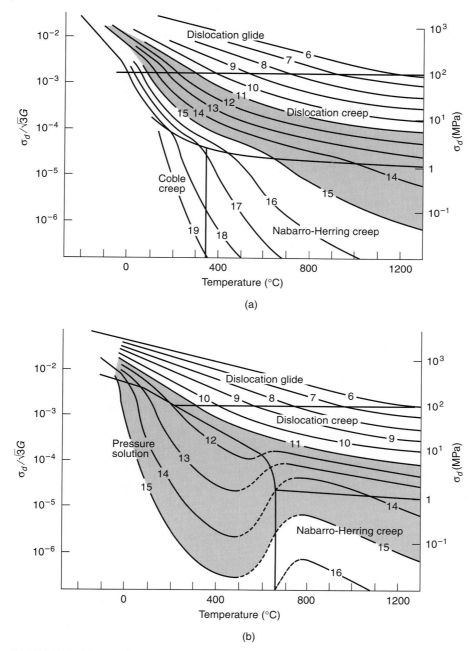

FIGURE 9.35 Deformation mechanism map for quartz (a) without and (b) with a pressure solution field for a grain size of 100 μm. Contours of −log strain rate are shown; σ_d is differential stress; G is shear modulus; the σ_d-scale on the right is for a shear modulus, G, at 900°C. The stress exponent, n, in the power-law field is 4. The region of dashed strain-rate contours represents the inhibition of pressure solution through decrease in pore water concentration.

the highest strain rate is the dominant deformation mechanism. Fields in deformation mechanism maps are defined by calculating the strain rate for each mechanism. Then, the mechanism giving the fastest rate is taken as representative for a field (i.e., the mechanism dominating flow). For example, the field for dislocation creep represents the range of conditions for which dislocation glide creates a strain rate faster than

any other mechanism, even though such other mechanisms may be operating. This means that at a boundary the two adjacent deformation mechanisms are equally important.[10] Let's give a practical example. During

---

[10]To emphasize this aspect, deformation mechanism maps are also called *deformation regime maps.*

mylonitization of a quartzite, dynamic recrystallization may dominate; yet diffusional flow may occur simultaneously if the grain size and the strain rate are sufficiently small. Consequently, the map will indicate that we are in the power-law field, but we also see microstructural evidence for diffusional flow.

A general pattern is common to all deformation mechanism maps, which we illustrate with the mineral olivine (Figure 9.36). Instead of homologous temperature we plot depth in Earth, based on a thermal gradient that exponentially decreases from 300K at the surface to 1850K at a depth of 500 km. This enables us also to take into account any effects of pressure, which play a role in the mantle by increasing the flow strength and decreasing the strain rate. From the olivine deformation mechanism map you see that cataclastic flow and exponential-law creep are restricted to relatively large differential stresses (here $\approx 8 \times 10^2$ MPa), meaning that these mechanisms are limited to shallow crustal levels. With depth, we pass from exponential-law creep to power-law creep to diffusional creep, given a constant geologic strain rate (say, 10^{-14}/s). In the latter regime, we may pass from grain-boundary diffusion (Coble creep) to volume diffusion (Nabarro-Herring creep), given further temperature or strain-rate change.

The value of a deformation mechanism map lies in its ability to predict the mechanism that dominates a flow under natural conditions. For example, if we assume that the Earth's upper mantle consists mainly of olivine,

we predict that at strain rates greater than 10^{-11}/s dislocation glide and climb dominate flow in the upper 100 km, given a grain size of 100 μm. If the strain rate is less, diffusional creep will be more important, especially if the grain size is small. The latter point, the effect of grain size, may not be clear from any of the deformation mechanism maps shown thus far, because grain size was taken as constant value. So how do we know the role of grain-size variation? Consider the flow laws for diffusional creep (Equation 9.15), which state that strain rate is inversely proportional to the square or cube of the grain size. Reducing grain size by, say, one order of magnitude will increase strain rate by two to three orders of magnitude, which will move the field of reasonable geologic strain rates into the regime of diffusional flow. Similarly, if we construct a map for a grain size of, say, 1 mm (1000 μm) or larger, the field of geologic strain rates moves into the regime of power-law creep. With grain sizes for upper-mantle olivine in the range of 100–1000 μm, the microstructures of mantle rocks generally support the predictions that we obtain from olivine deformation-mechanism maps.

9.10.1 How to Construct a Deformation Mechanism Map

The concept of deformation mechanism maps is best understood and appreciated when you construct your own. In Table 9.5, therefore, we list constitutive equations for various deformation mechanisms in natural limestones and marbles, which will allow you to construct a deformation mechanism map.

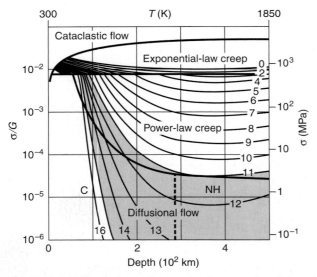

FIGURE 9.36 Deformation mechanism map for olivine with a grain size of 100 μm. Variables are the same as in Figure 9.33, except that depth is substituted for temperature, using an exponentially decreasing geothermal gradient with 300K at the surface and 1850K at 500 km depth. *From Ashby and Verrall, 1978.*

TABLE 9.5	EXPERIMENTALLY DERIVED CONSTITUTIVE EQUATIONS USED FOR THE CONSTRUCTION OF FIGURE 9.36
Exponential-law creep:	$\dot{e} = 10^{5.8}\, e^{(-62,000/RT + \sigma/114)}$
Power-law creep regime *a*:	$\dot{e} = 10^{-5.5}\, e^{(-75,000/RT)}\, \sigma^{6.0}$
Power-law creep regime *b*:	$\dot{e} = 10^{3.8}\, e^{(-86,000/RT)}\, \sigma^{2.9}$
Superplastic creep:	$\dot{e} = 10^{5.0}\, e^{(-51,000/RT)}\, \sigma^{1.7}\, d^{-3}$

$\dot{e}$ = strain rate [s^{-1}]
σ = differential stress (bar)
T = absolute temperature (K)
R = gas constant
d = grain size (μm)
Source: Rutter (1974), Schmid et al. (1977), Schmid (1982).

First, choose the axes of the plot. Let's decide to plot differential stress versus temperature. Now calculate the corresponding strain rate from each of the four constitutive equations at a specific stress and temperature condition (i.e., a point in the diagram). You recall that the mechanism producing the highest strain rate is dominant, so from the four solutions the one with the highest strain rate is dominant at that particular point in the diagram. Using individual points to fill the diagram is an unnecessarily slow and cumbersome approach. Instead, we calculate the stress–temperature curves at a given strain rate for each equation and plot these four curves in the diagram. Because some of these curves intersect, the final strain rate curve is composed of segments of the four curves for which the differential stress is smallest. When using different strain rates you will see that the positions of intersection points change. Also, the dominant deformation mechanism may change. Connecting these intersection point where mechanisms change defines the boundary between fields. A worked-out example is shown in Figure 9.37, in which differential stress is plotted as a function of grain size for $T = 475°C$. You can vary environmental conditions, such as stress, temperature, and grain size, and calculate the corresponding map using fairly simple spreadsheet calculations on a personal computer.

9.10.2 A Note of Caution

Deformation mechanism maps evaluate all types of material behavior, which is not restricted to rocks. They also permit predictions of the creep of metal in a nuclear reactor, thereby aiding their safe design; they determine the lifespan of lightbulb filaments; they explain the creep of ice sheets; and they provide crucial information on the sagging rate of ancient marble benches in parks (Figure 9.38). So, deformation mechanism maps are a powerful approach to understanding and exploring the rheology of materials, and to applying this information to significant earth science problems such as mantle convection or lithosphere subduction. However, these maps are not without limitations. First, extrapolation over several orders of magnitude is needed to move from experimentally derived flow laws ($\dot{e} > 10^{-8}$/s) to geologic conditions ($\dot{e} < 10^{-11}$/s), introducing a major source of uncertainty. Second, the maps assume steady-state flow (i.e., that stress is strain-independent), which may not be sufficiently representative of geologic conditions. Moreover, evolving microstructures affect the dominant deformation mechanism; for example, dynamic recrystallization tends to reduce the grain size, which in turn enhances the importance of diffusional creep and weakens the material. In spite of these limitations, deformation mechanism maps are a handy and powerful tool to evaluate and predict deformation mechanisms and ambient conditions for deforming materials. Comparison of natural deformation structures with predictions based on deformation mechanism maps offers a natural test, which indeed indicates that these maps provide reliable estimates about the conditions of

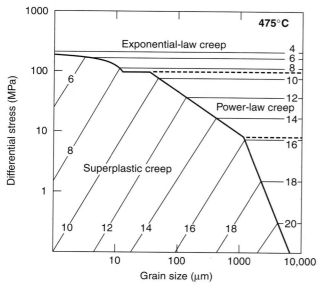

FIGURE 9.37 Deformation mechanism map for calcite at $T = 475°C$, constructed from the constitutive equations listed in Table 9.5. The thin lines represent strain rates (marked as −log), whereas the thick lines separate deformation mechanism fields.

FIGURE 9.38 Sharply dressed man with deformed marble bench that plastically sagged and locally fractured under the influence of gravity (and users).

TABLE 9.6	DEFORMATION MECHANISMS AND PROCESSES			
Process (P) or Mechanism (M)	Atomic-Scale Process	Diagnostic Microstructures	Rheological Implications	Common Minerals
Bulk rotation (M or P)	Physical rotation of whole or part of mineral grains	Helical inclusion trails, bending of crystals, delta and sigma porphyroclasts		Any
Climb (M)	Diffusional addition or removal of atoms at dislocation line			Any; more at high T
Diffusive mass transfer (M or P)	"Long range" diffusion of atoms	Veins, pressure shadows, porphyroblasts		Any, especially quartz and calcite
Dislocation glide (M)	Rearrangement of interatomic bonds	Deformation lamellae, deformation bands, undulose extinction	$\dot{e} \propto \sigma^3$, also a hardening with finer grain size	Any; more at low stress and high T
Fracturing (M)	Breaking of inter-atomic bonds	Gouge, breccias, boudinaged grains		Any; more at high stress and low T
Frictional sliding (M)	Frictional sliding on surfaces	Gouges, breccias, pseudotachylytes, domino grains	$\tau \propto \sigma^n$	Any; more at high stress and low T
Grain boundary migration (P or M)	Local diffusion and reorientation of atoms or atomic clusters	Irregular grain boundaries, pinning microstructures, orientation families; lattice-preferred orientations with strong point maxima, non-120° triple junctions	Produces low dislocation density; material softer	Any; more at high T; especially quartz, olivine, feldpar
Grain boundary sliding (M)	Dislocation movement on "clean" grain boundaries, shearing on "dirty" ones			Any

[From Jessell and Bons, 2002]

deformation. Now return to Figures 9.34 to 9.36, showing deformation mechanism maps for several common minerals, and use them to think about the interplay between deformation mechanism, strain rate, temperature, and stress, and the associated microstructures. These maps also offer a useful way to absorb and appreciate the various mechanisms and processes of ductile deformation that were the focus of this chapter.

9.11 CLOSING REMARKS

Modern structural geology interpretations are relying increasingly on a synthesis of observations on all scales, including microscopic. An analysis of microstructures plays a growing role in unraveling the deformation histories of rocks and regions. For example, mechanical twin-

ning can be used to unravel the early stress and strain history of fold-and-thrust belts; mylonitic microstructures allow us to estimate the conditions of T; σ_d, and $\dot{e}$ during deformation, and olivine fabrics in xenoliths, give us information on conditions of mantle flow. One additional consequence of crystal plastic deformation, crystallographic-preferred fabrics, will be introduced in Chapter 12. Meanwhile, we will explore common ductile structures in outcrop, namely folds, foliations, and lineations, in the next chapters. We close this chapter with a summary table (Table 9.6) that groups mechanisms and processes, and serves as a handy reference.

ADDITIONAL READING

Ashby, M. F., and Verrall, R. A., 1978. Micromechanisms of flow and fracture, and their relevance to

| TABLE 9.6 | (CONTINUED) |

Process (P) or Mechanism (M)	Atomic-Scale Process	Diagnostic Microstructures	Rheological Implications	Common Minerals
Kinking (M)	Dislocation glide on single slip system	Kink bands		Micas, low T quartz, kyanite
Lattice diffusion (M)	Diffusional movement of vacancies and interstitials	New crystal void of preexisting impurities (hard to prove in nature)	$\dot{e} \propto \sigma/d^2$ (Nabarro-Herring creep); $\dot{e} \propto \sigma/d^3$ (Coble creep)	Any; more at low stress and high T
Lattice rotation (P)	Dislocation glide and/or bulk rotation of grains	Lattice-preferred orientations	Well-developed fabrics may be stronger or weaker than random fabrics	Any; more at low stress and high T
Phase change (M or P)	Changed crystal structure without change in bulk chemistry	Phase boundaries in minerals	Often associated with volume change	Quartz, calcite-aragonite, olivine
Recovery (P)	Climb, mutual annihilation of dislocations of opposite signs, formation of sub-grain walls	Polygonization, foam texture, 120° triple junctions	Produces low dislocation density; material softer	Any; more at high T
Rotation recrystallization (P)	Progressive addition of dislocations of same sign to subgrain wall	Mortar texture or core-and-mantle texture, bimodal grain size	Change in grain size can strengthen or weaken material	Any; more at low stress and high T, especially quartz, feldspar, olivine
Twinning (M)	Rearrangement of interatomic bonds and reorientation of lattice site	Twins (sharp-nosed, narrow, parallel to rational twin planes)		Calcite (low T and low strain), plagioclase, quartz, amphibole

the rheology of the upper mantle. *Philosophical Transactions of the Royal Society of London,* Series A, 288, 59–95.

Frost, H. J., and Ashby, M. F., 1982. *Deformation-mechanism maps. The plasticity and creep of metals and ceramics.* Pergamon Press: Oxford.

Groshong, R. J., Jr., 1972. Strain calculated from twinning in calcite. *Geological Society of America Bulletin,* 83, 2025–2038.

Hayden, H. W., Moffatt, W. G., and Wulff, J., 1965. *The structure and properties of materials: Volume III, Mechanical behavior.* J. Wiley and Sons: New York.

Hull, D., and Bacon, D. J., 1984. *Introduction to dislocations.* Pergamon Press: Oxford.

Jamison, W. R., and Spang, J. H., 1976. Use of calcite twin lamellae to infer differential stress. *Geological Society of America Bulletin,* 87, 868–872.

Jessell, M., and Bons, P., 2002. *On-line short course in microstructures.* http://www.earth.monash.edu.au/Teaching/mscourse/index.html.

Loretto, M. H., and Smallman, R. E., 1975. *Defect analysis in electron microscopy.* Chapman and Hall: London.

Mercier, J.-C., Anderson, D. A., and Carter, N. L., 1977. Stress in the lithosphere: inferences from steady-state flow of rocks. *Pure and Applied Geophysics,* 115, 199–226.

Nicolas, A., and Poirier, J.-P., 1976. *Crystalline plasticity and solid state flow in metamorphic rocks.* John Wiley and Sons: London.

Ord, A., and Christie, J. M., 1984. Flow stresses from microstructures in mylonitic quartzites of the Moine thrust zone, Assynt area, Scotland. *Journal of Structural Geology,* 6, 639–654.

Poirier, J.-P., 1985. *Creep in crystals: high-temperature deformation processes in metals, ceramics and minerals.* Cambridge University Press: Cambridge.

Ross, J. V., Ave Lallemant, H. G., and Carter, N. L., 1980. Stress dependence of recrystallized-grain and subgrain size in olivine. *Tectonophysics, 70,* 39–61.

Rutter, E. H., 1974. The influence of temperature, strain rate and interstitial water in the experimental deformation of calcite rocks. *Tectonophysics, 22,* 311–334.

Schmid, S. M., Boland, J. N., and Paterson, M. S., 1977. Superplastic flow in finegrained limestone. *Tectonophysics, 43,* 257–291.

Schmid, S. M., Paterson, M. S., and Boland, J. N., 1980. High temperature flow and dynamic recrystallization in Carrara marble. *Tectonophysics, 65,* 245–280.

Spang, J. H., 1972. Numerical method for dynamic analysis of calcite twin lamellae. *Geological Society of America Bulletin, 83,* 467–472.

Turner, F. J., 1953. Nature and dynamic interpretation of deformation in calcite of three marbles. *American Journal of Science, 251,* 276–298.

Twiss, R. J., 1986. Variable sensitivity piezometric equations for dislocation density and subgrain diameter and their relevance to quartz and olivine. In Hobbs, B. E., and Heard, H. C., eds., *Mineral and rock deformation: laboratory studies (the Paterson volume),* American Geophysical Union, Geophysical Monograph 36, pp. 247–261.

Urai, J. L., Means, W. D., and Lister, G. S., 1986. Dynamic recrystallization of minerals. In Hobbs, B. E., and Heard, H. C., eds., *Mineral and rock deformation: laboratory studies (the Paterson volume),* American Geophysical Union, Geophysical Monograph 36, pp. 161–200.

Wenk, H.-R. (ed.), 1985. *Preferred orientation in deformed metals and rocks: an introduction to modern texture analysis.* Academic Press: Orlando.

White, S., 1976. The effects of strain on the microstructures, fabrics, and deformation mechanisms in quartzites. *Philosophical Transactions of the Royal Society of London,* Series A, 283, 69–86.

APPENDIX: DISLOCATION DECORATION

The principle behind optical imaging by dislocation decoration of olivine is that iron oxides preferentially precipitate along defects in olivine. In order to decorate dislocations in olivine, a sample with one polished surface is heated in air for approximately one hour at 900°C. A standard petrographic thin section is then prepared with the previously polished surface in contact with the glass slide. For most crystallographic directions, dislocation lines as far as 50 μm from the polished surface are decorated. Under the optical microscope, screw dislocations generally appear as long and straight lines. Using a microscope that is equipped with a universal stage, the crystallographic relationship can be determined. Decoration is most effective in samples with relatively low dislocation densities ($<10^8/cm^2$), such as mantle xenoliths in volcanic flows.

Folds and Folding

10.1 INTRODUCTION

Ask a structural geologist, or any other geologist for that matter, about their favorite structure and chances are that they will choose folds. If you have seen a fold in the field you will have marveled at its appearance (Figure 10.1). Let's face it, it is pretty unbelievable that hard rocks are able to change shape in such a dramatic way. In simple terms, a fold is a structural feature that is formed when planar surfaces are bent or curved. If such surfaces (like bedding, cleavage, inclusions) are not available you will not see a fold even though the rock was deformed. Folding is a manifestation of ductile deformation because it can develop without fracturing, and the deformation is (heterogeneously) distributed over the entire structure. Rather than fracturing, processes such as grain sliding, kinking, dissolution, and crystal plasticity dominate. Looking at a fold from a kinematic perspective, you realize that strain in this structure cannot be the same everywhere. We recognize distinct segments in a fold, such as the hinge area and the limbs, the inner and the outer arc, each of which reflect different strain histories, regardless of scale.

Why do folds exist, how do rocks do it, and what does folding mean for regional analysis? These and other questions were first asked quite some time ago and much of what we know today about folds and folding was well established before the 1980s. The geometry of folds tells us something about, for example, the degree and orientation of strain, which in turn provides critical information about the deformation history of a region. Much of the work in recent years represents refinements of some of the earlier work; we can apply increasingly sophisticated numerical and experimental approaches. Yet, the fundamental observations remain essentially intact. Therefore, in this chapter we will mainly look at some of the first principles of folding

FIGURE 10.1 Large-scale recumbent fold in the Caledonides of northeast Greenland. The height of the cliff is about 800 m and the view is to the Northwest. *(Kildedalen)*

and their application to structural analysis. First, however, we discuss the basic vocabulary needed to communicate about folds and fold systems.

10.2 ANATOMY OF A FOLDED SURFACE

The schematic illustration in Figure 10.2 shows the basic geometric elements of a fold. The **hinge area** is the region of greatest curvature and separates the two limbs. The line of greatest curvature in a folded surface is called the **hinge line.** You may think of a **limb** as the less curved portion of a fold. In a limb there is a point where the sense of curvature changes, called the **inflection point.** Folds with a straight hinge line (Figure 10.3a) are called **cylindrical folds** when the folded

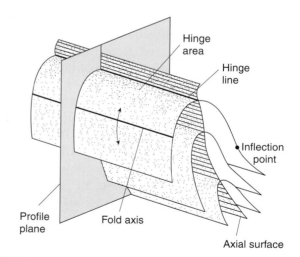

FIGURE 10.2 The terminology of a fold.

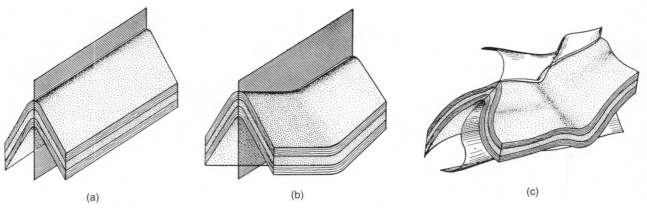

FIGURE 10.3   A cylindrical fold (a) is characterized by a straight hinge line and a noncylindrical fold (b) by a curved hinge line. The axial surface may be planar, as in (a) and (b), or curved (c).

surface can wrap partway around a cylinder. If this is not the case and the hinge line curves, the folds are called **noncylindrical** (Figure 10.3b). In reality the lateral extent of cylindrical folds is restricted to the outcrop scale or even less, because over greater distances the hinge line of folds typically curves. Nevertheless you will find that we may conveniently treat natural folds as cylindrical by dividing them into segments with straight hinge lines.

A cylindrical surface consists of an infinite number of lines that are parallel to a generator line. This generator line is called the **fold axis,** which, when moved parallel to itself through space, outlines the folded surface. In the case of cylindrical folds the fold axis is of course parallel to the hinge line.[1] The topographically highest and lowest points of a fold are called the **crest** and **trough,** respectively, and these do not necessarily coincide with hinge lines. The surface containing the hinge lines from consecutive folded surfaces in a fold is the **axial surface** (Figures 10.2 and 10.3). The term **axial plane** is loosely used by some, but the surface is not necessarily planar as seen in Figure 10.3c (recall the distinction between surface and plane). Moreover, the axial surface does not necessarily divide the fold into equal halves that are mirror images of one another. The reference plane used to describe fold shape is called the **fold profile plane,** which is perpendicular to the hinge line (Figure 10.2). Note that the profile plane is not the same as a cross section through the fold, which is any vertical plane through a body, much like the sides of a slice of layered cake. If the hinge line is not horizontal, then the profile plane is not parallel to

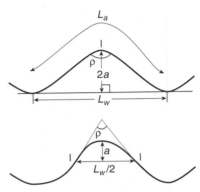

FIGURE 10.4   The interlimb angle (ρ), the wavelength (L_w), the amplitude (a), and the arc length (L_a) of a fold system in profile.

the cross-sectional plane, which has implication for the fold geometry description.

The angle between fold limbs as measured in the profile plane is called the **interlimb angle** (Figure 10.4). Intuitively you realize that the interlimb angle offers a qualitative estimate of the intensity of folding; the smaller the interlimb angle, the greater the intensity of folding. Finally, we recognize the **amplitude, wavelength,** and **arc length** of a fold in profile. These terms are used in the same manner as they are in wave physics, because folds tend to look a bit like harmonic functions (such as a sine curve). The wavelength is defined as the distance between two hinges of the same orientation, while the arc length is this distance measured over the folded surface; the amplitude is half the height of the structure measured from crest to trough (Figure 10.4). These and other terms associated with folds are summarized in Table 10.1.

When successive layers in a folded stack have approximately the same wavelength and amplitude, the

---

[1]Sometimes fold axis is used as a synonym for hinge line, but this is not correct.

TABLE 10.1	VOCABULARY OF A FOLD
Amplitude	Half the height of the structure measured from crest to trough
Arc length	The distance between two hinges of the same orientation measured over the folded surface
Axial surface	The surface containing the hinge lines from consecutive folded surfaces
Crest	The topographically highest point of a fold, which need not coincide with the fold hinge
Cross section	A vertical plane through a fold
Culmination	High point of the hinge line in a noncylindrical fold
Cylindrical fold	Fold in which a straight hinge line parallels the fold axis; in other words, the folded surface wraps partway around a cylinder
Depression	Low point of the hinge line in a noncylindrical fold
Fold axis	Fold generator in cylindrical folds
Hinge	The region of greatest curvature in a fold
Hinge line	The line of greatest curvature
Inflection point	The position in a limb where the sense of curvature changes
Limb	Less curved portion of a fold
Noncylindrical fold	Fold with a curved hinge line
Profile plane	The surface perpendicular to the hinge line
Trough	The topographically lowest point of a fold, which need not coincide with the fold hinge
Wavelength	The distance between two hinges of the same orientation

folds are called **harmonic.** If some layers have different wavelengths and/or amplitudes, the folds are **disharmonic** (Figure 10.5). In extreme circumstances, a series of folded layers may be totally decoupled from unfolded layers above or below. When this happens, a **detachment** horizon exists between folded and unfolded layers.

10.2.1 Fold Facing: Antiform, Synform, Anticline, and Syncline

Take a deep breath. We have already sprung a sizable array of terms on you, but before we explore the significance of folding, we have yet to add a few more. Maybe you will find comfort in the knowledge that generations of students before you have plowed their way through this terminology, happily discovering that in the end it really is important for the description and interpretation of regional deformation. Having said this, now draw a fold on a piece of paper. Chances are that you place the hinge area at the top of the structure, outlining something like a sharp mountain. In fact, a

psychological study among geologists found this invariably to be true (just look at your neighbor's sketch).[2] This particular fold geometry is called an **antiform.** The opposite geometry, when the hinge zone is at the bottom (outlining a valley), is called a **synform.** The explanation for the modifiers "anti" and "syn" is that the limbs dip away from or toward the center of the fold, respectively. You will find that many geologists use the terms anticline and syncline as synonyms for antiform and synform, but this is incorrect. The terms anticline and syncline imply that the stratigraphic younging direction in the folded beds is known. This is an important distinction for regional analysis, so let's look at this in some detail.

Imagine a sequence of beds that is laid down in a basin over a period of many millions of years. Obviously, the youngest bed lies at the top while the oldest bed is at the bottom of the pile (this is Steno's **Law of Superposition**). When we fold this sequence into a

---

[2]This is not (yet) linked to any criminal behavior.

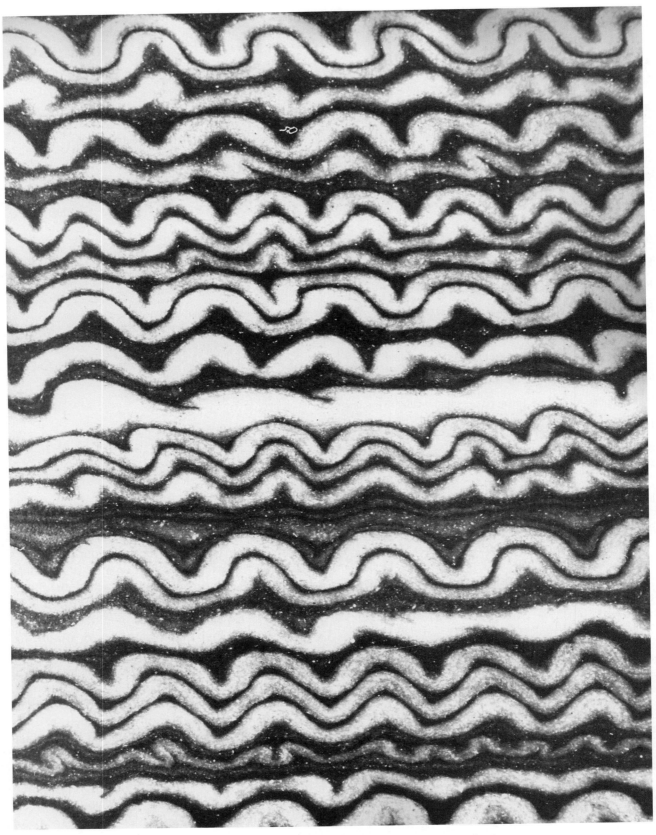

FIGURE 10.5 Small-scale disharmonic folds in anhydrite of the Permian Castile Formation in the Delaware Basin of Texas. White layers are anhydrite; dark layers consist of calcite that is rich in organic material (hence the dark color). Note that detachments occur in the organic-rich calcite layers and that the fold shapes (including box folds) in the anhydrite vary as a function of layer thickness.

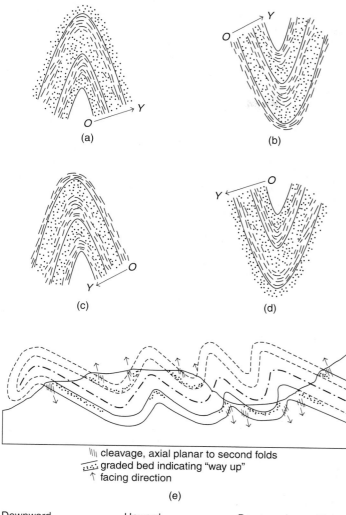

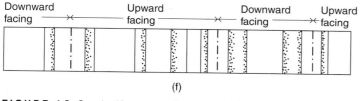

III cleavage, axial planar to second folds
∴∴ graded bed indicating "way up"
↑ facing direction

(e)

Downward facing →←× Upward facing →←× Downward facing →←× Upward facing

(f)

FIGURE 10.6 Antiforms, synforms, and fold facing. An upward-facing antiform (a) is also called an anticline and an upward-facing synform (b) is called a syncline. Downward-facing antiforms (c) and downward-facing synforms reflect an early history that placed the beds upside down prior to folding. These forms may occur in a region containing two generations of folding (e). The corresponding facing in map view across this area is shown in (f). Younging direction is indicated by $O \rightarrow Y$ arrow.

structures **upward-facing folds.** Now turn the original sequence upside down: the oldest bed now lies at the top and the youngest bed at the bottom (Figure 10.6c and 10.6d). While we generate the same geometry of antiforms and synforms, the younging direction is opposite to what we had before. In this antiform, the beds young toward the core, while in the synform the beds young away from the core. Both cases are **downward-facing folds,** and an antiform with this younging characteristic is therefore called a **downward-facing antiform;** analogously, we recognize a **downward-facing synform.**[3] Remember that when you find downward-facing folds in the field, you immediately know that some secondary process has inverted the normal stratigraphic sequence; that is, we cannot violate the Law of Superposition. Downward-facing folds are not as uncommon as one might guess. They are typically found in areas containing an early "generation" of regional folds with horizontal axial surfaces, which are quite common in collisional mountain belts. Subsequent folding of these early structures generates a series of upward- and downward-facing folds, as shown in Figure 10.6e. But we must not get ahead of ourselves; the principle of superposed folding is discussed in a later section of this chapter.

10.3 FOLD CLASSIFICATION

Now that we have established a basic vocabulary we can further classify folds. The classification of folds is based on four components:

1. Fold shape in three dimensions, primarily distinguishing between cylindrical folds and noncylindrical folds (Figure 10.3).
2. Fold facing, separating upward-facing folds and downward-facing folds (Figure 10.6).
3. Fold orientation.
4. Fold shape in the profile plane.

The first two components, three-dimensional fold shape and fold facing, have already been introduced (Figures 10.3 and 10.6, respectively). In this section

series of antiforms and synforms, we see that the oldest bed lies in the core of the antiform and the youngest bed lies in the core of the synform (Figure 10.6a and 10.6b). Under these circumstances we call them anticlines and synclines, respectively. In an **anticline** the beds young away from the core; in a **syncline** the beds young toward the core. In both cases the younging direction points (or faces) upward, so we call these

---

[3]The antiform and synform have the younging characteristics of a syncline and anticline, respectively, and are therefore also called *antiformal syncline* and *synformal anticline,* respectively.

FIGURE 10.7   An asymmetric, plunging fold (the Sheep Mountain Anticline in Wyoming, USA).

we concentrate on the other two components of fold classification: fold orientation and fold shape.

10.3.1   Fold Orientation

Looking at the curved surface of a natural fold makes one wonder if there is any one representative measurement for the structure (Figure 10.7). Taking your compass to the folded surface will give you a large number of different readings for dip, and dip and strike (or dip direction) if the fold is noncylindrical. In folds with limbs that are relatively straight, you will find that all the measurements in a single limb are pretty much alike (Figure 10.8), but in folds with curving limbs this will not be the case. So what do we measure if we want to give the orientation of a fold to another geologist? The first measurement we take is the orientation of the hinge line (Figure 10.9). On the scale of an outcrop the hinge

line is typically fairly straight, and we determine its plunge (say, 20°) and direction of plunge (say, 190°). We now say that the fold is shallowly plunging to the South. Secondly, we measure the orientation of the axial surface. We measure a dip direction/dip of 270°/70° for the axial surface, which completes our description of the fold: *a shallowly south-plunging, upright fold*. Remember that the hinge line always lies in the axial surface. Test your measurements in a spherical projection to see whether this relationship holds. What constrains terms like **shallow** and **upright?** As a practical convention we use the angular ranges shown in Table 10.2.

In Figure 10.10 we show some representative combinations of hinge line and axial surface orientations with their terminology. A fold with a horizontal axial surface by definition must have a horizontal hinge line, and is called a **recumbent fold** (Figure 10.1). In the European Alps, for example, large-scale recumbent

FIGURE 10.8 Chevron folds in Franciscan chert of California, USA (Marin County).

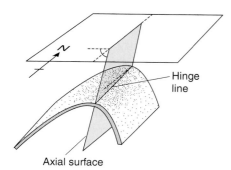

FIGURE 10.9 Fold orientation. Note that the axial surface is a *plane* whose orientation is given by dip and strike (or dip direction), whereas the hinge is a *line* whose orientation is given by plunge and bearing.

TABLE 10.2	FOLD CLASSIFICATION BY ORIENTATION	
Plunge of Hinge Line	**Dip of Axial Surface**
Horizontal: 0°–10°	Recumbent: 0°–10°
Shallow: 10°–30°	Inclined: 10°–70°
Intermediate: 30°–60°	Upright: 70°–90°
Steep: 60°–80°	
Vertical: 80°–90°	

folds are often associated with thrust faulting, and they are called **nappes** (see Figure 9.1). A term that is used for a steeply plunging, inclined fold is a **reclined fold.** In all cases remember that your field measurements will be no more accurate than ±2° (compass accuracy), but that the feature you measure will probably vary over an even greater angle of ±5°–10°. Thus, the values in Table 10.2 serve as a guide and should not be applied too strictly; they are not carved in stone.

10.3.2 Fold Shape in Profile

The profile plane of a fold is defined as the plane perpendicular to the hinge line (Figure 10.2). The fold shape in profile (as viewed, by convention, down the plunge) allows further classification of folds. Because the profile plane is perpendicular to the hinge line, we need not concern ourselves with the orientation of the fold. Fold shape in profile describes the interlimb angle and any changes in bed thickness. The *interlimb*

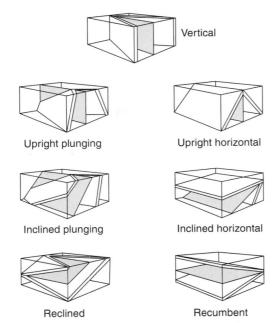

FIGURE 10.10 Fold classification based on the orientation of the hinge line and the axial surface (shaded).

Vertical

Upright plunging

Upright horizontal

Inclined plunging

Inclined horizontal

Reclined

Recumbent

TABLE 10.3	FOLD CLASSIFICATION BY INTERLIMB ANGLE
Isoclinal	0°–10°
Tight	10°–60°
Open	60°–120°
Gentle	120°–180°

angle of a fold is the angle between the limbs. We assume that the limbs are relatively planar or we use the tangent at the inflection points (Figure 10.4). The values corresponding to the various terms are listed in Table 10.3. As with those in Table 10.2, they serve only as a rough guide.

The second characteristic of fold shape in profile is any change in *bed thickness* across the structure. If you look at Figure 10.11a, you will notice that the bed thickness does not change appreciably as we go from one limb of the fold to the other. In contrast, the fold in Figure 10.11b has thin limbs and a relatively thick hinge area. We quantify these observations by using a method called **dip-isogon analysis.** Dip isogons connect points on the upper and lower boundary of a folded layer where the layers have the same dip relative to a reference frame (Figure 10.12). The construction method is explained step by step in all structural geology laboratory manuals, to which you are referred.[4] Three classes are recognized: **convergent dip isogons** (Class 1), **parallel dip isogons** (Class 2), and **divergent dip isogons** (Class 3). The terms "convergence" and "divergence" are used with respect to the core of the fold;[5] when the dip isogons intersect in a point in the core of the fold, the fold is called con-

vergent, and vice versa. The two geometries shown in Figure 10.11 are special cases. Dip isogons that are perpendicular to bedding throughout the fold define a **parallel fold,** whereas dip isogons that are parallel to each other characterize a **similar fold.** This terminology (especially the use of "parallel") may be confusing, but remember that parallel and similar describe the geometric relationship between the top and bottom surfaces of a folded layer; they do not describe the relationship between individual dip isogons in a fold. Parallel and similar folds anchor the finer fivefold subdivision that is used mainly for detailed description. In the field, loosely using the terms similar (representing Class 2 and 3) and parallel (representing Class 1A, 1B, and 1C) is usually sufficient to describe the fold shape in profile.

So we added two more components to our description of a fold. Now as a test, sketch a shallowly plunging, upright, tight, similar, downward-facing synform in the margin of the text. Hopefully these terms have become sufficiently clear that the task, unlike its description, is relatively simple. The only parameter we have excluded in our classification is **fold size.** To specify this we can use terms like microfold (microscopic size; up to millimeter scale), mesofold (hand specimen to small outcrop size; centimeter to meter scale), and macrofold (mountain size and larger; hundreds to thousands of meters). Although the lengthy description above is certainly not pretty, it ends up being very informative and complete. Remember that the goal of any good description is first to recall the characteristics for yourself, and secondly to relay this information in an understandable and unequivocal fashion to someone else.

10.4 FOLD SYSTEMS

Our treatment of folds so far has concentrated mostly on single antiforms and synforms. When we have a series of antiforms and synforms, we call this a **fold**

---

[4]Graphically, the dip isogon classification plots angle α versus normalized distance between the two tangents defining a dip isogon.

[5]The same terminology returns with cleavage (Chapter 12).

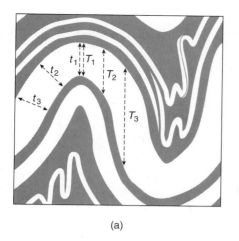

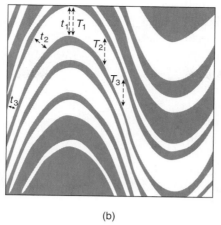

(a) (b)

FIGURE 10.11 Parallel folds (a) maintain a constant layer thickness across the folded surface, meaning, $t_1 = t_2 = t_3$, but the layer thickness parallel to the axial surface varies $(T_1 < T_2 < T_3)$. Note that parallelism must eventually break down in the cores of folds because of space limitation, which is illustrated by the small disharmonic folds in (a). In similar folds (b), the layer thickness parallel to the axial surface remains constant, so, $T_1 = T_2 = T_3$, but the thickness across the folded surface varies $(t_1 > t_2 > t_3)$. Similar folds do not produce the space problem inherent in parallel folds.

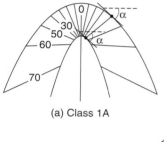

(a) Class 1A

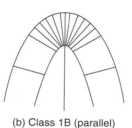

(b) Class 1B (parallel)

(c) Class 1C

(d) Class 2 (similar)

(e) Class 3

FIGURE 10.12 Fold classification based on dip isogon analysis. In Class 1A (a) the construction of a single dip isogon is shown, which connects the tangents to the upper and lower boundary of the folded layer with equal angle (α) relative to a reference frame; dip isogons at 10° intervals are shown for each class. Class 1 folds (a–c) have convergent dip isogon patterns; dip isogons in Class 2 folds (d) are parallel; Class 3 folds (e) have divergent dip isogon patterns. In this classification, parallel (b) and similar (d) folds are labeled as Class 1B and Class 2, respectively.

system. The information we can obtain from fold systems provides some of the most powerful information for the interpretation of regional structure, and involves such elements as fold symmetry, fold vergence, and the enveloping surface. We start with the last of these.

10.4.1 The Enveloping Surface

Draw an imaginary plane that is tangential to the hinge zones of a series of small folds in a layer (surface *A*, Figure 10.13). We call this surface the **enveloping surface.** It contains all the antiformal or synformal hinges.[6] Figure 10.13 also shows that we can draw an additional enveloping surface (surface *B*) when we connect the

hinges of the curved enveloping surface *A*. We call the enveloping surface for the largest folds the first-order enveloping surface (here surface *B*). The enveloping surfaces of successively smaller structures have a higher order (second-order enveloping surface, third-order enveloping surface, and so on). The first-order enveloping surface is typically of regional scale, while higher-order enveloping surfaces may go down as low as the thin-section scale. But what is the point of determining the enveloping surface? With decreasing order, the enveloping surfaces reduce the structural information of a folded area into increasingly simple patterns. For example, the second-order enveloping surface in Figure 10.13 shows that the small-scale folds define a larger-scale fold pattern consisting of antiforms and synforms. These large-scale structures have also been designated by the terms **anticlinorium** and **synclinorium,** respectively. Unfortunately the use of these terms suggests upward-facing structures, which is not

---

[6]In the case of folds with horizontal or shallowly dipping axial surfaces, we refer to these as crests and troughs, respectively.

always intended. The presence of anticlinoria and synclinoria implies that many small folds are somehow related, even though they vary in shape and position in the larger structure. It is important to realize that the orientations of these small folds (both hinge line and axial surface) are often the same, and also that these parameters approximately parallel that of the anticlinorium and synclinorium. For that reason, these small folds are sometimes called **parasitic folds,** because they are related to a larger structure.

The geometric relationship between parasitic folds and regional structures offers a powerful concept in structural analysis, which states that the orientation of small (high-order) structures is representative of the orientation of regional (low-order) structures.[7] Thus, the orientations of the hinge line and the axial surface of a small fold can predict these elements for a large regional fold that is otherwise not exposed; in Figure 10.13 you indeed see that this is the case. Obviously, this "rule" serves only as a convenient working hypothesis, but it has proven to be very robust in regional mapping. Use this rule on a field trip and surprise friend and foe with your quick insight into regional structure.

10.4.2 Fold Symmetry and Fold Vergence

The relationship between the enveloping surface and the axial surface of folds also enables us to describe the symmetry of folds. If the enveloping surface and the axial surface are approximately perpendicular (±10°), we have **symmetric folds** (Figure 10.14a); otherwise the folds are **asymmetric** (Figure 10.14b). In the case of an isolated fold, an enveloping surface cannot be defined. To determine if a fold is symmetric or asymmetric we use the median surface, which is the surface that passes through the inflection points of opposing limbs. If the axial surface is perpendicular to the median surface, then the fold is symmetric; otherwise the fold is asymmetric. There are other definitions of fold symmetry that involve, for example, the relative steepness of limbs, but these descriptions of fold symmetry are often ambiguous and should not be used.

Now let's look at a practical application of fold symmetry. The second-order enveloping surface defined by the small folds of Figure 10.13 outlines a large antiform-synform pair. The small folds (the *parasitic folds*) show characteristic shapes and asymmetries as we move along the second-order enveloping surface. On the west limb of the large antiform, the minor folds are asymmetric and have what we call a clockwise asymmetry when looking down the plunge of the fold. In the hinge area, the folds are symmetrical, because the axial surface is perpendicular to the enveloping surface. In fact, as we move from the limb toward the hinge area, the clockwise asymmetry becomes progressively less, until the fold is symmetrical. As we move into the east limb of the antiform, the fold asymmetry returns, but now with the opposite sense to that in the west limb; in the east limb the asymmetry is counterclockwise. Note that clockwise and counterclockwise

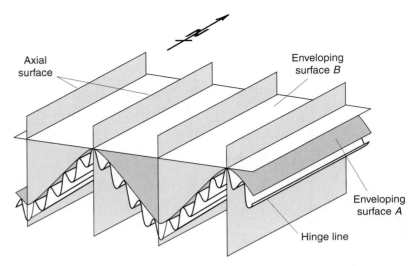

FIGURE 10.13 The enveloping surface connects the antiform (or synform) hinges of consecutive folds (surface *A*). If this imaginary surface appears to be folded itself, we may construct yet a higher-order enveloping surface (surface *B*). Note that the orientations (hinge line and axial surface) of the small folds and the large-scale folds are very similar.

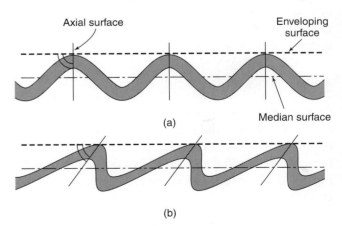

FIGURE 10.14 Symmetric (a) and asymmetric folds (b) are defined by the angular relationship between the axial surface and the enveloping surface.

---

[7]This is otherwise known as *Pumpelly's Rule,* after the nineteenth-century American geologist Raphael Pumpelly.

are defined by the rotation of the axial surface relative to a hypothetical symmetrical fold (Figure 10.15). In the past, parasitic folds were erroneously given the genetic name "drag folds," because it was assumed that the apparent rotation of the axial surfaces reflects drag between the layers during folding. Rather, these minor folds are probably symmetrical in the incipient stages of regional folding and become more asymmetrical when the large folds tighten. So, across a large fold the **vergence**[8] of parasitic folds changes in a characteristic manner that allows us to predict the location of the hinge area of large antiforms and synforms (Figure 10.16). Using parasitic folds, we may even predict the orientation of these large regional folds (see previous section).

There is need for some caution when folds are not plunging; that is, in horizontal upright folds where down-plunge observations cannot be defined. If we view the structure in Figure 10.16 from the opposite side, the clockwise folds become counterclockwise (just hold the page with Figure 10.16 facing the light, viewing it from the back)! This situation is not as confusing as it seems at first. Imagine an antiform that is cut by a road perpendicular to the axial surface. The asymmetry of parasitic folds in this large structure appears clockwise or counterclockwise on opposite sides of the road. However, in both cases they make the same prediction for the location of the hinge area. As long as you define the direction in which you view the minor structure, there is no problem using fold vergence as a mapping tool in an area. A practical tip is to copy the geometry of Figure 10.16 into the back of your field notebook. Matching field observation with asymmetries in your sketch, which may require some rotation of your notebook, will ensure a reliable application of this mapping tool. In any case, remember that a pattern of fold vergence opposite to that in Figure 10.16 (a "Christmas-tree" geometry) cannot be produced in a single fold generation (Figure 10.17).[9]

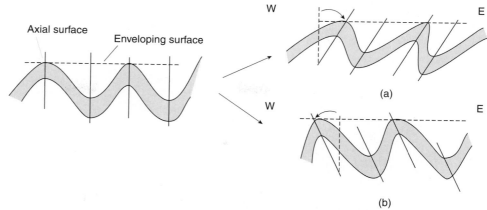

FIGURE 10.15 Fold vergence, clockwise (a) and counterclockwise (b), is defined by the apparent rotation of the axial surface from a hypothetical symmetric fold into the observed asymmetric fold, without changing the orientation of the enveloping surface. In a given geographic coordinate system we may also say east-verging (a) and west-verging (b) folds for clockwise and counterclockwise folds, respectively (for this example). In all descriptions we are looking down the plunge of the fold.

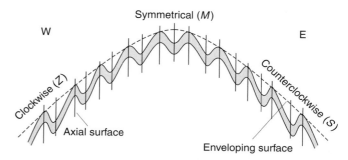

FIGURE 10.16 Characteristic fold vergence of parasitic folds across a large-scale antiform. Looking down the plunge, the parasitic fold changes from clockwise asymmetry (east-verging in the geographic coordinate system) to symmetric to counterclockwise asymmetry (west-verging) when going from W to E. You may also find that some geologists use the terms Z-, M-, and S-folds for this progression (for obvious reasons).

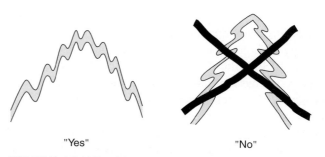

FIGURE 10.17 Right and wrong in fold vergence. It is useful to copy the images in Figures 10.16 and 10.17 in your field notebook for reference.

---

[8]Do not confuse fold vergence with fold facing.

[9]In fact, this geometry is diagnostic of the presence of at least two fold generations.

FIGURE 10.18 Monocline in the Big Horn Mountains (Wyoming, USA).

10.5 SOME SPECIAL FOLD GEOMETRIES

We end the descriptive part of this chapter with a few special fold geometries that you may encounter in your field career. **Monoclines** are fold structures with only one tilted limb; the beds on either side of the tilted limb are horizontal. Monoclines typically result from a vertical offset in the subsurface near the tilted portion of the structure. The fault uplifts a block of relatively rigid igneous or metamorphic rock, and the overlying sedimentary layers drape over the edge of the uplifted block to form the monocline (Figure 10.18). Spectacular examples are found in the Colorado Plateau of the western United States. **Kink folds** are small folds (less than a meter) that are characterized by straight limbs and sharp hinges. Typically they occur in finely laminated (that is, strongly anisotropic) rocks, such as shales and slates (Figure 10.19). Sharply bending a deck of cards is a good analogy for the kinking process, because kink folds are formed by displacements between individual laminae (individual cards in the analogy). **Chevron folds** (Figure 10.8) are the larger-scale equivalent of kink folds.

The term **box fold** describes a geometry that is pretty self-explanatory (Figure 10.5). In order for a box fold to form, a layer must be detached from the underlying and overlying layers. They are therefore common in areas with weak basal layers, such as in the Jura Mountains of Switzerland. **Ptygmatic folds** are irregular and isolated

FIGURE 10.19 Kink folds in mica-rich portion of greywackes of the Cantabrian Mountains (northern Spain).

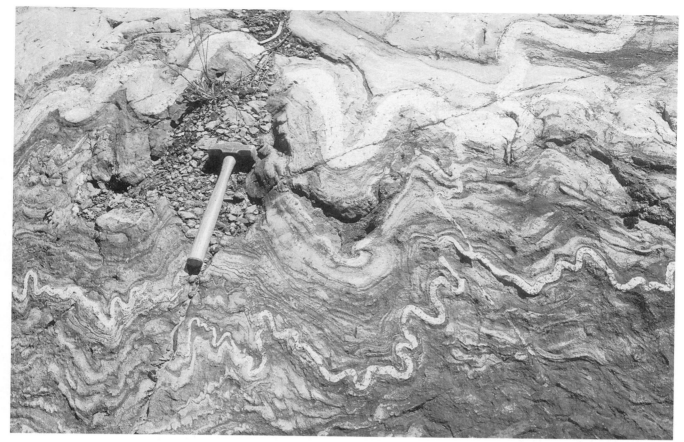

FIGURE 10.20 Ptygmatic folds in the Grenville Supergroup (Ontario, Canada); hammer for scale. Note the wavelength variation as a function of layer thickness.

fold structures that typically occur as tightly folded veins or thin layers of strongly contrasting lithology (and, thus, contrasting competency; Figure 10.20). Most metamorphic regions around the world contain ptygmatic folds, which, unglamorously, resemble intestines.

Doubly plunging folds are structures with hinge lines that laterally change curvature. Along the trend of plunge the folds may die out or even change from antiforms to synforms. The high point of the hinge line in a doubly plunging fold is called the **culmination** and the low point along the same hinge line is called a **depression.** The change in plunge angle is normally less than 50°. When additional folds are present, changes in plunge may result in *en echelon* **folds,** in which a gradually opening fold is replaced by a neighboring, gradually tightening fold of opposite form. Such a geometry occurs on all scales, from hand specimens (Figure 10.21) to the size of mountain ranges (such as the Valley-and-Ridge of the central Appalachians). Note that doubly plunging folds are, by definition, noncylindrical. **Sheath folds**[10]

show extreme hinge line curvature, to the extent that hinge line curvature approaches parallelism (change in plunge up to 180°!). What is typically found in outcrop is the elliptical cross section of the nose of the fold (see Figure 12.28); however, such a pattern itself does not necessarily imply a section through a sheath fold. Any doubly plunging fold may give the same outcrop pattern, but only when a highly curved hinge line is visible can

FIGURE 10.21 *En echelon* folds on the scale of centimeters; coin for scale.

---

[10]They resemble the sheath of a sword; the more imaginative term is *condom fold.*

sheath folds be recognized as such. Sheath folds are produced by taking a mildly doubly plunging fold and "pulling" at its crest, as occurs in zones of high shear strain. We return to the formation and significance of these structures in Chapter 12, which deals with ductile shear zones.

Finally, two additional fold types have been studied extensively in recent years because of their association with hydrocarbon potential in fold-thrust belts, namely, fault-bend folds and fault-propagation folds. Because of their intimate association with thrusting, we further examine their formation and significance in Chapter 18 ("Fold-Thrust Belts"). At this point we merely include them for completeness: **fault-bend folds** are formed as thrust sheets move over irregularities in the thrust plane (such as ramps), whereas **fault-propagation folds** are accommodation structures above the frontal tip of a thrust.

10.6 SUPERPOSED FOLDING

Structural geologists use the term **fold generation** to refer to groups of folds that formed at approximately the same time interval and under similar kinematic conditions. Commonly we find several fold generations in an area, which are labeled by the letter F (for $Fold$) and a number reflecting the relative order of their formation: F_1 folds form first, followed by F_2 folds, F_3 folds, and so on. Several fold generations may in turn form during an **orogenic phase** (such as the Siluro-Devonian "Acadian" phase in the Appalachians or the Cretaceous-Tertiary "Laramide" phase in the North American Cordillera), which is noted by the letter D (for $Deformation$). In any mountain belt several phases may be present, which are labeled D_1, D_2, and so on, each containing one or more generations of folds. For example, the Appalachian Orogeny of the eastern United States contains three main deformation phases (Taconic, Acadian, and Alleghenian). From the onset it is important to realize that neither a deformation phase nor each of the individual fold generations have to be present everywhere along the orogen, nor do they occur everywhere at the same time. On a regional scale *deformation is irregularly distributed and commonly diachronous.* You can imagine that fold generations and deformation phase can rapidly become pretty complex. So we'll stick to two fold generations to examine the principles of superposed folding, which allows us to unravel the sequence (that is, relative timing) of folding.

Generation is a relative time concept and only implies "older than" or "younger than"; you are the younger generation in the eyes of your parents. There are methods to determine the absolute ages of folds, such as dating of minerals that formed during folding, but we will not get into them here. The relative time principle of **superposed folding**[11] is simple: folds of a later generation are superimposed on folds of an earlier generation. The determination of this temporal sequence, however, is not straightforward and requires careful spatial analysis. Superposed folding is a widespread phenomenon that is not restricted to high-grade metamorphic areas. Even in regions below the **greenschist facies** (temperatures below ~300°C), superposed folding is found. It is worthwhile therefore to give attention to this topic, after which we close this section with the concept of fold style that is used to place our findings on fold generations in a regional context.

10.6.1 The Principle of Fold Superposition

Figure 10.22 shows a field photograph of a complex fold geometry that contains two fold generations. How do we know this from looking at the picture and how can we separate F_1 and F_2 folds in this pattern? Fold superposition is simple at its root, but the concept requires the ability to visualize and analyze sometimes very complex three-dimensional geometries. Let us first start with the rule: *a superposed fold must be younger than the structure it folds.* This merely restates the Law of Superposition such that it applies to folding. Unless a fold was present previously, it cannot be modified by a younger fold. How do we determine the criteria by which one obtains this temporal relationship? We begin with an example.

Figure 10.23a shows a sequence of recumbent folds that we will call F_A; the associated axial surface is called S_A. We now superimpose a series of upright folds of approximately the same scale (F_B with axial surface S_B; Figure 10.23c). The superimposition of F_B on F_A produces the interference pattern shown in Figure 10.23b. Elements of both fold generations are preserved; for example, the recumbent nature of F_A is still there, but its limbs are now folded. Similarly, the upright F_B folds remain visible, but they are superposed on a pattern that repeats and inverts bedding (from the recumbent F_A folds). The way to determine the temporal relationship from our interference pattern is to invoke the rule of superposition. Both the bedding

---

[11]Also called fold superimposition or fold overprinting.

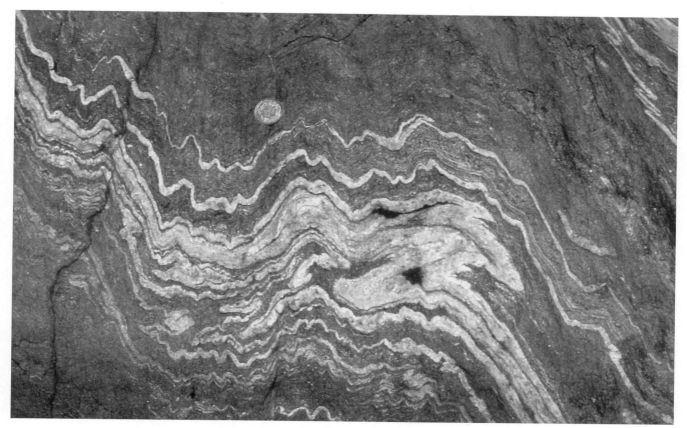

FIGURE 10.22 Fold interference pattern of Type 3 ("refolded fold") geometry.

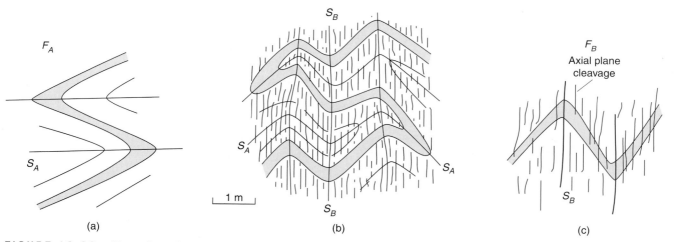

FIGURE 10.23 Dissecting a fold interference pattern. The cross sections show F_A recumbent folds (a) that are overprinted by F_B upright folds (b), producing the fold interference pattern in (c).

and S_A are folded, but S_B is essentially planar. So, bedding and S_A were already present before S_B; consequently, the upright folds must be F_2, which are younger than the recumbent F_1 folds. The axial surfaces may not be always visible in the field (although axial plane foliations are common; Chapter 11), but you can always use an imaginary axial surface to eval-

uate these complex folding patterns. Now examine this pattern yourself with an analog. Take a piece of paper and fold it in two (our F_1 fold) and orient it into a recumbent orientation. Then fold the paper again to create an upright fold (our F_2 fold) whose hinge parallels the hinge of the recumbent fold. *Voilà,* you get the pattern of Figure 10.23b. If you are comfortable with this example, we will proceed with the four basic fold interference patterns.

10.6.2 Fold Interference Patterns

Four basic patterns are recognized from the superimposition of upright F_2 folds on F_1 folds of variable orientation (Figure 10.24). Looking at these **fold inter-** **ference types** you will notice that we produced Type 3 using the piece of folded paper above. Types 0 and 2 can equally well be examined by folding a piece of paper, but Type 1 requires additional crumpling. Instead of describing these patterns in confusing words, look at Figure 10.24 and reproduce the geometries with a piece of paper and your hands. An approach that may offer further insight is to make fold

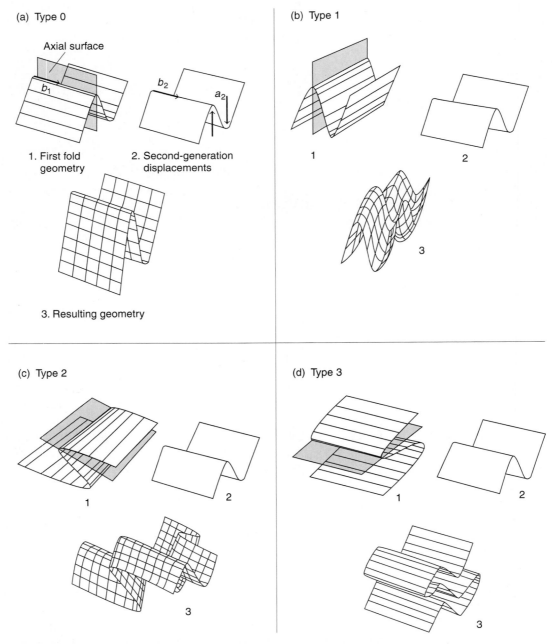

FIGURE 10.24 The four basic patterns arising from fold superposition. The analysis assumes that F_2 shear folds (a_2 is the relative shear direction and b_2 is the hinge line) are superimposed on a preexisting F_1 fold of variable orientation. Shear folds are modeled by moving a deck of cards. The shaded surface is the S_1 axial surface.

interference patterns using a few thin layers of colored modeling clay, which can then be cut with a knife to see the effect of intersecting surfaces. Open a refreshing drink and start your experimentation. . . .

Welcome back from spending time with clay, paper, and Figure 10.24. We now turn to some important properties of the various fold interference types. Type 0 is a special condition, because the hinge lines and the axial surfaces of both fold generations are parallel. As a consequence, F_1 is merely tightened by the superimposition of F_2. You realize that, practically, Type 0 cannot be recognized in the field as an interference type by geometry alone (that's why we use the number 0). Type 1 is also called a "dome-and-basin" structure and resembles an egg carton. Both the axial surfaces and the hinge lines of the two generations are perpendicular, producing this characteristic geometry (Figure 10.24a). Type 2 is perhaps the most difficult geometry to visualize, but folding a piece of paper helps enormously. In outcrop, we often see a section through this geometry that resembles a "mushroom" pattern (Figure 10.24b). Note that this outcrop pattern is only generated in the horizontal surface that intersects Type 2; if we take another cut, say vertical, the outcrop pattern is quite different. Finally, Type 3 (Fig-ure 10.24c) is sometimes referred to as the "refolded fold" pattern, which is a misnomer because all four types are refolded folds. We just present the name so that you have heard it, and because very few people are otherwise able to remember the corresponding numbers of the types. We recommend that you use the descriptive terms "dome-and-basin," "mushroom," and "refolded-fold," however flawed, instead of the abstract Type 1, Type 2, and Type 3, respectively.

Interference patterns are a function of the spatial relationship between hinge lines and axial surfaces of the fold generations, as well as the sectional surface in which we view the resulting patterns. Thus, the analysis of fold superposition is a three-dimensional problem. The four types that are shown in Figure 10.24 are only end-member configurations in an infinite array of possibilities. Figure 10.25 is a summary diagram that shows patterns from varying the spatial relationships between fold generations, as well as the observation surface (or, the outcrop). Even more so than before, understanding these patterns requires self-study. Ultimately, interference patterns reward you with complete information on the sequence and orientation of fold generations. So, again take your time.

The fold interference patterns we have analyzed are produced when fold generations of similar scale are superimposed. If the scales are very different there may be no interference pattern visible on the outcrop scale, and only through regional structural analysis does the large-scale structure appear. After some field work it is therefore not uncommon to find one or more additional fold generations that only show on the map scale. A reexamination of some puzzling field notes and outcrop sketches may all of a sudden be explained by recognizing this missing fold generation.

The presence of multiple fold generations has major implications for the interpretation of the deformation history of your area. First, it implies that the kinematic conditions have changed to produce a fold generation with different orientation than before (except Type 0); so the deformation regime must somehow have changed. Secondly, folds of the first generation will have variable orientations depending on where they are measured in the fold superposition pattern. Orientation, therefore, is *not* a characteristic of fold generations in multiply deformed areas and should be used carefully as a mapping tool (see below). That leaves a final question: How do we recognize folds of a certain generation in the absence of interference patterns at each and every locality in our area? For this we turn to the powerful concept of fold style.

10.6.3 Fold Style

When we encounter a number of folds in our field area, the logical question of their significance arises. Are they part of the same generation or do they represent several generations? Say that, at one locality in our area, we are actually able to determine a sequence of F_1 and F_2 folds, so we know that there are at least two generations. From our experience with superposed folding, we are also aware that only F_2 folds have an orientation that may persist over any distance, and that the orientation of F_1 folds depends entirely on their position in the fold interference pattern (we measure their orientation nonetheless because the distribution should "fit" the pattern). We now are at an outcrop where we only find one fold, which is not in the exact same orientation as either F_1 or F_2 in the previous outcrop. Nonetheless we wish to predict to which generation it belongs, and for that we use characteristics for each fold generation that are grouped under the term **fold style.** The fold style characteristics are listed in Table 10.4.

The four elements of Table 10.4, parallel/similar, interlimb angle, cylindrical/noncylindrical, and foliations/lineations, are used to describe the style of a fold. The first three have been discussed in detail and need no further clarification. Foliations and lineations will have more meaning after you read Chapter 11, but this fourth characteristic is included here because of its discriminatory ability. For example, an axial plane

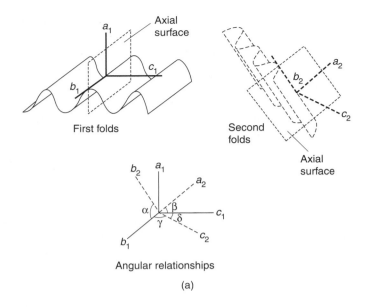

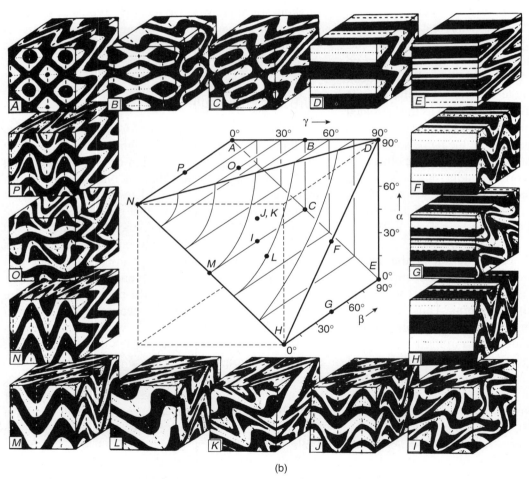

FIGURE 10.25 Geometric axes describing the orientation of fold generations F_1 and F_2 (a), and corresponding interference patterns (b). In all patterns, the layering was initially parallel to the front face of the cube. F_1 resembles case D; F_2 is similar to the folding in case D, but with different orientations. Axial surface S_1 is shown with dotted lines and axial surface S_2 with dashed lines.

TABLE 10.4	THE CHARACTERISTIC ELEMENTS OF FOLD STYLE

- In profile plane, is the fold classified as parallel or similar (or a further refinement)?
- What is the interlimb angle in profile?
- In three dimensions, is the fold cylindrical or noncylindrical?
- Is there an associated axial plane foliation and/or lineation present, and of what type are they?

Note that orientation and symmetry are not style criteria.

crenulation cleavage may be a characteristic of F_2 folds, and the presence of a mineral lineation may reflect special metamorphic conditions that only occurred during the first fold generation. Notably absent in our list are fold orientation and fold symmetry, which are not style criteria. Discriminating a fold generation on its orientation may only work for the last fold generation; the older ones most likely have become variably oriented. Secondly, we already learned that fold symmetry may change within a single-generation, large fold (Figure 10.16). So, just like orientation, symmetry is not a style criterion.

10.6.4 A Few Philosophical Points

We close the section on superposed folding with a few considerations. You will often find that it is not possible in any single outcrop to determine the complete sequence of fold generations, because discriminatory interference patterns may not be exposed, or one or more generations may not be visible at all. However, by combining information from several outcrops as well as using fold style you should eventually be able to obtain a reasonable fold sequence. As you map, you should continue to test this hypothesis, and after a thorough job the foundation on which you base the folding sequence will be firm. Only then will the time have come to place your findings in a regional kinematic picture. For example, you may find that the first generation of folds is recumbent, which we quite commonly associate with nappe style, and that the second fold generation is upright, reflecting folding of the thrust sequence. Maybe yet a third fold generation reflects a very different shortening direction, possibly related to a different orogenic phase. We may also find

small kink folds in well-foliated rocks that complete the deformation sequence. Although the possibilities seem limitless, reasonable interpretations are not.

There is sometimes a tendency to recognize too many fold generations by structural geologists. In the end, the number of fold generations should be based *only* on interference patterns, on either local or regional scales. Practically, in any one outcrop you may be able to see two or three fold generations, and regionally perhaps a couple more. Remember that structural analysis is not helped by proposing an unnecessarily long and complex sequence of fold generations, because each generation must reflect a corresponding deformation regime. One can reasonably expect only so many different tectonic patterns. With these musings and a closing field example of a Type 1 fold interference pattern (Figure 10.26), we leave the descriptive part of folds and field analysis to turn our attention to the mechanics of folding.

10.7 THE MECHANICS OF FOLDING

Why does folding occur? After spending so much time on the description of folds, this is a question whose time has come. Well, if you ever saw a car collision you do not need to be reminded that forces can cause folding.[12] Indeed, forces applied to rocks cause folding, but we will see that forces alone are not sufficient to form folds. Consider a block of clay that is reshaped by external forces (from your hands). The block will change form, but in doing so the internal structure does not show any folds (Figure 10.27a). After we add irregularly shaped layers of different color but with the same material properties to the block, we get folds when the irregularities are amplified (Figure 10.27b). If we add straight, thin sheets of rubber to our clay block and a force is applied, folds also appear (Figure 10.27c). These experiments provide us with a fundamental subdivision of folding based on the mechanical role of layers: passive folding and active folding.

10.7.1 Passive Folding and Active Folding

During **passive folding** the layering of a material has no mechanical significance. In the color-banded clay block, folds are formed by the amplification of small

---

[12]This recurring analogy does not reflect the authors' personal experiences.

FIGURE 10.26 Fold interference pattern of Type 1 ("dome-and-basin") geometry; compass for scale.

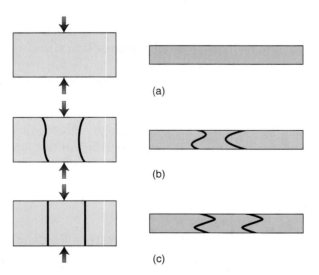

(a)

(b)

(c)

FIGURE 10.27 Compression of a clay block of uniform color (a), with irregularly shaped layers of different colors (b), or with uniform colored layers separated by thin sheets of rubber (c).

perturbations in the bands, but the strain pattern in the block is unaffected by the presence of these layers. Squeezing multicolor toothpaste on a counter top is a fresh and minty experiment where complex folding patterns are visible only because of the color contrast; if you do the same experiment with single-color toothpaste you will not see folds, even though the internal structure of the two blobs is similar. We find such toothpaste-like behavior in nature where rocks have little or no competency contrast between layers. Elevated temperatures can produce the right conditions for passive folding and it is common to find toothpaste-like structures in deformed metamorphic rocks. Rocks that were deformed at or near their melting temperature (that is, a high homologous temperature, $T_h = T/T_m$) are called **migmatites,** which often contain wonderfully complex fold structures (see Figure 2.22). Similarly, passive folding occurs in glaciers that deform close to their melting temperature. Passive folds are the amplification of natural irregularities in the layers, or are a consequence of differential flow in a volume of rock. But don't think that passive folds have to be chaotic in appearance because of this. Sheath folds in a shear zone are another natural example of passive folding and typically show very consistent orientation and style.

During **active folding,** also called **flexural folding,** the layering has mechanical significance. This means that the presence of layers with different competency directly affects the strain pattern in the deforming body and that there is contrasting behavior between layers. There are two dynamic conditions that we distinguish for active folding: bending and buckling. In *bending,* the applied force is oriented at an oblique angle to the layering (Figure 10.28a). In nature this may occur during basin formation or loading of a lithospheric plate (also called flexural loading), or during the development of monoclines over fault blocks. In *buckling,* the force is oriented parallel to the mechanical anisotropy (Figure 10.28b), the most common situation for folding.

Let's see what happens in a series of analog experiments of active folding, in this case buckling (Figure 10.29). We surround a band of rubber with foam in a plastic box that is open at one end, at which we place a plunger. As we push on the plunger, the band of rubber forms a series of folds. The surrounding foam (the "matrix") accommodates the shape of the rubber band by filling the gaps that would otherwise be present. We repeat the experiment with a thinner band of rubber and the same foam. Now we produce several more folds and the weaker foam again accommodates this new pattern. So, in spite of applying the same force and producing the same bulk shortening strain, the folding patterns are different as a function of the thickness of the rubber band. In other words, the rubber band introduces a **mechanical anisotropy.**

10.7.2 Buckle Folds

We return to the above rubber band–foam experiments, where we saw that the thickness of the band somehow affects the fold shape, to explore some systematic properties of buckle folds. The applied force (and thus stress), the bulk strain (the distance the plunger moves into our box), the strain rate (the speed at which the plunger moves), and the ambient conditions (room temperature and pressure) are assumed to be the same in all experiments; only thickness, *t*, of the layer varies. We notice that with increasing thickness the wavelength and arc length become larger. Secondly, if we use bands of equal thickness but with different

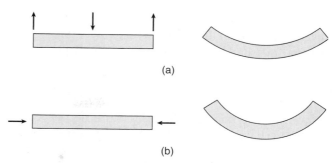

FIGURE 10.28 Bending (a) and buckling (b) of a layer.

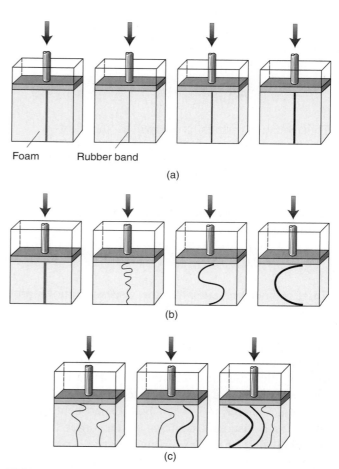

Foam   Rubber band

(a)

(b)

(c)

FIGURE 10.29 Line drawings of deformation experiments with transparent boxes containing foam, and rubber bands. In (a), four starting settings are shown that contain, from left to right, foam only (with marker line added), a thin, a medium, and a thick rubber band. When applying the same displacement, shown in (b), the setups respond differently. The foam-only box shows thickening of the marker line, but no folding. The boxes with rubber bands show folds with arc lengths varying as a function of thickness of each band. When using more than one band (c), the behavior depends on the combination of bands and their thicknesses, with the effect of the thicker bands being dominant.

stiffness, we find that arc length of the stiffer layer is larger than that of the weaker layer. So, we find that both thickness and the parameter "stiffness" increase the wavelength. In nature, we are not really dealing with elastic layers. Folds are permanent strain features, so it is more useful to consider this problem in terms of viscous behavior or even more complex rheologic models (such as elastico-viscous or non-linear viscosity; Chapter 5). For our current purposes we assume simple, Newtonian (linear) viscous behavior.

Using Newtonian viscosity, the theoretical arc length–thickness relationship for a layer with viscosity η_L surrounded by a matrix with viscosity η_M is:

$$L = 2\pi t \, (\eta_L/6\eta_M)^{1/3} \qquad \text{Eq. 10.1}$$

This equation, which is known as the **Biot-Ramberg equation,**[13] tells us that arc length (L) is directly proportional to thickness and to the cube root of the viscosity ratio. Therefore, if we know the arc length/thickness ratio of a layer, we can obtain the viscosity ratio. Reorganizing Equation 10.1, we get

$$\eta_L/\eta_M = 0.024 \, (L/t)^3 \qquad \text{Eq. 10.2}$$

This formulation states that the viscosity ratio is proportional to the cube of the L/t ratio. The measurements of folded sandstone layers in sedimentary rocks, shown in Figure 10.30, give a viscosity ratio of about 475. We intentionally say "about," because Figure 10.30 is a log–log plot, meaning that a small change in L/t ratio will result in a large change in viscosity ratio. Note that the same analysis for our box experiments above produces a viscosity contrast on the order of 1000, meaning that these experiments are a reasonable approximation of low-grade sandstone deformation.

There is an important consequence when the viscosity ratio of layer and matrix is small (say, <<100). Again we return to our box experiments. This time we place only foam in our plastic box. On the foam we draw a vertical line with a marker pen (Figure 10.29) to represent a layer with a small viscosity ratio (in this case, of course, η_L/η_M = 1). As we compress the foam we do not get any folds, but we find that the layer thickens. So, low viscosity contrasts result in a pronounced component of strain-induced *layer thickening.* We simplify this effect in our

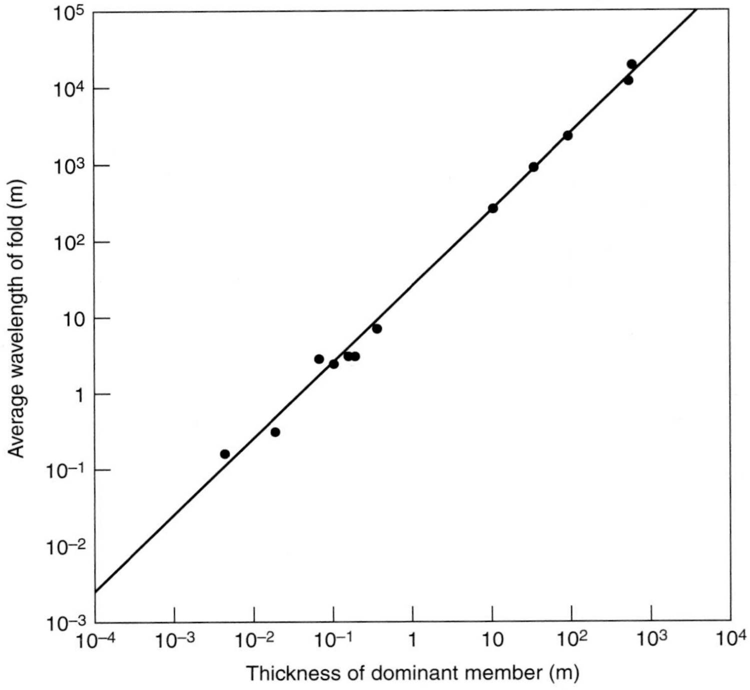

FIGURE 10.30 Log–log plot of wavelength versus layer thickness in folded sandstone layers.

analysis by inferring that a component of layer thickening occurs before folding instabilities arise. Recalling the effect of layer thickness on L (Equation 10.1), we therefore include a strain component in Equation 10.1:

$$L = 2\pi t \left[\frac{\eta_L(R_S - 1)}{6\eta_M \cdot 2R_S^{\,2}} \right]^{1/3} \qquad \text{Eq. 10.3}$$

where R_S is the strain ratio X/Z. This **modified Biot-Ramberg equation**[14] gives a reasonable prediction for the shape of natural folds in rocks with low viscosity contrast, such as one finds in metamorphic regions.

The respective roles of viscosity contrast and layer thickening during shortening are also well illustrated in numerical models of folding. The advantage of mathematical models is their ability to vary parameters with ease. Figure 10.31 shows the results of one series of computer simulations of single-layer folding using a finite-element method. As we already saw in the physical experiment, with decreasing viscosity ratio, the arc length becomes less and the layer increasingly thickens. Another advantage of computer modeling is that we can also track the strain field in our system. Note, for example, the strain pattern in and immediately surrounding the folded layer, which is shown by the

---

[13]Biot and Ramberg independently carried out this analysis in the early 1960s.

[14]Also known as the *Sherwin-Chapple equation,* after its authors.

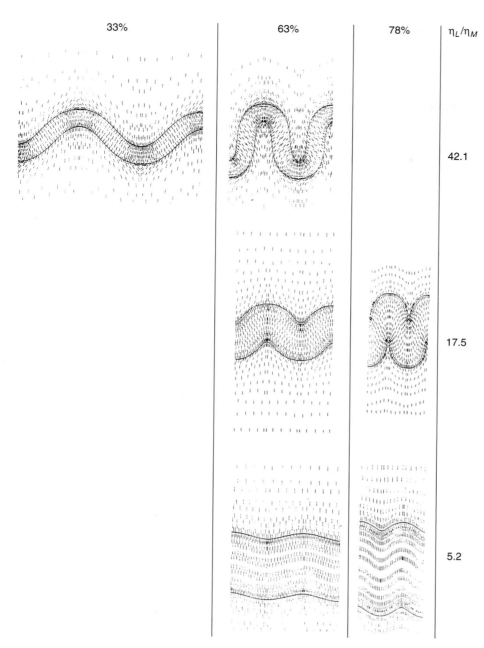

33%	63%	78%	η_L/η_M
			42.1
			17.5
			5.2

FIGURE 10.31 Finite-element modeling of single-layer buckling for various viscosity contrasts between layer (η_L) and matrix (η_M), and shortening strains (%). The short marks represent the orientation of the long axis of the strain ellipse in profile plane.

orientation of the x-axis in Figure 10.31. The strain pattern is increasingly homogeneous with decreasing viscosity contrast. Indeed, no viscosity contrast (η_L/η_M = 1) reflects the situation in which there is no more mechanical significance to the layer (see the foam-only box experiment in Figure 10.29).

These analyses of folding are entirely based on Newtonian-viscous behavior of both the layer and the matrix. However, it is likely that folding under elevated conditions of pressure and temperature involves

nonlinear rheologies, in which the viscosity is stress dependent (Chapter 5). Without going into any details, we only mention that appropriate nonlinear rheologies result in lower values of L/t as well as small to negligible components of layer-parallel shortening. Natural folds in metamorphic terrains typically have low L/t ratios ($L/t < 10$), suggesting that nonlinear viscous rheology, and therefore crystal-plastic processes, are indeed important during folding under these conditions.

10.7.3   Folded Multilayers

Our discussion so far has focused on folding of a single layer in a weaker matrix. While this situation occurs in nature, it is not representative. What happens when several layers are present? Again we turn to our simple experimental setup (Figure 10.29), but now use two rubber bands of equal thickness to see what folds develop. Can you explain why the results differ from those of our previous single-layer experiments? The wavelength in the multilayer experiment is greater and the two layers seem to act as a thicker single layer. In another experiment we combine a thick and a thin layer (Figure 10.29). In this case, the thin layer does not at all give the fold shape predicted from single-layer theory, while the thick layer behaves pretty much the same as predicted from the single-layer experiment. In fact, the thin layer mimics the shape of the thick layer, indicating that its behavior is dominated by that of the thick layer. We can try many other combinations, which all show that the behavior of a multilayer system is sensitive to the interaction between layers. The previous Biot-Ramberg buckling theory, therefore, only applies when layers in natural rocks are sufficiently far removed from one another to avoid interaction. We can theoretically determine the region over which the effect of a buckled layer in a weaker matrix dies off to negligible values. This is known as the **contact strain zone (CSZ).** The width of this contact strain zone ($2d_{CSZ}$, where d_{CSZ} is the distance measured from the midpoint of the buckled layer) is a function of the arc length:

$$d_{CSZ} = 2/\pi L \qquad \text{Eq. 10.4}$$

Practically this means that the width of the CSZ is slightly greater than the arc length of a fold, which seems supported by field observations on natural folds.

Interacting layers require a relatively simple extension of the theory of folding, if we assume a stack consisting of several superposed thin layers that are free to move relative to one another (that is, no coupling). The corresponding equation for this multilayer case is

$$L_a = 2\pi t \, (N\eta_L/6\eta_M)^{1/3} \qquad \text{Eq. 10.5}$$

where L_a is the arc length of the multilayer, N is the number of layers, and the space between the layers is infinitely small. Note that this is not the same as the equation for a single layer with the same thickness ($N \cdot t$). If we calculate the ratio of the arc length of a system with N superposed multilayers (L_m) and a system with a single layer of thickness Nt (L_s) by dividing Equation 10.1 (with $t = Nt$) and Equation 10.5, we get

$$L_s/L_m = N^{2/3} \qquad \text{Eq. 10.6}$$

This shows that the arc length of N multilayers of thickness t is less than that of a single layer with thickness Nt. This scenario is certainly not applicable to all geologic conditions. Commonly we find that layers of one viscosity alternate with layers of another viscosity. For example, a turbidite sequence contains alternating layers of sandstone and shale. In this case the analysis is considerably more complex and, because increasingly restrictive assumptions have to be made (including the spacing between layers), we stop here. The main point is that multilayers behave much like a thick single layer, but the resulting arc length in a multilayer system will be less than that of a single layer.

10.8   KINEMATIC MODELS OF FOLDING

The distinction we made earlier between active and passive folds describes the mechanical role of layers under an imposed stress, but this says little about the inner workings of the folded layer and the associated strain pattern. To this end, we differentiate between three fundamental models, flexural folding, neutral-surface folding, and shear folding, as well as modification of folds by superimposed strain; each of these have distinct properties and characteristics. These models are compared with natural examples in the final section.

10.8.1   Flexural Slip/Flow Folding

Take a phone book or a deck of playing cards, and bend it into a fold.[15] The ability to produce the fold is achieved by slip on the surfaces of the cards relative to one another, without appreciable distortion within the surface of any individual card (they remain of the same size). If we place small marker circles on the top surface as well as on the sides of the card stack for use as a strain gauge, we see that strain only accumulates in surfaces that are at an angle to the individual cards

---

[15]Computer punch cards are well suited, but remain only in the possession of old structural geology instructors.

(that is, the sides) when we fold the deck. The circles become ellipses in the profile plane (Figure 10.32) and on the other side of the folded deck parallel to the hinge line. Within the plane of each card, however, there is no strain, as seen from the fact that the circles on the top card in the deck remain circles. Folds that form from slip between layers are called **flexural slip folds.** The amount of slip between the layers increases away from the hinge zone and reaches a maximum at the inflection point. Moreover, the amount of slip is proportional to limb dip: slip increases with increasing dip. The card deck analogy highlights three important characteristics of flexural slip folding. First, at each point in the profile plane the strain ratio and orientation differ. Second, in three dimensions the strain state of the fold is plane strain ($X > Y = 1 > Z$), with the orientation of the intermediate strain axis (Y) parallel to the hinge line. Third, a geometric consequence of the flexural slip model is that the fold is cylindrical and parallel (Class 1B); the bed thickness in flexural slip folds does not change. The geometric consequences of flexural slip folding are not diagnostic of the model, because they also occur in other models (see neutral-surface folding to follow). The strain pattern, however, is. Chevron folds (Figure 10.8) and kink folds (Figure 10.19) are examples of flexural slip folding in natural rocks that form because of a strong layer anisotropy. Slip that occurs on individual grains within a layer, without the presence of visible slip surfaces, we call *flexural flow folding.*[16] Although they differ in a few details, the geometric and kinematic consequences of both flexural slip and flexural flow folding are alike.

A diagnostic feature of flexural slip folding that can be used in field analysis is that any original angular relationship in the slip surface before folding (say, flute casts in the bedding surfaces of turbidites) will maintain this angular relationship across the fold, because there is no strain on the top and bottom of the folded surface. Consequently, a lineation at an angle to the hinge line will distribute as a cone around the hinge line with that angle (a small-circle pattern in spherical projection).

10.8.2 Neutral-Surface Folding

When we bend a layer of clay or a metal bar we obtain a fold geometry that seems identical to one produced by flexural folding, but with a distinctly different strain

---

[16]Flexural flow folding has also been used to describe the migration of material from the limbs to the hinge area of folds, but we do not adopt this usage here.

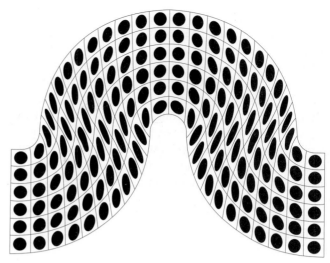

FIGURE 10.32 The characteristic strain pattern of flexural folding in the fold profile plane (the plane perpendicular to the hinge line).

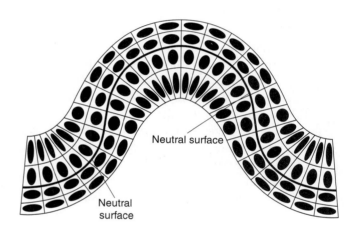

FIGURE 10.33 The strain pattern of neutral-surface folding in the fold profile plane.

pattern. This is illustrated by tracking the distortion of circles drawn on the sides of the undeformed layer (Figure 10.33). On all three surfaces we find that circles have become ellipses, including the folded top and bottom surface. On the top folded surface, the long axis of each ellipse is perpendicular to the hinge line, but on the bottom the long axis is parallel to the hinge line. In the profile plane the long axis is parallel or perpendicular to the top and bottom surfaces of the folded layer, depending on where we are in that plane (Figure 10.33). There must, therefore, be a surface in the fold where there is no strain. This zero-strain surface gives the model its name, **neutral-surface fold.** Trying to mimic this behavior with the deck of cards used earlier will require some muscle power, because the cards in the outer arc need to stretch while those in the inner arc are compressed.

The fold shape from neutral-surface folding is parallel and cylindrical, with the intermediate bulk strain axis parallel to the fold axis. These geometric characteristics and plane strain conditions also hold for flexural slip folding, so they are not diagnostic for either model; their strain patterns, however, are. Because strain accumulates in the folded surfaces during neutral-surface folding, the orientation of any feature on these surfaces changes with position in the fold. In the outer arc an initial angle with the hinge line increases, while in the inner arc this angle decreases. So, the angle of flute marks with the hinge line on the top surface of the folded layer increases, while on the bottom surface this angle decreases. In both cases the orientations in any individual surface describe neither a cone nor a plane (or small circle and great circle in spherical projection, respectively). Only in the neutral surface is the angle unchanged and does the linear feature describe a cone around the hinge line. It is not easy to use this criterion as a field tool unless one measures lineations from individual top and bottom layers in a fold. The strain pattern in the profile plane, however, is more characteristic, and is (cumbersomely) called **tangential longitudinal strain.** Note that the position of the neutral surface is not restricted to the middle of the fold, nor does it necessarily occur at the same relative position across the fold. In extreme cases the neutral surface may coincide with the outer arc, in which case the long axis of each strain ellipse in the profile plane is perpendicular to the folded surface. Is it possible to have the neutral surface at the inner arc? Answer this question to test your understanding of neutral-surface folding before moving on to the third and final folding model.

10.8.3 Shear Folding

To represent shear folding, we again turn to a deck of cards, but now we draw a layer on the sides of the deck. When we differentially move the cards relative to one another we produce a fold by a mechanism called **shear folding** (Figure 10.34). The fold shape varies with the amount and sense of displacement between individual cards and the layer has no mechanical significance; that is, shear folds are passive features. While slip occurs on individual cards, the slip surface and the slip vector are not parallel to the folded surface, as they are with flexural folding. Circles drawn on the deck before we shear the cards show that there is no strain in the surface of individual cards, but there is strain in the other surfaces. The overall strain state is

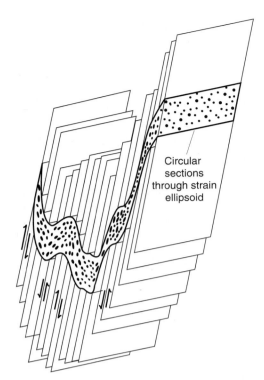

FIGURE 10.34   The strain pattern of shear folding.

plane strain, but the hinge line is not by definition perpendicular to the displacement vector, nor parallel to the intermediate (Y) strain axis. Most notably, the folds we produce have a distinct shape, because the trace of the layer on each card remains equal in length after shearing. As a consequence, we produce similar folds (Class 2). It is about time that we produced this class of folds, which are common in the field, while our models so far only generated parallel folds, which are not. Much of the charm of shear folding lies in the ability to form similar folds, which are only formed by the other two mechanisms after additional modification (see Section 10.8.4).

In nature, there are no playing cards slipping past one another. Axial plane cleavage (which is discussed in Chapter 11) may act as shear planes,[17] but rocks mostly flow as a continuum. Shear folds can be formed in regions where the flow field is heterogeneous, such as in glaciers (see fold geometry in Figure 10.1). In shear zones, relatively narrow regions with high shear strains (Chapter 12), a mylonitic foliation is common and may act as a shear plane for shear folding. In fact, sheath folds are a spectacular example of the development of such passive folds.

---

[17]Axial plane cleavage cannot exactly parallel the finite XY principal plane of strain, as this is not a shear plane.

10.8.4 Fold Shape Modification

The appeal of shear folding is the formation of similar (Class 2) folds, whereas both flexural folding and neutral-surface folding produce parallel (Class 1B) folds. What happens in the latter scenarios when we allow modification of the fold shape? Experiments and geometric arguments place a limit on the degree of strain that can be accumulated by fold tightening in parallel folds. You may have noticed this when folding the card deck, where the inner arc region increasingly experiences space problems as the fold tightens (Figure 10.11a). Material that occupies the inner arc region may be able to accommodate this space crunch, much like the foam does in our box experiments. However, as the fold tightens, the later strain increments will increasingly affect the entire fold structure (limbs and hinge equally), resulting in **superimposed homogeneous strain.** This strain component has a pronounced effect on the fold shape.

Figure 10.35 shows the effect of superimposed homogeneous strains (constant volume, plain strain) with 20% shortening ($X/Z = 1.6$) and 60% shortening ($X/Z = 6.3$) on shape and corresponding strain distributions in a flexural fold and a neutral-surface fold. You see that initially parallel folds change shape by thinning of the limbs relative to the hinge area, resulting in a geometry that approaches similar folds (Class 1C). Perfectly similar (Class 2) folds are only achieved at an infinite X/Z ratio, which is obviously unrealistic. A reminder on strain superimposition. In Chapter 4 you learned that strain is a second-rank tensor, and that tensors are not commutative; that is, $\mathbf{a}_{ij} \cdot \mathbf{b}_{ij} \neq \mathbf{b}_{ij} \cdot \mathbf{a}_{ij}$. This implies that we get different finite strains when we reverse the sequence of parallel folding and homogeneous strain, or simultaneously add homogeneous strain during folding. The implications of these scenarios are explored in more advanced texts, as they require tensor calculations.

In all models we fail to produce Class 3 folds, which pose the final challenge. One likely scenario for the formation of Class 3 folds lies in the interaction between layers of different competency in a multilayered system. Consider a sequence of shale and sandstone layers (Figure 10.36). As we shorten the sequence, the strong (competent) sandstone layers form Class 1B to 1C folds (as outlined above). The weaker (less competent) intervening shale layers accommodate the shape of the folded sandstone layers when they are in the contact strain zone. This means that folds of Class 3 will be formed in intervening shale when the sandstone layers are closely spaced, and folds of Class 1C to 2 when the sandstone layers are farther apart.

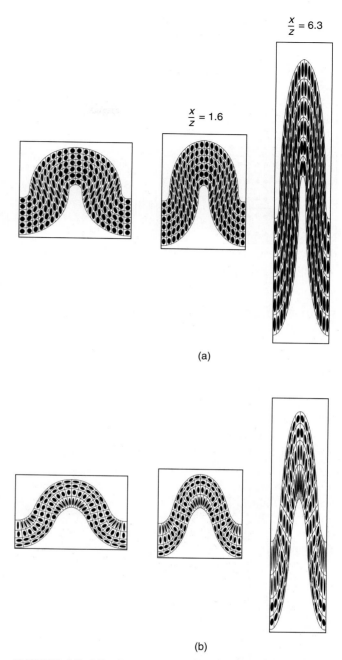

$\frac{x}{z} = 6.3$

$\frac{x}{z} = 1.6$

(a)

(b)

FIGURE 10.35 The effect of superimposed homogeneous strain on (a) a flexural and (b) a neutral-surface fold. Constant volume, plane strain with $X/Z = 1.6$ (20% shortening), and $X/Z = 6.3$ (60% shortening) are shown.

10.8.5 A Natural Example

How well does all this theory apply to nature? To answer this, we look at an example of a parallel fold in a limestone-pebble conglomerate. Strain values that were measured from pebbles across the folded layer give an estimate of the overall strain pattern (Figure 10.37a). When comparing this pattern with those predicted for flexural folding (Figure 10.37c) and

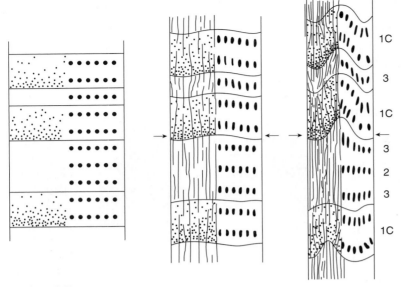

FIGURE 10.36 Folding of a multilayer consisting of sandstone (stippled) and shale layers. The incompetent shale layers accommodate the strong sandstone layers. This results in Class 3 and Class 2 folds in shale when the sandstone layers are closely and more widely spaced, respectively. The sandstone layer forms Class 1 folds.

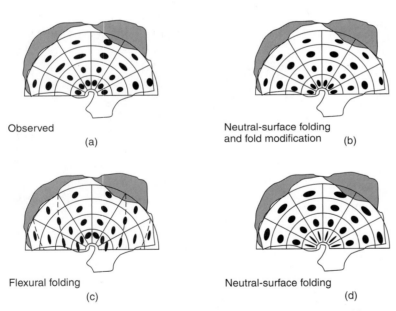

Observed
(a)

Neutral-surface folding
and fold modification
(b)

Flexural folding
(c)

Neutral-surface folding
(d)

FIGURE 10.37 Strain pattern in a natural fold of limestone-pebble conglomerate (a). This pattern more closely resembles the strain predicted in neutral-surface folding (d) than in flexural folding (c). With further modification, which consists of initial compaction and material transport away from the inner arc region, a strain pattern much like that observed in the natural sample can be produced (b).

neutral-surface folding (Figure 10.37d), we see that the natural pattern most resembles that predicted by neutral-surface folding. The x-axis is parallel to the folded surface in the outer arc and perpendicular to this surface in the inner arc. Yet the magnitude of the strain ratios predicted by neutral-surface folding is too low in the natural pattern, so the pattern requires additional modification to match the natural fold. One solution is shown in Figure 10.37b, where prefolding compaction (nonconstant volume, layer-perpendicular shortening) is followed by neutral-surface folding, during which material is preferentially removed from the inner arc. Dissolution and material transport during folding are quite common in natural rocks. Quartz and calcite veins in folded rocks are examples of this transport. Other natural folds show that flexural folding and shear folding are the dominant mechanisms, so all our folding models offer reasonable first-order approximations to the inner workings of a folded layer.

The example in Figure 10.37 emphasizes that strain in folds is highly heterogeneous, but we can nonetheless estimate the bulk shortening strain from the fold shape by adapting Equation 4.1. The bulk strain is given by

$$\mathbf{e} = (W - L)/L \qquad \text{Eq. 10.7}$$

where L is the arc length and W is the wavelength. Applying this to our example we obtain a value for $\mathbf{e}$ of approximately -0.35, or ~35% bulk-shortening strain.[18]

10.9 A POSSIBLE SEQUENCE OF EVENTS

We close the chapter on folding by an interpretive sequence of events in the formation of a single-layer fold. Immediately follow-

---

[18]If compaction does not lengthen the layer, this longitudinal strain estimate is tectonic strain.

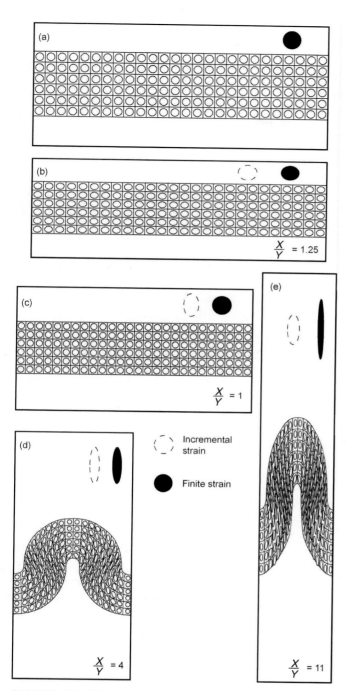

FIGURE 10.38 Folding scenario with the corresponding strain states at each step. An initial layer (a) undergoes 20% compaction (b). This is followed by layer-parallel shortening (c) and buckling (d). The final stage is a homogeneous shortening strain which transforms the parallel fold into a similar fold. The finite strain in the layer is indicated at each step; the incremental strain (dashed) and finite strain (solid) ellipses of the system are shown in the upper right. Strain ratios are shown for each step. It is assumed that volume loss occurs only at the compaction stage (b).

ing deposition of the bed, compaction reduces its thickness. We use an intermediate value of 20% layer-perpendicular shortening strain for this first component in our example (i.e., $X_c/Z_c = 1.25$), which represents 20% area loss (Figure 10.38b); in nature, compactional strains range from 0% to 50%. During the first stage of buckling, the competent layer changes dimensions by layer-parallel shortening (lps). You recall that layer thickening is more important at low viscosity ratios, so in our example we assume 20% layer-parallel shortening. This constant-volume, homogeneous strain component (conveniently) restores the finite strain ellipse to a circle ($X_f/Z_f = 1$), but the corresponding lps strain ratio X_{lps}/Z_{lps} is 1.25 (Figure 10.38c). Continued shortening results in the initiation and growth of a parallel fold by flexural folding, with a characteristic arc length (L) as a function of the thickness (t) of the layer and the linear viscosity ratio (η_L/η_M). We can estimate the viscosity ratio of the system by measuring the ratio L/t and the strain ratio X_{lps}/Z_{lps} of the layer. Until this buckling stage, the strain has been homogeneous, but after fold initiation the strain is heterogeneous, with coaxial strain dominating the hinge area and non-coaxial strain dominating in the limbs of the fold (Figure 10.38d). The bulk finite strain of the system is represented by the ratio X_b/Z_b (i.e., 4). When resistance to further folding has been reached, continued shortening is achieved by superimposed homogeneous strain (Figure 10.38e). The end result of these stages produces a similar fold with a strain pattern that varies as a function of the degree of compactional strain, the operative fold mechanism, the viscosity ratio, and the degree of superimposed homogeneous strain. The finite strain ratio X_f/Z_f for our history is 11, which corresponds to a total layer-parallel shortening strain of ~70%.

One can vary this scenario in many ways by simply changing the values of the strain ratio at each step, but also by introducing volume loss during the stages of buckling and superimposed homogeneous strain. Examining alternative scenarios will give you a good idea of folding and strain distributions under different conditions; modern imaging programs for personal computers offer a simple way to experiment. For example, in metamorphic rocks there is little competency contrast between the layers, and layer-parallel thickening may be much more important than in low-grade sedimentary beds. Moreover, elevated temperature conditions promote grain dissolution and transport of material. It is therefore particularly instructive to examine the history of a system experiencing 50%

FIGURE 10.39 Large recumbent fold in Nagelhorn, Switzerland, showing the characteristic nappe style of deformation in this part of the European Alps.

volume loss ($\Delta = -0.5$) during the buckling and superimposed homogeneous strain stages. This represents a geologically reasonable condition we will return to in the chapter on rock fabrics (Chapter 11).

10.10 CLOSING REMARKS

Hopefully you did not lose sight of the natural beauty of folds after learning so much about description, terminology, and mechanics. Folds are simply fascinating to look at, and the bigger the better (Figure 10.39). We study them to understand the conditions and significance of deformed rocks and regions. Folding patterns are good representations of the orientation of regional strain, and we can follow regional strain changes with time through the analysis of fold superposition. Superposed folding presents a particular challenge that awakens the puzzler in us. Fold generations should be based only on fold superimposition patterns, perhaps aided by fold style for correlation between outcrops. As a "tongue-in-cheek" rule, the number of fold generations should always be much less than the number of folds you have encountered in the field.

Folds are strain indicators, but the mechanical contrast between neighboring layers and with the matrix implies that strain data is representative of the folded layer, and not necessarily of the bulk rock. This is not a crippling limitation for regional analysis, nor is the problem unique to folds; it holds for all strain markers. Strain within a fold is markedly heterogeneous and the local pattern may be very different from regional conditions. The relationship of folds to regional stress is even less straightforward because of the mechanical interaction between layers with contrasting competency. Nonetheless, in many circumstances, folds may also tell us about these and other rheologic properties of rocks.

Although most of this chapter focuses on single-layer folding, the material should give you sufficient insight to tackle the literature on multilayer systems and other advanced topics on folding (such as nonlinear material rheologies). You are hopefully itching to go out into the field and test some of these concepts and ideas; otherwise a few additional laboratory experiments may satisfy your appetite until summer comes around. Finally, you will find that folds are commonly associated with other structural fabrics, such as foliations and lineations; these features are the topic of the next chapter.

ADDITIONAL READING

Biot, M. A., 1961. Theory of folding of stratified viscoelastic media and its implication in tectonics and orogenesis. *Geological Society of America Bulletin,* 72, 1595–1620.

Currie, J. B., Patnode, H. W., and Trump, R. P., 1962. Development of folds in sedimentary strata. *Geological Society of America Bulletin,* 73, 655–674.

Dietrich, J. H., 1970. Computer experiments on mechanics of finite amplitude folds. *Canadian Journal of Earth Sciences,* 7, 467–476.

Hudleston, P. J., and Lan, L., 1993. Information from fold shapes. *Journal of Structural Geology,* 15, 253–264.

Latham, J., 1985. A numerical investigation and geological discussion of the relationship between folding, kinking and faulting. *Journal of Structural Geology,* 7, 237–249.

Price, N. J., and Cosgrove, J. W., 1990. *Analysis of geological structures.* Cambridge University Press: Cambridge.

Ramberg, H., 1963. Strain distribution and geometry of folds. *Bulletin of the Geological Institute, University of Uppsala,* 42, 1–20.

Ramsay, J. G., and Huber, M. I., 1987. *The techniques of modern structural geology, volume 2: folds and fractures.* Academic Press.

Sherwin, J., and Chapple, W. M., 1968. Wavelengths of single layer folds: a comparison between theory and observation. *American Journal of Science,* 266, 167–179.

Thiessen, R. L., and Means, W. D., 1980. Classification of fold interference patterns: a reexamination. *Journal of Structural Geology,* 2, 311–316.

CHAPTER ELEVEN

Fabrics: Foliations and Lineations

11.1 INTRODUCTION

In everyday language, we commonly use the word "fabric." When talking about fabrics that are used to make garments, we mean a patterned cloth made by weaving fibers in some geometric arrangement. But the word "fabric" is not used only to refer to material products. In a philosophical moment we might consider the "fabric of life," by which we mean the underlying organization of life. As we found to be the case with many terms, the word fabric has a related yet somewhat different meaning in geology. To a structural geologist, the **fabric** of a rock is the geometric arrangement of component features in the rock, seen on a scale large enough to include many samples of each feature. The features themselves are called **fabric elements.** Examples of fabric elements include mineral grains, clasts, compositional layers, fold hinges, and planes of parting. Fabrics that form as a consequence of tectonic deformation of rock are called **tectonic fabrics,** and fabrics that form during the formation of the rock are called **primary fabrics** (Chapter 2). It may sound picky, but structural geologists also make a distinction between microstructure and texture. Although texture is sometimes used as a synonym for microstructure, for example igneous texture, here we restrict the term **texture** to crystallographic orientation patterns in an aggregate of grains (see Chapter 13) and **microstructure** to their geometric arrangement.

Tectonic fabrics provide clues to the strain state of the rock, the geometry of associated folding, the processes involved in deformation, the kinematics of deformation, the timing of deformation (if the fabric is defined by an arrangement of datable minerals), and ultimately about the tectonic evolution of a region. The purpose of this chapter is to explore two common tectonic fabric elements in rocks, **foliations** and **lineations,** and to introduce you to the characteristics and interpretation of these elements.

11.2 FABRIC TERMINOLOGY

Let's start by developing the inevitable vocabulary to discuss tectonic fabrics (see Table 11.1). If there is no preferred orientation (i.e., alignment) of the fabric elements, then we say that the rock has a **random fabric**

TABLE 11.1	TECTONIC FABRIC TERMINOLOGY
Axial plane cleavage	Cleavage that is parallel or subparallel to the axial plane of a fold; it is generally assumed that the cleavage formed roughly synchronous with folding and is subparallel to the *XY*-plane of the bulk finite strain ellipsoid.
Cleavage	A secondary fabric element, formed under low-temperature conditions, that imparts to the rock a tendency to split along planes.
Cross-cutting cleavage	Cleavage that is not parallel to the axial plane of a fold (also *nonaxial plane cleavage*); the term *transecting cleavage* is used when cleavage and folding are considered roughly synchronous in a transpressional regime.
Fabric	The geometric arrangement of component features in a rock, seen on a scale large enough to include many samples of each feature.
Foliation	The general term for any type of "planar" fabric in a rock (e.g., bedding, cleavage, schistosity).
Gneissosity	Foliation in feldspar-rich metamorphic rock, formed at intermediate to high temperatures, that is defined by compositional banding; the prefixes "ortho" and "para" are used for igneous and sedimentary protoliths, respectively.
Lineation	A fabric element that is best represented by a line, meaning that one of its dimensions is much longer than the other two.
Migmatite	Semichaotic mixture of layers formed by partial melting and deformation.
Mylonitic foliation	A foliation in ductile shear zones that is defined by the dimensional preferred orientation of flattened grains; the foliation tracks the *XY* flattening plane of the finite strain ellipsoid and is therefore at a low angle to the shear-zone boundary.
Phyllitic cleavage	Foliation that is composed of strongly aligned fine-grained white mica and/or chlorite; the mineralogy and fabric of phyllites give the rock a distinctive silky appearance, called *phyllitic luster*.
Schistosity	Foliation in metamorphic rock, formed at intermediate temperatures, that is defined by mica (primarily muscovite and biotite), which gives the rock a shiny appearance.
Texture	The pattern of crystallographic axes in an aggregate of grains; also *crystallographic fabric*.

(Figure 11.1a). Undeformed sandstone, granite, or basalt are rocks with random fabrics. Deformed rocks typically have a **nonrandom fabric** or a **preferred fabric,** in which the fabric elements are aligned in some manner and/or are repeated at an approximately regular spacing (Figure 11.1b). There are two main classes of preferred fabrics in rock. A planar fabric, or **foliation** (Figure 11.1c), is one in which the fabric element is a planar or tabular feature (meaning it is shorter in one dimension than in the other two), and a linear fabric, or **lineation** (Figure 11.1d), is one in which the fabric element is effectively a linear feature (i.e., it is longer in one dimension relative to the other dimensions). Structural geologists may use the word "fabric" alone to imply the existence of a preferred fabric (as in, "that rock has a strong fabric"), but you should use appropriate modifiers if your meaning is not clear from context alone.

We're not quite done with terminology yet! Fabrics are complicated features, and there are lots of different adjectives used by structural geologists to describe them. For example, if you can keep splitting the rock into smaller and smaller pieces, right down to the size of the component grains, and can still identify a preferred fabric, then we say that the fabric is **continuous** (Figure 11.2a). In practice, if the fabric elements are closer than 1 mm (that is, below the resolution of the eye), the fabric is continuous. When there is an obvious spacing between fabric elements, we say that the fabric is **spaced** (Figure 11.2b).

Rocks with a **penetrative** tectonic fabric are also called **tectonites.** When linear fabric elements dominate, the rock is called an **L-tectonite,** whereas a rock with dominantly planar fabrics is called an **S-tectonite,** and, not surprisingly, rocks with both types of fabric

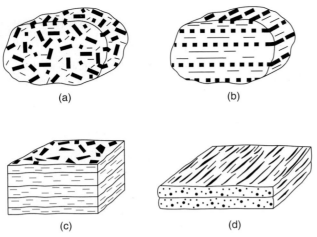

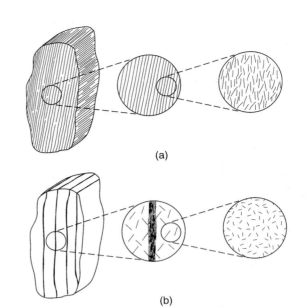

FIGURE 11.1   The basic categories of fabrics. (a) A random fabric. The fabric elements are dark, elongate crystals. The long axes of these crystals are not parallel to one another. (b) A (1-dimensional) preferred fabric, in which the long axes of elongate crystals are aligned with one another. (c) A foliation. The fabric elements are planar and essentially parallel to one another, creating a 2-dimensional fabric. (d) A lineation. The fabric elements are linear; in this example, we show the alignment of fabric elements in a single plane.

FIGURE 11.2   The distinction between continuous and spaced fabrics. (a) A continuous fabric. The lines represent a planar fabric element that continues to be visible no matter how small your field of view (at least down to the scale of individual grains). (b) A spaced fabric. The rock between the fabric elements does not contain the fabric. The circled areas represent enlarged views.

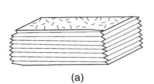 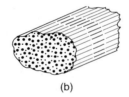 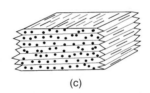

FIGURE 11.3   The nature of tectonites. (a) An S-tectonite. This fabric is dominantly a foliation, and the rock may tend to split into sheets parallel to the foliation. Within the planes of foliation, linear fabrics are not aligned, or are not present at all. (b) An L-tectonite. The alignment of linear fabric elements creates the dominant fabric, so the rock may split into rod-like shapes. In L-tectonites, there is not a strong foliation. (c) An L/S-tectonite. The rock possesses a strong foliation and a strong lineation.

elements are called **LS-tectonites**[1] (Figure 11.3). Why create such jargon? Simply to highlight rocks whose internal structure has been substantially changed during deformation. Typically, the deformation that leads to the formation of a tectonite is accompanied by metamorphism, so the fabric is defined by grains that have been partially or totally recrystallized, and/or by new minerals that have grown during deformation (called neomineralization). Because most rocks are deformed, tectonites are among the most

common rocks you will see (Figure 11.4); some examples are slates, schists, and mylonites, all of which will be discussed in a later section.

11.3   FOLIATIONS

A **foliation** is any type of planar fabric in a rock. We are admittedly a bit loose in our use of the term planar. Since, strictly speaking, a plane does not contain any curves or changes in orientation, the terms curviplanar or surface would be more appropriate. Although foliations are generally not perfectly planar, structural geol-

---

[1]"L" stands for "lineation"; "S" stands for "surface" in English, "schis- tosité" in French, or "Schieferung" in German.

FIGURE 11.4 Disjunctive cleavage in mica-rich rock (Rhode Island, USA). Note the variation in cleavage spacing between steeply dipping beds. Width of view is ~ 4 m.

ogists nonetheless are in the habit of talking about planar fabrics. Thus, bedding, cleavage, schistosity, and gneissosity all qualify as foliations. Fractures, however, are not considered to be foliations, because fractures are breaks through a rock and are not a part of the rock itself. A rock may contain several foliations, especially if it has been deformed more than once. To keep track of different foliations, geologists add subscripts to the foliations: S_0, S_1, S_2, and so on. S_0 is used to refer to bedding, S_1 is the first foliation formed after bedding, and S_2 forms after S_1. The temporal sequence of foliation development is defined by cross-cutting relationships, but in complexly deformed areas, it may be quite a challenge to determine which foliation is which unless independent constraints on (relative) time are available.

There are many types of tectonic foliations that are distinguished from one another simply on the basis of what they look like. The physical appearance of a foliation reflects the process by which it formed, and the process, in turn, is controlled partly by the composition and grain size of the original lithology, and partly by the metamorphic conditions under which the foliation formed. Different names are used for the different types of foliations. In Table 11.2 and the following discussion, we will introduce different types of foliation roughly in sequence of increasingly higher metamorphic conditions—cleavage first, then schistosity, then gneissic layering.

11.3.1 What Is Cleavage?

Cleavage in rocks has been defined in different ways by different people, so the use of this term in the literature is confusing. We advocate a nongenetic definition,

TABLE 11.2	FOLIATION CLASSIFICATION SCHEME
Spaced cleavage	(a) Disjunctive cleavage (e.g., stylolitic cleavage) (b) Crenulation cleavage
Continuous cleavage	(a) Coarse cleavage (e.g., pencil cleavage) (b) Fine cleavage (e.g., slaty cleavage)
Phyllitic cleavage	Continuous cleavage with a distinctive silky luster in low-grade metamorphic rock (lower greenschist facies)
Schistosity	Mica-rich foliation with a distinctive high sheen in low- to medium-grade metamorphic rock (greenschist facies)
Gneissic layering or gneissosity	Coarse compositional banding in high-grade metamorphic rock (upper amphibolite and granulite facies)

in which cleavage is defined as *a secondary fabric element, formed under low-temperature conditions, that imparts on the rock a tendency to split along planes.* The point of this definition is to emphasize

1. That cleavage is a feature of the rock that forms subsequent to the origin of the rock.
2. That the term "cleavage" is used, in practice, for tectonic planar fabrics formed at or below lower greenschist facies conditions (i.e., $\leq 300°C$). The term "cleavage" is not used when referring to the fabric in schists or in gneiss.
3. That in a rock with cleavage, there are planes of weakness across which the rock may later break when uplifted and exposed at the surface of the earth, even though cleavage itself forms without loss of cohesion. By this definition, an array of closely spaced fractures is not a cleavage.

We recognize four main categories of cleavage that are differentiated from one another by their morphological characteristics (or, by how they look in outcrop). These are disjunctive cleavage, pencil cleavage, slaty cleavage, and crenulation cleavage.

11.3.2 Disjunctive Cleavage

Disjunctive cleavage is a foliation that forms mostly in sedimentary rocks[2] that have been subjected to a tectonic differential stress under sub–greenschist facies metamorphic conditions. It is defined by an array of subparallel fabric elements, called **cleavage domains,** in which the original rock fabric and composition have been markedly changed by the process of pressure solution. Domains are separated from one another by intervals, called **microlithons,** in which the original rock fabric and composition are more or less preserved (Figure 11.5). The adjective "disjunctive" implies that the cleavage domains cut across a preexisting foliation in the rock (usually bedding), without affecting the orientation of the preexisting foliation in the microlithons. Because pressure solution is always involved in the formation of a disjunctive cleavage, other terms such as **(pressure-)solution cleavage** and **stylolitic cleavage** have been used for this structure. If the context is clear, some geologists may refer to the structure simply as spaced cleavage, though, as we see later, crenulation cleavage is also a type of spaced cleavage. Earlier in this century, many geologists incorrectly considered cleavage domains to be brittle fractures formed by loss of cohesion. The old term "fracture cleavage" should therefore be avoided when referring to disjunctive cleavage, because a cleavage cannot be composed of fractures. Such arrays of closely spaced fractures should be called a fracture or joint array.

Now that we've gotten through the cleavage terminology, let's see how disjunctive cleavage forms. Consider a horizontal bed of argillaceous (clay-rich) limestone or sandstone that is subjected to a compressive stress (σ_1 in Figure 11.6). Dissolved ions created by pressure solution are transported away from the site of dissolution through a water film that adheres to the grain surfaces. The ions may then precipitate at crystal faces where compressive stress is less, precipitate nearby in the **pressure shadows** adjacent to rigid grains, or enter the pore fluid system to be carried out of the local rock environment entirely. Note that in order for pressure solution to occur, a thin layer of water molecules must be chemically bonded to grain surfaces. If the water is not bonded, a grain in the water can only sustain isotropic stress, because fluids cannot support shear stresses, and pressure solution won't occur.

[2]There are descriptions of disjunctive cleavage in igneous rocks, but it is most typical of sedimentary rocks.

(a)

(b)

(c)

S_0

S_1

Microlithon   Cleavage domain

FIGURE 11.5 Spaced disjunctive cleavage (or solution cleavage) in limestone from the Harz Mountains of Germany. The cleavage is marked by the narrow dark bands which cut across the original, lighter-colored argillaceous limestone. (a) Outcrop view; (b) close-up view of central portion. In (c) the cleavage domain and microlithon of spaced cleavage are illustrated in gently left-dipping beds.

The distribution of clay in rocks is not uniform. Pressure solution occurs more rapidly where the initial concentration of clay is high. How clay affects pressure solution rates remains enigmatic. Perhaps swelling clay, which contains interlayers of bonded water, increases the number of diffusion pathways available for ions and thus multiplies the diffusion rate. Alternatively, the highly charged surfaces of clay grains may act as a chemical catalyst for the dissolution reaction. Field observations suggest that a rock must contain >10% clay in order for solution or sty-lolitic cleavage to develop. Where it occurs, the process preferentially removes more soluble grains. Thus, in an argillaceous limestone, calcite is removed, and clay and quartz are progressively concentrated. In argillaceous sandstone, the process is effectively the same as that in argillaceous limestone, except that quartz is the mineral that preferentially dissolves, and clay alone is concentrated. As the framework grains of carbonate and quartz are removed, the platy clay grains collapse together like a house of cards. Concentration of clay in the domain further enhances the solubility of

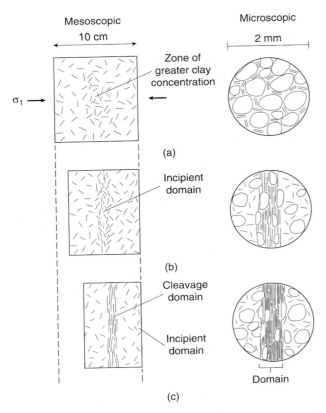

Mesoscopic
10 cm

Microscopic
2 mm

Zone of greater clay concentration

$\sigma_1 \rightarrow$

(a)

Incipient domain

(b)

Cleavage domain

Incipient domain

Domain

(c)

FIGURE 11.6 Evolution of spaced disjunctive cleavage. (a) Precleavage fabric of the rock. In the area indicated by the arrow in the mesoscopic image, there happens to be a greater initial concentration of clay. The microscopic image indicates that the clay flakes are randomly oriented. (b) As shortening and pressure solution occur, the zone in which there had initially been a greater clay concentration evolves into an incipient cleavage domain. At this stage, grains are being preferentially dissolved on the faces perpendicular to S_1, and the clay flakes are collapsing together. (c) Ultimately, a clearly defined cleavage domain is visible. In the domain, the clay flakes are packed tightly together and only small relicts of the soluble mineral grains are visible.

the rock, so there is positive feedback. Eventually, a discrete domain develops in which there is a **selvage,** the material filling the domain, composed of mostly clay (and quartz) with some relict corroded calcite grains. In the selvage, the clay flakes are packed together with a dimensionally preferred orientation. If deformation continues, the domain continues to thicken as pressure solution continues along its edges. As a result, compositional contrast between cleavage domains and microlithons becomes so pronounced that it defines a new stratification in the rock that nearly obscures the original bedding. From this description

you see that spaced cleavage formation is identical to the processes by which bedding-parallel stylolites form as a consequence of compaction loading; hence the use of the term stylolitic cleavage.

The **spacing** of cleavage domains in a rock depends on the initial clay content. If the clay content is high, the domains are closely spaced. Spacing also changes with progressive deformation. As strain increases, more cleavage domains nucleate and thus the domain spacing decreases. Thirdly, domain spacing may also be related to the magnitude of differential stress, for experiments suggest that the rate of pressure solution is proportional to the magnitude of differential stress. During the process of domain formation, the fabric of the microlithons remains relatively unaffected, though high-resolution microscopic analysis (such as transmission electron microscopy) may indicate the presence of incipient pressure solution features or newly crystallized grains of the soluble mineral in microlithons.

Before we leave the topic of disjunctive cleavage, we need to present the vocabulary that geologists use to describe disjunctive cleavage in the field. The classification and description of disjunctive cleavage is based on the characteristics of surface morphology of domains, and on domain spacing (Figure 11.7). If the clay content in the host rock is low, domain surfaces are severely pitted. In cross section, such domains resemble the toothlike or jagged sutures on a skull; this type of cleavage domain is called a **sutured domain.** If the clay content is high, cleavage domains tend to have thick selvages that have smooth borders; these domains are called **nonsutured.** When viewed under high magnification with a microscope, you will see that thick domains, in some cases, are composed of a dense braid of threadlike sutured domains. Cleavage domains can be either *wavy,* if the domain undulates, or *planar,* if the domain does not. If wavy domains are closely spaced, such that the domains merge and bifurcate and give the fabric the appearance of braided hair, the cleavage is called **anastomosing.**

The average spacing between domains is also a useful criterion for classification of cleavage. Figure 11.7e shows a simple field classification based on spacing. If the spacing between domains is greater than about 1 m, then the rock really doesn't have an obvious fabric. Such isolated pressure-solution domains, when formed in response to tectonic stress, are called **tectonic stylolites.** The adjective "tectonic" distinguishes these structures from bedding-parallel stylolites formed by compaction. In general practice, if domains are spaced

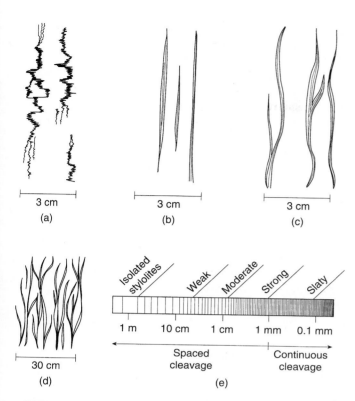

FIGURE 11.8   Pencil cleavage in shale, with horizontal bedding trace (Knobs Formation, Virginia, USA).

FIGURE 11.7   Cross-sectional sketches of morphological characteristics used for cleavage description and classification. (a) Sutured domains; (b) planar domains; (c) wavy domains; (d) an anastomosing array of wavy domains. In (e) the description of spaced cleavage based on domain spacing is shown.

between about 10 cm and 1 m apart, we call the feature a **weak cleavage;** a spacing of 1–10 cm defines a **moderate cleavage,** and a spacing of less than 1 cm denotes a **strong cleavage.** When cleavage domains are less than 1 mm apart, the cleavage is **continuous** (see slaty cleavage, below). We caution you again that different authors use these adjectives differently, so our definitions are only generalizations.

11.3.3   Pencil Cleavage

If a fine-grained sedimentary rock (shale or mudstone) breaks into elongate pencil-like shards because of its internal fabric, we say that it has a **pencil cleavage** (Figure 11.8). Typically, pencils are 5–10 cm long and 0.5–1 cm in width. In outcrop, pencil cleavage looks as if it results from the interaction of two fracture sets (and in some locations, it is indeed merely a consequence of the intersection between a fracture set and bedding), but actually the parting reflects an internal alignment of clay grains in the rock.

Pencil cleavage forms because of the special characteristics of clay. The strong shape anisotropy of clay flakes creates a preferred orientation parallel to bedding when they settle out of water and are compacted. This preferred orientation imparts the tendency for clay-rich rocks to break on bedding planes that is displayed by *shale.* Shale, mudstone, and slate are all names that are used to describe clay-rich rocks; mudstone is the general term for these rocks, whereas shale and slate are foliated clay-rich rocks. After compaction, a strain ellipsoid representing the state of strain in the shale would look like a pancake parallel to bedding (Figure 11.9). Now, imagine that the shale is subjected to layer-parallel shortening. Cleavage formation processes begin to take place: large detrital phyllosilicates *fold and rotate,* while fine grains undergo *pressure solution* along domains perpendicular to the shortening direction, and new clay *crystallizes. Microfolding* may occur during this stage of the process, but because of the fine grain size these microfolds are only visible under the microscope. In addition, quartz may begin to dissolve, and as these framework grains are removed, clay flakes collapse so that their basal planes are perpendicular to the plane of shortening. The plane defined by new or rotated clay grains is roughly perpendicular to the shortening direction, so it forms a tectonic foliation at high angles to the original bedding (Figure 11.9c). At an early stage during this process, the new tectonic fabric is comparable in degree of development to the initial bedding-parallel fabric. At this stage, the strain ellipsoid representing this state would look something like a big cigar, and the rock displays pencil cleavage (Figure 11.9b). In sum, pencil

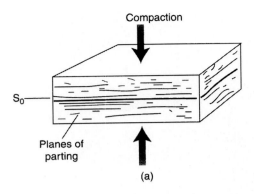

Compaction

S_0

Planes of
parting

(a)

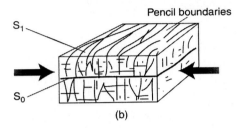

S_1

Pencil boundaries

S_0

(b)

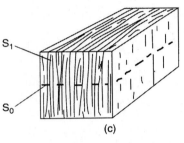

S_1

S_0

(c)

FIGURE 11.9 Sketches illustrating the progressive development of slaty cleavage via the formation of pencil structure. (a) Compaction during burial of a sedimentary rock produces a weak preferred orientation of clay parallel to bedding. A representative strain ellipsoid is like a pancake in the plane of bedding. (b) Shortening parallel to layering creates an incipient tectonic fabric. Superposition of this fabric on the primary compaction fabric leads to the formation of pencils. The representative strain ellipsoid is elongate, with the long axis parallel to the pencils. (c) Continued tectonic shortening leads to formation of slaty cleavage at a high angle to bedding. The phyllosilicates are now dominantly aligned with the direction of cleavage, and a representative strain ellipsoid is oblate and parallel to cleavage.

cleavage is a fabric found in weakly deformed shale in which the tendency of the shale to part on bedding planes is about the same as the tendency for it to part on an incipient tectonic cleavage that is at a high angle to bedding. Deformation in most areas produces a fabric that is much stronger than the original bedding parting, so pencils are not as common as the more evolved stage, represented by slaty cleavage.

11.3.4 Slaty Cleavage

Pencil cleavage may be considered a snapshot of an early stage in the process by which slaty cleavage develops. As shortening perpendicular to cleavage planes accumulates, clay throughout the rock develops a preferred orientation at an angle to the original sedimentary fabric and this orientation dominates over the primary fabric (Figure 11.9c). The finite strain ellipsoid for cleavage development at this stage has the shape of a pancake that parallels the tectonic fabric. Formation of slaty cleavage occurs by much the same process as did the formation of disjunctive cleavage in argillaceous sandstone or limestone, but the resulting domains are so closely spaced that effectively there are no uncleaved microlithons, and the entire rock mass displays the tectonically induced preferred orientation. When a rock has this type of continuous fabric, we say that it has **slaty cleavage** (Figure 11.10). In other words, slaty cleavage is defined by strong dimensionally preferred orientation of phyllosilicates in a very clay-rich rock, and the resulting rock, which is considered to be a low-grade metamorphic rock, is called a *slate*. Slaty cleavage tends to be smooth and planar. This characteristic, coupled with the penetrative nature of slaty cleavage, means that slates can be split into thin sheets, which made them popular roofing materials in the nineteenth century and early parts of the twentieth century.

At high magnifications the detailed character of slaty cleavage becomes apparent. In many cases only a network of anastomosing surfaces around partially dissolved quartz or feldspar grains is preserved (Figure 11.11a), but in some instances remnants of the original bedding orientation can be found (Figure 11.11b). Because slaty cleavage forms under temperature conditions that mark the onset of metamorphism (250°C–350°C), the mineralogy of slate tends to resemble that of shale. However, there is a notable decrease in the amount of interlayered water in clays; that is, smectite, the water-bearing clay, transforms to illite. As metamorphic temperatures increase, the illite becomes more micalike in structure, though it is still fine grained. Thus, rocks with higher-grade slaty cleavage display a distinct sheen on cleavage planes. They are also significantly harder than shale, so they ring when hit with a hammer. These are the slates that provide excellent material for the construction of pool tables.

11.3.5 Phyllitic Cleavage and Schistosity

When metamorphic conditions reach the lower greenschist facies, the clay and illite in a pelitic rock react to

FIGURE 11.10 An overturned syncline with a well-developed axial plane slaty cleavage (southern Appalachians, USA).

form white mica[3] and chlorite. If reaction occurs in an anisotropic stress field, these phyllosilicates grow with a strong preferred orientation. Rock that is composed of strongly aligned fine-grained white mica and/or chlorite is called **phyllite**[4] and the foliation that it contains is called **phyllitic cleavage.** The mineralogy and

fabric of phyllites give the rock a distinctive silky luster. Phyllitic cleavage is intermediate between slaty cleavage and schistosity.

When metamorphic conditions get into the middle greenschist facies, the minerals in a pelitic lithology react to form coarser-grained mica and other minerals. When these reactions again take place in an anisotropic stress field, the mica has a strong preferred orientation. The resulting rock is a **schist,** and the foliation it displays is called **schistosity.** The specific assemblage of minerals that forms depends not only on the pressure and temperature conditions at which metamorphism

---

[3]"White mica" is fine-grained muscovite. It contains more potassium and aluminum than illite, and is better ordered.

[4]Phyllite is not the same as *phyllonite,* which is a low-grade mylonitic rock.

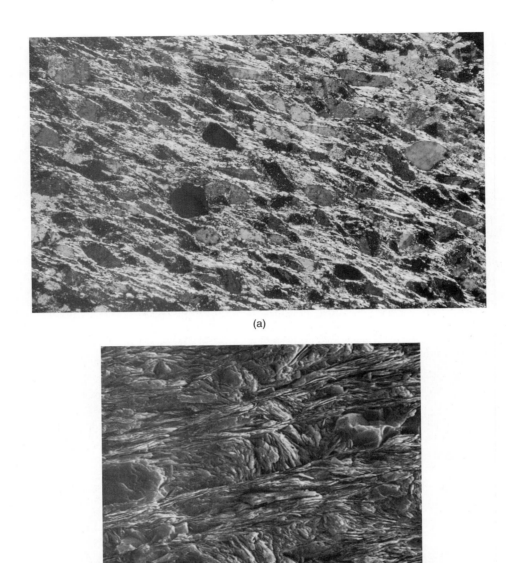

(a)

(b)

FIGURE 11.11 (a) Photomicrograph of a continuous cleavage from Newfoundland (Canada); width of view is 2 mm. (b) Scanning electron micrograph from a slate in the Rheinische Schiefergebirge (Germany). Note the microcrenulation of bedding-parallel micas, creating distinct microlithons with mica roughly parallel to bedding and cleavage domains with mica parallel to cleavage. Width of view is 100 μm.

takes place, but also on the chemical composition of the protolith and on the degree to which chemicals are added or removed from the rock by migrating fluids. Conveniently, a schist is named by the assemblage of metamorphic index minerals that it contains (e.g., a garnet-biotite schist). In schist that contains **porphyroclasts** (relict large crystals) or **porphyroblasts** (newly grown large crystals), the schistosity tends to be wavy, as the micas curve around the large crystals.

11.3.6 Crenulation Cleavage

A lithology containing a closely and evenly spaced foliation that is shortened in a direction at a low angle to this foliation will crinkle like the baffles in an accordion (Figure 11.12). In fine-grained lithologies like slate or phyllite, these microfolds are closely spaced and the spacing tends to be very uniform. The axial planes of the crenulations define a new foliation called

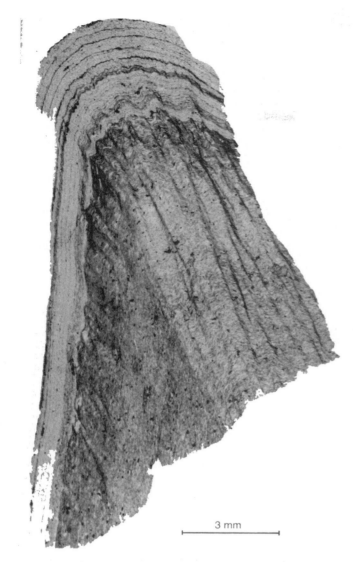

FIGURE 11.12 Photomicrograph of crenulation cleavage in a phyllite (Vermont, USA). The spaced crenulation cleavage deforms an earlier continuous cleavage, with cleavage spacing and intensity that are less in the sandy layers (white).

3 mm

crenulation cleavage (Figure 11.13). Like mesoscopic folds, crenulations can be symmetric (Figure 11.13b) or asymmetric (Figure 11.13c). In a given outcrop, both symmetric and asymmetric crenulation cleavages may occur. For example, on the limbs of a fold, asymmetric crenulation occurs, whereas in the hinge zone of the fold, the crenulation is symmetric (Figure 11.12).

A prerequisite for the formation of crenulation cleavage is the existence of a preexisting strong lamination or foliation. Crenulation cleavage won't form in a sandstone, but may form in a micaceous shale with a strong bedding-plane foliation, or in a rock that already contains a slaty or phyllitic cleavage (Figure 11.14). Crenulation cleavage in outcrop is typically an S_2 foliation that has been superimposed on an earlier (S_1) foliation.

Crenulation cleavage forms under conditions that are also amenable to the occurrence of pressure solution. When the starting rock contains a mixture of quartz and clay or fine-grained mica, the quartz is preferentially removed from the limbs of the microfolds and precipitates in the hinges as the crenulations form (Figure 11.15). Gradually, phyllosilicates concentrate on the limbs and quartz is concentrated in the hinges. This mineralogical differentiation can be so complete that the old foliation disappears entirely and is replaced entirely by a new foliation, that is defined not only by preferred orientation of the phyllosilicates, but also by microcompositional layering (Figure 11.15b). This process is a type of **transposition** (see Chapter 10), by which a preexisting foliation is transposed into a new orientation. If quartz is largely removed by progressive pressure solution, a new foliation eventually develops and the crenulated appearance of the rock fades (Figure 11.15c). If this happens, all traces of the original fabric, the one predating the crenulation, may be destroyed. Thus, crenulation cleavage forms in a rock

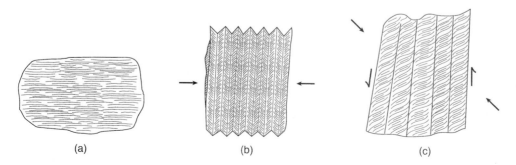

(a)   (b)   (c)

FIGURE 11.13 The two basic categories of crenulation cleavage. (b) Symmetric crenulation cleavage; (c) asymmetric (sigmoidal) crenulation cleavage. The arrows indicate a possible component of shear associated with this crenulation geometry.

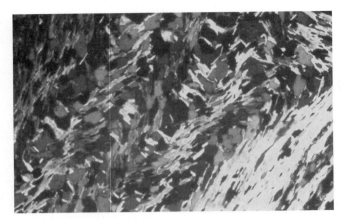

FIGURE 11.14 Photomicrograph of incipient differentiation in crenulation cleavage (Pyrenees). Width of view is ~0.5 mm

with a preexisting cleavage, but, paradoxically, it can evolve into a rock with seemingly only one foliation.[5]

11.3.7 Gneissic Layering and Migmatization

Foliated **gneiss** is a metamorphic rock in which the foliation is defined by compositional banding (Figure 11.16). Commonly, light and dark bands of felsic and mafic mineralogy alternate. The light-colored layers are rich in feldspar and quartz, whereas darker layers contain more of the minerals amphibole/pyroxene (and/or biotite). This color banding in gneiss is called **gneissosity.** Under the metamorphic condition for gneiss formation

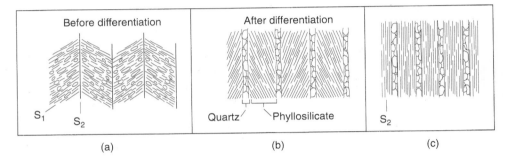

FIGURE 11.15 Differentiation during the formation of crenulation cleavage. (a) Fairly homogeneous composition, before migration of the quartz. (b) Quartz accumulates in the hinges of the crenulations, and the phyllosilicates are concentrated in the limbs; the result is the formation of compositionally distinct bands in the rock. (c) Complete transposition of the S_1 foliation into a new S_2 cleavage or schistosity.

FIGURE 11.16 Fold hinges in transposed gneiss near Parry Sound, Grenville Orogen (Ontario, Canada).

---

[5]To some extent, crenulation cleavage is a matter of scale of observation; many slates show evidence of microfolds in detrital mica flakes.

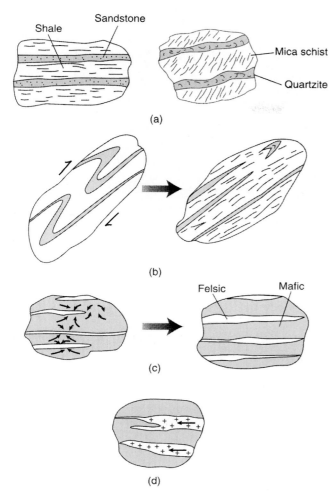

FIGURE 11.17 Mechanisms of formation of a gneiss. (a) Inheritance from an original lithology; (b) creation of new compositional banding via transposition; (c) metamorphic differentiation; (d) *lit-par-lit* intrusion.

(amphibolite to granulite facies), muscovite reacts to form feldspar, so the rock contains no schistosity. Gneiss can be derived from a sedimentary protolith, in which case it is called a **paragneiss,** or an igneous protolith, in which case it is called an **orthogneiss.** It is often difficult to decide whether a particular rock is an orthogneiss or a paragneiss, however; this requires careful field and petrologic analysis. A special type of gneiss, called **augen gneiss,** contains relatively large feldspar clasts floating in a finer-grained matrix.[6]

How does the compositional banding in gneiss form? There may be several processes involved in the formation of gneissic layering (Figure 11.17). First, it may occur by *inheritance* from original compositional contrasts. If the protolith (the rock from which the gneiss

formed) was a stratified sequence with layers of different composition (such as alternating sandstone and shale), then metamorphism will transform this sequence into a compositionally banded metamorphic rock. Secondly, it may result from *transposition* via folding of an earlier layering. Transposition is a common process during deformation under high-temperature conditions, and is discussed in several places in this book (see also Chapters 10 and 12). Rock containing compositional layering that is subjected to intense, high-grade deformation may develop isoclinal folds. If the hinges of the fold are detached and a new sequence of compositional layers has formed, then we say that the new layering is a type of transposed foliation. The layering in a rock with a transposed foliation does not represent the original stratigraphy of the rock, though it may have been derived from it. In other words, the sequence of compositional layers in a transposed rock does not represent the original stratigraphic succession. Thirdly, gneissic layering may be formed by *metamorphic differentiation* when the thermodynamics governing diffusion during metamorphism causes certain ions to be excluded from the formation of new metamorphic minerals in a layer. The excluded ions accumulate to form different minerals in an adjacent layer. Thus, minor differences in the original composition of successive layers may be amplified into major compositional changes after metamorphism. The resulting rock with alternating layers of different composition is a gneiss. Finally, gneissic banding may originate from an igneous process called **lit-par-lit intrusion** (French for "layer-by-layer"). Melts inject as thin sills along many weak planes in the protolith, and this interlayering of sills and host rock defines the gneiss. Usually the process of injection is accompanied by passive folding, so the igneous nature of the contacts is often not evident.

When metamorphic temperatures are sufficiently high, a rock begins to melt; but not all minerals melt at the same temperature. Quartz, some feldspar, and muscovite melt at lower temperatures than mafic minerals like amphibole, pyroxene, and olivine. Therefore, when a rock of intermediate composition begins to melt, certain minerals become liquid while others remain solid. The minerals that stay solid until higher temperatures are achieved are called **refractory minerals.** When only part of a rock melts, we say that it has undergone *partial melting* or **anatexis.** A rock that is undergoing partial melting is a mixture of pockets of melt and lenses of solid, both of which are quite soft. Shortening will cause the mass to flow, and the contrasting zones of melt and solid fold and refold much like chocolate and vanilla batter in marble cake. When this happens in rock, the resulting semichaotic mixture of light and dark layers is called a **migmatite.** Because

---

[6]Because the altered feldspar grains look like eyes, or "Augen" in German.

of their origin and chaotic nature, the analysis of structures in migmatite may provide little or no information about the kinematics of regional deformation.

11.3.8 Mylonitic Foliation

Mylonites are fine-grained, foliated and/or lineated fault rocks that form by crystal-plastic deformation processes. They are discussed in detail in the chapter on shear zones (Chapter 13), but we briefly want to include mylonitic foliation here. The foliation in mylonites is defined by shape-preferred orientation of flattened grains (usually quartz and calcite, but also feldspar and olivine) and mica. Typically, transposition via folding occurs during mylonitization, so the lithologic banding in a mylonite is not the original layering of the rock. **Mylonitic foliation** is associated with these folds, but fold hinges vary in orientation or are strongly curved in the foliation plane as a consequence of differential flow. Like other fine laminations, mylonitic foliation is susceptible to the formation of crenulations, giving rise to structures that aid recognizing the kinematic history of shear zones. Mylonites rival foliations in their importance for structural analysis, so we direct you to Chapter 13 for much more on mylonitization.

11.4 CLEAVAGE AND STRAIN

Does the study of a foliation in the field provide any constraints on the nature of strain in the region? Unfortunately, we can only offer a wishy-washy answer: sometimes yes, sometimes no. In order to determine the relationship of cleavage to strain, we need to look at strain markers in a cleaved rock. Red slate is a good lithology for such strain studies, because it may contain reduction spots. **Reduction spots** (Figure 11.18) are small zones where iron in the rock is reduced and therefore is greenish in color. Assuming these spots start out as early diagenetic spheres around an inclusion, they make ideal strain markers because they behave in a totally passive manner, meaning that they have no mechanical contrast with the host rock. In studies of reduction spots in slate, geologists find that the deformed spots are flattened ellipsoids, and that the plane of flattening (i.e., the XY principal plane of the finite strain ellipsoid) is essentially parallel to the slaty cleavage. In these cases, therefore, cleavage appears to approximate the orientation of a principal plane of strain, with the shortening direction being perpendicular to cleavage. But cleavage is probably not strictly

FIGURE 11.18 Reduction spots in slate that are ellipsoidal in shape (elliptical in section); Appalachians, USA. They can be used as strain markers if they were formed as spherical regions around a reducing phase prior to deformation; coin for scale.

parallel to the XY principal plane of strain in all situations. For example, when cleavage occurs in flexural slip folds, or adjacent to fault zones, there may be a component of shear on the cleavage planes themselves, and when this happens the cleavage by definition is not a principal plane. Moreover, a spaced cleavage may initiate as a principal plane of strain, but subsequent folding of the bed in which cleavage occurs may rotate it away from the bulk principal strain directions. Reduction spot studies give estimates of a total shortening strain (e_1) for the formation of slaty cleavage in the range of 50–60%.

The most controversial issue about low-grade cleavage formation is the question of whether cleavage formation is a volume-constant strain or a volume-loss strain process. Imagine a cube of rock that is 10 cm long on each edge. If it is shortened in one direction, but does not stretch in the other directions ($X = Y = 1 > Z$), it must lose volume during deformation (Figure 11.19). Alternatively, if shortening in one direction causes it to expand by an equal proportion in one other direction ($X > Y = 1 > Z$ or plane strain) or in two other directions ($X > Y > Z$), then the strain may be volume-constant. Remember that formation of low-grade cleavage involves significant activity in the way of dissolution and new growth of grains. If the ions of soluble minerals enter the pore water system, they can be carried out of the local rock system by the movement of groundwater. Thus, volume loss during cleavage formation is a likely possibility; in fact, studies of deformed markers (such as the fossil graptolites, whose regular protrusion spacing can be used to measure finite strain) and geochemical studies demonstrate that rock volume may have decreased by up to 50% during cleavage formation

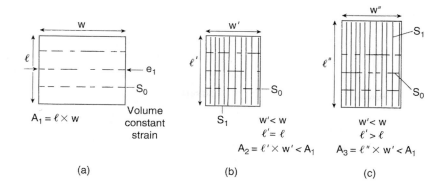

FIGURE 11.19 To lose or not to lose volume; that's the question. (a) A block of height l and width w is shortened and forms a cleavage. If there is volume loss strain, then $w' < w$, but $l' = l$ (we are assuming no change in the third dimension; that is, the intermediate strain axis (Y) equals 1). If there is volume constant strain, then $l'' > l$, as in (c).

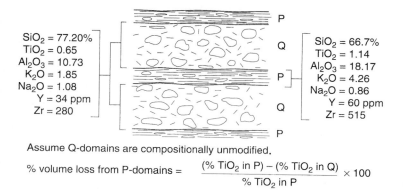

Assume Q-domains are compositionally unmodified.

% volume loss from P-domains = $\dfrac{(\% \text{ TiO}_2 \text{ in P}) - (\% \text{ TiO}_2 \text{ in Q})}{\% \text{ TiO}_2 \text{ in P}} \times 100$

FIGURE 11.20 Volume loss may occur by preferential removal of certain elements. If the composition of the microlithons (Q-domains) is assumed to be constant, then the amount of volume loss can be calculated (see equation); based on relatively immobile elements TiO_2, Y, and Zr, the volume loss is calculated as ~45%.

(Figure 11.20). Interestingly, these values of bulk strain and volume loss suggest that plane strain conditions are representative for cleavage formation, and that flattening strains are apparent rather than real (field of apparent flattening; Figure 4.16). But not all strain analyses of cleaved rocks yield indications of large volume loss. Probably the degree of volume loss is affected by the degree to which the rock is an open or closed geochemical system during deformation, and whether the fluids passing through the rock during deformation are saturated or undersaturated with respect to the soluble mineral phase.

Volume-loss scenarios also have an important implication concerning the nature of **rock–water interactions** during deformation. If we assume that most of the volume loss reflects dissolution and removal of certain minerals, then we can calculate the minimum volume of fluid required for a given amount of volume loss if we know the solubility of the mineral and the amount that can be held in solution (saturation). Such calculations suggest that the amounts of fluid needed for the proposed amounts of volume loss are very large. The results of these calculations are

usually expressed in terms of the ratio of fluid volume to rock volume (called the **fluid/rock ratio**). In the example in Figure 11.20[7] this ratio is >>100, but whether such large ratios are reasonable remains a matter of debate.

11.5 FOLIATIONS IN FOLDS AND FAULT ZONES

We have so far discussed tectonic foliations mainly from a descriptive point of view, without examining how they are related to other structures. Tectonic foliations do not occur in isolation; rather, they are integral components of the suite of structures that represent the manifestation of deformation in a region. In this section we will look at how cleavage develops during progressive deformation, and introduce a number of powerful field characteristics of foliations.

---

[7]Given the solubility of quartz is 0.019 kg/l and a 20% undersaturated fluid.

First we consider the geometric relationship between low-grade cleavage (slaty, spaced disjunctive, and crenulation cleavage) and folds. Low-grade cleavage forms in regions where metamorphic conditions are low (hence the name), a condition that is prevalent in the foreland region of collisional orogens or in convergent plate margins. Such regions display deformation of the fold-thrust belt style, in which a part of the upper crust several kilometers thick is shortened above a basal detachment. Shortening is partitioned among several mechanisms: thrust faulting, folding (which is generally upright to inclined), and the formation of cleavage and intragranular strains (calcite twinning, quartz deformation bands, kinked micas, and so on). These different strain mechanisms are geometrically related because they all accommodate the same regional shortening.

Imagine a region of flat-lying strata that is subjected to layer-parallel shortening (Figure 11.21a). At stresses that are insufficient to initiate faults, pressure solution is activated. As a consequence of pressure solution, spaced disjunctive cleavage domains are formed in argillaceous sandstone and limestone, and pencil cleavage develops in shale. Because of the orientation of the shortening direction, the cleavage domains are oriented approximately perpendicular to bedding. The initial spacing of the domains, you will recall, depends on the original clay content, so in units that are more clay-rich, the cleavage domains tend to be more closely spaced (Figure 11.21a inset). With continued deformation, individual domains get thicker, and new domains are formed. As a consequence, cleavage gets stronger (i.e., domains are more closely spaced). In shales, an incipient slaty cleavage develops. If a detachment fault occurs at depth, deformation of the strata may be decoupled from the deformation of rock below the detachment. In fact, there may be movement on the detachment simply to accommodate the required shortening and cleavage formation in the overlying strata.

Eventually, if stresses get high enough for faulting to begin, the package of strata containing the cleavage is carried in the hanging wall of a thrust and is folded as it navigates a ramp. During folding, flexural slip occurs between layers, with weaker shale layers caught in-between more competent sandstone or limestone layers. As a consequence, the cleavage rotates and becomes inclined to bedding in shales (Figure 11.21b), while in the more rigid layers it maintains its original orientation at a high angle to bedding. At the ramp, cleavage in the more rigid layers fans around the folds, whereas cleavage in the weaker (shaly) horizons is roughly parallel to the axial plane of the fold (Figure 11.21b inset). During the final phases of deformation, the fold as a whole is flattened and the limbs are squeezed together. In steep to overturned limbs, late-stage tectonic cleavage forms, that is just about parallel to the steeply dipping beds.

Because cleavage fans change from convergent in competent beds (sandstone, limestone) to divergent or axial planar in incompetent beds (shale, marl), cleavage changes orientation from bed to bed, a pattern that is called **cleavage refraction** (Figure 11.21b inset and Figure 11.22). Cleavage refraction is the change in cleavage attitude that occurs where cleavage domains cross from one lithology into another of different competency, and reflects variation in the local strain field between beds. In graded beds, this change in cleavage orientation occurs gradually across the bed, producing curved cleavage surfaces (Figure 11.22). Changes in domain spacing typically accompany cleavage refraction, as this is also controlled by lithology.

If the entire unit being deformed is dominantly shale, then the regional slaty cleavage forms approximately

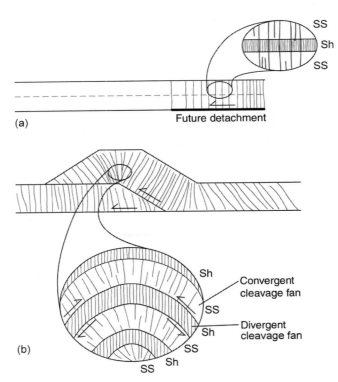

FIGURE 11.21 Evolution of cleavage in a foreland fold-thrust belt. (a) Initiation of cleavage during layer-parallel shortening. Cleavage starts by being nearly perpendicular to bedding. Inset shows that cleavage is parallel in the sandstone (SS) and shale (Sh) beds, but that the initial spacing is different. After formation of a ramp anticline (b), regional shear has caused cleavage to be inclined with respect to bedding. In the fold itself, cleavage refracts as it passes from lithology to lithology. Cleavage in the shale beds is axial planar with respect to the folds, but fans around the folds in the sandstone layers.

FIGURE 11.22 Cleavage refraction in a sandstone—shale sequence (Eifel, Germany). Note that the spacing and nature of cleavage changes between rock types; width of view is ~15cm.

parallel to the axial plane of regional folds. Often, in such tectonic settings, folds are overturned in the direction of regional tectonic transport, so the regional cleavage dips toward the hinterland. In cases where the cleavage is not parallel to the axial plane, but cuts obliquely across the folds, we say that the cleavage is **cross-cutting.** The occurrence of cross-cutting cleavage may indicate that the cleavage was superimposed on preexisting folds, or that there were local complexities in the strain field. For example, rotation of a thrust sheet during folding may cause fold hinges to become oblique to the regional shortening direction. **Transecting cleavage** (Figure 11.23) is a term for cross-cutting cleavage that forms in transpressional environments (meaning there are components of both pure and simple shear). Counterclockwise transecting cleavage, in which the cleavage cuts counterclockwise relative to roughly synchronous fold hinges (Figure 11.23), indicates a component of dextral shear (dextral transpression). Similarly, a component of sinistral shear may produce a clockwise transecting cleavage. However, the use of transecting cleavage as a shear-sense indicator is controversial.

Besides the special situation above, cleavage-bedding relationships provide powerful clues to the relative position of an outcrop with respect to a large regional fold, and cleavage refraction can help you determine the facing ("younging") of folds. Take the example of an initially recumbent fold (F_1) with an S_1 axial planar cleavage that is folded by a second folding generation

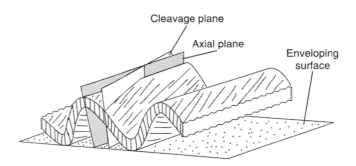

FIGURE 11.23 The relationship between cleavage, axial plane, and enveloping surface in folds with transecting cleavage. The counterclockwise cleavage transection that is illustrated may be indicative of dextral transpression. Note the obliquity of the bedding-cleavage intersection lineation to the hinge line.

(F_2), creating an upright, open fold (Figure 11.24). As a consequence, a portion of the F_2 fold faces up, while another portion of the fold faces down. If the strata in this fold still contain the S_1 cleavage, and if this cleavage is axial planar to the F_1 fold in clay-rich horizons, then, on the overturned limb of the upward-facing part of the F_2 fold (A), bedding dips more steeply than cleavage, and on the upright limb (B) cleavage dips more steeply than bedding. On the upright limb of the downward-facing part of the large fold (C), bedding dips more

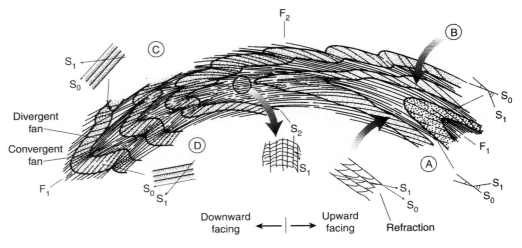

FIGURE 11.24 Cleavage-bedding relationships and cleavage refraction in upright and overturned limbs of upward-facing and downward-facing folds.

steeply than cleavage, whereas on the overturned limb cleavage is steeper (D). You might also find a crenulation cleavage (S_2) that is axial planar to the F_2 fold, especially in its hinge region. The geometry of cleavage refraction further indicates the younging direction (i.e., upright and overturned limbs) in the large fold. Figure 11.24 is another one of those rich diagrams that explore a variety of geometries and relationships that will prove to be quite helpful in the field; that is, once you have figured it out. After taking a suitable amount of time to study this final illustration, let's now turn to another element of deformed rocks: lineations.

11.6 LINEATIONS

A **lineation** is any fabric element that can be represented by a line, meaning that one of its dimensions is much longer than the other two. There are many types of lineations. Some are associated with other structures (such as folds or boudins), some are visible only on specific surfaces in a rock body, while others reflect the arrangement and shape of mineral grains or clasts within the rock. Some lineations reflect strain axes or kinematic trajectories, while others appear to have no kinematic significance. For their description, we broadly group lineations into three categories: form lineations, surface lineations, and mineral lineations.

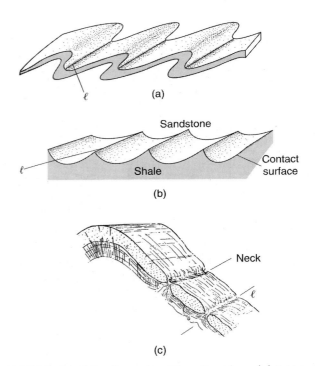

FIGURE 11.25 Examples of form lineations. (a) Fold and crenulation hinges, (b) mullions, and (c) boudins.

11.6.1 Form Lineations

The hinge of any fold is a linear feature. If folds are closely spaced, the fold hinges effectively define a rock fabric that we can measure as a **fold hinge lineation** (Figure 11.25a). Similarly, a **crenulation lineation** is defined by the hinge lines of the microfolds

FIGURE 11.26 Elongate pebbles (arrows) in a stretched conglomerate (Narragansett, Rhode Island, USA); knife for scale.

in a crenulated rock. Why bother measuring crenulation lineations if a crenulation cleavage is present in the rock? The reason is that sometimes the interval in which crenulations occur is not thick enough for cleavage domains to be measurable. For example, deformation and metamorphism of thick sandstone beds separated by thin interbeds of shale will create quartzite beds separated by thin layers of phyllite. Even if the phyllite layer is so thin that a crenulation cleavage plane cannot be measured, it may be possible to see and measure the crenulation lineation.

When intense deformation detaches the limbs of folds, as occurs during fold transposition, isolated fold hinges may be left in the rock. Such hinges are called **rods** and typically occur in a multilayer composed of phyllite (or schist) and quartzite; the quartz layers are relatively rigid and define visible folds, the limbs of which may be thinned so severely that they pinch out, and the quartz flows into the hinge zone, where it is preserved as a rod. Rodding may also occur in mylonites, because the progressive folding in mylonites may generate rootless isoclinal folds whose limbs are detached and whose hinges (the rods) have rotated into parallelism with the shear direction (see Chapter 13). **Mullions** are cusplike corrugations that form at the contact between units of different competencies in a deformed multilayered sequence (Figure 11.25b); the axes of mullions are a lineation. Typically, the more rigid lithology occurs in convex bulges that protrude into the ductile lithology, and the bulges connect in pointed troughs.

Because of their mechanical origin, mullions cannot be used as a facing indicator.

Boudins are tablet-shaped lenses of a relatively rigid lithology, embedded in a weaker matrix, that have collectively undergone layer-parallel stretching (Figure 11.25c). In the third dimension, these long tabular bodies are separated by narrow **boudin necks** that are linear objects. In rare cases we find boudins that record extension in two directions, lovingly called chocolate-tablet boudinage. Other elongate objects in rock that are useful lineations include elongate pebbles and elongate pumice fragments. Again, when the long axes of such elongate objects in a rock are aligned, then they define a measurable lineation. An example of stretched-pebble conglomerate is shown in Figure 11.26. The elongation of these embedded objects is generally a manifestation of deformation in some form, which we used in Chapter 4 for strain quantification. When using these objects for structural analysis, however, it is important to make sure that their alignment is a tectonic and not a primary feature (i.e., alignment during deposition in flowing water or air).

11.6.2 Surface Lineations

An **intersection lineation** is a linear fabric element formed by, as the name suggests, the intersection of two planar fabric elements (Figure 11.27a). An intersection lineation that structural geologists often use in the field is the **bedding-cleavage intersection,** which

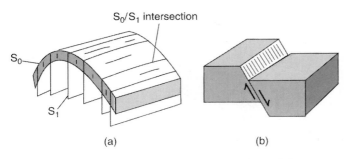

FIGURE 11.27 Sketches of surface lineations. (a) Intersection lineation of bedding (S_0) and (axial plane) cleavage (S_1) in a fold, and (b) slip lineation on a (normal) fault surface.

is manifested by the traces of cleavage domains on a bedding plane (or vice versa). When cleavage is parallel to the axial plane of a fold, the bedding-cleavage intersection must parallel the hinge line in mostly cylindrical folds. The field application is powerful, as it allows one to predict regional fold geometry in areas with otherwise sparse outcrop.

Slip lineations form on surfaces that move in opposite directions (Figure 11.27b). This occurs, for example, on fault surfaces, but also at the interface between beds in flexural slip folds. There are two basic types of slip lineations: **groove lineations,** formed by plowing of surface irregularities, and **fiber lineations** that are formed when vein mineral fibers precipitate along a sliding surface (see Chapter 8). Slip lineations, of course, are parallel to the slip direction and their roughness may indicate slip sense.

11.6.3 Mineral Lineations

When geologists say that a rock contains a **mineral lineation,** they mean that the fabric element defining this lineation is the size of a mineral grain or a cluster of mineral grains (Figure 11.28). Mineral lineations commonly occur in the foliation plane of metamorphic rocks, on shear surfaces, or in the plane of mylonitic foliation. There are several types of mineral lineations. Not all of these have the same tectonic significance, so it is important to determine what type of mineral lineation is present in the rock.

Some minerals, such as kyanite and amphibole, grow such that they are very long in one direction relative to the other two directions. If the long axes of the crystals are aligned in a rock, they create a mineral lineation. The alignment may be due either to growth of the crystal in a preferred direction (con-

trolled by differential stress or by flow-controlled diffusion) or because elongate grains are rotated toward a principal strain direction during deformation. This type of linear fabric is taken to indicate the direction of stretching in the rock and is therefore called a **stretching lineation.** It is quite common in deformed metamorphic terranes to find a consistently oriented mineral lineation that reflects the regional transport direction of the rocks; for example, in areas of regional thrusting.

We earlier mentioned elongate objects, such as stretched pebbles in a conglomerate. Elongate grains or grain clusters produce the same phenomenon, only on a smaller scale. Quartz, which deforms quite readily under metamorphic conditions, may deform into long ribbons that define a distinct lineation in outcrop.

11.6.4 Tectonic Interpretation of Lineations

In structural analysis, the **bedding-cleavage intersection lineation** is widely used, because this lineation offers a clue to the orientation of folds in a region where the hinges themselves may not be exposed. Other types of lineations, however, may be more difficult to interpret, because there are at least two alternative interpretations for their origin. First, a lineation can parallel a principal strain; specifically, the direction of stretching or elongation. When we talk about **stretching lineations,** as defined by elongate mineral grains or pebbles, we are implying that the lineation is roughly parallel to the direction of maximum elongation (the X-axis of the finite strain ellipsoid). Other lineations, like boudin necks, are roughly perpendicular to the stretching direction. Second, a lineation can be parallel to a shear direction, meaning that it is parallel to a vector defining the motion of one part of a rock with respect to another. Slip lineations (fibers or grooves) are good examples of shear-direction lineations. However, the shear direction is not parallel to the stretching direction, except in special cases. Only in zones of high shear strain does the shear direction approach the finite elongation direction; in other words, grains are stretched and mineral clusters are smeared out in the direction of shear. Clearly, you must be careful when interpreting lineations, and take into account that strain and kinematic indicators can be different.

FIGURE 11.28 A steeply plunging mineral lineation that is defined by hornblende (Gotthard Massif, Switzerland); coin for scale.

11.7 OTHER PHYSICAL PROPERTIES OF FABRICS

We focused in this chapter primarily on dimensional and, to a lesser extent, crystallographic aspects of rock fabrics, but these characteristics also impart an anisotropy that may be manifested by other physical properties. It is beyond the scope of this text to explain the underlying principles of these properties or to discuss details of these methods, so we merely mention two of these properties in closing: seismic properties and magnetic anisotropy. **Seismic velocity** varies as a function of material characteristics (such as density and elasticity), so we can use seismic velocity to obtain increasingly more detailed images of the Earth's deep structure (see Chapter 14). Seismic velocity also varies with direction in a single crystal or aggregate of crystals, because the spacing of atoms in the crystal lattice varies in different directions. Thus, when a foliation is defined by the preferred orientation of crystals, the seismic velocity of the sample will vary (by a few percent) in different directions (this phenomenon is called wave splitting). Similarly, the **magnetic properties** of rocks vary as a function of direction, because the ability of a rock to be magnetized depends on the geometric arrangement of atoms in individual crystals and/or the orientation of its mineral constituents. The geometric anisotropy of a foliated rock produces a magnetic fabric that is anisotropic. **Anisotropy of magnetic susceptibility** (AMS) is one measure of a rock's magnetic fabric that is easy to obtain in the laboratory, and is a quick indicator of even the weakest fabric element. In Figure 11.29, we show the distribution of the maximum and minimum susceptibility axes in a progressively deformed sequence of rocks. Notably the distribution of the minimum axes is a sensitive indicator of cleavage intensity. While the orientation of the strain ellipsoid and the ellipsoid representing directional variability of the magnetic fabric are often parallel, the relative magnitudes of these ellipsoids depend strongly on the magnetic phases and processes involved. The relationships between finite strain and magnetic fabric are therefore not straightforward. Nevertheless, the orientation and intensity of magnetic fabrics offer a sensitive measure of fabric elements in natural rocks.

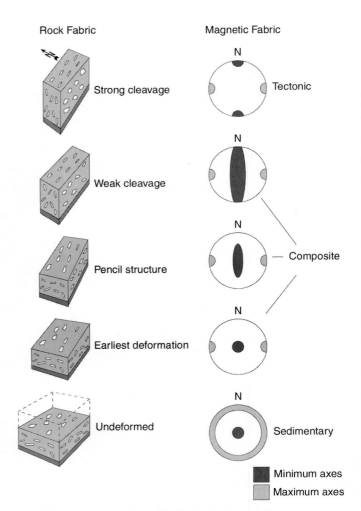

FIGURE 11.29 Magnetic fabrics in progressively cleaved mudrocks. The degree of foliation development, from compaction fabric to strong cleavage, is characterized by the distributions of the maximum and minimum susceptibility axes.

11.8 CLOSING REMARKS

You may find it surprising that after nearly 150 years of research on foliations and lineations (dating from the early studies of Sorby, Darwin, and their contemporaries in the nineteenth century), many questions about these fabric elements remain unresolved. Perhaps one reason for this uncertainty is our inability to create these fabric elements in the laboratory under conditions and processes similar to those in nature; fluid migration and material transfer, especially, pose experimental limitations. So, many of our ideas about tectonic fabrics are based on field evidence, which by its nature is often circumstantial. In the past few decades, however, new progress has been made in our understanding of fabric formation using electron microscopy (scanning and transmission electron microscopy). These high-resolution studies, accompanied by microchemistry, reveal the detailed structure of mineral grains (Figure 11.30). Complementary data obtained from X-ray techniques (texture goniometry) further support a general view of fabric formation based on strain magnitude and metamorphic grade. And all scientists agree? Well, it remains common for a group of geologists to stand on an outcrop and argue about the origin and the meaning, or even the very existence of a fabric. We have attempted to curtail our own biases, but they have undoubtedly crept into the text and may be criticized by your instructor. No matter, debate is what keeps this a vibrant and exciting field of study!

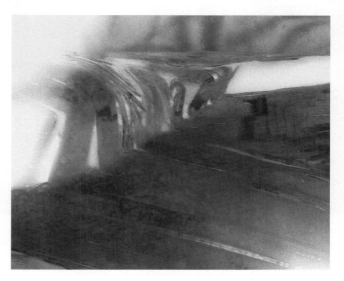

FIGURE 11.30 Transmission electron micrograph of microfolds in mica from some of the earliest studied slates in southern Wales (UK). Width of view is ~ 0.4 μm.

ADDITIONAL READING

Borradaile, G. J., 1978. Transected folds: a study illustrated with examples from Canada and Scotland. *Geological Society of America Bulletin,* 89, 481–493.

Borradaile, G. J., 1988. Magnetic susceptibility, petrofabrics and strain. *Tectonophysics,* 156, 1–20.

Engelder, T. E., and Marshak, S., 1985. Disjunctive cleavage formation at shallow depths in sedimentary rocks. *Journal of Structural Geology,* 7, 327–343.

Etheridge, M. A., Wall, V. J., and Vernon, R. H., 1983. The role of the fluid phase during regional metamorphism and deformation. *Journal of Metamorphic Geology,* 1, 205–226.

Gray, D. R., 1979. Microstructure of crenulation cleavages: an indicator of cleavage origin. *American Journal of Science,* 279, 97–128.

Hobbs, B. E., Means, W. D., and Williams, P. F., 1982. The relationship between foliation and strain: an experimental investigation. *Journal of Structural Geology,* 4, 411–428.

Housen, B. A., and van der Pluijm, B. A., 1991. Slaty cleavage development and magnetic anisotropy fabrics. *Journal of Geophysical Research (B),* 96, 9937–9946.

Knipe, R. J., 1981. The interaction of deformation and metamorphism in slates. *Tectonophysics,* 78, 249–272.

Powell, C. M. A., 1979. A morphological classification of rock cleavage. *Tectonophysics,* 58, 21–34.

van der Pluijm, B. A., Ho, N.-C., Peacor, D. R., and Merriman, R. J., 1998. Contradictions of slate formation resolved? *Nature,* 392, 348.

Weber, K., 1981. Kinematic and metamorphic aspects of cleavage formation in very low-grade metamorphic slates. *Tectonophysics,* 78, 291–306.

Williams, P. F., 1972. Development of metamorphic layering and cleavage in low grade metamorphic rocks at Bermagui, Australia. *American Journal of Science,* 272, 1–47.

Wood, D. S., Oertel, G., Singh, J., and Bennett, H. F., 1976. Strain and anisotropy in rocks. *Philosophical Transactions of the Royal Society of London,* 283, 27–42.

Wright, T. O., and Platt, L. B., 1982. Pressure dissolution and cleavage in the Martinsburg shale. *American Journal of Science,* 282, 122–135.

Ductile Shear Zones, Textures, and Transposition

12.1 INTRODUCTION

Imagine a cold and wet day in northern Scotland, which is not a far stretch of the imagination if you've visited the area. While mapping part of the Scottish Highlands you are struck by the presence of highly deformed rocks that overlie relatively undeformed, flat-lying, fossiliferous sediments. This relationship is even more startling because the overlying unit has experienced much higher grade metamorphism than the underlying sediments, and it contains no fossils. When you arrive at the contact between these two rock suites, you notice that they are separated by a distinctive layer of light-colored, powdery-looking rock. The regional relationships of the suites and their superposition already suggest that the contact is a low-angle reverse (or, thrust) fault. So, what is the distinctive fine-grained rock at the contact? In your mind you envision the incredible forces associated with the emplacement of the hanging wall unit over the footwall, and you surmise that the rock at the contact was crushed and milled, like what happens when you rub two bricks against each other. Using your class in ancient Greek, you decide to coin the name **mylonite** for this fine-grained layer, because "mylos" is Greek for milling.

Something like this happened over a hundred years ago in Scotland where the late Precambrian Moine Series ("crystalline basement") overlie a Cambro-Ordovician quartzite and limestone ("platform") sequence along a Middle Paleozoic low-angle reverse fault zone, called the Moine Thrust. This classic Caledonian area was mapped by Charles Lapworth of the British Geological Survey in the late nineteenth century. Anecdote has it that Lapworth became convinced that the Moine Thrust was an active fault and that it would ultimately destroy his nearby cottage and maybe take his life; Lapworth's later years were spent in great emotional distress.

In most areas around the world you will find zones in which the deformation is markedly concentrated (Figure 12.1). Deformation in these zones is manifested by a variety of structures, which may include isoclinal folds, disrupted layering, and particularly well-developed foliations and lineations, among other deformation features. These zones, called ductile shear zones, often preserve crucial information about the tectonic history of an area, so they merit careful study. We begin their description with a definition. A **ductile shear zone** is a tabular band of definable width in which there is considerably higher strain than in the surrounding rock. The total strain within a shear zone typically has a large component of

(a)

(b)

FIGURE 12.1 (a) The Parry Sound shear zone in the Grenville Orogen (Ontario, Canada) displays characteristics that are common in ductile shear zones, including mylonitic foliation, mineral lineation (stretching lineation) and rotated clasts. (b) A close-up of the zone highlights the mylonitic foliation.

mulates relative displacement of rock bodies, but, unlike a fault, displacement in a ductile shear zone occurs by ductile deformation mechanisms and no throughgoing fracture is formed. This absence of a single fracture is a consequence of movement under relatively high temperature conditions or low strain rates, as we saw in the deformation experiments of Chapter 5.

Consider a major displacement zone that cuts through the crust and breaches the surface (Figure 12.2). In the first few kilometers below Earth's surface, **brittle processes** occur along the discontinuity, which result in earthquakes if the frictional resistance on discreet fracture planes is overcome abruptly. Displacement may also occur by movement on many small fractures, a ductile process called *cataclastic flow* (Chapter 9). In either case, frictional processes dominate the deformation at upper levels of the discontinuity, and this crustal segment is, therefore, called the **frictional regime.** Mechanically, this region is pressure sensitive. With depth, crystal-plastic and diffusional processes, such as recrystallization and superplastic creep, become increasingly important, primarily because temperature increases. Where these mechanisms are dominant, typically below a depth of 10–15 km for normal geothermal gradients (20°C/km–30°C/km) in quartz-dominated rocks, we say that displacement on the discontinuity occurs in the **plastic regime.** Rather than being pressure sensitive, deformation in this region is mostly temperature sensitive. Not surprisingly, the transitional zone between a dominantly frictional and dominantly plastic regime is called the **frictional-plastic transition,** or, more commonly, the **brittle-plastic transition.**[1] While

simple shear, where rocks on one side of the zone are displaced relative to those on the other side. In its most ideal form, a shear zone is bounded by two parallel boundaries, outside of which there is no strain. In real examples, however, shear zone boundaries are gradational. The adjective "ductile" is used because the strain accumulates by ductile processes, which can range from cataclasis to crystal-plasticity to diffusion (Chapter 9). So, a shear zone is like a fault in the sense that it accu-

---

[1]*Frictional-plastic transition* avoids confusion surrounding the terms brittle, ductile, and plastic (see Chapter 5), but brittle-ductile and brittle-plastic are more widely used synonyms.

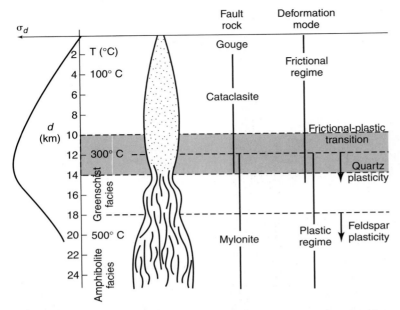

FIGURE 12.2 Integrated model for a displacement zone that cuts deep into the crust, showing the frictional and plastic regimes, the frictional–plastic transition, and crustal strength [σ_d]; this is sometimes called the Sibson-Scholz fault model. Fault rocks typically found at crustal levels are indicated.

this term is in common use, it is technically not correct to call this the *brittle-ductile transition,* because ductile processes (such as cataclasis) may occur in the frictional regime. For example, we see that a crustal-scale displacement zone is a brittle fault at the surface and is a ductile shear zone at depth. Based on this contrast in deformation processes, we predict that the displacement zone will change from a relatively narrow fault zone to a broader ductile shear zone with increasing depth, because the strength contrast between deformation zone and host rock decreases.

Mylonites are dominated by the activity of crystal-plastic processes, which produce another characteristic of deformed rocks: **crystallographic-preferred fabrics** or **textures.** Textures are discussed in this chapter, although the topic logically follows the material on crystal-plasticity that is presented in Chapter 9. Secondly, rocks within ductile shear zones are typically intensely folded and the original layering is transposed into a tectonic foliation. A description of transposition will therefore close this chapter, but we emphasize that it is not unique to ductile shear zones (see also Chapter 11). We'll see that shear zones may produce more than one foliation and one or more lineations, and that shear-zone rocks commonly contain rotated fabric elements, grain-shape fabrics, and a grain size that is characteristically smaller than that of the host rock. Arguably, ductile shear zones are the most varied structural feature, and perhaps the most

informative for tectonic analysis. Enough hype; let's look at these things.

12.2 MYLONITES

The inference made by Charles Lapworth concerning the formation of mylonites at the Moine Thrust is incomplete and, mechanically, incorrect. As mentioned earlier, the derivation of the name mylonite suggests that cataclastic flow is the process of grain-size reduction. In fact, microstructures in samples that record the change from relatively undeformed to mylonitized rock show that **crystal-plastic processes,** and **dynamic recrystallization** in particular, are mainly responsible for grain-size reduction (Figure 12.3). Therefore, we restrict the use of the term mylonite to a fault rock type with a relatively fine grain size as compared to the host rock and resulting from crystal-plastic processes. When brittle fracturing produces a reduced grain size, these fault rock types are called cataclasites. The field terminology of fault rocks and the processes by which they generally form are summarized in Table 12.1.

Mylonites are associated with all kinds of ductile shear zones, whether they result in reverse displacement (such as the Moine Thrust in Scotland), normal displacement (such as those in core complexes and the Basin-and-Range of the western USA), or strike-slip displacement (such as the Alpine fault in New Zealand). In each of these structural settings mylonites have one element in common: conditions that promote crystal-plastic deformation mechanisms. Such conditions are reached at various values of temperature, strain rate, and stress, depending on the dominant mineral in the rock (Chapter 9). For example, marble mylonites and quartzite mylonites form at temperatures that are lower than those under which feldspathic mylonites form, because the onset of dynamic recrystallization occurs at different temperatures in calcite (>250°C), quartz (>300°C), and feldspar (>450°C). Rocks that contain a variety of minerals show mixed behavior; for example, quartz grains in a sheared granite may dynamically recrystallize, while feldspar grains deform predominantly by fracturing. If the rock displays this contrast, we surmise that it was sheared at temperatures greater than 300°C, but less than 500°C (the greenschist facies of metamorphism). However, mylonite textures should not be used as a quantitative indicator of the conditions of temperature, stress, or

FIGURE 12.3 Clastomylonite containing relatively rigid clasts of varied lithologies in a fine-grained, crystal-plastically deformed marble matrix (Grenville Orogen, Ontario, Canada); hammer for scale.

TABLE 12.1	FAULT ROCKS AND PROCESSES
Breccia	Incohesive fault rock with randomly oriented fragments that make up >30% of the rock mass, formed by brittle processes. Because breccia is also a sedimentary rock, the adjective "tectonic" is often included for these fault rocks.
Cataclasite	Cohesive fault rock generally with randomly oriented fabric, formed by brittle processes.
Gouge	Fault rock formed under brittle conditions with randomly oriented fragments that make up <30% of the rock mass. Clay-rich gouge is typically more cohesive than clastic gouge, because of the presence of sticky, swellable clays.
Mylonite	Cohesive, foliated fault rock, formed dominantly by crystal-plastic processes.
Pseudotachylite	Dark, glassy fault rock along fractures, formed by melting of the host rock from heat generated during frictional sliding; they are likely associated with earthquake activity. *Note:* Tachylite is an old name for an igneous rock with a glassy texture.

strain rate, because of the uncertainties surrounding these parameters and their mutual dependence; at best we can make semiquantitative estimates such as those above. Quantitative metamorphic petrology and isotope geology, in many instances, provide reliable methods to determine the past conditions of temperature and pressure of deformed rocks (see Chapter 13), so we do not have to rely on microstructures only.

12.2.1 Types of Mylonites

The study of mylonites, which occur widespread in crustal rocks, has led to the development of several prefixes to distinguish among different types of mylonites. These terms are listed in Table 12.2.

Protomylonite and **ultramylonite** are used to describe mylonites in which the proportion of matrix is <50% and 90–100%, respectively. In a protomylonite, only part of the rock is mylonitized, whereas pervasive mylonitization has occurred in an ultramylonite. Mylonites containing 50–90% matrix are known simply as *mylonite*. **Blastomylonite** and **clastomylonite** are used to describe mylonites containing large grains surrounded by a fine-grained matrix that grew during mylonitization or that remained from the original rock, respectively. The terms derive from the Greek words "blastos," meaning growth and "klastos," meaning

TABLE 12.2	TYPES OF MYLONITES
Blastomylonite	Mylonite that contains relatively large grains that grew during mylonitization (e.g., from metamorphic reactions or secondary grain growth).
Clastomylonite	Mylonite that contains relatively large grains or aggregates that remain after mylonitization reduced the grain size of most of the host rock (e.g., relatively undeformed feldspar grains or clumps of mafic minerals).
Phyllonite	Mica-rich mylonite.
Protomylonite	Mylonite in which the proportion of matrix is <50% (i.e., rocks in which only a minor portion of the minerals underwent grain-size reduction).
Ultramylonite	Mylonite in which the proportion of matrix is 90–100% (i.e., rocks in which mylonitization was nearly complete).

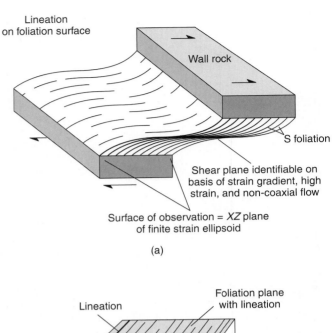

(a)

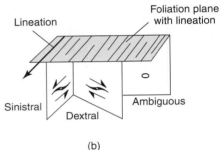

(b)

FIGURE 12.4 (a) Schematic, right-lateral ductile shear zone showing mylonitic foliation (S) and lineation. The optimal surface for study is the *XZ*-plane of the finite strain ellipsoid. (b) Apparent difference in shear sense, which results from observing the shear-sense indicator in different surfaces. Note that the surface perpendicular to the lineation and the foliation gives no shear-sense information.

broken. These prefixes are used for microstructural description of mylonites, but they are also used as field terms. Clastomylonite is used to describe mylonites that contain coarse fragments of less deformed host rock or exotic lithologies; for example, mafic clasts in a marble mylonite (Figure 12.3).

12.3 SHEAR-SENSE INDICATORS

Ductile shear zones concentrate displacement at deeper levels in the crust, where recognizable markers that determine offset, such as bedding, are often absent. Consider a greenschist-facies shear zone in a large granitic body. The granitic rocks on either side of the mylonite are indistinguishable, so there is nothing at first glance to predict the sense of displacement, let alone the magnitude of displacement. **Sense of displacement** describes the relative motion of opposite sides of the zone (left-lateral or right-lateral, up or down, and so on), whereas **magnitude of displacement** is the distance over which one side moves relative to the other. The solution to this challenge is to look for **shear-sense indicators**[2] in ductile shear zones.

---

[2]The term *kinematic indicators* is often used as a synonym, but this suggests information about the strain state rather than sense of displacement.

12.3.1 Plane of Observation

The recognition and interpretation of shear-sense indicators require that we examine a shear zone in a particular orientation. Most mylonites contain at least one foliation and a lineation, which we use as an **internal reference frame.** In the field we look for outcrop surfaces (or cut an oriented sample in the lab) that are perpendicular to the mylonitic foliation and parallel to the lineation (Figure 12.4a). We make the reasonable assumption in this case that the lineation coincides with the movement direction of the shear zone. When two foliations are present, this surface is also generally perpendicular to their intersection. This plane, which parallels the *XZ*-plane of the finite strain ellipsoid, maximizes the expression of the rotational component of the deformation; in all other surfaces this component is less. Then we must place the orientation of our surface in the context of the region. Say, we find that a right-lateral displacement is the surface of observation. Were we to look at this same surface from the opposite side, the displacement would appear to be left-lateral (Figure 12.4b). This is not a paradox, but simply a matter of reference frame; we encountered the same situation with fold vergence (Chapter 9). Because the displacement sense is the same in geographic coordinates, it is a good habit to analyze surfaces in the same orientation across the field area to avoid confusion. If this is not possible, make careful field notes of the orientation of the surface in which you determined shear sense. Back in the laboratory the rock saw offers complete control, provided you oriented the sample prior to removing it from the field. Having cautioned you sufficiently about orientation, let us now look at types of shear-sense indicators. They fall into five main groups: (1) grain-tail complexes, (2) disrupted grains, (3) foliations, (4) textures (or crystallographic fabrics), and (5) folds.

12.3.2 Grain-Tail Complexes

Mylonites commonly contain large grains or aggregates of grains that are surrounded by a finer matrix; for convenience, we use the term *grain* in a general sense to describe both large single grains and coherent grain aggregates. These grains may have *tails* of material with a composition and/or grain shape and size that differ from the matrix, such that they are distinguishable. For example, large feldspar grains connected by thin layers of finer-grained feldspathic material are common in gneiss (Figure 12.5). The tail may represent highly attenuated, preexisting mineral grains, it may be a consequence of dynamic recrystallization of material at the rim of the grain, or it may be material formed by synkinematic metamorphic reactions (neocrystallization). During deformation, the grains act as rigid bodies and we may be able to determine the sense of displacement from the tails. Based on their relationship with the shear-zone foliation, we recognize two types of grain-tail complexes: σ-type and δ-type. A third type, θ-complexes, has been proposed, but their interpretation is equivocal and we do not further discuss them.

Grain-tail complexes of the **σ-type** are characterized by wedge-shaped tails that do not cross the reference plane when tracing the tail away from the grain (Figure 12.6a). Sometimes the tail is flat at the top and the other side is curved toward the reference plane. Overall it has a stair-stepping geometry in the direction

of displacement. This grain-tail geometry looks like the Greek letter σ (at least in the case of right-lateral displacement), hence the name σ-type. Figure 12.5a is a field example of a σ-type complex in a quartzofeldspathic shear zone. Obviously, in the case of left-lateral displacement the geometry is a mirror image.

In **δ-type** grain-tail complexes the tail wraps around the grain such that it cross cuts the reference plane when tracing the tail away from the grain (Figure 12.6b). If you rotate the Greek letter δ over 90° you will see why we use this symbol. Figure 12.5b is a field example of a δ-type complex in another feldspathic shear zone. The rotation on the δ-type complex is counterclockwise for left-handed and clockwise for right-handed displacement. The stair-stepping geometry that we find in σ-type grain-tail complexes is no longer a characteristic of displacement in δ-type complexes.

It is common to find both σ- and δ-types in one surface; even complexes that have characteristics of both σ- and δ-types may occur, because they are related (Figure 12.6c). One reason for this mixed occurrence is a varying relationship between the rate of recrystallization/neocrystallization and rotation of the grain. If tail formation is fast relative to rotation, the tails are of the σ-type. If, on the other hand, the rotation of the grain is faster, the tail will simply be dragged along and wrap around the grain (δ-type). The case of preexisting tails, which often occur with pegmatites that are incorporated into a shear zone, falls in the latter category.

The presence of both σ- and δ-type grain-tail complexes may indicate different rates of tail growth, different initial grain shape, different times of tail formation, or different coupling (see later section). From these variables it is clear that we should use grain-tail complexes with considerable caution for strain quantification purposes, but there is no doubt about their power as shear-sense indicators. In practice, geometries of the σ-type may be difficult to recognize, but δ-type complexes offer intuitively obvious and unequivocal information on shear sense.

12.3.3 Fractured Grains and Mica Fish

Minerals such as feldspar may deform by fracturing (Figure 12.7a) along crystallographic directions or parallel to the shortening direction (extension cracks). As long as the approximate orientation of these fractures before shear is known, we can determine the shear sense from their displacement (Figure 12.8). Fractures oriented at low angles to the mylonitic foliation have a

(a)

(b)

FIGURE 12.5 Grain-tail complexes. (a) A K-feldspar clast with a tail of fine-grained plagioclase of the σ-type complex (California, USA). (b) A δ-type complex in a feldspathic gneiss from the Parry Sound shear zone of the Grenville Orogen (Canada). Width of view is ~10 cm.

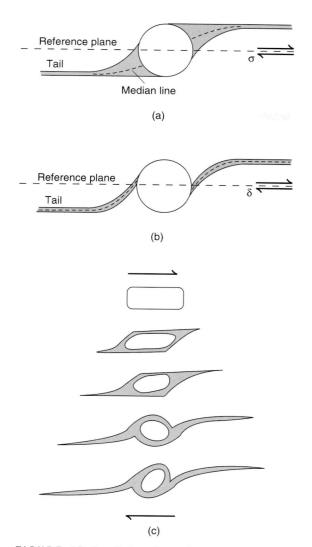

FIGURE 12.6   Grain-tail complexes as shear-sense indicators. (a) σ-type complex, (b) δ-type complex, and (c) the evolution of a σ-type complex into a δ-type grain-tail complex.

(a)

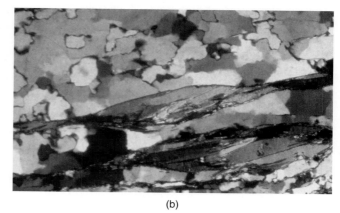

(b)

FIGURE 12.7   (a) Photomicrograph of fractured feldspar grain showing bookshelf- or domino-type, antithetic displacement. (b) Mica fish in a quartz mylonite. Right-lateral displacement in both images; width of view ~0.5 mm.

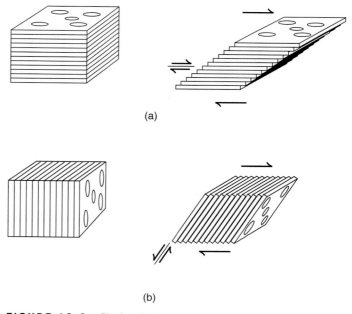

FIGURE 12.8   Placing dominos between a pair of hands to demonstrate sense of shear from fractured grains, if the fractures are at a low angle (a) or high angle (b) to the shear plane. Note that rotation according to the "domino model" of individual segments of fractured grains in (a) is the same as that for rotated grains.

displacement sense that is consistent with the overall shear sense of the zone; these fractures are called **synthetic fractures** (Figure 12.8a). Fractures at angles greater than ~45° to the foliation show an opposite sense of movement; these are called **antithetic fractures** (Figure 12.8b). The opposite motion is not contradictory, as we can see from a simple experiment. Place a series of dominos upright between your hands, and move your hands in opposite directions. You will notice that as long as the angle of the dominos with your hands remains greater than ~45°, the displacement between individual dominos is opposite (antithetic) to the relative movement direction of your hands. At lower angles you will find that displacement of the dominos has a motion that is the same (synthetic) to the motion of your hands. Fractured minerals

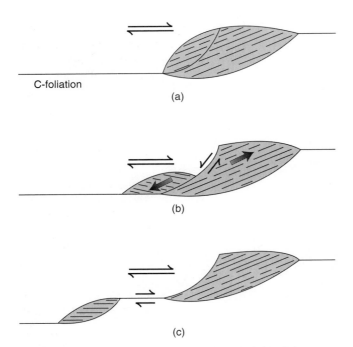

C-foliation

(a)

(b)

(c)

FIGURE 12.9 The formation of mica fish; (a) to (c) show successive stages in mica-fish development, with C-foliation marking the shear plane.

and clasts behave similarly in shear zones, and some therefore call this the *domino model*[3] for shear sense.

Feldspar and quartz are not the only minerals useful for determining shear-sense in mylonites. It is quite common to find large phyllosilicate grains, such as mica and biotite in quartzo-feldspathic rocks, and phlogopite in marbles, that display a characteristic geometry. The micas are connected by a mylonitic foliation, and their basal (0001) planes are typically oriented at an oblique angle to the mylonitic foliation, such that they point in the direction of the instantaneous elongation axis. In this orientation they show a stair-stepping geometry in the direction of shear (Figures 12.7b and 12.9), which is similar to σ-type grain-tail complexes that also step up in the shear direction. When phyllosilicates are large enough to be seen in hand specimens they look like scales on a fish (hence they are called **mica fish**) and you can use a simple field test to determine their approximate orientation. The basal planes of phyllosilicates are excellent reflectors of light, so when you turn the foliated shear-zone sample in the sun you encounter an orientation that is particularly reflective. When this happens and the sun is behind you, you are looking in the direction of shear. The method is affectionately known as the "fish flash,"

and was obviously developed by those fortunate geologists who work in sunnier parts of the world.[4]

12.3.4 Foliations: C-S and C-C' Structures

Most mylonites show at least one well-developed foliation that is generally at a low angle to the boundary of the shear zone. Previously this foliation was called the **mylonitic foliation,** and it is otherwise known as the **S-foliation;** S is derived from the French word for foliation, "schistosité." Its angle with the shear zone boundary may be as little as a few degrees, at which point it is hard to distinguish from a foliation that parallels the shear zone boundary, called the **C-foliation;** C comes from "cisaillement," which is French for shear. A third foliation showing discrete shear displacements that is oblique to the shear zone boundary is called the **C'-foliation.** These are the three most

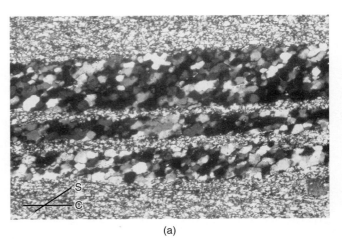

(a)

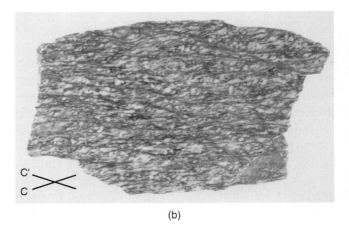

(b)

FIGURE 12.10 Photomicrograph of a C-S structure in a quartzite mylonite (a) and C-C' structure in a micaceous mylonite (b); width of view is ~1 mm and ~15 cm, respectively.

---

[3]More literate, less playful geologists prefer the term *bookshelf model.*

[4]Less fortunate geologists resort to the fish flash*light* method.

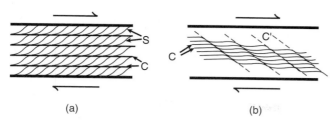

FIGURE 12.11 Characteristic geometry of (a) C-S and (b) C-C' structures in a dextral shear zone. The C-surface is parallel to the shear zone boundary and is a surface of shear accumulation (i.e., not parallel to a plane of principal finite strain). The S-foliation is oblique to the shear-zone boundary and may approximate the *XY*-plane of the finite strain ellipsoid. The C'-foliation in (b) displaces an earlier foliation (C or composite C/S).

common foliations in shear zones and may reflect grain-shape fabrics or discrete shear surfaces. At first, these many foliations may seem confusing, but when correctly identified they become very powerful shear-sense indicators.

Thin-section study of well-developed mylonites in quartzites, granites, and marbles often shows the presence of a foliation that is defined by elongate grains (Figure 12.10a). This foliation reflects the activity of crystal-plastic processes that tend to elongate grains toward the extension axis of the finite strain ellipsoid, and is called the S-foliation. It is uncertain whether the S-foliation exactly tracks the *XY*-plane of the finite strain ellipsoid; nor is it certain whether S- and C-foliations form simultaneously or sequentially. These distinctions are only important when **C-S structures** are used to determine the degree of non-coaxiality or the kinematic viscosity number (Chapter 4). For their purpose as a shear-sense indicator, these questions are somewhat academic, because in all cases the geometry of a C-S structure gives the same shear sense. The long axis of elongate grains in the S-surface points up in the direction of shear, and the shear direction is perpendicular to the intersection line of the S- and C-foliations (Figure 12.11a).

Another common shear-sense indicator is a series of oblique, discrete shears that are present in strongly foliated mylonites. These small shears, called C'-surfaces because they accumulate shear strain, are particularly common in phyllosilicate-rich mylonites and crenulate or offset the mylonitic foliation (Figure 12.10b). C'-foliations are, therefore, also called **shear bands** or extensional crenulations. The offset on C'-surfaces is in the same direction as the overall displacement in the shear zone (i.e., displacement along C). The C'-surfaces contrast with S-surfaces that do not appear to displace the C-surface, suggesting that C'-surfaces form late in

the mylonitic evolution. As with C-S structures, the strain significance of **C-C' structures** is incompletely understood, but their formation reflects a component of extension along the main anisotropy (the C-surface) of the mylonite. Thus, the shear sense on C-C' structures is *synthetic* to the sense of shear of the zone as a whole (Figure 12.11b).

12.3.5 A Summary of Shear-Sense Indicators

Shear sense in ductile shear zones is only reliably determined when two or more different indicators give a consistent sense of displacement. So we close this section with a summary diagram (Figure 12.12) showing common shear-sense indicators that may be encountered in a ductile shear zone. Of all indicators, C-S and δ-clasts are most readily interpretable. Planar objects may have synthetic and antithetic shear, and C' surfaces may reflect opposing motions; for example, boudins may show antithetic bookshelf faulting or synthetic displacement shear. The occurrence of a particular indicator will vary within a zone and even within the same outcrop, as a function of the dominance and

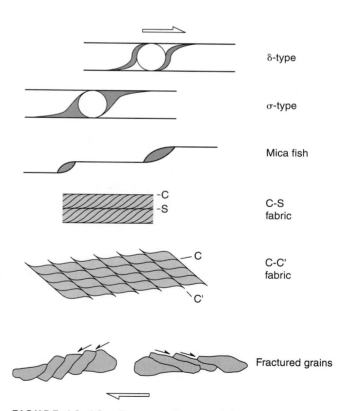

FIGURE 12.12 Summary diagram of shear-sense indicators in a dextral shear zone. A copy of this figure on a transparency (for left- and right-lateral shear) makes a handy inclusion in your field notebook.

mineralogy of grains and presence of foliations. Except for fractured grains, the shear sense can be determined from any of these indicators without knowing their (original) relationship to the shear-zone boundary. Figure 12.12 is another handy reference tool for your field notebook.

12.4 STRAIN IN SHEAR ZONES

In Chapter 4 we introduced a number of methods to quantify strain in deformed rocks, including the use of grain shapes. But do any of these methods apply to shear zones? Well, even if we accept that grain shape or degree of grain alignment reflects strain, at best we may be able to measure an incremental part of the strain history from the activity of dynamic recrystallization in mylonites. However, we have little or no idea how much of the finite strain this increment represents, nor to what extent the increment coincides with the orientation of the finite strain ellipsoid. Two approaches can

be used to give at least a first-order estimate of the amount of finite strain in shear zones, namely, rotation of grains, and deflection of foliations; but, as you will see, these methods are not without their limitations. It is exactly because of the uncertainties surrounding strain analysis in shear zones that we have left a description of these methods until this chapter.

12.4.1 Rotated Grains

The formation of a ductile shear zone involves an internal rotation (the internal vorticity; Chapter 4) that may be recorded in rotated grains. So, from grains that preserve evidence of rotation we can determine this component of strain. We saw this earlier with δ-type grain-tail complexes, but this is also found in minerals that grow and incorporate matrix grains during rotation. In particular the mineral garnet shows this behavior, in which "trapped" matrix grains eventually produce a spiraling trail. Such garnets are often called **snowball garnets,** for obvious reasons (Figure 12.13). Let's use another simple analog experiment. Place a

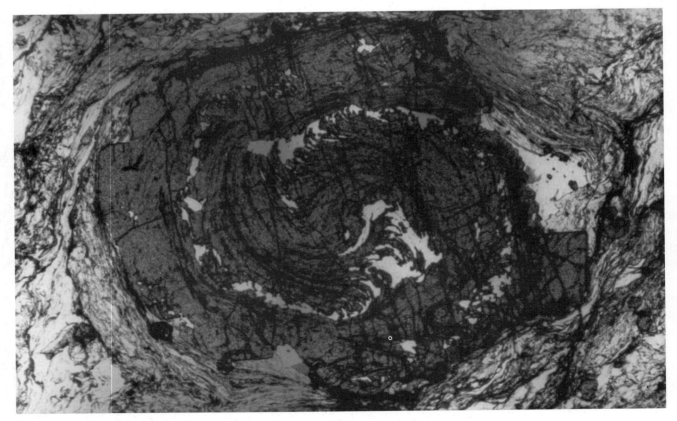

FIGURE 12.13   Snowball garnet (Sweden); width of view is ~7 mm.

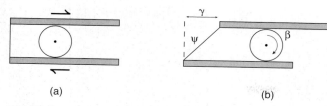

FIGURE 12.14 (a) A simple ball-bearing experiment (b) that illustrates the relationship between rotation [β] and shear (ψ, γ).

ball bearing or a marble between oppositely sliding hands and you will see that its rotation is directly related to the motion of your hands (Figure 12.14). In fact, the amount of rotation of the marble is proportional to the relative displacement of your hands (i.e., the amount of simple shear); mathematically, this relationship is straightforward:

$$\beta = \tan \psi = \gamma \qquad \text{Eq. 12.1}$$

where β is the rotation angle in radians (1 radian is 180°), ψ is the angular shear, and γ is the shear strain (see Chapter 4). However, if the ball bearing is greasy, the rotation angle may be less, because there is some slip between your hands and the ball bearing. This adds to Equation 12.1 a factor that describes the *coupling* between matrix and grain:

$$\beta = \Omega \tan \psi = \Omega \gamma \qquad \text{Eq. 12.2}$$

where the parameter Ω describes the mechanical coupling between the ball bearing and your hands. The value of Ω is equal to 1 for full coupling (clean ball bearing), less than 1 for partial coupling (greasy ball bearing), and 0 for no coupling. In kinematic terms, coupling describes the degree by which internal vorticity is converted to spin. There is no single value for Ω that is unique for natural rocks, but if we assume that grains rotate in a viscous (Newtonian) fluid, considerable slippage will occur at the contact between matrix and grain, and we obtain a theoretical value for Ω of 0.5. Thus, by measuring the rotation angle of the spiraling snowball garnet we can determine the shear strain given some assumption about coupling. Regardless of the coupling assumption, using Ω = 1 gives us an estimate of the minimum shear strain.

12.4.2 Deflected Foliations

Now let's take a closer look at the strain distribution in a shear zone by deforming the middle of three blocks in Figure 12.15 under conditions of non-coaxial strain (simple shear). In Figure 12.15a the strain in the sheared block is homogeneous, because the circles become ellipses with the same axial ratio and orientation. The strain pattern in Figure 12.15b, on the other hand, is heterogeneous as shown by the ellipses of different axial ratio and orientation. Comparing these two patterns with shear zones in a natural rock (e.g., Figure 12.16), we find that inhomogeneous strain dominates. The shear zone in our example is characterized by a mylonitic foliation (S-foliation) that is at ~45° to the shear-zone boundary at the edges of the zone. The S-foliation becomes increasingly parallel to the shear-zone boundary as we approach the center of the zone. If we assume that the trace of the foliation tracks the X-axis of the finite strain ellipsoid (Figure 12.17a), the shear strain is determined by the equation

$$\gamma = 2/\tan 2\phi' \qquad \text{Eq. 12.3}$$

where φ' is the angle between the foliation and the shear-zone boundary. Note that small differences in the

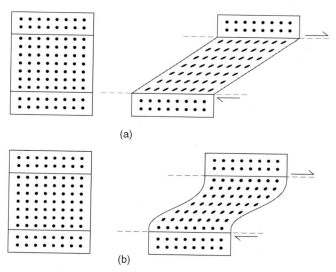

FIGURE 12.15 Homogeneous (a) and heterogeneous (b) strain in shear zones. The similarly shaped and oriented ellipses in (a) show that strain in the zone is homogeneous, whereas variably oriented and elongated ellipses in (b) show heterogeneous strain across the zone.

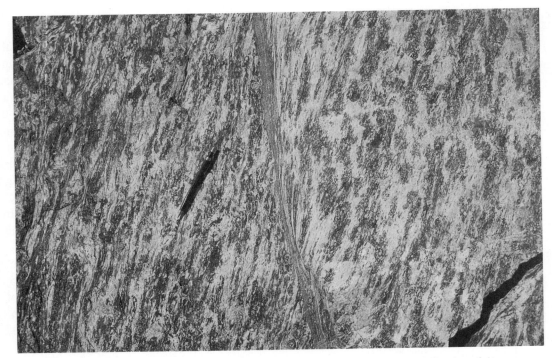

FIGURE 12.16 Small-scale, left-lateral shear zone in anorthosite, showing deflection of the mylonitic foliation (Grenville Orogen, Ontario, Canada). Width of view is ~20cm.

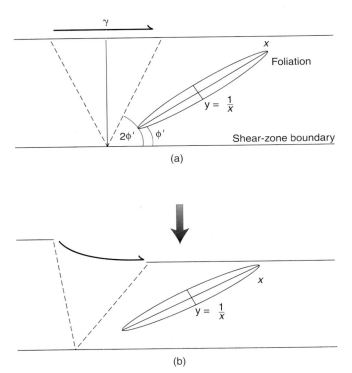

FIGURE 12.17 Angular relationship, ϕ', between foliation and shear-zone boundary, and shear strain, γ, in (a) a perfect shear zone (kinematic vorticity number, W_k, is 1), and in (b) a zone with a component of shortening perpendicular to the shear-zone boundary $(0 < W_k < 1)$. Dashed lines are traces of circular sections of the strain ellipse.

value of ϕ' (which occur in the center of the zone) significantly change the calculated magnitude of the shear strain. Just compare the results for ϕ' values of, say, 10° and 5°. Again, in practice, we use this method to obtain a minimum shear strain.

There is yet another issue. Strain determination using Equation 12.3 is more complicated when a component of shortening or extension perpendicular to the shear zone boundary is present (Figure 12.17b); that is, when the strain is nonperfect simple shear (or general shear; Chapter 4). General shear with a shortening component is called **transpression** and with an extensional component is called **transtension.** This pure shear component will change the angle of ϕ', and thus invalidate Equation 12.3, which is based on progressive simple shear. The combination of coaxial and non-coaxial strain cannot be modeled by simply superimposing the two components, because the order of addition matters (recall that tensors are noncommutative). Moreover, simultaneously adding these components produces yet another finite strain, which is shown schematically in Figure 12.18. Deriving the associated equations requires an understanding of strain tensors and matrix operations that is beyond the scope of this book. We leave you simply with the observation that the angle ϕ' will be small even for low shear strains in the presence of a shear zone–perpendicular shortening component and, analogously, the angle may be large with a zone-perpendicular

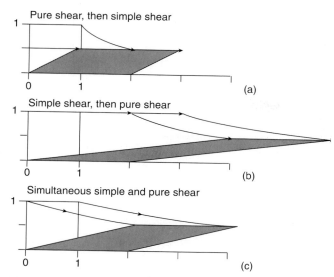

Pure shear, then simple shear

Simple shear, then pure shear

Simultaneous simple and pure shear

FIGURE 12.18 Comparison of (a) superimposing simple shear on pure shear, (b) superimposing pure shear on simple shear, and (c) simultaneously adding simple and pure shear. The magnitudes of the simple shear and pure shear components are equal in all three examples, but they produce distinctly different finite strains because of the noncommutative nature of tensors.

extensional component. Shear strains obtained from Equation 12.3 will be overestimates and underestimates, respectively, under these conditions. Clearly, strain analysis in shear zones is a complex problem because of the many assumptions it requires, and the results are often rough estimates at best.

12.5 TEXTURES OR CRYSTALLOGRAPHIC-PREFERRED FABRICS

Microstructures describe the geometric relationship between the various constituents of a rock, give an indication of the operative deformation mechanism, and often enable us to form an estimate of the ambient conditions (see Chapter 9). Crystal-plastic deformation mechanisms also provide us with a tool for the interpretation of deformation that is particularly useful in shear zones. Hence we have waited until this point to bring it up. Recall that dislocation glide is the strain-producing crystal-plastic process that occurs on specific crystallographic planes in a crystal (Chapter 9; Table 9.1 lists these glide planes in common rock-forming minerals). The properties of dislocation movement hold the key to interpreting the significance of **textures** or **crystallographic-preferred fabrics** in rocks. It turns out that the type of texture may be indicative of

the dominant deformation mechanism and can provide valuable information on the rheologic conditions.

Yet another type of fabric in deformed rocks! You have already learned about dimensional-preferred fabric and now we add crystallographic-preferred fabric to our vocabulary. Let's first see how they differ. Dimensional- and crystallographic-preferred fabrics describe different properties of rocks. A **dimensional-preferred fabric** is the quantification of grain shapes in a rock; it is, in essence, a geometric parameter. Aligned hornblende crystals provide an example of a dimensional-preferred fabric. A **crystallographic-preferred fabric** describes the collective crystallographic orientation of grains that make up the rock. In other words, crystallographic-preferred fabrics represent the degree of alignment of crystallographic axes.

We start by looking at the principles governing the development of crystallographic-preferred orientation by intracrystalline slip. The square $ABCD$ in Figure 12.19a marks the schematic cross section of a single crystal that is deformed by homogeneous shortening perpendicular to the top and bottom sides of the square. Stated more specifically, the infinitesimal shortening strain (Z_i) is parallel to AD (heavy arrows). The grain only deforms by dislocation glide along specific crystallographic planes, which are shown by the thin lines (parallel to the diagonal BD). Let's say that these planes coincide with the basal plane of the crystal with the indices (0001) for hexagonal minerals, so that the crystallographic c-axis is oriented perpendicular to these glide planes. We add one other restriction to the deformation: the faces AB and CD and the infinitesimal strain axes are held in a constant orientation relative to the external framework, which defines a nonspinning deformation. Because dislocation glide is a volume constant mechanism, the dimensions of the grain measured parallel to the glide plane do not change during deformation. At first glance this may not appear to be the case, but measure the length of the same glide plane in each step (a to c) to prove this to yourself.

During deformation, shortening is accomplished by slip on the glide planes; this is accompanied by simultaneous extension, as shown by the finite strain axis (X_f) in Figure 12.19b. The critical ingredient is that if neither length nor spacing of the glide planes are to change, these planes have to rotate to accommodate the strain. As a consequence, the c-axis (perpendicular to the glide plane) rotates progressively *toward* the infinitesimal shortening direction. Meanwhile, the grain continues to elongate, with the long axis of the finite strain ellipsoid increasingly approaching AB (or CD). Thus, the finite strain axes and infinitesimal strain axes rotate relative to each other, which defines

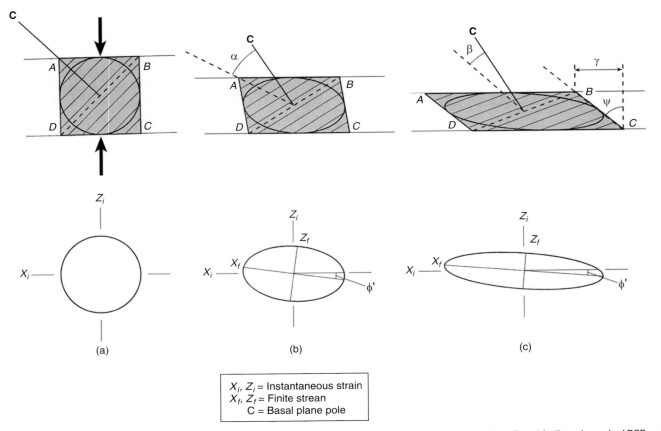

X_i, Z_i = Instantaneous strain
X_f, Z_f = Finite stream
C = Basal plane pole

FIGURE 12.19 The development of a crystallographic-preferred orientation by dislocation glide. The thin lines in grain *ABCD* are crystallographic glide planes, which here coincide with the basal plane (0001). The c-axis is indicated with the heavy line, labeled **C**, and the instantaneous shortening direction by heavy arrows. Strain arises from glide on the crystal planes; note that the length of individual planes remains the same, requiring that the glide planes rotate to accommodate the distortion. Various angular relationships are indicated: (α) angle of shear along the glide plane; (β) rotation angle of the c-axis with respect to an external reference system (e.g., shear-zone boundary); (ψ) rotation angle of material line *BC* with respect to an external reference system (angular shear); (ϕ') angle of finite extension axis (subscript "*f*") with respect to an external reference system. Note that the instantaneous strain axes (subscript "*i*") define the external reference system, and that they do not change in orientation during the progressive history (i.e., there is no spin).

non-coaxial strain with simultaneous nonspinning deformation. In this generalized scenario, we recognize two important consequences: (1) the c-axes rotate toward the instantaneous shortening axis (Z_i) and (b) the finite strain ellipsoid elongates toward the instantaneous shortening direction (X_i), which means that they are only parallel at very large strains. In Nature, matters may be more complex as we are dealing with three-dimensional space in which several glide planes may be active simultaneously. Computer modeling of the development of crystallographic fabrics in these more complex situations, however, shows overall behavior similar to that in our simple model. We can use these characteristics of texture development for the determination of shear sense, but, for this, we first must be familiar with the Symmetry Principle.

12.5.1 The Symmetry Principle

Symmetry is a common and fascinating aspect of the world around us. Look in the mirror or at your neighbor and you see that a person's face is symmetric (at least in general) around a single plane. This symmetry plane cuts between the eyes, and divides the nose and mouth in halves (bilateral or mirror symmetry). Most objects, living or inanimate, display some degree of symmetry, as do many of the geometric concepts basic to mathematics and physics. A cube has a higher symmetry than the human face, in that it contains three mutually perpendicular **mirror planes.** We say that a cube has a higher symmetry than a human face, because it contains more symmetry elements. In mineralogy, you have learned about the symmetry of various types of crystals; for example, triclinic (only a center of symmetry or a two-fold axis),

TABLE 12.3	CRYSTAL SYSTEMS IN ORDER OF INCREASING SYMMETRY	
System	Symmetry	Crystal Axes[1]
Triclinic	1 one-fold axis or center of symmetry	$a \neq b \neq c, \alpha \neq \beta \neq \gamma \neq 90°$
Monoclinic	1 two-fold axis or 1 symmetry plane	$a \neq b \neq c, \alpha = \gamma = 90°, \beta \neq 90°$
Orthorhombic	3 two-fold axes or 3 symmetry planes	$a \neq b \neq c, \alpha = \beta = \gamma = 90°$

[1]a, b, and c describe the lengths of the crystal axes; α is the angle between b and c; β is the angle between a and c; γ is the angle between a and b.

monoclinic (one mirror plane), and orthorhombic (three mutually perpendicular planes with intersections of unequal length). Any introductory mineralogy textbook will refresh your memory on crystal symmetry and in Table 12.3 we list the characteristics of these systems that are most pertinent to our discussion.

Let's explore the **Symmetry Principle** (or Curie principle[5]) with a simple example (Figure 12.20). Imagine that we apply strain to a cube such that the resulting shape is a rectangular block. The symmetry of the rectangular block is orthorhombic (three perpendicular symmetry planes or three perpendicular two-fold axes; Figure 12.20a). The symmetry of the simplest strain path causing this shape change is also orthorhombic, meaning that the incremental and finite

strain ellipsoids differ only in shape, not in orientation (i.e., coaxial strain). We restate this by distinguishing the cause (the strain path) and the effect (the rectangular block): the effect has a symmetry that is equal to or greater than the cause. This is called the Symmetry Principle. Now we distort our cube, say by shear, such that it becomes a block with a lower symmetry (Figure 12.20b). In this case the resulting symmetry is monoclinic, because we can only recognize one mirror plane and one two-fold axis. Using the Symmetry Principle we therefore predict that the strain path (meaning, the cause) must have monoclinic or lower symmetry. Thus, the strain that caused the distortion must have been non-coaxial, as coaxial strain would have produced higher symmetry.

If the Symmetry Principle says that the effect is of equal or greater symmetry than the cause, is the reverse also true? No, the cause cannot have a symmetry that is

[5]After the French scientist Pierre Curie (1859–1906).

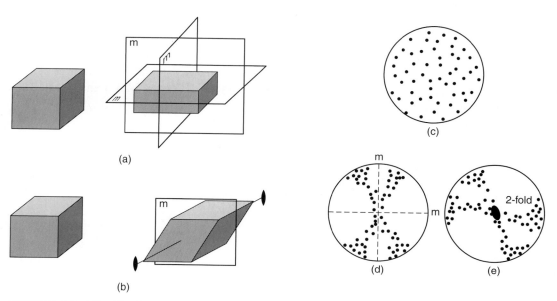

FIGURE 12.20 The Symmetry Principle. A cube is distorted into a body with (a) orthorhombic and (b) monoclinic symmetry. (c) A random distribution of c-axis pattern reorganizes in response to these deformations into (d) a high-symmetry (orthorhombic) and (e) a low-symmetry (monoclinic) pattern. The symmetry of the c-axes patterns enables us to predict the strain path.

higher than the effect. This is all fine and good, but what does it have to do with crystallographic-preferred fabrics? Figure 12.20c shows a random pattern of crystallographic c-axes in lower hemisphere projection, say from a statically recrystallized quartzite. In two separate deformation experiments we form characteristic c-axis patterns (Figure 12.20d and e). The symmetry of the two patterns is different; the pattern in Figure 12.20d has orthorhombic symmetry, whereas the pattern in Figure 12.20e is monoclinic. What can we say about the strain that produced these crystallographic patterns? Using the Symmetry Principle, the pattern in Figure 12.20d must have been caused by a strain path with a symmetry equal to or less than orthorhombic, whereas the symmetry of the strain path that produced the pattern in Figure 12.20e must have been monoclinic or lower. Thus, Figure 12.20e ("the effect") can only have been formed by non-coaxial strain ("the cause"). Note that the pattern in Figure 12.20d, however, cannot be uniquely interpreted; it could have formed by either coaxial or non-coaxial strain according to the Symmetry Principle. With this information we can use textures to indicate whether the rocks were deformed in a non-coaxial or coaxial strain regime. Perhaps more importantly, the Symmetry Principle provides a means to determine shear sense in rocks.

12.5.2 Textures as Shear-Sense Indicators

Imagine an aggregate of grains that each slip on the same single glide plane, but that initially is randomly oriented in our sample (Figure 12.21). This situation is slightly more complex than our previous single-grain condition (Figure 12.19), because neighboring grains will affect the ability of grains to slip.[6] When shearing the aggregate, a pattern emerges in which the majority of c-axes rotate toward an orientation perpendicular to

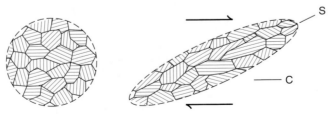

FIGURE 12.21 The relationship between shape, crystallographic fabric, mylonitic foliation (S) and shear plane (C) in a grain aggregate with a single operative glide plane for each grain.

---

[6]This is called the *compatibility problem*.

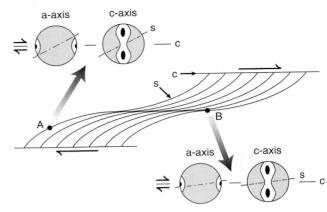

FIGURE 12.22 Schematic illustration of foliation in shear zone and associated crystallographic fabrics. The angular relationship between S and C decreases with increasing shear strain (as recorded by S-foliation deflection), producing characteristic c-axis patterns (compare regions A and B). The corresponding a-axis patterns show no change.

the bulk shear plane (C in Figure 12.21). At the same time a dimensional-preferred fabric is formed that defines the mylonitic foliation (S-foliation), which is oblique to the shear zone boundary (C-surface). Thus, the c-axis fabric is oblique to the S-foliation in a sense similar to that of the shear sense in the zone (region A in Figure 12.22); in other words, the c-axis girdle tilts in the direction of shear. With increasing shear, the obliquity is less because the S and C surfaces approach parallelism, which means that in the center of the shear zone the interpretation of textures is restricted (region B in Figure 12.22). The solution to this situation is measurement of a traverse across the shear zone.

The natural c-axis fabric for quartz shown in Figure 12.23b is consistent with our model. But is this pattern a rule that can be applied to all minerals? Unfortunately, the answer is no. Look at the calcite c-axis fabric (Figure 12.23a), in which slip occurred on

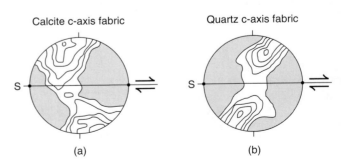

(a)   (b)

FIGURE 12.23 Asymmetric c-axis fabrics for (a) calcite and (b) quartz from natural shear zones, showing contrasting patterns resulting from the operation of different glide planes. In quartz, basal slip occurred, whereas e-twinning dominated calcite deformation. Note that the mylonitic foliation (S) is used for reference (compare with Figure 12.22).

the e-plane of calcite (calcite twinning; Table 9.1). The displacement is again right-lateral, but here the c-axes lie in the opposite quadrant from those of the quartz fabric in Figure 12.23b. This is not a paradox, but reflects the operative slip system. Quartz deformed predominantly by slip on the basal plane [(0001)-plane], whereas calcite slip occurred by e-twinning. Calcite crystallographic fabrics that are formed by slip on other glide planes will, in fact, produce c-axis patterns that are similar to that for quartz. These observations from natural rocks highlight the importance of knowing the operative slip system before crystallographic patterns can be used as a shear-sense indicator. Determining these patterns requires complementary electron microscopy and measuring the orientation of other crystallographic axes in the same sample (such as the a-axis; Figure 12.22).

The measurement of crystallographic fabrics requires careful preparation of oriented thin sections and tedious measurement of crystallographic orientations with an optical microscope that is equipped with a universal stage or optical sensors. Other techniques, such as scanning electron microscopy, and X-ray and neutron-source methods, are rapidly becoming available for more automated analysis. Crystallographic fabric analysis is most commonly applied to monomineralic rocks consisting of relatively simple minerals such as quartz, calcite, and olivine, but more complex minerals, such as feldspar, can also be studied. For each grain the orientation of a particular crystallographic axis is measured and plotted in spherical projection. Typically, the orientation of the mylonitic foliation (S) or the shear zone boundary (C) is taken as E-W (East–West) and vertical; it is critical for the interpretations to label these foliations carefully (recall their relationship as a function of shear sense; see Figure 12.11). A reliable crystallographic-preferred fabric measurement involves at least 100–150 grains. A crude estimate of the degree of crystallographic-preferred orientation may be obtained with the optical microscope by inserting the gypsum plate (1/4-wavelength) under crossed polarizers and slowly rotating the sample. If there is an obvious fabric, the color of most grains will change at the same time; for example, from blue to yellow in quartz. Otherwise, you will see an irregular pattern of colors that is a mix of red (the extinction color), yellow, and blue.

A final note: Crystallographic fabrics mainly reflect the *latest stages* of the deformation history, because recrystallization processes tend to erase the early history. Nonetheless, crystallographic fabrics offer a valuable tool to unravel the deformation history of rocks and regions, as you will see by looking at the modern structural geology literature.

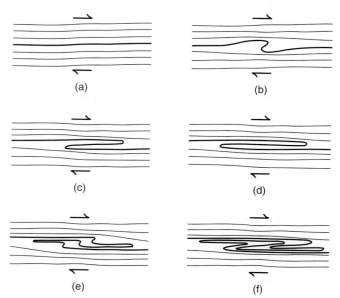

FIGURE 12.24 Fold transposition in a layered rock that undergoes non-coaxial, layer-parallel displacement. An asymmetric fold develops at a perturbation (a–d), which in turn gets refolded (e–f). The end result is a superimposed fold structure that is essentially parallel to the dominant layering. Note that if the second fold had formed at another location, the two would have been indistinguishable, although formed at different times.

12.6 FOLD TRANSPOSITION

At first glance, folds appear to be less common in ductile shear zones than one would expect; but upon closer inspection they are found to occur with a peculiar appearance. In Figure 12.24 an asymmetric fold is formed from a small instability by foliation-parallel shear. With increasing shear, the oblique (short) limb of the asymmetric fold rotates back into a foliation-parallel orientation (Figure 12.24a–d). The resulting perturbation gives rise to a new fold that is superimposed on the original structure. Continued shear reorients the fold pattern back into a layer-parallel orientation (Figure 12.24e–f), leaving behind a complex geometry that can only be interpreted by carefully tracing the layering (Figure 12.25). This scenario, called **fold transposition,**[7] highlights two aspects of folds in shear zones. (1) Fold asymmetry may be representative for the sense of shear; that is, Z-vergence occurs in right-lateral shear zones while S-vergence occurs in left-lateral shear zones. We state this only as a possibility and not as a rule, because at high shear strains the vergence of small folds may actually reverse

[7]Originally described as "Umfaltung" by the early-twentieth-century German geologist Bruno Sander, who pioneered modern structural analysis.

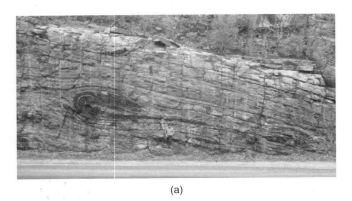

(a)

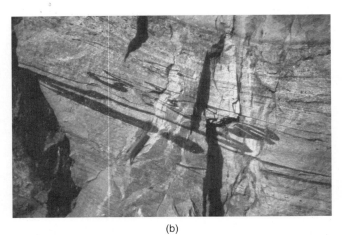

(b)

FIGURE 12.25 Transposed mafic layer in a granitic gneiss (Grenville Orogen, Ontario, Canada). The mafic (dark) layer can be traced as a single bed that is refolded numerous times in the outcrop (a). A detail of the large structure, which is affectionately known as the "snake outcrop," highlights the complexity of folding; person for scale.

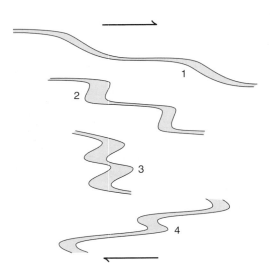

FIGURE 12.26 Schematic illustration of reversal in fold vergence (from "S-shape" to "Z-shape"), with increasing shear strain in a right-lateral shear zone.

(Figure 12.26). (2) Folding is a progressive process, resulting in complex patterns of folding and refolding. Fold transposition occurs at all scales, from microfolds to kilometer-scale folds, and is, in fact, not restricted to zones of non-coaxial strain.

Folds in areas of high strain are often disrupted, preserving only isolated fold hinges or fold hooks (so called because of their resemblance to fishing hooks). Consider shortening that occurs parallel to competent layers in a less competent matrix, say sandstone layers in a shale matrix. The result is folds with an upright axial surface (Figure 12.27a); the **fold enveloping surface** is drawn for reference. Progressive shortening produces thinning of limbs and, locally, hinges become detached (Figure 12.27b; a natural example of this stage is shown in 12.27d). Eventually, we are left with a rock texture that is characterized by a single layering with lenses of more competent material (Figure 12.27c). Without knowledge of the history, the fabric in Figure 12.27c shows little or no indication of the original geometry, especially if the structures are very large (say kilometers). A cursory examination would suggest that overall bedding is up-down, until you notice the presence of the small fold hooks and maybe minor isoclinal folds. Only then do you realize that fold transposition has occurred. Transposed fabrics are even more pronounced when an axial-plane foliation is formed, which will be nearly parallel to transposed layering (Figure 12.27c). As a consequence, pretty much all layering in the rock seems parallel, with the exception of preserved fold hinges. Transposition in non-coaxial shear zones similarly results in a geometry in which all layering is roughly parallel to the shear-zone boundary. Try to sketch the evolution of folds in a non-coaxial environment, using Figure 12.26 as a starting point, and track the orientation of the enveloping surface for comparison with the coaxial strain scenario in Figure 12.27.

Rather than being the exception, transposition is more likely the rule in deformed metamorphic areas, where it is promoted by the overall weakness of layers at elevated pressures and temperatures (such as the rocks in Figure 12.25). In addition to unraveling the deformation history, identifying transposition is important if you wish to erect a regional stratigraphy for an area, because it poses the question of whether repetition of a particular lithology is primary or structural in origin. So, are there *criteria* to recognize transposition? One clue is a regular repetition of lithologies; for example, repeated marble layers in an area of mafic gneisses may, on a regional scale, outline large transposed folds. When a sequence of lithologies exists, say gabbro-gneiss-marble, its repetition in reversing order is suggestive of transposition. Parallelism between foliation and bedding is not diagnostic by itself, but it should

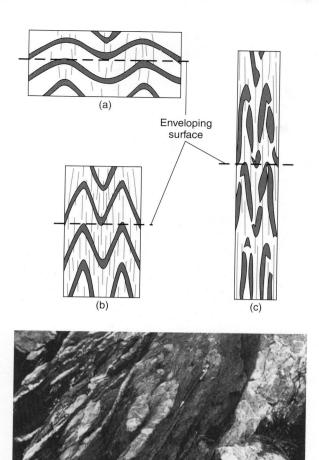

FIGURE 12.27 Fold transposition. Progressive coaxial strain (a–c) results in isoclinal folds that become disrupted by limb attenuation and boudinage. The fold enveloping surface is shown for reference. (d) A natural example of an early stage of transposition (Newfoundland, Canada); pencil for scale.

make you suspicious. The occurrence of minor isoclinal folds and fold hooks (especially if the foliation is axial-planar to these structures) adds weight to the case. You wrap it up by showing reversals in the direction of younging in layers across the area changes. Unfortunately, younging evidence is hard to find in medium- to high-grade metamorphic areas, but in low-grade metamorphic areas (greenschist facies and down) the observant geologist often finds evidence for younging. Your efforts to prove a case of transposition will bear fruit. Transposition results in a radical reinterpretation of stratigraphic relationships by showing that "stratigraphy" in the area is, in fact, structural in origin, and that the real stratigraphy is simpler. Secondly, a previously unrecognized early stage of isoclinal folding has major implications for the structural evolution of the area. In the section on regional geology (Chapters 21 and 22) you will see that exhumed rocks in many mountain

belts show early stages of large-scale isoclinal folding, typically associated with thrusting, which is manifested in outcrop by fold transposition.

12.6.1 Sheath Folds

In closing this chapter, we describe one particular type of fold that is restricted to regions of high shear: **sheath folds.** In contrast to the ambiguity of fold vergence in shear zones already described (Figure 12.26), sheath folds can define shear-sense in ductile shear zones. They are typically exposed as eye-shaped outcrop patterns that represent a section through the nose of the fold (Figure 12.28). Sheath folds are a special type of doubly plunging folds (see Chapter 10), where the hinge line is bent around by as much as 180°. In other words, layering in a sheath fold is everywhere at a high angle to the profile plane, which gives the characteristic eye-shaped outcrop pattern. Sheath folds are formed when the hinge line of a fold rotates passively into the direction of shear, while the axial surface rotates toward the shear plane (Figure 12.29a). Note that sheath folds differ from "dome-and-basin" type fold interference patterns, because they form by progressive strain under the same overall deformation regime, rather than reflecting two discrete deformation regimes (see superposed folding; Section 10.6.2). Since hinge rotations require high amounts of shear strain, the occurrence of sheath folds is limited to ductile shear zones. The degree of hinge rotation is a function of shear strain and the angular relationship of hinge line with the shear plane. However, sheath folds cannot be used as a strain gauge beyond the general comment of "large amount of shear," unless these initial angular relationships are known.

The location of the nose of sheath folds points in the direction of movement, but this can be determined only when the folds are fully exposed (including the nose!). In less-rotated portions of a sheath fold, the shear sense may also be derived from hinge line rotation (Figure 12.29a).[8] Most commonly, sheath folds define the direction of shear rather than shear sense, with the hinge line approximately parallel to the shear direction.

12.7 CLOSING REMARKS

It is fairly certain that you will encounter shear zones (and transposition) on a field trip or in your study area; you may already have done so. There is no better way to learn about the variety of deformation

---

[8]Similar to the Hansen slip line method.

FIGURE 12.28 A sheath fold in amphibolite gneiss showing the characteristic eye-shaped outcrop pattern (Grenville Orogen, Ontario, Canada); hammer for scale.

features in shear zones than by direct observation. In many instances you will be unable to convince yourself of important concepts, such as sense of shear or "stratigraphy," unless you allow ample time for outcrop examination. As a rule, it is more useful to understand one outcrop well, than many outcrops in a cursory fashion; one single outcrop may, in fact, hold the key to the interpretation of an entire area. Unfortunately, it is another rule that this special outcrop is invariably to be found in the most remote part of the area, on the highest peak, and on the rainiest day. Until you find that Rosetta Stone[9] of structural analysis, any well-studied outcrop offers you a working hypothesis that can be tested in other places. Only later, after many outcrops, will you learn which of them is your area's "Rosetta Stone."

Once mylonitization begins (protomylonite stage), the resulting microstructures further weaken the zone relative to the host rock. Thus, strain will continue to be localized in a zone and the mylonite will evolve. A variety of potential shear-sense indicators are

[9]This slab of basalt, discovered in the late eighteenth century, enabled linguists to decipher Egyptian hieroglyphs because it contained the same text in three languages, one of which was Greek.

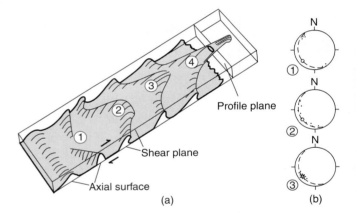

FIGURE 12.29 Sheath folds are formed when weakly doubly-plunging folds are modified in a zone of high shear strain. Several stages of fold modification occur, with the most evolved stage producing the characteristic conical geometry of a sheath fold (fold 4). The progression in (a) has the lowest amount of shear at the left and the highest shear strain at the right. Note the progressive curvature of the hinge line in folds 1 to 3, which typically is accompanied by a (stretching) lineation in outcrop. In (b), the corresponding lower-hemisphere projections of hinge lines in folds 1, 2, and 3 are shown. The great circle represents the shear plane and the open circle the shear direction; small dots are hinge-line measurements. When fully exposed, sheath folds can be used as a shear-sense indicator with the nose pointing in the direction of transport; otherwise, sheath folds indicate the direction of shear only.

available for ductile shear zones, but their kinematic significance is not fully understood. Most of these indicators are based on empirical observations, and one should not interpret shear sense based on a single indicator in one location only, no matter how classic an example it is. Reliable predictions are based on several independent indicators, which each support the same shear sense. Even then, caution should be exercised with tectonic interpretations, because a shear zone may be subsidiary to a larger system, so that the shear sense is only representative of the zone in question (remember the analogous situation with changing fold vergence in a large fold; Section 10.4.2). After addressing all caveats, you will find that ductile shears will deliver on their promise. To end this sequence of chapters on ductile deformation, we close with some tools from related Earth science disciplines that further aid the modern structural geologist in unraveling the history of deformed rocks and regions.

ADDITIONAL READING

Fossen, H., and Tikoff, B., 1993. The deformation matrix for simultaneous simple shearing, pure shearing and volume change, and its application to transpression and transtension tectonics. *Journal of Structural Geology,* 15, 413–422.

Hanmer, S., and Passchier, C., 1991. Shear-sense indicators: a review. *Geological Survey of Canada,* Paper 90–17, 72 pp.

Law, R. D., 1990. Crystallographic fabrics: a selective review of their applications to research in structural geology. In Knipe, R. J., and Rutter, E. H., eds., *Deformation mechanisms, rheology and tectonics. Geological Society Special Publication,* 54, 335–352.

Lister, G. S., and Snoke, A. W., 1984. S-C mylonites. *Journal of Structural Geology,* 6, 617–638.

Passchier, C. W., and Simpson, C., 1986. Porphyroclast systems as kinematic indicators. *Journal of Structural Geology,* 8, 831–843.

Ramsay, J. G., and Graham, R. H., 1970. Strain variation in shear belts. *Canadian Journal of Earth Sciences,* 7, 786–813.

Ramsay, J. G., and Huber, M. I., 1983/1987. *The techniques of modern structural geology, volumes 1 and 2.* Academic Press, 700 pp.

Sibson, R. H., 1977. Fault rocks and fault mechanisms. *Journal of the Geological Society of London,* 133, 191–213.

Simpson, C., 1986. Determination of movement sense in mylonites. *Journal of Geological Education,* 34, 246–261.

Simpson, C., and De Paor, D. G., 1993. Strain and kinematic analysis in general shear zones. *Journal of Structural Geology,* 15, 1–20.

CHAPTER THIRTEEN

Deformation, Metamorphism, and Time

13.1 INTRODUCTION

Unraveling the deformation history of a region is somewhat like a "whodunit" mystery. Well, it perhaps isn't quite as exciting as some of the stuff in the movies, but a mystery it is. Just like any modern detective you need tools in the forensics lab to solve the geologic mystery. While it is impossible to be expert in all of today's methods (ranging from materials science to chemistry), a basic knowledge is required to help you decide what may be useful for your particular case. In this essay we look at some approaches, mainly from the fields of metamorphic petrology and isotope geochemistry, that are integral parts of many studies of deformed regions. We concentrate on medium- to high-grade metamorphic areas, which are representative of processes that are active in the deeper levels of the crust, to complement the emphasis on shallow-level deformation earlier in the book. We use a hypothetical area with hypothetical (i.e., perfect?) rocks,[1] in order to show the complementary nature of other approaches to "traditional" structural analysis. As we go along, you will find that this chapter remains far from comprehensive; hence we call it an essay.

First we need to collect, organize, and examine the evidence, and then chart our course of action. Some of the associated terminology and concepts that will follow are listed in Table 13.1 for reference.

13.2 FIELD OBSERVATIONS AND STUDY GOALS

We are studying the tectonic evolution of a metamorphic terrane containing rocks in greenschist to granulite facies. Our first field summer was used to map the lithologies and structure of the area. Now that we have a basic understanding of the structural geometry (i.e., locations of shear zones, nature of fabrics, and so on), we select samples for laboratory analysis (Figure 13.1a). The goal of our study is to determine the spatial and temporal conditions at which these rocks were deformed. Specifically, we try to determine the crustal

---

[1]While based on real geographic names and natural rocks, any resemblance to reality is purely coincidental.

TABLE 13.1	SOME TERMS AND CONCEPTS RELATED TO DEFORMATION, METAMORPHISM, AND TIME
(Geo)barometry	Determination of pressure during metamorphism.
(Geo)chronology	Isotopic age determination of a mineral or rock; these ages are given in years. Sometimes called "absolute" age.
Clockwise *P-T-t path*	Burial history plotted in *P-T* space of rapidly increasing temperature relative to pressure followed by uplift history of rapidly decreasing pressure relative to temperature. *Note:* Clockwise is defined in a *P-T* plot with *P* increasing downward; petrologists use the opposite terminology because they plot *P* increasing upward.
Closure temperature	Temperature at which a system becomes closed to loss of the radiogenic daughter isotope.
Cooling (rate)	Temperature decrease (with time).
Counterclockwise *P-T-t* path	Burial history plotted in *P-T* of rapidly increasing pressure relative to temperature followed by uplift history of rapidly decreasing temperature relative to pressure. Counterclockwise is defined with *P* increasing downward.
Exhumation (rate)	Displacement relative to the surface (with time); also *denudation, unroofing*.
Geothermal gradient	Temperature change per depth unit (typically °C/km or K/km).
P-T-t path	History of rock or region in pressure(*P*)-temperature(*T*)-time(*t*) space.
Paragenesis	Mineral assemblage in a metamorphic rock.
Peak metamorphism	Condition of peak-temperature and corresponding pressure. Note that peak temperature generally does *not* coincide with peak pressure (and vice versa).
Porphyroblastesis	Relative timing of mineral growth with respect to deformation (strictly: mineral growth).
Postkinematic growth	Metamorphic mineral growth after deformation.
Prekinematic growth	Metamorphic mineral growth before deformation.
Prograde metamorphism	Metamorphic history before peak-temperature, characterized by dehydration reactions and density increase.
Retrograde metamorphism	Metamorphic history after peak-temperature condition, which may involve hydration reactions in the presence of an H_2O-rich fluid.
Synkinematic growth	Metamorphic mineral growth during deformation.
Tectonite	General term for a deformed rock; the adjectives L, S, and L-S indicate dominance of a lineation, a foliation, or a combination, respectively.
(Geo)thermobarometry	Temperature–pressure determination using equilibrium reactions.
(Geo)thermochronology	Temperature–time history of rocks.
(Geo)thermometry	Determination of metamorphic temperature.
Uplift (rate)	Displacement relative to a fixed reference frame, such as the geoid (with time).
(Mineral) zonation	Presence of compositionally distinct regions in a mineral grain; this records changing conditions of chemical equilibrium during growth or subsequent modification at times of chemical disequilibrium. Typically zonation is found from core to rim in such minerals as garnet and feldspar.

FIGURE 13.1 Field photograph of mylonitic gneiss used in our analysis and corresponding lower hemisphere, equal-area projection of spatial data (inset).

depth at which these rocks were deformed, the timing of metamorphic and deformational events, and their burial and exhumation history in the context of tectonic evolution.

Our rocks are called *paragneiss* and *marble,* which are metamorphic rocks of sedimentary origin (as opposed to *orthogneiss,* which is a metamorphic rock of igneous origin). Prior to removing samples, we characterize the outcrop as a whole and carefully orient our specimens by marking dip and strike or dip direction on the samples. These **tectonites** have several field characteristics. Most obvious is a well-developed foliation that is defined by compositional variation or color banding, and a lineation that is defined by oriented hornblendes in the paragneiss and black graphitic stripes in the marble. The orientation of these fabric elements in outcrop is shallowly SE-dipping and SE-plunging, respectively. Upon detailed inspection of the gneiss, we see a fine matrix that surrounds millimeter-size grains with distinct shapes and textures. In particular, the asymmetry of tails on feldspar grains in the gneiss draws our attention, because it indicates a left-lateral sense of displacement along the foliation when looking to the northeast

(δ-type clasts; Section 12.3.2). Based on the presence of a strongly foliated and lineated fabric, fine grain size of the rocks relative to neighboring outcrops, and the occurrence of shear-sense indicators, we conclude that the rock is a mylonite; that is, the rock was subjected to significant shear strain and deformed by crystal-plastic processes. We plot the spatial information in spherical projection (Figure 13.1 inset) and, based on the collective information, we conclude that the outcrop is part of a low-angle, SE-dipping shear zone with a reverse sense of displacement.

The paragneiss contains the minerals hornblende, garnet, quartz, feldspar, biotite, muscovite, and kyanite, and accessory phases that cannot be conclusively identified with the hand lens. This mineral assemblage suggests that the rock was metamorphosed in the amphibolite facies, indicating a temperature range of 500°C to 700°C and a pressure range of 400–1200 MPa (Figure 13.2). Staining of the marble with a solution of HCl and red dye indicates the presence of dolomite in addition to calcite; the main accessory phase is graphite. The assemblage of minerals in the marble does not allow us to determine the metamorphic grade in the field, but the coexistence of calcite and dolomite will be useful in the

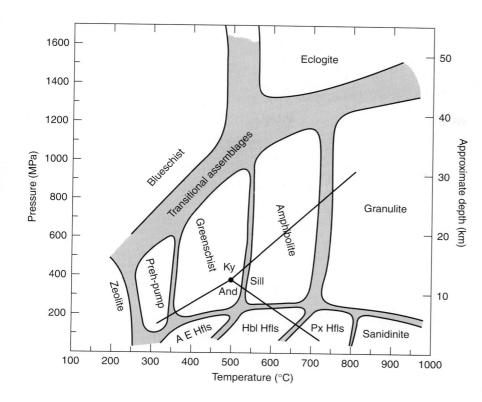

FIGURE 13.2 Pressure-temperature diagram showing the fields of the various metamorphic facies and the kyanite (Ky)-andalusite (And)-sillimanite (Sill) triple point. The approximate crustal depths that correspond to lithostatic pressures are also shown. Abbreviations used are: A-E = albite-epidote, Hbl = hornblende, Hfls = hornfels, Preh-Pump = prehnite-pumpellyite, Px = pyroxene.

laboratory as a geothermometer. After taking some photographs of the scene and making some sketches of geometric relationships, the samples are removed from the outcrop, labeled, and taken to the laboratory for further analysis.

Upon returning to the bug-free environment of the laboratory, we cut oriented samples and prepare two mutually perpendicular thin sections. The sections are oriented such that one is perpendicular to foliation and parallel to lineation, and the other is perpendicular to both foliation and lineation. These two surfaces provide a three-dimensional description of the microscopic characteristics of each sample. Thin sections of the gneiss reveal that our field conclusions about metamorphic grade were generally correct. The main minerals in the rock are plagioclase and alkali feldspar, quartz, hornblende, biotite, muscovite, garnet, and kyanite, and the accessory phases are rutile, ilmenite, and titanite (also called sphene). The foliation is defined by alternating light and dark bands, whose color is a consequence of the relative proportion of light-colored quartz and feldspar, and dark-colored hornblende and biotite in these layers. The marble consists of calcite, dolomite, and minor amounts of graphite, monazite, and titanite. The foliation in the marble is defined by alternating layers of different grain size and by concentrations of opaques (mainly graphite) and other accessory phases.

13.3 PRESSURE AND TEMPERATURE

We first constrain the metamorphic conditions of our samples. **Geothermobarometry** is the quantification of temperature (T) and pressure (P) conditions to which a rock was subjected during its metamorphic history. It contributes three very important pieces to our geologic puzzle: (1) the peak temperature condition, (2) the approximate depth to which the rocks were buried (from pressure determination), and (3) (part of) the prograde and retrograde history of the rock. We will look at each piece in turn.

Experimental and thermodynamic work in metamorphic petrology, supported by empirical observations, have identified a large number of mineral reactions that can be used to estimate the P and T conditions of metamorphism. In making such estimates, it is assumed that the mineral assemblage or parts of individual minerals represent *equilibrium conditions,* which means that at a given condition of pressure and temperature certain minerals have specific compositions. Mineral assemblages make good geobarometers if the equilibrium reaction is relatively insensitive to temperature (Figure 13.3). Similarly, a good geothermometer is largely pressure independent (Figure 13.3); thermodynamically, this means there is

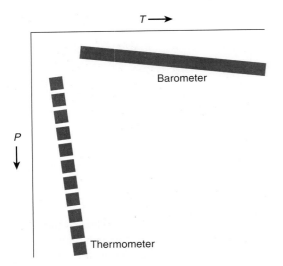

FIGURE 13.3 Reliable geobarometers (solid band) are largely pressure insensitive, whereas good geothermometers (dashed band) are largely temperature independent; they are schematically plotted in *P-T* space.

<table>
<tr><td colspan="2">**TABLE 13.2**   **SELECTED GEOBAROMETERS AND GEOTHERMOMETERS**</td></tr>
</table>

Geobarometers:

Garnet-plagioclase [reactions involving Ca exchange between garnet and plagioclase, for example, garnet-plagioclase-quartz-Al_2SiO_5 (GASP); garnet-plagioclase-rutile-ilmenite-quartz (GRIPS); garnet-rutile-ilmenite-Al_2SiO_5-quartz (GRAIL)]

Pyroxene-plagioclase-quartz (Na or Ca exchange)

Phengite (phengite content in muscovite)

Hornblende (Al content)

Geothermometers:

Garnet-biotite (Fe-Mg exchange)

Garnet-pyroxene (Fe-Mg exchange)

Two-feldspar (Na exchange between alkali feldspar and plagioclase)

Two-pyroxene (Ca-Mg exchange)

Calcite-dolomite (Mg exchange)

Calcite-graphite (^{13}C isotopic fractionation)

In part after Essene (1989).

a large entropy[2] change, but little volume change in geothermometers. For example, consider the following reaction involving the exchange of ions:

$$A(i) + B(j) = A(j) + B(i) \qquad \text{Eq. 13.1}$$

where A and B are two minerals in equilibrium (e.g., garnet and biotite) and i and j are two cations (e.g., Fe^{2+} and Mg). Because only cation exchange occurs in this reaction, there is little or no change in volume involved, and consequently these types of reactions are pressure insensitive.

Now let's look at a geobarometer. The mineral garnet is common in amphibolite facies rocks, and is formed when the rocks contain sufficient amounts of the respective elemental constituents of that mineral. An example of a common garnet-forming reaction in metasediments is

$$3\ CaAl_2Si_2O_8 = Ca_3Al_2Si_3O_{12} + $$
$$2\ Al_2SiO_5 + SiO_2 \qquad \text{Eq. 13.2}$$

plagioclase (anorthite) = garnet (grossular) + aluminosilicate (e.g., kyanite) + quartz

Garnet is formed by the breakdown of plagioclase with increasing pressure. This particular reaction (also

known by the acronym GASP[3]) is a geobarometer because it involves significant volume change, but is relatively insensitive to temperature. Thus, in a plot of pressure versus temperature (Figure 13.3), we find that lines with a shallow slope are good barometers and those with steep slopes are good geothermometers. Intermediate slopes indicate sensitivity to both pressure and temperature (one reaction, two unknowns), so the reaction can be used only if one parameter is determined independently.

A single mineral reaction describes a line in the *P-T* diagram; for example the kyanite-sillimanite boundary in Figure 13.2. Two mineral reactions define a single point in a *P-T* diagram, provided that the lines intersect; the greater the angle between the two intersecting lines the more reliable the *P-T* estimate. Examples of some widely used geothermometers and geobarometers are listed in Table 13.2; for details of these systems

---

[2]Entropy, *S,* is the thermodynamic parameter describing the degree of randomness (or disorder) in a system.

[3](G)arnet + (A)luminosilicate + (S)ilica = (P)lagioclase (silica is quartz).

and their applications you are referred to an extensive petrologic literature (see "Additional Reading").

In theory, the application of thermobarometric systems should give unique, internally consistent values for pressure and for temperature, but in practice this is rarely the case. Nevertheless, with multiple reactions you are generally able to constrain the pressure conditions within ±100 MPa (i.e., ±1 kbar) and temperature conditions within ±50°C. In all our future results we assume these practical error estimates.

After considering the mineral assemblage in our samples, we decide to apply the GASP and GRIPS geobarometers, and two-feldspar and garnet-biotite geothermometers to the paragneiss. Recall that the most reliable results are obtained when more than one system is used to determine P or T. The marble from the same outcrop does not contain an assemblage that allows pressure determination, but the calcite-dolomite and calcite-graphite systems are good geothermometers. After many hours in a darkened microprobe room and some stable isotopic work using a mass spectrometer (neither of which we describe here), we obtain an internally consistent pressure of 700 MPa (7 kbar) and a temperature of 600°C for our samples. These results place the rocks firmly in the amphibolite facies, just above the kyanite-sillimanite boundary (Figure 13.2).

What do these values of temperature (T) and pressure (P) really mean? In theory, thermobarometry measures the conditions at the time of mineral formation or when chemical exchange between phases ceases to occur. If the mineral formed during **prograde metamorphism** (i.e., there is a history of increasing metamorphic temperature), we assume that the temperature represents the peak temperature of metamorphism, because at these conditions the rate of mineral growth is highest. However, the chemical composition of minerals formed at peak temperature conditions may be modified during cooling of the rock, which is called **retrograde metamorphism.** Retrograde metamorphism may be preserved in zoned minerals, where only part of the mineral reequilibrates, or it may lead to complete compositional resetting. The latter is like erasing with one hand what you write on a board with the other. The rock achieves equilibrium for certain temperature conditions until this thermally activated process is too slow to allow further exchange.

The pressures that we derive from geobarometry do not describe the stress state that caused deformation of the rock. This is an important point. Rather, metamorphic pressures are lithostatic pressures and measure the weight of the overlying rock column (recall

Chapter 3). So, if we know the density of the rock column, we can translate metamorphic pressures into values for depth. For every kilometer of depth in the average crust the pressure increases by about 27 MPa (assuming an average rock density of 2700 kg/m³; Chapter 3). Thus, the previously determinate pressure estimate of 700 MPa for our sample implies that these rocks were buried to a depth of about 26 km.

Zoned minerals are particularly useful for deciphering P-T history, because zonation records changes in conditions during prograde and/or retrograde metamorphism. Often the rims of minerals reset during retrograde P-T conditions. Zonation is sometimes visible in the optical microscope, but is usually determined using an electron microprobe, which recognizes small compositional differences. Mineral zonation stems from the incomplete equilibration of previously formed minerals. Garnets, which are particularly prone to zonation, are relatively fast-growing minerals that act as a memory of the metamorphic history. Figure 13.4 shows an example of a large garnet that preserves a complex history of growth, based on the irregular Ca pattern, and retrograde exchange, established from the decreasing Mg/Fe ratio (Mg decreases and Fe increases) and increasing Mn content toward the rim. From our garnet sample we are able to determine pressure and temperature for both core and rim of the mineral. The analysis of garnet cores and inclusions were given earlier (700 MPa, 600°C) and we obtain rim values of 500 MPa and 550°C.

13.3.1 Status Report I

What have we learned so far? After the gneiss and marble were deposited at the surface (they are metasediments!), they were buried by about 25 km where they became exposed to a peak temperature of 600°C ("core" P-T). Today these rocks are back at the surface, because we were able to sample them, and indeed they preserve a record of lower temperatures and pressures, during their ascent, as the "rim" P-T data (550°C at ~19 km depth). There is no record of shallower conditions because the kinetics (or rate) of the geothermobarometric reactions was too low. Thus, we are beginning to unravel a detailed burial and exhumation history, or the **P-T-t path,** of our rocks, which we eventually integrate with the structural evolution of our area. But before we can do this, we need to establish absolute and relative temporal relationships of deformation and metamorphism.

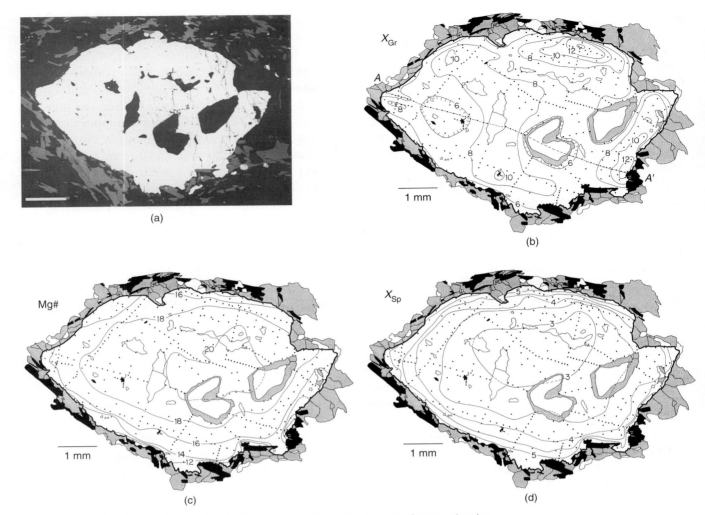

FIGURE 13.4 Complexly zoned, inclusion-rich garnet from the Grenville Orogen, showing (a) back-scattered electron image, and (b) Ca [$X_{Gr} = 100 \times Ca/(Ca + Mn + Mg + Fe)$], (c) Mg# [$= 100 \times Mg/(Mg + Fe)$], and (d) Mn [$X_{Sp} = 100 \times Mn/(Ca + Mn + Fe)$] composition based on detailed electron microprobe analysis. Each dot in garnet represents an analytical point; black is biotite, gray is plagioclase, and white is quartz. Scale bar is 1 mm; zonation is given in mol%.

13.4 DEFORMATION AND METAMORPHISM

To understand the conditions during deformation, we must determine the relative timing of metamorphism with respect to deformation. **Prekinematic growth** means that mineral growth occurs before deformation, **synkinematic growth** means that growth occurs during deformation, and **postkinematic growth,** you guessed it, implies that minerals grow after deformation. The shape and internal geometry of minerals and their relationship to external fabric elements, in particular foliations, help us to determine this relative temporal relationship. Such newly grown minerals are

called **porphyroblasts,** derived from the Greek word "blastos" for growth, and the relationship between growth and deformation is called **porphyroblastesis.**

The beautiful crystals that decorate geodes may come to mind when thinking of mineral growth, but rocks at depths beyond a few kilometers rarely have any open spaces. Mineral growth under these conditions mostly occurs by the replacement of preexisting phases. For example, garnet grows by the consumption of another mineral, and phases that are not involved in the reaction (accessory phases such as zircon, monazite), or phases that are left over because the rock does not contain the right mix of ingredients, may become inclusions. These inclusions may form

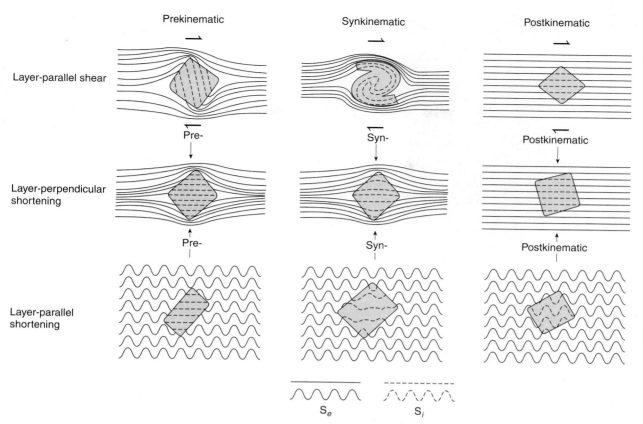

FIGURE 13.5 Schematic diagram showing the diagnostic forms of porphyroblasts that grow before (prekinematic), during (synkinematic), and after (postkinematic) layer-parallel shear, layer-perpendicular shortening, and layer-parallel shortening. The temporal relationship is based largely on the relationship between the internal foliation (S_i; dashed lines) and the external foliation (S_e; solid lines).

ordered trails that define an **internal foliation** (S_i) in the porphyroblast; for contrast, the foliation outside the blast is called the **external foliation** (S_e). The relationship between S_i and S_e allows us to determine the relative timing of mineral growth and deformation (Figure 13.5). Prekinematic growth is characterized by an internal foliation (S_i) whose shape is unrelated to S_e, and typically the internal and external foliations do not connect. At the other end of the spectrum we find postkinematic growth, which shows an external foliation that continues into the grain seemingly without any disruption. Figure 13.6 shows natural examples of these relationships. In the intermediate case (synkinematic growth), the timing of growth and deformation coincide. The evidence for synkinematic growth is an external foliation that can be traced into a blast containing an internal foliation with a pattern different but connected to S_e. A classic example is a snowball garnet (Figure 12.13), where S_i spirals around the core of the garnet until it connects with S_e, which generally displays a simpler pattern.

How do we preserve a spiral pattern in garnet and what can it tell us? Imagine a rock that becomes metamorphosed in an active shear zone. After a small porphyroblast nucleates, it rapidly grows by consumption and inclusion of other minerals (Figure 13.7). The shear zone foliation is incorporated into the blast as an inclusion trail. Because the mineral is affected by the shearing motion, like ball bearings between your hands (Section 12.4.1), continued mineral growth produces an internal foliation that appears to be rolled up, producing the characteristic snowball geometry. Now consider the alternative: the same pattern can be generated if the matrix rotates around a stationary, growing clast. Relative rotation of blast versus matrix is a matter of some debate and it has been suggested that blasts are located in non-deforming pods surrounded by bands that concentrate deformation. Most work, however, indicates that blasts rotate. Returning to our rocks, we are fortunate to find rotated garnets, so we conclude that metamorphism is synkinematic. Furthermore, the counterclockwise rotation sense supports the

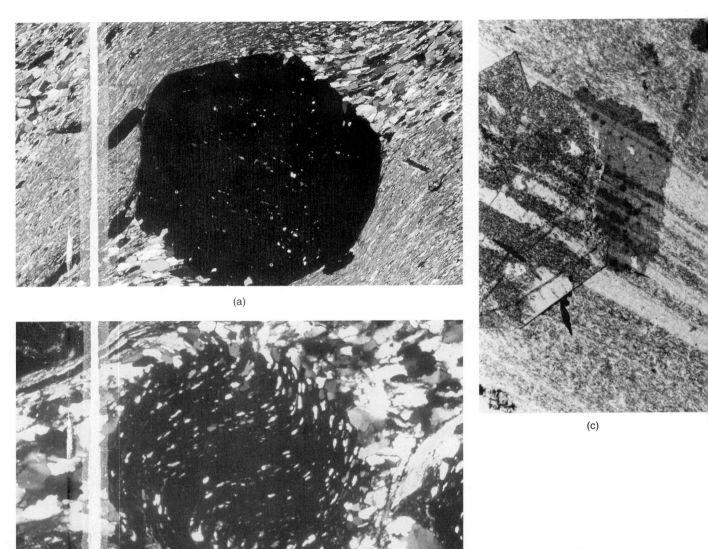

(a)

(b)

(c)

FIGURE 13.6 Photomicrographs of (a) prekinematic, (b) synkinematic and (c) postkinematic growth of garnet (a and b) and staurolite (c). Long dimension of view is ~1.5 mm.

reverse displacement in our mylonite zone that was also determined from other shear-sense indicators (see previous discussion).

In many instances we may not find evidence for synkinematic growth as compelling as that of snowball garnets. Nevertheless, it seems generally true that mineral reactions are triggered by deformation, because the stress gradients that exist during deformation promote material transport and grain growth (e.g., by fluid-assisted or "wet" diffusion; Chapter 9). Thus, mineral assemblages and mineral compositions may be metastable until deformation triggers the reactions that produce equilibrium assemblages for the ambient metamorphic conditions. While mineral growth dur-

ing deformation is common, it is not a rule. Environments of contact metamorphism, for example, are one exception.

13.4.1. Status Report II

From our latest observations we learn that previously determined *P* and *T* values represent the metamorphic conditions during deformation. In other words, deformation in the shear zone occurred under amphibolite-facies conditions. Much of the information necessary to solve our geologic mystery has been obtained, but we still don't know *when* these conditions of pressure and temperature prevailed. This has long been a diffi-

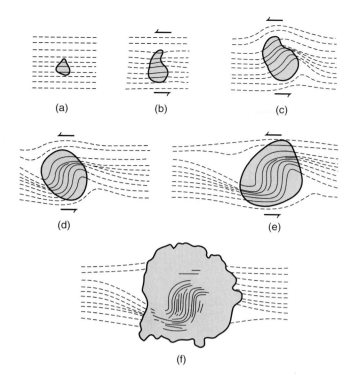

FIGURE 13.7 The progressive development of snowball textures in counterclockwise rotating, growing porphyroblast (a–f). Compare the rotation with that in the garnet in Figure 13.6b.

cult question to answer, but huge progress in radiogenic isotope geochemistry comes to the rescue.

13.5 TIME

Minerals consist of a specific arrangement of atoms (crystal structure). Atoms, in turn, consist of protons and neutrons that make up the nucleus, which is surrounded by a cloud of electrons. The number of protons, the **atomic number** (symbol Z), defines an element; for example, Si contains 14 protons and its atomic number is therefore 14 ($_{14}$Si). This atomic number specifies the order in the periodic table of the elements. The protons and neutrons combined define the mass of an atom, which is therefore called the **mass number** or **isotopic number** (symbol A). The mass number is the second value for an element that you find in the periodic table; for example Si has a mass of 28, which we place on the upper left of the element symbol: $_{14}^{28}$Si. In contrast to the atomic number, a single element can have several mass numbers; that is, different numbers of neutrons. For example the element rubidium has the isotopes $_{37}^{85}$Rb and $_{37}^{87}$Rb. Some isotopes are unstable and they come apart spontaneously. Of the two Rb isotopes, $_{37}^{87}$Rb is

unstable and decays to $_{38}^{87}$Sr. In this process we call rubidium the **parent isotope** and strontium the **daughter isotope.**[4]

The number of isotopes that decay per unit time is proportional to the number of parent isotopes; this proportionality factor is a constant for each unstable isotope and is called the **decay constant, λ**. Notice that isotope geologists use the same symbol that we previously defined as the quadratic elongation (Chapter 4); they are unrelated. Perhaps you are more familiar with the concept of **half-life** ($t_{1/2}$) of an isotope, which is the time required for half of a given number of parent isotopes to decay. For the transformation of Rb to Sr the decay constant is 1.4×10^{-11}/y, which means a half-life of 48.8×10^9y, based on the relationship

$$t_{1/2} = 0.693/\lambda \qquad \text{Eq. 13.3}$$

Table 13.3 lists common radiogenic systems with their corresponding half-lives and decay constants.

The principle behind isotopic dating is simple. We measure the amount of parent (P) and daughter (D) in a mineral today. The sum of these equals the original amount of parent (P_0). The ratio D/P_0 is then proportional to the age of the mineral. A useful device to illustrate the fundamentals of geochronology is an hourglass. Starting with one side of the hourglass full (containing the "parent") and the other side empty (representing the "daughter"), we only need to know the rate at which sand moves from one chamber to the other (the "decay constant") and the amount of sand in the daughter chamber (or the amount of parent remaining) to determine how much time has passed. In mathematical terms

$$-dP/dt = \lambda P \qquad \text{Eq. 13.4}$$

where dP/dt is the rate of change of the parent. This has a negative sign because the amount of the parent decreases with time. In practice, the process is a little more complex, as we will see next.

13.5.1 The Isochron Equation

The first complication arises at the time the radiogenic clock starts, because the rock already contains some daughter; in other words, some sand already exists in the daughter chamber before we start the timer. This amount of daughter is referred to as the **initial daughter.** When we measure the amount of daughter product in our

---

[4]There are no sons in this dating game.

TABLE 13.3

COMMONLY USED LONG-LIVED ISOTOPES IN GEOCHRONOLOGY

Radioactive Parent (P)	Radiogenic Daughter (D)	Stable Reference (S)	Half-life, $t_{1/2}$ $(10^9\ y)$	Decay constant, λ (y^{-1})
^{40}K	^{40}Ar	^{36}Ar	1.25	0.58×10^{-10}
^{87}Rb	^{87}Sr	^{86}Sr	48.8	1.42×10^{-11}
^{147}Sm	^{143}Nd	^{144}Nd	106	6.54×10^{-12}
^{232}Th	^{208}Pb	^{204}Pb	14.01	4.95×10^{-11}
^{235}U	^{207}Pb	$^{204}Pb^1$	0.704	9.85×10^{-10}
^{238}U	^{206}Pb	$^{204}Pb^1$	4.468	1.55×10^{-10}

[1] ^{204}Pb is not stable, but has an extremely long half-life of $\sim 10^{17}$ years.
From Faure (1986).

specimen we are actually combining the amounts of daughter from decay of the parent and initial daughter. The latter amount needs to be subtracted for age determination. The solution to this problem involves using minerals of the same sample containing different isotopic ratios, which allows us to define an **isochron.** The details of this approach lie outside the scope of this chapter, but you will find them in every textbook on geochronology. We merely give you the **isochron equation** without more theory:

$$D/S = D_0/S + P/S\ (e^{\lambda t} - 1) \qquad \text{Eq. 13.5}$$

where D is total amount of daughter present at time t (radiogenic and initial daughter), D_0 is common daughter, P is parent present at time t, S is a reference isotope (a stable, non-daughter isotope), λ is the decay constant, and t is time. You notice that this equation has the general form $y = a + bx$, which is the equation of a straight line. So, plotting the values of D/S and P/S for several minerals in a single rock defines a line (the isochron) whose slope represents the age of the sample; the slope is defined by $(e^{\lambda t} - 1)$ (Equation 13.5). An illustration is given in Figure 13.8 for the Rb-Sr system.

The situation is more complicated for the U/Pb system, because here we are dealing with two parents and two daughters (Table 13.3); however, this added complexity also results in great reliability for U/Pb ages. We simultaneously solve two isochron equations. If the ages are equal within error (typically within 2–3 m.y.), the age is called **concordant;** otherwise we have a **discordant** age (Figure 13.9). Discordant mineral ages are common and reflect loss of Pb after the mineral was formed, but they are not useless. Tech-

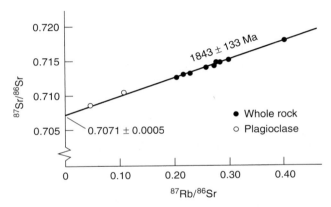

FIGURE 13.8 Rb-Sr isochron based on a combination of whole-rock analyses and plagioclase concentrates; each point represents an analysis. The different ratios lie on a straight line (the isochron) whose slope gives the age of the rocks.

niques to extract the original age of the mineral as well as the time of Pb loss have been developed. These properties and the high closure temperature for several minerals (see the following section) make U/Pb dating a very sensitive (standard error around 2 m.y.) and informative method that has become widely used in tectonics.

The $^{40}Ar/^{39}Ar$ method of dating is another geochronologic approach that we use in our study. The advantage of this method over traditional K-Ar determinations is that all measurements are carried out on the same sample (after neutron irradiation), allowing us to use a stepwise incremental heating technique, during which Ar is progressively released. In theory, the age obtained at each step (derived from the $^{40}Ar/^{39}Ar$ ratio) should be the same, but in practice this is generally not the case. Some Ar may have escaped

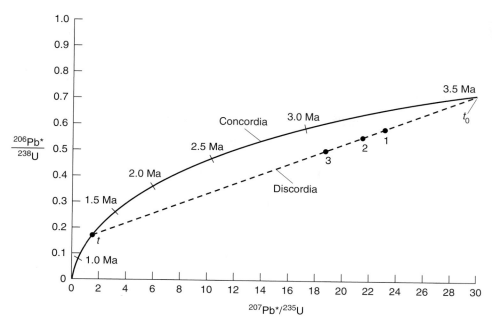

FIGURE 13.9 U-Pb concordia diagram showing three discordant zircon ages that reflect Pb loss at time $t = 1250$ Ma; the original zircon growth age is derived from the intersection between the line passing through the discordant zircon ages and concordia ($t_0 = 3500$ Ma). Modern U/Pb analysis uses portions or regions of individual grains to determine discordia. For example, the grain is progressive abraded to remove reset rims.

(called Ar-loss) or have been gained (called excess-Ar) in part of a mineral, which can be examined by stepwise release patterns (Figure 13.10); the details are outside the scope of this book. A reliable $^{40}Ar/^{39}Ar$ age for the mineral is aided by the recognition of a "plateau," comprising several steps of equal age within some error limit. But before interpreting the radiometric data we need to understand the concept of closure temperature of a geochronometer.

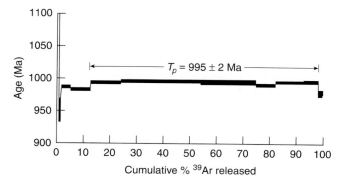

FIGURE 13.10 $^{40}Ar/^{39}Ar$ stepwise release spectrum of hornblende, with a plateau age, T_p, of 995 ± 2 Ma that is indistinguishable from the total gas age (also called the fusion age).

13.5.2 The Isotopic Closure Temperature

What does an isotopic age mean, or when does the radiogenic clock start? Previously you saw that the age of a mineral is a function of the ratio D/P. Now consider that the daughter chamber in our hourglass has a leaky bottom, so that the sand will not stay put until we close the chamber. Likewise, in nature, closure of the system for daughter product does not automatically coincide with the time of mineral growth nor does it occur at the same temperature for each isotopic system. This critical characteristic is addressed by the concept of the **closure temperature,** T_c. Below its closure temperature a mineral retains the amount of daughter produced; in other words, the radiogenic clock starts to tick upon reaching T_c. Above the closure temperature the daughter can escape. Table 13.4 lists closure temperatures for common minerals and their isotopic systems. Clearly, T_c differs for different minerals and isotopic systems. The closure temperature of a mineral is not a set value, but changes as a function of grain size (larger grains have higher T_c), cooling rate, and other parameters. The values listed in Table 13.4 are therefore based on common grain sizes, as well as on cooling rates that are typical of metamorphic rocks.

TABLE 13.4	SELECTED CHRONOLOGIC SYSTEMS AND APPROXIMATE CLOSURE TEMPERATURES

System-Mineral	$T_c{}^1$
U/Pb, zircon	>900°C
U/Pb, garnet	>800°C
U/Pb, monazite	725°C
U/Pb, titanite	600°C
U/Pb-rutile	400°C
^{40}Ar/^{39}Ar, hornblende	480°C
^{40}Ar/^{39}Ar, muscovite	350°C
^{40}Ar/^{39}Ar, biotite	300°C
^{40}Ar/^{39}Ar, alkali feldspar	350–200°C
Fission track-apatite	110°C
(U-Th)/He, apatite	~80°C

[1]For a cooling rate of 1–10°C/m.y. and a common grain size; larger/smaller grain sizes have higher/lower T_c.
After Mezger (1990) and other sources.

A high value of T_c implies that the radiogenic clock in a mineral is not easily reset by regional metamorphism, which explains, for example, the popularity of zircon U/Pb dating for formation ages of igneous rocks. Thus, dating porphyroblasts with a closure temperature that exceeds the metamorphic temperature (e.g., monazite and sphene in greenschist facies rocks) gives the *time of mineral growth*. Note that these minerals can grow at temperatures higher, equal to, or lower than T_c! In general, analysis of minerals gives **growth ages** if the metamorphic temperature is less than T_c; otherwise we obtain **cooling ages.** The difference in T_c for various minerals has advantages that are used in various ways to understand the history of a rock or region, as we'll see later.

13.5.3 Dating Deformation

Using our understanding of isotopic closure, growth, and cooling ages, we can finally turn to the question of when our rocks were deformed. Unfortunately, most "dateable" minerals are not involved in microstructures that reflect deformation. One solution is to bracket deformation by dating rocks that intrude the area; U/Pb ages on zircons from granites and peg-

matites are especially useful to this end. Deformed intrusive rocks give ages that predate the deformation, whereas undeformed intrusives are postkinematic. With some luck you constrain the deformation within tens of millions of years. In old rocks these constraints are often very loose, resulting in broad age brackets. Careful mapping may offer a second approach, involving intrusives that are restricted in their occurrence to deformation zones. This suggests that intrusion occurred due to deformation, so these synkinematic intrusives date the deformation. U/Pb zircon ages from pegmatites in Precambrian rocks are widely used for this purpose.

In our field area we find no dateable intrusions to bracket deformation, but we do have a range of minerals present that can be individually dated. Monazites and titanites in our mylonite show no evidence that they were derived elsewhere (i.e., eroded and transported). The crystals are euhedral and look "fresh," and have distinctly different morphologies (color, size) from those in relatively undeformed samples elsewhere in the area. This indicates that they grew in our shear zone; they are synkinematic. Dating monazite and titanite using the U/Pb method gives us concordant ages of 1010 ± 2 Ma and 1008 ± 2 Ma, respectively. The closure temperature for monazite (725°C; Table 13.4) is well above peak metamorphic temperatures determined from thermometry (600°C), so this mineral grew below its closure temperature and, thus, dates the deformation. The titanite age is slightly younger, but the difference is not statistically significant (although it might reflect some resetting). Because the monazite represents a growth age, we conclude that the age of mylonitization is 1010 Ma.

Both approaches (dating of intrusives and dating of accessory phases) have drawbacks: (1) they do not date the mineral(s) involved in the mineral reactions on which we base the P and T conditions, and (2) they do not date minerals that are synkinematic based directly on microstructural relationships (as in Figure 13.5). In greenschist facies rocks this can be overcome by looking at mineral reactions involving hornblende, muscovite, and other phases that can be dated with the ^{40}Ar/^{39}Ar technique. In amphibolite facies rocks, however, this method only yields cooling ages, because T_c for these minerals (Table 13.4) is less then T_{peak}. As an option for high-grade rocks, garnets may be used for dating.[5] This common mineral is involved in many geothermobarometric reactions (Table 13.1) and often

[5]There is some question as to whether U resides in the garnet or in inclusions; this ambiguity creates problems with high T_c inclusions.

allows determination of the relative timing of deformation and metamorphism (Figure 13.6). After considerable effort, U/Pb dating of synkinematic garnets in our samples produces an age of 1011 ± 2 Ma. This age is within the error of the monazite age and supports our previous determination for the age of deformation.

With our geochronologic work we have determined when deformation occurred, but we are also able to learn something about the subsequent history by dating other minerals. The range in T_c values for various systems gives an opportunity to constrain the cooling history of the area, an approach called **thermochronology.** If we date several minerals within a single rock we obtain the times at which the rock passes through the respective closure temperatures, provided that the minerals grew at or above their closure temperatures. In other words, we are able to calculate the temperature change per time unit (Figure 13.11). We apply the $^{40}Ar/^{39}Ar$ method to our gneiss sample, which yields ages for hornblende, muscovite, and biotite of 1000 Ma, 984 Ma, and 976 Ma, respectively. The **cooling rate** is the temperature difference divided by the age difference:

$$\text{cooling rate} = \delta T/\delta t \qquad \text{Eq. 13.6}$$

In our sample, the cooling rate from the time of peak metamorphic temperature to the closure temperature for hornblende is:

$$(T_{\text{peak}} - T_{c,\text{hornblende}})/(\text{age}_{\text{monazite}} - \text{age}_{\text{hornblende}}) = 120°C/10 \text{ m.y.} = 12°C/\text{m.y.}$$

Furthermore, if we calculate the cooling rate based on minerals with different ages, it appears that the cooling rate changes with time:

$$(T_{c,\text{hornblende}} - T_{c,\text{muscovite}})/(\text{age}_{\text{hornblende}} - \text{age}_{\text{muscovite}}) = 130°C/16 \text{ m.y.} = 8°C/\text{m.y.}$$
$$(T_{c,\text{muscovite}} - T_{c,\text{biotite}})/(\text{age}_{\text{muscovite}} - \text{age}_{\text{biotite}}) = 50°C/8 \text{ m.y.} = 6°C/\text{m.y.}$$

The decreasing cooling rate of our rocks places constraints on the uplift history of our area, as will be discussed after the third and final status report.

13.5.4 Status Report III

Our database has considerably expanded in size and scope since we hammered our samples in the field. Let's summarize the key information:

- Low-angle mylonite zone with reverse sense of displacement; based on field data and microfabrics.

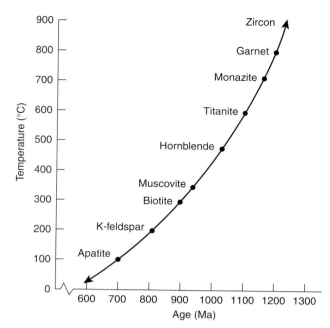

FIGURE 13.11 Schematic cooling history of a rock based on determining the ages of various minerals with different closure temperatures (see Table 13.4).

- Metamorphosed sediments; thus, originally deposited at the surface.
- Peak temperature of 600°C (middle amphibolite facies).
- Peak pressure of 700 MPa and burial depth of 26 km; thus, metamorphosed in the middle crust.
- 1010 Ma age of peak temperature and deformation (i.e., mid-Proterozoic orogenic activity) based on U/Pb ages of monazite and garnet.
- Retrograde metamorphic conditions of 500 MPa and 550°C based on garnet rim analyses.
- Cooling ages of 1000 Ma at 480°C, 984 Ma at 350°C, and 976 Ma at 300°C, based on hornblende, muscovite, and biotite $^{40}Ar/^{39}Ar$ dating, respectively.
- Cooling rate of 12°C/m.y. during early retrogression that slows down to 6°C/m.y.

13.6 *D-P-T-t* PATHS

Status Report III provides quite a lot of information to digest, but we are now able to reconstruct a remarkably detailed history of our rocks, from deposition to burial and back to exhumation. This history is described by the pressure-temperature-time (*P-T-t*) path, which makes predictions about the deformation setting of the area. In Figure 13.12 the *P, T,* and *t* data that we collected (solid dots) and inferences based on them (open dots) are summarized in three related diagrams: a

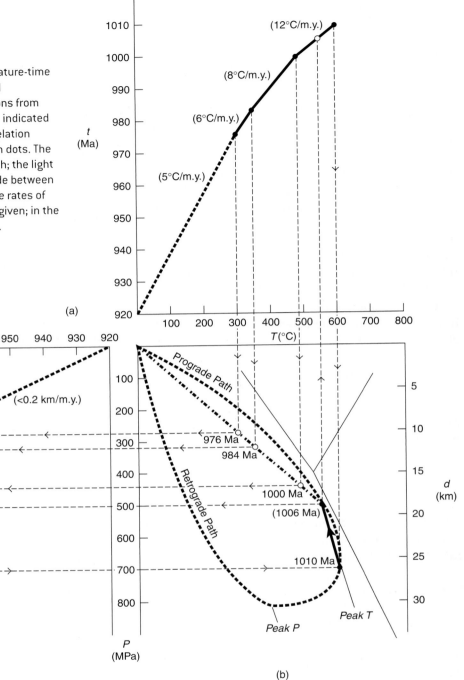

FIGURE 13.12 Evolution in temperature-time (a), pressure/depth-temperature (b), and pressure/depth-time space. Determinations from thermobarometry and geochronology are indicated with solid dots; inferences based on correlation between the diagrams are shown by open dots. The solid arrows indicate the sense of the path; the light arrows indicate how inferences were made between the various diagrams. In each diagram the rates of corresponding segments of the path are given; in the P-T plot the age along the path is marked.

(a)

(b)

(c)

temperature-time diagram, a pressure (or depth)–temperature diagram, and a pressure (or depth)–time diagram. Each is explored separately in the following discussion.

13.6.1 Temperature-Time (T-t) History

Our careful geochronologic and thermochronologic work offers quite a few data points from which we may construct a reliable T-t diagram. Each mineral age is plotted against the corresponding peak T or T_c, depending on whether it is a growth or cooling age, respectively. From this plot we see that the curve is not linear, but that it decreases in slope toward the surface temperature (for convenience we place the surface temperature at 0°C). Each segment of the curve has a different cooling rate that changes from 12°C/m.y. to 8°C/m.y. to 6°C/m.y.; the youngest segment is unconstrained other than that it must intersect the T-axis at 0°C. We use a gradient of 5°C/m.y., so that the rocks are at the surface around 920 Ma (heavy dashed line in Figure 13.12a).

13.6.2 Pressure-Temperature (P-T) History

In P-T space we have determined two points from thermobarometry: 700 MPa and 600°C, and 500 MPa and 550°C. We also know from the occurrence of retrograde rims on garnet that the change in P and T trends toward the lower values, so we should have a counterclockwise sense for this part of the P-T path. In addition, we know that the path should intersect the origin after 976 Ma (dashed line in Figure 13.12b), because the rocks are now at the surface. With the temporal information we can place ages at points where the T-t curve intersects the P-T curve. This exercise highlights an important characteristic of P-T paths: there is no simple relationship between length of a P-T path segment and length of time. A P-T segment early in the retrograde history represents only a few m.y., whereas in the late history a segment of equal length represents tens of millions of years. The time at which we reach the P-T point at 500 MPa and 550°C is predicted from the T-t curve to be 1006 Ma. Note that the path from peak temperature to surface exposure, the **retrograde path,** is most likely smooth rather than kinked, as in Figure 13.12b (marked by the dashed line).

We have no direct information on the **prograde path,** before the rocks reached peak metamorphic conditions, other than that the path starts at the origin (when the rocks were deposited as sediments). We

simply connect the high-temperature end of our retrograde path with the origin (Figure 13.12b) to complete the P-T path.

13.6.3 Pressure-Time (P-t) History

Only one point is constrained in P-t space by barometric and chronologic work: 700 MPa at 1010 Ma. Nevertheless we are able to construct a P-t path by combining the T-t and P-T curves (Figure 13.12c). The resulting path is a pattern of decreasing pressure change with time, which means a decreasing rate of removal of the overlying rock column with time (or decreased **exhumation rate**).[6] For our purposes it is most informative to convert pressure to depth (1 km = 27 MPa); this shows that the exhumation rate changes by at least one order of magnitude, from approximately 2 km/m.y. to <0.2 km/m.y.

13.6.4 The Geothermal Gradient

The diagrams in Figure 13.12 highlight a second property of the crust's thermal structure in deformational settings: the geothermal gradient is not a constant. This gradient of temperature (T) change with depth (d) equals the ratio of cooling rate (Equation 13.6; Figure 13.12a) over exhumation rate (Figure 13.12b):

$$\text{geothermal gradient} = \delta T/\delta d \qquad \text{Eq. 13.7}$$

Using Equation 13.7 for various time intervals constrained previously, we get

1010–1000 Ma	12°C/km
1000–984 Ma	26°C/km
984–976 Ma	30°C/km
976–(920) Ma	30°C/km

In other words, the value of the geothermal gradient changes with time, especially during the early retrograde history. Thus, no single geothermal gradient is representative for the entire orogenic history. Using the metamorphic P and T peak values only approximates the average geothermal gradient; therefore, the pattern is also called the **metamorphic field gradient.** In our area the average gradient is 600°C/26 km, or 23°C/km, which lies

---

[6]*Uplift* is sometimes used, but this describes displacement relative to a fixed coordinate system. Here we uplift the rock and remove its cover.

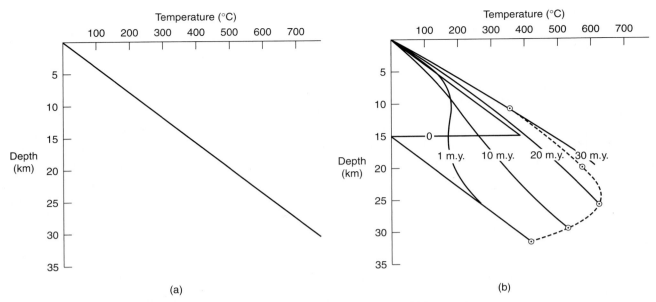

FIGURE 13.13 Thermal evolution after instantaneous doubling of a 15 km thick crustal section. A simplified geothermal gradient of 25°C/km is assumed before thickening of the crust (a). In (b) we show the geotherms at the time of instantaneous thrusting (0 m.y.), and after 1, 10, 20 and 30 m.y. After a few m.y. the irregularity in the thermal perturbation is largely removed, but the geothermal gradient remains depressed. After 10 m.y. the gradient is <20°C/km; after 20 m.y. it is approximately back to the original value (~25°C/km), and after 30 m.y. the gradient is 30°C/km. Although these patterns are schematic, the shapes of the curves are based on thermal modeling.

in a reasonable range and agrees with values determined previously.

But why does the geothermal gradient change with time? Consider a slice of regular continental crust with an undisturbed geothermal gradient of 25°C/km (Figure 13.13a). This gradient is not really linear, because heat-producing elements are preferentially concentrated in the upper crust, but for our purposes this simplification will do (in Chapter 14 we return to the thermal structure of Earth). At time t_0, the top 15 km of the crust becomes doubled by thrusting. This results in a disturbed thermal gradient that steadily increases toward the thrust boundary, but at this boundary, sharply jumps back to the surface temperature, after which temperature again increases. This thermal pattern is called the **sawtooth model.** Note that the development of a sawtooth shape implies that deformation (here thrusting) is instantaneous relative to thermal equilibration. We can evaluate this.

The time over which a thermal perturbation decays by conduction is given by the "rule of thumb" equation for heat flow:

$$t = h^2/K \qquad \text{Eq. 13.8}$$

where h is the distance over which thermal conduction occurs and K is the thermal diffusivity (5×10^{-7} m^2/s

for average crustal rock). Substituting $h = 15$ km gives an equilibration time on the order of 14 m.y. Although thrusting is not really instantaneous, it occurs at much faster rates than changes in thermal structure, permitting the simplification of geologically instantaneous deformation we just used. Equation 13.8 quantifies the change of a sawtooth thermal gradient with time, which we explore with a comforting analogy. Placing a hot blanket over your cold body creates a strong temperature gradient at the contact between blanket and skin. Pretty soon the heat of the blanket affects your body surface and the temperature difference becomes less. After a while there is little or no temperature contrast between blanket and skin, and you are comfortably warm. The actual thermal evolution is a bit more complex, because it also involves heat loss to the air and heat production by your body. But the analogy describes the situation in Nature pretty well. Before dozing off, we return to the cold reality of Figure 13.13b, where you see that after about 20 m.y. the original gradient is restored, and that after 30 m.y. it even exceeds the original value (~30°C/km). Thus, we conclude that deformation significantly affects the thermal structure of the crust for time periods of up to tens of millions of years. Note that adding contemporaneous erosion to our history further affects the thermal evolution.

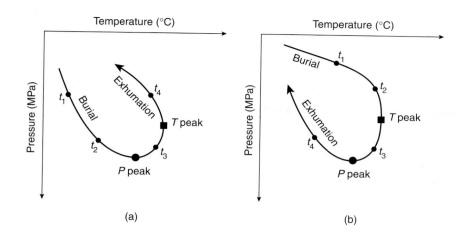

FIGURE 13.14 Counterclockwise (a) and clockwise (b) *P-T-t* paths may reflect characteristic deformation histories. Peak temperature (square) and peak pressure (circle) do not occur at the same time. Because metamorphic reactions proceed faster at peak temperatures, metamorphic pressures obtained from geothermobarometry are generally less than peak pressures. Note that the time steps (t_i) do not represent equal time increments.

13.6.5 The Deformational Setting

At the end of a chapter that seems dominated by metamorphic petrology and geochronology, we return to the "*D*" in *D-P-T-t* paths: deformation. The patterns in Figures 13.12 and 13.13 can be constructed and analyzed using numerical models involving input parameters describing heat production and heat transfer, and familiar parameters such as density, time, and depth. The theoretical approach, the details of which remain outside the scope of our discussion, can give critical insights into orogenic evolution. Observations on natural rocks, such as in our study, combined with modeling have produced characteristic relationships between *P-T-t* paths and the deformation history. For example, crustal thickening by thrusting results in rapid burial of rocks (deformation is relatively instantaneous; see previous discussion), a geothermal gradient that changes from a sawtooth shape to elevated values, and peak pressures that precede peak temperatures (Figure 13.14a). As a consequence of the latter, rocks record pressure at the time of peak temperature, which may differ from peak pressure. Our combined burial (prograde) and exhumation (retrograde) history describes a *counterclockwise path*[7] (Figure 13.14a), which fits well with the analysis of our field observations and *P-T-t* data.

Paths from other tectonic environments have similarly characteristic geometries; for example, a clockwise path (Figure 13.14b) may reflect increasing heat from below due to extension and crustal thinning, or adding a heat source at the base of the crust (a process called **metamorphic underplating**). While numerical models do not substitute for observations, they can constrain part of the history that is not preserved in

natural rocks. Modeling, however, requires a critical attitude, because the input parameters may vary considerably and the equations are quite sensitive to uncertainties.

13.7 CLOSING REMARKS

Modern structural analysis integrates a variety of approaches. This essay offers only a glimpse of the possibilities that move structural geology into the full realm of orogenic evolution and tectonic processes of the deeper crust. Different disciplinary approaches have their strengths and weaknesses, but combined they provide internally consistent information to determine a more complete scenario. A crucial component in all this remains good field observations. There is no sense in analyzing a sample with highly sophisticated techniques and numerical models if the results are not representative for the area. Once we have met these requirements, an integrated approach, such as illustrated in this essay, will greatly expand our understanding of the tectonic history of a region and provide fundamental information on lithospheric processes. With these remarks we close the part of the book on ductile structures and move to the fundamentals of plate tectonics and lithospheric structure.

ADDITIONAL READING

Barker, A. J., 1990. *Introduction to metamorphic textures and microstructures.* Blackie: Glasgow.

Bell, T. H., Johnson, S. E., Davis, B., Forde, A., Hayward, N., and Wilkins, C., 1992. Porphyroblast inclusion-tail orientation data: eppure non son girate! *Journal of Metamorphic Geology,* 10.

---

[7]Petrologists often plot pressure up, thereby creating a clockwise path; this generates terminology confusion with structural geologists plotting pressure/depth down.

England, P. C., and Thompson, A. B., 1984. Pressure-temperature-time paths of regional metamorphism I. Heat transfer during the evolution of regions of thickened continental crust. *Journal of Petrology,* 25, 894–928.

Essene, E. J., 1989. The current status of thermobarometry in metamorphic rocks. In Daly, J. S., Cliff, R. A., and Yardley, B. W. D., eds., *Evolution of metamorphic belts.* Geological Society Special Publication, 43, 1–44.

Faure, G., 1986. *Principles of isotope geology* (2nd edition). J. Wiley & Sons; New York.

Hodges, K. V., 1991. Pressure-temperature-time paths. *Annual Reviews of Earth and Planetary Sciences,* 19, 207–236.

Mezger, K., 1990. Geochronology in granulites. In Vielzeuf, D., and Vidal, P., eds., *Granulites and crustal evolution.* Kluwer, pp. 451–470.

Passchier, C. W., Meyers, J. S., and Kroner, A., 1990. *Field geology of high-grade gneiss terranes.* Springer Verlag.

Passchier, C. W., Trouw, R. A. J., Zwart, H. J., and Vissers, R. L. M., 1992. Porphyroblast rotation: eppur si muove? *Journal of Metamorphic Geology,* 10, 283–294.

Spear, F. S., and Peacock, S. M., 1989. Metamorphic pressure-temperature-time paths. *American Geophysical Union, Short Course in Geology,* 7.

Spry, A., 1969. *Metamorphic textures.* Pergamon: Oxford.

Vernon, R. H., 1976. *Metamorphic processes.* Murby: London.

Villa, I. M., 1998. Isotopic closure. *Terra Nova,* 10, 42–47.

Wood, B. J., and Fraser, D. G., 1977. *Elementary thermodynamics for geologists.* Oxford University Press.

Zwart, H. J., 1962. On the determination of polymetamorphic mineral associations, and its application to the Bosost area (central Pyrenees). *Geologische Rundschau,* 52, 38–65.

PART D

TECTONICS

and Plate Tectonics

14.1 INTRODUCTION

Up to this point in the book, we have focused on understanding the nature and origin of specific structures (such as faults, folds, and fabrics), and on characterizing relations among deformation, stress, and strain. This discussion has emphasized features that we see at the microscopic to outcrop scale. Now, let's change our focus to the *regional* scale by discussing tectonics.[1] Geologists use the term **tectonics,** in a general sense, to refer to the sum of physical processes that yield regional-scale geologic features (Figure 14.1). Studies in tectonics consider such issues as the origin of mountain belts, the growth of continents, the formation of the ocean floor, the development of sedimentary basins, and the causes of earthquakes and volcanoes. In the next six chapters (Part D), we introduce these "big-picture" issues, and show how rock deformation occurs hand-in-hand with many other geologic phenomena as the Earth evolves.

This chapter covers two topics that provide the foundation for Part D of this book. We begin by describing **whole Earth structure,** meaning, the internal layering of the Earth. Geologists subdivide the Earth's insides into layers in two different ways. First, based on studying the velocity of seismic waves that pass through the Earth, we subdivide the *entire* planet into the **crust,** the **mantle,** and the **core,** named in sequence from surface to center. Second, based on studying rock rheology (the response of rock to stress), we subdivide the *outer several hundred kilometers* of the Earth into the lithosphere and asthenosphere. The **lithosphere** is the outermost of these **rheologic layers** and forms Earth's rigid shell. The lithosphere, which

---

[1]The word "tectonics" was derived from the Greek *tektos,* meaning "builder."

FIGURE 14.1 Landsat image of tectonically active region, the Zagros Mountains of Iran. The dome shapes are relatively recent developed anticlines. The topography clearly reflects the underlying structure.

consists of the crust *and* the outermost mantle, overlies the **asthenosphere,** a layer that behaves plastically (meaning that, though solid, it can flow). The asthenosphere lies entirely within the mantle. A background on whole Earth structure sets the stage for introducing **plate tectonics theory** (or simply, "**plate tectonics**"). According to this theory, the lithosphere consists of about 20 discrete pieces, or **plates,** which slowly move relative to one another. In our discussion, we describe the nature of plates, the boundaries between plates, the geometry of plate movement, and the forces that drive plate movement.

14.2 STUDYING EARTH'S INTERNAL LAYERING

Before the late nineteenth century, little was known of the Earth's interior except that it must be hot enough, locally, to generate volcanic lava. This lack of knowledge was exemplified by the 1864 novel *Journey to the Center of the Earth,* in which the French author Jules Verne speculated that the Earth's interior contained a network of caverns and passageways through which intrepid explorers could gain access to the planet's very center. Our picture of Earth's insides changed in the late nineteenth century, when researchers compared the gravitational pull of a mountain to the gravitational pull of the whole Earth and calculated that our planet has a mean density about 5.5 g/cm³, more than twice the density of surface rocks like granite or sandstone. This fact

meant that material inside the Earth must be much denser than surface rocks—Verne's image of a Swiss-cheese-like Earth could not be correct.

Once researchers realized that the interior of the Earth is denser than its surface rocks, they worked to determine how mass is distributed inside the Earth. First, they assumed that the increase in density occurred gradually, due entirely to an increase in **lithostatic pressure** (the pressure caused by the weight of overlying rock) with depth, for such pressure would squeeze rock together. But calculations showed that if density increased only gradually, so much mass would lie in the outer portions of the Earth that centrifugal force resulting from Earth's spin would cause our planet to flatten into a disc. Obviously, such extreme flattening hasn't happened, so the Earth's mass must be concentrated toward the planet's center. This realization led to the image of a **layered Earth,** with a very dense central region called the "core," surrounded by a thick "mantle" of intermediate density. The mantle, in turn, is surrounded by a very thin skin, the relatively low density "crust." Eventually, studies showed that the density of the core reaches 13 g/cm³, while crustal rocks have an average density of 3–6 g/cm³.

In the twentieth century, geoscientists have utilized a vast array of tools to provide further insight to the mystery of what's inside the Earth. Our modern image of the interior comes from a great variety of data sources, some of which are listed in Table 14.1. Standard textbooks on geophysics (see Additional Reading) will explain these in greater detail. Here, we focus on constraints on Earth's structure provided by seismic data.

14.3 SEISMICALLY DEFINED LAYERS OF THE EARTH

Work by seismologists (geoscientists who study earthquakes) in the first few decades of the twentieth century greatly refined our image of the Earth's interior. **Seismic waves,** the vibrations generated during an earthquake, travel through the Earth at velocities ranging from about 4 km/s to 13 km/s. The speed of the waves, their **seismic velocity,** depends on properties (e.g., density, compressibility, response to shearing) of the material through which the waves are traveling. When waves pass from one material to another, their velocity changes abruptly, and the path of the waves bends. You can see this phenomenon by shining a flashlight beam into a pool of water; the beam bends when it crosses the interface between the two materials.

TABLE 14.1 | SOURCES OF DATA ABOUT THE EARTH'S LAYERS

Drilling	Deep holes have been drilled in the continents and in the ocean floor (Figure 14.2a). Samples extracted from these holes, and measurements taken in these holes, provide constraints on the structure of the uppermost part of the Earth. But even the deepest hole on Earth, drilled in Russia, penetrates only about 15 km deep, a mere 0.2% of the Earth's radius.
Electrical conductivity	The manner in which electricity conducts through subsurface Earth materials depends on rock composition and on the presence of fluids (water and oil near the surface, magma at depths). Electrical conductivity measurements can detect the presence of partially molten rock (rock that is starting to melt, so that it consists of solid grains surrounded by films of melt) in the mantle and crust.
Earth's density	By measuring the ratio between the gravitational force generated by a mountain (of known dimensions and composition) and the gravitational force generated by the whole Earth, geoscientists determined the mass of the Earth and, therefore, its average density. Knowledge of density limits the range of possible materials that could comprise the Earth.
Earth's shape	The Earth is a spinning sphere. Thus, Earth must be largely solid inside, for if its insides were liquid, its spin would cause the planet to be disc-shaped. Similarly, if density were uniform through the planet, Earth's spin would cause the planet to be more flattened than it actually is. Thus, the shape of the Earth requires that it have a core that is denser than the surrounding mantle and crust.
Exposed deep crust	In ancient mountain belts, the combined process of faulting, folding, and **exhumation** (removal of overlying rock) have exposed very deep crustal levels. In fact, outcrops in the interior of large mountain ranges contain rocks that were once at depths of 20 to 50 km in the crust. In a few localities geologists have even found rocks brought up from about 100 km depth (these rocks contain an ultrahigh-pressure mineral named coesite). Study of these localities gives a direct image of the geology at depth.
Geochemistry	Rocks exposed at the surface were, ultimately, derived by extraction of melt from the mantle. Thus, study of the abundance of elements in rocks at the surface helps to define the range of possible compositions of rocks that served as the source of magma at depth.
Gravity field	Measurements of variations in the strength of Earth's gravity field at the surface give a clue to the distribution of rocks of different density below the surface, for denser rocks have more mass and thus cause greater gravitational attraction. Gravity measurements can indicate where the lithosphere is isostatically compensated and where it is not (see our discussion of isostasy later in this chapter for a further explanation).
Lab experiments	Lab studies that determine the velocity of seismic waves as a function of rock type, under various conditions of pressure and temperature, allow geoscientists to interpret the velocity versus depth profile of the Earth, and to interpret seismic-refraction profiles. Sophisticated studies of elastic properties of minerals squeezed in diamond anvils and heated by lasers even allow the study of materials at conditions found in the lower mantle or core.
Lithospheric flexure	The lithosphere, the outer relatively rigid shell of the Earth, bends ("flexes") in response to the addition or removal of a surface load. For example, when a huge glacier spreads out over the surface of a continent during an ice age, the surface of the continent bends down, and when the glacier melts away, the surface slowly rises or "rebounds." When the lithosphere bends down, the underlying asthenosphere must flow out of the way, and when rebound occurs, the underlying asthenosphere flows back in. Thus, the rate of sinking or rebound depends on the rate at which the asthenosphere moves, and thus on the viscosity (resistance to flow) of the asthenosphere. As another example, the shape of a lithosphere where it bends down into the mantle at convergent plate boundaries (subduction zone) provides insight into flexural strength ("bendability") of the plate. Thus, studies of flexure give insight into the rheology of crust and mantle (e.g., they tell us if it is it elastic, viscous, or viscoelastic).
Magnetic anomalies	We say that a magnetic anomaly occurs where the measured strength of the Earth's magnetic field is greater or lesser than the strength that would occur if the field were entirely due to the Earth's internal field (caused by the flow of iron alloy in the outer core). Anomalies occur because of the composition of rock in the crust, or due to the polarity of the magnetic field produced by tiny grains of iron-bearing minerals in a rock.

TABLE 14.1	SOURCES OF DATA ABOUT THE EARTH'S LAYERS
Meteorites	Meteorites are chunks of rock or metal that came from space and landed on Earth. Some are relict fragments of the material from which planets first formed, while others are fragments of small planets that collided and broke apart early in the history of the solar system. Thus, some meteorites may be samples of material just like that which occurs inside the Earth today.
Ophiolites	An ophiolite is a slice of oceanic crust that was thrust over continental crust during collisional orogeny, and thus is now exposed on dry land for direct examination by geologists. Study of ophiolites gives us an image of the structure of the oceanic crust.
Seismic reflection	Geoscientists have developed methods for sending artificial seismic waves (vibrations generated by explosions or by large vibrating trucks) down into the crust and upper mantle. These waves reflect off boundaries between layers in the subsurface and then return to the surface. Sophisticated equipment measures the time it takes for this process to occur and, from computer analysis of this data, geoscientists can create cross sections of the subsurface that reveal formation contacts, folds, faults, and even the Moho.
Seismic refraction	When a seismic wave reaches the boundary between two layers, some of the energy reflects, or bounces off the boundary, while some refracts, meaning that it bends as it crosses the boundary. Studies of refracted waves can be used to define the velocity of seismic waves in a layer. Such studies provide insight into the composition and dimensions of subsurface layers.
Seismic-wave paths	By studying the paths that earthquake waves follow as they pass through the Earth, and the time it takes for the waves to traverse a distance, geoscientists can identify subsurface layer boundaries and layer characteristics.
Seismic tomography	New computer techniques, similar to those used when making medical "CAT scans," allow geoscientists to create a three-dimensional picture of seismic velocity as a function of location in the crust, mantle, and inner core (see Figure 14.9). These images can be interpreted in terms of variation in material properties controlled by temperature and/or chemistry.
Xenoliths	Xenoliths (from the Greek "xeno-," meaning foreign or strange) are preexisting rocks that have been incorporated in a magma and brought to or near the Earth's surface when the magma flows upward. Some xenoliths are fragments of the deep crust and/or upper mantle, and thus provide samples of these regions for direct study (Figure 14.2b).

(a)

(b)

FIGURE 14.2 (a) Photograph of the *Joides Resolution,* a drilling ship used to drill holes in the sea floor for research purposes. The ship has drilled hundreds of holes all over the Earth, allowing geologists to understand the layering of the ocean crust. (b) Photo of a block of basalt containing xenoliths of dunite. Dunite, a variety of mantle peridotite, consists almost entirely of small, green olivine crystals.

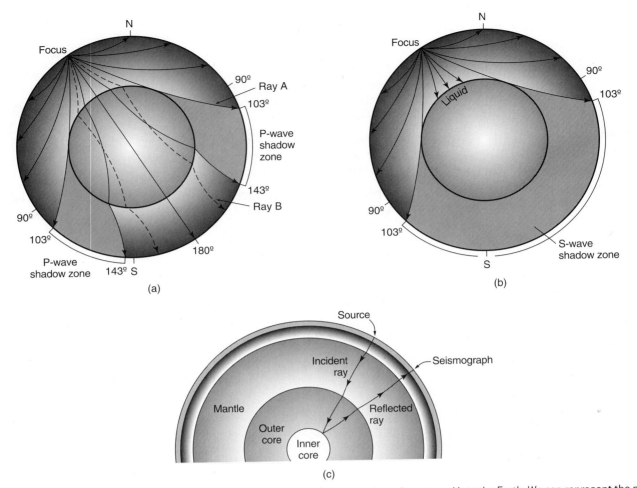

FIGURE 14.3 (a) Earthquake P-waves (compressional waves) generated at a focus travel into the Earth. We can represent the path that they follow by the use of "rays." A "ray" is an arrow drawn perpendicular to the wave front at a point. Note that waves bend or "refract" as they pass downwards. This happens because material properties of the Earth change with depth, so that wave velocity increases. If the change in material properties is gradual, the bending of a ray is gradual. At abrupt changes, the ray bends abruptly. P-waves bend abruptly at the mantle/core boundary. As a result, there is a P-wave shadow zone, where P-waves from a given earthquake do not reach the Earth's surface. The presence of this shadow zone proves the existence of the core. (b) S-waves (shear waves) also travel into the Earth. But shear waves cannot pass through a liquid, and thus cannot pass through the outer core. This creates a large S-wave shadow zone. It is the presence of this shadow zone that proves the existence of liquid in the core. (c) The existence of a solid inner core was deduced from studies that showed that earthquake waves bounce off a boundary within the core.

By studying the records of earthquakes worldwide, seismologists have been able to determine how long it takes seismic waves to pass through the Earth, depending on the path that the wave travels. From this data, they can calculate how seismic-wave velocity changes with depth. The data reveal that there are specific depths inside the Earth at which velocity abruptly changes and waves bend—these depths are called **seismic discontinuities** (Figure 14.3a and b). Seismic discontinuities divide the Earth's interior into distinct shells; within a shell, seismic wave velocity increases gradually, and waves bend gradually, but at the boundary between shells, wave velocity changes suddenly and the waves bend. The change in seismic velocity at a seismic discontinuity can be a consequence of a **compositional change,** meaning a change in the identity or proportion of atoms, and/or a **phase change,** meaning a rearrangement of atoms to form a new mineral structure. Phase changes occur because of changes in temperature and pressure.

Figure 14.4a illustrates an average **seismic velocity versus depth profile** for the Earth. Seismic discontinuities define layer boundaries, as shown in Figure 14.4b. Discontinuities led geologists to subdivide the earlier-mentioned major layers—crust, mantle, and core. We now recognize the following **seismically defined layers: crust, upper mantle, transition zone** (so named because it contains several small discontinuities), **lower mantle, outer core,** and **inner core.** Let's now look at the individual layers more closely.

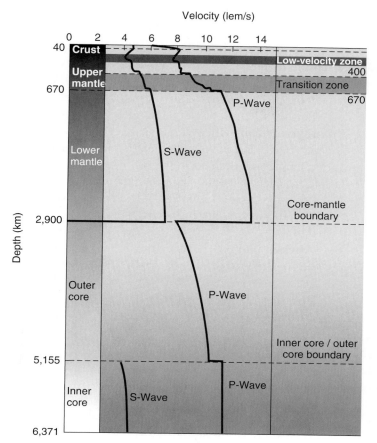

Velocity (lem/s)

FIGURE 14.4 (a) Graph of the variation in P-wave and S-wave velocities with depth in the Earth. The side of the graph shows the correlation of these boundaries with seismically defined layering in the Earth. (b) Diagram illustrating the internal layering of the Earth. Note the relationship between seismically defined layering and rheologic layering.

(a)

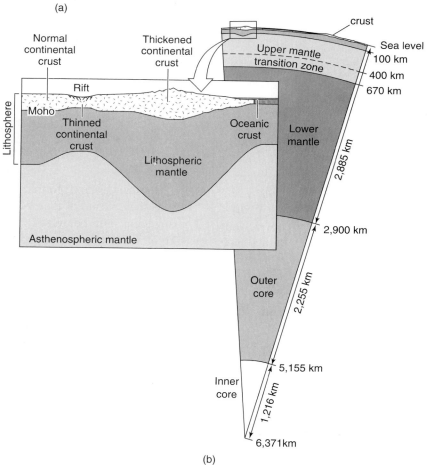

(b)

14.4 THE CRUST

The **crust** of the Earth refers to the outermost layer of our planet. All of the geologic structures that we have discussed in this book so far occur in rocks and sediments of the crust. Geologists distinguish between two fundamentally different types of crust on Earth: continental crust, which covers about 30% of the Earth's surface, and oceanic crust, which covers the remaining 70%. These types of crust differ from each other in terms of both thickness and composition.

The proportions of the two types of crust can be easily seen on a graph, called a **hypsometric curve,** that shows the percentage of Earth's surface area as a function of elevation or depth, relative to sea level (Figure 14.5a). Most continental crust lies between sea level and 1 km above sea level; mountains reach a maximum height of 8.5 km, and continental shelves (the submerged margins of continents) lie at depths of up to 600 m below sea level. Most oceanic crust lies at depths of between 3 km and 5 km below sea level. The greatest depth of the sea occurs in the Marianas Trench, whose floor lies over 11 km below sea level. The bottom of the crust is a seismic discontinuity called the **Moho.**[2] Interestingly, the elevation pattern of Earth is different from that of other bodies in the inner Solar System, as shown by histograms of elevation (Figure 14.5b). This distinction can be used as one piece of evidence that our planet's evolution has been different from that of the Moon, Mars, or Venus; Earth is probably the only planet whose surface has been affected by plate tectonics.

14.4.1 Oceanic Crust

Earth's oceanic crust is 6–10 km thick, and consists of mafic igneous rock overlain by a sedimentary blanket of varying thickness. Field studies of **ophiolites** (slices of oceanic crust emplaced on land by thrusting), laboratory studies of drill cores, and seismic-refraction studies (Table 14.1) indicate that oceanic crust has distinct layers. These layers, when first recognized in seismic-refraction profiles, were given the exciting names Layer 1, Layer 2a, Layer 2b, and Layer 3 (Figure 14.6). Layer 3, at the base, consists of **cumulate,** a rock formed from mafic (magnesium- and iron-rich) minerals that were the first to crystallize in a cooling magma and then settled to the bottom of the magma chamber. The cumulate is overlain in succession by a layer of **gabbro** (massive, coarse-grained mafic igneous

rock), a layer of basaltic **sheeted dikes** (dikes that intrude dikes), a layer of **pillow basalt** (pillow-shaped blobs extruded into sea water), and a layer of **pelagic sediment** (the shells of plankton and particles of clay that settled like snow out of sea water). We'll explain how the distinct layers of crust form in Chapter 16.

Earth's ocean floor can be divided into four distinct **bathymetric provinces,** regions that lie within a given depth range and have a characteristic type of submarine landscape, defined in the following list. We indicate the plate-tectonic environments of these features, where appropriate (Figure 14.5c). These environments will be discussed later in the chapter.

- **Abyssal plains:** These are the broad, very flat, submarine plains of the ocean that lie at depths of between 3 km and 5 km. They are covered with a layer of pelagic (deep-sea) sediment.
- **Mid-ocean ridges:** These are long, submarine mountain ranges that rise about 2 km above the abyssal plains. Their crests, therefore, generally lie at depths of about 2–3 km. Mid-ocean ridges are roughly symmetric relative to a central axis, along which active submarine volcanism occurs. Mid-ocean ridges mark the presence of a divergent plate boundary, at which seafloor spreading occurs.
- **Oceanic trenches:** These are linear submarine troughs in which water depths range from 6 to 11 km. Trenches border an active volcanic arc and define the trace of a convergent plate boundary at which subduction occurs. The volcanic arc lies on the overriding plate.
- **Seamounts:** Seamounts are submarine mountains that are not part of mid-ocean ridges. They typically occur in chains continuous along their length with a chain of oceanic islands. The island at the end of the chain may be an active volcano. A seamount originates as a hot-spot volcanic island, formed above a mantle plume. When the volcano drifts off of the plume, it becomes extinct and sinks below sea level.
- **Guyots:** Guyots are flat-topped seamounts. The flat top may have been formed by the erosion of a seamount as the seamount became submerged, or it may be the relict of a coral reef that formed as the seamount became submerged.
- **Submarine plateaus:** These are broad regions where the ocean is anomalously shallow. Submarine plateaus probably form above large hot-spots.

14.4.2 Continental Crust

The continental crust differs in many ways from the oceanic crust (Table 14.2). To start with, the thickness

---

[2]Named for Andrija Mohorovičić, the seismologist who discovered it in 1909.

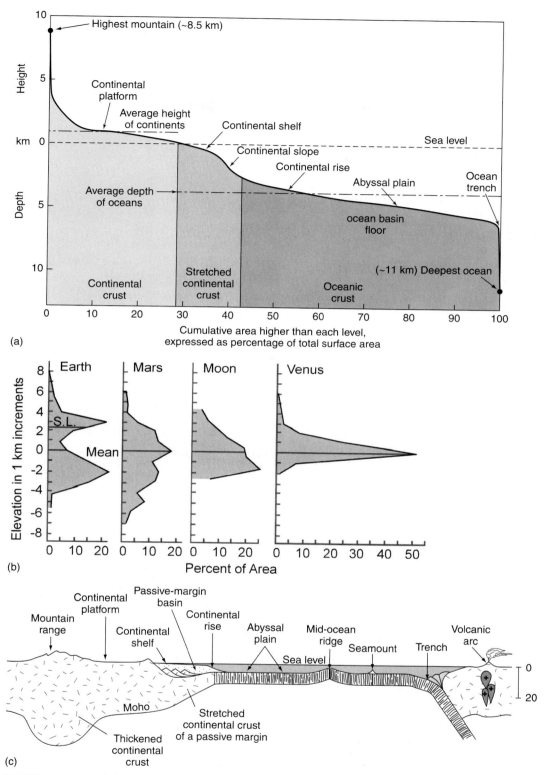

FIGURE 14.5 (a) Hypsometric curve of the Earth, showing elevation as a function of cumulative area. From this diagram, 29% of the Earth's surface lies above sea level; the deepest oceans and highest mountains comprise only a small fraction of the total area. The total surface area of the Earth is 510×10^6 km². Note that most of the continental shelf regions of the Earth coincide with passive margins, underlain by stretched continental crust. (b) Histograms that compare the elevation distribution of Earth, Mars, Venus, and Moon, showing that the bimodality of Earth's surface elevation (due to the presence of oceans and continents is not found on other planets). (c) Cross section of Earth's crust, showing various bathymetric features of the sea floor.

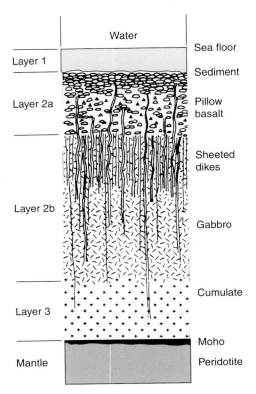

FIGURE 14.6 Columnar section of the oceanic crust and upper mantle. This section schematically represents the layers of the crust between the sea floor and the Moho.

of continental crust is, on average, four times that of oceanic crust. Thus, continental crust has an average thickness of about 35–40 km. In addition, continental crustal thickness is much more variable than oceanic crustal thickness. For example, beneath mountain belts formed where two continents collide and squash together, the crust can attain a thickness of up to 70 km, while beneath active rifts, where plates are being stretched and pulled apart, the crust thins to less than 25 km (Figure 14.7a). Next, if you were to calculate the average relative abundances of elements comprising continental crust, you would find that it has, overall, a silicic to intermediate composition (i.e., a composition comparable to that of granite to granodiorite). Thus, it is less mafic, overall, than oceanic crust. Finally, continental crust is very heterogeneous. In contrast, oceanic crust, which all forms in the same way at mid-ocean ridges, has the same overall structure everywhere on the planet. Specifically, continental crust consists of a great variety of different igneous, metamorphic, and sedimentary rocks formed at different times and in different tectonic settings during Earth history. The varieties of rock types have been deposited in succession, interleaved by faulting, corrugated by folding, or juxtaposed by intrusion (Figure 14.7b).

In the older part of continents can we find gross layering. Here, the upper continental crust has an average chemical composition that resembles that of granite

TABLE 14.2	DIFFERENCES BETWEEN OCEANIC AND CONTINENTAL CRUST
Composition	Continental crust has a mean composition that is less mafic than that of oceanic crust.
Mode of formation	Continental crust is an amalgamation of rock that originally formed at volcanic arcs or hot spots, and then subsequently passes through the rock cycle. Mountain building, erosion and sedimentation, and continued volcanism add to or change continental crust. Oceanic crust all forms at mid-ocean ridges by the process of seafloor spreading.
Thickness	Continental crust ranges between 25 km and 70 km in thickness. Most oceanic crust is between 6 km and 10 km thick. Thus, continental crust is thicker than oceanic crust.
Heterogeneity	Oceanic crust can all be subdivided into the same distinct layers, worldwide. Continental crust is very heterogeneous, reflecting its complex history and the fact that different regions of continental crust formed in different ways.
Age	Continental crust is buoyant relative to the upper mantle, and thus cannot be subducted. Thus, portions of the continental crust are very old (the oldest known crust is about 4 Ga). Most oceanic crust, gets carried back into the mantle during subduction, so there is no oceanic crust on Earth older than about 200 Ma, with the exception of the oceanic crust in ophiolites that have been emplaced and preserved on continents.
Moho	The Moho at the base of the oceanic crust is very sharp, suggesting that the boundary between crust and mantle is sharp. The continental Moho tends to be less distinct.

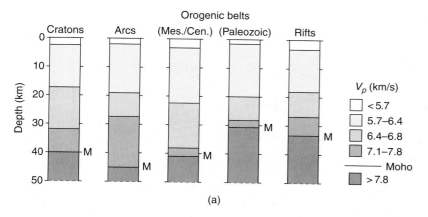

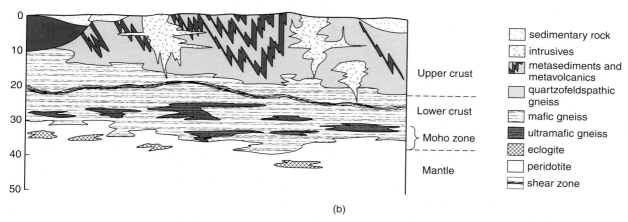

FIGURE 14.7 (a) Vertical sections through continental crust of different tectonic settings, based primarily on seismic data. The P-wave velocities are shown; M is Moho. Note the varying thickness of the highest velocity layer in the lower crust, which ranges from 20 km in arc regions to only a few kilometers in young orogenic belts. (b) Schematic cross section of the continental crust, illustrating its complexity.

(a felsic igneous rock) or granodiorite (an intermediate composition igneous rock), while the lower continental crust has an average chemical composition that resembles that of basalt (a mafic igneous rock). The boundary between upper crust and lower crust occurs at a depth of about 25 km. Contrasts in composition between upper and lower continental crust probably resulted, in part, from the differentiation of the continental crust during the Precambrian. Perhaps, partial melting of the lower crust created intermediate and silicic magmas which then rose to shallow levels before solidifying, leaving mafic rock behind. Mafic rock in the lower crust may also have formed from magma that formed by partial melting in the mantle, that rose and pooled at the base of the crust or intruded to form sills near the base of the crust. The process by which mafic rock gets added to the base of the crust is called **magmatic underplating.** Note that the compositional similarity between the lower continental crust and the oceanic crust does *not* mean that there is oceanic crust at the base of continents.

The heterogeneity of the continental crust reflects its long and complicated history. Most of the atoms forming the present continental crust were extracted from the Earth's mantle by partial melting and subsequent rise of magma in Archean and Paleoproterozoic time, between about 3.9 Ga and 2.0 Ga. Early continental crust grew by the amalgamation of volcanic arcs formed along convergent plate boundaries, as well as of oceanic plateaus formed above hot spots. Once formed, continental crust is sufficiently buoyant that it cannot be subducted, but it can be recycled through the stages of the rock cycle.

Because of its heterogeneity, geologists have found that it is useful to distinguish among several categories of continental crust (Figures 14.7a and 14.8a). Differences among these categories are based on: the age and type of rock making up the crust; the time when the crust was last involved in pervasive metamorphism and deformation; and the style of tectonism that has affected the crust.

Precambrian shields are broad regions of continents in which Precambrian rocks are presently exposed at the Earth's surface (Figure 14.8b). Most of these rocks are plutonic or metamorphic, but locally, shields contain exposures of Precambrian volcanic and sedimentary rocks. Some shields (e.g., the Canadian and Siberian shields) have relatively low elevations and subdued topography, but others (e.g., the Brazilian shield) attain elevations of up to 2 km above sea level and have been deeply disected by river erosion. Fault-bounded slices of Precambrian rocks incorporated in Phanerozoic mountain belts are not considered to be shields.

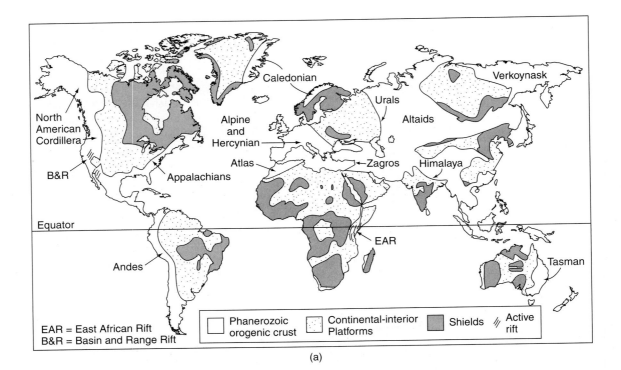

(a)

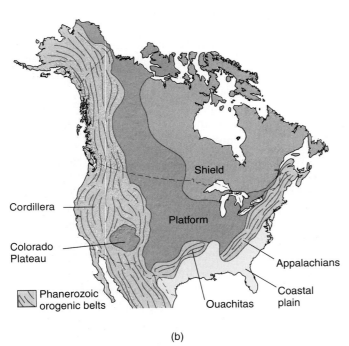

(b)

FIGURE 14.8 [a] Map of the continents, showing the distribution of different kinds of continental crust. Ancient rifts and details of mountain belts are not shown. [b] Detail of North America, illustrating the location of the platform and shield. The coastal plain is the region where the Appalachian Orogen has been buried by Mesozoic and Cenozoic strata.

Continental platforms are regions where Precambrian rocks are currently covered by a relatively thin veneer of unmetamorphosed and generally flat-lying latest Precambrian (Neoproterozoic) or Phanerozoic sedimentary strata. Geologists commonly refer to the rock below the sedimentary veneer as "basement" and the sedimentary veneer itself as "cover." Cover strata of platforms is, on average, less that 2 km thick, but in intracratonic basins, it may reach a maximum thick-

ness of 5 to 7 km. The Great Plains region of the American Midwest is part of North America's continental platform (Figure 14.8b).

Cratons are the old, relatively stable portions of continental crust. In this context, we use the word "stable" to mean that cratonic crust has not been pervasively deformed or metamorphosed during at least the past one billion years. As such, cratons include both continental platforms and pre-Neoproterozoic Precam-

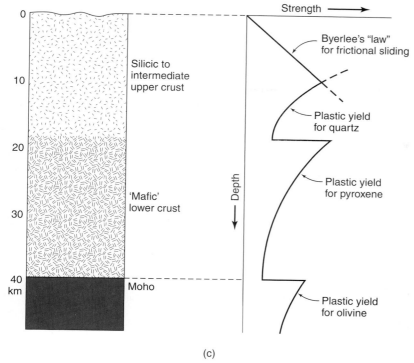

FIGURE 14.8 (continued) (c) Schematic graph illustrating the variation of strength with depth beneath the continental crust. Strength (measured in arbitrary units of stress) is controlled by frictional sliding in the upper crust, and by plastic deformation mechanisms below. The composition of the crust determines which mineral's flow law determines crustal strength.

brian shields (Figure 14.8b). Because of their age, cratons tend to be relatively cool (i.e., they are regions with low heat flow), and thus quite strong. Beneath cratons, the crust has a thickness of 35 to 40 km.

Younger orogenic belts are regions of continental crust that have been involved in orogeny during the Neoproterozoic or Phanerozoic. This means that they contain igneous rocks, metamorphic rocks, and geologic structures that are younger than about one billion years old. The crust in younger orogenic belts can be either "juvenile" (meaning that it formed by extraction of new melt from the mantle at the time of orogeny), or "recycled" (meaning that it is composed of older rock that was remelted, was intensely metamorphosed, or was transformed into sediment and then lithified at the time of orogeny). Geologists subdivide younger orogenic belts into two subcategories: **active or recently active orogens** (e.g., the Alps, the Andes, and the Himalays) are those in which topography and crustal thickness still reflect the consequences of mountain-building processes, while **inactive orogens** (e.g., the Appalachians, the Caledonides, and the Brasiliano/Pan African orogens in South America and Africa) are those in which tectonic activity ceased so long ago that erosion has exhumed deep levels of the orogen and the crust is not necessarily anomalously thick. Notably, in inactive orogens, current topography is not necessarily a relict of the uplift accompanying the original orogeny, but may be due to a much younger, independent uplift event. Because crust slowly cools through time, younger orogenic belts are warmer than cratons. Not surprsingly, heat flow in active orogens is greater than that of inactive orogens. Younger orogens generally have thicker crust than do cratons. In fact, beneath some active orogens (e.g., the Himalayan orogen), the crust reaches a thickness of 65 to 70 km.

Rifts are regions where the continental lithosphere has been stretched and thinned. As we will see, rifting leads to the formation of normal faults, formation of deep sedimentary basins, and volcanic activity. Geologists subdivide rifts into **active rifts,** where stretching continues today and active faulting is occurring (e.g., the Basin and Range of the western United States, and the East African Rift of Africa), and **inactive rifts,** where stretching ceased long ago (e.g., the Mesozoic rifts of the eastern United States). Crust in active rifts may be only 25 km thick, and is very warm—heat flow in active rifts may be many times that of cratons. Crustal thickness in inactive rifts is variable, as it reflects the history of tectonic events that the rift has endured subsequent to its original formation.

Passive margins are regions of continental crust that were stretched during a rifting event that succeeded in breaking apart a supercontinent to form two pieces that are now separated by an ocean basin. When rifting ceased and sea-floor spreading began, the relict of the rift became the tectonically inactive transition between the continent and the ocean basin (hence the name "passive margin"). As time passed, the stretched continental crust of passive margins slowly subsided (sank), and became buried by a very thick wedge of sediment derived from erosion of the adjacent continent or from the growth of reefs. The top surface of this sedimentary wedge forms a broad continental shelf (e.g., the region of relatively shallow ocean bordering the east coast of North America). Crustal thickness (including the wedge of sediment) and heat flow of a passive margin depend on the age of the margin. Older margins are thicker and cooler.

The strength of continental crust depends on its composition, thickness, and temperature. Temperature, as represented by heat flow, is particularly important because warmer crust is weaker than cooler crust. This relationship between temperature and strength simply reflects the fact that warmer rock is more ductile than cold rock. Continental crustal types can be ranked in order of relative strength. Active rifts are the weakest. The strength of younger orogenic belts varies greatly depending on the age of the belt, for their strength depends on their warmth. Specifically, Cenozoic orogens are warmer and weaker than Early Paleozoic orogens. In cratons, old shields are the strongest, while continental platforms are weaker, but both types of crust are stronger than younger orogenic belts. The strength of crust determines its behavior during tectonic activity. For example, the Cenozoic collision of India with Asia had little effect on the strong, old craton of India, but transformed southern Asia, an amalgamation of weak younger orogenic belts, into a very broad mobile belt. The interaction somewhat resembles a collision between an armored bank truck and a mountain of jelly. Similarly, it is much easier to rift apart a supercontinent along a recent orogen, than it is to rift apart an Archean craton.

Because the strength of continental crust varies with composition and temperature, strength varies with depth. The graph of Figure 14.8c illustrates this behavior. In the upper, cooler crust, strength is controlled by the frictional behavior of rocks as defined by Byerlee's law (see Chapter 6). Since frictional strength is linearly proportional to the normal stress across the sliding surface, strength increases with depth, following a straight-line relation. At greater depth, below the brittle–plastic transition, temperature and confining pressure become great enough that the rock becomes plastic. At these depths, the strength of rock is controlled by the plasticity of quartz, the dominant mineral of upper crustal rocks. Plastic strength decreases exponentially with increasing depth, because strength depends on temperature and temperature increases with depth. At the boundary between the upper and lower crust, rock composition changes and olivine becomes the dominant mineral. Olivine is stronger than quartz under the same conditions of pressure and temperature, so we predict an abrupt increase in strength at this boundary. Below the boundary, strength again decreases exponentially, due to the plastic behavior of olivine.

14.4.3 The Moho

The **Moho,** short for *Mohorovičić discontinuity,* is a seismic discontinuity marking the base of the crust.

Above the Moho, seismic P-waves travel at about 6 km/s, while below the Moho, their velocities are more than 8 km/s (Figure 14.7a). Beneath the ocean floor, this discontinuity is very distinct and probably represents a very abrupt contact between two rock types (gabbro above and peridotite below). Beneath continents, however, the Moho is less distinct; it is a zone that is 2–6 km thick in which P-wave velocity discontinuously changes. Locally, continental Moho appears to coincide with a distinct reflector in seismic-reflection profiles, but this is not always the case. Possibly the Moho beneath continents represents different features in different locations. In some locations, it is the contact between mantle and crustal rock types, whereas elsewhere it may be a zone of sill-like mafic intrusions or underplated layers in the lower crust.

14.5 THE MANTLE

14.5.1 Internal Structure of the Mantle

The mantle comprises the portion of the Earth's interior between the crust and the core. Since the top of the mantle lies at a depth of about 7–70 km, depending on location, while its base lies at a depth of 2,900 km beneath the surface, the mantle contains most of the Earth's mass. In general, the mantle consists of very hot, but solid rock. However, at depths of about 100–200 km beneath the oceanic crust, the mantle has undergone a slight amount of partial melting. Here, about 2–4% of the rock has transformed into magma, and this magma occurs in very thin films along grain boundaries or in small pores between grains. Chemically, mantle rock has the composition of **peridotite,** meaning that it is an ultramafic igneous rock (i.e., it contains a very high proportion of magnesium and iron oxide, relative to silica).

Researchers have concluded that the chemical composition of the mantle is roughly uniform throughout. Nevertheless, the mantle can be subdivided into three distinct layers, delineated by seismic-velocity discontinuities. These discontinuities probably mark depths at which minerals making up mantle peridotite undergo abrupt phase changes. At shallow depths, mantle peridotite contains olivine, but at greater depths, where pressures are greater, the olivine lattice becomes unstable, and atoms rearrange to form a different, more compact lattice. The resulting mineral is called β-phase. The phase change from olivine to β-phase takes place without affecting the overall chemical composition. The three layers of the mantle, from top to bottom, are as follows:

- **Upper mantle.** This is the shell between the Moho and a depth of about 400 km. Beneath ocean basins, the interval between about 100 km and 200 km has anomalously low seismic velocities; this interval is known as the **low-velocity zone.** As noted earlier, the slowness of seismic waves in the low-velocity zone may be due to the presence of partial melt; 2–4% of the rock occurs as magma in thin films along grain surfaces, or in small pores between grains. Note that the low-velocity zone constitutes only part of the upper mantle and that it probably does not exist beneath continents.
- **Transition zone.** This is the interval between a depth of 400 km and 670 km. Within this interval, we observe several abrupt jumps in seismic velocity, probably due to a succession of phase changes in mantle minerals.
- **Lower mantle.** This is the interval between a depth of 670 km and a depth of 2,900 km. Here, temperature, pressure, and seismic velocity gradually increase with depth.

As we noted above, the mantle consists entirely of solid rock, except in the low-velocity zone. But mantle rock is so hot that it behaves plastically and can flow very slowly (at rates of no more than a few centimeters per year). Given the slow rate of movement, it would take a mass of rock about 50–100 million years to rise from the base to the top of the mantle

Because of its ability to flow, the mantle slowly undergoes **convection** (Table 14.3). Convection is a mode of heat transfer during which hot material rises, while cold material sinks. (You can see convection occurring as you warm soup on a stove.) Convective movement takes place in the mantle because heat from the core warms the base of the mantle. Warm rock is less dense than cool rock, and thus feels a buoyant force pushing it upwards. In other words, warmer rock has "positive buoyancy" when imbedded in cooler rock. In the mantle, this buoyant force exceeds the strength of plastic peridotite, and thus buoyant rock originating deep in the mantle can rise. As it does so, it pushes aside other rock in its path, like a block of wood rising through water pushes aside the water in its path. Meanwhile, cold rock, at the top of the mantle, is denser than its surroundings, has "negative buoyancy" and sinks, like an anchor sinking through water. Geologists do not know for sure whether the upper mantle and lower mantle convect independently, and remain as separate chemical reservoirs, or if they mix entirely or at least partially during convection.

Because of convection, the onion-like image of the mantle that we provided above is actually an oversimplification. **Seismic tomography** (a technique for generating a three dimensional image of the Earth's interior; Table 14.1 and Figure 14.9) suggests that, in fact, the mantle is heterogeneous within

TABLE 14.3	TYPES OF HEAT TRANSPORT INSIDE THE EARTH
Conduction	This phenomenon occurs when you place one end of an iron bar in a fire; heat gradually moves along the bar so that the other end eventually gets hot too. We say that heat flows along the bar by conduction. Note that iron atoms do not physically move from the hot end of the bar to the cool end. Rather, what happens is that the atoms nearest the fire, when heated, vibrate faster, and this vibration, in turn, causes adjacent atoms further from the fire to vibrate faster, and so on, until the whole bar becomes warm.
Convection	This phenomenon occurs when you place a pot of soup on a hot stove. Conduction through the base of the pot causes the soup at the bottom to heat up. The heated soup becomes less dense than the overlying (cold) soup; this density inversion is unstable in a gravity field. Consequently, the hot soup starts to rise, to be replaced by cold soup that sinks. Thus, convection is driven by density gradients that generate buoyancy forces and occurs by physical flow of hot material. Convection occurs when (1) the rate at which heat is added at the bottom exceeds the rate at which heat can be conducted upward through the layer, and (2) the material that gets heated is able to flow.
Advection	This phenomenon occurs when water from a boiler flows through pipes and the pipes heat up. The water carries heat with it, and the heat conducts from the water into the metal. Thus, advection is the process by which a moving fluid brings heat into a solid or removes heat from a solid. In the Earth, heating by advection occurs where hot water or hot magma passes through fractures in rock, and heats the surrounding rock. Cooling by advection occurs where cold seawater sinks into the oceanic crust, absorbs heat, and then rises, carrying the heat back to the sea.

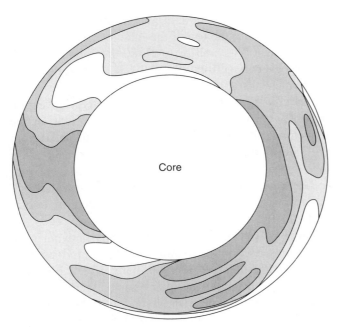

FIGURE 14.9 Seismic tomographic image of the mantle. The shading indicates differences in the velocity of seismic waves, with the region of highest velocity (coldest material) indicated by the darkest color and the region of lowest velocity (hotter material) by white. Note that, in places, these patterns suggest mantle convection, in which relatively hot material rises from the core-mantle boundary region to the upper mantle.

Geologists speculate that hot-spot volcanoes form above **mantle plumes,** columns of hot mantle rising from just above the core-mantle boundary. According to this model, mantle plumes form because heat conducting out of the core warms the base of the mantle, creating a particularly hot, positively buoyant layer.[3] At various locations, the surface of this layer bulges up, and a dome of hot rock begins to rise diapirically. This dome evolves into a column or "plume." Beneath a hot spot volcano, a plume has risen all the way to the base of the lithosphere. Because of the decompression that accompanies this rise, the peridotite at the top of the plume partially melts, producing basaltic magma. This magma eventually rises through the lithosphere to erupt at the Earth's surface.

14.6 THE CORE

The core of the Earth has a diameter of about 3,481 km; thus, the core is about the size of the Moon. Because of its great density, geologists have concluded that the core consists of iron alloy. As indicated in Figure 14.3a, a seismic discontinuity divides the core into two parts. Note that seismic P-waves (compressional waves) can travel through both the outer core and the inner core, while seismic S-waves (shear waves) travel through the inner core but not through the outer core. Shear waves cannot travel through a liquid, so this observation means that the outer core consists of liquid (molten) iron alloy, while the inner core consists of solid iron alloy. Convective flow of iron alloy in the outer core probably generates Earth's magnetic field. Of note, recent studies indicate that the inner core does not rotate with the same velocity as the mantle. Rather, the inner core spins slightly faster. Similar studies show that the inner core is anisotropic, so that seismic waves travel at different speeds in different directions through the inner core.

the layers. Specifically, tomographic studies reveal that, at a given depth, the mantle contains regions where seismic velocities are a little faster, and regions where seismic velocities are a little slower, than an average value. These variations primarily reflect variations in temperature (waves slow down in hotter, softer mantle). An overall tomographic image shows blobs and swirls of faster-transmitting mantle interspersed with zones of slower-transmitting mantle. The slower-transmitting mantle is warmer, and less dense, and thus is rising, while the faster-transmitting mantle is colder, and denser, and thus is sinking.

14.5.2 Mantle Plumes

A map of active volcanoes on the Earth reveals that most volcanoes occur in chains (which, as we see later in the chapter, occur along the boundaries between plates), but that some volcanoes occur in isolation. For example, several active, or recently active, volcanoes comprise the Cascades chain in Oregon and Washington, but the big island of Hawaii erupts by itself in the middle of the Pacific Ocean, far distant from any other volcano. Geologists refer to isolated volcanic sites as hot spots. There are about 100 **hot spots** currently active on the surface of the Earth.

14.7 DEFINING EARTH LAYERS BASED ON RHEOLOGIC BEHAVIOR

In the early years of the twentieth century, long before the discovery of plate tectonics, geologists noticed that the thickness of the crust was not uniform, and that

---

[3]Seismologists refer to this thin hot layer as the D″ layer (D″ is pronounced "D double prime").

where crust was thicker, its top surface lay at higher elevations. This behavior reminded geologists of a bathtub experiment in which wood blocks of different thickness were placed in water—the surface of the thicker blocks lay at a higher elevation than the surface of thinner blocks. Could the crust (or the crust together with the topmost layer of mantle) be "floating" on a weaker layer of mantle below? If so, then the mantle, below a certain depth, would have to be able to respond to stress by flowing. Geologists also noticed that when a large glacier grew on the surface of the Earth, it pushed the surface of the Earth down, creating a broad dimple that was wider than the load itself. (To picture this dimple, imagine the shape of the indentation that you make when you stand on the surface of a trampoline or a mattress.) Could the outermost layer of the Earth respond to loads by flexing, like a stiff sheet of rubber?

To explain the above phenomena, geologists realized that we needed to look at earth layering, at least for the outermost several hundred kilometers, by considering the way layers respond to stress. In other words, we had to consider **rheologic layering** (Figure 14.10a). By doing this, we can divide the outer several hundred kilometers into two layers based on the way that layers respond to stress. This perspective is different from that provided by studying seismic discontinuities in the Earth.

We refer to the outer, relatively rigid shell of the Earth that responds to stress by bending or flexing, as the **lithosphere.** The lithosphere, which consists of the crust and outermost mantle, overlies an interval of the mantle that responds to stress by plastically flowing. This plastic layer is called the **asthenosphere.** When plate tectonics theory came along in the 1960s, it proved convenient to picture the plates as large pieces of lithosphere. In fact the term "lithospheric plate" has become standard geologic jargon.[4] The plates could move over the asthenosphere because the asthenosphere is weak and plastic.

14.7.1 The Lithosphere

The **lithosphere,** derived from the Greek word "lithos" for rock (implying that it has strength, or resistance to

stress), is the uppermost rheologic layer of the Earth. As noted above, lithosphere consists of the crust and the uppermost (i.e., coolest and strongest) part of the mantle; the mantle part of the lithosphere is called the **lithospheric mantle.** As we've noted already, the lithosphere can be distinguished rheologically from the underlying layer, the asthenosphere, because, overall, the lithosphere behaves rigidly on geologic timescales. If you place a load on it, the lithosphere either supports the load or bends—it does not simply flow out of the way. In technical terms, we say that lithosphere has "flexural rigidity."

To picture what we mean by flexural rigidity, imagine a stiff rubber sheet resting on a layer of honey (Figure 14.11a). The sheet has flexural rigidity, so if you place an empty can (a small load) on the sheet, the sheet supports it, and if you place a concrete block (a large load) on the sheet, the sheet bends. If we were to do the same experiment with a material that does not have flexural rigidity, the results would be different. For example, if we place a can directly on the surface of a pool of honey, a material without flexural rigidity, the can sinks in a little and then floats, while if we place the concrete block in the honey, the block sinks (Figure 14.11b). The honey is able to flow out of the way of the block as the block sinks and can then flow over the top of the block as the block passes. In sum, **flexural rigidity** is the resistance to bending (flexure) of a material. A steel beam has a relatively high flexural rigidity, but a sheet of rubber has low flexural rigidity. Totally nonelastic materials, like honey or lava, have no flexural rigidity at all. Not all lithosphere on the Earth has the same flexural rigidity. Old, cold cratons have greater flexural rigidity than younger and warmer orogens.

Significantly, the rheologic behavior of the lithosphere affects the way in which heat can be transported through it. Because the lithosphere cannot flow easily, heat moves through the lithosphere only by conduction or advection (see Table 14.3), not by convection. In contrast, as we see later, heat is transported in the asthenosphere primarily by convection.

As we noted above, the lithosphere includes all of the crust and the uppermost part of the mantle. For simplicity, the base of the lithosphere can be defined by an isotherm, meaning an imaginary surface on which all points have the same temperature. Geologists often define the 1280°C isotherm in the mantle as the base of the lithosphere, because at approximately this temperature, olivine, the dominant silicate mineral in mantle peridotite, becomes very weak; this weakness happens because dislocation glide and climb become efficient deformation mechanisms at high temperatures.

---

[4]Oceanic plates indeed seem to be composed only of lithosphere. They are decoupled from the asthenosphere at the very weak low-velocity zone. However, recent work suggests that continental plates have roots that extend down into the asthenosphere, in that some of the asthenosphere, perhaps down to a depth of 250 km, moves with the continental lithosphere when a plate moves. This thicker entity—lithosphere plus coupled asthenosphere—has been called the "tectosphere." The coupled asthenosphere may have subtle chemical differences that make it buoyant relative to surrounding asthenosphere. For simplicity in this book, we will consider plates to be composed of only lithosphere.

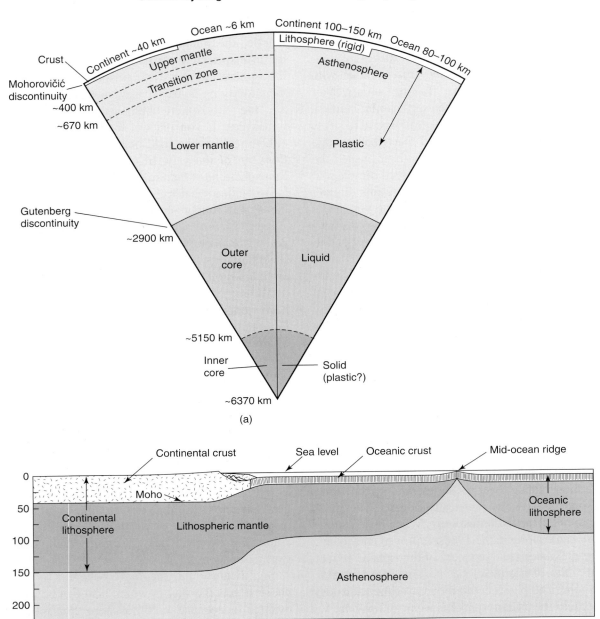

FIGURE 14.10 (a) Comparison of seismic layering and rheologic layering of the Earth. (b) Cross section of the lithosphere showing variations in lithosphere thickness.

In other words, we can view the boundary between the lithosphere and underlying asthenosphere as a **thermal boundary.** In contrast, the boundary between the crust and the upper mantle (which collectively make up the lithosphere) is a **compositional boundary,** meaning it is due to a change in chemical makeup.

The depth of the 1280° C isotherm is not fixed in the Earth, but varies depending on the thermal structure of the underlying mantle and on the duration of time that the overlying lithosphere has had to cool. Thus, the base of the lithosphere is not at a fixed depth everywhere around the Earth. Moreover, at a given location, the depth of the base of the lithosphere can change with time if the region is heated or cooled. For example, directly beneath the axis of mid-ocean ridges, the lithosphere is very thin, because the 1280° C isotherm rises almost to the base of the newly formed crust. Yet beneath the oceanic abyssal plains, oceanic lithosphere

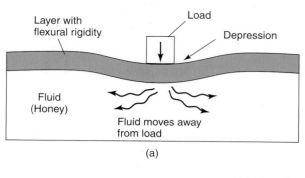

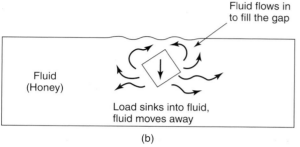

FIGURE 14.11 The concept of flexural rigidity. (a) A sheet of rubber has flexural rigidity and bends when a load is placed on it. The underlying honey flows out of the way. (b) A load placed in a material with no flexural rigidity simply sinks into it, if the load has negative buoyancy.

is old and has had time to cool; here it reaches a thickness of about 100 km. In other words, the lithospheric mantle of an oceanic plate thickens with time as seafloor spreading occurs, and the plate moves away from a mid-ocean ridge; this thickening occurs as heat conducts upward through the lithosphere to the surface of the Earth. Beneath continental cratons, the lithosphere is very ancient and may be more than 150 km thick.

14.7.2   The Asthenosphere

The **asthenosphere,**[5] which is the layer of the mantle that underlies the lithosphere, is so warm and plastic that it behaves somewhat like a viscous fluid. It does not have flexural rigidity over geologic time scales. (Though over very short timescales, it behaves elastically, for it can transmit seismic waves.) Thus, if a load is placed on the asthenosphere, the load sinks as the asthenosphere in the path of the load flows out of the way (Figure 14.11b). Because the asthenosphere has no flexural rigidity, lateral density differences that result from variations in temperature and/or composition can cause the asthenosphere to flow. Specifically, warmer and less dense parts of the asthenosphere rise,

while denser parts sink; in other words, heat transfer in the asthenosphere occurs by convection, as we've already noted (Table 14.3). Keep in mind that even though the asthenosphere is able to flow, it is almost entirely composed of solid rock—it should *not* be viewed as a subterranean sea of magma. Indeed, the only magma in the asthenosphere occurs either in the low-velocity zone, where a slight amount of melt is found as films on the surface of grains, or in regions beneath volcanic provinces, where blobs of magma accumulate and rise. Plastic flow of the asthenosphere takes place largely by dislocation movement and diffusion.

As we've also noted, the top of the asthenosphere can be defined approximately by the 1280° C isotherm, for above this thermal boundary, peridotite is cool enough to behave rigidly, while below this boundary, peridotite is warm enough to behave plastically. There appears to be no significant contrast in chemical composition across this boundary. Defining the base of the asthenosphere is a matter of semantics, because all mantle below the lithosphere is soft enough to flow plastically. For convenience, some geoscientists consider the asthenosphere to be equivalent to the low-velocity zone in the upper portion of the upper mantle. Others, equate it with the interval of mantle between the base of the lithosphere and the top of the transition zone. Still others equate it with the interval between the base of the lithosphere and the top of the lower mantle, or even with all the mantle below the lithosphere. There really isn't a consensus on this issue.

14.7.3   Isostasy

Now that we have addressed the concept of lithosphere and asthenosphere, we can consider the principle of isostasy (or, simply, "isostasy"), which is an application of Archimedes' law of buoyancy to the Earth. Knowledge of isostasy will help you to understand the elevation of mountain ranges and the nature of gravity anomalies. To discuss isostasy, we must first review Archimedes' law.

Archimedes' law states that when you place a block of wood in a bathtub full of water, the block sinks until the mass of the water displaced by the block is equal to the mass of the whole block (Figure 14.12a). Since wood is less dense than the water, some of the block protrudes above the water, just like an iceberg protrudes above the sea. When you place two wood blocks of different thicknesses into the water, the surface of the thicker block floats higher than the surface of the thinner block, yet the proportion of the thick block above the water is the same as the proportion of the

---

[5]The term is derived from the Greek *asthenes,* meaning "weak."

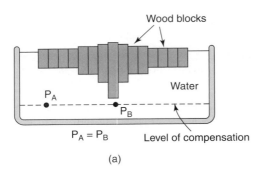

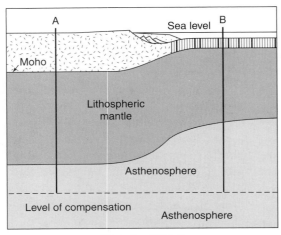

FIGURE 14.12 The concept of isostasy. (a) Blocks of wood of different thicknesses float at different elevations when placed in water. Therefore, the pressure at point A is the same as the pressure at point B. (b) In the Earth, isostasy requires that the mass of a column drilled down to the level of compensation at A equals the mass of a column drilled at B, if isostatic equilibrium exists at both locations.

thin block above the water. Thus, the base of the thick block lies at a greater depth than the base of the thin block. Now, imagine that you put two wood blocks of the same thickness but of different density in the water; this will be the case if one block is made of oak and the other is made of pine. The dense oak block floats lower than the less-dense pine block does. Note that, in the experiment, the pressure in the water at the base of the tub is the same regardless of which block floats above. Also, if you push down on or pull up on the surface of a block, it will no longer float at its proper depth.

If we make an analogy between the real Earth and our bathtub experiment, the lithosphere plays the role of the wooden blocks and the asthenosphere plays the role of the water. For a given thickness of lithosphere, the surface of more buoyant lithosphere floats higher than the surface of less buoyant lithosphere, if the lithosphere is free to float. Further, the pressure in the

asthenosphere, at a depth well below the base of the lithosphere, is the same regardless of thickness and/or density of the lithosphere floating above (if the lithosphere is floating at the proper depth). We call a depth in the asthenosphere at which the pressure is the same, regardless of location, a **depth of compensation.**

With this image of floating lithosphere in mind, we can now state the **principle of isostasy** more formally as follows: When free to move vertically, lithosphere floats at an appropriate level in the asthenosphere so that the pressure at a depth of compensation in the asthenosphere well below the base of the lithosphere is the same. Where this condition is met, we say that the lithosphere is "isostatically compensated" or in "isostatic equilibrium."

Another way to picture isostatic equilibrium is as follows. If a location in the ocean lithosphere and a location in the continental lithosphere are both isostatically compensated, then a column from the Earth's surface to the depth of compensation at the ocean location has the same mass as a column of the same diameter to the same depth in the continental location (Figure 14.12b). Ocean basins exist because ocean crust is denser and thinner than continental crust, and thus ocean lithosphere sinks deeper into the asthenosphere than does continental lithosphere. Low-density water fills the space between the surface of the oceanic crust and the surface of the Earth.

Note that with isostasy in mind, we see that changing the relative proportions of crust and mantle within the lithosphere will change the depth to which the lithosphere sinks and thus will change the elevation of the lithosphere's surface. This happens because crustal rocks are less dense than mantle rocks. For example, if we increase the proportion of buoyant crust (by thickening the crust beneath a mountain range or by underplating magma to the base of crust), the surface of the lithosphere lies higher, and if we remove dense lithospheric mantle from the base of the plate, the plate rises.

If the lithosphere doesn't float at an appropriate depth, we say that the lithosphere is "uncompensated." Uncompensated lithosphere may occur, for example, where a relatively buoyant piece of lithosphere lies embedded within a broad region of less buoyant lithosphere. Because of its flexural rigidity, the surrounding lithosphere can hold the buoyant piece down at a level below the level that it would float to if unimpeded. The presence of uncompensated lithosphere causes gravity anomalies. Positive anomalies (gravitational pull is greater than expected) occur where there is excess mass, while negative anomalies (gravitational pull is less than expected) occur where there is too little mass.

14.8 THE TENETS OF PLATE TECTONICS THEORY

While lying in a hospital room, recovering from wounds he received in World War I, Alfred Wegener (1880–1930), a German meteorologist and geologist, pondered the history of the Earth. He wondered why the eastern coastlines of North and South America looked like they could cuddle snugly against the western coastlines of Europe and Africa. He wondered why glacial tills of Late Paleozoic age crop out in India and Australia, land masses that now lie close to the equator. And he wondered how a species of land-dwelling lizard could have evolved at the same time on different continents now separated from one another by a vast ocean. Eventually, Wegener realized that all these phenomena made sense if, in the past, the continents fit together like pieces of a jigsaw puzzle into one "supercontinent," a vast landmass that he dubbed **Pangaea.** According to Wegener, the supercontinent of Pangaea broke apart in the Mesozoic to form separate, smaller continents that have since moved to new locations on the Earth's surface. Wegener referred to this movement as **continental drift.** Before the breakup of Pangaea, Wegener speculated, the Atlantic Ocean didn't exist, so lizards could easily have walked from South America to Africa to Australia without getting their feet wet, and land that now lies in subtropical latitudes instead lay near the South Pole where it could have been glaciated.

Though Wegener's proposal that continents drift across Earth's surface seemed to explain many geologic phenomena, the idea did not gain favor with most geoscientists of the day because Wegener (who died while sledging across Greenland in 1931) could not provide an adequate explanation of *why* continents moved. Such an explanation did not appear until 1960, when an American geologist named Harry Hess circulated a manuscript in which he proposed a process known as **seafloor spreading** (a concept also credited to Robert S. Dietz). During seafloor spreading, new ocean floor forms along the axis of a submarine mountain range called a mid-ocean ridge. Once formed, the new ocean floor moves away from the ridge axis. Two continents drift apart when seafloor spreading causes the ocean basin between them to grow wider.

In the 1960s, researchers from around the world rushed to explore the implications of the seafloor spreading hypothesis and to reexamine the phenomenon of continental drift. The result of this work led to a broad set of ideas, which together comprise the theory of plate tectonics[6] (or simply "plate tectonics"). According to this theory, the lithosphere, Earth's relatively rigid outer shell, consists of discrete pieces, called **lithosphere plates,** or simply **plates,** which move relative to one another. Continental drift and seafloor spreading are manifestations of plate motion.

Plate tectonics is a **geotectonic theory.** It is a comprehensive set of ideas that explains the development of regional geologic features, such as the distinction between oceans and continents, the origin of mountain belts, and the distribution of earthquakes, volcanoes, and rock types. Acceptance of plate tectonics represented a revolution in geology, for it led to the inevitable conclusion that Earth's surface is mobile— the map of the planet constantly changes (though very slowly).[7]

A plate can be viewed geometrically as a cap on the surface of a sphere. The border between two adjacent plates is a **plate boundary.** During plate movement, the **plate interior** (the region away from the plate boundary) stays relatively coherent and undeformed. Thus, most plate movement is accommodated by deformation along plate boundaries, and this deformation generates earthquakes. A map of earthquake epicenters defines seismic belts, and these belts define the locations of plate boundaries (Figure 14.13).

A closer look at the map of earthquake-epicenter locations shows that not all plate boundaries are sharp lines. Along some, such as the boundary that occurs between India and Asia, and the boundary that occurs between parts of western North America and the Pacific Plate, earthquakes scatter over a fairly broad area. Such **diffuse plate boundaries** typically occur where plate boundaries lie within continental crust, for the quartz-rich continental crust is relatively weak and

---

[6]Recall that a "hypothesis" is simply a reasonable idea that has the potential to explain observations. A "theory" is an idea that has been rigorously tested and has not yet failed to explain relevant observations. Nevertheless, a theory could someday be proven wrong.

[7]Prior to the proposal of plate tectonics theory, most geologists had a "fixist" view of the Earth, in which continents were fixed in position through geologic time. In this context, geologists viewed mountain building to be a consequence predominantly of vertical motions. Pre–plate tectonics ideas to explain mountain building included: (1) The **geosyncline hypothesis,** which stated that mountain belts formed out of the deep, elongate sedimentary basins (known as "geosynclines") that formed along the margins of continents. According to this hypothesis, mountain building happened when the floor of a geosyncline sank deep enough for sediment to melt; the resulting magma rose and in the process deformed and metamorphosed surrounding rock; (2) The **contracting Earth hypothesis,** which stated that mountains formed when the Earth cooled, shrank, and wrinkled, much like a baked apple removed from the oven. Both ideas have been thoroughly discredited.

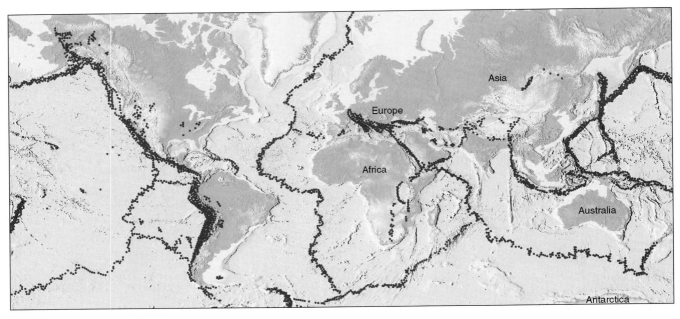

FIGURE 14.13 A map of earthquake epicenters (dots) outlines the locations of plate boundaries. Intraplate earthquakes occur away from plates.

continental crust contains many preexisting faults, so the deformation is not confined to a narrow zone. Some seismic activity and deformation does occur entirely away from a plate boundary. Such activity, called **plate-interior deformation,** probably indicates the presence of a particularly weak fault zone within a plate, capable of moving in response to the ambient stress state within a plate. Plate-interior deformation currently occurs along the New Madrid Fault System in the Mississippi Valley of the central United States.

Geoscientists distinguish three types of plate boundaries (Figure 14.14a–d). (1) At **divergent plate boundaries,** defined by mid-ocean ridges (also called oceanic ridges, because not all occur in the middle of an ocean), two plates move apart as a consequence of seafloor spreading. Thus, these boundaries are also called "spreading centers." The process of seafloor spreading produces new oceanic lithosphere (see Chapter 16). (2) At **convergent plate boundaries,** one oceanic plate sinks into the mantle beneath an overriding plate, which can be either a continental or an oceanic plate (see Chapter 17). During this process, which is called **subduction,** an existing oceanic plate gradually disappears. Thus, convergent plate boundaries are also called "destructive boundaries" or "consuming boundaries." Volatile elements (water and carbon dioxide) released from the subducted plate trigger melting in the overlying asthenosphere. The resulting magma rises and erupts in a chain of volcanoes, called

a **volcanic arc,** along the edge of the overriding plate. "Arcs" are so named because many, though not all, are curved in map view. The actual plate boundary at a convergent boundary is delineated by a deep ocean **trench.** Convergence direction is not necessarily perpendicular to the trench. Where convergence occurs at an angle of less than 90° to the trend of the trench, the movement is called **oblique convergence.** (3) At **transform plate boundaries,** one plate slides past another along a strike-slip fault. Since no new plate is created and no old plate is consumed along a transform, such a boundary can also be called a "conservative boundary." Transform plate boundaries can occur either in continental or oceanic lithosphere. At some transform boundaries, there is a slight component of convergence, leading to compressive stress. Such boundaries are called **transpressional boundaries.** Similarly, at some transform boundaries there is a slight component of extension, leading to tensile stress. Such boundaries are called **transtensional boundaries.**

In addition to the plate boundaries just discussed, geologists recognize two other locations where movement of lithosphere creates structures. (1) At **collision zones,** two buoyant blocks of crust converge (Figure 14.15a). The buoyant blocks, which may consist of continental crust, island arcs, or oceanic plateaus, are too buoyant to be subducted, so collisions result in broad belts of deformation, metamorphism, and crustal

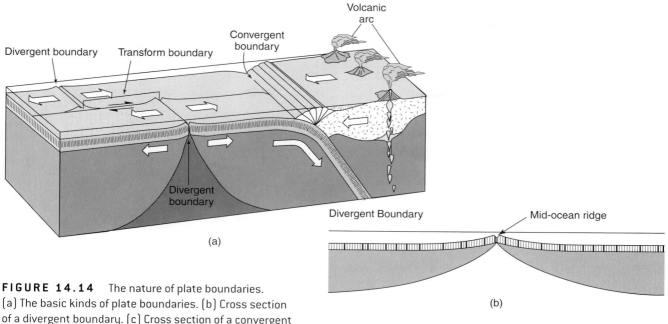

(a)

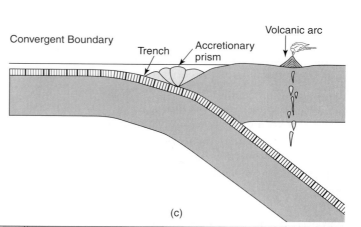

Divergent Boundary — Mid-ocean ridge

(b)

Convergent Boundary — Trench — Accretionary prism — Volcanic arc

(c)

FIGURE 14.14 The nature of plate boundaries.
(a) The basic kinds of plate boundaries. (b) Cross section
of a divergent boundary. (c) Cross section of a convergent
boundary. (d) The seven major and several minor plates of
the Earth. Ocean ridges (double lines), trenches (lines with
teeth on the overriding plate), and transform faults (single
lines) mark the plate boundaries. In a few places, plate
boundaries are ill-defined. These boundaries are marked by
dashed lines.

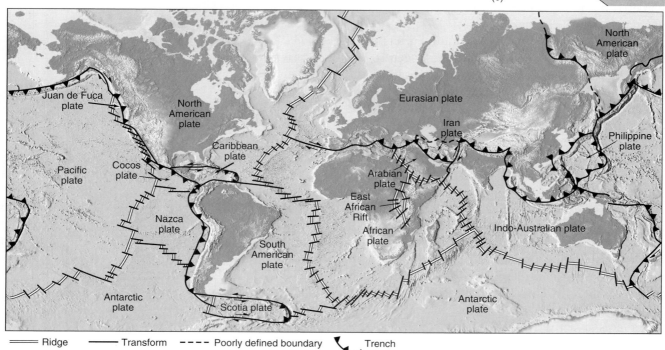

═══ Ridge   ─── Transform   - - - Poorly defined boundary   ⌣ Trench

(d)

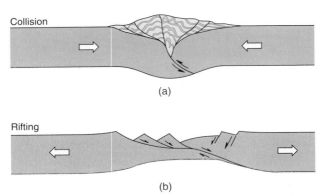

FIGURE 14.15 (a) Schematic cross section of a collision zone, where two buoyant continents converge, creating a broad belt of deformation and crustal thickening. (b) Schematic cross section of a rift, where stretching of the crust causes thinning and normal faulting.

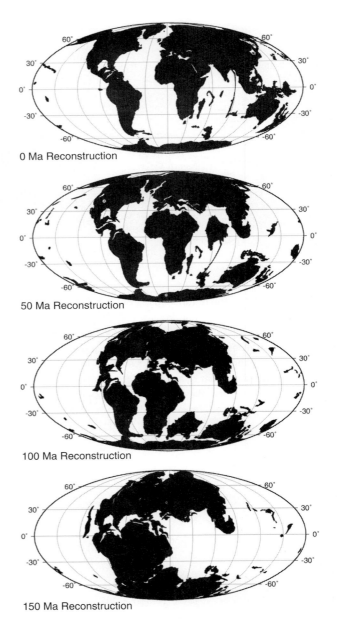

0 Ma Reconstruction

50 Ma Reconstruction

100 Ma Reconstruction

150 Ma Reconstruction

FIGURE 14.16 A sequence of maps showing the motion of continental elements on Earth, following the Early Mesozoic breakup of Pangaea (Molweide projection).

thickening (see Chapter 17). At the end of a collision, two continents that were once separate have become stuck together to form one continuous continent; the boundary here is called a **suture.** Large continents formed when several smaller continents have sutured together are called **supercontinents.** (2) At a **rift,** an existing continent stretches and starts to split apart (Figure 14.15b). At a **successful rift,** the continent splits in two and a new mid-ocean ridge forms. The stretched continental crust along the boundary of a successful rift evolves into a passive continental margin. At an **unsuccessful rift,** rifting stops before the split is complete so the rift remains as a permanent scar in the crust. It is usually marked by an elongate trough, bordered by normal faults and filled with thick sediment and/or volcanic rock.

As shown in Figure 14.13, the Earth currently has seven major plates (Pacific, North American, South American, Eurasian, African, Indo-Australian, and Antarctic Plates) and seven smaller plates (e.g., Cocos, Scotia, and Nazca Plates). In addition, Earth has several microplates. A plate can consist entirely of oceanic lithosphere (such as the Nazca Plate), but most plates consist of both oceanic and continental lithosphere (Figure 14.15). For example, the North American Plate consists of the continent of North America *and* the western half of the Atlantic Ocean. Thus, not all of today's continental margins are plate boundaries, and for this reason, we make the distinction between **active continental margins,** which are plate boundaries, and **passive continental margins,** which are not plate boundaries. The western margin of Africa is a passive margin, while the western margin of South America is an active margin. As noted earlier, passive margins form from the stretched continental crust left

by the rifting that led to the successful formation of a new mid-ocean ridge.

Continents on either side of a mid-ocean ridge move apart as seafloor spreading causes the intervening ocean basin to grow. Continents separated from each other by a subduction zone passively move together as subduction consumes intervening seafloor. Thus, plate motion causes the map of the Earth's surface to constantly change through time, so the process of **continental drift,** as envisioned by Alfred Wegener, does indeed occur (Figure 14.16).

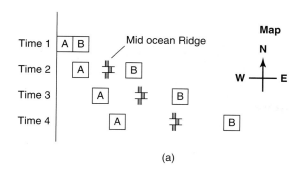

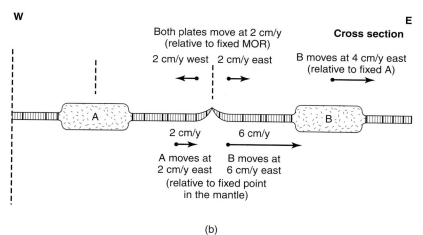

(b)

FIGURE 14.17 Describing plate velocity.
(a) A sequence of map-view drawings illustrating the movement of two continents on the surface of the Earth, as seafloor spreading occurs between them. Both continents are drifting east, in an absolute reference frame (the line on the left). At the same time, B is moving to the east relative to A, which is equivalent to saying that A is moving to the west relative to B. (b) This cross section illustrates the velocities of plates A and B in different reference frames. First we consider relative velocities. Plate A is moving west at 2 cm/y relative to the mid-ocean ridge (MOR), and plate B is moving east at 2 cm/y relative to the MOR. Thus, plate B moves east at 4 cm/y relative to a fixed plate A, and plate A moves at 4 cm/y west relative to a fixed plate A. Now we consider absolute velocities (i.e., velocities relative to a fixed point in the mantle). Plate A moves east at 2 cm/y relative to a fixed point in the mantle, while plate B moves east at 6 cm/y relative to a fixed point in the mantle. Note that the relative motion of plate B with respect to plate A occurs because plate B's absolute velocity to the east is faster than that of plate A.

14.9 BASIC PLATE KINEMATICS

When we talk of **plate kinematics,** we are referring to a description of the rates and directions of plate motion on the surface of the Earth. Description of plate motion is essentially a geometric exercise, made somewhat complex because the motion takes place on the surface of a sphere. Thus, the description must use the tools of **spherical geometry,** and to do this, we must make three assumptions: First, we must assume that the Earth is, indeed, a sphere. In reality, the Earth is slightly flattened at the poles (the Earth's radius at the poles is about 20 km less than it is at the equator), but this flattening is not sufficiently large for us to worry about. Second, we must assume that the circumference of the Earth remains constant through time. This assumption implies that the Earth neither expands nor contracts, and thus that the amount of seafloor spreading worldwide is equivalent to the amount of subduction worldwide. Note that this assumption does *not* require that the rate of subduction on one side of a specific plate be the same as the rate of spreading on the other side of that plate, but only that growth and consumption of plates are balanced for all plates combined over the whole surface of the Earth. Third, we must assume that plates are internally rigid, meaning that all motion takes place at plate boundaries. This assump-

tion, as we noted earlier, is not completely valid because continental plates do deform internally, but the error in plate motion calculations resulting from this assumption is less than a few percent.

Geoscientists use two different reference frames to describe plate motion (Figure 14.17). When using the **absolute reference frame,** we describe plate motions with respect to a fixed point in the Earth's interior. When using the **relative reference frame,** we describe the motion of one plate with respect to another. To understand this distinction, imagine that you are describing the motion of two cars cruising down the highway. If we say that Car A travels at 60 km/h, and Car B travels at 40 km/h, we are specifying the *absolute* velocity of the cars relative to a fixed point on the ground. However, if we say that Car A travels 20 km/h faster than Car B, we are specifying the *relative* velocity of Car A with respect to Car B. We will next look at how we specify absolute plate velocity and then we will examine relative plate velocity.

14.9.1 Absolute Plate Velocity

We mentioned earlier that not all volcanoes occur along plate boundaries. Some, called hot-spot volcanoes, erupt independently of plate-boundary activity;

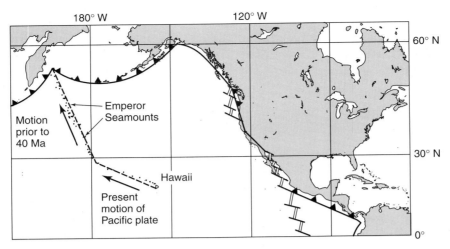

FIGURE 14.18 The Hawaiian-Emperor hot-spot track. The bend shows that the movement of the Pacific Plate relative to the Hawaiian plume changed abruptly at 40 Ma.

they form above a mantle plume. Since mantle plumes appear to be independent of plate boundaries, they can be used as the "fixed" reference points for calculating absolute plate velocities.

To see how this works, picture a lithosphere plate moving over a plume. At a given time, a hot-spot volcano forms above the plume. But as the plate moves, it carries the volcano off of the plume, and when this happens, the volcano stops erupting. Eventually, a new volcano forms above the plume, but as more time passes this volcano will also be carried off the plume. Thus, over time, a chain of volcanic islands and seamounts develops on the lithosphere. As long as the plume exists, one end of this chain will be an active volcano. The age of the volcanoes in the chain increases progressively along the chain away from the plume. These chains are called **hot-spot tracks.** The Hawaiian-Emperor chain in the Pacific Ocean serves as a classic example of a hot-spot track (Figure 14.18). Some plumes are long-lived, lasting 100 million years or more, whereas others are short-lived, lasting less than 10 million years. The orientation of a hotspot track gives the direction of plate motion, and the rate of change in the age of volcanic rocks along the length of the track represents the velocity of the plate.

Note that a distinct bend in a hot-spot track indicates that there has been a sudden change in the direction of absolute motion. For example, the bend in the Hawaiian-Emperor chain reveals that the Pacific Plate moved north-northwest before 40 Ma, but has moved northwest since 40 Ma. Geologists determined the age of this shift by radiometrically dating volcanic rocks from Midway Island, the extinct volcanic island that occurs at the bend.

The hot-spot frame of reference gives a reasonable *approximation* of absolute plate velocity, but, in reality, it's not perfect because hot spots actually do move with respect to one another. Nevertheless, the velocity of hot-spot movement is an order of magnitude less than the velocity of plate movement, so the error in absolute plate motions based on the hot-spot reference frame is only a few percent.[8]

Figure 14.19 shows the absolute velocity of plates. You will notice that absolute plate rates today range from less than 1 cm/y (for the Antarctic and African Plates) to about 10 cm/y (for the Pacific Plate). North America and South America are moving west at about 2–3 cm/y. For comparison, these rates are about the same at which your hair or fingernails grow. This may seem slow, but remember that a velocity of only 2 cm/y can yield a displacement of 2000 km in 100 million years! There has been plenty of time during Earth's history for large oceans to open and close many times.

14.9.2 Relative Plate Velocity

The motion of any plate with respect to another can be defined by imagining that the position of one of the plates is fixed. For example, if we want to describe the motion of plate A with respect to plate B, we fix

---

[8]To get more accurate measurements of absolute plate motion, geoscientists use the **no net torque calculation.** In this calculation, we assume that the sum of all the plate movements shearing against the underlying asthenosphere creates zero torque on the asthenosphere. The no net torque calculation is well suited for computer calculations. See Cox and Hardt (1986) for details.

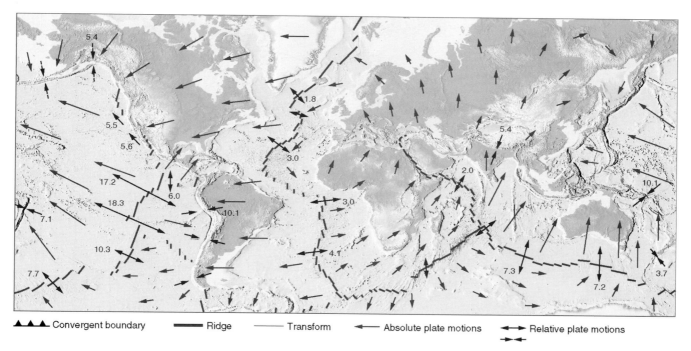

FIGURE 14.19 Directions and rates in mm/y of plate motion. Relative motions of the major plates are given for selected points along their boundaries by pairs of inward-pointing (convergent) or outward-pointing (divergent) arrows. The numbers next to these arrows give the relative velocity in cm/y.

plate B on the surface of the Earth and see how plate A moves (Figure 14.17). Recall that in plate kinematic calculations, the Earth is considered to be a sphere, so such a movement is described by a rotation at a specified angular velocity around an imaginary axis that passes through the center of the Earth. The intersection between this imaginary rotation axis and the surface of the Earth is called an **Euler pole** (Figure 14.20a–c). Note that Euler poles are merely geometric elements and that they are not related to the Earth's geographic poles (the points where the Earth's spin axis intersects the surface), nor are they related to the Earth's magnetic poles (the points where the Earth's internal dipole intersects the surface).

To avoid confusion, it is useful to distinguish between two types of Euler poles. An **instantaneous Euler pole** can be used to describe relative motion between two plates at an instant in geologic time, whereas a **finite Euler pole** can be used to describe total relative motion over a long period of geologic time. For example, the *present-day* instantaneous Euler pole describing the motion of North America with respect to Africa can be used to determine how fast Chicago is moving away from Casablanca today. We could calculate a single finite Euler pole to describe the motion between these two locations over the past

80 million years, even if the instantaneous Euler pole changed at several times during this interval.

14.9.3 Using Vectors to Describe Relative Plate Velocity

Let's now see how to describe relative plate motion with the use of vectors. Imagine two plates, A and B, that are moving with respect to each other. We can define the relative motion of plate A with respect to plate B by the vector $_A\Omega_B$, where

$$_A\Omega_B = \omega\mathbf{k} \qquad \text{Eq. 14.1}$$

In this formula, ω is the angular velocity and $\mathbf{k}$ is a unit vector parallel to the rotation axis.

For most discussions of plate kinematics, however, it is easier for us to describe motion in terms of the **linear velocity, v,** as measured in centimeters per year at a point on a plate. For example, we can say that New York City, a point on the North American Plate, is moving west at 2.5 cm/y with respect to Paris, a point on the Eurasian Plate. Note that we can *only* describe **v** if we specify the point at which **v** is to be measured. If we know the value of $_A\Omega_B$ defining the relative plate motion, we can calculate the value of **v** at a point.

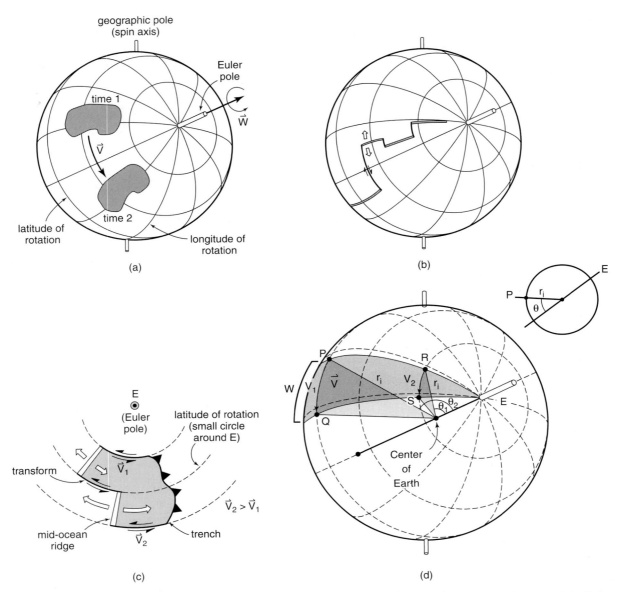

FIGURE 14.20 Describing the motion of plates on the Earth's surface. (a) Latitudes and longitudes around the Euler pole (E) are small circles and great circles, respectively. The motion of a plate between time 1 and time 2 is a **rotation (ω)** around the Euler axis. The linear velocity (v) is measured on the surface of the Earth. (b) Transform faults are small circles (latitudes of rotation) relative to the Euler pole. (c) We can also see the relation of transforms relative to the Euler pole in a projection looking down on the Euler pole. Note that spreading velocity and rate of slip along transform faults increase with increasing distance away from the Euler pole. (d) Linear velocity increases as the angle θ increases. r_i is the radius vector to a point on the Earth's surface. Thus, since $\theta_2 > \theta_1$, $V_2 > V_1$.

Specifically, **v** is the vector cross product of $_A\Omega_B$ and the radius vector ($\mathbf{r}_i$) drawn from the center of the Earth to the point in question. This relation can be represented by the equation

$$\mathbf{v} = {}_A\Omega_B \times \mathbf{r}_i \qquad \text{Eq. 14.2.}$$

As in any vector cross product, this equation can be rewritten as

$$\mathbf{v} = \mathbf{r}_i \sin\theta \qquad \text{Eq. 14.3}$$

where θ is the angle between $\mathbf{r}_i$ and the Euler axis (Figure 14.20d).

We see from the preceding equations that **v** is a function of the distance along the surface of the Earth between the point at which **v** is determined and the Euler pole. As you get closer to the Euler pole, the value of θ becomes progressively smaller, and at the pole itself, $\theta = 0°$. Since $\sin 0° = 0$, the relative linear velocity (**v**) between two plates *at* the Euler pole is 0 cm/y. To picture this relation, think of a Beatles record playing on a stereo. Even when the record spins at a constant angular velocity, the linear velocity at the center of the record is zero and increases toward the edge of the record. Thinking again of plate motions, note that the maximum relative linear velocity occurs where $\sin \theta = 1$ (i.e., at 90° from the Euler pole). What this means is that the relative linear velocity between two plates changes along the length of a plate boundary. For example, if the boundary is a mid-ocean ridge, the spreading rate is greater at a point on the ridge at 90° from the Euler pole than it is at a point close to the Euler pole. Note that in some cases, the Euler pole lies on the plate boundary, but it does not have to be on the boundary. In fact, for many plates the Euler pole lies off the plate boundary.

Because the relative velocity between two plates can be described by a vector, plate velocity calculations obey the **closure relationship,** namely

$$_A\Omega_C = {_A\Omega_B} + {_B\Omega_C} \qquad \text{Eq. 14.4}$$

Using the closure relationship, we can calculate the relative velocity of two plates even if they do not share a common boundary. For example, to calculate the relative motion of the African Plate with respect to the Pacific Plate, we use the equation:

$$_{Africa}\Omega_{Pacific} = {_{Africa}\Omega_{S.\,America}} +$$
$$_{S.\,America}\Omega_{Nasca} + {_{Nasca}\Omega_{Pacific}} \qquad \text{Eq. 14.5}$$

An equation like this is called a **vector circuit.**

At this point, you may be asking yourself the question, how do we determine a value for $_A\Omega_B$ in the first place? Actually, we can't measure $_A\Omega_B$ directly. We must calculate it by knowing ω, which we determine, in turn, from a knowledge of **v** at various locations in the two plates or along their plate boundary. We can measure values for **v** directly along divergent and transform boundaries.

As an example of determining **v** along a divergent boundary, picture point P on the Mid-Atlantic Ridge, the divergent boundary between the African and the North American Plates. To determine the instantaneous Euler pole and the value for **v** at point P, describing the motion between these two plates, we go through the following steps:

- First, since **v** is a vector, we need to specify the *orientation* of **v,** in other words, the spreading direction. To a first approximation, the spreading direction is given by the orientation of the transform faults that connect segments of the ridge, for on a transform fault, plates slip past one another with no divergence or convergence. Geometrically, a trans-

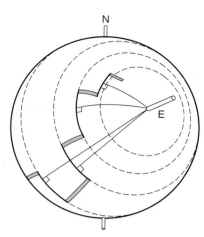

FIGURE 14.21 The location of an Euler pole can be determined by drawing a great circle (longitude of rotation) perpendicular to each of the transforms. The point at which the great circles meet is the Euler pole (E).

form fault describes a small circle around the Euler pole (Figure 14.20), just as a line of latitude describes a small circle around the Earth's geographic pole. Therefore, the direction of **v** at point P is parallel to the nearest transform fault.

- Second, we need to determine the location of the Euler pole describing the motion of Africa with respect to North America. Considering that transform faults are small circles, a great circle drawn perpendicular to a transform fault must pass through the Euler pole (Figure 14.21), just as geographic lines of longitude (great circles perpendicular to lines of latitude) must pass through the geographic pole. So, to find the position of the Euler pole, we draw great circles perpendicular to a series of transforms along the ridge, and where these great circles intersect is the Euler pole.

- Finally, we need to determine the magnitude of **v.** To do this, we determine the age of the oceanic crust on either side of point P.[9] Since velocity is distance divided by time, we simply measure the distance between two points of known, equal age on either side of the ridge to calculate the spreading velocity across the ridge. This gives us the magnitude of **v** at point P.

On continental transform boundaries, like the San Andreas Fault in California, the magnitude of **v** can be determined by matching up features of known age that

---

[9]We can determine the age of the crust by dating the fossils directly at the contact between the pillow basalt layer and the pelagic sediment layer, or we can use the age of known "marine magnetic anomalies." We will discuss these anomalies in Chapter 16.

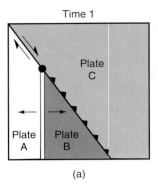
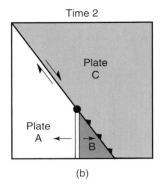

Time 1   Time 2

Plate C

Plate A   Plate B

Plate C

Plate A   B

(a)   (b)

FIGURE 14.22 Geometry of triple junctions. (a) This ridge-trench-transform triple junction is stable. (b) With time the ridge-trench-transform triple junction location changes, but the geometry stays the same.

formed across the fault after the fault formed, but are now offset by the fault. We cannot directly determine relative motion across a convergent boundary, so the value of **v** across convergent boundaries must be determined using vector circuits.

Until the end of the twentieth century, relative plate motion could not be confirmed by direct observation. But now, the **global positioning system** (GPS), which uses signals from an array of satellites to determine the location of a point on Earth's surface, makes this possible. By setting up a network of GPS stations, it is possible to define the location of a point to within a few millimeters, and with this accuracy, plate movements over a period of a few months to years can indeed be detected.

14.9.4 Triple Junctions

At this point, you should have a fairly clear idea of what we mean by a "plate boundary" between two plates. We can represent the intersection of a plate boundary with the surface of the Earth by a line on a map. The point where three plate boundaries meet is called a **triple junction.** You can see several examples of triple junctions in Figure 14.14d. We name specific types of triple junction by listing the plate boundaries that intersect. For example, at a ridge-ridge-ridge triple junction, three divergent boundaries intersect (e.g., in the southern South Atlantic and in the southern Indian Ocean), and at a ridge-trench-transform triple junction, a divergent boundary, a convergent boundary, and a transform boundary intersect (e.g., off the west coast of northern California).

Geoscientists distinguish between stable and unstable triple junctions. The basic configuration of a **stable**

triple junction can exist for a long time, even though the location of the triple junction changes. For example, the ridge-trench-transform junction in Figure 14.22a is a stable triple junction, because even though the location of the triple junction (point T) migrates to the southeast with time, the geometry stays the same. In contrast, the basic geometry of an **unstable triple junction** must change rapidly to create a new arrangement of plate boundaries.

The migration of a triple junction along a plate boundary can lead to the transformation of a segment of a plate boundary from one type of boundary to another. Such a change occurred along the coast of California, an active margin, during the Cenozoic. Through most of the Mesozoic, and into the Early Cenozoic, the western North American margin was a convergent plate boundary. Toward the end of this time interval, convergence occurred between the North American and Farallon Plates. At around 30 Ma, the Farallon-Pacific Ridge (the divergent boundary between the Farallon and Pacific Plates) began to be subducted. When this occurred, the Pacific Plate came into contact with the North American Plate and two triple junctions formed, one moving north-northwest, and the other moving south-southeast. The margin between the triple junction changed from being a convergent boundary into a transform boundary, the San Andreas Fault.

14.10 PLATE-DRIVING FORCES

Alfred Wegener was unable to convince the geologic community that continental drift happens, because he could not explain *how* it happens. The question of what drives the plates remains controversial to this day. In the years immediately following the proposal of plate tectonics, many geoscientists tacitly accepted a **convection-cell model,** which stated that convection-driven flow in the mantle drives the plates. In this model, plates were carried along on the back of flowing asthenosphere, which was thought to circulate in simple elliptical (in cross section) paths; **upwelling** (upward flow) of hot asthenosphere presumably occurred at mid-ocean ridges, while **downwelling** (downward flow) of hot asthenosphere occurred at the margins of oceans or at subduction zones. In this model, the flowing asthenosphere exerts **basal drag,** a shear stress, on the base of the plate, which is sufficient to move the plate. This image of plate motion, however, was eventually discarded for, while it is clear that the mantle does convect,[10] it is impossible to devise a global geometry of convection cells that can explain the observed geometry of plate boundaries that now exist on Earth. Subsequent calculations showed that two other forces, ridge push and slab pull, play a major role in driving plates.

Ridge-push force is the outward-directed force that pushes plates away from the axis of a mid-ocean ridge. It exists because oceanic lithosphere is higher along mid-ocean ridges than it is in the abyssal plains at a distance away from the ridge. In Earth's gravity field, the difference in elevation means that the lithosphere along the ridge has more gravitational potential energy than the lithosphere of the abyssal plain, and this energy provides an outward push perpendicular to the ridge axis. To picture this push, imagine a thin pool of molasses on a table top. If you pour more molasses onto the center of the sheet, a mound of molasses builds up in the center. The new molasses flows outward and pushes the molasses that was already on the table in front if it, causing the diameter of the molasses sheet to increase.

Slab-pull force is the force that pulls lithosphere into a convergent margin. It exists because old, cold ocean lithosphere is negatively buoyant relative to the underlying asthenosphere, so if given a chance, the oceanic lithosphere sinks downwards. Once the cold subducting plate (also called a subducted "slab") descends into the mantle, phase transformations change basalt to much denser eclogite, so a subducted plate is even denser than a plate at the Earth's surface. Thus, like a sinking anchor, a subducted slab "pulls" the rest of the plate down with it. To picture the process, imaging that you smooth a sheet of wet tissue paper onto the ceiling. If you peel one end of the paper off, and let go, the paper slowly moves down and peels the rest of the paper off with it, until the whole sheet eventually separates from the ceiling and falls.

Let's recap and consider all the forces that act on plates (Figure 14.23). Ridge-push forces drive plates away from mid-ocean ridges, and slab-pull forces drive them down subduction zones. Thus, plate motion is, to some extent, a *passive* phenomenon, in that it is

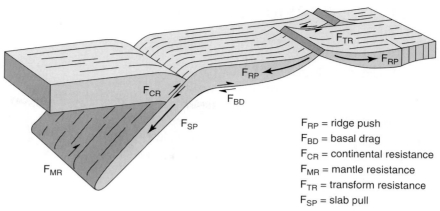

F_{RP} = ridge push
F_{BD} = basal drag
F_{CR} = continental resistance
F_{MR} = mantle resistance
F_{TR} = transform resistance
F_{SP} = slab pull

FIGURE 14.23 Simplified diagram illustrating the forces acting on a plate.

the gravitational potential energy of the plate itself that causes a plate to move. Does basal drag due to mantle convection play a role at all in driving plates? Probably yes, but not in the simple way that researchers first envisioned. Convection in the mantle, according to seismic tomography studies, seems to be accommodated *globally* by a few upwelling zones and a few downwelling zones. These zones do not exactly correspond to the present configuration of ridges and trenches. The flow of asthenosphere due to convection probably does create some basal drag at the base of plates, but the basal drag force can assist *or* retard motion. Specifically, where the asthenosphere flows in the same direction as the plate motion caused by ridge push and/or slab pull, basal drag may accelerate the motion. In contrast, where the asthenosphere flows in a direction opposite to the plate motion caused by ridge push and/or slab pull, it may retard the motion. And if the asthenosphere flows at an angle to the plate motion caused by ridge push and/or slab pull, it may change the direction of motion. Calculations show that the size of the basal-drag force caused by asthenosphere flow is less than that of ridge push or slab pull, and thus basal drag is not the dominant force. The forces that drive plates are resisted by frictional forces between plates.

Note that even though circulation of the asthenosphere alone does not drive plate motion, the mobility of the asthenosphere does make plate motion possible. If the asthenosphere could not flow up at mid-ocean ridges to fill space created by seafloor spreading, and if the asthenosphere could not move out of the way of subducting slabs, then buoyancy forces (negative or positive) would not be sufficient to cause plates to move. Further, the formation of new hot lithosphere at mid-ocean ridges and the eventual sinking of this lithosphere, once it has cooled, at trenches is itself a form of convection.

---

[10]Seismic tomography shows variations in temperature in the mantle, and submarine measurements show that heat flow is higher at mid-ocean ridges than elsewhere.

14.11 THE SUPERCONTINENT CYCLE

While Wegener's view that continental drift happens ultimately proved correct, his view that drift *only* happened after the breakup of Pangaea was just part of the story. J. Tuzo Wilson, a Canadian geophysicist, noted that formation of the Appalachian-Caledonide Orogen involved closure of an ocean. Thus, there must have been another ocean (not the Atlantic) on the east side of North America *before* the formation of Pangaea. Wilson envisioned a cycle of tectonic activity in which an ocean basin opens by rifting, grows by seafloor spreading, closes by subduction, and disappears during collision and supercontinent formation and, further, that more than one such cycle has happened during Earth history. The successive stages of rifting, seafloor spreading, convergence, collision, rifting, recorded in a single mountain range came to be called the **Wilson cycle.** Because of the Wilson cycle, we cannot find oceanic lithosphere older than about 200 Ma on Earth—it has all been subducted. Very old continental crust, however, can remain because it's too buoyant to be subducted; that's why we can still find Archean rocks on continents.

Eventually, geologists realized that Wilson cycles were part of a global succession of events leading to formation of a supercontinent, breakup of a supercontinent, dispersal of continents, and reassembly of continents into a new supercontinent. This succession of events has come to be known as the **supercontinent cycle** (Figure 14.24). At various times in the past, continental movements and collisions have produced supercontinents, which lasted for tens of millions of years before eventually rifting apart. The geologic record shows that a supercontinent, Pangaea, had formed by the end of the Paleozoic (~250 Ma), only to disperse in the Mesozoic. Similarly, a supercontinent formed at the end of the Precambrian (1.1 Ga, called Rodinia), only to disperse at about 900 Ma to form a new supercontinent, Panotia, which itself broke up about 600 Ma (i.e., at the beginning of the Paleozoic). And there is growing evidence that supercontinents also formed even earlier in Earth history. Thus, it seems that supercontinents have formed, broken up, formed, and broken up at roughly 200–500 m.y. intervals.

Modeling suggests that the supercontinent cycle may be related to long-term convection patterns in the mantle. In these models, relative motion between plates at any given time—the "details" of plate kinematics—is determined by ridge-push and slab-pull forces. But over long periods of time (about 200–500 m.y.), continents tend to accumulate over a zone of major mantle

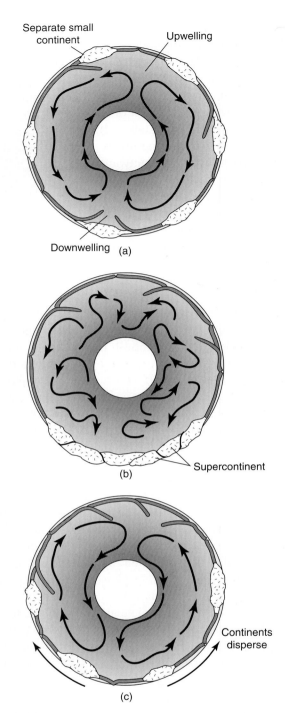

FIGURE 14.24 Stages in the supercontinent cycle. (a) Continents move relative to one another, but gradually aggregate over a mantle downwelling zone. (b) While the supercontinent exists, large-scale convection in the mantle reorganizes. (c) Upwelling begins beneath the supercontinent and weakens it. Eventually, rifting occurs and the supercontinent breaks up, forming smaller continents that drift apart.

downwelling to form a supercontinent (Figure 14.18). However, supercontinents don't last forever; once a supercontinent forms, the thermal structure of the mantle beneath it changes. This change happens because a supercontinent acts like a giant insulator that does not allow heat to escape from the mantle. Over 80% of the Earth's internal heat escapes at mid-ocean ridges, where seawater circulating through the hot crust and upper mantle cools the lithosphere much like a coolant cools an automobile engine; however, within the area of a supercontinent, there are no ridges, so heat cannot easily be lost. As a consequence, the mantle beneath the supercontinent eventually heats up, and can no longer be a region of downwelling. When this happens, a new downwelling zone develops elsewhere, and hot asthenosphere must begin to upwell beneath the supercontinent. Upwelling causes the continental lithosphere of the supercontinent to heat up and weaken. Ultimately, the supercontinent rifts apart into smaller continents separated by new oceans.

If the supercontinent-cycle model works, then we can say that Earth is currently in the dispersal stage of the cycle. During the next 200 million years, the assembly stage will begin and the floors of the Atlantic and Pacific Oceans will be subducted. After a series of collisions, the continents will once again coalesce to form a new supercontinent.[11]

14.12 CLOSING REMARKS

A complete description of whole-Earth structure and plate-tectonics theory would each fill an entire book; in fact, many such books have been written (see Additional Reading). Here, our purpose was simply to remind you of some of the key features of whole-Earth structure, and some of the key tenets of the theory. This information sets the stage for our discussion of structures and other geologic features that develop in specific kinds of tectonic settings. After reviewing basic geophysical methods, we introduce extensional tectonics, meaning the geology of rifts and passive margins. You will see that understanding the architecture of rifts and passive margins is a prerequisite for understanding the consequences of collision and convergence. You will find that many of the topics discussed in this and subsequent chapters will resurface in the essays on representative orogens in Chapter 22.

---

[11]The hypothetical supercontinent formed by the predicted collision of the Americas and Asia has been called "Amasia."

ADDITIONAL READING

Anderson, D. L., 1989. *Theory of the earth.* Blackwell Scientific Publications: Boston.

Bolt, B. A., 1982. *Inside the earth.* Freeman: San Francisco.

Brown, G. C., and Musset, A. E., 1992. *The inaccessible earth* (second edition). Chapman and Hall: London.

Butler, R. J., 1992. *Paleomagnetism: Magnetic domains to geologic terranes.* Blackwell: Boston.

Condie, K. C., 1997. *Plate tectonics and crustal evolution* (fourth edition). Butterworth: Oxford.

Condie, K. C., 2001. *Mantle plumes & their record in earth history.* Cambridge University Press: Cambridge.

Cox, A., and Hart, R. B., 1986. *Plate tectonics: how it works.* Blackwell Scientific Publications: Oxford.

Davies, G. F., 1992. Plates and plumes: dynamos of the earth's mantle. *Science, 257,* 493–494.

Fowler, C. M. R., 1990. *The solid Earth: an introduction to global geophysics.* Cambridge University Press, Cambridge.

Keary, P., and Vine, F. J., 1990. *Global tectonics.* Blackwell Scientific Publications: Oxford.

Lillie, R. J., 1999. *Whole earth geophysics.* Prentice Hall: Upper Saddle River.

Marshak, S., 2001. *Earth: portrait of a planet.* W. W. Norton & Co.: New York.

McFadden, P. L., and McElhinny, M. W., 2000. *Paleomagnetism: continents and oceans* (second edition). Academic Press.

Moores, E. M. (ed.), 1990. *Shaping the earth—tectonics of continents and oceans.* Freeman: New York, 206 pp.

Moores, E. M., and Twiss, R. J., 1995. *Tectonics.* Freeman: New York.

Musset, A. E., and Khan, M. A., 2000. *Looking into the earth.* Cambridge University Press: Cambridge.

Nance, R. D., Worsley, T. R., and Moody, J. B., 1986. Post-Archean biogeochemical cycles and long-term episodicity in tectonic processes. *Geology, 14,* 514–518.

Oreskes, N. (ed.), 2003. *Plate tectonics: an insider's history of the modern theory of the earth.* Westview Press: Boulder, CO.

Tackley, P. J., 2000. Mantle convection and plate tectonics: toward an integrated physical and chemical theory. *Science, 288,* 2002–2007.

Turcotte, D. L., and Schubert, G., 2001. *Geodynamics* (second edition). Cambridge University Press: Cambridge.

Windley, B. F., 1995. *Evolving continents* (third edition). Wiley: Chichester.

Geophysical Imaging of the Continental Lithosphere

An essay by Frederick A. Cook

15.1 INTRODUCTION

The lithosphere beneath the continents is a vast and largely unexplored region of the Earth. Because it is inaccessible to normal geologic observations, various geophysical tools, including measurements of magnetism, gravity, electricity, subsurface temperatures, and earthquake waves, have been used to gather information about it for more than 100 years. Since about 1975, however, application of controlled-source **seismic reflection methods** has produced images of the deep subsurface that are visually similar to geologic cross sections and are therefore readily accessible to most earth scientists. When analyzed in conjunction with other **geophysical imaging methods,** as well as with the known geology, these data provide the most detailed information on subsurface structure presently available.

In many areas of the continents, preliminary reconnaissance has been accomplished and a few major features have been discovered: some of these are extensions of the surface geologic context; others resemble features we know but cannot be directly linked to our geologic base; still others are unusual and outside our frame of reference. Overall, we have only begun the search.

15.2 WHAT IS SEISMIC IMAGING?

Seismic imaging methods use vibrational (elastic) energy generated on or near the Earth's surface as a source of waves that propagate into the subsurface, reflect from or refract through interfaces at depth, and then return to the surface where they are digitally recorded for subsequent data enhancement. Refracted waves are valuable to delineate regional characteristics of the crust and upper mantle, including the base of the crust, whereas reflected waves are useful for mapping structural detail. The source of energy may be produced artificially, as with an explosion or vibratory signal, or it may be natural, as with an earthquake. In general, artificial sources do not have as much energy as earthquakes do, and hence do not penetrate as deeply, but the results derived from artificial sources have much finer detail and are easier to relate to known geologic features. Seismic reflection profiling is not new. It was first employed by the petroleum industry about 75 years ago in oil exploration. However, many refinements, particularly following the advent of digital technology in the 1960s and 1970s, have made it possible to obtain images from greater depths and with greater precision than was considered feasible even just a few years ago.

The **seismic reflection method** is conceptually simple but logistically intensive (Figure 15.1). In most applications today, the method includes four essential components:

1. A vibrational energy source, usually several (3–5) synchronized, truck-mounted vibrators (to +/– 0.001 s or less). The elastic energy radiates into the subsurface, reflects off boundaries at depth, and returns to the Earth's surface where the resulting ground motion is measured by a series of miniseismometers (geophones). Once a vibration point is completed, the vibrator trucks move a few tens of meters to the next point and repeat the process.

2. A line of geophones. Each geophone is typically slightly larger than a 35-mm film container, and there are usually several thousand geophones spread over a single line to receive the signals from a single vibration location. The line of receivers is commonly several kilometers long (10–12 km or longer for most modern lithosphere-scale reflection surveys). As the sources (e.g., vibrator trucks) move for each source point, the receiving line also moves. For a survey that is 500 km long, there may be 10,000 vibration points.

3. A recording system, usually a box with computers and recording media (e.g., digital tapes) mounted on a truck. The signals are transmitted to the recording truck where they are stored for later computer enhancement. The electronics in the recording truck also provide the radio control that is sent to the vibrator trucks to initiate the vibrator signal.

4. A data processing system, which includes a suite of computer software that is applied in a series of steps to enhance the desired signal at the expense of unwanted noise. Some testing can be accomplished in the field to examine the data, but most of the intensive data processing requires weeks or months of effort to optimize the results.

The final product is a two-dimensional cross section of the earth that is displayed in terms of the signal transit time (the time required for a signal to follow a path from the surface source to a reflecting boundary at depth, and then to return to the surface). The display of the image in terms of transit time is the primary difference between a seismic cross section and a geologic cross section, as the "time section" needs to be converted to depth for an effective comparison to geologic boundaries. Once this is accomplished, the most fundamental result of all seismic profiles is an outline of the subsurface geometric framework.

Conversion to depth requires knowledge of the **seismic wave velocities** in the rocks. Although average velocities can be estimated for a general understanding of a cross section (we know that seismic waves generally travel faster in crystalline rocks than in sedimentary rocks), the more accurately the variations in wave velocity are known, the greater will be the accuracy of the image. Nevertheless, a good rule-of-thumb estimate for average seismic velocities in crystalline rocks of the crust is about 6.0 km/s; hence the depth in kilometers may be estimated by multiplying one-half of

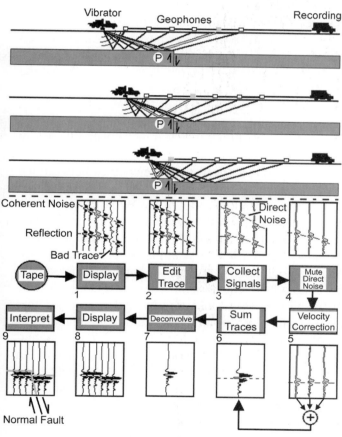

FIGURE 15.1 Schematic diagram illustrating the principles of seismic reflection profiling. At the top, three steps in field acquisition are shown to indicate how a reflection from a single subsurface point (P) is recorded by several different positions of the vibrator sources and the geophone receivers (gray geophone). This process, known as a common midpoint, or CMP, acquisition allows the recorded traces from P to be analyzed in the data processing steps (below) for the purposes of improving the signal.

the transit time by a factor of 6.0. In this conversion a transit time of 5.0 s corresponds to a depth of about 15.0 km. Below the Mohorovičić discontinuity, the seismic wave velocity increases to about 8.0 km/s.

15.3 HOW ARE DATA INTERPRETED?

Great strides have been made in using seismic reflection profiling for mapping the internal structure of the lithosphere. Reflections from the deep crust (to a depth of about 30–40 km) were considered curiosities until the 1970s; reflections from the subcrustal lithosphere to twice this depth, or even more, are not unusual now. However, with technological advancements and the resulting improvements in signal quality and image depth, the challenges to interpretations have been great, because, as we obtain images from structures that are far removed from our geologic reference on the surface, the ability to relate them to what we know is increasingly limited. In many cases, ancillary geophysical methods that respond to different physical properties may be helpful in limiting possible interpretations, because the seismic images provide a geometric framework of structures that are found at many scales in the lithosphere.

Interpretation of the resultant seismic sections is in many ways analogous to interpreting a geologic cross section; indeed, the goal of the data processing is to provide an image that closely approximates a geologic cross section. However, careful interpretations require extensive knowledge of seismic wave propagation in rocks, geologic principles, and the regional geologic and geophysical context. Interpretation is often an iterative process: new images of subsurface geometry from the reflection profiles commonly spawn new geologic (or other geophysical) projects to test some of the ideas. Data resulting from these projects can then be used to review and often reinterpret the subsurface images. It is not unusual to rework data that may be 10 or 15 years old as new data processing techniques and new geologic information become available.

Although this essay is not intended to provide a complete review of image types or all major discoveries, it is helpful to consider images of a variety of common geologic features.

15.4 SOME EXAMPLES

Many of the criteria that are commonly used to identify faults in layered sedimentary rocks are not applicable in most crystalline rocks. For example, in layered strata, faults are usually delineated by offset layers, whereas reflections from a fault plane are rare. In generally unlayered crystalline rocks, however, offsets are difficult to observe (because it is not easy to correlate from one side of a fault to the other) and reflections from fault planes or fault zones are common.

In Figure 15.2a, for example, prominent layered reflections from Precambrian sills outline an anticline, the right (east) side of which is faulted by a west-dipping normal fault, the Rocky Mountain Trench in southwestern Canada. In this case prominent layering can be correlated across the fault to provide some estimate of the type (listric normal) and amount of displacement (about 10 km) along the fault. On the other hand, in Figure 15.2b reflections are visible from the Wind River fault zone in western Wyoming. Near the surface (to about 4 s travel time, or about 12 km depth), Precambrian crystalline rocks on the east are thrust westward over sedimentary strata of the Green River Basin on the west. Accordingly, the seismic velocity contrast between the crystalline rocks and the sediments is quite large, and a prominent reflection from the boundary is produced. The Wind River Thrust can then be followed as a series of subparallel reflections that downdip eastward to more than 7.0 s travel time (about 21 km depth).

Here then is a dilemma. If the contrast between the sedimentary rocks of the Green River Basin and the overlying crystalline rocks of the Wind River uplift is what produces the reflection along the shallow portion of the fault, then what is the cause of the reflection at greater depths where crystalline rocks are juxtaposed with crystalline rocks? Although this is discussed later in more detail, the answer in this case appears to be that the reflections are cause by mylonitic rocks that were formed as a result of the faulting. The process of **mylonitization** causes a very strong preferred orientation of crystals that, in turn, produces a contrast with the more randomly oriented crystals above and below. Thus, even where crystalline rocks are seismically homogeneous, the faulting process may produce surfaces that appear as subparallel reflections, particularly if faulting occurs below the brittle–plastic transition.

Seismic reflections from relatively homogeneous igneous rocks (e.g., plutons, basalt flows) are generally not easily observed. Exceptions are rocks that are deformed so that reflections may arise from the deformation surfaces (as described above for the Wind River uplift), and igneous intrusions (e.g., sills) that are sufficiently thin that there is a measurable contrast with surrounding rocks such as sediments or other crystalline rocks (Figure 15.2c).

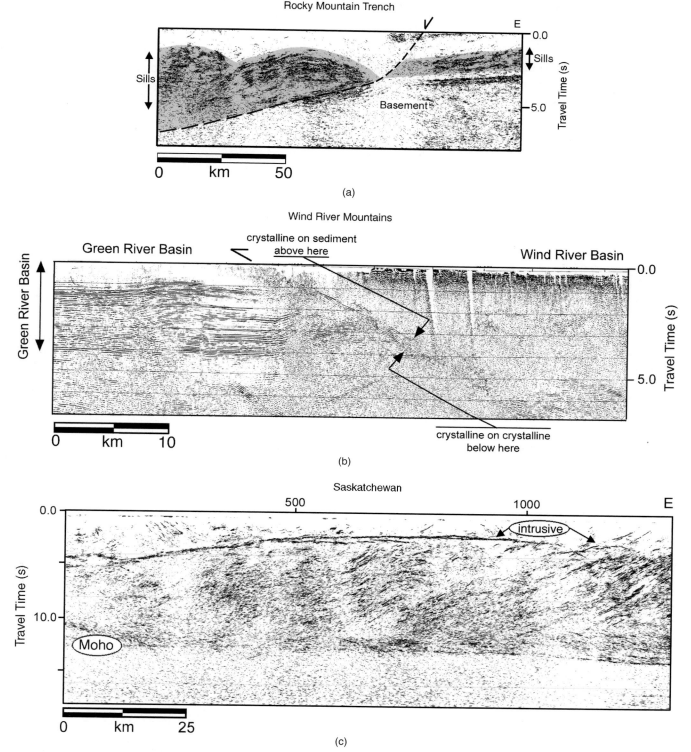

FIGURE 15.2 (a) Seismic reflection profile across the southern Rocky Mountain Trench near the Canada-U.S. border. Note that the prominent layering, which is drilled on the west and is known to be dominantly Proterozoic sills, is offset along a west-dipping listric normal fault that has about 10 km of dip-slip displacement. Data were recorded by Duncan Energy of Denver, Colorado. (b) Seismic profile from the Wind River Mountains in Wyoming (USA). The Wind River fault juxtaposes crystalline rocks of the Wind River Mountains with sedimentary rocks of the Green River Basin along a moderately east-dipping fault, and this provides a simple explanation for the prominent reflection. Below a travel time of about 3.5–4.0 s, however, the fault zone places crystalline rocks onto crystalline rocks and the reflections must be caused by other mechanisms. Data recorded by COCORP (Consortium for Continental Reflection Profiling) in 1977. (c) Seismic profile from the Proterozoic Trans-Hudson Orogen in northern Saskatchewan (Canada) illustrating prominent subhorizontal reflections that have been interpreted as intrusive rocks. Note that the reflector appears to cross cut several dipping reflections. Note also the prominent Moho on these data.

15.5 THE CRUST–MANTLE TRANSITION

The transition from the crust to the mantle is generally considered to be a relatively simple surface that has mafic rocks such as gabbro or mafic granulites above, and ultramafic rocks below (see Chapter 14). Indeed, much research in the 1950s to 1970s attempted to address the question of whether the transition is a compositional change (i.e., gabbro or granulite in the crust to peridotite in the mantle) or whether it could be a change in phase (as from mafic granulites in the crust to eclogites in the mantle). Central to the discussion was the observation from regional seismic refraction

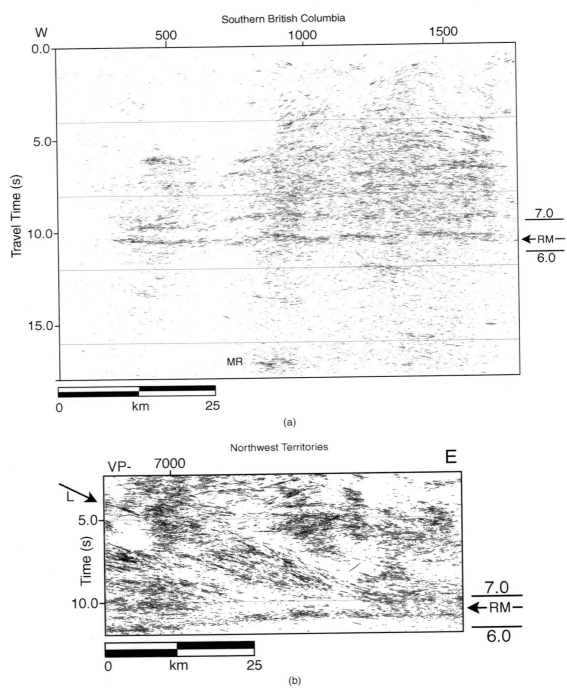

(a)

(b)

FIGURE 15.3 Some reflection characteristics of the crust–mantle transition. (a) Profile from south-central portion of the Canadian Cordillera illustrating a relatively simple, single reflection from near the transition. On the right side of the figure, the numbers 6.0 and 7.0 represent the positions of the Moho, as identified from adjacent seismic refraction data, for average velocities of 6.0 and 7.0 km/s, respectively. RM represents the preferred position of the Moho using the crustal velocity structure determined from the refraction profile.

and earthquake data that there is almost always a prominent seismic velocity increase at 40–50 km beneath the continents and about 10 km beneath the oceans. On these grounds, the boundary appears to be relatively simple and globally significant, and, as such, is known as the **Mohorovičić discontinuity,** or **Moho.**

As the information obtained from reflection profiling becomes increasingly detailed, however, we see that the crust–mantle transition is clearly not a uniform boundary because lateral variations in its geometry and reflection characteristics are common (Figure 15.3). It can be structurally complex or simple, multilayered or single

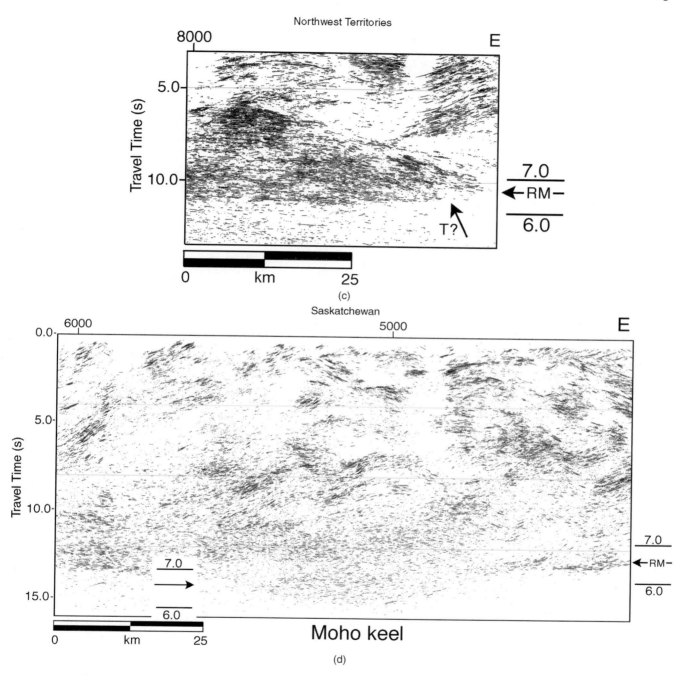

FIGURE 15.3 (Continued) Note that the Moho appears to be located at the base of crustal reflectivity, and that the underlying mantle has fewer reflections (e.g., MR). Data were recorded by LITHOPROBE in 1988. (b) Portion of a seismic profile that illustrates listric structures into the crust–mantle transition. Data were recorded by LITHOPROBE in 1996. This segment is from beneath the Great Bear arc region on the regional profile (Figure 15.5). (c) Portion of a seismic profile that illustrates many lower crustal layers that are parallel to the Moho as well as a possible truncation (T?). Data were recorded by LITHOPROBE in 1996. (d) Portion of a seismic profile from northern Saskatchewan (Canada) that illustrates a local deepening of the crust–mantle transition (Moho keel). Note that although there is not a prominent reflection near the transition, the reflectivity does diminish near it. In this figure, two locations for estimates of travel time to the reflection Moho are indicated from adjacent refraction profiles.

surface, flat or dipping; and any of these variations may be present along a single profile, sometimes changing over distances of only a few kilometers.

On the other hand, one of the most obvious characteristics of many reflection profiles is the transition from reflective crust to relatively nonreflective mantle (Figure 15.3). Indeed, this effect is so pervasive on a global scale that it is commonly used as a means to identify the crust–mantle transition; the "reflection Moho" is generally interpreted to be at the base of prominent crustal reflectivity. Thus, on one hand the detailed structural and reflection characteristics of the transition are complex and variable (Figure 15.3), while on the other there are regional, large-scale differences in the reflectivity of the crust and the upper mantle. Until the advent of crustal reflection profiling, and particularly high-quality detailed images of the lower crust and upper mantle, these characteristics were not observable. As a result, any future interpretation of the crust–mantle transition must account for geometric complexities at relatively small scales (kilometers to tens of kilometers), and relative uniformity when viewed at larger scale with lower resolution. This may ultimately be one of the most fundamental results of these kinds of data as it will lead to new concepts of how the crust and mantle interact.

One of the more heated debates in the interpretation of deep-crustal reflection profiles has been the cause (or causes) of reflectivity. Some aspects are well understood. For example, a reflection must result from a change in seismic velocity and/or rock density and the magnitude of the reflection (amplitude) is related to the magnitude of the contrast. Hence, a contrast between a rock with relatively low seismic velocity, such as a sandstone, and another with a relatively high seismic velocity, such as a gneiss, will produce a substantial reflection. At great depth, however, seismic velocities tend to be somewhat more homogenized than they are near the surface, because microcracks and pores are closed within a few kilometers of the surface, so that differences in seismic velocity from one rock type to another tend to be diminished. Coupled with the fact that boundaries are not often easily traced to known interfaces at the surface or in drill holes, the causes of deep reflections are not always clear. They may be from metasedimentary rocks, mylonite zones, layered intrusions, fluids, or combinations of these.

Where such boundaries can be related to known features, it has been found that any feature in the preceding list may explain the reflections, so that without some additional information, it is difficult to uniquely identify the lithology of specific reflectors.

Nevertheless, whether or not geologic causes of specific reflectors can be determined, the patterns of reflectivity provide first-order geometric frameworks for interpretation.

15.6 THE IMPORTANCE OF REGIONAL PROFILES— LONGER, DEEPER, MORE DETAILED

In order to provide valuable information on the regional structure of the lithosphere, **seismic profiles** must be hundreds of kilometers long. Imagine trying to look into a dark room through a small hole with the illumination on your side of the hole. As the hole is increased in size, more of the light can penetrate through the aperture, and more of the reflected light from the objects inside the room is then visible from the vantage point outside the hole. Furthermore, as higher energy (brighter) light is used, the features within the darkened room become more visible. The situation is similar with seismic profiling: as longer profiles (apertures) are used, large-scale features are more likely to be seen, and as more energy (truck vibrator sources) is used, the input signal is larger, and more reflections are usually visible. By analogy, therefore, long seismic lines with many truck vibrators will be the most beneficial for mapping large structures of the lithosphere. Of course, these kinds of surveys require efforts that are correspondingly more expensive.

In addition to long profiles with large energy sources, it is desirable to obtain the most detailed geologic information possible. In order to accomplish this, the signal must include the widest possible range of frequencies. When much of the seismic profiling was initiated on regional scales across North America by the COCORP (COnsortium for COntinental Reflection Profiling) project from 1975–1980, technological limitations (long travel time for signals from the upper mantle) precluded acquisition of deep data with sufficiently high frequencies to provide much detail. Now there are close to 20,000 km that have been recorded in North America alone, and over the past 20 years technological developments have allowed acquisition of such extensive and detailed data, often with remarkable results. A data set from the LITHOPROBE program in Canada serves as an example to illustrate the approach to interpretation as well as some of the information that can be obtained.

15.7 AN EXAMPLE FROM NORTHWESTERN CANADA

A nearly 700-km long profile of the lithosphere in northwestern Canada recorded in 1996 and processed in 1996–1997 was acquired in an effort to map the deep structure of the western portion of the Canadian Shield, both where it is exposed on the east end of the profile and then where it projects beneath younger sedimentary rocks of the Western Canada Sedimentary Basin to the west (Figure 15.4). In this region, the Canadian Shield consists of the Archean Slave Province on the east, and younger, Proterozoic rocks on the west. The Proterozoic rocks are primarily associated with an orogen, the Wopmay Orogen, that has been interpreted from surface geologic measurements to represent remnants of tectonic accretion associated with subduction on the west margin of the Slave craton at about 1.85–2.1 Ga. On the west, the profile ended east of the Cordillera, although three more profiles that cross the Cordillera have since been recorded to provide nearly 3,000 km of data that extend from some of the oldest rocks in the world (Slave Province) to the modern active margin near Alaska.

The most dramatic features of this profile are the reflectivity throughout the crustal section, the regionally subhorizontal Moho, and the extensive, but comparatively sparse, upper mantle reflections (Figure 15.5). It is commonly observed that the crust is more reflective than the mantle and this profile is a nice illustration. This makes sense geologically because the crust is lithologically heterogeneous at many scales, including scales of tens to hundreds of meters, in which the seismic waves are most responsive. In contrast, the mantle tends to be more lithologically, and thus seismically, homogeneous.

The difficulties of interpreting the causes of crustal reflectivity are, however, amplified for mantle reflections because (1) mantle reflections cannot be linked directly to outcrop, and (2) the relatively homogeneous lithology of the mantle is not usually expected to have sufficient contrasts in properties to produce reflections. Nevertheless, reflections are present from within the mantle, and the large lateral extent of them implies that they are related to major regional features. Furthermore, even though the large-scale features are visible and mappable along this section, many smaller features, from the size of a sedimentary basin down to a few hundred meters, can also be delineated and are, indeed, most helpful in interpreting the large-scale structures.

It has been suggested that the regional patterns along this profile are related to Proterozoic subduction, with East-dipping mantle reflectors as images of a remnant subduction zone, and many of the crustal structures associated with this subduction and accretion process. The ability to image structures at various

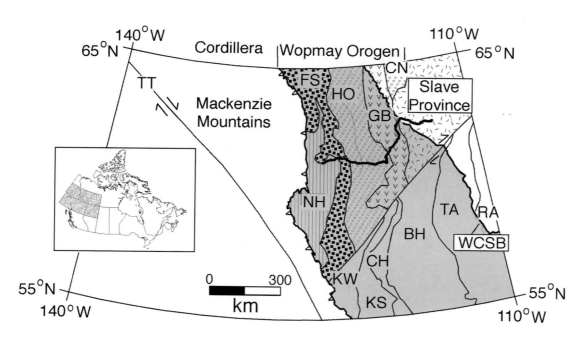

FIGURE 15.4 Map of northwestern Canada showing the division of major geologic domains and the location of a ~700-km long reflection profile (prominent dark line). Precambrian domains HO, GB, CN, RA, and TA are all defined on the basis of regional gravity and magnetic anomaly patterns that can be correlated to outcrops in the exposed Canadian Shield to the north and east. Domains NH, FS, KW, KS, CH, and BH are entirely covered by the sedimentary rocks of the Western Canada Sedimentary Basin (gray), the eastern edge of which is labeled WCSB.

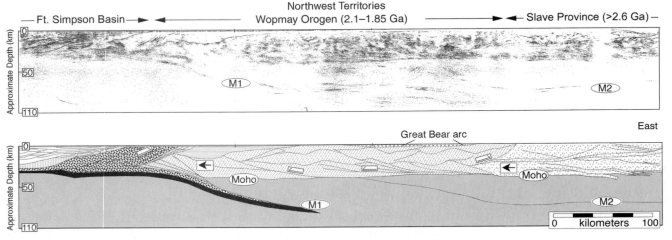

FIGURE 15.5 (upper) Regional seismic profile from ancient (>2.6 Ga) rocks of the Slave Province on the east, across the Proterozoic (2.1–1.85 Ga) Wopmay Orogen in the center, and then the younger Proterozoic (~1.74–0.55 Ga) Fort Simpson Basin on the west. The data are plotted to 32.0 s travel time, or about 120 km depth. Note the prominent crustal reflectivity, the crust–mantle transition, and sparse, but important reflections from within the upper mantle (M1 and M2). A general interpretation is shown (lower) to illustrate that the accretion of the Proterozoic rocks to the Slave Province probably resulted from subduction, the remnants of which are probably the dipping mantle reflections.

scales further allows the large and regionally significant boundaries to be related to more local features that, in turn, can be correlated with geologic observations (e.g., outcrop patterns, drill holes, and so on). For example, a key factor in the interpretation of the mantle reflections is that they can be followed to structures in the crust that can be approximately dated.

Consider the relationship between reflections M1 (Figure 15.5) and the crustal geometry previously discussed. On the west (left) side of the section, a series of layers thickens westward between the surface and 30 km depth (Figure 15.6a). These are almost certainly the expression of a westward thickening Proterozoic basin (the Fort Simpson Basin). They are overlain by shallow, more or less flat-lying, Paleozoic sedimentary rocks, and they are underlain by west-dipping surfaces that can be followed updip eastward to where they subcrop at the base of the Paleozoic. Drill holes have intersected the crystalline rocks that underlie the east flank of the basin and samples dated with radiometric techniques yield ages of about 1.845 Ga. Thus the basin layers overlying these crystalline rocks must be younger than about 1.845 Ga, but older than the Paleozoic.

Within the basin, even finer scale structures and stratigraphy may be discerned (Figure 15.6b). Near the surface, the unconformity between the base of the Paleozoic and the Proterozoic is evident as a truncation of dipping layers at ~0.2 s (about 500 m depth). Near here, drill holes penetrated from the Paleozoic sedi-

mentary rocks into Proterozoic argillaceous rocks, thus establishing that the uppermost layers of the Proterozoic are indeed of sedimentary origin. Note also that the lower Paleozoic layers appear to be arched slightly into an anticline at the position of the truncation (Figure 15.6b). This anticline must have formed after the Paleozoic strata were deposited, and was probably associated with the uplift of the Cordillera, the eastern front of which is located about 50 km west of the profile. At greater reflection times, unconformities are visible within the Proterozoic layering, thus indicating that these deep layers are indeed also of sedimentary origin and filled a deep basin during this time (Figure 15.6a). Although the depth of the basin is not certain, the thickness of the layering indicates it may be as much as 20 km (Figure 15.6a). A prominent reflection crosses the stratified basin layers at a relatively high angle (Figure 15.6b). Although it is not known with certainty what this is, because it does not outcrop, its cross-cutting geometry is characteristic of dike intrusions, and there are such intrusions known within the Proterozoic sedimentary rocks of this region.

Thus, even though there are no direct observations (drill holes or outcrops) of the layering to 20 km depth, the large-scale geometry of a basin shape, the geometric relationships between the layers indicating unconformities and stratigraphic thickening (Figure 15.6a and b), and regional relationships that indicate Proterozoic sedimentary rocks are very thick to the west,

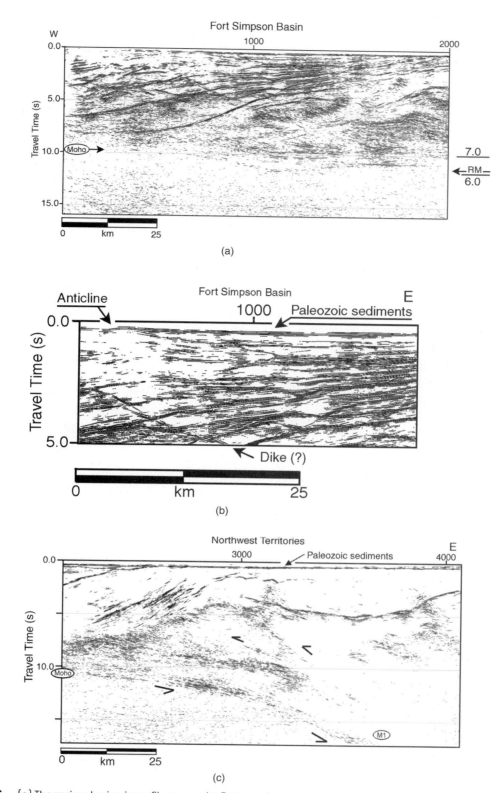

FIGURE 15.6 (a) The regional seismic profile across the Proterozoic basin illustrating the huge thickness of strata on the west and the associated shallowing of the Moho. (b) Enlargement of the regional profile in the upper part of the Proterozoic Fort Simpson Basin on the west. Note the sedimentary features such as the unconformity at the base of the Paleozoic sediments, unconformities in the eastward-thinning Proterozoic layers, and the prominent cross-cutting reflection that may be an igneous dike. (c) Enlargement of the regional profile across a feature that has been interpreted as the remnants of an accretionary complex. Note that the mantle reflections, M1, can be followed westward where they correlate with the Moho and that dipping layers above M1 tend to steepen eastward (upper arrows) as is common in accretionary wedges.

all lead to the conclusion that the western 100 km or so of this profile provides an image of a large, deep, and previously unknown sedimentary basin.

Reflections from layered sediments are well known in petroleum industry exploration. Thus, mapping such reflections from a large basin, even though it is Proterozoic, are not surprising. However, most of the deep continent includes crystalline metamorphic and igneous rock, and the common belief 20 to 25 years ago was that the velocity and density contrasts in these rocks were insufficient to produce reflections. If properly recorded and processed, however, data from the crystalline crust can have reflections that are just as prominent as those from a sedimentary basin.

To the east of the sedimentary basin discussed above, the crust beneath the thin Paleozoic cover consists of Proterozoic igneous and metamorphic rocks. This is known because the Paleozoic rocks thin to zero eastward (Figure 15.4), with crystalline rocks exposed on the surface east of there, and because drill holes intersect Precambrian crystalline rocks where the Paleozoic cover is present. Throughout this region, however, reflections are visible between the surface and about 35 km depth (Figure 15.5), which must all be within the crystalline basement.

The complex reflections from the crystalline crust can be interpreted by applying the same principles as with the Proterozoic basin: Drill holes provide direct evidence for the lithology and ages of rocks near the surface; the regional geology is incorporated to the extent possible, and detailed geometric relationships (i.e., truncations) are utilized to establish structural patterns and age relationships. Although there are too many details to address them completely here, three important characteristics stand out when the reflections are interpreted:

1. The reflectivity pattern delineates a series of complex structures associated with the accretion of middle Proterozoic rocks to an older Archean craton (Slave craton).
2. These structures are for the most part confined to the crust.
3. The base of the layering is remarkably abrupt at about 10–11 s (about 30–33 km) beneath both the Proterozoic and the Archean regions.

The first of these is significant because it provides key evidence on how subduction and accretion occurred during the Proterozoic (about 1.85–2.1 Ga in this region). From the geometric and geologic information, it appears that the products (what is visible today) of the tectonic process acting at that time are nearly identical to structures in modern subduction accretion zones. For example, east of the Fort Simpson Basin and above the mantle subduction reflections, the crustal geometry is nearly identical to that of an accretionary wedge and associated structures (Figure 15.6c).

The second and third characteristics emphasize the apparent structural (or at least reflection) differences between the crust and the mantle. The base of the crustal reflectivity (the "reflection Moho") is nearly horizontal along most of the profile east of the Fort Simpson Basin and is at a travel time that corresponds to the Moho identified from collocated regional seismic refraction data. This means that the crust–mantle transition here is either a zone of late intrusives (e.g., sills) that underlies the crust, or that it is a structural detachment zone. Examination of some detailed features with travel times near 10–11 s (about 30–33 km depth) provides information to distinguish between these possibilities (Figure 15.3b), as many of the layers in the lower crust beneath the Great Bear arc (Figure 15.5) of the Wopmay Orogen are listric (flatten) into the horizontal reflections near the Moho. Thus, the crust–mantle transition at this location is almost certainly a structural detachment rather than a layered intrusion zone. There are many profiles around the world that have images of lower crustal structures listric into the Moho, but the data must be of sufficiently high quality and have sufficiently fine detail for such subtle structures to be observed.

At some locations, in contrast to the listric structures just noted, the crust–mantle transition may have characteristics appropriate for an interpretation of sill-like intrusions; an example is visible beneath the Slave Province (Figure 15.7). Here, reflections project from the lower crust to below the Moho, and horizontal reflections at the Moho appear to cross cut them.

Other characteristics of the variable crust–mantle transition on this profile include the following: At two locations, reflections dip from the lower crust into the mantle; one of these corresponds to the interpreted Proterozoic subduction zone, and the other is located beneath the Archean craton. In some parts of the profile, the reflectivity of the crust–mantle transition is weak or nonexistent; whereas in others it is flat and prominent. At some locations the reflection Moho is flat, whereas beneath the Proterozoic basin it shallows by a few kilometers (Figure 15.6b). Thus, many of the variable characteristics of the crust–mantle transition that occur on deep reflection data around the world are visible along this single profile over relatively short lateral distances.

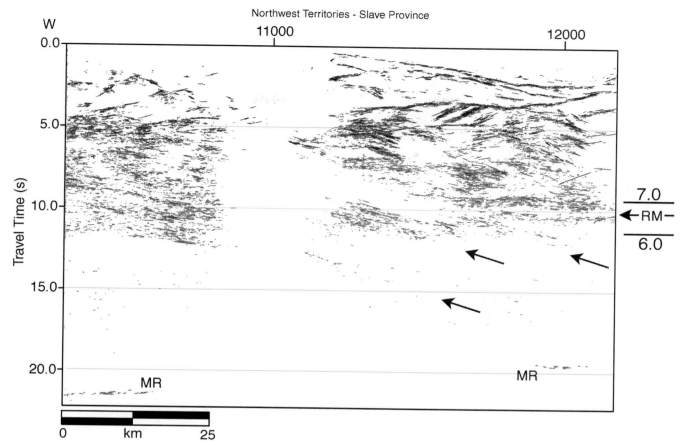

FIGURE 15.7 Enlargement of a segment of the regional profile from the Slave Province (Figure 15.5). Here, the Moho appears to have a series of dipping surfaces (arrows) that are cross cut by horizontal reflections (RM). One possible interpretation is that these horizontal reflections represent intrusives.

15.8 OTHER GEOPHYSICAL TECHNIQUES

Intracrustal structures such as the Proterozoic basin are sufficiently large to be compared with other regional data. The basin geometry on the west is observed on several other seismic profiles that are parallel to this one but that are located over a lateral distance of more than 500 km. They can be correlated from one to another with the application of other geophysical data; seismic profiles provide regional cross sections, but it is often difficult to project far away from the two-dimensional sections without additional information. In many areas, geophysical data such as gravity and magnetics, can provide such information.

The map in Figure 15.8a shows the isostatic gravity variations in northwestern Canada, and Figure 15.8b is an enlargement of the central portion of the map in the vicinity of the regional seismic profile. To produce this map, known characteristics of the Earth's shape and other effects have been estimated and removed from the measured values. The residual values, or anomalies, were contoured, and ideally represent variations due to rocks in the near subsurface. The contoured values were then plotted as a pseudotopographic image. As **gravity anomalies** result from variations in rock mass, in principle we should be able to determine the relative positions of different masses at depth. In practice, however, there is a fundamental problem underlying the interpretations of these results, as well as other geophysical anomalies such as magnetics. Because neither the subsurface structure nor the values of mass (or magnetism in the case of magnetic anomalies) of the rocks is known, an anomaly may be caused by small regions with large contrasts in properties, or large regions with small contrasts in properties. There are some limits (spatially), of course, because the anomalies are located according to map position (e.g., anomaly FS in Figure 15.8b must be due to something in the subsurface beneath it), but it is difficult to determine much more detail without some additional information from other techniques.

In the context of regional variations of continental structure, however, the patterns of large-scale anomalies can be extremely valuable in delineating patterns of continental structures. For example, in Figure 15.8a, the gravity patterns exhibit prominent, but relatively subdued, anomalies in the eastern part of the map (FS in Figure 15.8b); and more random and higher frequency patterns in the west, which are crossed by some major northwest oriented features (TT in Figure 15.8a). This change occurs where the sedimentary rocks of the Western Canada Sedimentary Basin give way to the complexly deformed rocks of the Mackenzie Mountains in the northern Cordillera. It is logical to interpret the change in gravity anomalies as being related to the large-scale geologic transition from the basin to the Cordillera.

On the other hand, the causes of the patterns beneath the Western Canada Basin are less clear. In this area, the sedimentary rocks are relatively flat and thin (Figure 15.6a), hence they should not exhibit major changes in gravity signature. The observed gravity variations

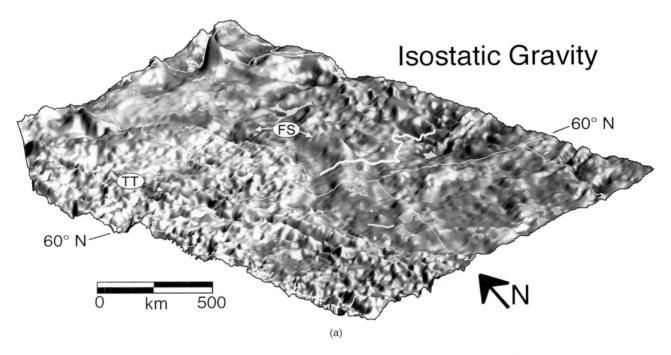

Isostatic Gravity

(a)

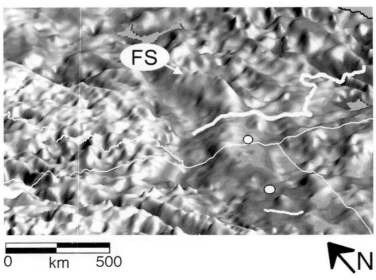

(b)

FIGURE 15.8 (a) Isostatic gravity map of northwestern Canada plotted with shaded relief (artificial illumination from the west, view toward the northeast). The position of the regional seismic profile is shown by the thick white line. TT represents the Tintina Fault, a late strike-slip fault within the Cordillera, and FS represents the Fort Simpson Trend associated with the Fort Simpson Basin. The gridded digital gravity data were provided by the Canadian Geophysical Data Centre, and the original version of this figure was made by Kevin Hall. (b) Enlargement of the map in the vicinity of the seismic profile to emphasize the relationship of the profile to the FS anomaly. The smaller white line near the bottom right is the location of a second profile across the southern portion of the FS trend, and the white circles represent locations of drill holes that penetrated crystalline rocks below the Western Canada Sedimentary Basin strata.

must therefore be associated with structure and lithology beneath the sedimentary basin, such as the large-scale structures observed on the reflection profile.

The interpretation of these observations is facilitated by the fact that the anomalies can be followed eastward into the Canadian Shield, where they are correlated with regional structures on the surface in the ancient (1.8–3.5 Ga) rocks. Accordingly, the patterns observed in the basin provide an image of structural patterns that project westward from the Canadian Shield, beneath the basin, to the eastern part of the Cordillera. Thus, while this approach does not necessarily provide us with much detail on the nature of the causes of individual anomalies, it does provide important information on the orientation and extent of subsurface structures in a region where there are no exposures.

All of the other seismic profiles that cross anomaly FS (Figure 15.8b) have essentially the same geometry; the Proterozoic basin previously described occurs everywhere to the west of FS. In all locations, the gravity signature indicates that the rocks at depth (below the Paleozoic sediments) are low density west of FS because the gravity anomalies are low, and this information is consistent with a deep, but old, basin within the crust. In applications for regional crustal and lithospheric imaging, therefore, one of the most valuable contributions of potential field maps is to project information over long distances away from the much higher resolution seismic cross sections.

15.9 CLOSING REMARKS

Geophysical imaging techniques have become standard tools for mapping the subsurface structure of the continental lithosphere. The most successful results derive from seismic reflection data that can be linked to known geologic features, either in outcrop or in drill holes. Improvements in field acquisition methods and signal processing have led to remarkable images of the crust and mantle lithosphere. Thus, as more, longer, and increasingly detailed profiles are acquired, the ability to compare, contrast, and link the results with other geologic and geophysical data will continue to foster new concepts on the origin and tectonic development of the continental lithosphere.

ADDITIONAL READING

Cook, F., van der Velden, A., Hall, K., and Roberts, B, 1999. Frozen subduction in Canada's Northwest Territories: lithoprobe deep lithospheric reflection profiling of the western Canadian Shield. *Tectonics,* 18, 1–24.

Lucas, S., Green, A., Hajnal, Z., White, D., Lewry, J., Ashton, K., Weber, W., and Clowes, R., 1993. Deep seismic profile of a Proterozoic collision zone: surprises at depth. *Nature,* 363, 339–342.

Mandler, H., and Clowes, R., 1997. Evidence for extensive tabular intrusions in the Precambrian shield of western Canada: a 160-km long sequence of bright reflections. *Geology,* 25, 271–274.

Smithson, S., Brewer, J., Kaufman, S., Oliver, J., and Hurich, C., 1978. Nature of the Wind River thrust, Wyoming, from COCORP deep reflection data and from gravity data. *Geology,* 6, 648–652.

van der Velden, A., and Cook, F., 1996. Structure and tectonic development of the southern Rocky Mountain trench. *Tectonics,* 15, 517–544.

van der Velden, A., and Cook, F., 1999. Proterozoic and Cenozoic subduction complexes: a comparison of geometric features. *Tectonics,* 18, 575–581.

Rifting, Seafloor Spreading, and Extensional Tectonics

16.1 INTRODUCTION

Eastern Africa, when viewed from space, looks like it has been slashed by a knife, for numerous ridges, bounding deep, gash-like troughs, traverse the landscape from north to south (Figure 16.1a and b). This landscape comprises the East African Rift, a region where the continent of Africa is quite literally splitting apart—land lying on the east side of the rift moves ever so slowly to the east, relative to land lying on the west side of the rift. If this process were to continue for another 10 or 20 million years, a new ocean would form between the two land masses. Initially, this ocean would resemble the Red Sea, which now separates the Arabian Peninsula from Africa. Eventually, the ocean could grow to be as wide as the Atlantic, or wider.

The drama that we just described is an example of rifting. Simply put, **continental rifting** (or simply "**rifting**") is the process by which continental lithos-phere undergoes regional horizontal extension. A **rift** or **rift system** is a belt of continental lithosphere that is currently undergoing extension, or underwent extension in the past. During rifting, the lithosphere stretches with a component roughly perpendicular to the trend of the rift; in an oblique rift, the stretching direction is at an acute angle to the rift trend.

Geologists distinguish between active and inactive rifts, based on the timing of the extension. **Active rifts,** like the East African Rift, are places on Earth where extension currently takes place (Figure 16.1). In active rifts, an array of recent normal faults cuts the crust, earthquakes rumble with unnerving frequency, and volcanic eruptions occasionally bury the countryside in ash and lava. The faulting taking place in active rifts yields a distinctive topography characterized by the occurrence of linear ridges separated by nonmarine or shallow-marine sedimentary basins (Figure 16.2). In **inactive rifts,** places where extensional deformation ceased

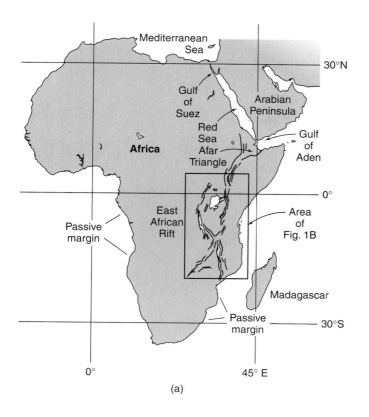

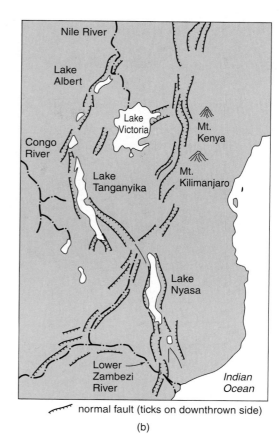

normal fault (ticks on downthrown side)

(b)

FIGURE 16.1 (a) Map of Africa, showing the East African Rift, the Red Sea, and the Gulf of Aden. As we see in East Africa, a continental rift consists of a belt of normal faults, bounding deep troughs. Some of the troughs may fill with water to become lakes. If the rift is "successful," a narrow ocean basin, like the Red Sea or Gulf of Aden, develops. The Afar Triangle lies at the triple junction between the Red Sea, the Gulf of Aden, and the East African Rift. If seafloor spreading continues for a long time, the narrow ocean could grow to become a large ocean, like the Atlantic. The passive continental margins bordering the ocean basin are underlain by stretched continental crust. (b) Detail of the East African Rift, illustrating the approximate fault pattern.

FIGURE 16.2 Photo of the Gregory Rift in southwest Kenya. Note the escarpments. These are the eroded footwalls of normal faults. Floods bring in sediment to bury the surface of the hanging-wall block. Magmatic activity associated with rifting has led to the growth of the volcano on the left.

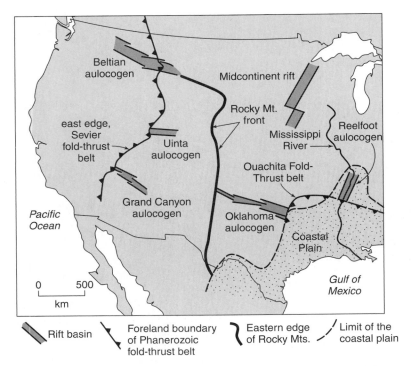

Rift basin | Foreland boundary of Phanerozoic fold-thrust belt | Eastern edge of Rocky Mts. | Limit of the coastal plain

FIGURE 16.3 Sketch map of the western two-thirds of the United States showing inactive Proterozoic rifts that trend at a high angle to North America's Phanerozoic orogens and continental margins. Because these rifts cut into the continent at a high angle, they are called aulacogens. They may be relics of unsuccessful rifting events that occurred prior to the successful rifting that generated the margins of North America at the beginning of the Phanerozoic. Late Proterozoic successful rifts of the western United States trended north-northeast, forming a north-northeast trending, passive margin. Thus, Phanerozoic convergent-margin tectonism, along this margin, generated orogens at high angles to the aulacogens.

some time ago, we find inactive normal faults and thick sequences of redbeds, conglomerates, evaporites, and volcanics. A preserved inactive rift can also be called an **unsuccessful rift,** in that its existence reflects the occurrence of a rifting event that stopped before it succeeded in splitting a continent in two. Unsuccessful rifts that cut into cratonic areas of continents, at a high angle to the continental margin, are known as **aulacogens** (from the Greek for "furrow"; Figure 16.3).

A **successful rift** is one in which extensional deformation completely splits a continent into two pieces. When this happens, as we have already noted, a new **mid-ocean ridge** (oceanic spreading center) forms between the now separate continent fragments, and seafloor spreading produces new oceanic lithosphere. Typically, 20 to 60 million years pass between inception of a rift and the time (called the **rift–drift transition**) at which active rift faulting ceases and seafloor spreading begins. The amount of lithospheric stretching that takes place prior to the rift–drift transition is variable. Typically, the continental lithosphere stretches by a factor of 2 to 4 times before separation, meaning that

the lithosphere of the rift region eventually becomes 2 to 4 times its original width and about one half to one quarter of its original thickness (Figure 16.4). The amount of stretching prior to rifting at a given locality, as well as the overall width of the rift, depends largely on the pre-rift strength of the lithosphere. Rifts formed in old, cold, strong shields tend to be narrow, while rifts formed in young, warm, soft orogens tend to be wide.[1] Once active faulting ceases in a successful rift, the relics of the rift underlie the continental margins on either side of a new ocean basin. Since no tectonic activity happens along such continental margins after they have formed, we refer to them as **passive margins.** The inactive, rifted crust of passive margins slowly subsides (sinks) and gets buried by sediment eroded from the bordering continent.

---

[1] For example, the Midcontinent Rift, a 1.1 Ga rift that ruptured the old, cold craton of the central United States, is much narrower than the Basin and Range Province, a Cenozoic rift that formed in the warm, soft North American Cordillera immediately following a protracted period of Mesozoic and Cenozoic convergent tectonism and associated igneous activity.

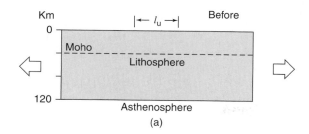

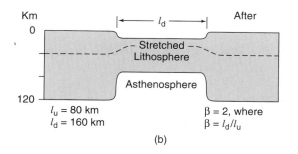

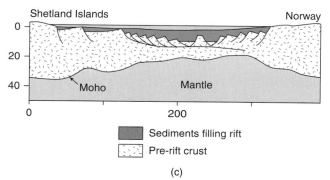

FIGURE 16.4 Illustration of the concept that rifting leads to stretching of the lithosphere. (a) This cross section shows that before rifting, the unstretched lithosphere is 120 km thick. The region that will evolve into the rift is 80 km wide. We call this the undeformed length (l_u). (b) After rifting, the lithosphere in the rift has stretched and thinned, so its deformed length (l_d) is 160 km, and it is 60 km thick. We can represent the strain that results from rifting by a number, β, called the **stretching factor**. In this example, $\beta = l_d/l_u = 2$. The larger the value of β, the greater the amount of stretching and thinning. A value of $\beta = 2$ means that the lithosphere has thinned by 50%. (c) In reality, the crust portion of the lithosphere stretches by developing normal faults. This simplified cross-sectional sketch through the Viking Graben (North Sea) shows the faulting in the crust, and the thinning of the crust. (Note: The base of the lithosphere is not shown.) The depression formed by rifting has filled with sediment.

Rifts and passive margins are fascinating regions, and geologists have worked for decades to decipher the complex structural, stratigraphic, and igneous assemblages that they contain. They are also important from a practical standpoint, because they contain significant petroleum resources. In this chapter, we

survey the principal features of rifts and passive margins (i.e., of extensional tectonism) and will introduce current speculations about how and why rifting occurs. Table 16.1 summarizes basic rift terminology, for quick reference.

16.2 CROSS-SECTIONAL STRUCTURE OF A RIFT

Let's start our discussion of rifts by developing a cross-sectional image of an active rift. Rifts evolve with time, so the geometry of a rift at a very early stage in development differs from that of a rift just prior to the rift–drift transition. We'll examine rift evolution later, but for now, we'll focus on the geometry of a rift that is at an intermediate stage in its development.

16.2.1 Normal Fault Systems

Rifts are regions in which extensional tectonics (i.e., stretching of continental lithosphere) takes place. Recall from Chapter 6 that, in the brittle field, extensional strain can be accommodated by slip on a "system" (group or array) of normal faults. On pre–1970s cross sections of rifts, geologists implied that normal-fault systems are symmetric, in that the borders of the rift were defined by normal faults that dipped toward the interior of the rift. In this **symmetric rift model,** pairs of normal faults dipping toward each other outline **grabens,** while pairs of normal faults dipping away from each other outline **horsts** (Figure 16.5a). Note that horsts and grabens are bounded on *both* sides by faults and that in old symmetric rift models, horsts formed mountain ridges ("ranges"), and the grabens underlay sediment-filled troughs ("basins"). At depth, in old cross sections, faults simply die out, either in a zone of distributed strain or in a cluster of question marks.

Modern seismic-reflection surveys of rifts do not agree with the above image, but rather show that most rifts are asymmetric. In this **asymmetric rift model,** upper-crustal extension is accommodated by displacement on arrays of subparallel normal faults, most of which dip in the same direction. These faults merge at depth with a regional subhorizontal **basal detachment** (Figure 16.5b). The fault that defines the boundary of the rift, where the basal detachment curves up and reaches the ground surface, is the **breakaway fault.**

TABLE 16.1	TERMINOLOGY OF EXTENSIONAL TECTONICS
Accommodation zones	Normal fault systems are not continuous along the length of a rift. Rather, rifts are divided into segments, whose axes may be offset from one another. Further, the faults of one segment may dip in the opposite direction to the faults of another segment. An accommodation zone is the region of complex structure that links the ends of two rift segments. Accommodation zones typically include strike-slip faults.
Active margin	A continental margin that coincides with either a strike-slip or convergent plate boundary, and thus is seismically active.
Aulacogen	An unsuccessful rift that cuts across a continental margin at a high angle to the margin. Typically, aulacogens transect the grain of an orogen that borders the margin. Aulacogens may represent failed arms of three-armed rifts, or they may simply be older rifts (formed long before the development of the continental margin, during an earlier episode of rifting at a different orientation) that were cut off when the margin formed (Figure 16.3).
Axis (of rift or MOR)	The center line along the length of a rift or a mid-ocean ridge (MOR). The trend of the axis is the overall trend of the rift.
Breakaway fault	The normal fault that forms the edge of the rift. (A breakaway fault forms the boundary between stretched and unstretched crust).
Graben	A narrow, symmetric trough or basin, bounded on both sides by normal faults that dip toward the center of the trough.
Half graben	An asymmetric basin formed on the back of a tilted fault block; one border of the basin is a normal fault.
Horst	An elongate, symmetric crustal block bordered on both sides by normal faults; both faults dip away from the center of the horst.
Listric normal fault	A normal fault whose dip decreases with depth, thereby making the fault surface concave upward.
Midocean ridge	The elongate submarine mountain range that is the bathymetric manifestation of a divergent plate boundary. Though some midocean ridges (e.g., the Mid-Atlantic Ridge) do lie in the center of ocean basins, some (e.g., the East Pacific Rise) do not. Therefore, some geologists use the term "oceanic ridge" in place of "mid-ocean ridge" for these features.
Oblique rifting	Rifting that occurs where the stretching direction is at an acute angle to the rift axis.
Passive margin	A continental margin that is not a plate boundary and, therefore, is not seismically active. It is underlain by the relict of a successful rift. The rift relict subsides and is buried by a thick wedge of sediment.
Planar normal fault	A normal fault whose dip remains constant with depth.
Nonrotational normal fault	A normal fault on which slip does not result in rotation of the hanging-wall block.
Rift (rift system)	A belt of continental lithosphere that is undergoing, or has undergone, extensional deformation (i.e., stretching); also called a continental rift.
Rift–drift transition	The time at which active rift faulting ceases and seafloor spreading begins (i.e., the time at which a mid-ocean ridge initiates, and the relicts of a rift become the foundation of a passive margin).
Rifting	The process by which continental lithosphere undergoes extensional deformation (stretching) by the formation and activity of normal faults.
Rotational normal fault	A normal fault whose hanging wall block rotates around a horizontal axis during slip.
Subsidence	The sinking of the surface of the lithosphere. Subsidence produces sedimentary basins. For example, the relict of a successful rift subsides to form a passive-margin basin.
Successful rift	A rift in which stretching has proceeded until the continent cut by the rift ruptures to form two pieces separated by a new mid-ocean ridge.
Transfer fault	A dominantly strike-slip fault that links two normal faults that are not coplanar; some transfer faults serve as accommodation zones.
Unsuccessful rift	A rift in which extensional deformation ceased prior to rupture of the continent that was cut by the rift.

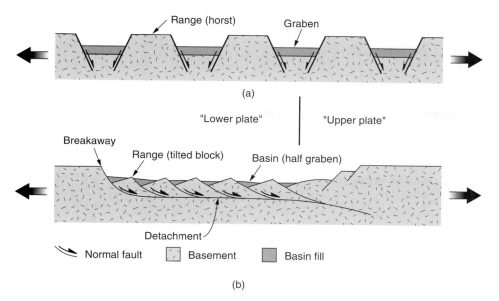

FIGURE 16.5 Contrasting models of rift faulting. (a) Cross section illustrating the old concept of symmetric horsts and grabens. (b) Cross section illustrating the contemporary image of tilted fault blocks and half grabens, all above a detachment.

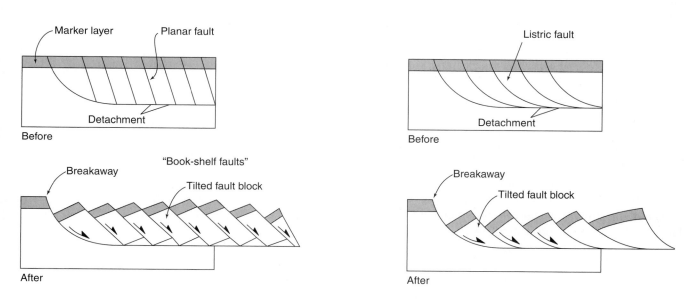

FIGURE 16.6 Geometry of normal fault arrays, in cross section. (a) Parallel rotational faults. Before faulting (top), the faults are parallel and not curved. After faulting (bottom), fault blocks are tilted. In reality, crushing and small-scale faulting at the base of the blocks fill the gaps. (b) Listric faults and associated unfaulted rollover. Before faulting (top), the faults shallow with depth and merge with a detachment. After faulting (bottom), the blocks have moved. The block to the right curves down to maintain contact with the footwall, forming a rollover anticline.

Rifts contain both **planar normal faults,** meaning faults whose dip remains constant with depth (Figure 16.6a), and **listric normal faults,** meaning faults whose dip decreases with depth, thereby making the fault surface concave upward (Figure 16.6b). Move-ment on listric normal faults results in rotation of the hanging-wall block, so that this block progressively tilts during regional extension, with the amount of tilting proportional to the amount of displacement on the fault. Because of the curvature of listric faults,

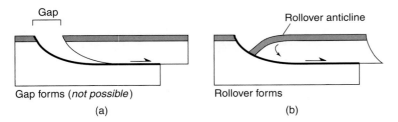

Gap

Gap forms (*not possible*)

(a)

Rollover anticline

Rollover forms

(b)

FIGURE 16.7 Formation of a rollover anticline above a listric normal fault. (a) Cross section showing that if the hanging-wall block moves to the right, without bending over, a gap develops. (b) Gravity pulls the lip of the hanging-wall block down, to maintain contact with the footwall, forming a rollover anticline.

movement of the hanging-wall block along a subhorizontal detachment creates a gap (Figure 16.7a). In Earth's gravity, rock isn't strong enough for such gaps to remain open, so either the hanging wall breaks up into fault slices (Figure 16.6b), or it sinks downward and curves to form a rollover anticline (Figure 16.7b).

Though slip on planar normal faults can occur without rotation (i.e., making the faults **nonrotational normal faults**), planar faults typically occur in arrays above a basal detachment; so as slip on the system takes place, the blocks between the faults do progressively rotate. As a result, the blocks tilt, yielding a geometry that resembles a shelf-full of tilted books. As displacement increases, the tilt increases. Faults in such arrays are sometimes referred to as **book-shelf faults.** Any normal fault (listric or planar) on which tilting accompanies displacement can be considered to be a **rotational normal fault.**

Individual normal faults may consist of subhorizontal segments (**flats**) linked by steeply dipping segments (**ramps**), producing a staircase geometry in profile. During displacement on a listric or stair-step normal fault, strata of the hanging wall can bend to form a **rollover fold** (either a rollover anticline or a rollover syncline, depending on the geometry of the underlying fault; Figure 16.8a). Synthetic and antithetic faults may also develop in the hanging-wall block (Figure 16.8b). Recall that a **synthetic fault** is a secondary fault that dips in the same direction as a major fault and that an **antithetic fault** is one that dips in the direction opposite to that of the major fault. In an **imbricate array of normal faults,** adjacent faults die out updip or break the ground surface, whereas in an **extensional duplex,** adjacent ramps merge with the same upper and lower flats (Figure 16.8c). In real rifts, combinations of stair-step faults, imbricate arrays, duplexes, rollovers, antithetic faults, and synthetic faults yield a very complex geometry. This geometry can become even more complex in regions where deposition occurs during faulting (because different stratigraphic levels

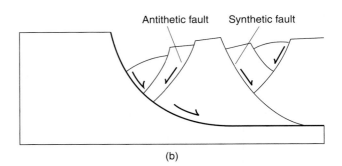

Rollover anticline   Rollover syncline

(a)

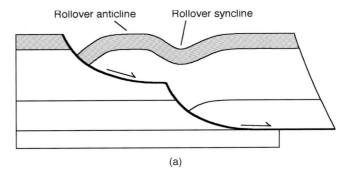

Antithetic fault   Synthetic fault

(b)

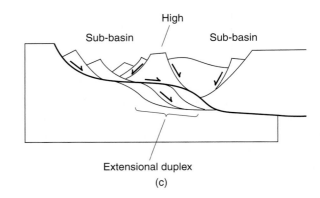

High

Sub-basin   Sub-basin

Extensional duplex

(c)

FIGURE 16.8 Complex fault systems and related folds found in rifts. (a) Cross section showing a rollover anticline above a listric normal fault, and a rollover syncline forming at the intersection of a ramp and flat. (b) Here, antithetic faults (dipping toward the main fault) and synthetic faults (dipping in the same direction as the main fault) break up the hanging-wall block. (c) Complex fault system underlain by an extensional duplex. Note the sub-basins and the high block between them.

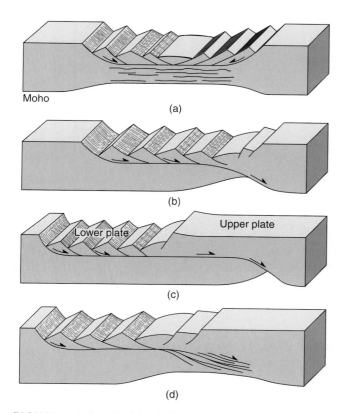

FIGURE 16.9 Models of rifting at the crustal scale.
(a) Pure-shear model; (b) simple-shear model;
(c) delamination model; (d) hybrid model (simple shear plus broad zone of distributed shear at depth).

are offset by different amounts), and/or where salt movement accompanies rifting (because salt may rise diapirically into overlying strata).

16.2.2 Pure-Shear versus Simple-Shear Models of Rifting

We've concentrated, so far, on the structural geometry of rifts at shallow levels in the continental crust. What do rifts look like at greater depths? In the **pure-shear model** (Figure 16.9a), the detachment defining the base of upper-crustal normal faulting lies at or near the brittle–plastic transition in the crust. Beneath this detachment, the crust accommodates stretching across a broad zone, either by development of penetrative plastic strain, or by movement on an array of anastomosing (braided) shear zones. Geologists apply the name "pure-shear model" to this geometry because a square superimposed on a cross section of the crust prior to extension becomes a rectangle after rifting, due to shortening in the vertical direction and stretching in the horizontal direction.

In the **simple-shear model** (Figure 16.9b), by contrast, the basal detachment cuts down through the crust,

and perhaps deeper, as a discrete shear zone. In some versions of this model, the detachment may be subhorizontal for a substantial distance beyond the edge of the rift before bending down, so that the region of upper-crustal extension does not lie directly over the region of deeper extension (this is sometimes called the "delamination model"; Figure 16.9c). Geologists refer to the portion of the detachment that traverses the crust as a **transcrustal extensional shear zone,** or **translithosphere extensional shear zone,** depending on how deep it goes. Because the deeper portions of the shear zone involve portions of the crust where temperatures and pressures are sufficiently high for plastic deformation mechanisms to operate, movement in these zones yields mylonite. In a "combination model," the transcrustal extensional shear zone of the simple-shear model spreads at depth into a diffuse band of anastomosing shear zones and disappears in a zone of distributed strain in the lower crust, and lithosphere beneath the detachment stretches penetratively (Figure 16.9d).

Simple-shear models have the advantage of readily explaining the asymmetry of many rifts; one side of the rift, called the **lower plate,** undergoes extreme stretching, while the other side, called the **upper plate,** undergoes less stretching (Figures 16.9 and 16.10). Quite likely, some combination of pure-shear and simple-shear models represents the actual geometry of rifts.

16.2.3 Examples of Rift Structure in Cross Section

To conclude our discussion of rift structure in cross section, let's look at two documented examples of rifts. In the first example, we look at the Viking Graben in the North Sea Rift. This graben separates the United Kingdom from the mainland of Europe. The second example, the Gulf of Suez, lies between Egypt and the Sinai Peninsula.

The North Sea Rift formed during the Mesozoic and Early Cenozoic, and has been studied intensively because it contains valuable petroleum reserves. A cross-sectional interpretation of the northern portion of the rift, a region called the Viking Graben (Figure 16.11a), shows that the base of the rift has a stair-step profile. Numerous tilted fault blocks underlie half grabens, which have filled with sediment. Some of the deposition is syntectonic, meaning that it occurred as faulting progressed. Note that layers of **syntectonic strata** thicken toward the normal fault that bounds one edge of the half graben and that they have themselves been tilted as fault blocks rotated during later stages of extension. Because the faults grow updip as more sediment accumulates, they are also called **growth faults.**

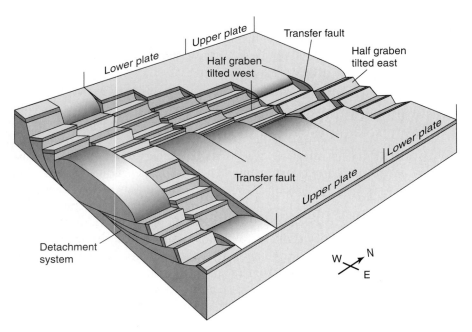

FIGURE 16.10 Block diagram illustrating the concept of upper-plate and lower-plate parts of rifts. Note how the asymmetry of rifts changes at transfer faults (accommodation zones).

In places, hanging-wall blocks contain rollover folds, synthetic faults, and antithetic faults. The magnitude of extensional strain is not uniform across the rift. Thus, this rift includes more than one **rift sub-basin.** A rift sub-basin is a portion of a rift separated from another adjacent portion by a **basement high,** a horst in which the crust has not been stretched as much. The Bergen High in Figure 16.11a is an example of a basement high. The Horda basin is a rift sub-basin.

The Gulf of Suez formed in the Cenozoic during a phase of northward propagation of the Red Sea Rift. According to the cross section interpretation of Figure 16.11b, the rift is very asymmetric, in that most of the faults dip eastward at the latitude of the cross section. Extension lies to the east of a distinct breakaway fault. In the Gulf of Suez, it appears that after the original breakaway had been active for some time, it became inactive as a new breakaway developed to the southwest. This younger breakaway cuts down to a deeper crustal level.

16.3 CORDILLERAN METAMORPHIC CORE COMPLEXES

The simple-shear model of rifting provides a possible explanation for the origin of enigmatic geologic features now known as **Cordilleran metamorphic core**

complexes (sometimes referred to as "metamorphic core complexes" or, simply, "core complexes"). Cordilleran metamorphic core complexes are found in a belt that rims the eastern edge of the Cenozoic Basin and Range Province (Figure 16.12), a broad rift that occurs in portions of Idaho, Utah, Nevada, California, Arizona, and Sonora in the North American Cordillera. They were originally named because they are distinct domes of metamorphic rock that occur in the "core" (or hinterland) of the orogen (Figure 16.13). More recently, the phrase "metamorphic core" has been used by geologists in reference to the metamorphic rocks exposed at the center in the domes. An idealized Cordilleran metamorphic core complex includes the following features:

- The interior consists of nonmylonitic country rock (gneiss, sedimentary strata, or volcanics), locally cut by Cenozoic or Mesozoic granite.
- A carapace of mylonite, formed by intense shearing, surrounds the nonmylonitic interior (Figure 16.13). This rock is very fine grained, has very strong foliation and lineation, and contains rootless isoclinal folds whose axial planes are parallel to the foliation.
- Regionally, the mylonite carapace arches into a gentle dome, shaped somewhat like a turtle shell. Notably, the mineral lineation in the mylonite has the same bearing regardless of the dip direction of the foliation. Commonly, shear-sense indicators in the mylonite (e.g., rotated porphyroclasts, C-S

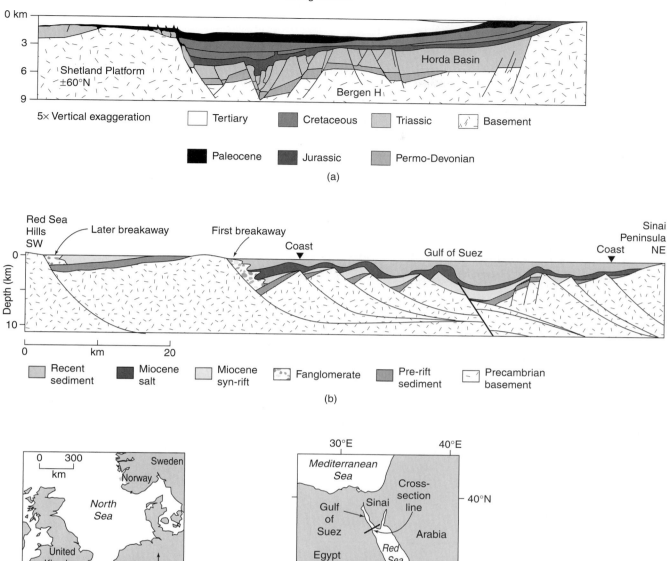

FIGURE 16.11 (a) Cross section of the Viking Graben, in the North Sea Rift. Note that the cross section is at 5 × vertical exaggeration, so the faults look much steeper than they are in nature. Note the basement highs that separate the graben into sub-basins. (b) Cross section of the Gulf of Suez. Note the asymmetry of the rift, in that most faults dip to the northeast; also, note how the original breakaway was abandoned when a later one formed further to the southwest. (c) Location maps showing the North Sea and the Gulf of Suez.

fabric, σ- and δ-tails; see Chapter 12) indicate the same regional displacement sense everywhere in the carapace.

- The basal contact of the mylonite carapace is gradational, such that the degree of mylonitization diminishes progressively downward into the nonmylonitic rocks of the core complex's interior.

- A layer of chloritic fault breccia locally occurs at the top of the mylonite carapace. Rocks above this breccia zone do not contain the mylonitic foliation. The boundary between the chlorite breccia zone and the mylonite can be quite abrupt.

- Movement on rotational normal faults cuts the brittle hanging wall above the chlorite breccia zone,

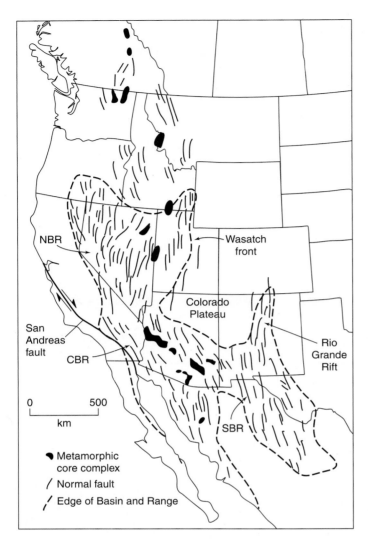

FIGURE 16.12 Simplified geologic map of the Basin and Range Province of the North American Cordillera. The province can be subdivided into three parts: NBR = Northern Basin and Range, CBR = Central Basin and Range, and SBR = Southern Basin and Range. The Rio Grande Rift is an arm of extensional strain that defines the eastern edge of the Colorado Plateau. Metamorphic core complexes are shown in black.

Map labels: NBR, Wasatch front, Colorado Plateau, San Andreas fault, CBR, Rio Grande Rift, SBR

Legend:
● Metamorphic core complex
/ Normal fault
/ Edge of Basin and Range

Scale: 0 — 500 km

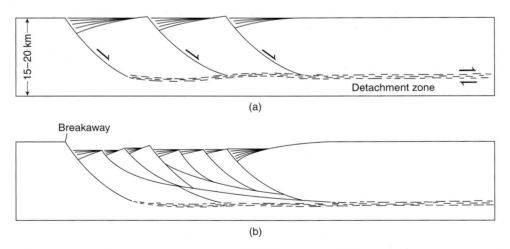

15–20 km

Detachment zone

(a)

Breakaway

(b)

FIGURE 16.13 Idealized cross sections (a–d) showing stages in the development of a metamorphic core complex. (a) An initially subhorizontal, midcrustal ductile detachment zone is formed beneath an array of steeply dipping normal faults in the upper plate; (b) additional normal faults have formed, increasing the geometric complexity; (c) as a result of unloading and isostatic compensation, the lower plate bows upward; (d) extreme thinning of the hanging wall exposes the "metamorphic core" (an exposure of the mylonitic shear zone of the detachment). Some of the hanging-wall blocks have rotated by 90°. (e) Photograph of the contact between tilted fault blocks composed of Tertiary volcanics of the hanging wall and mylonitized basement of the footwall. Whipple Mountains, California (USA).

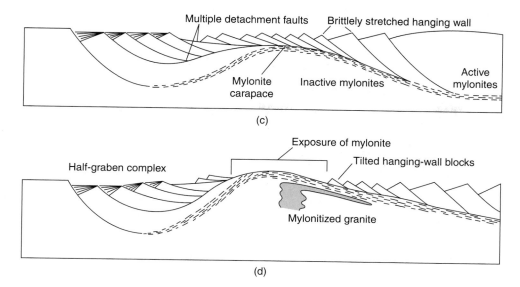

Multiple detachment faults   Brittlely stretched hanging wall

Mylonite carapace   Inactive mylonites   Active mylonites

(c)

Exposure of mylonite

Half-graben complex   Tilted hanging-wall blocks

Mylonitized granite

(d)

(e)

FIGURE 16.13   (Continued)

resulting in large extensional strains in the hanging wall. In fact, locally, faulting has isolated individual blocks of the hanging wall, so that they appear as islands of unmylonitized rock floating on a "sea" of mylonite. In places, stratification in these blocks has been tilted by almost 90°, and thus intersects the mylonite/breccia zone at almost a right angle.

An explanation of how Cordilleran metamorphic core complexes formed remained elusive for many years, until the process of mylonitization, the tools for interpreting shear-sense indicators, and the implications of asymmetric rifting became clear. In the context of these new concepts, geologists consider the complexes to be exposures of the regional detachment at the base of the normal fault system in a rift (Figure 16.13). Specifically,

Cordilleran metamorphic core complexes form where normal faulting in a rift has stretched the hanging wall above the rift's basal detachment by so much that it locally thins to zero, so that the footwall beneath the detachment becomes exposed. In this interpretation, the nonmylonitic interior of the core complex consists of footwall rock below the rift's basal detachment, and the mylonitic carapace is an exposure of the detachment itself. Note that the mylonite of the carapace forms at depth in the crust (along a transcrustal extensional shear zone) and then moves up to shallower crustal levels because of the progressive normal-sense motion on the shear zone. The chlorite breccia zone represents the boundary between fault rocks deformed in the brittle field and the mylonite; it forms at shallower depths where deformation leads to brittle failure of rock. Chlorite in this breccia results from retrograde metamorphism accompanying fluid circulation as hot rocks of the footwall are brought up to shallower crustal levels. The mylonitic shear zone arches to form a turtleback-shaped carapace as a consequence of isostatic uplift in response to unloading, when the hanging wall is tectonically thinned and removed. Rocks in the middle crust must flow plastically to fill in the area under the dome.

In sum, Cordilleran metamorphic core complexes represent regions where crustal-scale normal-sense shear on a rift's detachment brings mylonitized footwall rocks up from depth and juxtaposes them beneath brittly deformed rocks of the hanging wall. The exposed "metamorphic core" represents an exposure of the mylonitic detachment zone, revealed by extreme stretching of the hanging wall.

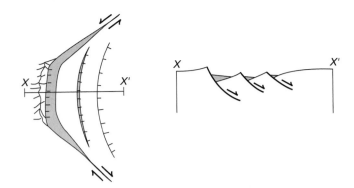

FIGURE 16.14 Map view and cross section of a "C-shaped" half graben. Note that the basin tapers toward its ends as the displacement on the fault changes from dip-slip to strike- or oblique-slip.

16.4 FORMATION OF A RIFT SYSTEM

Extension along the entire length of a continent-scale rift system does not begin everywhere at the same time. In a given region, the rift system begins as a series of unconnected **rift segments,** 100–700 km long, each containing a set of normal faults, which die out along their length. The numerous normal faults that comprise an individual rift segment dip dominantly in the same direction. In many locations, individual faults within segments have a C-shaped trace (i.e., are spoon-shaped in three dimensions); at the center of the "C," displacement on the fault is dominantly downdip, whereas at the ends of the C, displacement is strike-slip or oblique-slip (Figure 16.14). The thickness of sediment in the half graben developed above the down-dropped hanging-wall blocks diminishes toward the ends of the C. As displacement increases, segments grow along strike until adjacent segments overlap and interact. When this happens, segments link end-to-end to form a continuous zone of extension. Note that segments that ultimately link to form the continuous rift are not necessarily aligned end-to-end.

Regions where two rift segments interact and connect are called **accommodation zones,** and in these zones, you'll find complex deformation involving strike-slip, dip-slip, and oblique-slip faulting (Figure 16.15a and b). The geometry of an accommodation zone depends on whether the faults that it links dip in the same or different directions, on whether the faults were initially aligned end-to-end, on how much the fault traces overlap along strike, and on whether regional stretching is perpendicular or oblique to the traces of faults. As a consequence, accommodation-zone geometry is quite variable. In some cases, accommodation zones consist entirely of strike-slip faults (called **transfer faults**) oriented at a high angle to the regional trend of the rift (Figure 16.c). If the rift is successful, *transfer* faults evolve into the *transform* faults that connect segments of mid-ocean ridges (see Chapter 19 for strike-slip fault terminology).

Let's now look more closely at the sequence of events during which rift segments link up to form a long rift system. At an early stage in rifting, segments are separated from one another along their length by unfaulted crust (Figure 16.16a). As displacement on the faults in the segments increases, the length of the segments also increases. Eventually, the individual segments interact along strike, with one dip direction dominating—at which time opposite-dipping faults cease to be so active; they are preserved as the faults of the "upper plate" (Figure 16.16b). The underlying

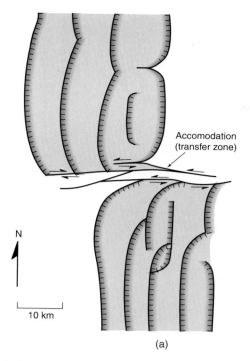

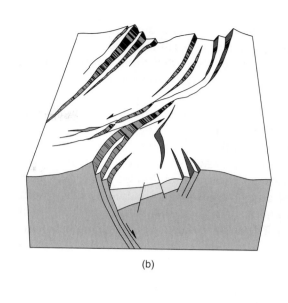

Accomodation
(transfer zone)

N

10 km

(a)

(b)

(c)

FIGURE 16.15 (a) Map-view geometry of an idealized accommodation zone, formed because the axes of two rifts are not aligned along strike. (b) Three-dimensional model of an accommodation zone. (c) Three-dimensional model of an accommodation zone that is at right angles to rift segments. This could evolve into a transform fault.

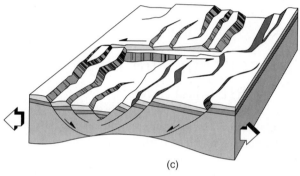

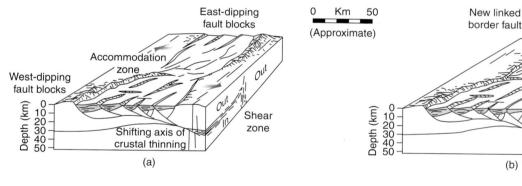

East-dipping
fault blocks

Accommodation
zone

West-dipping
fault blocks

Depth (km)

0
10
20
30
40
50

Out

In

Out

Shear
zone

Shifting axis of
crustal thinning

(a)

0   Km   50

(Approximate)

New linked
border fault   Abandoned fault

Depth (km)

0
10
20
30
40
50

Out

In

Out

(b)

Relict basin

Depth (km)

0
10
20
30
40
50

Planar faults   Axial dikes

(c)

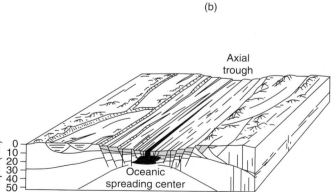

Axial
trough

Depth (km)

0
10
20
30
40
50

Oceanic
spreading center

(d)

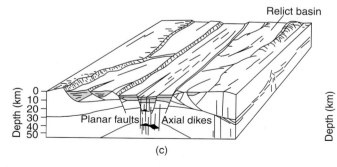

FIGURE 16.16 Rift evolution. (a) After initial rifting, two north-south-trending rift segments that face in opposite directions begin to interact. (b) A new breakaway propagates from the southern rift segment northwards, across the less active segment of the other rift. The original breakaway of the northern fault becomes inactive. (c) A swarm of dikes begins to intrude along the axis of the rift, and the rift–drift transition begins. (d) A new midocean ridge initiates, and seafloor spreading begins.

detachment also grows during this time, and propagates to greater depths in the crust.

As a rift evolves, and the lithosphere continues to stretch, the **rift–drift transition** takes place (Figure 16.16c). When this happens, discrete elongate troughs, in which heat flow is particularly high, form in places along the rift. These troughs are sites of nascent seafloor spreading. As stretching continues, the troughs widen until the continental lithosphere breaks apart, a mid-ocean ridge forms (Figure 16.16d), and a new ocean is born. Geologists can examine the evolution of rifting to seafloor spreading in the Red Sea. The northern part of the Red Sea is a mature continental rift, the central part of the sea is currently undergoing the rift–drift transition, while the southern part of the sea is a narrow ocean underlain by young oceanic crust.

16.5 CONTROLS ON RIFT ORIENTATION

Two fundamental factors, preexisting crustal structure (e.g., preexisting fabrics, faults, and sutures) and the syn-rift stress field, appear to control the regional orientation of a rift. Examination of rift geometry worldwide suggests local correlation between rift-axis strike and/or accommodation-zone strike and the orientation of basement foliations, faults, and shear zones. In the Appalachian Mountains, for example, the trend of Mesozoic rift basins (Figure 16.17a) closely follows the trend of Paleozoic fabrics and faults, and in the South Atlantic Ocean, the Romanche Fracture Zone aligns with a major Precambrian shear zone in the adjacent Brazilian Shield. Your intuition probably anticipates such relationships, because planes of foliation are weak relative to other directions in a rock. Preexisting orogens are particularly favored sites for later rifting, for not only do orogens contain preexisting faults that can be reactivated, but orogens are warmer than cratonic areas of continents, and thus are weaker. Orogens are, effectively, the weakest link between continental blocks that have been connected to form a larger continent.

Though the extension direction is commonly perpendicular to the rift axis, **oblique rifting** takes place where the direction of regional stretching lies at an acute angle to the rift axis (Figure 16.17b). In such rifts, oblique-slip displacement occurs on faults. Oblique rifting may happen where fault geometry was strongly controlled by preexisting fabric and the faults

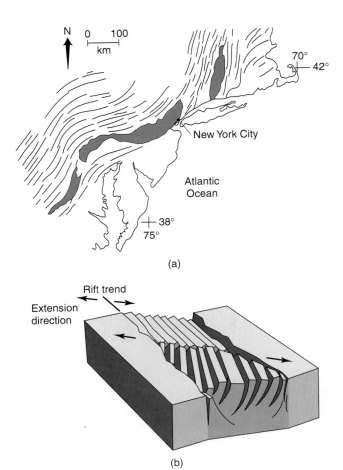

(a)

(b)

FIGURE 16.17 (a) Map of Triassic/Jurassic rift basins along the east coast of the United States. These basins formed during the early stage of rifting that separated North America from Africa, as Pangaea broke up. Note how the axes of these basins parallel the Paleozoic structural grain (defined by the trends of thrusts, folds, and fabrics) of the Appalachians. (b) Oblique rifting, as modeled in a sandbox. Here, the extension direction is oblique to the rift axis.

happen to be at an angle to the regional stretching direction. It may also develop where stretching directions change once a rift has initiated, so that the rift axis no longer parallels a principle axis of stress.

Preexisting structures do not always control the geometry of rifting. For example, in the interior of Africa, contemporary faults of the East African Rift system obliquely cut across a preexisting rift system. At this location, it appears that the syn-rifting stress field plays a more important role in controlling rift geometry than does preexisting structure. In East Africa, stress measurements suggest that faults in the rift are perpendicular to the contemporary σ_3 direction, as predicted from the theory of brittle failure (Chapter 6).

16.6 ROCKS AND TOPOGRAPHIC FEATURES OF RIFTS

Perhaps the first geologic feature that comes to mind when you think about a rift is the system of normal faults that crack the crust and accommodate regional extensional strain. But faults are not the only geologic features of rifts. Rifts also contain distinct assemblages of sedimentary and igneous rock, and they have a distinct topography.

16.6.1 Sedimentary-Rock Assemblages in Rifts

Rifts are low areas, or "depressions," relative to the **rift margins** (the borders of the rift). Within the regional depression, movement on individual normal faults results in the development of numerous narrow, elongate **basins** (or **rift basins**), each separated from its neighbor by a **range.** These basins are grabens or half grabens, which fill with sediment, while the ranges are horsts or tilted fault blocks. Of note, the Basin and Range Province, which we noted earlier is a broad rift in the western United States, lies between two high rift margins (the Sierra Nevada Mountains on the west and the Colorado Plateau on the east); it derives its name from the many basins and ranges it contains. Geographers refer to the central portion of the Basin and Range Province as the "Great Basin;" this portion has **interior drainage,** meaning that streams from surrounding highlands flow into it, but it has no outlet.

Subsidence (sinking) of rifted crust, overall, can be relatively rapid, so that a succession of sediment several kilometers thick can accumulate in the basins of a rift during just a few million years. The depositional sequence in a rift basin reflects stages in the evolution of the basin. When a rift first forms, the floor of the rift lies above sea level (Figure 16.18a). At this stage, sediment deposited in the rift basin consists entirely of nonmarine clastic debris supplied by the erosion of rift margins or by the erosion of ranges in the rift. This sediment accumulates in overlapping alluvial fans fed by streams draining ranges or rift margins. Streams may flow down the axis of the basin, providing an environment in which fluvial deposits accumulate. Water may collect in particularly low areas of basins to form a lake; examples of such lakes include Lake Victoria in Africa, and the Great Salt Lake of Utah. If the rift lies in a desert climate, the water of the lake may evaporate to leave a salt-encrusted playa, but in more humid climates the water lasts all

year. Fine-grained mud settles out in the quiet water of such lakes. Turbidites (graded beds) may also accumulate, if the lake is deep enough, for rift-related earthquakes trigger subaqueous avalanches. Thus, the sedimentary sequence deposited during the early stages of rifting typically contains coarse gravels interstratified with red sandstone and siltstone, and lacustrine sediments.

As a potentially successful rift continues to get wider, its floor eventually subsides below sea level (Figure 16.18b). At this stage, the seawater that covers the rift floor is very shallow, so evaporation rates are high. In fact, if the basin floor temporarily gets cut off from the ocean (either due to tectonic uplift at the ends of the rift or to global sea level drop) seawater in the rift basin becomes isolated from the open ocean. The trapped seawater evaporates and deposits of salt (dominantly halite ± gypsum ± anhydrite) precipitate, burying the continental clastic deposits that had formed earlier. The resulting evaporite sequence may become very thick, because global sea level may rise and fall several times, so that the rift repeatedly floods and then dries up. At this stage, the crests of ranges may be buried by sediment, so the whole rift contains only a single, wide rift basin.

With continued extension, a successful rift broadens and deepens, and evolves into a narrow ocean (Figure 16.18c). At this stage, marine strata (carbonates, sandstones, and shales) bury the evaporites. As we will see in our later discussion of passive margins, if the rift is successful, this marine sequence continues to accumulate long after rifting ceases and seafloor spreading begins, forming a very thick passive-margin sedimentary wedge. If the rift is unsuccessful, a very thick marine sequence does not cover the rift. However, as the lithosphere beneath the central part of the rift cools, thickens, and subsides, the rift margins eventually sink. As a result, a thin tapering wedge of sediment covers the rift margins, so that a vertically exaggerated profile of the rift basin and its margins resembles the head of a longhorn bull. Basins with this geometry are informally called **steerhead basins** (Figure 16.19).

16.6.2 Igneous-Rock Assemblage of Rifts

Stretching during rifting decreases the thickness of the lithosphere, and thus decreases lithostatic pressure in the asthenosphere directly beneath the rift. The asthenosphere is so hot (>1280° C) that, when such decompression occurs, partial melting takes place. Partial melting of the ultramafic rock (peridotite) comprising the asthenosphere, yields a mafic (basaltic)

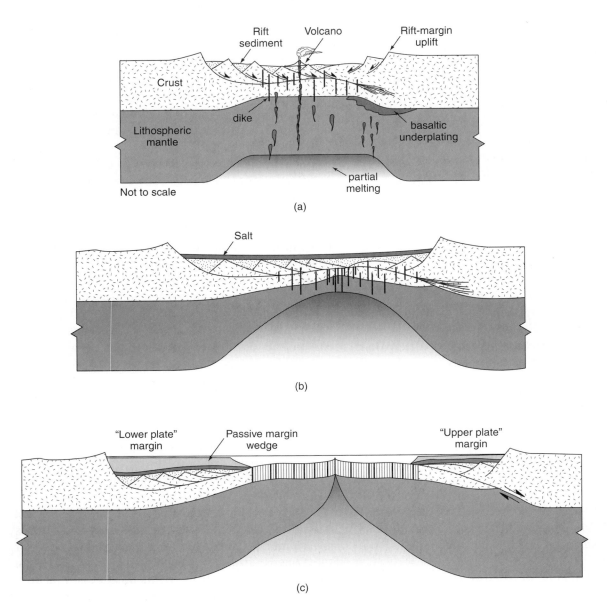

FIGURE 16.18 Evolution of a rift system into paired passive margins and sedimentary environment. (a) Rift stage with nonmarine basins; (b) rift–drift transition with evaporite deposition; (c) drift stage, with seafloor spreading occurring and passive-margin basins evolving. Marine deposition occurs in the basins.

magma.[2] The amount of partial melting at a given pressure depends on temperature, so if there is a mantle plume beneath the rift, making the asthenosphere hotter than normal, large amounts of magma will form.

Magma is less dense than the overlying lithosphere, and thus rises into the lithosphere. In some cases, a portion of the magma gets trapped and solidifies at the base of the crust, a process called **magmatic underplating.** Magmatic underplating is an important process that thickens the crust by adding mass to its base. The magma that does not solidify at the base of the crust rises higher, through cracks and along faults, and intrudes into the crust. Some magma even makes it to the Earth's surface and erupts at volcanoes. Mt. Kilimanjaro in east Africa is an example of a rift volcano.

[2]By **partial melting,** we mean that only the minerals with lower melting temperatures in the rock melt. Minerals with a higher melting temperature remain in the solid state. Typically, only 2–4% of the asthenosphere actually melts during the formation of rift magma. Partial melting results in a magma that is less mafic than its source, because mafic minerals (such as olivine and pyroxene) have higher melting temperatures than felsic minerals (such as quartz and K-feldspar).

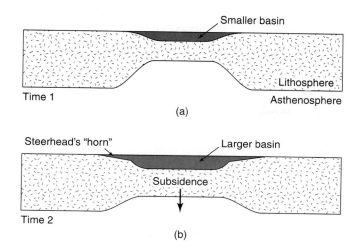

FIGURE 16.19 Thermal subsidence of a rift basin. (a) Just after rifting, the lithosphere is very thin, and the surface of the lithosphere has sunk to form a rift basin that rapidly fills with sediment. The details of the rift (faults) are not shown. (b) After time has passed, the lithosphere cools and thickens, and therefore sinks to maintain isostasy. The basin therefore gets deeper. The margins of the basin may warp down, so that sediments lap onto the unthinned lithosphere. This geometry is called a steerhead basin, because it resembles a head-on view of a bull with horns.

In rifts, which are zones of horizontal extension, the maximum principal stress (σ_1) in the crust is due to gravity and must be vertical, while the least principal stress (σ_3) trends roughly perpendicular to the rift axis (except where oblique rifting occurs). Thus, basalt that rises into the crust in rifts forms subvertical **dike swarms,** arrays of many parallel dikes, that strike parallel to the rift axis. At very shallow crustal levels, however, magma pressure may be sufficient to lift up overlying layers of rock. Here, basalt intrudes between layers and forms **sills.**[3] Basalt erupting at the ground surface in rifts has low viscosity and thus spreads laterally to create flows covering a broad area.

In a few locations, such as the Columbia Plateau in Washington and the Parana Basin in Brazil, immense quantities of basaltic lava erupted burying the landscape in immense sheets; the resulting rock is aptly named **flood basalt.** The formation of flood basalts probably occurs where a mantle plume underlies the rift, for the peridotite in the plume is significantly hotter than the peridotite comprising normal asthenosphere, and thus undergoes a greater degree of partial melting than does normal asthenosphere during

---

[3]A large sill, called the Palisades Sill, intruded the redbeds of the Mesozoic Newark rift basin in New Jersey. Rock of this sill forms the high cliffs that border the west bank of the Hudson River, opposite New York City.

decompression. Because there is more melting, more magma is produced. Further, this lava is hotter and less viscous than normal rift lava. When rifting opens conduits, the magma formed at the top of the mantle plume rushes to the surface and spews out in flows up to several hundred kilometers long.

Mafic magma is so hot ($>1000°$ C) that when it becomes trapped in the continental crust, it conducts enough heat into the adjacent crust to cause partial melting of this crust. This melting takes place because the rock comprising the continental crust contains minerals with relatively low melting temperatures ($<900°$ C). Partial melting of continental crust yields silicic magma that then rises to form granite plutons and rhyolite dikes at depth, or rhyolite flows and ignimbrites (sheets of welded tuff formed when hot ash flows blast out of volcanoes) at the surface. The common association of silicic and mafic volcanism in rifts is called a **bimodal volcanic suite** (Figure 16.20).

16.6.3 Active Rift Topography and Rift-Margin Uplifts

Because of their fault geometry, continental rifts typically display **basin-and-range topography** (such as occurs in the Basin and Range Province of the western United States). As noted earlier, some of the **ranges** (high ridges) are horsts, but most are the unburied tips of **tilted fault blocks** (i.e., hanging-wall blocks). Similarly, some of the **basins** (low, sediment-filled depressions) are grabens, but most are **half grabens,** depressions bounded on one side by a normal fault and on the other side by the surface of a tilted fault block.

Figure 16.21 shows profiles across two active rifts, the Basin and Range Province in Utah and Nevada, and the Red Sea rift in Egypt. Note that in both examples, the axis of the rift is a low area relative to the rift margin. As we mentioned earlier, the highlands bordering the rift are known as **rift-margin uplifts,** and they can reach an elevation of 2 km above the interior of the rift. In the Basin and Range, an early- to intermediate-stage rift, the interior of the basin is itself substantially above sea level, whereas in the Red Sea, which is a late-stage rift, the interior of the basin is below sea level. Clearly, the topography of rifts evolves through time in concert with the development of strain in the rift. During most of its evolution, a rift is actually relatively high. It is only during its old age, long after it has ceased being active, or when it enters the rift–drift transition, that rift floors subside to sea level or deeper.

Why are many rifts high during the early stages of their evolution? One cause may be preexisting elevation

FIGURE 16.20. Photograph of a basalt flow on top of a rhyolitic ignimbrite, in the Basin and Range Rift of the western United States.

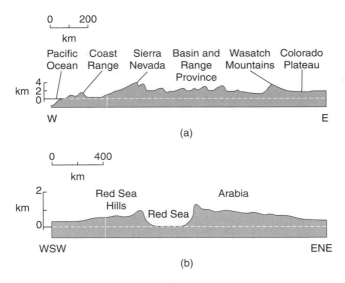

FIGURE 16.21 Topographic profiles across two rifts. (a) The Basin and Range Rift of western United States; (b) the Red Sea Rift of northeastern Africa.

of the region undergoing extension. Rifts forming in regions that previously were orogenic belts are high to start with. For example, rifting is currently occurring in Tibet, a region that is several kilometers above sea level (Chapter 21). But even if rifting begins in a region at relatively low elevation, the process of rifting can cause uplift, because rifting heats the lithosphere and makes it more buoyant. Let's see how this works.

Heating in rifts occurs for two reasons. First, the magma that rises through the lithosphere during rifting carries heat, and this heat radiates into the surrounding rock. Second, thinning of the lithosphere that accompanies rifting causes the 1280° C isotherm defining the lithosphere-asthenosphere boundary to move closer to the Earth's surface. Bringing the isotherm closer to the surface is much like placing a hot plate beneath the lithosphere, so the lithosphere at shallower depths gets warmer.[4] Heating the lithosphere causes it to become less

[4]Heat-flow studies demonstrate that heating of the lithosphere does occur in rifts. For example, the average heat flow for the Canadian shield is ~1 HFU whereas the average heat flow in the East African Rift is about 2.5 HFU. (The abbreviation "HFU" means "Heat Flow Unit." It is a measure of the heat passing through an area in a given time. 1 HFU = 40mW/m^{-2} where mW = megawatt).

dense because rock expands when heated; so, to maintain regional isostatic compensation, the surface of the lithosphere in the rift rises when stretching takes place.

As rifting progresses, stretching leads to normal faulting and graben and half-graben formation. This deformation causes the interior of the rift to sink, relative to its margins. In addition, normal faulting decreases the vertical load (weight) pressing down on the rift margins, so they rebound elastically and move up even farther, much like a trampoline surface moves up when a gymnast steps off of it. The combination of subsidence within the rift and elastic rebound of the rift margins can lead to a situation in which rift margins evolve into mountain ranges that tower over the internal portion of the rift.

After rifting ceases, rifts eventually subside. This subsidence happens because, when rifting ceases (i.e., lithospheric stretching ceases), the stretched lithosphere begins to cool. As a consequence, the 1280° C isotherm migrates down to lower depth in the mantle that, by definition, means that the lithosphere thickens. As the lithosphere cools and becomes thicker, its surface sinks in order to maintain isostatic equilibrium. In some cases, the rift margins also subside.

Significantly, not all rift margins sink after rifting ceases—some remain elevated long afterwards. Examples of **long-lived rift-margin uplifts** include the Serra do Mar along the southeast coast of Brazil (Figure 16.22a), the Blue Mountains of Australia, and the Transantarctic Mountains of Antarctica (Figure 16.22b). Note that the elevation of such uplifts is not a relict of collisional orogeny. For example, the rocks that comprise the famous high peaks of Rio de Janeiro formed during a collisional orogeny in the Proterozoic, but the present uplift developed during the Late Mesozoic and Cenozoic initially in association with opening of the South Atlantic. Similarly, metamorphism of rocks in the Transantarctic Mountains took place during a collisional orogeny in Late Proterozoic to Early Paleozoic times, but this orogen was beveled flat and was covered by Jurassic strata and volcanics. The uplift we see today developed in Late Mesozoic and Cenozoic times, in association with the rifting that formed the Ross Sea. Jurassic strata of the Transantarctic Mountains are still nearly flatlying, even though they now lie at elevations of over 4 km.

The cause of long-lived rift-margin uplifts remains a matter of debate. Two processes may contribute to their development (Figure 16.23):

- First, the uplifts may have gone up because of basaltic underplating. This process thickens the less-dense crust relative to denser (ultramafic) man-

(a)

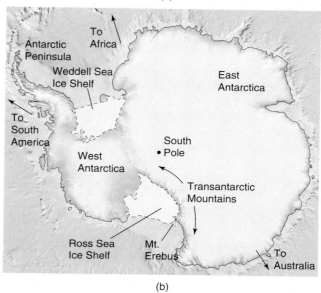

(b)

FIGURE 16.22 (a) Photo of the Serra do Mar, along the east coast of Brazil. Here, mountains exposing Precambrian granite and gneiss rise directly out of the ocean, along South America's passive margin. (b) Shaded relief map of Antarctica, showing the Transantarctic Mountains along the edge of the East Antarctic craton. The Ross Sea and Wedell Sea were formed by Mesozoic rifting.

tle lithosphere and leads to isostatic uplift, which may be long-lived. Thickening the crust, relative to the lithospheric mantle, is like adding a float to a block of wood floating in a bathtub.

- Second, the uplifts may have gone up because rifting decreased the strength of the lithosphere. When this happened, the presence of isostatically uncompensated buoyant loads (e.g., large granitic plutons from an earlier episode of intrusion) could have caused the lithosphere containing the loads to rise to attain isostatic equilibrium. Prior to rifting, these loads had been held down by the strength of the pre-rift lithosphere. As an analogy, picture a balloon trapped under a floating piece of plywood. The strength of the plywood can hold the balloon completely under water. But if the wood is replaced by

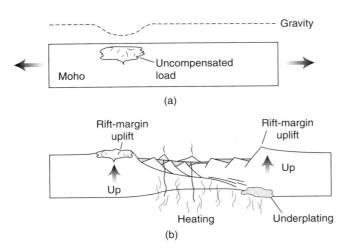

FIGURE 16.23 Hypotheses to explain long-lived rift-margin uplifts. (a) The crust, before rifting, containing an uncompensated buoyant mass (e.g., a large pluton); (b) during rifting, the strength of the lithosphere decreases, so the buoyant crust rises to achieve isostatic equilibrium. Underplating of basalt thickens the crust and can also cause uplift.

a thin sheet of rubber, a much weaker material, the buoyancy of the balloon can cause the surface of the rubber to rise and stay high.

16.7 TECTONICS OF MIDOCEAN RIDGES

In December, 1872, the H.M.S. *Challenger* set sail from England with a staff of researchers to begin the world's first oceanographic research voyage. Over the next four years, the ship traversed 127,500 km of the Atlantic and Pacific Oceans, during which, among other tasks, it made depth soundings at 492 different locations. Each measurement took many hours, for it required sailors to let out enough cable for a lead weight to reach the seafloor. Nevertheless, the crew obtained enough measurements to suggest that elongate submarine mountain ranges, known as midocean ridges, rose from the abyssal plains of the oceans. Sonar surveys in the twentieth century provided much more data and, by the 1950s, had provided our modern image of midocean ridges. Midocean ridges occur in all ocean basins, rising about 2 km above the floor of adjacent abyssal plains. Stretched out end-to-end, the midocean ridges of Earth today would make a chain 40,000 km long (Figure 16.24). Ridges are not continuous, but rather

consist of segments, ranging from ten to several hundred kilometers in length, linked at their ends by transform faults. (We'll discuss these transform faults in Chapter 19). The transform faults are roughly perpendicular to the ridge axis.

In 1960, Harry Hess suggested that midocean ridges mark divergent plate boundaries at which new oceanic lithosphere forms as seafloor spreading occurs. The new lithosphere moves away from the ridge axis, and the ocean basin becomes wider. At the ridge axis, seafloor spreading involves extensional deformation, producing structural features that are similar to those found in continental rifts.

Oceanic lithosphere consists of two components (Chapter 14). The upper part of the lithosphere, the oceanic crust, ranges from 6 to 10 km in thickness and consists of five distinct layers named, from bottom to top: the cumulate layer, the massive-gabbro layer, the sheeted-dike layer, the pillow-basalt layer, and the pelagic-sediment layer. The lower part of the oceanic lithosphere consists of lithospheric mantle; this layer varies in thickness from 0 km, beneath the ridge, to over 90 km, along the margins of a large ocean. In Chapter 14, we noted that oceanic crust forms by magmatic processes at the midocean ridge. Here, we delve into these processes in greater detail, and add information about bathymetry and the structures of midocean ridges.

Figure 16.25 provides a simplified cross section of a midocean ridge. Hot asthenosphere rises beneath the ridge axis. The resulting decrease in pressure (decompression) in the asthenosphere at this location causes partial melting, producing basaltic magma, which rises to fill a large magma chamber in the crust beneath the ridge axis. As time passes, the magma cools, and crystals of olivine and pyroxene begin to grow within it. These crystals are denser than the magma and sink to the bottom of the magma chamber to form a layer of cumulate (**cumulate,** by definition, is a rock formed from dense minerals that sank to the bottom of a magma chamber). Meanwhile, crystallization at the sides of the magma chamber produce gabbro, a coarse-grained mafic igneous rock. As seafloor spreading progresses, this gabbro moves away from the axis of the chamber, creating more room for magma. Spreading also cracks the rock overlying the magma chamber, creating conduits for magma to rise still further. Some of this magma freezes in the cracks to form the dikes of the sheeted dike layer, while some reaches the surface of the Earth and extrudes on the seafloor to form the pillow basalt layer. Continued seafloor spreading moves new-formed crust away from the ridge axis. Eventually, sediment (plankton shells and clay) snow-

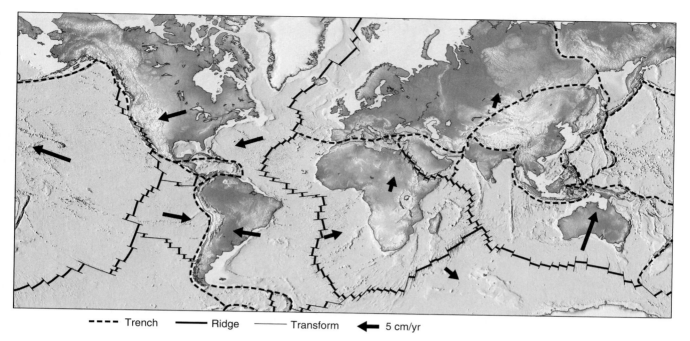

FIGURE 16.24 The midocean ridges of Earth. Note that ridges, in map view, consist of short segments, linked by transform faults. The lengths of arrows are proportional to the velocity.

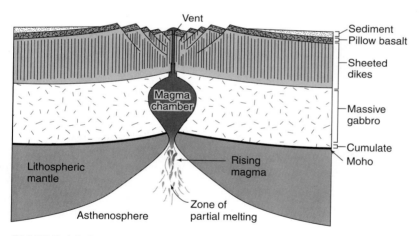

FIGURE 16.25 Schematic cross section (not to scale) through the oceanic crust and upper mantle at a fast-spreading ridge. This illustrates a model for the formation of the distinct layers of oceanic crust.

ing down from the overlying sea buries the pillow basalt.

The bathymetry and structure that occur at a midocean ridge reflect the rate of seafloor spreading (Figure 16.26a). Geoscientists distinguish between **slow ridges** (e.g., the Mid-Atlantic Ridge), where plates move apart at rates of <4 cm/y, and **fast ridges** (e.g., the East Pacific Rise), where plates move apart at rates of > 8 cm/y. (There are also intermediate-spreading ridges, with spreading rates of between 4 and 8 cm/y).

Slow ridges are relatively narrow (hundreds of kilometers wide), and have deep (>500 m) axial troughs bordered by steplike escarpments (Figure 16.26b and c). In contrast, fast ridges are broad (up to 1500 km wide), relatively smooth swells, with no axial trough.

Why are fast ridges wider than slow ridges? The answer comes from a consideration of isostasy and how it affects the level at which oceanic lithosphere floats in the sea (Figure 16.26a; see Chapter 14). Beneath a midocean ridge, the lithospheric mantle is

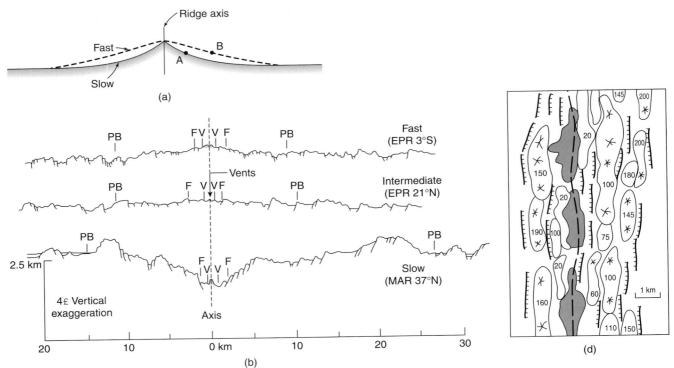

(a)

(b)

2.5 km

4£ Vertical exaggeration

Axis

20   10   0 km   10   20   30

(d)

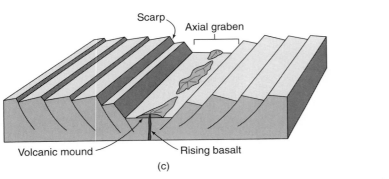

Scarp

Axial graben

Volcanic mound

Rising basalt

(c)

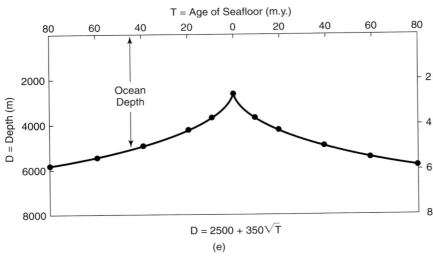

T = Age of Seafloor (m.y.)

Ocean Depth

D = Depth (m)

D = 2500 + 350√T

(e)

FIGURE 16.26 Morphology of a midocean ridge axis. (a) Simplified profiles of a fast ridge versus a slow ridge. Notice that fast ridges are wider. That's because the depth to which the seafloor sinks depends on its age. The slow plate at point A is the same age as the fast plate at point B, but both are at the same depth. (b) Profiles contrasting fast and slow ridge axes. The top profile is across the East Pacific Rise (3° S latitude), while the lower profile is across the Mid-Atlantic Ridge (37° N latitude). Note that an axial trough only occurs at slow ridges, not at fast ridges. "V" marks the position of volcanic vents. (c) Three-dimensional block diagram emphasizing the axial graben of a ridge. Mounds of pillow basalt build up over individual vents. Fault scarps border the graben. (d) A detailed map of the axial graben of the Mid-Atlantic Ridge, showing dated mounds of basalt. Note that the entire ridge is not active at any given time. The barbed lines are faults, and the stars are vents. The heavy dashed line is the plate boundary. Ages are in hundreds of thousands of years. (e) The age versus depth curve for seafloor.

very thin (in fact, at the axis of the ridge, hot asthenosphere lies just below the crust, so there is no lithospheric mantle). As the seafloor drifts away from the ridge axis, cooling takes place, and the 1280°C isotherm sinks, making the lithospheric mantle thicker. As a consequence, the ratio of less-dense crust to denser lithospheric mantle decreases as the lithosphere gets older. This makes the base of the older lithosphere sink deeper into the asthenosphere, relative to the base of the younger lithosphere, to maintain isostatic compensation.[5] Thus, the depth of the seafloor increases with age, as shown by Figure 16.26e. With this concept in mind, we see that fast ridges are wider than slow ridges because young seafloor (with a thin lithospheric mantle) exists at a greater distance from the axis of a fast ridge than it does from the axis of a slow ridge. Put another way, because seafloor spreading takes place more slowly at a slow ridge, lithosphere at a given distance from the ridge axis, has aged more, and thus has subsided more, than lithosphere at the same distance from a fast-ridge axis.

The morphology at the axis of a midocean ridge reflects the faulting that accompanies horizontal extension. Specifically, the stretching of newly formed crust breaks the crust and forms normal faults, which generally dip toward the ridge axis. Studies using research submersibles, like the *Alvin,* demonstrate that fault escarpments expose fault breccias whose matrix includes the mineral serpentine; this serpentine forms when the olivine of ocean-crust basalt reacts with hydrothermal fluids circulating in the crust of the ridge.[6] Submarine debris flows, formed from broken-up rock that breaks free and falls down fault scarps, accumulate at the base of the scarps. Submarine mapping shows that, on a timescale of thousands of years, the entire length of a ridge is not active simultaneously, for geologists can identify different mounds of lava formed at different times (Figure 16.26d).

The contrast between the morphology of the ridge-axis zone at fast ridges and that of the ridge-axis zone at slow ridges may reflect, in part, a balance between the amount of magma rising at the axis and the rate at which stretching of the crust occurs. At slow ridges, relatively little magma forms, and the magma chamber beneath the ridge axis (see Chapter 15) periodically freezes solid. As seafloor spreading at slow ridges takes place, the brittle crust stretches and breaks, and a graben develops over the axis. This graben comprises the axial trough. At fast ridges, in contrast, a nearly steady-state magma chamber may exist in places beneath the ridge axis, meaning the supply of new magma added to the chamber from below roughly balances the volume of solid gabbro formed along the margins of the chamber in a given time. Thus, the magma chamber is always inflated with magma. Magma pressure in this chamber may keep the crust high at the axis, preventing formation of an axial graben. Also, the voluminous extrusion of basalt may fill what would have been the central graben. At a distance from the axis of any midocean ridge, fault scarps disappear, at least in part because pelagic sediment progressively buries the structures once they become inactive. Some researchers have suggested, however, that slip on faults reverses away from the ridge axis, so that hanging-wall blocks rise back up to higher elevation. The nature of this movement, if it occurs, is not well understood.

16.8 PASSIVE MARGINS

As noted earlier, geologists distinguish between two basic types of continental margins. **Active margins** are continental margins that coincide with either transform or convergent plate boundaries, and thus are seismically active, while **passive margins** are not plate boundaries and thus are not seismically active. Passive margins develop over the edge of the inactive rift relict that remains after the rift–drift transition has taken place and a new midocean ridge has formed. Examples of present-day passive margins include the eastern and Gulf Coast margins of North America (Figure 16.27), the eastern margin of South America, both the eastern and the western margins of Africa, the western margin of Europe, the western, southern, and eastern margins of Australia, and almost all margins of Antarctica.

Once the drift phase begins, the rift relict (i.e., thinned continental lithosphere) underlying a passive margin gradually subsides. This subsidence happens because, when stretching ceases, thinned lithosphere, which had been heated during rifting, cools. As cooling takes place, the lithospheric mantle thickens, and thus the lithosphere as a whole sinks to maintain isostatic

---

[5]As an analogy, think of a tanker ship being loaded with crude oil. Before loading, the tanker contains air and floats high, with its deck well above sea level. As the tanker fills, it sinks into the water, and the deck becomes lower.

[6]Hydrothermal fluids cause extensive alteration of ocean-crust basalt. These fluids form when seawater sinks into the crust via cracks, is warmed by heat brought into the crust by magma, and then rises back to the seafloor. Magmatic heat effectively drives convection of water through the crust. The vents at which hot seawater is released back into the ocean are called **black smokers,** because the hot water contains dissolved sulfide minerals that precipitate when the hot water mixes with cold seawater. Chimneys of sulfide minerals accumulate around the vents.

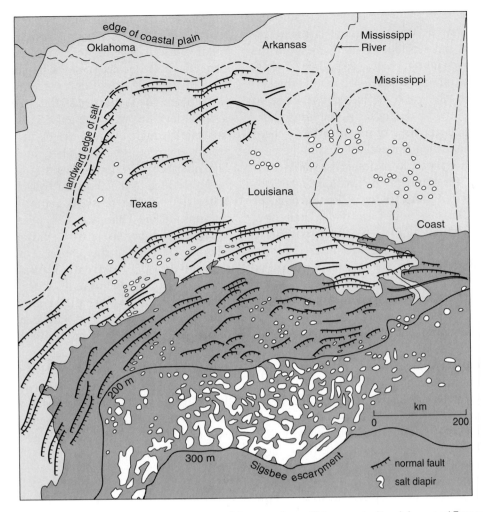

FIGURE 16.27 Simplified map of the Gulf Coast region, off the coast of Louisiana and Texas (USA), illustrating the normal faults that dip down to the Gulf of Mexico, and the abundant salt diapirs. Part of the region lies within the Coastal Plain Province, a region submerged by the sea in the Cretaceous and early Tertiary. The edge of the salt (zero isopach) is shown. Offshore, we show the 200-m and 3000-m bathymetric contours. Thrusts at the toe of the passive-margin wedge emerge at the Sigsbee Escarpment.

compensation. The process of sinking to maintain isostatic equilibrium during cooling is called **thermal subsidence.**

Effectively, thermal subsidence of a passive margin creates a space that fills with sediment eroded off of the adjacent continent and carried into the sea by rivers. This "space" is a **passive-margin basin,** and the sediment pile filling the basin is a **passive-margin sedimentary wedge.** The surface of the landward portion of a passive-margin sedimentary wedge comprises the **continental shelf,** a region underlain by an accumulation of shallow-water marine strata. The **continental slope** and **continental rise** form the transition between the shelf and the abyssal plain, and are underlain by

strata deposited in deeper water.[7] Passive-margin basins are typically segmented along strike into discrete sub-basins separated by basement highs; sediment is thinner over the basement highs. Individual sub-basins probably correspond to discrete rift segments, while the highs correspond to accommodation zones. Note that loading by sediment causes the floor of the basin to sink even more than it would if no sediment was present. Specifi-

---

[7]In pre–plate tectonic literature, a passive-margin sedimentary wedge was referred to as a "geosyncline." Geosynclines, in turn, were subdivided into a "miogeosyncline," which consisting of shallower-water facies of the continental shelf, and a "eugeosyncline," consisting of deep-water facies of the slope and rise.

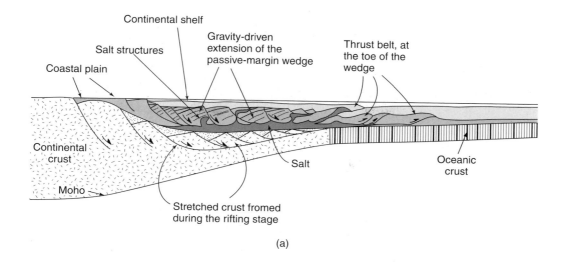

(a)

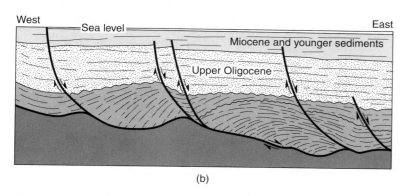

(b)

FIGURE 16.28 Cross section of a passive-margin wedge. This cross section resembles that of the Gulf Coast (USA), but has been simplified, and is not an exact representation. Note that there are two levels of extensional faulting. The lower level formed during the original rifting of the continental crust that led to the formation of this passive margin. The upper level developed within the passive-margin wedge sediments and is due to seaward gravity-driven slip of the wedge. The detachment for this upper level probably lies in the salt near the base of the section. The salt has risen to form diapirs, especially in the footwalls of normal faults, due to buoyancy and to differential vertical loads. A thrust belt has developed at the toe of the wedge. The inset shows a detail of bedding in Cenozoic strata. Note that the faults are listric growth faults and that strata in the hanging-wall blocks thicken toward the fault.

cally, the weight of 1 km of sediment causes the floor of the basin to sink by about 0.33 km.

Stretching and thinning of continental lithosphere ceases after the rift–drift transition. Normal faults, however, do develop in a passive-margin sedimentary wedge, long after the drift stage has begun (Figure 16.28). This faulting, which involves only the strata deposited above the older, rifted continental crust, is a consequence solely of gravitational force. It can occur because the seaward face of the passive-margin wedge is a sloping surface separating denser sediment below from less-dense water above. Thus, strata of the passive margin are able to slowly slip seaward, much like a slump in weak sediment of a hill slope. As it slips, the sedimentary wedge stretches; the stretching

is accommodated by normal faulting. Typically, the evaporite horizon near the base of the sedimentary section serves as the sliding surface (i.e., basal detachment) for this normal faulting (Figure 16.28).

Seaward sliding of passive-margin wedge sediment leads to the development of a system of half grabens, grabens, and rollover folds in the wedge. Many of these faults can be called **growth faults,** in the sense that they are growing (displacement is taking place and the fault cuts updip) as sedimentation continues. The geometry of the normal-fault system that forms in a passive-margin wedge may be modified substantially by salt tectonics (see Chapter 2). Salt diapirs, salt pillows, and salt walls rise in response to the unloading that accompanies movement on normal faults. In some cases, rising salt flows spread laterally to form an **allochthonous** salt sheet at a higher stratigraphic level, and this sheet acts as a basal detachment for a new generation of normal faults, affecting only strata above the allochthonous salt sheet. Because of the complex interplay between gravity-driven faulting, deposition, and salt movement, the structural architecture of passive-margin basins tends to be extremely complex. Oil exploration companies have invested vast sums in trying to understand this architecture because passive-margin basins contain large oil reserves.

Just as the body of a slump on a hill slope rises up and thrusts over the land surface at the downhill toe of the slump, the seaward-moving mass of sediment in a passive-margin wedge rides up on a system of submarine thrusts at the base of the continental slope. Thus, a small fold-thrust belt develops along the seaward toe of the passive-margin wedge, while normal faults continue to develop in the landward part of the wedge (Figure 16.28).

Considering that passive-margin basins evolve from continental rifts, and that rifts tend to be asymmetric, we find two different endmember classes of passive margins. These classes depend on whether the margin evolved from the upper-plate side of the rift or the lower-plate side of the rift (see Figure 16.10). In **upper-plate margins,** which evolve from the upper-plate portion of a rift, the lithosphere has not been stretched substantially, and there is relatively little subsidence; upper-plate margins typically develop small passive-margin basins and correspond to narrow continental shelves. In **lower-plate margins,** which evolve from the lower-plate portion of a rift, the lithosphere has been stretched substantially, so extreme subsidence takes place, leading to wide passive-margin basins that may contain a succession of sediment up to 20 km thick; these correspond to wide continental shelves. Figure 16.18c schematically illustrates the basic differences between these two classes. Researchers have been able to document a few examples in which a lower-plate margin lies on the other side of an ocean from an upper-plate margin. Notably, both upper-plate and lower-plate margins may occur along the same continental margin. In such cases they link along a strike at accommodation zones.

We conclude our brief description of passive margins by noting that not all of these form by stretching perpendicular to the margin. Some margins have evolved from regions where transtensional faulting took place. For example, the northeast margin of South America began as a transtensional system as South America began to pull away from Africa by rifting of the South Atlantic. Eventually, the extensional component on this system was sufficient to break South America and Africa apart. As a consequence of the transtensional phrase of its history, the northeast margin of South America has a fairly narrow continental shelf, locally underlain by "flower structure" (see Chapter 19).

16.9 CAUSES OF RIFTING

Up to this point, we've focused on the architecture and evolution of rifts. Now we turn our attention to the question of why rifting occurs in the first place. Given that convection occurs in the asthenosphere, is it correct to picture all rifts as places where the continent lies above the upwelling part of large mantle convective cell? No—because it is impossible to devise a geometry of simple convective cells that is compatible with the present-day geometry of rifts. (For example, if upwelling takes place at the South Atlantic Ridge and at the Indian Ocean Ridge, how can there also be upwelling along the East African Rift?) Shear stresses applied to the base of plates by convective flow of the asthenosphere may contribute to rifting in some places, but several additional processes appear to play a role as well. Let's consider various reasons for rifting (Figure 16.29):

- In places where mantle plumes rise, forming hot spots at the base of the lithosphere, the lithosphere heats up and rises. Thus, it must undergo stretching. As a consequence of this stretching, normal faults form in the upper part of the lithosphere. Rifts that form in response to the rise of hot mantle are called **thermally activated rifts.** Three rift arms, each at an angle of 120° to its neighbor, may nucleate above a single hot spot. In classic models of thermally activated rifting, a long continuous rift develops when the arms radiating from one hot spot link with the arms radiating from a neighboring hot spot. In such situations, one of the arms of the original three-armed rift shuts off. This unsuccessful rift, called a **failed arm,** becomes an aulacogen. Whether thermally activated rifts can be successful, and whether large rifts really form by linkage of rift arms radiating from hot spots, remains unclear.
- Rifting may be caused by changing the radius of curvature of a plate. Such **flexure-related rifting** occurs where the lithosphere bends just prior to descending beneath a collisional boundary. As a consequence of the lithosphere's bending, a series of normal faults that parallel the boundary develops in the descending plate. Flexure-related rifting may also occur because the Earth is not a perfect sphere.

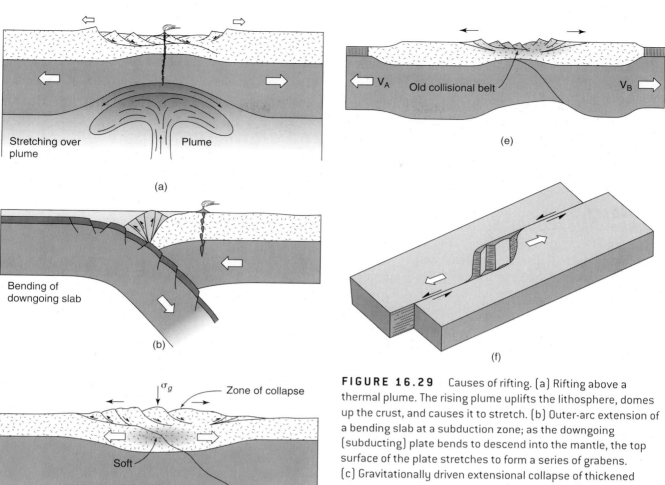

FIGURE 16.29 Causes of rifting. (a) Rifting above a thermal plume. The rising plume uplifts the lithosphere, domes up the crust, and causes it to stretch. (b) Outer-arc extension of a bending slab at a subduction zone; as the downgoing (subducting) plate bends to descend into the mantle, the top surface of the plate stretches to form a series of grabens. (c) Gravitationally driven extensional collapse of thickened crust at an orogen; the crust becomes soft at depth, and gravitational potential energy causes it to spread laterally, even if convergence is continuing. The upper crust breaks up by normal faulting. (d) Backarc extension associated with convergence. If the overriding plate is not moving in the same direction as the rollback of the subducting plate, the overriding plate breaks up at, or just behind the volcanic arc. (e) If plates are moving apart, perhaps dragged by slab pull, an intervening continent may be stretched and broken apart, like a loaf of bread that you break apart in your hands. Rifting localizes at a weak old orogen; (f) a pull-apart basin formed at a releasing bend along a strike-slip fault.

Specifically, the radius of curvature of an elastic plate changes as the plate moves from one latitude to another. The issue of whether such **membrane stresses** are sufficiently large to break plates also remains controversial.

- As discussed in Chapter 14, plates feel ridge-push and slab-pull forces, and these contribute to driving plates, independent of asthenosphere flow. Some rifts may form when the two ends of a continent are pulled in opposite directions by such plate-driving forces. If the continent contains a weak zone (e.g., a young orogen), these forces may be sufficient to pull the continent apart.

- Some rifts develop in regions of thickened and elevated crust in convergent or collisional orogens, even as contractional deformation continues. This observation implies that zones of extreme crustal thickening become zones of extension. Researchers suggest that this relationship makes sense because the quartz-rich rocks of the continental crust are not very strong, especially where heated in an orogenic belt. So, when continental crust thickens and rises relative to its surroundings during a collisional or convergent orogeny, gravitational potential energy causes the thickened and elevated zone to spread laterally under its own weight. This process is known as **extensional collapse** (or **gravitational collapse,** or **orogenic collapse**). To visualize extensional collapse, take a block of soft cheese that has a hard rind. When you put the block in the sun, the cheese warms, weakens and flows slowly outwards. Eventually the rind splits along discrete "faults" to accommodate the overall displacement.

- Rifting occurs at "releasing bends" along continental strike-slip faults. Here, the strike-slip fault trace makes a jog in map view creating a geometry that requires the crust to pull apart in order to accommodate regional movement. Normal faults develop parallel to the jog, and thus oblique to the regional trend of the strike-slip system. Normal faulting along a releasing bend can lead to formation of a pull-apart basin (see Chapter 19).

- As we will see in Chapter 17, a zone of extension may develop in back-arc regions at convergent margins. In the case of continental arcs, the resulting backarc extensional zone becomes a rift.

- Finally, rifting may develop in the foreland of collisional orogens as a consequence of indentation. As further discussed in Chapter 17, if the lateral margins of the foreland are unconstrained, the collision of a rigid continental block with another continent may cause portions of the continent to squeeze sideways, in order to get out of the way, a process called lateral escape. In the region between the escaping blocks, the lithosphere stretches and rifts develop. These rifts trend roughly perpendicular to the trace of the orogen.

16.10 CLOSING REMARKS

This chapter was devoted to a discussion of rifting, the process by which continental breakup occurs. In some cases, rifts evolve into divergent plate boundaries, at which a new oceanic basin, bordered by passive margins, begins to grow (Figure 16.28). Rifting and seafloor spreading comprise the divergent end of the plate-tectonic conveyor. In the next chapter, we jump to the other end of the plate-tectonic conveyor and consider convergent tectonics, which is the process by which the oceanic lithosphere sinks back into the mantle, and collisional tectonics, which is the process by which buoyant crustal blocks merge and squeeze together.

ADDITIONAL READING

Bosworth, W., 1994. A model for the three-dimensional evolution of continental rift basins, northeast Africa. *Geologische Rundschau,* 83, 671–688.

Coney, P. J., 1980. Cordilleran metamorphic core complexes: An overview. *Geological Society of America Memoir,* 153, 7–31.

Dixon, T. H., Ivins, E. R., and Franklin, B. J., 1989. Topographic and volcanic asymmetry around the Red Sea: constraints on rift models. *Tectonics,* 8, 1193–1216.

Eaton, G. B., 1982. The Basin and Range Province: origin and tectonic significance. *Annual Review of Earth and Planetary Science,* 10, 409–440.

Ebinger, C. J., 1989. Tectonic development of the western branch of the East African Rift. *Geological Society of America Bulletin,* 101, 885–903.

Gibbs, A. D., 1984. Structural evolution of extensional basin margins. *Journal of the Geological Society of London,* 141, 609–620.

Lister, G. S., and Davis, G. A., 1989. The origin of metamorphic core complexes and detachment faults during Tertiary continental extension in the northern Colorado River region. *Journal of Structural Geology,* 11, 65–94.

Macdonald, K. C., 1982. Mid-ocean ridges: fine-scale tectonic, volcanic and hydrothermal processes within the plate boundary zone. *Annual Review of Earth and Planetary Sciences,* 10, 155–190.

McClay, K. R., and White, M. J., 1995. Analogue modelling of orthogonal and oblique rifting. *Marine and Petroleum Geology,* 12, 137–151.

McKenzie, D. P., 1978. Some remarks on the development of sedimentary basins. *Earth and Planetary Science Letters,* 40, 15–32.

Parsons, B., and Sclater, J. G., 1977. An analysis of the variation of ocean floor bathymetry and heat flow with age. *Journal of Geophysical Research,* 82, 803–827.

Rosendahl, B. R., 1987. Architecture of continental rifts with special reference to East Africa. *Annual Review of Earth and Planetary Sciences,* 15, 445–503.

Rosendahl, B. R., Kaczmarick, K., and Kilembe, E., 1995. The Tanganyika, Malawi, Rukwa, and Turkana rift zones of East Africa; an inter-comparison of rift architectures, structural styles, and stratigraphies. *Basement Tectonics,* vol. 10. Kluwer, Dordrecht, 139–146.

Wernicke, B., 1985. Uniform-sense normal simple shear of the continental lithosphere. *Canadian Journal of Earth Sciences,* 22, 108–125.

Wernicke, B., 1992. Cenozoic extensional tectonics of the U.S. Cordillera. In: *The Geology of North America,* vol. G-3, The Cordilleran Orogen: Conterminous U.S. Geological Society of America, Boulder, 553–581.

Wernicke, B., and Burchfiel, B. C., 1982. Modes of extensional tectonics. *Journal of Structural Geology,* 4, 105–115.

Convergence and Collision

17.1 INTRODUCTION

The Andes Mountains, a 6,000-km long rampart of rugged land speckled with several peaks over 6 km high, rim the western edge of South America (Figure 17.1a). Along the range, powerful volcanoes occasionally spew clouds of ash skyward. Halfway around the world, Mt. Everest, the highest mountain in the Himalayan chain (and in the world), rises 8.5 km above sea level (Figure 17.1b). At its peak, air density is so low that climbers use bottled oxygen to stay alive. Why did the immense masses of rock comprising such mountains rise to such elevations? Before the 1960s, geologists really didn't know. But plate tectonics theory provides a ready explanation—the Andes Mountains formed where the Pacific Ocean floor subducts beneath South America along a convergent plate boundary, while the Himalayan Mountains rose when India rammed into Asia, forming a collisional orogen.

Complex suites of structures (involving thrust faults, folds, and tectonic foliations) develop at convergent plate boundaries and collisional orogens. As a consequence, the crust shortens and thickens. In the process, metamorphism and, locally, igneous activity takes place. And, though it may seem surprising at first, gravity can cause the high regions of convergent and collision orogeny to collapse and spread laterally, yielding extensional faulting. In this chapter, we describe both the structural features and the rock assemblages that develop during convergent-margin tectonism and continental collision.

(a)

(b)

FIGURE 17.1 (a) Photo of the southern Andes Mountains of Chile. Rocks exposed on these peaks include relicts of old accretionary prisms, as well as granitic intrusions of a continental volcanic arc. (b) Photo of the central Himalaya Mountains, Nepal. The highest peak, which appears to be nucleating a cloud, is Mt. Everest, the highest mountain on Earth.

17.2 CONVERGENT PLATE MARGINS

When oceanic lithosphere first forms at a mid-ocean ridge, it is warm and relatively buoyant. But as lithosphere moves away from the ridge axis, it cools and the lithospheric mantle thickens, so that when lithosphere has aged more than 10 or 15 million years, it becomes negatively buoyant. In other words, old oceanic lithosphere is denser than underlying hot asthenosphere, and thus can **subduct** or sink into the asthenosphere, like an anchor sinks through water. Such subduction occurs at a **convergent plate boundary** (which may also be called a **subduction zone** or a **convergent margin**). Here, oceanic lithosphere of the **downgoing plate** (or **downgoing slab**) bends and sinks into the mantle beneath the **overriding plate** (or **overriding slab**). An overriding plate can include either oceanic crust or continental crust, but a downgoing plate can include only oceanic crust, because continental crust is too buoyant to subduct.

Exactly how the subduction process begins along a given convergent plate boundary remains somewhat of a mystery. Possibly it is a response to compression across a preexisting weakness such as may occur at a contact between continental and oceanic lithosphere along a passive continental margin, at a transform fault, or at an inactive mid-ocean ridge segment. Conceivably, compression causes thrusting of the overriding plate over the soon-to-be subducting plate (Figure 17.2a and b). Once the subducting plate turns down and enters the asthenosphere, it begins to sink on its own because of its negative buoyancy. The subducting plate pulls the rest of the oceanic plate with it and gradually draws this plate into the subduction zone. In other words, because of its negative buoyancy, the subducted plate exerts a **slab-pull force** on the remaining plate and causes subduction to continue.

Note that the process of subduction resembles the peeling of a wet piece of paper off the bottom of a table when you pull on one end. During the process of subduction, the position of the bend in the downgoing slab migrates seaward with time, relative to a fixed reference point in the mantle; this movement is called **rollback** (Figure 17.3). When the subducting slab reaches a depth of about 150 km, it releases volatiles (H_2O and CO_2) into the overlying asthenosphere, triggering partial melting of the asthenosphere. The melt rises, some making it to the surface, where it erupts to form a chain of volcanoes called a **volcanic arc.**

Convergent plate boundaries presently bound much of the Pacific Ocean. In fact, the volcanism along these boundaries led geographers to refer to the Pacific rim as the "ring of fire." Other present-day convergent plate boundaries define the east edge of the Caribbean Sea, the east edge of the Scotia Sea, the western and southern margin of southeast Asia, and portions of the northern margin of the Mediterranean Sea (see Figure 14.14d). In the past, the distribution of convergent plate boundaries on the surface of the Earth was much different. For example, during most of the Mesozoic, the west coast of North America and southern margins of Europe and Asia were convergent plate margins, but convergence at these localities ceased during the Ceno-

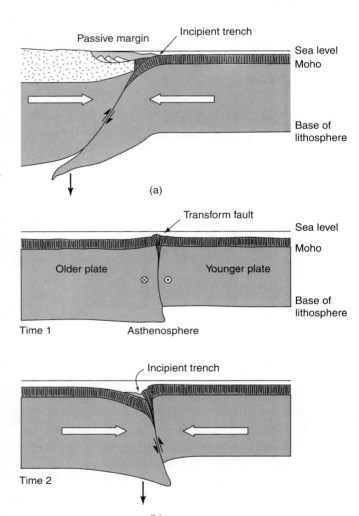

FIGURE 17.2. (a) Cross section that illustrates how a convergent margin may initiate along a passive continental margin. Here, thrusting of the margin over the denser oceanic plate has just begun. (b) Two stages during the evolution of an oceanic transform fault into a convergent plate boundary. At Time 1, two plates of different age are in contact along a transform fault. At Time 2, compression has developed across the fault, and it has become a thrust fault. The older plate has just begun to bend and sink into the asthenosphere.

zoic. Some convergent plate boundaries mark localities where oceanic lithosphere subducts under oceanic lithosphere (e.g., along the Mariana Islands and Aleutian Islands), and others mark localities where oceanic lithosphere subducts under continental lithosphere (e.g., along the Andes).

If you were to make a traverse from the abyssal plain of an ocean basin across a convergent plate boundary, you would find several distinctive tectonic features (Figure 17.4). A deep trough, the **trench,** marks the actual boundary between the downgoing and overriding plates. An **accretionary wedge** or **accretionary prism,** consisting of a package of intensely

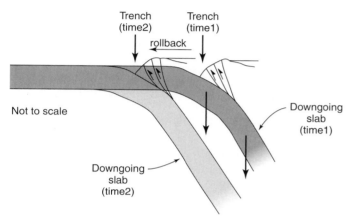

FIGURE 17.3 The concept of rollback. As a subducting slab sinks into the asthenosphere, the position of the trench relative to the a fixed point in the mantle migrates. This movement is called rollback.

deformed sediment and oceanic basalt, forms along the edge of the overriding plate adjacent to the trench. Undeformed strata of the **forearc basin** buries the top of the accretionary prism and, in some cases, trapped seafloor or submerged parts of the volcanic arc. This basin lies between the exposed accretionary prism and the chain of volcanoes that comprises the **volcanic arc.** For purposes of directional reference, we refer to the portion of a convergent margin region on the trench side of a volcanic arc as the **forearc region,** while we refer to the portion behind the arc as the **backarc region.** These and related terms are summarized in Table 17.1.

To get an overview of what a convergent plate margin looks like, we now take you on a brief tour from the ocean basin across a convergent plate margin. We start on the downgoing slab, cross the trench, climb the accretionary wedge, and trundle across the forearc basin and frontal arc into the volcanic arc itself. We conclude our journey by visiting the backarc region.

17.2.1 The Downgoing Slab

The first hint that oceanic lithosphere is approaching a subduction zone occurs about 250 km *outboard* of the trench (i.e., in the seaward direction, away from the trench). Here, the surface of the lithosphere rises to form a broad arch called the **outer swell** or **peripheral bulge** (Figure 17.5a). The elevation difference between the surface of an abyssal plain of normal depth and the crest of the swell itself, is about 500–800 m. Outer

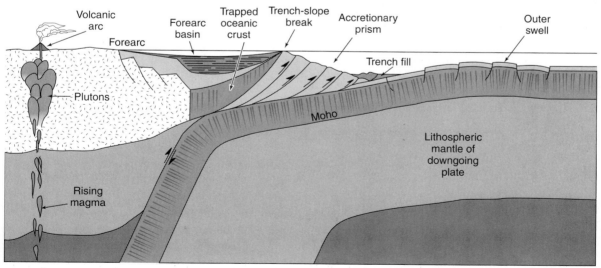

FIGURE 17.4 Idealized cross section of a convergent plate margin and related terminology. In this case, the margin occurs along the edge of a continent, bordered by a sliver of trapped oceanic crust.

TABLE 17.1	TERMINOLOGY OF CONVERGENT PLATE BOUNDARIES
Accretionary prism	A wedge of deformed sediment, and locally deformed basalt, that forms along the edge of the overriding slab; the material of the accretionary prism consists of pelagic sediment and oceanic basalt scraped off the downgoing plate, as well as sediment that has collected in the trench.
Arc-trench gap	The horizontal distance between the axis of the volcanic arc and the axis of the trench.
Backarc basin	A narrow ocean basin located between an island arc and a continental margin.
Backarc region	A general term for the region that is on the opposite side of the volcanic arc from the trench.
Continental arc	A volcanic arc that has been built on continental crust.
Convergent plate boundary	The surface between two plates where one plate subducts beneath another; as a consequence of subduction, oceanic lithosphere is consumed.
Coupled subduction	A type of subduction in which the overriding plate is pushing tightly against the downgoing plate.
Décollement	Synonym for detachment; the term is French.
Detachment	A basal fault zone of a fault system; in accretionary prisms it marks the top of the downgoing plate.
Downgoing slab (plate)	Oceanic lithosphere that descends into the mantle beneath the overriding plate.
Forearc basin	A sediment-filled depression that forms between the accretionary prism and the volcanic arc. The strata of a forearc basin buries the top of the accretionary wedge and/or trapped oceanic crust, and in general, are flat-lying.
Forearc region	A general term for region on the trench side of a volcanic arc.
Island arc	A volcanic arc built on oceanic lithosphere; the arc consists of a chain of active volcanic islands.
Marginal sea	Synonym for a backarc basin that is underlain by ocean lithosphere.
Mélange	A rock composed of clasts of variable origin distributed in a muddy matrix.
Oceanic plateau	A broad region where seafloor rises to a shallower depth. It is underlain by anomalously thick oceanic crust, probably formed above a large mantle plume.
Offscraping	The process of scraping sediment and rock off the downgoing slab at the toe of the accretionary prism.
Outer swell	A broad arch that develops outboard of the trench, in response to flexural bending of the lithosphere at the trench.
Overriding plate (slab)	The plate beneath which another plate is being subducted at a convergent plate boundary.
Peripheral bulge	Synonym for outer swell.
Retroarc basin	Synonym for backarc basin.
Rollback	The seaward migration of the bend in the downgoing plate as subduction progresses.
Subduction	The process by which one plate sinks into the mantle beneath another.
Subduction complex	Synonym for accretionary wedge.
Underplating	In the context of subduction, this is the process of scraping of material from the downgoing slab beneath the accretionary prism, so that the material attaches to the base of the prism.
Trench	A deep marine trough that forms at the boundary between the downgoing and overriding slabs; the trench may be partially or completely filled with sediment eroded from the volcanic arc or continental margin.
Trench slope break	Topographic ridge marking a sudden change in slope at the top of the accretionary wedge.
Uncoupled subduction	A process of subduction in which the downgoing plate is not pushing hard against the overriding plate, so the subduction system is effectively under horizontal tension.
Volcanic arc	A chain of subduction-related volcanoes.

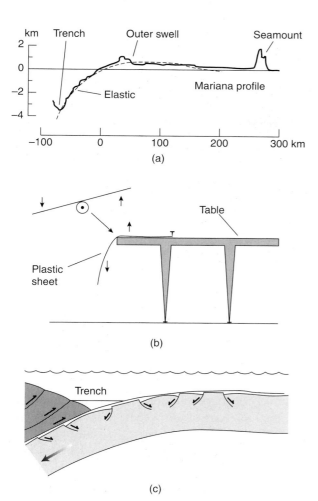

FIGURE 17.5 The peripheral bulge of the downgoing plate. (a) Example of the peripheral bulge just east of the Mariana Trench in the western Pacific. The solid line is a profile of the actual surface of the downgoing slab, while the dashed line is a calculated profile assuming the plate behaves like a sheet with flexural rigidity. (b) Table-top model of a peripheral bulge, with the inset showing lever concept. (c) Stretching along the outer swell produces horsts and grabens at the surface of the downgoing slab.

swells form because of the flexural rigidity of the lithosphere. Specifically, downward bending of the lithosphere at a convergent plate boundary levers up the lithosphere outboard of the trench and causes it to rise; you can illustrate this phenomenon by bending a sheet of plastic over the edge of a table (Figure 17.5b). Oceanic crust stretches to accommodate development of an outer swell and, as a result, an array of trench-parallel normal faults develops along the crest of the outer swell (Figure 17.5c).

Not surprisingly, as a downgoing plate shears along the base of the overriding plate, rock ruptures abruptly, to cause earthquakes. But even after it has descended into the mantle, the downgoing plate remains seismi-

cally active. Earthquakes in subducting lithosphere define an inclined belt, called the **Wadati-Benioff zone,**[1] that reaches a maximum depth of around 670 km (Figure 17.6a, b, and c). In fact, it is the distribution of earthquakes in the Wadati-Benioff zone that defines the location of the subducted plate. Based on the shape of the Wadati-Benioff zone, researchers find that not all subducted plates dip at the same angle. In fact, dips vary from nearly 0°, meaning that the slab shears along the base of the overriding slab, to 90°, meaning that it plunges straight down into the mantle. Subducted-slab dip may be controlled, in part, by the age of the subducting lithosphere, for older oceanic plate is denser and may sink more rapidly. It may also be controlled by **convergence rate,** the horizontal rate at which plates are converging across the trench, for if we assume that sinking velocity is constant, an increase in convergence velocity decreases the dip of the subducting plate. The angle may also be affected by the flow direction and velocity of the asthenosphere into which the lithosphere sinks.

Because the lithosphere is cooler then the asthenosphere, the subducting plate perturbs the thermal structure of the mantle (Figure 17.7). The internal part of the downgoing plate remains relatively cool down to significant depths because rock has such low thermal conductivity. Under the pressure and temperature conditions found in the subducting plate, basalt of the oceanic crust undergoes a phase transition to become a much denser rock called **eclogite;** formation of eclogite may increase the slab-pull force.

The type of stress associated with earthquakes changes character with depth along the Wadati-Benioff zone (Figure 17.8). Beneath the outer swell, earthquakes result from tension caused by plate bending, whereas in the region beneath the accretionary wedge, earthquakes result from compression; thrust movements are due to shear between the overriding and downgoing slab. Huge, destructive earthquakes, such as the 1964 "Good Friday" earthquake of southern Alaska, result from ruptures in this zone. At depths of about 150–300 km, earthquakes of the Wadati-Benioff zone occur in a tensional stress field. Perhaps slab pull by the deepest part of the subducting plate stretches the plate in this interval. At deep levels, earthquakes of the Wadati-Benioff zone indicate development of compression, perhaps caused by shear between the deep downgoing plate and the asthenosphere. Seismologists do not understand why deep-focus earthquakes of the

---

[1]Named after its discoverers, K. Wadati (in Japan) and H. Benioff (in the USA) who worked independently and several decades apart.

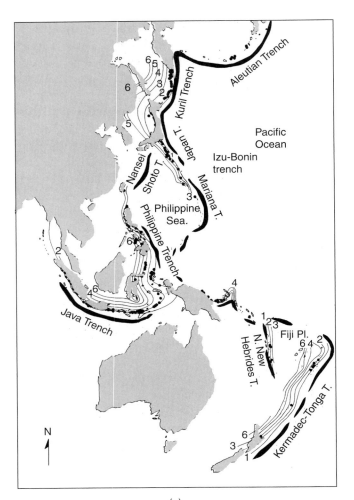

(a)

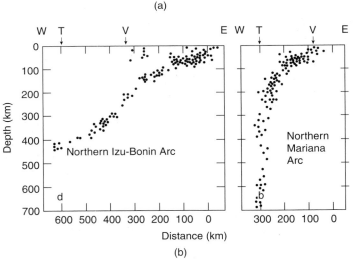

(b)

Wadati-Benioff zone can occur, because at great depths, the downgoing slab should be warm enough to be ductile. Some researchers suggest that deep-focus earthquakes happen when sudden mineralogical phase transition or sudden dehydration reactions take place in rock comprising the downgoing plate, and that these cause an abrupt change in the volume of the rock; this change generates vibrations.

The deepest earthquakes of the Wadati-Benioff zone occur near the boundary between the seismically defined transition zone of the mantle and the lower mantle. Earthquakes from depths greater than 670 km have not been detected. But the deepest earthquakes do not necessarily define the greatest depth to which the downgoing slab sinks. Seismic tomography studies show that downgoing plates, because they are relatively cool, show up as bands of anomalously fast velocity. Some bands continue downwards into the lower mantle, suggesting that the downgoing slab flows downwards into the lower mantle. In fact, subducted plates may eventually sink almost to the base of the mantle, accumulating in "slab graveyards" near the core-mantle boundary (Figure 17.9). If this image is correct, then the base of the Wadati-Benioff zone does not mark the base of subducted slabs, but merely the depth at which earthquakes no longer occur because the fracturing and/or phase changes that produce seismic energy in slabs can no longer take place.

17.2.2 The Trench

Trenches are linear or curvilinear troughs that mark the boundary, at the Earth's surface, between the downgoing slab and the accretionary prism of the overriding plate (Figure 17.4). Trenches exist because the subducted portion of the downgoing slab pulls the slab downwards to a

FIGURE 17.6 (a) Map of the western Pacific, showing trenches (heavy lines), and volcanoes (black dots) related to subduction in the western Pacific. The depth to Wadati-Benioff zones is shown by contour lines. The contours are given in multiples of 50 km (e.g., "2" means 2 × 50 = 100 km depth below the surface.) (b) Cross section showing earthquake foci defining the moderately dipping Wadati-Benioff zone of the northern Izu-Bonin arc. (c) Cross section showing earthquake foci defining the steeply dipping Wadati-Benioff zone of the northern Mariana Arc. T = location of trench; V = location of volcanic arc; the distance between T and V is the arc-trench gap.

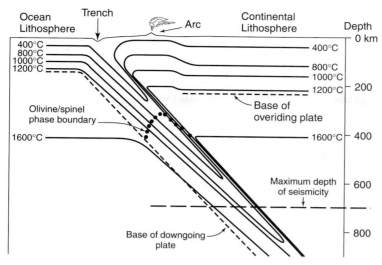

FIGURE 17.7 A simplified model of the thermal structure of the downgoing plate. (The thermal effects of mineral phase changes are not shown.) Note that within the downgoing plate, relatively low temperatures are maintained to great depth. For example at a depth of 400 km, the asthenosphere is at about 1600°C, while the interior of the downgoing plate may be as cool as 750°C.

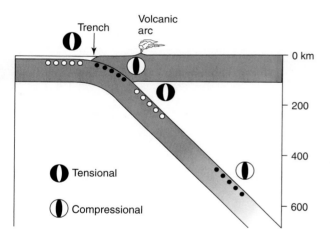

FIGURE 17.8 Schematic cross section illustrating the different types of earthquakes that occur at different depths within the downgoing plate.

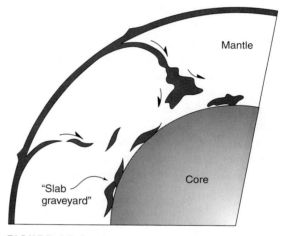

FIGURE 17.9 Schematic cross section of the Earth illustrating the concept of a slab graveyard in which masses of subducted oceanic lithosphere may accumulate near the base of the mantle.

depth greater than it would be if the lithospheric plate were isostatically compensated.[2]

The deepest locations in the oceans occur at trenches. In fact, the floor of the Mariana Trench in the western Pacific (Figure 17.6a) reaches a depth of over 11 km, deep enough to swallow Mt. Everest (nearly 9 km high) without a trace. But not all trenches are so deep. For example, the Juan de Fuca Trench in the Pacific, off the coast of Oregon and Washington

(northwestern USA), is not much deeper than the adjacent abyssal plain of the ocean floor. Trench-floor depth reflects two factors: (1) the age of the downgoing slab (the floor of older oceanic lithosphere is deeper than the floor of younger oceanic lithosphere), and (2) the sediment supply into the trench (if a major river system from a continent spills into a trench, the trench fills with sediment). To see the effect of these parameters, let's compare the geology of the Mariana Trench and that of the Oregon-Washington Trench. The great depth of the Mariana Trench is a result of its location far from a continental supply of sediment and the fact that the plate being subducted at the Mariana Trench is relatively old (Mesozoic). In contrast, the

---

[2]The resulting mass deficit from this depression at trenches produces a large negative gravity anomaly, which is a signature of subduction zones.

trench along the Pacific northwest margin of the United States has filled with sediments carried into the Pacific by the Columbia River, and the downgoing slab beneath the trench is quite young (Late Cenozoic).

Even though the thickness of sediments in trenches is variable, all trenches contain some sediment, called **trench fill.** Typically, trench fill consists of flat-lying turbidites and debris flows that decended into the trench via submarine canyons (Figure 17.10). The sediment comes from the volcanic arc and its basement, from the forearc basin, and from older parts of the accretionary wedge. Eventually, the trench fill becomes incorporated into the accretionary prism, where it becomes deformed.

17.2.3 The Accretionary Prism

During the process of subduction, the surface of the downgoing plate shears against the edge of the overriding plate. As we have already noted, the shear between the two plates produces an **accretionary prism** (or **accretionary wedge**). This is a wedge consisting of deformed pelagic sediment and oceanic basalt, which were scraped off the downgoing plate, and of deformed turbidite that had been deposited in the trench. Researchers have described two different accretionary prism geometries. Figure 17.11 illustrates these differences for the case of a convergent margin near a continent. In Figure 17.11a, the prism forms seaward of a trapped sliver of oceanic crust, whereas in 17.11b, the edge of the continent comes directly in contact with the surface of the downgoing plate.

Traditionally, the process of forming an accretionary prism has been likened to the process of forming a sand pile in front of a bulldozer (Figure 17.11c). The blade of the bulldozer can be called a **backstop,** in the sense that it is a surface that blocks the movement of material that had been moving with the downgoing

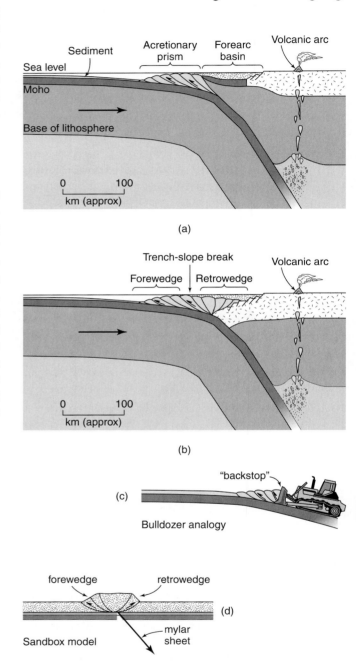

FIGURE 17.11 Two possible configurations of accretionary prisms (roughly to scale). (a) An accretionary prism caught beneath the lip of trapped ocean lithosphere. (b) An accretionary prism being scraped off the edge of a continent. Note that the prism, in this case, is bivergent. (c) The bulldozer analogy for the formation of an accretionary prism. The blade acts as the backstop. (d) Sandbox model showing the formation of a bivergent wedge.

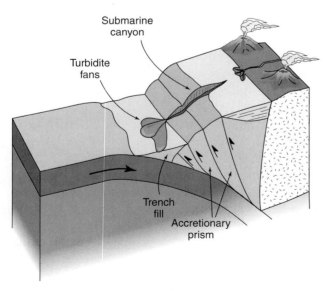

FIGURE 17.10 Trenches fill with turbidites. Much of this sediment flows down submarine canyons and then accumulates in turbidite fans on the floor of the trench.

plate. Another way to visualize accretionary prism formation process comes from a simple sand-box model. In this model, a sand layer buries a sheet of mylar (thin plastic) that can be pulled through a slit in the base of the box; the slit represents the contact between the overriding and downgoing plates. As the mylar sheet moves down through the slit, the nonmoving sand on the "overriding plate" acts as a backstop, so the sand brought into the subduction zone piles up (Figure 17.11d). Note that in this configuration, a **bivergent wedge** (or **bivergent prism**) forms. This means that the prism consists of a forewedge and retrowedge. In the **forewedge,** the portion of the accretionary prism closer to the trench, structures verge toward the trench (i.e., toward the downgoing plate), while the **retrowedge,** the portion of the prism closer to the arc, structures verge toward

the arc (i.e., toward the overriding plate). Note that the material of the wedge itself serves as the backstop.

Compressional deformation in the accretionary prism produces thrust faults, folds, and cleavage. But tectonic compressional stress is not the only cause of strain in a prism. Gravity sliding causes slumping of rock and sediment down the slope of the prism toward the trench. And once the prism has become very thick, it begins to undergo **extensional collapse** under its own weight, like soft cheese. This means that gravitational energy overcomes the strength of material at depth in the internal part of the prism, so this material spreads sideways, leading to horizontal stretching in the above prism. As a consequence of this stretching, the region near the surface of the prism undergoes normal faulting (Figure 17.12a and b).

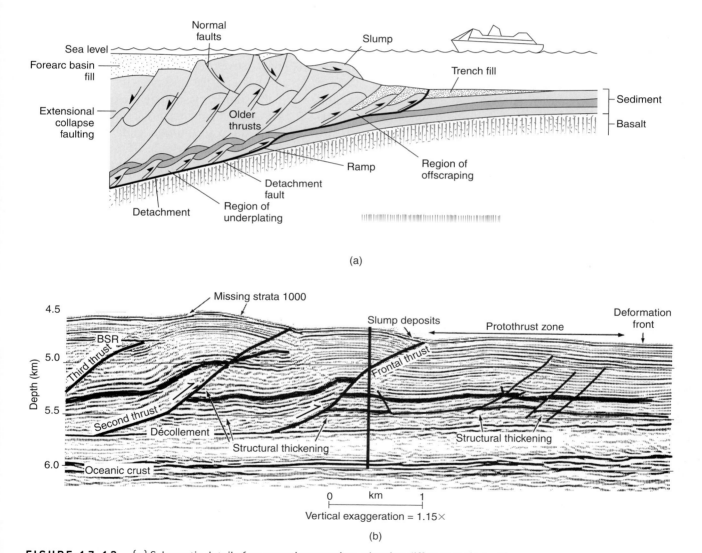

(a)

(b)

FIGURE 17.12   (a) Schematic detail of an accretionary prism, showing different regimes of deformation referred to in the text. (b) Interpreted seismic-reflection profile of the toe edge of an accretionary prism forming in the Nankai trough off Japan. Several faults can be imaged.

Let's first look a little more closely at the consequences of gravity-driven sliding down the slope of the prism. Such downslope movement events, which may be triggered by the relatively frequent earthquakes that occur in accretionary prisms, may lead to displacement of slump blocks (Figure 17.12), ranging from meters to tens of kilometers across. Very large (100s to 1,000s of meters long) slump blocks that remain semi-coherent during displacement are called **olistostromes.** The movement of slump blocks may lead to the formation of penecontemporaneous folding (see Chapter 1). Slope failure may also generate submarine debris flows in which muddy slurries flow downslope, and turbidity currents in which sediment disaggregates into an avalanche-like cloud that settles out in graded beds in the trench (Figure 17.10). Material that reaches the toe of the wedge will be recycled back into the wedge by offscraping. Thus, in some accretionary prisms, geologists find blocks composed of conglomerate containing clasts of conglomerate.

Now let's look more closely at the structural development of an accretionary prism. Effectively, an accretionary wedge is a fold-thrust belt that forms where the subducting plate slides under (underthrusts) an overriding plate (or, viewed from different perspective, where the overriding plate slides over the subducting plate). A **detachment fault** (i.e., **décollement**) delineates the boundary between the top of the subducting plate and the base of the accretionary prism and defines the actual plate boundary. This detachment ramps to progressively shallower stratigraphic levels toward the trench. At the toe of the prism the décollement grows and ramps up into trench strata or pelagic strata. As soon as strata becomes part of the hanging wall of the fault, it becomes, by definition, part of the overriding plate. The processes of detachment growth and ramping continue as subduction brings more material into the plate boundary, so with time, more and more material adds to the toe of the prism. This process is called **offscraping** (Figure 17.12).

Meanwhile, at the base of the accretionary prism, a duplex forms as the basal detachment cuts trenchward into strata of the downgoing plate and then ramps up to merge with a higher-level detachment. When a new **horse** (lens of rock or sediment surrounded by faults) forms, the material of the horse, by definition, becomes part of the overriding plate. Thus, material can be transferred from the downgoing plate to the overriding plate at the base of the accretionary prism, a process called **tectonic underplating**[3] (Figure 17.12). While faulting

takes place, the trench fill continues to be deposited over the subducting ocean floor. With continued deformation, faults in the internal part of the accretionary prism progressively steepen and rocks become penetratively strained. These processes lead to thickening within the prism.

Throughout most of an accretionary prism, thrusts and associated folds verge toward the trench, but along the arc side of the trench, thrusting may verge toward the arc. In this regard, accretionary prisms can be **bivergent** (Figure 17.11b). As noted earlier, you can simulate the development of a bivergent accretionary prism with a simple sand-box experiment (Figure 17.11d). Take a wooden box and cut a slit (representing a trench) in the floor of the box, about one-quarter of the distance in from one end of the box. Now, place a sheet of mylar (thin plastic) on the base of the box, and run the end through the slit. Bury the floor of the whole box (including the mylar) with a layer of sand. As you begin to pull the mylar sheet through the slit, to simulate subduction, the sand above the sheet moves toward the slit. Sand piles up above the slit, because the moving sand on the mylar collides with the stationary sand on the overriding side of the slit, and as this happens, thrust faults develop on both sides of the slit to accommodate the sand buildup. Those thrusts formed on one side of the slit verge in the direction opposite to the thrusts on the other side, creating a bivergent wedge. Note that the material of the wedge itself serves as the backstop that causes material to be scraped off the downgoing plate.

Recall that thrusting is not the only type of faulting to occur in an accretionary prism. When an accretionary prism reaches a substantial thickness, the upper part may undergo extensional collapse leading to the development of normal faults. Slip on these faults results in exhumation (uplift and exposure) of deeper rocks (Figure 17.12a).

The formation of an accretionary prism, as we discuss further in Chapter 18, can be described by the concepts of **critical taper theory.** To picture the essentials of this theory, imagine a plow blade moving into a layer of sand. The blade acts as a rigid **backstop** that transmits stress into the sand. (At a convergent plate boundary, the edge of the overriding plate and, later, the already accreted portions of the wedge act as a backstop with respect to new material being added to the wedge.) According to critical taper theory, a dynamic balance develops to maintain the **critical-taper angle,** the angle between the seaward surface of the prism and the surface of the downgoing plate. To maintain this angle, processes that cause the prism to grow wider, relative to its height (e.g., offscraping,

---

[3]Not to be confused with magmatic underplating, the process of adding basalt to the base of the crust along the margins of rifts or above mantle plumes.

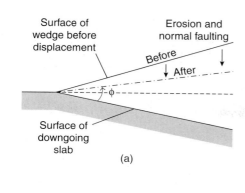

Surface of wedge before displacement

Erosion and normal faulting

Before

After

φ

Surface of downgoing slab

(a)

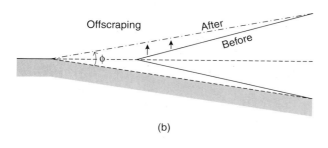

Offscraping

After

Before

φ

(b)

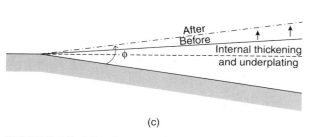

After

Before

Internal thickening and underplating

φ

(c)

FIGURE 17.13 Cross sections illustrating maintenance of critical taper angle in an accretionary wedge. The solid line represents the wedge when it does not have a critical taper (i.e., before adjustment) and the dot-dash line represents the wedge that does have a critical taper. (a) If the wedge slope is too steep, steepness can be decreased by erosion and normal faulting in the upper part of the wedge. (b) If the slope is too steep, steepness can also be decreased by offscraping and building out of the wedge at the toe. Note the surface of the downgoing slab sinks. (c) If the slope is not steep enough, steepness can be increased by underplating and internal thickening of the wedge.

normal faulting, and trenchward slumping), compete with processes that cause the prism to grow thicker, relative to its width (e.g., underplating and penetrative shortening). The taper angle of the wedge can also be decreased as a consequence of trenchward slumping or by normal faulting in higher parts of the wedge (Figure 17.13a). If internal shortening and underplating cause the internal part of the wedge to thicken, so that the surface slope becomes steeper, then the taper angle of the wedge becomes too large. At this time, the wedge as a whole slides seaward and new material is

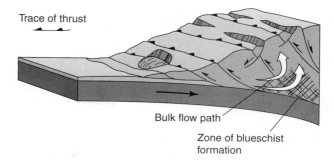

Trace of thrust

Bulk flow path

Zone of blueschist formation

FIGURE 17.14 Bulk flow path of sediment in an evolving accretionary wedge. The large arrows trace the average movement of a grain of material. Note that during its movement, the grain may enter the region of blueschist metamorphism.

added to the toe of the wedge by offscraping, the net result being that the taper angle decreases below the critical value again (Figure 17.13b). If, alternatively, the taper angle becomes less than the critical value, then internal strain of the wedge and underplating occurs, thickening the wedge and resulting in an increase of the taper angle (Figure 17.13c).

The overall consequence of deformation and mass-wasting processes in accretionary wedges results in long-term internal "circulation" of material within the wedge. During this process material first moves down to the base of the wedge and then moves back up toward the surface (Figure 17.14). Thus, sediment that accreted to the base of the wedge by subduction may later end up exposed at the surface of the wedge. This movement reflects both internal thrusting in the wedge that pushes material up, and normal faulting and slumping that strips away the overlying material of the wedge (Figure 17.14). Such net material flow within an accretionary wedge explains how blueschist, formed at the base of the wedge, can eventually be brought to the surface of the wedge.

The combination of tectonic deformation, gravity-driven slumping, and extensional collapse that takes place in prisms makes them structurally complex and heterogeneous. In some places, accretionary prisms consist of coherent sequences of strata containing parallel arrays of folds and faults and an axial-planar cleavage. Elsewhere, prisms consist of **broken formation** in which beds can be traced for only a short distance before they terminate at another lithology, or **mélange,** a chaotic mixture of different rock types (Figure 17.15).[4] In mélange, bedding cannot be traced very far at all, and rocks of radically different lithology and metamorphic grade are juxtaposed.

---

[4]From the French word for "mixture."

FIGURE 17.15 Photograph of Paleozoic mélange exposed in an outcrop in north-central Newfoundland. Note how layers are disrupted, strongly cleaved, and complexly folded, and that a variety of different rock types are visible.

On a traverse across an accretionary prism, you will find exposures of slate and lithic sandstone (derived from trench-fill turbidites), bedded chert (derived from pelagic silicic ooze), micrite (derived from pelagic carbonate ooze), greenstone (altered seafloor basalt), and **blueschist** (metamorphic rocks containing a blue amphibole called glaucophane). The origin of blueschist puzzled geologists for years, because glaucophane can only form under unusual conditions of very high pressure (as occurs at depths >20 km) and relatively low temperature. Because of the geothermal gradient (rate of increase of temperature with depth) characteristic of continental crust, glaucophane does not form in normal continental crust. But once geologists began to understand the process by which accretionary prisms formed, the location of blueschist formation became clear. Blueschists form at the base of the deepest part of the prism, where pressures are great, due to the overburden of 20 km of sediment, but temperatures are relatively low because the underlying oceanic lithosphere is relatively cold.

17.2.4 The Forearc Basin and the Volcanic Arc

As we continue our tour up the slope of the accretionary prism and toward the volcanic arc, we find that the top of the prism is defined by an abrupt decrease in slope. This topographic ridge is the **trench-slope break** (Figure 17.11). In a few locations around the world (e.g., Barbados, east of the Lesser Antilles volcanic arc along the east edge of the Caribbean), the trench-slope break emerges above sea level.

At many convergent margins, a broad shallow basin covers the region between the trench-slope break and the volcanic arc (Figure 17.11). This **forearc basin** contains flat-lying strata derived by erosion of the arc and the arc's substrate. Typically, strata of the forearc basin overlie older, subsided, portions of the prism. But locally, these strata overlie ocean lithosphere that had been trapped between the arc axis and the trench when subduction initiated. The strata may also overlie older parts of the volcanic arc and its basement.

The **volcanic arc** is the chain of volcanoes that forms along the edge of the overriding plate, about 100–150 km above the surface of the subducted oceanic lithosphere. As noted earlier, most of the magma that rises to feed the arc forms by partial melting in the asthenosphere above the surface of the downgoing slab. This partial melting takes place primarily because of the addition of volatiles (H_2O or CO_2) released from the downgoing plate into the mantle as the downgoing plate heats up. Some researchers argue that small amounts of melt may be derived locally from the downgoing plate.

Island arcs form where one oceanic plate subducts beneath another, or where the volcanic arc grows on a sliver of continental crust that had rifted from a continent; **continental arcs** grow where an oceanic plate subducts beneath continental lithosphere. Volcanism at island arcs formed on oceanic crust tends to produce mostly mafic and intermediate igneous rocks, whereas volcanism at continental arcs also produces intermediate and silicic igneous rocks, including massive granitic batholiths. The large volumes of silicic magmas in continental arcs form when hot mafic magmas rising from the mantle transfer heat into the surrounding continental crust and cause melting of the crust. While partial melting of mantle peridotite (an ultramafic rock) yields basaltic (mafic) magma, partial melting of mafic or intermediate continental crust yields intermediate to silicic magma.

The **arc-trench gap,** meaning the distance between the arc axis and the trench axis, varies significantly among convergent margins (Figure 17.6). Two factors control the width of the arc-trench gap at a given convergent margin: (1) *Dip of the downgoing slab:* Geometric principles dictate that if the downgoing slab dips very steeply, then the arc-trench gap must be narrow, but if the downgoing slab dips gently, then the arc-trench gap must be broad. (2) *Width of the accretionary prism:* Where subduction has continued for a long time, or where a large river fills the trench with sediment, the accretionary prism grows to be very large. When this happens, the prism acts as a weight that flexurally depresses the downgoing slab, and as the prism builds seaward, the trench location migrates seaward.

17.2.5   The Backarc Region

The **backarc region** refers to the region on the opposite side of the volcanic arc from the forearc basin. The structural character of backarc regions varies with tectonic setting (Figure 17.16). For the purpose of discussion, we define three types of backarc regions: (1) contractional, (2) extensional, and (3) stable.

In a **contractional backarc** (Figure 17.16a), a backarc basin does not form. Rather, crustal shortening generates a fold-thrust belt and/or a belt of basement-cored uplifts (see Chapter 18). Both types of deformation developed on the east side of the Andes during the Cenozoic, so a contractional backarc is commonly called an **Andean-type backarc.** The style of deformation that develops in a contractional backarc depends on the angle of subduction and on the nature of the crustal section in the overriding plate. If subduction angles are moderate to steep and the backarc region contains thick strata of a former passive margin, a fold-thrust belt develops. (Such a backarc fold-thrust developed during the Mesozoic Sevier Orogeny in western North America.) If, alternatively, subduction angles are shallow, so the subducted plate shears along the base of the overriding plate, stress activates preexisting basement-penetrating faults further to the foreland, and reverse-sense movement on these faults uplifts basement blocks. Overlying strata drape over the block uplifts and form monoclinal folds. (Such block uplifts, and associated monoclines, formed in the Rocky Mountain region of the United States during the 80–40 Ma Laramide Orogeny.)

In **extensional backarc** regions, crustal stretching takes place (Figure 17.16b). This stretching produces a **backarc basin.** If the stretching produces only a continental rift, then continental crust underlies the backarc basin, but if seafloor spreading takes place, then oceanic crust underlies the backarc basin. A clear example of a backarc basin formed by seafloor spreading occurs behind the Mariana Island Arc in the western Pacific Ocean (Figure 17.7), so extensional backarcs are commonly called **Mariana-type backarcs.** Backarc basins appear to initiate by rifting along the length of the volcanic arc. When the rift evolves into a new mid-ocean ridge, it splits off a slice of the volcanic arc, and then seafloor spreading separates this now-inactive slice of arc crust from the still active volcanic arc. A slice of arc crust, separated from the active arc by a new segment of seafloor, is called a **remnant arc.** Large backarc basins like the Philippine Sea contain several remnant arcs. In the Philippine Sea these appear to have been produced by a succession of separate rifting episodes, each of which yielded a short-lived mid-ocean ridge.

In a **stable backarc,** no strain accumulates (Figure 17.16c). Some stable backarcs may have once been contractional or extensional, but then later became stable when plate motions changed. Others are composed of oceanic lithosphere trapped behind the arc when a convergent margin developed far off the coast of the passive margin. The Bering Sea, for example, is

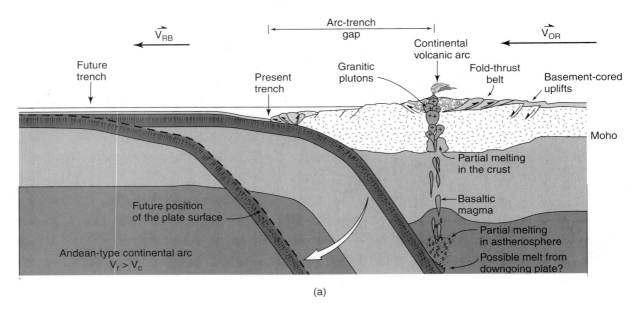

(a)

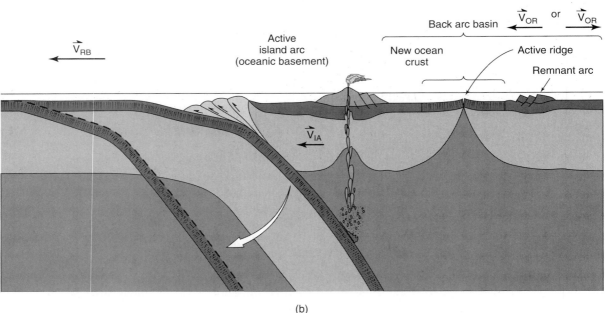

(b)

FIGURE 17.16 Different kinds of volcanic arcs and backarc regions. (a) An Andean-type continental arc, with a compressional backarc region. Here, the volcanic arc grows on continental crust, and compression has generated a fold-thrust belt and "Laramide-style" basement-cored uplifts. Large granitic plutons develop. This situation develops where the velocity of the overriding plate (v_{OR}) is in the same direction and exceeds the rollback velocity (v_{RB}). (b) A Mariana-type island arc, with an extensional backarc. Here, the volcanic arc grows on oceanic crust, and a backarc basin develops in which there is seafloor spreading. A remnant arc, composed of a rifted-off fragment of the arc may occur in the basin. This situation develops when $v_{OR} < v_{RB}$, or is in the opposite direction to v_{RB}. The island arc must move to keep up with the rollback. (c) A Japan-type volcanic arc, in which the island arc has continental basement that had rifted off a continent when a backarc basin grew. Here, the backarc spreading has ceased, and the backarc is stable. This is because $v_{OR} = v_{RB}$. A strike-slip fault could develop in the backarc region.

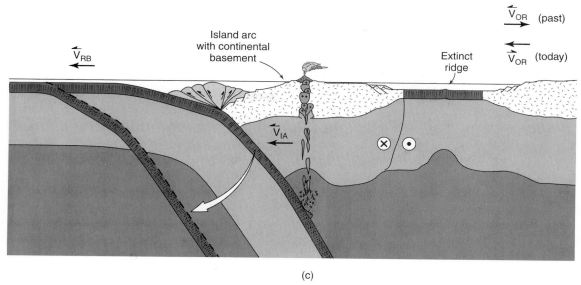

Island arc
with continental
basement

Extinct
ridge

V⃗_RB

V⃗_IA

⊗ ⊙

(c)

FIGURE 17.16 (Continued)

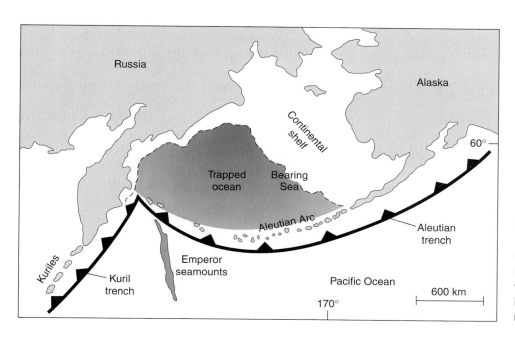

Russia

Alaska

Continental
shelf

60°

Trapped
ocean

Bearing
Sea

Aleutian Arc

Aleutian
trench

Kuriles

Emperor
seamounts

Pacific Ocean

600 km

Kuril
trench

170°

FIGURE 17.17 The Bering Sea is an example of a backarc region formed from trapped ocean. Here, the oceanic lithosphere that now lies behind the arc formed in the Mesozoic—it is not a consequence of Cenozoic backarc spreading.

underlain by Mesozoic-age trapped ocean floor that was isolated from the rest of the Pacific plate when the Aleutian volcanic arc formed (Figure 17.17).

Why do we observe such a wide range of kinematic behavior in backarc regions? The answer comes from examining the relative motion between the backarc region and the volcanic arc. As subduction progresses, the location of the bend in the downgoing plate rolls back, away from the backarc (Figure 17.18). The axis of the volcanic arc moves with the rollback. Thus, if the overriding plate is moving in the same direction but at a rate faster than rollback, a contractional backarc develops. If, however, the overriding plate is stationary or moves away from the trench, then rifting and a backarc basin develop. And if the overriding plate moves in the direction of rollback at the same rate as rollback, then the backarc region is stable. If there is a component of lateral motion between the overriding and downgoing plates, then some strike-slip faulting may also occur in the backarc.

Backarc regions can evolve with time. For example, the Japan Sea (Figure 17.7) started as an extensional backarc. In fact, Japan's basement consists of continental crust that originally linked to eastern Asia; the islands separated from the rest of Asia when seafloor spreading in the backarc produced the Japan Sea. Presently, however, the character of earthquakes indicates that shortening, accompanied by strike-slip

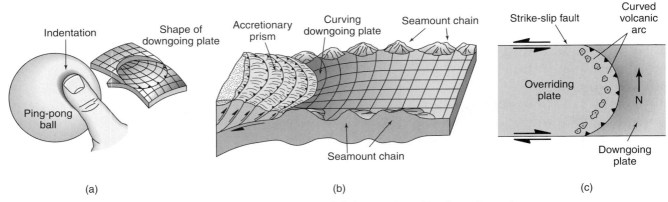

(a)   (b)   (c)

FIGURE 17.18 Reasons for the curvature of volcanic arcs. (a) If you make a thin slit on the surface of a ping-pong ball, and push it in, the piece must pucker into a curve, to maintain its surface area; (b) If a chain of seamounts collides with the arc, it may cause indentations because the lithosphere of the seamount is more buoyant and subducts at a shallower angle, and because the seamounts act as an obstacle to growth of the accretionary prism. (c) Strike-slip and drag origin for island-arc curvature. Shear on strike-slip faults may bend the volcanic arc in map view.

deformation, now takes place in the Japan Sea. Because of this motion, the Japan Sea may someday close, suturing Japan back to Asia.

17.2.6 Curvature of Island Arcs

The reason that island arcs are called *arcs* (e.g., Figure 17.17), is that many have a curved trace in map view. Let's now examine a few possible explanations for why such curvature exists.

First, a curve reflects the natural shape of an indentation on a sphere. Places where oceanic lithosphere subducts can be viewed as indentations on the surface of the spherical Earth; the downgoing plate must curve in order to maintain the same surface area for a given length of the convergent boundary, and if the downgoing plate curves in three dimensions, then the trench and volcanic arc must also curve on the Earth's surface (Figure 17.18a). To visualize this geometry, take a ping-pong ball and push in one side with your thumb—the trace of the indentation is a curve on the ball's surface.

Second, a curve forms where a seamount collides with an originally straight arc during subduction. To picture this geometry, look at a map of the western Pacific, and note that some of the major cusps in subduction zones coincide with sites of seamount subduction (Figure 17.18b). For example, the Emperor Seamount Chain subducts at the cusp between the Aleutian Trench and the Kuril Trench (Figure 17.17). This relationship may develop because a seamount acts as an obstacle that inhibits propagation of the accretionary wedge, and/or the buoyancy of a seamount decreases the rate of rollback, so that por-

tions of the downgoing plate away from the seamount roll back faster than the plate under the seamount.

Third, strike-slip faulting causes map-view shear at the end of an arc (Figure 17.18c). To picture this geometry, look at a map of the region encompassing the southern end of South America, the Scotia Sea, and the northern tip of the Antarctic Peninsula. Transform faults delineate both the northern and the southern boundaries of the Scotia Plate. Conceivably, shear along the northern fault bent the southern end of South America and shear along the southern fault bent the northern end of the Antarctic Peninsula. Shear on these faults also caused the Scotia arc to bend into an arc. Alfred Wegener noted the shape of the Scotia Arc and used the shape as evidence for the westward drift of South America. A similar geometry of faulting may explain the curvature of the Lesser Antilles volcanic arc at the eastern edge of the Caribbean Sea.

17.2.7 Coupled versus Uncoupled Convergent Margins

Taking into account the description of convergent margins that we've provided earlier in this section, we see that not all convergent margins display the same suite of rocks and structures. Based on the contrasts among various convergent margins worldwide, geologists distinguish between two end-member types.

In a **coupled convergent margin,** the downgoing plate pushes tightly against the overriding plate, so the plate boundary overall is under compression. As a consequence, large shear stresses develop across the contact, causing efficient offscraping and tectonic underplating, and therefore buildup of a large accretionary

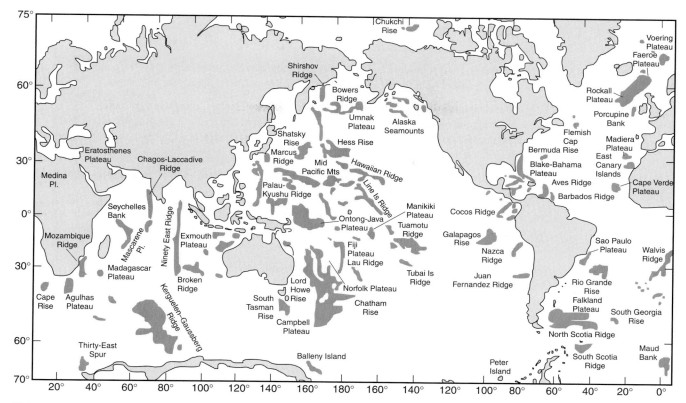

FIGURE 17.19 A map showing blocks of buoyant crust including volcanic arcs, oceanic plateaus, and continental fragments. Blocks that have darker shading rise above today's sea level. Blocks that have lighter shading are presently submerged. Convergence between any two blocks could cause collisional tectonics.

wedge. This shear stress also triggers devastating earthquakes, and development of a contractional backarc region. Perhaps because compression squeezes crustal fractures closed, magma rises slowly, and therefore has time to fractionate and/or cause partial melting of adjacent continental crust before intruding at shallow depth or erupting at the surface. Since partial melting of continental crust produces intermediate to felsic magma, and since fractionation removes mafic minerals from a melt, intermediate to felsic igneous rocks predominate at coupled convergent margins.

In an **uncoupled convergent margin,** the downgoing plate does not push tightly against the overriding plate, so compression across the margin is not great. As a consequence, shear stresses across the plate boundary are relatively small, thrust earthquakes at the boundary have smaller magnitude, and relatively little offscraping and underplating occurs. In uncoupled systems we find extensional backarcs, and cracks in the overriding plate remain somewhat open, so mantle-derived magmas rise directly to the surface before significant fractionation or crustal contamination occurs. Thus, mafic igneous rocks are more common at such convergent margins.

17.3 BASIC STAGES OF COLLISIONAL TECTONICS

As subduction consumes an oceanic plate, a piece of buoyant lithosphere attached to the downgoing plate may eventually be brought into the convergent boundary. Examples of buoyant lithosphere include large continents, small continental fragments, island arcs, oceanic plateaus (broad regions of anomalously thick oceanic crust, formed by hot-spot volcanism), and spreading ridges (Figure 17.19). Regardless of type, buoyant lithosphere generally cannot be completely subducted, and when it merges with the overriding slab, the boundary becomes a **collision zone.**[5] When the forces driving collision cease, the relative motion between the colliding blocks ceases, and when this happens, the once separate blocks of lithosphere have merged to become one. The shear surface that marks the boundary between these once-separate plates is a

---

[5]Here, we've described collision tectonics that follows convergent tectonics. Note that collision can also result from the closure of a rift, even if no oceanic lithosphere existed between the colliding blocks.

TABLE 17.2	TERMINOLOGY OF COLLISION
Accreted terrane	A piece of exotic crust that has been attached to the margin of a larger continent. (Note the spelling of *terrane,* which differs from that of the geographic term for a tract of land, spelled *terrain.*)
Basin inversion	The process whereby a region that had undergone crustal extension during basin formation subsequently undergoes crustal shortening during collisional tectonics; in the process, faults that began as normal faults are reactivated as reverse faults, and the strata of the basin thrusts up and over the former basin margin.
Collision	An event during which two pieces of buoyant lithosphere move toward each other and squash together, after the intervening oceanic lithosphere has been subducted.
Delamination	In the context of collisional tectonics, this refers to the separation of the basal portion of the thickened lithospheric mantle beneath a collisional orogen; this delaminated lithospheric mantle then sinks downward through the asthenosphere.
Exotic terrane	An independent block of buoyant crust that has been brought into a convergent margin during subduction, where it collides and docks against the continent (also called **accreted terrane**). The adjective "exotic" in this context is used simply to emphasize that the block in question did not originate as part of the continent to which it is now attached, but rather came from somewhere else.
Lateral escape	The process, accompanying collision, during which crustal blocks of the overriding plate slide along strike-slip faults in a direction roughly perpendicular to the regional convergence direction; effectively, the "escape" from the collision zone resembles the movement of a watermelon seed that you squeeze between your fingers.
Orogenic collapse	The process that occurs when thickened crust in a collisional orogen weakens and starts to sink under its own weight. Effectively, gravitational loads cause horizontal extensional strain to develop. In some cases, extension is coeval with thrusting (synorogenic collapse), in other cases it occurs after thrusting has ceased (postorogenic collapse). (Also called *extensional collapse.*)
Suspect terrane	A crustal block in an orogen whose tectonic origin is unclear. The block does not appear to correlate with adjacent crust in the orogen where it now resides and thus may be exotic. A block remains "suspect" only until its origin (i.e., whether it's exotic or not) has been determined.
Suture	The shear surface within an orogen that marks the boundary between once-separate continents; commonly, slivers of ophiolites occur along a suture.
Tectonic collage	A region of crust that consists of numerous exotic terranes that have been sutured together; in other words, a tectonic collage consists of accreted terranes that docked during a protracted period of convergent-margin tectonism.

suture; slivers of ophiolites (ocean crust thrust over continental crust) crop out locally along sutures. We define these terms, as well as others used in the discussion of collisional tectonics in Table 17.2.

The types of rocks and structures formed during a particular collisional orogeny depend on numerous variables, including:

- *Relative motion between the colliding blocks.* **Frontal collisions** yield thrust faults whose movement is perpendicular to the edge of the colliding blocks, while strain in **oblique collisions** may be partitioned between thrusting and strike-slip faulting (Figure 17.20a). In addition, where blocks col-

lide obliquely, the timing of collision may be diachronous along strike (Figure 17.20b), and the blocks merge together like the two sides of a zipper.
- *Shape of the colliding pieces.* The collision of broad, smooth continental margins yields fairly straight orogens, while the collision of irregular continental margins (with **promontories,** which are seaward protrusions of the continental margin, and **recesses,** which are indentations along the continental margin) yields sinuous orogens, in map view. Promontories act as indenters that push into the opposing margin, creating a localized region of high strain. In some cases, slices of crust may move sideways (relative to the frontal collision direction) along strike-slip faults

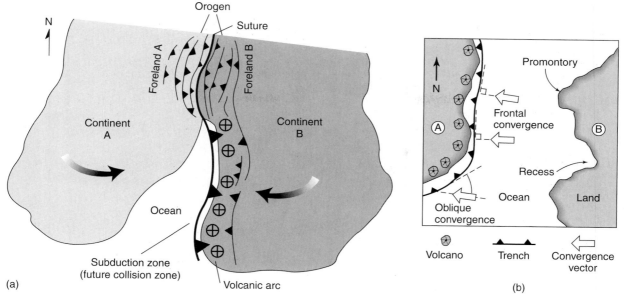

FIGURE 17.20 (a) A map showing a zipper-like collision between two continents. Here, the ocean between the two continents is closing progressively from north to south. In the collision zone, the boundary between what had originally been two separate continents. (b) A map showing the convergence of two continents. Promontories and recesses make the west coast of Continent B irregular. Because of the change in trend of the subduction zone along Continent A, frontal convergence (and, eventually, frontal collision) will occur to the north, while oblique convergence (and, eventually, oblique collision) will occur to the south.

to get out of the way of the colliding blocks. This movement is a type of **lateral escape,** and will be discussed further in Section 17.4.1.

- *Physical characteristics of the colliding pieces.* Physical characteristics, such as temperature, thickness, and composition, influence the way in which crustal blocks deform during collision. For example, warmer and, therefore, softer crust of a younger orogenic belt will develop greater strains during collision than will old, cold cratonic crust. During collision, a craton acts as a rigid indenter that pushes into the relatively soft, younger orogenic belt. The collision between India, an old craton, and southern Asia, a weak Phanerozoic orogen, illustrates such behavior. During this collision, much more deformation has happened in the weak southern margin of Asia than in strong India. In fact, a map of the collision zone (see section 17.4.1) shows that India has actually pushed into Asia, so that a transform fault now bounds each side of the Indian subcontinent.

Because so many variables govern the nature of a collisional orogeny, no two collisional orogenies are exactly the same. Nevertheless, we can provide a basic image of the collision process by outlining, in the following section, the various stages in an idealized collision between

two continents (A and B; Figure 17.21a, b, c). For reference, we call the portion of the orogen that is on the craton side of the collision the **foreland** and the internal part of the orogen the **hinterland.**

17.3.1 Stage 1: Precollision and Initial Interaction

Let's begin by setting the stage for the collision between two continents. In this scenario, Continent A moves toward Continent B as the oceanic lithosphere connected to Continent A subducts beneath the margin of Continent B (Figure 17.21a). Note that the margin of Continent A is a passive margin, along which a passive-margin sedimentary basin has developed; in contrast, the margin of Continent B is an Andean-type convergent margin along which a volcanic arc has developed.

Continent A remains oblivious to the impending collision until the edge of the continent begins to bend, prior to being pulled into the subduction system by the downgoing plate (Figure 17.21b). When this happens, flexure causes the surface of the continental margin to rise, so that the continental shelf rises above sealevel and undergoes erosion. The margin of Continent A also undergoes stretching, and as a result, normal faults trending parallel to the edge of the margin start to slip.

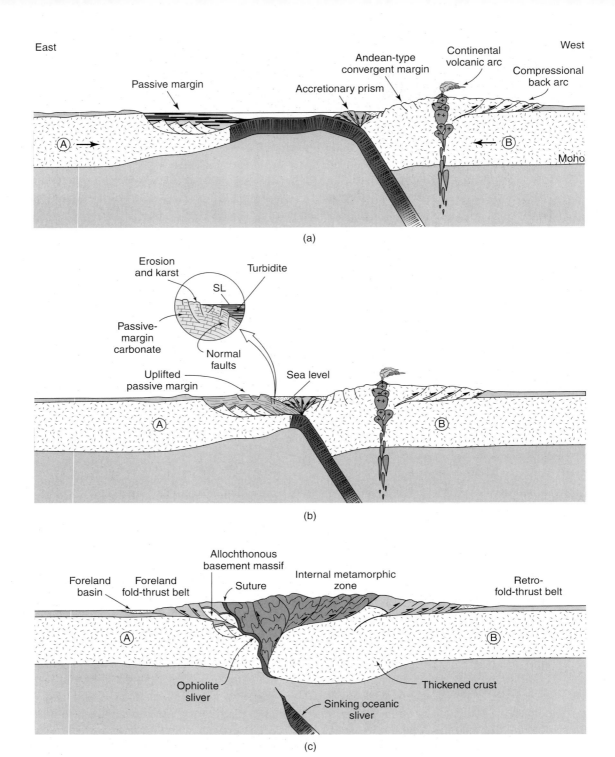

East **West**

FIGURE 17.21 Stages in an idealized continent–continent collision. (a) Precollision configuration. Continent A has a passive-margin basin on its east coast, while Continent B has a convergent margin on its west coast. (b) During the initial stage of collision, the passive margin is uplifted, and an unconformity (locally, with karst) develops. Turbidites derived from Continent B soon bury this unconformity (see inset). Normal faults break up the strata of the passive-margin basin, due to stretching. But soon, thrusts begin to develop, transporting the deeper parts of the basin over the shallower parts. (c) In a mature collision orogen, the subducting slab has broken off, a suture has formed, and metamorphic rocks are uplifted and exhumed in the interior of the orogen.

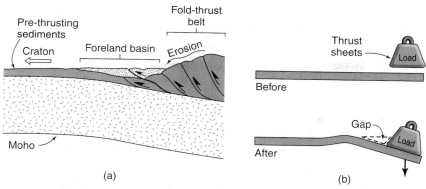

FIGURE 17.22 Origin of a foreland basin. (a) A schematic cross section showing how a stack of thrust slices, when emplaced on the edge of a continent, loads the continent. As a result, a depression develops on the continent. This depression fills with sediment eroded from the orogenic highlands and becomes the foreland basin. (b) You can envision the process of foreland-basin formation by imagining a load placed on the edge of a sheet. If the sheet has flexural rigidity, as does lithosphere, the load pushes the edge of the sheet down and creates a gap.

17.3.2 Stage 2: Abortive Subduction and Suturing

With continued convergence, the surface of Continent A's continental shelf becomes the floor of the trench (Figure 17.21b). When this happens, turbidites derived from the margin of Continent B and its volcanic arc bury the now-eroded surface of the shelf.[6] Thus, a major unconformity defines the contact between the passive-margin basin sedimentary sequence and the turbidites.

Prior to collision, formation of the accretionary prism progressed along the margin of Continent B as new thrusts cut seaward into the strata on the down-going oceanic plate. But during collision, the strata on the downgoing plate consist of thick, well-stratified sedimentary beds of the former passive-margin basin. Thus, thrusts propagate into these beds, producing a fold-thrust belt that, with time, grows toward the fore-land (i.e., in the direction of the continental interior) of Continent A. The stack of thrust slices acts as a load that depresses the surface of Continent A, yielding a **foreland sedimentary basin** that spreads out over the edge of Continent A's craton (Figure 17.22a and b).

Such basins are asymmetric—they are thickest along the margin of the orogen and become thinner toward the interior of the continent. Meanwhile the backarc fold-thrust belt on Continent B continues to be active.

Shortening during the collision also reactivates the normal faults that bound basement slices at the base of the passive margin; because reactivation occurs in response to compression, these faults now move as reverse faults. This new movement emplaces slices of basement closest to the hinterland part of the fold-thrust belt over strata of the former passive margin (Figure 17.21c). The overall process of transforming the passive-margin basin into a thrust belt is called **basin inversion.** We use the term to emphasize that, during this process, a region that had previously undergone extension and subsidence during basin formation, now telescopes back together by a reversal of shear sense on preexisting faults, and undergoes uplift.

Eventually, a slice of the oceanic lithosphere that had once separated Continent A from Continent B may thrust over the inverted passive-margin of Continent A. This slice, which appears in the orogen as a band of highly sheared mafic and ultramafic rock, defines the **suture**; rock on one side of the suture was once part of Continent A, while rock on the other side was once part of Continent B.

Meanwhile, in the internal part of the orogen, or its "hinterland," the crust thickens considerably, and ductile folding (creating large, tight to isoclinal folds), shearing (creating mylonites), and regional metamorphism (creating schists and gneisses) occur at

---

[6]In older literature, this sequence of turbidites is called **flysch,** which was defined as "synorogenic" strata. Deposition of flysch was thought to signal the onset of orogenic activity (see also Chapter 20). However, turbidites form in other settings as well, so "flysch" as a tectonic term is confusing and we discourage its use in that way.

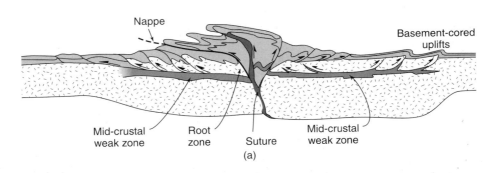

(b)

FIGURE 17.23 (a) An alternative cross section of a collisional orogen. Not all orogens are the same, and there are different ways to depict their geometry. (This version is modeled on the Alps.) Here, we do not show the erosional land surface, and we add features not shown in Figure 17.21. A mid-crustal weak zone serves as a basal detachment for faulting in the upper crust. Large recumbent nappes develop in the metamorphic interior of the orogen. (b) Photograph of a large nappe, transported on a subhorizontal thrust fault. This example is the Glarner Thrust of the Swiss Alps, placing Permian clastics over Tertiary flysch.

depth. With progressive deformation, the plastically deformed metamorphic rocks move upwards and toward the foreland. In some cases large recumbent folds develop. In the European literature, large sheets of such transported rock, locally containing recumbent folds, are called **nappes** (Figure 17.23a, b; see also Figure 18.15). Metamorphic rock of the hinterland eventually becomes exposed in the peaks of the mountain range due to **exhumation,** the combination of processes that strips off rock at the surface of the Earth to expose rock that had been deeper.

Eventually, the downgoing oceanic lithosphere breaks off the edge of Continent A and sinks slowly into the depths of the mantle. Without a source of new magma, the convergent-margin volcanic arc of Continent B shuts off. On Continent B, deformation styles are the same, but the vergence of structures is opposite to those that form on the edge of Continent A; rocks on the Continent B side of the orogen thrust toward the interior of Continent B. Thus, taken as a whole, the orogen is **bivergent,** meaning that opposite sides of the orogen, overall, verge in opposite directions.

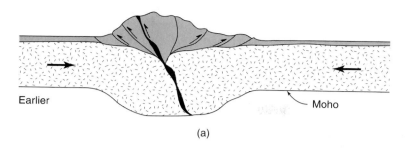

Earlier

Moho

(a)

Not to scale

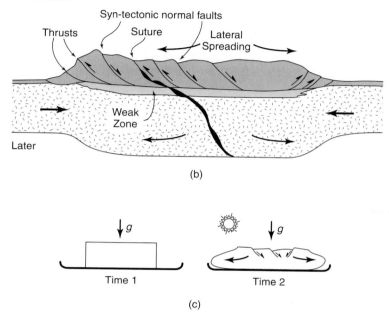

Syn-tectonic normal faults

Thrusts

Suture

Lateral
Spreading

Weak
Zone

Later

(b)

$\downarrow g$

Time 1

$\downarrow g$

Time 2

(c)

FIGURE 17.24 The concept of orogenic collapse. (a) A schematic cross section shows that during an early stage in a collision, the crust thickens by thrusting. (b) Later, as collapse occurs, extensional faults develop in the upper crust, while plastic flow occurs at depth. This process may contribute to development of a broad plateau. (c) The soft-cheese analogy for extensional collapse. A block of cold cheese can maintain its thickness. If the cheese warms up in the sun, it loses strength and spreads laterally. The rind of the cheese ruptures, and small faults develop.

17.3.3 Stage 3: Crustal Thickening and Extensional Collapse

So far, we've focused on the horizontal shortening that takes place in the crust during collisional tectonics. But keep in mind that as crust shortens horizontally, it also thickens (Figure 17.24a). In fact, the crust beneath collisional orogens may attain a thickness of 60–70 km, almost twice the thickness of normal crust. Shortening during collision also causes the lithospheric mantle to thicken substantially as well.

Thickening of the crust cannot continue indefinitely because, as the crust thickens, rock at depth becomes warm and, therefore, weaker. As a consequence, the differential stress developed in the orogen due to the weight of overlying rock (the "overburden") exceeds the yield strength of the rock at depth, and the rock begins to flow and develop horizontal extensional strain (Figure 17.24b). In other words, because of the force of gravity, very thick orogens collapse under their own weight. As we pointed out in Chapter 16, you can picture this process by imagining a block of cheese

that is heated in the sun (Figure 17.24c). Eventually, the cheese gets so soft that it spreads out, and the thickness of the block diminishes. This process is called **extensional collapse** (or, **orogenic collapse**).

Ductile extensional collapse at depth in an orogen causes stretching of the upper crust, where rock is cooler and still brittle. Therefore, during collapse, rock of the upper crust ruptures and normal faults form. Because collapse decreases the thickness of the uppermost crust, it causes decompression of the lower crust. This decompression may trigger partial melting of the deep crust, or even the underlying asthenosphere, producing magmas. Melting may also be caused by **lithosphere delamination.** This means that the deep keel of lithosphere that develops during thickening drops off (Figure 17.25). Warm asthenosphere rises to take its place and heats the remaining lithosphere. Some researchers argue that a broad portion of the mid-crust may actually become partially molten at this stage. What is certain is that magma does form and intrude the upper crust in many collisional orogens after

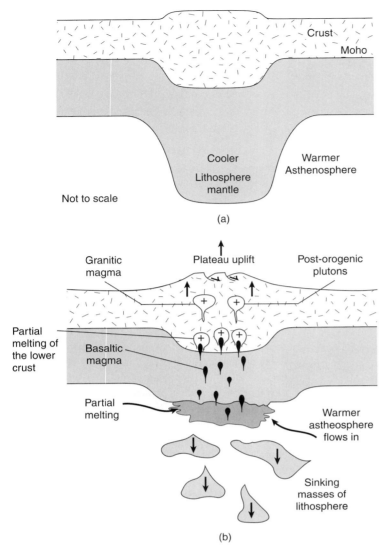

FIGURE 17.25 Postorogenic plutons and delamination.
(a) Thickening of lithosphere forms a keel-shaped mass of cool lithosphere to protrude down into the asthenosphere. (b) The keel drops off and is replaced by warm asthenosphere, causing partial melting and formation of anorogenic (postorogenic) plutons. The surface of the crust may rise as a consequence.

deformation has ceased. Because it intrudes after deformation, the granite has no tectonite fabric, so geologists refer to the granite as **postorogenic granite.**

The process of extensional collapse can occur while shortening and thrusting continue along the margins of collisional orogen, and it may continue after shortening has ceased. Extensional collapse, together with erosion, keeps mountain ranges from exceeding elevations of about 8 to 9 km, and contributes to the development of broad, high plateaus like the Tibetan Plateau of Asia (see Chapter 21).

17.4 OTHER CONSEQUENCES OF COLLISIONAL TECTONICS

In our description of the stages in an idealized continent–continent collision, we focused only on tectonic phenomena occurring adjacent to the colliding margins, and only on movements that can be illustrated by a two-dimensional cross-sectional plane. Here, we describe additional tectonic phenomena that may occur during collision, in specific situations, and consider movements in the third dimension.

17.4.1 Regional Strike-Slip Faulting and Lateral Escape

Collisional orogeny along a continental margin may lead to the development of regional-scale strike-slip faults that propagate far into the interior of the overriding plate. As an example, several large strike-slip faults start at the Himalayas and cut eastward, following curved trajectories across Asia to the Pacific margin (Figure 17.26a). Researchers suggest that some of the movement on these faults accommodates translation of large wedges of Asia relatively eastward, in response to the collision of India with Asia.

To simulate the development of these faults, researchers constructed models in which they push a wooden block into a cake of clay (Figure 17.26b). The wooden block represents the rigid craton of India, while the clay represents the relatively soft crust of Asia. (The crust of Asia is relatively soft because it formed during a protracted Phanerozoic orogeny; see Essay 21.2). In this model, a rigid wall, representing the mass of western Asia, constrains the left side of the clay cake. The right side, representing the Pacific margin, remains unconstrained. As the wooden block moves into the clay, wedges of clay, bounded by strike-slip faults, move laterally to the right. The pattern of faults resembles the trajectories of maximum shear stress predicted by the theory of elasticity; engineers refer to this pattern as a **slip-line field** (Figure 17.26c). So, based on the clay-cake experiments, geologists have suggested that the pattern of strike-slip faults in Asia represents a slip-line field resulting from stresses transmitted into the interior of Asia by the India–Asia

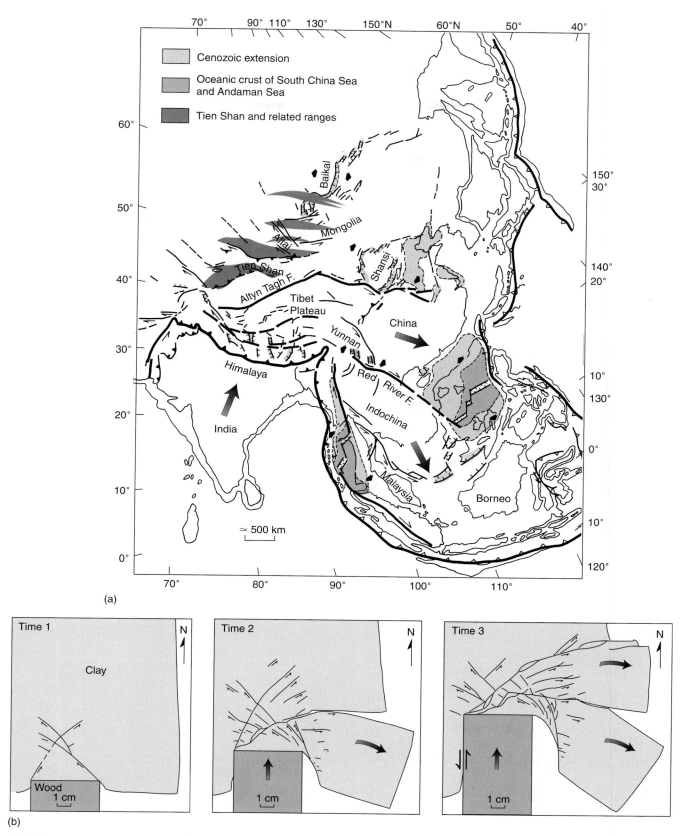

FIGURE 17.26 Lateral escape. (a) A sketch map of major structures in southeastern Asia. Note the major faults that slice across China. The large arrows indicate the motion of large crustal blocks. (b) Map-view sketch of a laboratory experiment to simulate lateral-escape tectonics. A wooden block (representing the Indian craton) is pushed northwards into a clay cake. The cake is restrained on the west, but not on the east. As the block indents, strike-slip faults develop in the clay, and large slices are squeezed eastwards. (c) Map view showing a theoretical model of a slip-line field caused by indentation of a rigid block into an elastic sheet. The lines represent the trajectories of maximum shear stress.

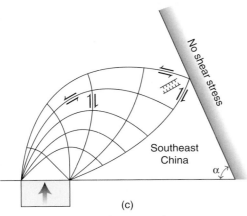

Southeast
China

α

(c)

FIGURE 17.26 (Continued)

sideways when you squeeze it between your fingers. This tectonic process is called **lateral escape.** Note that where lateral escape occurs, all strain resulting from collision cannot be depicted in a cross section, because of movement into or out of the plane of the section.

17.4.2 Plateau Uplift

Continental collisions cause the uplift of a linear belt of mountains, a "collisional range." But in some examples, collision also leads to the development of a broad region of uplifted land known as a **plateau.** As a case in point, consider the development of the Tibetan Plateau in southern Asia (Figure 17.26). Determining the time of uplift of this plateau has proven to be a challenging problem, but available evidence suggests that it rose up in concert with India's collision with Asia.

Researchers continue to debate the reason for the uplift of the Tibetan Plateau. Some consider it to be a consequence of thickening of the crust, either by plastic flow in the lower crust, or because the crust of India has been emplaced under the crust of Asia. Others suggest that it reflects

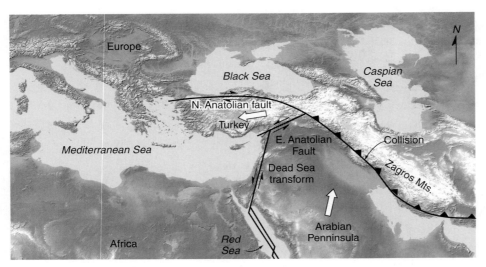

FIGURE 17.27 Sketch map of the eastern Mediterranean region, showing how Turkey is being squeezed to the west as the Arabian Peninsula moves northwards. This lateral escape of Turkey is accommodated by slip on the North Anatolian and East Anatolian Faults.

heating that occurred when the lower part of the lithospheric mantle detached (delaminated) from the base of the lithosphere and sank down into the mantle (Figure 17.25). Such **lithosphere delamination** would juxtapose hot asthenosphere at the base of the remaining lithosphere and cause the lithosphere to heat up. Thus, to maintain isostatic equilibrium, the surface of the lithosphere would have to rise.

17.4.3 Continental Interior Fault-and-Fold Zones

Look again at a Figure 17.26a. An *en echelon* band of short, but high thrust-bounded mountain belts, starting with the Tien Shan, tracks across the center of the continent. This band appears to be a zone of fault reactivation caused by stresses imparted to Asia by the collision of India. Stresses that occur in a continental

collision. Substantial debate continues about the significance and age of faulting in southern Asia, and thus the applicability of slip-line field theory to these faults remains uncertain. Possibly, the strike-slip faults are reactivated preexisting faults; the specific faults that reactivated were those that had the correct orientation to be part of the slip-line field.

A map of Turkey provides another example of strike-slip faulting resulting from a collision. Here, the northward movement of the Arabian plate has caught Turkey in a vise. The resulting stresses squeeze Turkey westward into the Mediterranean as the Saudi Arabian Plate moves north. This motion is accommodated by strike-slip displacement on along the North and East Anatolian Faults (Figure 17.27). This example illustrates how blocks of the overriding plate in a collisional orogen can be squeezed sideways out of the path of an indenter, much like a watermelon seed squirts interior as a result of tectonism along the continental margin are called **far-field stresses.** As India moves into Asia, it displaces the crust of southern Asia with respect to that of northern Asia. This displacement is accommodated by transpressional motion on faults in central Asia that are thousands of kilometers from the collision zone.

Similarly, the collision of Africa with North America during the late Paleozoic Alleghanian orogeny generated far-field stresses in the continental interior of North America. These stresses were sufficient to cause transpressional or transtensional movements on basement-penetrating faults of the region (Figure 17.28a). The reverse-sense component of motion on these faults led to the formation of monoclinal folds that drape platform strata over the uplifted blocks (see Essay 22.7). Sediments shed from the uplifts filled narrow, but deep, sedimentary basins. Exposures of the faults and associated structures (e.g., monoclinal folds, narrow sedimentary basins) can be found throughout the present-day Rocky Mountain region of the North American Cordillera in the western United States.

Geologists refer to the region in the western United States in which these Late Paleozoic faults, folds, uplifts, and basins formed as the **Ancestral Rockies,** because it occurs in the region now occupied by the present-day Rocky Mountains (a Cenozoic feature) but existed long before the present mountains formed (Figure 17.28a). Notably, structures very similar to those of the traditional Ancestral Rockies lie beneath the corn and wheat fields of the Midcontinent Region—these structures are called **Midcontinent fault-and-fold zones** (Figure 17.28b). Though they are not as well exposed as those of the west, and typically do not display as much displacement, they probably formed in much the same way, in that their movement probably occurred in response to far-field stresses generated during Paleozoic orogeny along the eastern and southern margins of North America. Movement on these faults displaced stratigraphic markers at depth and led to the development of monoclinal folds at shallower crustal levels.[7]

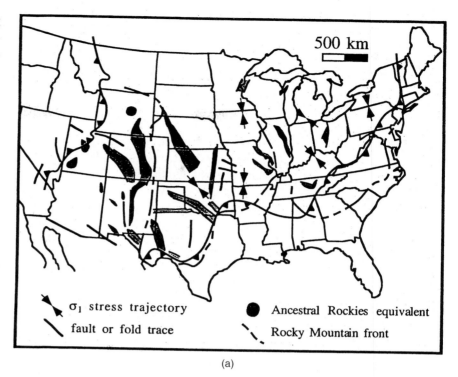

(a)

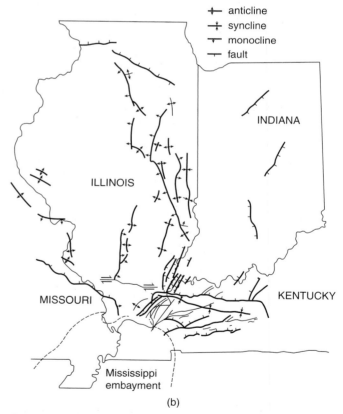

(b)

FIGURE 17.28 Continental-interior deformation of the United States. (a) Map of the Ancestral Rockies, and equivalent Midcontinent fault-and-fold zones. (b) Midcontinent fault-and-fold zones within the interior of the Illinois basin. These were active during the Alleghanian Orogeny (i.e., during the collision of Africa with North America).

---

[7]In addition to macroscopic folds and faults, the Alleghanian Orogeny also caused calcite twinning, a form of intragranular strain, to develop in the Midcontinent Region (see Figure 5.29). The strain due to this calcite twinning is only on the order of 1–6%.

The change in differential stress in continental interiors, and/or flexural loading (i.e., the emplacement of a load such as a stack of thrust slices) on a continental margin, due to collision may also trigger **epeirogeny,** meaning the gentle vertical displacement of broad regions, in continental interiors. These movements include the uplift of regional domes and arches as recorded by the development of unconformities, and the subsidence of intracratonic basins as recorded by sudden increases in the rate of basin subsidence. We discuss epeirogeny, and the factors that may cause it, in Essay 22.7.

17.4.4 Crustal Accretion (Accretionary Tectonics)

If you look at the present-day Pacific Ocean region, you will find many pieces of crust that differ in thickness and/or composition from the typical oceanic crust produced at a mid-ocean ridge (Figure 17.19). These pieces include:

- Small fragments of continental crust, such as Japan or Borneo, that rifted off larger continents in the past.
- Volcanic island arcs, such as the Mariana Arc, which formed along convergent plate boundaries.
- Seamount chains and oceanic island chains formed above hot spots.
- Oceanic plateaus, broad regions of anomalously thick crust, probably composed of basalts extruded at particularly productive hot spots.

All of these pieces are buoyant, relative to the asthenosphere, and thus cannot be subducted. Therefore, if subduction continues along the eastern margin of Asia for many more millions of years, the pieces would eventually collide with and suture to Asia. After each such **docking event,** a new convergent margin may form on the outboard (oceanic) side of each sutured piece. As a consequence, the continent grows. This overall process is called **crustal accretion** or **accretionary tectonics.**[8] The small crustal pieces that have been attached to a larger continental block by accretion are called **accreted terranes.**[9] In some cases,

the area of a continent grows by the addition of a broad belt of accreted terranes called a **tectonic collage** (Figure 17.29a and b).

How do we identify accreted terranes in an orogen? The first hint that a block of crust may be an accreted terrane comes from studying the geologic history preserved in a block. Do the rocks and structures of the block correlate with those of adjacent crust? If not, then the block is probably accreted. Geologists can test this proposal further by studying paleomagnetic data, for these may demonstrate that the accreted block and the continent to which it is now attached did not have the same apparent **polar-wander paths** prior to the time at which accretion occurred. Paleontological data may also indicate that the block originated at a different latitude from the host continent.

Based on mapping, paleomagnetic study, and paleontologic study, geologists have demonstrated that, during the Mesozoic, the North American Cordillera grew westward by crustal accretion. In fact, much of the vast tract of land that now comprises most of California, Oregon, Washington, and Alaska in the United States, and British Columbia and the Northwest Territories in Canada originated as crustal fragments outboard of North America that were accreted to the continental margin (Figure 17.29b). Much of this accretion occurred during oblique convergence, so strike-slip faults developed in the orogen and movement on these transported whole accreted terranes, or slivers of them, along strike to the north. In western North America, for example, bits and pieces of an accreted terrane known as Wrangelia crop out from Washington (maybe even Idaho) to Alaska ("W" in Figure 17.29b; see also Section 19.4). The Appalachian Mountains preserve a similar story. Thus, the eastern edge of what had been North America in the Precambrian lies well inboard of the continent's present coastline.

When two major continents collide along a margin that previously was the locus of terrane accretion, you can imagine that the resulting collisional orogen will be very complex. It will contain several sutures separating different blocks, and each block has its own unique geologic history. Most major collisional orogens involve accretion of exotic terranes prior to collision of the larger continents and final closure of the intervening ocean, so such complexity is the norm rather than the exception. Therefore, you should not assume that the history in one particular region represents that of the whole orogen, nor should you be surprised to find radically different geologic histories preserved in adjacent areas of crust in the orogen.

---

[8]An orogenic event during which a continent grows significantly larger by the addition of broad regions of accreted crust can be called an **accretionary orogeny.**

[9]They may also be called **exotic terranes,** to emphasize that they came from elsewhere (an "exotic" place) before attaching to the larger continent. A block of crust that might be accreted (or exotic) can be referred to as a **suspect terrane,** until its origin has been confirmed. Once geologists prove that a crustal block in an orogen is exotic, then the name "suspect" should no longer be used for it. Note the spelling of "terrane." This differs from the spelling of "terrain," which refers to a topographic surface.

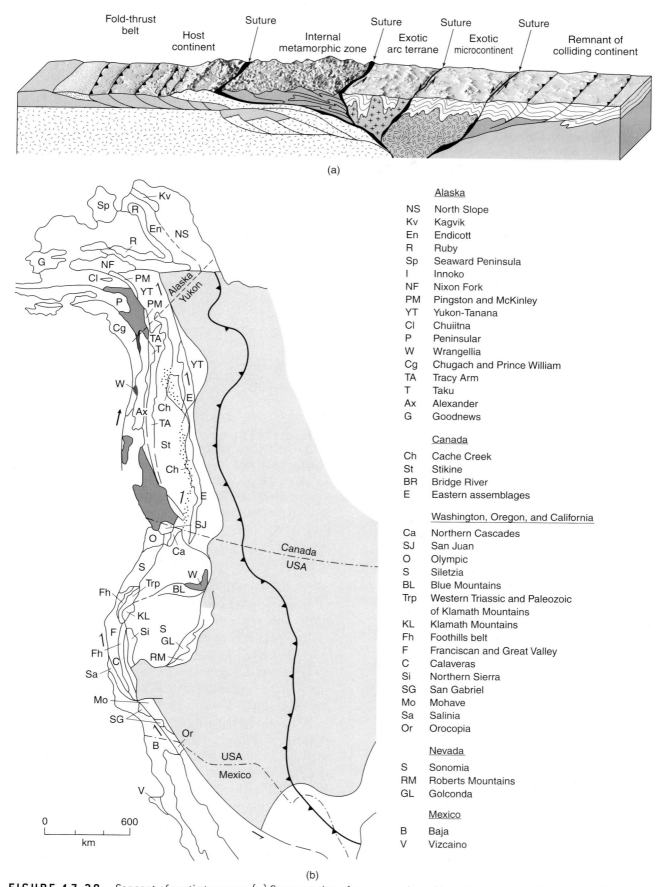

FIGURE 17.29 Concept of exotic terranes. (a) Cross section of an orogen, based loosely on the southern Appalachians, showing how the orogen includes an accreted volcanic arc, an accreted microcontinent, and crust that had been part of a large colliding continent. (b) Map of the North American Cordillera, showing the regions that consist of accreted crust. This crust can be divided into numerous exotic terranes.

(a)

Fold-thrust belt · Host continent · Suture · Internal metamorphic zone · Suture · Exotic arc terrane · Suture · Exotic microcontinent · Suture · Remnant of colliding continent

(b)

Alaska
NS   North Slope
Kv   Kagvik
En   Endicott
R   Ruby
Sp   Seaward Peninsula
I   Innoko
NF   Nixon Fork
PM   Pingston and McKinley
YT   Yukon-Tanana
Cl   Chuiitna
P   Peninsular
W   Wrangellia
Cg   Chugach and Prince William
TA   Tracy Arm
T   Taku
Ax   Alexander
G   Goodnews

Canada
Ch   Cache Creek
St   Stikine
BR   Bridge River
E   Eastern assemblages

Washington, Oregon, and California
Ca   Northern Cascades
SJ   San Juan
O   Olympic
S   Siletzia
BL   Blue Mountains
Trp Western Triassic and Paleozoic
     of Klamath Mountains
KL   Klamath Mountains
Fh   Foothills belt
F   Franciscan and Great Valley
C   Calaveras
Si   Northern Sierra
SG   San Gabriel
Mo   Mohave
Sa   Salinia
Or   Orocopia

Nevada
S   Sonomia
RM   Roberts Mountains
GL   Golconda

Mexico
B   Baja
V   Vizcaino

0   600
km

17.4.5 Deep Structure of Collisional Orogens

The depth in the crust to which faults penetrate during collision is not clear. Some researchers suggest that much of the deformation in a collisional orogen is confined to crust above a **mid-crustal detachment** or weak zone, at a depth of around 20 km (Figure 17.24b). This detachment may form at the lower boundary of discrete faulting—below the detachment, pressure, temperature, and stress conditions allow the crust to behave more like a viscous fluid that deforms independently of the crust above. In some cases, partial melting may occur in the vicinity of the mid-crustal detachment, creating magmas that seep along the detachment and perhaps follow faults to the surface, eventually crystallizing as granites during a late stage in the orogeny.

So far, we have mainly focused our attention on the crust, but it is important to keep in mind that the lithospheric mantle also thickens during collisional orogeny. What happens to this part of the lithosphere remains a subject of debate. Some geologists suggest that the lithospheric mantle separates from the crust during collision and sinks into the deeper mantle. Others suggest that the lithospheric mantle itself shortens and thickens during collision, leading to interesting consequences. For example, if the lithospheric mantle thickens by the same proportion as the crust during collision (i.e., doubles in thickness), then the base of the lithosphere will reach a depth of 200–300 km. Calculations show that the relatively cool rock composing the lithospheric mantle is denser than the rock composing the surrounding asthenosphere and, therefore, is negatively buoyant. Under such conditions, the lower part of the lithospheric mantle beneath a collisional orogen may peel off and sink into the mantle, a process we referred to earlier as **lithospheric delamination** (Figure 17.25). Lithospheric delamination may lead to postorogenic plutonism and to isostatic uplift of the overlying continent.

17.5 INSIGHT FROM MODELING STUDIES

In recent years, researchers have explored the process of convergent and collisional orogeny by means of sandbox models[10] and by computer models. This work allows

---

[10]Sand is a good laboratory-scale analog for continental crust. This means that, if continental crust were scaled down to the size of a typical laboratory sandbox (1 m × 1 m), the strength of crustal rock would be comparable to the strength of unconsolidated sand.

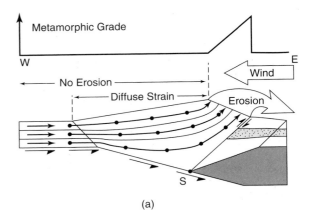

(a)

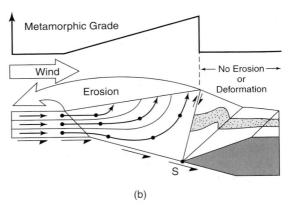

(b)

FIGURE 17.30 Results of a computer model study showing the flow path of points in an orogen under different weather conditions. (a) Wind comes from east, so exhumed metamorphic rocks occur in a narrow belt on the east side. (b) If the wind comes from the west, a broad wide metamorphic belt occurs on the west.

researchers to simulate the evolution of orogens in front of their eyes and to determine how a variety of variables affect this evolution. Variables that can be tested include detachment strength, subduction angle, crustal strength, and surface erosion rates. The effect of these variables on the geometry of an orogen can be understood in the context of critical taper theory (see Chapter 18) as applied to doubly vergent orogens.

With critical taper theory in mind, let's consider the effect of changing variables on an orogen. If there is an increase in the erosion rate (due to a change in climate), material transfers from the hinterland to the foreland, thereby decreasing the taper. This causes the orogen to grow wider and causes **exhumation** (erosion and removal of crust at the Earth's surface, thereby allowing the uplift of crust that had been at depth) in the interior of the orogen. The geometry of erosion depends on the prevailing wind direction, for this determines whether erosion attacks the retrowedge or the forewedge (Figure 17.30). The location of erosion, in turn, determines where exhumation takes place and thus where metamorphic rocks formed at depth eventually become exposed. As another example, if the foreland region contains a particularly weak detachment horizon, then the critical taper angle decreases, and orogen grows wider without becoming as thick. Again, this will affect metamorphic patterns in the interior of the orogen. If the geothermal gradient increases, crust at depth becomes weaker, which in turn decreases the taper angle. This decrease in crustal strength allows extensional collapse to take place so that the wedge achieves the lower taper angle.

17.6 CLOSING REMARKS

In the descriptions provided in Chapters 16 and 17, we've illustrated the start and finish of one loop in the cycle of opening and closing of ocean basins. Geologic mapping around the world emphasizes that patterns of structures created during continental breakup influence patterns of structures developed during collision, and vice versa. Rifts are superimposed on ancient collisional orogens, and in turn, rifts are the likely sites of later collision (so rift basins ultimately get inverted). Because rifts generally control the location of collisional orogens, and vice versa, rocks in a single orogen tend to record the effects of multiple phases of contraction and extension. For example, the eastern United States records a history that involves two principal phases of rifting (Late Precambrian and Middle Mesozoic) and two principal phases of continental collision (Late Precambrian and Paleozoic). Multiple reactivations have kept Phanerozoic collisional orogens weak relative to continental interiors (cratons). In a way, they are like weak scars that never heal, and thus protect cratons from major deformation in the Phanerozoic.

Our discussion of convergent and collisional tectonics is not yet complete. In the next chapter (Chapter 18), we narrow our focus and look into the world of fold-thrust belts, and in the final chapter of this book we will explore the geology of several collisional orogens in detail.

ADDITIONAL READING

Beaumont, C., Kooi, H., Willett, S., 2000. Coupled tectonic-surface process models with applications to rifted margins and collisional orogens. In M. S. Summerfield, ed., *Geomorphology and global tectonics*. John Wiley & Sons: Chichester. 29–55.

Ben-Avraham, Z., Nur, A., Jones, D., and Cox, A., 1981. Continental accretion: from oceanic plateaus to allochthonous terranes. *Science*, 213, 47–54.

Coney, P. J., 1989. Structural aspects of suspect terranes and accretionary tectonics in western North America. *Journal of Structural Geology*, 11, 107–125.

Dewey, J. F., 1980. Episodicity, sequence, and style at convergent plate boundaries. In Strangway, D.W., ed., *The continental crust and its mineral deposits*. *Geological Association of Canada Special Paper*, 20, 553–574.

Dewey, J. F., Hempton, M. R., Kidd, W. S. F., Saroglu, F., and Sengör, A.M.C., 1986, Shortening of continental lithosphere: the neotectonics of Eastern Anatolia—a young collision zone. In Coward, M. P., and Ries, A. C., eds., *Collision tectonics. Geological Society of London Special Publication*, 19, 3–36.

Dewey, J. F., 1988. Extensional collapse of orogens. *Tectonics*, 7, 1123–1139.

Dickinson, W. R., and Seely, D. R., 1979. Structure and stratigraphy of forearc regions. *American Association of Petroleum Geologists Bulletin*, 63, 2–31.

England, P., 1982. Some numerical investigations of large-scale continental deformation. In Hsü, K. J., ed., *Mountain building processes*. Academic Press: New York.

Hauck, M. L., Nelson, K. D., Brown, L. D., Zhao, W., Ross, A. R., 1998. Crustal structure of the Himalayan Orogen at approximately 90 degrees east longitude from Project INDEPTH deep reflection profiles. *Tectonics*, 17, 481–500.

Malavieille, J., 1993. Late orogenic extension in mountain belts: insights from the Basin and Range and the Late Paleozoic Variscan belt. *Tectonics*, 12, 1115–1130.

Moore, J. C., 1989. Tectonics and hydrogeology of accretionary prisms: role of the décollement zone. *Journal of Structural Geology*, 11, 95–106.

Oldow, J. S., Bally, A. W., Avé Lallemant, H. G., 1990. Transpression, orogenic float, and lithospheric balance. *Geology*, 18, 991–994.

Platt, J. P., 1986. Dynamics of orogenic wedges and the uplift of high pressure metamorphic rocks. *Geological Society of America Bulletin*, 97, 1106–1121.

Shen, F., Royden, L. H., Burchfiel, B. C., 2001. Large-scale crustal deformation of the Tibetan Plateau. *Journal of Geophysical Research*, 106, 4, 6793–6816.

Tapponnier, P., Peltzer, G., and Armijo, R., 1986. On the mechanics of the collision between India and Asia. In Coward, M. P., and Ries, A. C., eds., *Collision tectonics. Geological Society Special Publication*, 19, 115–157.

CHAPTER EIGHTEEN

Fold-Thrust Belts

An essay by Stephen Marshak and M. Scott Wilkerson

18.1 INTRODUCTION

Picture yourself hiking among the glaciated spires of the Canadian Rockies (Figure 18.1). Where did the beds of sedimentary rock forming these mountains come from, and how did they end up exposed on cliffs 2 km above sea level? The sediment composing the beds originally accumulated on the floor of a sea, tens of kilometers to the *west* of their present location. Eventually, the sediment was deeply buried until it lay several kilometers *below* the Earth's surface. Thus, these strata had to move large distances both horizontally and vertically to get to their present location in the Canadian Rockies. After uplift, erosion by rivers and glaciers carved the rugged cliffs on which the strata crop out today.

Formation of the Canadian Rockies involved displacement on **thrust faults,**[1] dip-slip faults on which hanging-wall blocks slide up the fault surface. Geologists refer to the bodies of rock that move during thrusting as **thrust sheets** or **thrust slices.** In the case of the Canadian Rockies, the stress driving thrust-sheet movement was probably generated by convergence and/or microplate collision along North America's western border during the Mesozoic and Early Cenozoic Sevier and Laramide Orogenies. As a consequence of thrusting, once-horizontal beds of sediment may become tilted or folded and may develop tectonic

---

[1]We'll introduce other names for such faults later in this chapter—unfortunately, fold-thrust belt jargon has become quite complex!

FIGURE 18.1 Photo of Mt. Kidd in the Canadian Rocky Mountain Front Ranges, Alberta (Canada). The folds affecting the Paleozoic strata exposed on these cliffs developed in association with transport on the Lewis and Rundle Thrusts. View is to the north.

cleavage. Geologic domains, such as the Canadian Rockies, in which regional horizontal tectonic shortening of the upper-crust yields a distinctive suite of thrust faults, folds, and associated mesoscopic structures, are called **fold-thrust belts** or **fold-and-thrust belts.**

Geologists have been struggling to understand the nature of fold-thrust belts since the 1820s, but it wasn't until the second half of the twentieth century that an integrated image of such belts evolved, because this image could not be developed without data from seismic-reflection profiling (see Chapter 15) and oil-well drilling. Work in the 1960s through 1980s led to a refinement of the geometric and kinematic rules governing the shape and evolution of **thrust-related folds** (folds formed as a result of the development of and slip on thrusts faults), and provided geologists with insight into the fundamental issue of *how* and *why* fold-thrust belts develop. Studies of fold-thrust belts continue in the twenty-first century, as geologists work to charac-

terize fold-thrust belts in three dimensions and try to predict the geometry of their component structures in the subsurface. Current work has practical applications because important oil reserves occur in fold-thrust belts.

In this chapter, for the sake of completeness, we first briefly review the tectonic setting of fold-thrust belts and their regional architecture—some of this material was covered in Chapter 10. We then focus on the geometry of thrust faults and thrust systems, and on the relationships between thrusts and folds. With this background, we can discuss methods for testing the reliability of cross sections depicting fold-thrust belts. We conclude by outlining the key ideas that have been proposed to explain the mechanics by which fold-thrust belts develop. You'll notice that this chapter uses a large number of terms. For better or worse, fold-thrust belt geologists are among the most "jargonistic" folks in geology (see Table 18.1)! Hopefully, you will find that the vocabulary simplifies discussion.

TABLE 18.1	FOLD-THRUST BELT TERMINOLOGY
Allochthon	A mass of rock, comprising a thrust sheet (i.e., a hanging-wall block), that has been displaced by movement on a thrust fault; commonly, use of the term implies that the mass has moved a considerable distance on a detachment from its point of origin.
Allochthonous	Adjective describing "out-of-place" rocks that have moved a large distance from their point of origin.
Autochthonous	Adjective describing rocks that are still at the site where they originally formed and have not been displaced by movement on a thrust fault or detachment.
Backarc	The region that lies behind the volcanic arc along a convergent plate boundary; the backarc and the trench are on opposite sides of the volcanic arc.
Backstop	A representation of the boundary load in the hinterland of a fold-thrust belt. The backstop generates horizontal compressional stress, which contributes to driving fold-thrust belt development. The backstop represents rock of the hinterland that is moving toward the foreland. As such, a backstop is like a snowplow pushing snow toward the foreland.
Backthrust	A thrust on which the transport direction is opposite to the regional transport direction.
Basal detachment	The lowest detachment of a thrust system; the regional basal detachment in a fold-thrust belt separates shortened crust above from unshortened crust below. In the foreland part of a fold-thrust belt, it typically lies at or near the basement-cover contact (also called a basal décollement).
Blind thrust	A thrust that, while it is active, terminates in the subsurface.
Branch line	The line of intersection between two fault surfaces, e.g., where a ramp branches (splays) off of a detachment, or where one ramp splays off another.
Break-forward sequence	A sequence of thrusting during which younger thrusts initiate to the foreland of older thrusts (also called a foreland-breaking sequence).
Break-thrust fold	A fold that initiates prior to thrusting, but later ruptures so that a thrust cuts through its forelimb.
Cutoff (cutoff line)	The line of intersection between a fault and a bedding plane.
Décollement	A subhorizontal fault (also called a detachment)
Detachment	A subhorizontal fault (also called a décollement)
Detachment fold	A fold that forms in response to slip above a subhorizontal fault, much like fold in a rug that wrinkles above a slick floor.
Duplex	A type of thrust system where a series of thrusts branch from a lower detachment to an upper detachment.
Fault-bend fold	A fold that forms in response to movement over bends in a fault surface.
Fault-propagation fold	A fold that forms immediately in advance of a propagating fault tip (also called a **tip fold**).
Floor thrust	The lower detachment of a duplex; it forms the base of the duplex.
Fold nappe	A thrust sheet that contains a regional-scale recumbent fold.
Fold-thrust belt	A geologic terrane in which upper-crustal shortening is accommodated by development of a system of thrust faults and related folds.
Footwall block	The body of rock beneath the fault.
Footwall cutoff	The intersection between bedding planes of footwall strata and a fault surface.
Footwall flat	The portion of the footwall where bedding surfaces parallel the fault.
Footwall ramp	The portion of the footwall where bedding surfaces truncate against the fault (i.e., the portion of the footwall along which there are **footwall cutoffs**).

TABLE 18.1	FOLD-THRUST BELT TERMINOLOGY
Forearc	The region to the trench side of the volcanic arc of a convergent plate boundary. The forearc is not the same as the foreland. The forearc lies on the ocean side of a continental volcanic arc.
Foreland	The part of the undeformed craton adjacent to an orogenic belt; some authors have used the term in a more general sense to include the portion of an orogenic belt closer to the undeformed continental interior.
Foreland basin	A sedimentary basin formed on the continent side of a fold-thrust belt that forms because the weight of the stack of thrust sheets in the belt depresses the lithosphere.
Forethrust	A thrust on which the transport direction is the same as the regional transport direction for the whole fold-thrust belt.
Frontal ramp	A ramp that strikes perpendicular to transport direction.
Hanging-wall block	The rock mass that has been transported above a fault surface.
Hanging-wall cutoff	The intersection between bedding planes of hanging-wall strata and the fault surface.
Hanging-wall flat	The portion of the hanging wall where bedding surfaces parallel the fault.
Hanging-wall ramp	The portion of the hanging wall where bedding surfaces truncate against the fault (i.e., the portion of the hanging wall where there are **hanging-wall cutoffs**).
Hinterland	The region closer to the high-grade core of an orogen; as a directional reference, it is the direction opposite to the foreland direction.
Horse	A body of rock in a duplex that is completely enveloped by faults.
Imbricate fan	A type of thrust system where a series of thrusts branch from a lower detachment without merging into an upper detachment horizon.
Inversion tectonics	The process by which a site of extension (e.g., a rift or passive margin basin) transforms into a site of shortening. During inversion, faults that had initiated as normal faults reactivate as thrust faults, and the sedimentary fill of the rift or passive-margin basin is shoved up and over the margins of the basin.
Klippe	An erosional outlier of a thrust sheet that is completely surrounded by footwall rocks; it is an isolated remnant of the hanging-wall block above a thrust.
Lateral ramp	A ramp that strikes parallel to transport direction.
Mechanical stratigraphy	The succession of rock types comprising the stratigraphy of a region, defined in terms of their relative strength.
Oblique ramp	A ramp that strikes oblique to transport direction.
Out-of-sequence thrust	A thrust that initiates to the hinterland of preexisting thrusts.
Out-of-plane strain	The strain due to movement outside the plane of cross section.
Regional transport direction	The dominant direction in which thrust sheets of a thrust belt moved during faulting. Some authors use the term **regional vergence direction** as a synonym.
Roof thrust	The upper detachment of a duplex.
Stair-step geometry	The geometry of a thrust that cuts upsection via a series of flats and ramps. The shape of the fault resembles a staircase in cross section. Typically, the ramps form in stronger units, and the flats in weaker units.
Tear fault	A nearly vertically dipping fault in a thrust sheet that that is parallel or subparallel to the regional transport direction. Motion on a tear fault is dominantly strike-slip and may accommodate differential displacement of one part of a thrust sheet relative to another (i.e., a tear fault is a nearly vertically dipping **oblique ramp** or **lateral ramp**).

TABLE 18.1	FOLD-THRUST BELT TERMINOLOGY
Tectonic inversion	The reactivation of preexisting faults by a reversal of slip direction on the faults.
Thick-skinned tectonics	The process of deformation that involves slip on basement-penetrating reverse faults; this movement uplifts basement and causes monoclinal forced-folds ("drape folds") to develop in the overlying cover.
Thin-skinned tectonics	The process of deformation in which folding and faulting are restricted to rock above a detachment. Some authors restrict the term to situations in which the detachment lies at or above the basement-cover contact. Others use the term even when basement occurs in thrust sheets, to imply that the basement has been transported or detached.
Thrust fault (thrust)	A shallowly to moderately dipping ($< 30°$) contractional fault with dip-slip reverse movement; in detail, thrusts may include several ramps and flats, and thus on a regional scale, do not necessarily have a uniform dip.
Thrust sheet	The hanging-wall block, above a thrust surface, that has been transported as a consequence of slip on the thrust (also called a **thrust slice**)
Thrust system	An array of related thrusts that connect at depth; a regional-scale thrust system may represent shortening above a specific regional detachment.
Tip line	The line along which displacement on the thrust becomes zero.
Triangle zone	A region in which a wedge of rock is bounded below by a forethrust and is bounded above by a backthrust.
Window (fenster)	An erosional hole through a thrust sheet that exposes the footwall (i.e., an exposure of the footwall completely surrounded by hanging wall rocks).

18.2 FOLD-THRUST BELTS IN A REGIONAL CONTEXT

18.2.1 Tectonic Settings of Fold-Thrust Belts

Fold-thrust belts occur worldwide in a variety of tectonic settings—basically, anywhere that a layer of the upper crust undergoes significant horizontal shortening under low-grade or submetamorphic conditions. To describe these settings, we first need to introduce a few terms. When specifying relative locations in fold-thrust belts that have formed on continental crust, we use the undeformed region of a continent outside of the fold-thrust belt as a point of reference. The **foreland direction** is toward the undeformed continental interior, whereas the **hinterland direction** is toward the orogen's more intensely deformed and metamorphosed internal zone (Figure 18.2a). Similarly, when referring to accretionary prisms formed on ocean lithosphere, geologists sometimes use the term "foreland" to refer to the less deformed side closer to the trench and the term "hinterland" to refer to the more intensely deformed and metamorphosed side of the prism closer to the volcanic arc. Portions of some fold-thrust belts on continents involve strata that had been deposited in a **foreland basin.** In accordance with our definition of "foreland," a foreland basin is a wedge of sediment deposited on the surface of the continent in the foreland of the orogen. Foreland basins form because the stack of thrust sheets in a fold-thrust belt acts as a weight that bends down the surface of the continent and creates a depression that collects sediment eroded from the orogen (Figure 18.2b). Let's now summarize six different settings in which fold-thrust belts develop.

1. *Foreland of an Andean-type convergent margin.* The Andes Orogen is currently developing at a convergent plate boundary, as the floor of the Pacific Ocean slides beneath the continent of South America. Because the South American Plate is moving westward while the Nazca Plate is moving eastward, the two plates squeeze together and crustal shortening takes place across the orogen. This shortening affects the backarc region. (Note that the "backarc region" occurs on the "foreland side" of the orogen—this terminology can be confusing, unless you keep in mind that the term "backarc" specifies

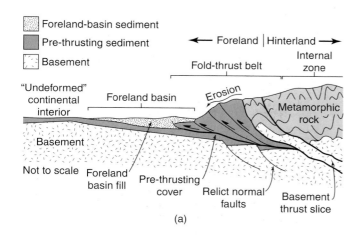

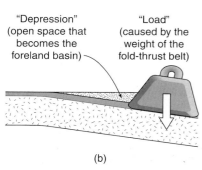

FIGURE 18.2 [a] Schematic cross section illustrating the location of a fold-thrust belt in an orogen. Note that the belt occurs between the foreland basin and the internal metamorphic region of the hinterland. Thrusts eventually cut across the strata of the foreland basin and incorporate the basin material into the fold-thrust belt. [b] A stack of thrust slices acts like a heavy load, pushing the surface of the crust down to create a depression that fills with sediment (i.e., to create the foreland basin).

location relative to the volcanic arc, whereas the term "foreland" specifies location relative to the continental interior.) As a consequence, sediment that accumulated along the western margin of South America before convergent tectonism began (Figure 18.3A) as well as strata formed from sediment eroded from the orogen and transported into the foreland basin, undergoes shortening and a fold-thrust belt evolves (Figure 18.3). The Sevier/Laramide fold-thrust belt of the North American Cordillera may be another example of an Andean-type fold-thrust belt.

2. *Accretionary prisms bordering a trench.* We've just focused on what happens in the backarc region of a convergent margin. Don't forget that on the ocean side of the arc, subduction scrapes rock and sediment from the surface of the downgoing plate and incorporates it, along with sediment filling the trench in an accretionary prism

(Chapter 17). Accretionary prisms are a type of fold-thrust belt because their development involves shortening and the formation of thrusts and folds (Figure 18.3b). Because the material incorporated in a prism tends to be poorly lithified at the time of incorporation, and because gravity-driven slumping takes place frequently as a prism grows, structures of accretionary prisms tend to be chaotic. Most contemporary examples of accretionary prisms now lie underwater. However, uplifted accretionary prisms do occur in some regions (e.g., Kodiak Island, Alaska, USA).

3. *Foreland sides of a collisional orogenic belt.* Eventually, subduction consumes the oceanic lithosphere between two continents and they collide. When this happens, strata that had originally accumulated in a passive-margin basin[2] on the downgoing plate get caught in a vise between the two continents and undergo tectonic shortening (Figure 18.3c). As a consequence, a fold-thrust belt evolves in which strata of the former passive margin undergo thrusting toward the foreland. During this process, strata of the deeper-water portion of the passive-margin basin may be placed on top of strata of the shallower-water part of the basin. In the hinterland portions of such fold-thrust belts, normal faults formed during the rifting that originally created the passive margin reactivate as thrust faults (i.e., undergo **inversion**), so that basement thrust slices move up and over sedimentary strata. Erosion attacks the rising collisional orogen, providing sediment that collects in a foreland basin. Foreland-basin sediment eventually becomes incorporated in the fold-thrust belt as well. The Valley and Ridge Province of the Appalachians, the Jura Mountains of Switzerland, and the Himalayan Mountains of Asia are examples of fold-thrust belts in collisional orogens.

4. *Inverted rift basins.* We've just described fold-thrust belts formed when an ocean is consumed and continents collide. Similar fold-thrust belts develop when large, sediment-filled rift basins later undergo compression and close. Geologists refer to this process as **rift inversion** to emphasize that a region that was once the site of crustal extension has become the site of crustal shortening.

---

[2]Recall that a passive-margin basin forms when rifting succeeds in forming a new ocean basin and the stretched lithosphere along the edges of the continents bordering this ocean slowly cools and subsides. See Chapter 16.

During rift inversion, faults that initiated as normal faults during rifting reactivate as thrust faults (Figure 18.4a and b). Movement on these reverse faults transports the contents of the rift basin up and over the rift's margins. The Pyrenees between Spain and France may represent such a fold-thrust belt.

5. *Seaward edge of passive-margin sedimentary basins.* Not all fold-thrust belts involving strata of passive-margin basins form in regional convergent or collisional settings. While the basin is still growing, gravity may drive the development of thrusts at the seaward edge of the basin, cre-

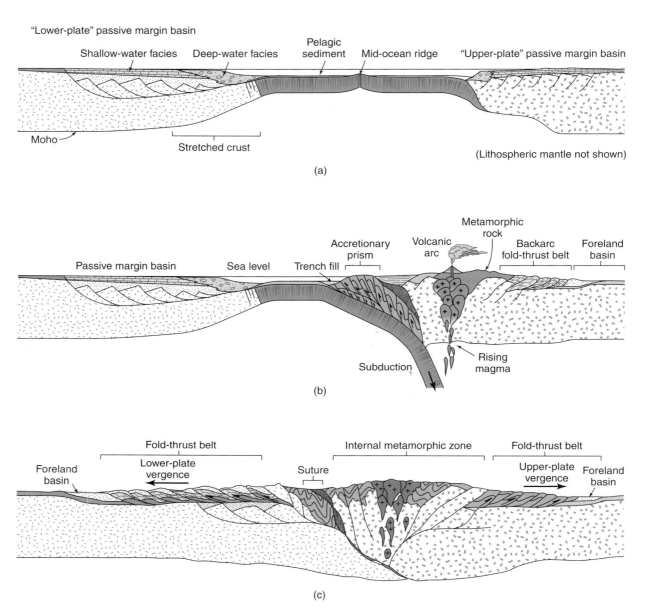

FIGURE 18.3 Regional cross sections depicting stages of fold-thrust belt development first during convergent-margin tectonism and then during continent-continent collision. (a) Passive-margin strata are deposited on thinned continental crust. In this sketch, basins on opposite sides of the margin do not have the same shape, because the basement beneath underwent different amounts of stretching. The so-called lower-plate margin underwent more stretching, whereas the so-called upper-plate margin underwent less stretching. (b) With the onset of convergence, an accretionary prism develops that verges towards the trench, and a backarc fold-thrust belt forms cratonward of the volcanic arc and verges towards the upper-plate craton. (c) Eventually the two continents collide. A fold-thrust belt forms in the foreland of the orogen on both sides of the orogen. Slivers of obducted ocean crust may separate lower-plate rocks from the metamorphic hinterland of the orogen and define the suture between the two plates.

ating an array of structures that resembles a fold-thrust belt. This process happens because gravity causes the strata of passive-margin basins to slump slowly seaward. In the continental shelf region, this movement results in a system of normal faults. But toward the seaward margin of the slumping pile, the mass of sediment slips up and over sediment of the deep ocean floor, just as the downhill edge of a slump on a hillslope rises up and over the ground surface (Figure 18.5a and b). The resulting movement creates a series of thrusts and related folds. Examples of such fold-thrust belts occur along the southern edge of the passive-margin basin of the United States Gulf Coast and along the western coast of Africa.

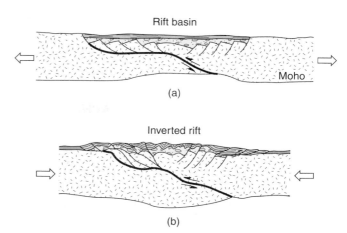

FIGURE 18.4 (a) Cross-sectional sketch of a rift basin just after it has formed. (b) Inversion of the rift occurs when the two margins are pushed towards each other. Note that faults which were originally normal faults turn into thrust faults.

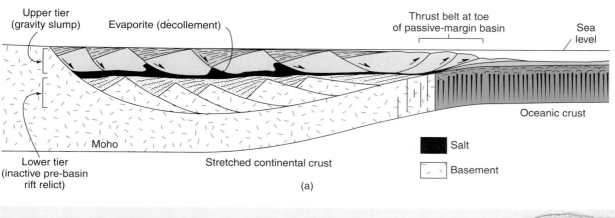

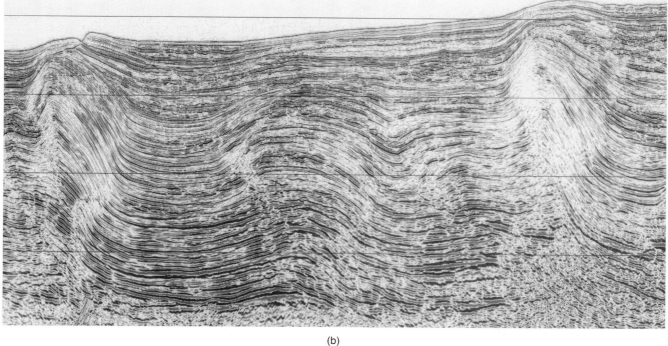

FIGURE 18.5 (a) Cross-section sketch of a fold-thrust belt forming at the seaward toe of a passive-margin basin. (b) Vertically exaggerated two-dimensional seismic-reflection profile illustrating an imbricate fan of thrust faults that has developed offshore of Nigeria.

6. *Restraining bends along large continental strike-slip fault.* Compression develops across a restraining bend along a strike-slip fault. As a result, strata bordering the fault undergo compression and shortening. This deformation may yield a fold-thrust belt bordering the strike-slip fault. Typically, such belts trend oblique to the strike-slip fault. Compression across a restraining bend in southern California has led to the rise of the Transverse Ranges of California adjacent to the San Andreas Fault. In some cases, the fold-thrust belt involves strata that had originally accumulated in a pull-apart basin (i.e., that had undergone transtension) along the fault, so thrusting represents inversion of the pull-apart basin. When this happens, the contents of the basin are thrust out over its former margins (Figure 18.6; see Chapter 19).

18.2.2 Mechanical Stratigraphy

The geometric characteristics of structures in a fold-thrust belt depend on the overall ductility of the rock sequence being deformed and the ductility contrast between layers within the sequence. In other words, fold-thrust belt characteristics depend on the **mechanical stratigraphy**—the succession of strong and weak rock layers—of the sequence being deformed. For example, a sequence consisting of massive layers of limestone behaves differently from one consisting of thin layers of sandstone interbedded with thick layers of shale. The former may break to form several large

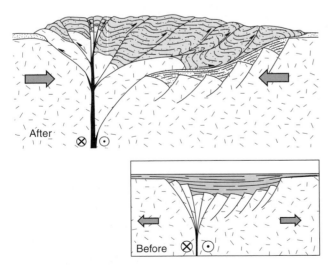

FIGURE 18.6 Cross-sectional sketch illustrating a narrow fold-thrust belt that formed by transpression at a restraining bend along a strike-slip fault. The inset shows the region before transpression began. Here, transtension formed a broad basin prior to inversion.

thrust slices or may flex to form large-amplitude folds, whereas the latter may buckle to form a train of short-wavelength folds. Because different mechanical stratigraphies develop in different tectonic settings, not all fold-thrust belts look the same.

For much of the remainder of this chapter, we focus discussion on so-called **classic fold-thrust belts.** These belts form where the mechanical stratigraphy subjected to deformation consists of laterally extensive layers that maintain coherence during deformation. In classic fold-thrust belts structures are well organized, in that thrust traces tend to be roughly parallel to one another and folds tend to have roughly the same wavelengths and amplitudes within a given stratigraphic interval. Such mechanical stratigraphy occurs in passive-margin basins and foreland basins. Thus, most of the examples that we use come from collisional and convergent-margin orogens, where deformation involves strata of passive-margin basins and foreland basins.

18.3 GEOMETRY OF THRUSTS AND THRUST SYSTEMS

18.3.1 A Cross-Sectional Image of a Thrust Fault

We've considered the regional setting in which fold-thrust belts occur, now let's focus on the individual structures, and arrays of structures, that occur within fold-thrust belts. To begin our discussion, we examine a cross section of the Pine Mountain Thrust in the southern Appalachians (Figure 18.7a and b). This fault formed in a sequence of Paleozoic strata during the Alleghanian Orogeny, when Africa collided with North America.

As shown in Figure 18.7c, thrust faults like the Pine Mountain Thrust cut *upsection* in the direction that the hanging-wall moves. Displacement on the fault puts older strata on top of younger strata. In the case of the Pine Mountain Thrust, we see that the fault lies at the base of the Cambrian strata in the southeast, cuts upsection to the northwest, and eventually flattens out in Siluro-Devonian strata. Movement on the Pine Mountain Fault placed Cambrian strata over Siluro-Devonian strata. Note that thrusting can duplicate a stratigraphic succession, so that a vertical hole drilled through the hanging wall, across the fault, and into the footwall could encounter the same stratigraphic units twice. Further, thrusting raises strata above its pre-faulting elevation. Strata in the hanging wall of the

Pine Mountain Thrust lie approximately 2.5 km above their original pre-faulting elevation!

Our example of the Pine Mountain Thrust also illustrates that some thrust faults resemble a flight of stairs, in that they consist of **flats** that lie approximately in the plane of bedding and **ramps** that cut across bedding (Figure 18.7c). The key to determining whether a fault segment is a ramp or a flat is to look for cutoffs. A **cutoff** is the intersection between a bedding plane and a fault surface along a ramp. Flats commonly exceed ramps in cross-sectional length, and typically lie within incompetent (weak) strata like shale and evaporite. Ramps tend to develop in competent (strong) rocks like sandstone, dolomite, and limestone.

Note that their placement depends on mechanical stratigraphy.

If you examine Figure 18.7c in detail, you will see that a ramp with respect to the footwall can be a flat with respect to the hanging wall. Thus, to be complete, a description of a given ramp should indicate whether it is a **hanging-wall ramp** that cuts across beds of the hanging wall, or a **footwall ramp** that cuts across beds of the footwall. Similarly, a description of a flat should indicate whether the flat is a **hanging-wall flat** that lies parallel to bedding in the hanging wall, or a **footwall flat** that lies parallel to bedding in the footwall. Note that some segments of a fault may display a "flat-on-flat relationship," meaning that the segment is a flat

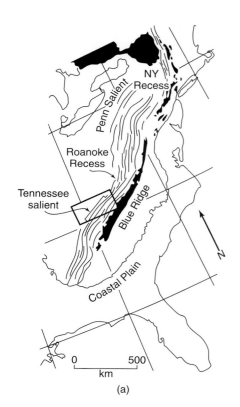

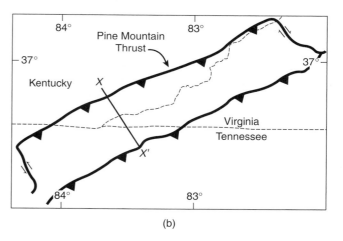

FIGURE 18.7 (a) The box shows the location of the Pine Mountain thrust sheet in the southern Appalachians (eastern United States). (b) Simplified map showing the trace of the Pine Mountain Thrust. (c) Cross section of Pine Mountain thrust sheet, southern Appalachians (simplified to illustrate terminology for describing hanging-wall and footwall structures). Location of the cross section shown by the *XX'* section line on map in (b). Barbs point toward hanging wall of the thrust faults. Northeast and southwest ends of the Pine Mountain thrust sheet are bounded by tear faults.

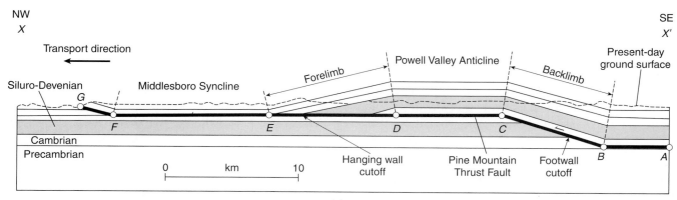

(c)

with respect to both the hanging wall and footwall. In Figure 18.7c, segment *FE* is a hanging-wall flat on a footwall flat, *ED* is a hanging-wall ramp on a footwall flat, and *CB* is a hanging-wall flat on a footwall ramp.

To better understand the cross-sectional geometry of a thrust, let's examine the relationships among ramps and flats on a fault both before and after slip. Figure 18.8a shows the trace of a stair-step thrust fault before movement has taken place. Before movement, hanging-wall ramps must lie adjacent to footwall ramps, and hanging-wall flats must lie adjacent to footwall flats. Thus, the number of hanging-wall flats and ramps *exactly* matches the number of footwall flats and ramps. After slip (Figure 18.8b), the hanging-wall block moves, so a hanging-wall ramp may end up on a footwall flat, and a hanging-wall flat may end up on a footwall ramp. Before slip on the fault, the hanging-wall ramp between *T* and *U* was adjacent to the footwall ramp between *V* and *W,* and the hanging-wall ramp between *P* and *Q* was adjacent to the footwall ramp between *R* and *S*. Note that after slip, the number of hanging-wall flats and ramps still matches the number of footwall flats and ramps.

Typically, ramps curve and change strike along their length. A ramp segment that strikes approximately perpendicular to the direction in which the thrust sheet moves is a **frontal ramp** (Figure 18.9), a ramp segment that cuts upsection laterally and strikes approximately parallel to the direction in which the thrust sheet moves is a **lateral ramp,** and a ramp segment that strikes at an acute angle to the transport direction is an **oblique ramp.** Note that dip-slip movement dominates on frontal ramps, oblique-slip movement dominates on oblique ramps, and strike-slip movement dominates on lateral ramps. **Tear faults** are lateral or oblique ramps that break a thrust sheet into segments that move by different amounts into the foreland.

While active, some thrust faults lie entirely in the subsurface, whereas others cut and displace the ground surface during movement. Consequently, geologists distinguish between **blind thrusts,**[3] which are faults that moved in the subsurface and do not intersect the "syntectonic ground surface" (the ground surface at the time of active faulting), and **emergent thrusts,** which are faults that intersected and displaced the syntectonic ground surface. After exhumation of a thrusted region by uplift and erosion, it is not always possible to determine if a given fault was emergent or blind while it was active.

---

[3]Notably, the disastrous 1994 Northridge earthquake in California occurred on a blind thrust fault; this fault had not been previously identified by any surface expression, and thus movement on it was a surprise.

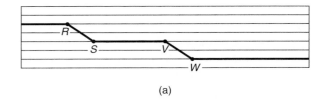

(a)

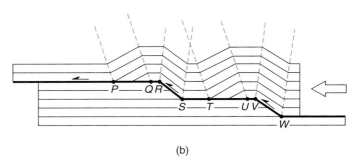

(b)

FIGURE 18.8 When a thrust sheet moves up a stair-step fault, ramp anticlines develop. (a) The cross-sectional trace of the fault before slip. Points at the tops and bottoms of ramps are labeled. (b) The cross-sectional trace of the fault after slip. Note that the number of hanging-wall ramps exactly matches the number of footwall ramps.

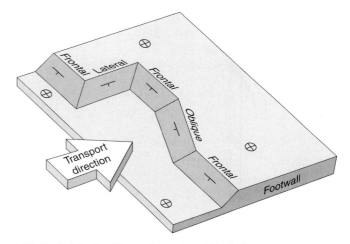

FIGURE 18.9 Three-dimensional block diagram illustrating different types of fault ramps (hanging wall removed). Tear faults are vertically dipping oblique ramps or lateral ramps.

If erosion produces a hole in a thrust sheet so that at the ground surface we see an exposure of the footwall completely surrounded by hanging-wall rocks, then the exposure is called a **window** or **fenster** (Figure 18.10a). Alternatively, if erosion removes most of a thrust sheet so that you can map a remnant of the thrust sheet that is completely surrounded by footwall strata, then the "island" of hanging-wall rock is called a **klippe** (Figure 18.10b).

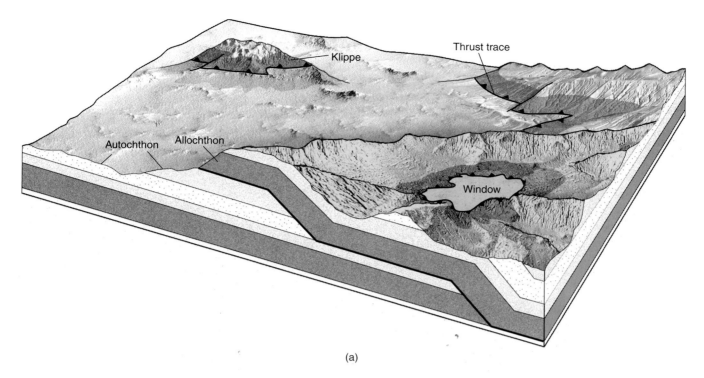

(a)

(b)

FIGURE 18.10 (a) Block diagram illustrating klippe, window, allochthon, and autochthon. (b) Photo of Crow's Nest Mountain Klippe, Alberta (Canada).

18.3.2 Thrust Systems

A **thrust system** refers to the family of related thrust faults that ramp up from a single **detachment fault** or **décollement** (e.g., Figure 18.11). The name "detachment" emphasizes that rock above the fault has detached or separated from rock below during movement. Detachments tend to develop in weak rock types, such as shale or evaporite.

There are two end-member types of thrust systems—imbricate fans and duplexes. Individual thrusts that make up an **imbricate fan** branch upsection from a common detachment and terminate updip without merging into an upper detachment (Figure 18.12a). The line (in three dimensions) along which the ramp connects to an underlying detachment is called a **branch line,**[4] and the line at which the fault terminates

and displacement decreases to zero is called the **tip line.** In a **duplex,** a series of thrusts branches upwards from a lower detachment and merges with a higher detachment (Figure 18.13). Sometimes geologists refer to the lower detachment of a duplex as the **floor thrust,** and the upper one as the **roof thrust** (Figure 18.13a and b). Note that adjacent thrust surfaces in a duplex completely surround bodies of rock; these fault-bounded bodies are called **horses.**[5] Depending on the spacing and relative displacement on thrusts in a duplex, the roof thrust of a duplex may be planar (Figure 18.13a) or corrugated (Figure 18.13c). The roof thrust and overlying strata of a corrugated-roof duplex is folded into a train of anticlines and synclines.

In both imbricate fans and duplexes, the faults comprising a thrust system do not all initiate at the same time. Generally they initiate in a **break-forward**

---

[4]The term "branch line" can also be used to define the line at which a ramp bifurcates to form two separate ramps, either updip or along strike.

[5]Note that a "horse" is not the same structure as a "horst."

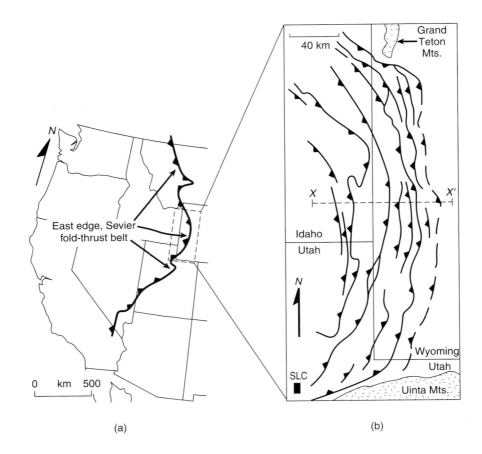

(a)

(b)

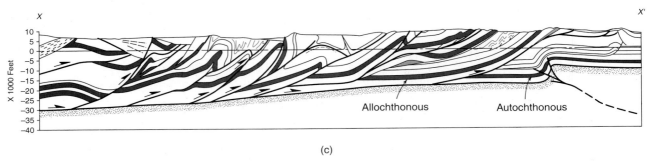

(c)

FIGURE 18.11   (a) Trace of the eastern edge of the Sevier fold-thrust belt in the western United States. (b) Map showing the traces of principal thrusts in the western Wyoming fold-thrust belt (a part of the Sevier fold-thrust belt). Note the thrusts curve so they are convex toward the foreland. (c) Cross section of the western Wyoming fold thrust belt. Approximate location of the cross section is indicated by the *XX'* section line in (b). (c) Cross section of the western Wyoming fold thrust belt.

sequence (Figure 18.12). This means that the faults of the system form one after the other, with each new fault forming to the foreland side of the previous one. Thus, the youngest fault in the system occurs on the foreland end of the system, whereas the oldest fault occurs at the hinterland end. There are exceptions to the break-forward sequence model. Slip at any given time may be partitioned among the youngest thrust and thrusts immediately behind it. Also, in some cases, existing thrusts of the system reactivate and/or new faults initiate to the hinterland of the preexisting faults.

These **out-of-sequence faults** can be recognized where they cross cut structures in the foreland.

Within a given fold-thrust belt, most thrust sheets move in the same overall direction, called the **regional transport direction** (Figures 18.11, 18.12, and 18.13).[6] In the case of classic collision-related or

<hr>

[6]Some geologists also use the term "vergence" to describe transport of thrust sheets as in, "Thrust sheets of the Ouachita Mountains verge to the north."

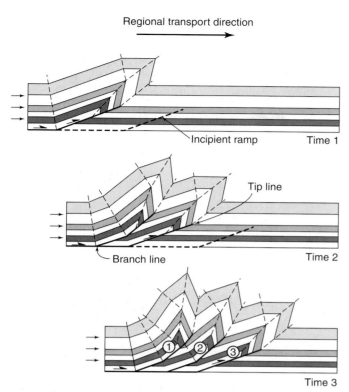

Incipient ramp   Time 1

Tip line

Branch line   Time 2

① ② ③

Time 3

FIGURE 18.12 An idealized imbricate fan that develops by progressive break-forward thrusting. Note that successively younger thrusts cut into the footwall, and older faults and folds become deformed by younger structures. The dashed lines are the traces of fold axial surfaces. In the cross section showing "Time 3," the sequence of thrusts is labeled. Fault 1 is the oldest and Fault 3 is the youngest. On this cross section, tip lines and branch lines are points; in three dimensions, they go into and out of the page.

convergent-margin fold-thrust belts, the regional transport direction carries rocks from the hinterland of an orogen toward the foreland. For example, the regional transport direction in the Appalachian fold-thrust belt is to the west, whereas the regional transport direction for the Canadian Rockies is to the east. Regional transport direction in accretionary prisms is generally toward the trench.[7] In discussion, it is convenient to use the term "vergence" to describe the movement of thrust sheets. For example, if the regional transport direction in a thrust system is to the east, we could also say that the thrusts "verge to the east."

In detail, however, not all thrust sheets of a fold-thrust belt move in the same direction, so we distinguish between **forethrusts** that verge in the regional transport direction and **backthrusts** that verge opposite to the regional transport direction. Backthrusts may form where the front of a thrust sheet wedges between layers of strata in the foreland as it moves up and over a footwall ramp and onto a footwall flat, or in the hanging wall as a thrust sheet rides up over a ramp (Figure 18.14a). A place where the thrust sheet wedges between layers is called a **triangle zone** (Figure 18.14b).

So far, we've focused on faults that branch from a detachment. It's important to point out that not all thrusts do so. Local thrusts may form during evolution of a fold-thrust belt, when synclines tighten so that their limbs move toward each other. Because of this tightening, there is no longer enough room in the hinge zone for the volume of rock within the fold. When such "room problems" develop, **out-of-the syncline faults** form to transport rock up and out of the hinge zone (Figure 18.14c). Note that out-of-the-syncline faults can be either forethrusts or backthrusts and that displacement decreases downdip, so that the faults die out within the fold.

18.3.3 Overall Fold-Thrust Belt Architecture

Let's now broaden our view to incorporate an entire classic fold-thrust belt. For the sake of discussion, we divide the belt into two parts, a foreland part and a hinterland part.

In the foreland portion of a classic fold-thrust belt, deformation is restricted to the rock above a regional fault called the **basal detachment** or **basal décollement** (Figure 18.11c). A basal detachment is the lowest detachment of a fold-thrust belt. The belt may contain higher level detachments and many imbricate fans and duplexes, as well as several detachment horizons, but they all lie above the basal detachment. Typically, the basal detachment of the foreland part of a fold-thrust belt lies in a weak shale or evaporite at or near the basement-cover contact.[8] Toward the foreland, the basal detachment may ramp upsection to higher stratigraphic levels. Rocks above the detachment are **allochthonous** in that they have been transported relative to their original location. In contrast, rocks that lie

---

[7]Some accretionary prisms, however, are bivergent, meaning that thrust sheets on the trench side of the prism move toward the trench, while those on the arc side move toward the arc.

[8]Commonly the "basement" consists of older crystalline rocks (e.g., Precambrian gneiss and granite) that form the substrate on which sedimentary beds were unconformably deposited. The term "cover" refers to the succession of overlying sedimentary beds.

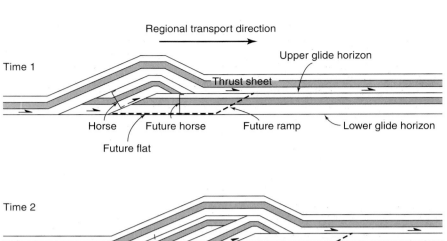

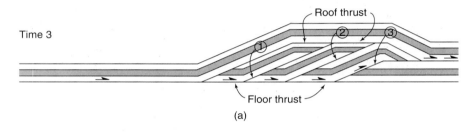

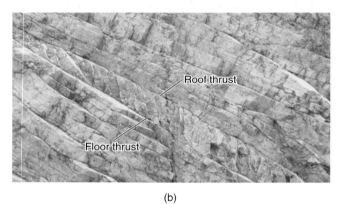

(a)

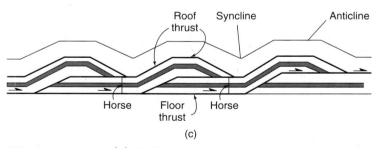

(b)

(c)

FIGURE 18.13 (a) Idealized flat-roofed duplex that develops by progressive break-forward faulting. Note that the roof thrust undergoes a sequence of folding and unfolding, and that formation of the duplex results in significant shortening. (b) Photo of a duplex involving a single bed at Crow's Nest Pass, Alberta (Canada). Duplex height is about 0.3 m. (c) Schematic sketch, using kink-style fold construction, of a corrugated roof duplex.

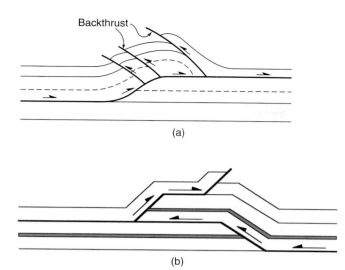

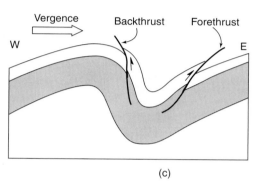

FIGURE 18.14 (a) Formation of backthrusts across the backlimb of a ramp anticline, and along bedding of the forelimb. (b) Cross section of a triangle zone. (c) Out-of-the-syncline forethrusts and backthrusts. These faults formed to accommodate a room problem that develops in the hinge zone of the syncline.

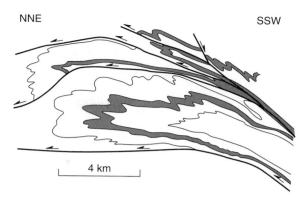

FIGURE 18.15 Profile sketch (based on down-plunge projection) of the Helvetic nappes in the Alps. Note that the limbs of these recumbent folds have been thinned during shearing.

below the detachment are **autochthonous,** in that they have not been transported by fault slip and thus lie in their original position.

In the hinterland portion of a fold-thrust belt, the basal detachment "roots" into basement (Figure 18.3b and c). By this we mean that basement rocks are incorporated in thrust sheets. In some cases, the basement slices have been transported long distances and end up on top of sedimentary rock. Basement slices in the hinterland of fold-thrust belts probably originated as the hanging-wall blocks above normal faults that formed during the rifting stage, prior to passive-margin basin formation. Thus, thrusting of these slices represents inversion of the rift that underlies the passive-margin basin. Further into the hinterland, deformation occurs under metamorphic conditions, so sedimentary rocks are able to deform plastically. Sedimentary layers incorporated in hinterland thrust slices may be folded into huge recumbent folds, underlain by a detachment.

Such folds are called **fold nappes.** The Swiss Alps contain excellent examples of fold nappes (Figure 18.15; see Section 21.1).

Geologists sometimes refer to the style of deformation in which faulting and folding occur only above a regional basal detachment as **thin-skinned tectonics.** Unfortunately, this term has not been used consistently in the literature. Some researchers use it to distinguish faulting that does not involve basement from faulting that does, and use the term "thick-skinned deformation" for the latter.[9] Others use the adjective "thin-skinned" in all cases where the deformed interval is underlain by a subhorizontal detachment, regardless of whether or not basement is involved. In general, the adjective "thick-skinned" is preferred for the style of deformation in which movement occurs on basement-penetrating reverse faults, which have fairly steep dips near the ground surface. These faults may die out up-dip in a monocline.

18.4 THRUST-RELATED FOLDING

Geologists refer to the Canadian Rockies and the Appalachian Valley and Ridge Province as "*fold*-thrust belts" because, in addition to the thrust systems that we have just described, these regions contain spectacular folds, with amplitudes ranging from millimeters up to a few kilometers (Figure 18.16a and b). We call these folds "thrust-related folds" because they form in

[9]As alternative terminology, one could distinguish between "basement-detached" deformation and "basement-involved" deformation. In the former, the detachment lies at or above the basement-cover contact, while in the later, basement rocks have been incorporated in thrust slices.

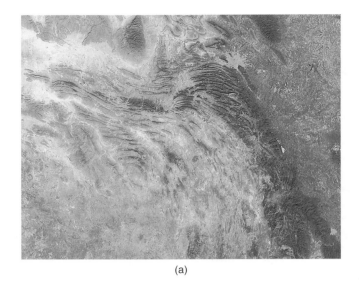

(a)

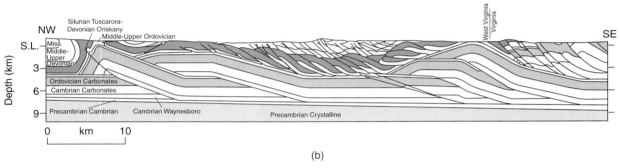

(b)

FIGURE 18.16 (a) Space photo of the Monterray Salient, Mexico, a prominent fold belt comprised predominantly of detached folds in Cretaceous carbonates. Note that the fold belt transitions to a fold-thrust. The features to the foreland (top of the photo) are salt diapirs. (b) Cross section of the Appalachian Valley and Ridge Province in Virginia and West Virginia. Observe that there are duplexes at two levels; the roof thrust of the Cambrian-Ordovician duplex is corrugated and is the floor thrust for a higher-level duplex.

association with displacement on thrust faults. We recognize four broad categories, based on the specific relationship between a fold and the underlying fault:

1. *Folding associated with break thrusts.* Picture the development of a fold resulting from the buckling of a stratified sequence. Initially, the layers bend, without rupturing, to form an open anticline-syncline pair (Figure 18.17a). The folds tighten and become more asymmetric as folding progresses. Eventually, strain can no longer be accommodated by folding alone, and rupturing produces a thrust fault that breaks through the overturning limb (the "forelimb") of the anticline. Because the thrust cuts through the limb of an already formed fold, it is called **break**

thrust[10] (Figure 18.17b). Note that a break thrust develops *after* folding. After a break thrust has ruptured the forelimb of an asymmetric anticline, it displaces the anticline of the hanging wall to the foreland, relative to the syncline of the footwall.

2. *Fault-bend folding:* Picture a simple thrust geometry in which a thrust that occurs as a flat in a weak layer bends, cuts upsection across a strong layer as a ramp, and then bends again and becomes a flat at the top of the strong layer (Figure 18.18a-b). If you push on the thrust sheet,

[10]Admittedly, this isn't an ideal name, as the term itself does not immediately bring to mind the process we've described, but researchers have ensconced the term in fold-thrust belt jargon.

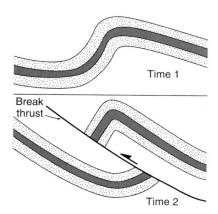

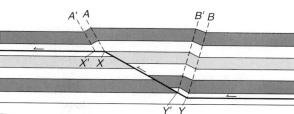

Time 1

Time 2

FIGURE 18.17 Formation of a "break-thrust fold." (a) An asymmetric fold begins to form and tighten. (b) Eventually, a fault breaks through the fold's forelimb.

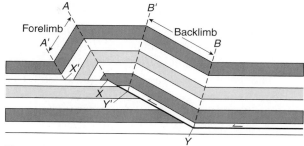

Time 1

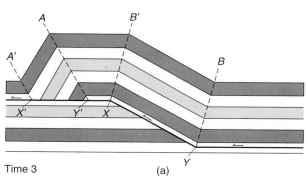

Time 2

FIGURE 18.18 (a) Cross-sectional model showing the progressive stages during the development of a fault-bend fold. The dashed lines are the traces of axial surfaces. (b) Photo of a fault-bend fold above the McConnell Thrust, near Seebe, Alberta (Canada). These Paleozoic strata have been displaced over 5 km vertically and 40 km horizontally, and now lie above Cretaceous foreland basin deposits. The foreland basin deposits have preferentially eroded and are now forested.

Time 3   (a)

(b)

the sheet must itself bend to climb the ramp. Because of Earth's gravity, the sheet cannot rise into the air at the top of the ramp. Rather, the layer bends again to form an anticline on the upper flat. Thus, by pushing strata up and over a preexisting stair-step thrust, strata of the thrust sheet must undergo folding. Note that this folding occurs *after* the formation of the thrust. The resulting fold is called a **fault-bend fold.**[11] In sum, a fault-bend fold forms where hanging-wall strata move up and over a stair-step in a fault; the strata deform in order to conform to changes in dip (bends) of the fault surface.

The geometry of a specific fault-bend fold depends on the geometry of the stair-step thrust beneath it. Note that if the bends in the fault surface are abrupt, the fold will have kink-style (i.e., not rounded) hinges. Also, note that once the fold has climbed over the ramp, the width of the anticline depends on the magnitude of displacement on the fault; as displacement increases, interlimb distance increases because the back limb (the limb closer to the hinterland) remains over the ramp (Figure 18.19). In an ideal fault-bend anticline, the fold's backlimb parallels the footwall ramp, the fold's forelimb is shorter and steeper than the backlimb, strata of the footwall remain flat-lying, and the kink-style hinges of the hanging-wall anticline directly reflect the shape of bends in the fault surface. In some cases, strata move through the kink hinges as the fold evolves.

3. *Fault-propagation folding.* In some cases, folding develops just in advance of the tip of a ramp as the ramp propagates updip. The resulting fold is called a **fault-propagation fold.** Such folds develop *concurrently* with thrust development. Typically, a fault-propagation fold develops only in front of the upper tip line as the ramp propagates upsection toward the ground surface. Fault-propagation folds are asymmetric and verge in the direction of thrusting (Figure 18.20a and b). Displacement along the fault dies out in the hinge zone of the fold.

In a geometric model of fault-propagation folds (Figure 18.20), the fault tip propagates upsection as the fold's backlimb and forelimb lengthen. The fault tip and the merge point of the

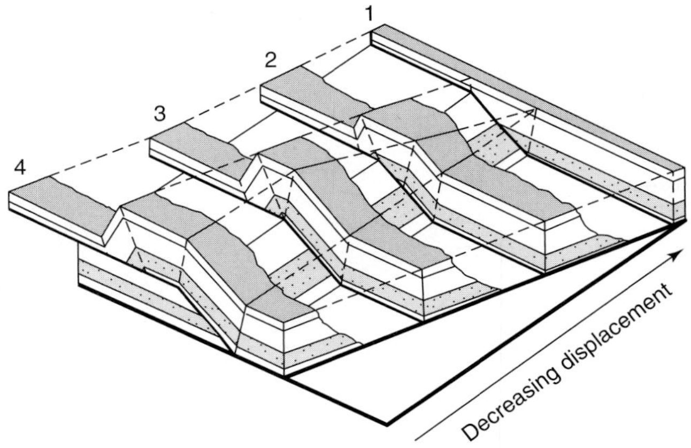

FIGURE 18.19 Cutaway block diagram depicting a possible along-strike termination geometry of a simple fault-bend fold that experiences an along-strike decrease in displacement, from a maximum on section 4, to zero on section 1. The dashed lines are the traces of axial surfaces, and the dotted lines are position reference lines.

fold's kink axes lie at the same stratigraphic horizon. As in the fault-bend fold model, the backlimb of a fault-propagation fold parallels the dip of the ramp, and strata in the footwall remain flat-lying. In some cases, strata move through the kink axes as the fold evolves.

In recent years, geologists have examined deformation in the region above the tip line of the fault, and have found that not all regions obey the classical geometric image of a fault-propagation fold with kink-style hinges as described above. Rather, a triangular (in profile) region of deformation develops beyond the fault tip. This region has come to be known as a **trishear** zone.[12] Figure 18.21 shows the geometric characteristics of a trishear deformation zone. You will note that because strain is distributed through the triangular region, folds have curved hinges, and beds undergo stretching and thinning.

4. *Detachment folding.* Folds may develop in fold-thrust belts above a detachment fault, even if no ramps develop. This happens where the strata above the detachment buckle or wrinkle up like a rug that has been shoved across a slick wooden

---

[11]The geometry of fault bend folds was first worked out rigorously by John Suppe of Princeton University, in classic papers published in the early 1980s.

[12]The concept of trishear deformation was proposed by Eric Erslev of Colorado State University. Richard Allmendinger of Cornell University has developed insightful computer models of trishear deformation.

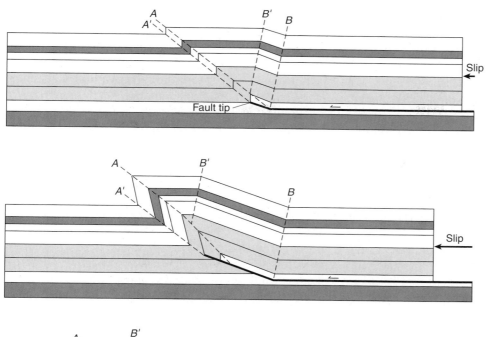

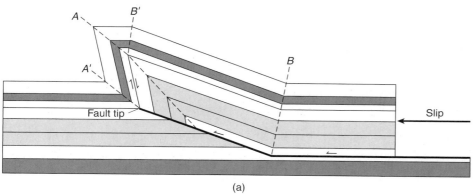

(a)

FIGURE 18.20 (a) Model for progressive development of a simple fault-propagation fold. (b) Exposure of a fold in the Lost River Range, Idaho, showing an asymmetric fold dying out updip in the core of a fold.

(b)

floor (Figure 18.22). Such **detachment folds** are particularly common in regions where detachments lie within thick shale or salt layers, as occurs in the Jura Mountains of Switzerland, for the weak rock can flow into the core of the fold as the structure develops. In some cases, a break thrust may develop at a late stage in the evolution of the fold; the fault cuts across the forelimb of the already formed detachment fold.

Division of folds in fold-thrust belts into the classes listed above is clearly an oversimplification. Today, geologists realize that an evolutionary continuum exists between these classes. For example, initial detachment folding may establish the spacing for folds. Further shortening causes fold amplification and tightening, with the initiation and propagation of a break thrust further modifying the overall fold geometry. If displacement along the fault is sufficiently high, the fault may break through the fold forelimb, transporting the fold along the fault. If the thrust merges updip with a flat, then continued displacement of the thrust sheet will cause the fold to evolve into a fault-bend fold. Alternatively, a thrust may "lock up" after some displacement has occurred. When this happens, the fault plane ceases to grow, and shortening in the area is accommodated by folding rather than by frictional sliding.

Before leaving our discussion of folds, we point out that the folds we've described above can be considered to be the "first-order folds" of fold-thrust belts (see Chapter 10). Typically, thrusting and folding also produce "second-order folds" as well, meaning smaller folds that form within larger folds. Folds in this category include parasitic intraformational folds formed due to flexural slip between beds on the limbs of larger folds, folds that develop within shear zones, and folds formed by buckling of thinner-bedded intervals in the hinge zones of larger folds.

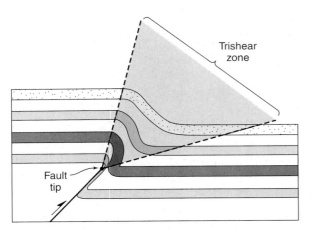

FIGURE 18.21 Cross section illustrating the concept of trishear deformation. Note that strain is distributed throughout a triangular zone in the region beyond the fault tip. The solid line is the fault trace and the dashed lines outline the region of trishear.

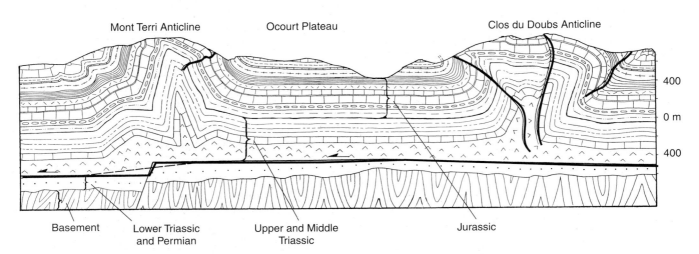

FIGURE 18.22 Cross section of detachment folds in the foreland of the Jura Mountains, Switzerland. Note that the folds do not involve the basement. The fold cores filled with ductile rock as the folds formed.

18.5 MESOSCOPIC- AND MICROSCOPIC-SCALE STRAIN IN THRUST SHEETS

Fold-thrust belts form in response to layer-parallel compression of the upper crust. This means that during the development of a fold-thrust belt, the maximum principal compressive stress (σ_1) is horizontal and has a bearing roughly perpendicular to map trace of folds and faults within the fold-thrust belt.[13] If you have a chance to study a fold-thrust belt up close, you'll discover that folds and thrusts are not the only structures that they contain. Compression also may cause a suite of mesoscopic- and microscopic-scale structures to form in fold thrust belts. Specific examples include the following:

1. *Tectonic cleavage.* Appropriate rock types (e.g., shale, argillaceous limestone, and argillaceous sandstone) tend to be susceptible to pressure-solution deformation and thus develop tectonic cleavage (Figure 18.23a-b). This cleavage varies from widely spaced cleavage to slaty cleavage. Generally, the strike of cleavage is approximately parallel to the trends of folds. Dip of the cleavage varies; slaty cleavage tends to be parallel to the axial planes of folds, whereas spaced cleavage tends to fan around folds. Adjacent to faults, cleavage becomes inclined at a low angle to the fault surface, for the cleavage domains rotate toward the direction of transport.

2. *Mesoscopic folds.* Stratigraphic units that consist of thin beds of relatively strong rock types interbedded with layers of relatively weak rock types may undergo buckling during the shortening that produces the fold-thrust belt. As a result, mesoscopic folds may develop within some thrust sheets (Figure 18.23a).

3. *Wedge thrusts.* If a thrust sheet contains a succession of strata in which strong beds are interlayered with weak beds, on the scale of centimeters to meters, deformation may generate single-bed ramps, meaning ramps that cut across a single strong bed and die out in the weak bed above or below (Figure 18.23b). Displacement on wedge thrusts is generally less than the thickness of the bed.

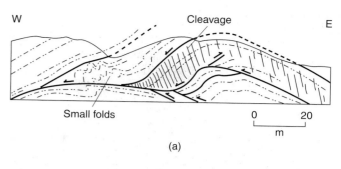

(a)

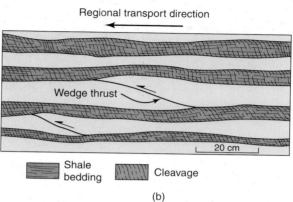

(b)

FIGURE 18.23 (a) Photo of a wedge thrust exposed on New World Island (Newfoundland). (b) Wedge thrusts forming in rigid limestone beds sandwiched between weak shale layers. In the shale layers, a cleavage has developed. The cleavage tips toward the hinterland.

4. *Grain-scale strain.* Rocks within thrust sheets locally undergo distortion at the scale of individual grains as a consequence of pressure solution between grains and/or plastic deformation within grains (e.g., by creating twins in calcite grains and deformation bands in quartz; see Chapter 9). The resulting strain ranges from a few percent to as much as 50%. Detection of grain-scale strain requires application of methods such as Fry analysis (see Chapter 4).

5. *Joints.* Fold-thrust belts typically contain systematic and non-systematic joint sets. The nonsystematic joints may develop in regions of tight folding, or adjacent to faults. Simplistically, we can divide systematic joints into two categories: strike-parallel joints and cross-strike joints. Chapter 7 provides a discussion of the origin of these joints.

18.6 FOLD-THRUST BELTS IN MAP VIEW

Faults are not surfaces with infinite dimensions; that is, the map trace of a fault must terminate along its length, either because the fault merges with or is truncated by

---

[13]Some authors refer to the traces of folds, faults, and fabrics as the **trend lines** of an orogenic belt. Further, some authors refer to the pattern of trend lines as the **structural grain** of the belt.

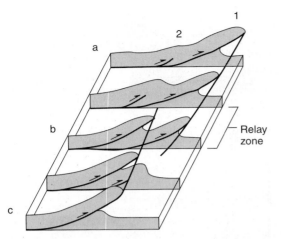

FIGURE 18.24 Concept of displacement transfer between thrusts in a simple relay zone. (a) The majority of displacement occurs on fault 1 and little occurs on fault 2. (b) Displacement is equally partitioned between faults 1 and 2. (c) The majority of displacement occurs on fault 2, little on fault 1.

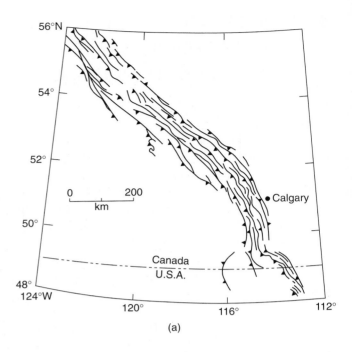

(a)

another fault, or because the magnitude of displacement across the fault decreases progressively along strike until, at a tip line, displacement is zero. Beyond the tip line of a fault, shortening may be accommodated instead by folding or by slip on neighboring faults. For example, as shown in Figure 18.24, a decrease in slip on fault 1 is matched by an increase in slip on fault 2. Slip transfers from one fault to another at a **relay zone** or **transfer zone,** much like a baton transfers from one racer to the next in a relay race (Figure 18.24).

Commonly, the map trace of a major thrust or thrust system is convex toward the foreland (Figure 18.25a). According to the "bow-and-arrow rule" of thrusting, if you connect the termination points of the bowed fault trace with a straight reference line, the regional transport direction on the fault lies in a direction roughly perpendicular to this line. Further, the displacement magnitude on the fault is largest where the distance between the reference line and the fault trace is greatest (Figure 18.25b).

As illustrated by the Appalachians of eastern North America, regional map traces of fold-thrust belts typically are sinuous (Figure 18.7a). A place where the belt bulges into the foreland is a **salient,** whereas a place where the belt has not propagated so far into the foreland is a **recess.** The curved traces of structures in fold-thrust belts stand out in satellite images (Figure 18.26).

The origin of fold-thrust belt curvature remains controversial. Curves may develop (Figure 18.27a–f)

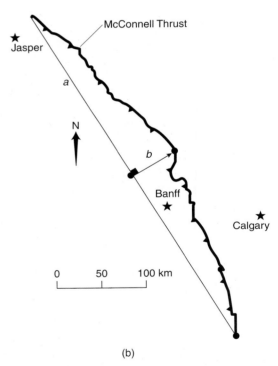

(b)

FIGURE 18.25 (a) Schematic map showing the traces of thrust faults in the southern Canadian Rockies. (b) The "bow-and-arrow rule," as applied to the McConnell Thrust, Alberta (Canada). For most foreland thrust belts, the ratio *b/a* is roughly 0.07–0.12.

FIGURE 18.26 Landsat satellite image of the curved Monterrey Salient (Mexico). The city of Monterrey is on the northern edge of the image. Long dimension of the photo is approximately 100 km.

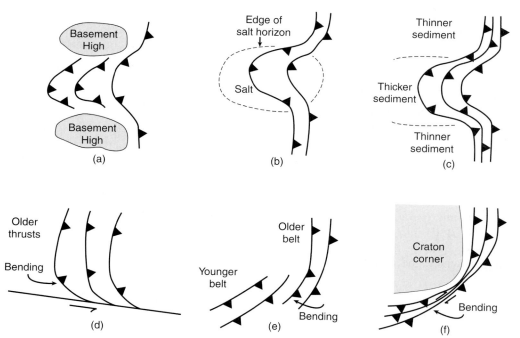

FIGURE 18.27 Map-view sketches illustrating possible processes leading to the formation of curved fold-thrust belts.
(a) Interaction with basement highs in the foreland. New thrusts forming between the two highs originate with curved traces.
(b) Lateral pinch-out of a stratigraphic glide horizon. The thrusts propagate further to the foreland over the weak salt horizon and the thrusts originate with curved traces. (c) Lateral variations in stratigraphic thickness. The thrusts propagate further to the foreland over the region where predepositional strata are thicker, and the thrusts originate with curved traces. (d) Interaction with a strike-slip fault. Motion on the strike-slip fault oroclinally bends the thrust traces. (e) Overprinting of two nonparallel thrust belts. Development of the younger belt bends the traces of the older belt. (f) Impingement against an irregular cratonic margin. Thrust sheets bend as they wrap around the corner.

where: (1) the belt interacts with basement highs in the foreland, for the basement highs retard propagation of thrusts; (2) the strength of the detachment horizon beneath changes along strike, because thrusts propagate further to the foreland where the detachment is weaker; (3) there are lateral variations in the thickness of the stratigraphic sequence being deformed, for thrust belts are wider where they involve thicker stratigraphic successions; (4) preexisting thrusts are bent by interaction with a strike-slip fault that cuts across the belt; (5) a second phase of thrusting overprints a preexisting belt at a high angle to the original belt; and (6) a fold-thrust belt impinges on the corner of a rigid craton.

Note that in some cases, the map-view curvature of fold-thrust belts develops when the thrusts initiate, so that right from the start the thrust has a curved trace. But in other cases, a preexisting straight thrust belt undergoes bending in map view during a second phase of deformation. A map-view curve that forms by the bending of a preexisting straight belt, such that the arms of the curve rotate around a vertical axis, is called an **orocline.**

18.7 BALANCED CROSS SECTIONS

In this chapter, we have presented several cross sections depicting the subsurface geometry of fold-thrust belts, right down to the basal detachment, even where these depths are not exposed. Perhaps you've asked yourself the fundamental question, "How do people draw such cross sections?" and "How reliable are they?" Well, to begin with, it is important to remember that a cross section is just an *interpretation* of the subsurface geology, and nothing more. We do not have access to outcrops several kilometers below the ground surface to let us see exactly where formation contacts, faults, and cutoffs are positioned. Cross-section interpretations are constrained by projecting surface geology into the subsurface, by interpreting seismic-reflection profiles, and by interpreting well data. Such data rarely provide a complete picture of subsurface geology, so we always must extrapolate when making cross sections. However, geologists have established a set of tests that permit us to evaluate cross sections to determine if the sections at least have a good chance of being correct. A cross section that passes these tests is said to be a **balanced cross section.** A balanced cross section has a reasonable chance of being correct, though we cannot guarantee it, whereas an unbalanced cross section is probably wrong (unless a good explanation can be provided for why the section does not balance).

Let's now look at four fundamental tests that help determine whether a cross section is balanced. Of note, these observations only apply when deformation does not result in movement in or out of the cross-section plane. Thus, they only apply to cross sections that have been drawn parallel to the transport direction on faults.

1. *The deformed-state cross section must be admissible.* Structures in the deformed-state cross section (i.e., the cross section depicting the way structures look today, after deformation) must resemble real structures that geologists have observed in outcrop or seismic profiles. For example, ramps should cut upsection, not downsection, unless they are out-of-sequence faults. We call cross sections that pass this test **admissible sections.** Figure 18.28a provides an admissible deformed-state cross section of the Lewis thrust sheet in Canada.

2. *Restoration of the cross section must yield reasonable geometries.* A **restored cross section** (Figure 18.28b) represents the predeformation configuration of strata and the predeformational location of faults in the region. "Restoration" of a cross section involves returning beds to horizontal by removing the effect of folding, by returning rocks to their original locations by removing the displacement on faults, and by undistorting thrust sheets by removing the effect of mesoscopic-scale and microscopic-scale deformation. The restored section must depict realistic-looking structures. For example, if a restored fault trace zigs and zags upsection, then there's likely something wrong with the section.

3. *The cross section must be "area balanced."* The area of rock shown on the restored cross section must equal the area shown on the deformed-state cross section unless pressure solution causes volume-loss strain. Cross sections that meet this criteria are **area balanced.** If volume-loss strain developed during deformation, this must be taken into account when restoring the section.

4. *The cross section must be kinematically reasonable.* It should be possible to create the deformed-state cross section from the restored cross section in a *kinematically reasonable way.* This means that you should be able to draw a series of cross sections depicting stages in the evolution of originally horizontal beds into the faulted and folded beds of the deformed-state cross section.

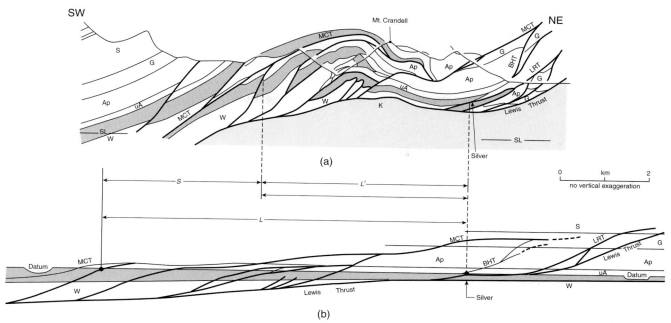

FIGURE 18.28 (a) Deformed-state cross section of a duplex within the Lewis Thrust Sheet, Waterton (Canada). No vertical exaggeration. (b) Restored version of the cross section. Hanging-wall strata are composed of the Precambrian Belt Supergroup (W = Waterton; shaded = lower Altyn; uA = upper Altyn; Ap = Appekunny; G = Grinnell; S = Siyeh; SL = sea level; MCT = McConnell thrust) that overlie footwall Cretaceous siliciclastics (K) across the Lewis Thrust. Shortening (S) is determined by comparison of the deformed and restored cross sections using the equation: $S = L - L'$. Note that this previously published cross section and restoration does not exactly balance in the foreland; to see why, try to match slices in the deformed and restored cross sections. Datuum is taken as the top of the shaded reference horizon.

The first two criteria in the above list ensure that the section doesn't depict impossible structural geometries. The third criterion ensures that the configuration of structures shown on the deformed-state section does not imply that undocumented volume change occurred during deformation. The fourth criterion emphasizes that you do not really understand the geometry of a complex structure until you can demonstrate how the structure formed from undeformed rock. "Balancing" a cross section involves the following steps. First, you carefully examine the deformed-state section for admissibility. Then, you construct a restored section and check it for admissibility and area balance. Finally, you think through a scenario that can explain the evolution of the deformed-state cross section from initially horizontal beds.

It is beyond the scope of this chapter to provide detailed guidelines for balancing cross sections; most structural geology laboratory manuals offer step-by-step instructions and exercises. But we do point out that, in some cases, you can use **quick-look techniques** to quickly scan a deformed-state cross section and determine if it has the potential to be balanced. To apply these techniques, first identify ramps and flats in each part of the cross section, and count them (Figure 18.29). Are there the same number of ramps and flats in the hanging wall as in the footwall? There should be, because in an admissible, restored cross section, the hanging wall fits over the footwall with no gaps or overlaps. Now, paying particularly close attention to the ramps, check to see if the same beds are truncated in the hanging wall as in the footwall. They must be, because the hanging-wall beds were originally adjacent to the footwall beds. These two simple tests will highlight the majority of common cross-section errors that lead to construction of unbalanced cross sections.

Let's apply the quick-look technique to the example depicted in Figure 18.28. First, note that the sliver of lower Altyn Formation (shaded), between faults *BHT* and *LRT* in the deformed section, has a short hanging-wall ramp and a long hanging-wall flat (Figure 18.28a). In the restored section (Figure 18.28b), however, this sliver only has a hanging-wall ramp. Further, in the restored section, the area of the sliver is much smaller than in the deformed-state section. Thus,

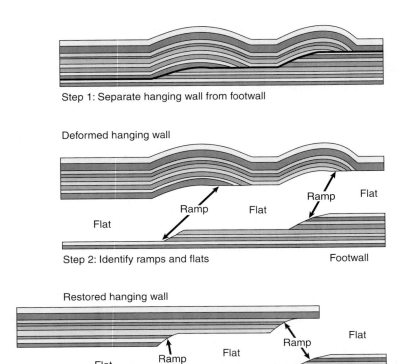

Step 1: Separate hanging wall from footwall

Deformed hanging wall

Ramp   Flat   Flat
Ramp   Flat

Flat

Step 2: Identify ramps and flats   Footwall

Restored hanging wall

Ramp   Flat
Flat   Ramp   Flat

Step 3: Match ramps and flats   Footwall

FIGURE 18.29 Diagram illustrating quick-look technique for checking a cross section for potential problems. The key is to recognize ramps and flats in the deformed-state section and realize that hanging-wall flats and ramps must exactly match footwall ramps and flats in number and in stratigraphic composition.

this part of the deformed-state cross section cannot be balanced. There are other mismatches on this section, as well, so the geologist constructing this section should check to determine if there has been movement out of the plane of section, if there is a drafting error, or if there is an alternative interpretation that can be balanced.

Before getting too carried away with the value of cross-section balancing, keep in mind that not all cross sections have to balance. Two-dimensional balancing techniques cannot deal with cross sections over lateral or oblique ramps, or across strike-slip faults. In such settings, movement takes place in or out of the plane of the section, and thus by definition, area balance in the cross-sectional plane is impossible. Similarly, in regions where deformation has occurred by flow of rock, units may be sheared and isoclinally folded, or may flow into or out of the section plane, to an extent that balance is again impossible. Even in low-grade rocks, pressure solution may cause significant volume change that may make area balance a challenge.

We conclude our discussion of balancing by once again pointing out that checking a cross section for balance does not automatically ensure a "correct"

interpretation or mean that the interpretation is unique (i.e., there may be other balanced interpretations that fit the data). Balancing procedures are simply meant to focus your attention on potentially problematic areas in the cross section that require geologic explanation and/or reinterpretation.

18.8 MECHANICS OF FOLD-THRUST BELTS

Now that we have a feel for the geometry of structures that occur in a fold-thrust belt, we can explore the mechanisms of fold-thrust belt development. The formation of fold-thrust belts has been particularly perplexing to geologists because the movement of large thrust sheets—tens of kilometers wide as measured in the transport direction, but less than a few kilometers thick—at first seems like a paradox. Picture a thrust sheet, in cross section, to be a rectangle resting on a surface (Figure 18.30a). If you assume a reasonable value for frictional resistance (σ_f) to sliding on the underlying detachment, assuming the detachment is dry, then the stress (σ_1) necessary to push the sheet over a horizontal surface or up a gentle incline greatly exceeds the failure strength of the intact rock comprising the thrust sheet (see Chapter 6). Thus, you would expect the hinterland end of the thrust sheet to crush or buckle before the sheet as a whole would move. Yet clearly, large thrust sheets do exist. So how do they move, and how do belts composed of many large thrust sheets develop? It took several decades for geologists to develop models that address these questions.

Geologists first attacked the issue of how to overcome the friction between solid surfaces that presumably provided the resistance for thrust-sheet movement. In the late 1950s researchers[14] realized that the force required to move a thrust sheet can be greatly diminished if hydrostatic pressure in the detachment zone increases to values approaching lithostatic loads (Figure 18.30b). "Hydrostatic pressure" refers to the pressure in the water that fills pores and cracks in rock, while "lithostatic pressure" refers to the pressure within solid grains generated where grains are in direct contact. A pore resembles a tiny balloon between

---

[14]M. King Hubbert and William Rubey provided some of the key contributions to this idea.

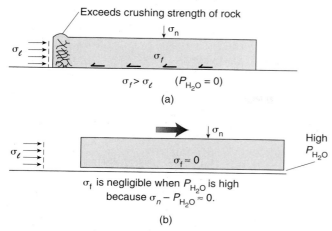

$\sigma_f > \sigma_\ell \quad (P_{H_2O} = 0)$

(a)

σ_f is negligible when P_{H_2O} is high because $\sigma_n - P_{H_2O} \approx 0$.

(b)

FIGURE 18.30 [a] When pushed from the rear, a rectangular thrust sheet sliding on dry rock would be crushed before overcoming frictional resistance. [b] If there is high fluid pressure at the basal detachment, the effective stress decreases and allows the thrust sheet to move under very small applied load. σ_l is the stress resulting from horizontal loading, σ_f is frictional resistance, σ_l represents the boundary load at the end of the thrust sheet, and P_{H_2O} represents the pore pressure. Other terms are defined in the text.

grains—if you force water into the pore, the water pushes outward, just like the air in a balloon pushes outward if you blow up the balloon. In effect, the fluid pressure in a detachment zone "lifts up" the thrust sheet so that it can glide over the detachment.

By looking once again at the Mohr-Coulomb criterion for failure (Chapter 8), we can gain further insight into the role of fluid pressure in thrust-sheet movement. For sliding to occur, the shear stress (σ_s) applied to the fault surface must exceed frictional resistance (σ_f). When rock is dry, the equation relating the shear stress (σ_s) necessary to cause sliding on a detachment surface to the normal stress (σ_n) squeezing the two sides of the detachment together can be written

$$\sigma_s = C + \mu\sigma_n \qquad \text{Eq. 18.1}$$

where C is the cohesion and μ is the coefficient of internal friction. This equation shows that as the normal stress (representing the vertical load due to the weight of the thrust sheet) increases, the shear stress needed to cause movement on the detachment increases. If, however, the rock in the detachment horizon is wet, water creates pore pressure. The pore pressure pushes up the load, and thus counteracts σ_n. Thus, the frictional resistance to sliding depends instead on the **effective normal stress** (σ_n^*) across the surface. We define σ_n^* by the equation

$$\sigma_n^* = \sigma_n - P_{H_2O} \qquad \text{Eq. 18.2}$$

where P_{H_2O} is the fluid pressure. Substituting effective stress (σ_n^*) back into the Mohr-Coulomb criterion yields

$$\sigma_s = C + \mu\sigma_n^* \qquad \text{Eq. 18.3}$$

Equation 18.2 indicates that as fluid pressure increases, the effective normal stress across the detachment decreases, because fluid in the detachment zone partially supports the weight of the thrust sheet. Equation 18.3 shows that as the effective normal stress decreases, the shear stress needed to cause sliding decreases. Therefore, the thrust sheet can be moved easily when fluid pressure is high, even when the boundary load at the hinterland edge of the sheet is significantly lower than the failure strength of rock comprising the sheet.

But what initiates thrust motion? Initially geologists assumed that thrust sheets slid toward the foreland in response to gravity when the basal detachment dipped toward the foreland, and thus that thrust sheets moved like slumps on a hillslope. Such **gravity sliding** models became very popular as a cause for thrusting, and in the 1960s, most structural geologists envisioned that development of fold-thrust belts occurred as thrust sheets glided down an incline created by uplift of the hinterland during orogeny (Figure 18.31a and b). However, petroleum exploration of many fold-thrust belts provided seismic-reflection data showing that basal detachments beneath almost all classic fold-thrust belts dip toward the hinterland, not the foreland! To account for this contradiction, some structural geologists suggested that fold-thrust belt formation was a consequence of **gravity spreading,** meaning that fold-thrust belts form when the thickened crust of an orogen "collapses" and spreads laterally under its own weight, much like a continental ice sheet spreads away from the region where snow accumulates (Figure 18.31c). The main difference between the gravity-spreading model and the gravity-sliding model is that the former implies that the direction of dip of the topographic surface of the fold-thrust belt, not the dip of the basal detachment, drives thrust movement.

The next step in formulating an understanding of how fold-thrust belts develop came in the 1970s, when researchers began to study laboratory models that simulated the development of the belts. Models involving the formation of a sand wedge building in front of a plow proved to be particularly informative (Figure 18.32). That's because sand is a **Coulomb material,** meaning an aggregate composed of grains that can frictionally slide past one another, and at the scale of a mountain range, rock of the upper crust behaves

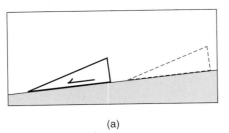 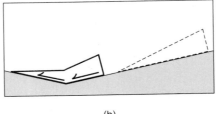 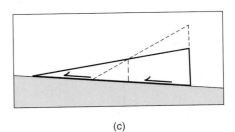

|(a)|(b)|(c)|

FIGURE 18.31 (a) The concept of gravity sliding. Here a block slides down a foreland-tilted slope. The dashed lines show the original position. (b) Gravity sliding partly downslope and partly upslope. The dashed lines show the original position. (c) The concept of gravity spreading. Before spreading, the wedge had the shape indicated by the dashed line.

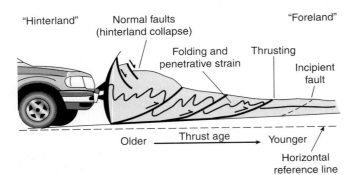

FIGURE 18.32 Snowplow analogy for fold-thrust belt development. The wedge of snow widens with continued shortening; younger thrusts generally initiate in a hinterland to foreland progression. While new thrusts are adding material at the toe of the wedge, the hinterland portions are developing penetrative strain, normal faults, and slump features.

essentially like a Coulomb material. In typical sand-wedge models (or, more generally, Coulomb-wedge models), the plow pushes into a sand layer underlain by a hard material. The boundary between the sand and the underlying hard material represents the basal detachment. Note that the wedge depicted in Figure 18.32 has a foreland-dipping topographic surface and a hinterland-dipping basal detachment—this shape is similar to the overall cross-sectional shape of a fold-thrust belt.

Two sources of stress drive the development of a Coulomb wedge. One source is a result of the displacement of the plow blade toward the foreland. This stress is called a **horizontal boundary load.** Another source is the result of the gravitational potential energy that develops when a foreland-dipping topographic slope develops (i.e., when the hinterland portion of the wedge rises and becomes higher than the foreland portion). Note that the gravitational potential energy, due to the elevation of the hinterland, creates both vertical and horizontal stresses. In Figure 18.33a, we represent

these stresses as follows: σ_{bs} is the horizontal boundary load caused by movement of the backstop toward the foreland; σ_{gv} is the vertical component of stress caused by gravity; σ_{gh} is the horizontal component of stress caused by gravity.

Let's look more closely at the evolution of a Coulomb wedge (e.g., a sand wedge). As the backstop moves toward the foreland, the wedge deforms internally (by forming folds, faults, and grain-scale distortion) and, as a consequence, its surface slope increases. When the wedge reaches a certain **critical taper angle,** ϕ_c (defined as the surface slope angle, α_1, plus the detachment dip, β), the wedge as a whole slides toward the foreland along the weak detachment (Figure 18.33a). Slip occurs on the detachment because the coefficient of sliding friction on the detachment is less than the coefficient of internal friction in the wedge. If the taper angle becomes too large (Figure 18.33b), processes take place within the wedge to cause the surface slope of the wedge to decrease (Figure 18.33c). Several processes can cause the slope angle to decrease, including the addition of new thrust slices at the toe (a process called offscraping), the erosion of the higher portions of the wedge, or the development of extensional faulting (extensional collapse) within the wedge. If these processes continue until the taper angle become less than the critical taper angle (Figure 18.33d), sliding of the wedge stops, and deformation within the wedge occurs once again to thicken the wedge internally and increase the surface slope (Figure 18.33e). This internal thickening can involve reactivation of thrusts, formation of out-of-sequence faults, formation of duplexes at the base of the wedge (a process also called underplating), or formation of penetrative strain and folding within thrust sheets. Internal thickening increases the topographic slope angle until the wedge achieves the critical taper angle again. Then, the wedge again starts sliding toward the foreland and new thrusts again form at the toe.

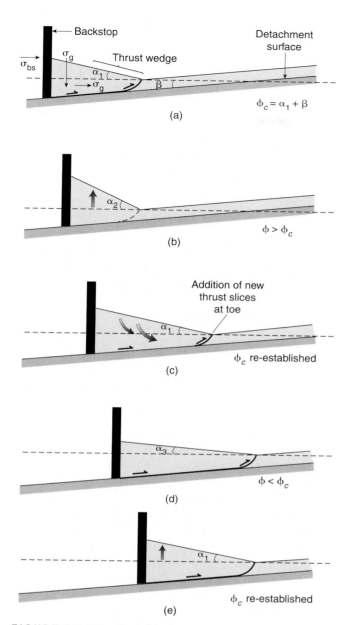

FIGURE 18.33 The critical taper theory of fold-thrust belt mechanics. The critical taper (ϕ_c) is defined as the sum of the surface slope angle (α_1) and the detachment slope angle (β). (a) Stress acting on a wedge is partly a horizontal boundary load caused by the backstop (σ_{bs}) and is partly caused by gravity (σ_g). (b) If the backstop moves, the wedge thickens, so the surface slope increases, and the taper (ϕ) eventually exceeds ϕ_c. (c) The wedge slides toward the foreland and new material is added to the toe, and extension of the wedge occurs so that surface slope decreases. (d, e) If the surface slope becomes too small, thrusting at the toe stops, and the wedge thickens by penetrative strain or out-of-sequence thrusting.

The observed critical taper of a given fold thrust belt depends on the material strength of the wedge, the resistance to sliding across the basal detachment, and the ratio of fluid pressure to overburden pressure both in the wedge and across the detachment. If the effective strength of the wedge is increased (either by increasing rock strength or by decreasing fluid pressure), then the critical taper angle decreases. If the resistance to sliding on the basal detachment is increased (either by increasing the coefficient of sliding friction or by decreasing the fluid pressure), then the critical-taper angle increases. Fold-thrust belts whose basal detachments lie in salt, a very weak lithology, have a critical taper angle as low as 1° to 2°, whereas fold-thrust belts that have detachments in stronger rocks may have critical taper angles as high as 8° to 10°.

Overall, we see that in the Coulomb-wedge model of fold-thrust belt development, now known formally as **critical taper theory,** the belt evolves and grows by maintaining a dynamic equilibrium between (1) addition of new material at the toe of the wedge, which decreases the surface slope, (2) internal deformation within the wedge, which increases the surface slope, and (3) thinning of the wedge due to extensional deformation (i.e., hinterland collapse) and/or erosion of the wedge.

Can critical taper theory explain the break-forward sequence of thrusting in a fold-thrust belt? Yes. Imagine a plow at the end of a snow-covered concrete driveway (the "basement") that slopes gently toward the street (Figure 18.32). As soon as the plow begins to move, a wedge-shaped pile of snow forms in front of the blade, and the surface of this pile slopes toward the foreland. A detachment forms beneath the pile, and a thrust develops at the front of the pile. A few meters in front of the plow, however, the snow remains totally unaffected. During the next increment of movement, the detachment propagates farther beneath the snow and above the concrete "basement" toward the foreland, and a new thrust develops. The thrust system now consists of two imbricate thrusts cutting upward from the basal detachment, with the imbricates forming in a break-forward sequence. As the process continues, an imbricate fan of thrusts develops, with the youngest thrust furthest to the foreland. In other words, Coulomb wedge models demonstrate that thrusting occurs in a break-forward sequence. Note that, because Coulomb materials are not very strong, the wedge cannot thicken indefinitely. Eventually, the hinterland of the wedge collapses under its own weight by slumping or by formation of normal faults within it. This hinterland collapse results in a net thinning of the thicker part of the wedge, thereby maintaining a dynamic

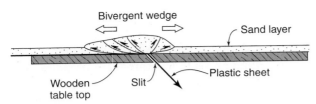

FIGURE 18.34 Simplified cross section of a sandbox model illustrating development of a bivergent thrust wedge. The sand was laid down on a mylar (plastic) sheet, and the sheet is pulled through a slit in the wooden table beneath it. Note that the wedge is not symmetrical.

balance between wedge thickening and the proper surface slope for wedge translation.

In recent years, geologists have modified Coulomb-wedge model design to make the models more realistic. In the newer models, there is no rigid backstop. Rather, sand is placed on a thin plastic sheet, and the sheet is then pulled through a slit at the bottom of the sand box. Such a model configuration resembles what happens where one plate slides beneath another at a convergent or collisional boundary. Note that the sand itself serves as the backstop. As motion progresses, the sand on the plastic sheet (i.e., the sand on the downgoing or underthrust plate) pushes against the stationary sand on the overriding plate. As a result, a **bivergent wedge** of sand develops (Figure 18.34). This consists of two thrust belts on opposite sides of the slit, each of which verges away from the slit. The "forewedge," which verges toward the interior of the downgoing plate, tends to be wider than the "retrowedge," which verges toward the interior of the overriding plate. The bivergent-wedge model explains the geometry of thrusting at some accretionary prisms, in which thrusting toward the sea occurs at the toe of the prism, while thrusting toward the arc occurs at the hinterland portion of the wedge. This model also provides an analog for the gross geometry of a collisional orogen, for in many such orogens, thrust belts form on both sides of the metamorphic hinterland.

18.9 CLOSING REMARKS

Fold-thrust belts are inherently fascinating geologic terranes. They contain all the components that make for a good scientific puzzle—intriguingly complex features and potentially quantifiable relationships. In addition, they yield beautiful mountain ranges that are exciting settings for field work, and which may contain valuable resources. No wonder fold-thrust belts have been the focus of such intense research for so long, and will undoubtedly challenge the talents of geologists for years to come.

ADDITIONAL READING

Boyer, S., and Elliott, D, 1982. Thrust systems. *American Association of Petroleum Geologists Bulletin*, 66, 1196–1230.

Dahlstrom, C., 1969. Balanced cross sections. *Canadian Journal of Earth Science*, 6, 743–758.

Davis, D., Suppe, J., and Dahlen, F., 1983. Mechanics of fold-and-thrust belts and accretionary wedges. *Journal of Geophysical Research*, 88, 1153–1172.

Marshak, S., and Woodward, N., 1988. Introduction to cross-section balancing. In Marshak, S., and Mitra, G., eds. *Basic methods of structural geology, Part II.* Prentice Hall: Upper Saddle River, 303–332.

McClay, K., ed., 1992. *Thrust tectonics.* Chapman & Hall.

McClay, K., ed., 2003. *Thrust tectonics and petroleum systems,* American Association of Petroleum Geologists Memoir, in press.

Price, R., 1981. The Cordilleran foreland thrust and fold belt in the southern Canadian Rocky Mountains. In McClay, K. R., and Price, N. J., eds., *Thrust and nappe tectonics,* Geological Society of London Special Publication 9, 427–448.

Rich, J., 1934. Mechanics of low-angle overthrust faulting as illustrated by Cumberland thrust block, Virginia, Kentucky, and Tennessee. *American Association of Petroleum Geologists Bulletin,* 18, 1584–1596.

Suppe, J., 1983. Geometry and kinematics of fault-bend folding. *American Journal of Science,* 283, 684–721.

Suppe, J., and Medwedeff, D., 1990. Geometry and kinematics of fault-propagation folding, *Eclogae Geologica Helvetica,* 83, 409–454.

Tearpock, D., and Bischke, R., 1991. *Applied subsurface geological mapping.* Prentice Hall.

Wilkerson, M. S., and Dicken, C. L., 2001. Quick-look techniques for evaluating two-dimensional cross-sections in detached contractional settings. *American Association of Petroleum Geologists Bulletin,* 95, 1759–1770.

Wilkerson, M. S., Fischer, M. P., and Apotria, T., eds., 2002. Fault-related folds: The transition from 2-D to 3-D. *Journal of Structural Geology (special issue),* 24, 4.

Strike-Slip Tectonics

19.1 INTRODUCTION

Every year, a few earthquakes startle the residents of California. Some tremors do little more than rattle a few dishes, but occasionally the jolting of a great earthquake tumbles buildings, ruptures roads, and triggers landslides. Most earthquakes in California signify that a sudden increment of slip has occurred somewhere along the San Andreas fault zone. This strike-slip fault zone, which contains countless individual faults, accommodates northward motion of the Pacific Plate relative to the North American Plate (Figure 19.1a and b). Most North American residents have heard of the San Andreas Fault—but it's not the only major strike-slip fault zone on this planet! Strike-slip faults cut both continental and oceanic crust in many places. Examples in continental crust include the Queen Charlotte Fault of western Canada, the Alpine Fault of New Zealand (Figure 19.2a), the faults bordering the Dead Sea (Figure 19.2b), the Anatolian Faults of Turkey, the Chaman Fault in Pakistan, and the Red River and Altyn Tach Faults of China. Strike-slip faults in oceanic crust most commonly occur along mid-ocean ridges, where they trend perpendicular to the ridge axis and link segments of the ridge. But there are important examples that link nonaligned segments of trenches as well.

As we noted in Chapter 8, a **strike-slip fault,** in the strict sense, is a fault on which all displacement occurs in a direction parallel to the strike of the fault (i.e., slip lineations on a strike-slip fault are horizontal). Thus, strict strike-slip displacement does not produce uplift or subsidence. In the real world, strike-slip movement is commonly accompanied by a component of shortening or extension. Specifically, **transpression** occurs where there is a combination of strike-slip movement and shortening, and can produce uplift along the fault. **Transtension** occurs where there is a combination of strike-slip movement and extension, and can produce subsidence along the fault.

In this chapter, we discuss the nature of deformation within strike-slip fault zones (both oceanic and continental), and we review the tectonic settings in which

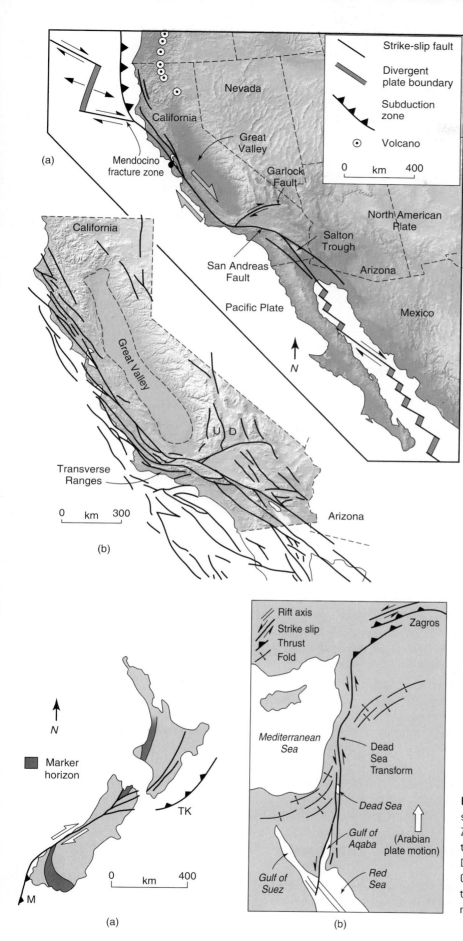

FIGURE 19.1 (a) Regional map of the plate boundary between the North American and Pacific Plates. The San Andreas Fault is the strike slip fault zone that defines this boundary in California. (b) An enlargement of California, showing the major strike slip faults. J & F = Juan de Fuca Plate.

FIGURE 19.2 Examples of major strike-slip faults. (a) The Alpine Fault in New Zealand links the Macquarie Trench (M) with the Tonga-Kermadec Trench (TK). (b) The Dead Sea transform (DST) runs from the Gulf of Aqaba north to the western end of the Zagros Mountains. It accommodates northward movement of the Arabian Plate.

TABLE 19.1	TERMINOLOGY FOR STRIKE-SLIP FAULTS
Flower structure	An array of faults in a strike-slip fault zone that merges at depth into a near-vertical fault plane, but near the ground surface diverges so as to have shallower dips. In a positive flower structure, there is a component of thrusting on the faults, and in a negative flower structure, there is a component of normal faulting.
Horsetail	A group of fault splays at the endpoint (fault tip) of a fault; the splays diverge to define a fanlike array in map view.
Lateral escape	The strike-slip movement of a crustal block in a direction perpendicular to the regional convergence direction in a collisional orogen; the block is essentially squeezed laterally out of the way of the colliding blocks.
Lateral ramp	A surface connecting two non-coplanar parts of a thrust fault or normal fault (i.e., a ramp that is roughly parallel to the transport direction). Strike-slip movement occurs on a lateral ramp.
Non-coplanar	An adjective used to describe a geometric arrangement of two planes in which the planes have the same attitude, but are not aligned end-to-end to form a single plane.
Pull-apart basin	A sedimentary basin that forms at a releasing bend along a strike-slip fault.
Releasing bend	In the context of describing a strike-slip fault, it is a map-view bend, in the fault plane, whose orientation is such that the block on one side of the fault pulls away from the block on the other, causing transtensional deformation.
Restraining bend	In the context of describing a strike-slip fault, it is a map-view bend, in the fault plane, whose orientation is such that the block on one side of the fault pushes against the block on the other, causing transpressional deformation.
Stepover	A geometry that occurs where the end of one fault trace overlaps the end of another, non-coplanar fault trace.
Strike-slip duplex	An arrangement of sigmoidal-shaped fault splays (in map view) at a stepover, a restraining bend, or a releasing bend, whose geometry in map view resembles the cross-sectional geometry of the duplexes that occur in thrust-fault or normal-fault systems.
Strike-slip fault	Any fault on which displacement vectors are parallel to the strike of the fault, in present-day surface coordinates. The use of this term is purely geometric, and has no genetic, tectonic, or size connotation.
Tear fault	A traditional term for strike-slip faults that occur in a thrust sheet and accommodate differential displacement of one part of a thrust sheet relative to an adjacent part; in more recent jargon, a tear fault is a vertical lateral ramp.
Transcurrent fault	A strike-slip fault that has the following characteristics: it dies out along its length; the displacement across it is less than the length of the fault; the length of the fault increases with time and continued movement; displacement on the fault is greatest at the center of the fault trace and decreases toward the endpoints (tips) of the fault.
Transfer fault	A fault trending roughly perpendicular to the axis of a rift that links together two non-coplanar parts of the rift; transfer faults accommodate map-view offset of the locus of extension. Kinematically, transfer faults are a type of transform fault. If a rift succeeds in evolving into a mid-ocean ridge, the transfer fault becomes an oceanic transform fault.
Transform fault	A strike-slip fault that has the following characteristics: once formed, displacement across it can be constant along the length of the fault; displacement across it can be much greater than the length of the active fault; its length can be constant, increase, or decrease with time; it terminates at another fault. Transform faults can be lithosphere plate boundaries, in which case they terminate at intersections with other plate boundaries.
Transpression	A combination of strike-slip and compressional deformation; transpression occurs where a fault is not parallel to the map projection of regional displacement vectors, so that there is a component of compression across the fault (leading to shortening).
Transtension	A combination of strike-slip and tension; transtension occurs where a fault is not parallel to the map projection of regional displacement vectors, so that there is a component of tension across the fault (leading to extension).
Wrench fault	A traditional synonym for a strike-slip fault. It was commonly used in the petroleum geology literature, typically in reference to a regional-scale continental strike-slip fault. We abandon the term, as it tends to be used inconsistently.

these faults develop. You will find that a wide variety of complex subsidiary structures develop in strike-slip fault zones. We begin the chapter by explaining the kinematic distinction between transfer faults and transcurrent faults, the two basic classes of strike-slip faults. Basic terms used in discussing strike-slip faults are provided in Table 19.1.

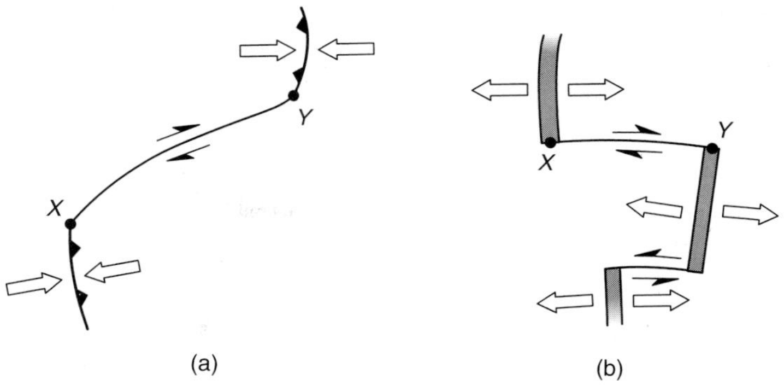

19.2 TRANSFORM VERSUS TRANSCURRENT FAULTS

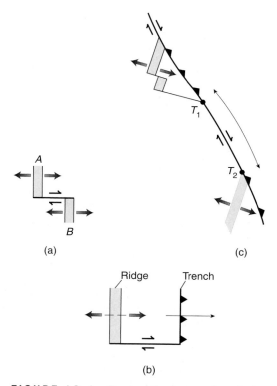

FIGURE 19.3 (a) Map sketch showing a transform fault linking two convergent plate boundaries. The fault trace occurs only between X and Y. (b) Map sketch showing a transform fault linking two ridge segments.

When examining the role that strike-slip faults play in crustal deformation, we find it convenient to distinguish between two kinematic classes of faults—transform faults and transcurrent faults.[1] A fault in one class differs from a fault in the other class in terms of the geometry of its endpoints (the locations along strike where the fault terminates), the way that slip magnitude varies along the fault's length, and the way that the fault's geometry evolves through time. To make this statement more concrete, let's look at the specific characteristics in more detail.

19.2.1 Transform Faults

J. Tuzo Wilson introduced the term "transform faults" to the geologic literature in the early 1960s, in reference to the third category of plate boundary (distinct from convergent and divergent boundaries). While some strike-slip faults are, indeed, plate boundaries, the term **transform fault** can be applied more broadly to describe any strike-slip fault that has the following characteristics:

- The active portion of a transform fault terminates at discrete endpoints. At the endpoints, the transform intersects other structures (Figure 19.3). For example, the transform can terminate at a shortening structure (e.g., convergent boundary, thrust fault, or stylolite) or at an extensional structure (e.g., divergent boundary, normal fault, or vein).
- The length of a transform fault can be constant, or it can increase or decrease over time. For example, in Figure 19.4a, the spreading rate on ridge segment A is the same as the spreading rate on ridge segment B. As a consequence, the length of the transform fault connecting these ridge segments remains constant

FIGURE 19.4 The length of a transform fault can change with time. (a) Transform length stays constant if spreading rates on ridge segments at both endpoints are the same. (b) Transform length decreases if the spreading rate at one endpoint is less than the subduction rate at the other. (c) Transform length increases as the two triple junctions (T_1 and T_2) defining the endpoints move apart.

over time. In contrast, Figure 19.4b shows a transform fault where one end terminates at a ridge segment, but the other end terminates at a trench. If the rate of subduction at the trench exceeds the rate of spreading on the ridge, then the length of the transform fault connecting them decreases over time. In Figure 19.4c, the length of the transform fault

---

[1]This discussion is based on a paper by R. Freund (1974).

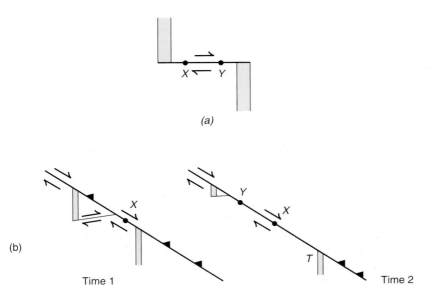

FIGURE 19.5 (a) The amount of displacement remains the same along the length of a transform if the length of the transform stays constant or decreases. Displacement at X is the same as the displacement at Y. (b) If the transform length changes with time, then the amount of slip varies along the length. At time 1, the fault is fairly short. At time 2, the length of the fault is longer. Displacement at X, in the middle of the fault, is greater than displacement at point Y, near an endpoint.

between two triple junctions labeled (T_1 and T_2) increases with time if the triple junctions move apart.

- In cases where the fault length is constant or decreasing, the amount of displacement along the length of a transform fault is constant. For example, the displacement at point X on the fault in Figure 19.5a is the same as the displacement at point Y. If the length of a transform fault increases over time, however, the amounts of displacement on the younger portions of the fault are less than the amounts on the older portions (Figure 19.5b).

- Displacement across a transform fault can be greater than the length of the fault itself. For example, consider a 10-km long transform fault that links two ridge segments. If more than 10 km of spreading occurs on the ridge segments, then there will be more than 10 km of slip on the transform fault (Figure 19.6).

Transform faults occur in a variety of settings and at a variety of scales (from mesoscopic to regional). At a mesoscopic scale, transform faults can link non-coplanar stylolites or non-coplanar veins. In Chapter 16, we saw how transfer faults link segments of rifts. Transfer faults are oriented at a high angle to the rift segments and terminate at normal faults. Kinematically, transfer faults behave like transform faults.

Where rifting is successful, so that a new divergent plate boundary forms, transfer faults in the rift evolve into classic transform faults that link spreading seg-

ments of a divergent plate boundary (Figure 19.7a and b). Note that such transform faults are plate boundaries, as first recognized by J. Tuzo Wilson.[2] Let's see how such transform faults display the kinematic characteristics we described above by looking at an example, a 10-km long transform fault linking two segments of the Mid-Atlantic Ridge (Figure 19.7c). Each end of the transform fault terminates abruptly at a ridge segment (i.e., at a zone of extension) trending at a high angle to the transform. If the spreading rate on each ridge segment is the same, then the length of the transform stays constant over time, regardless of the amount of slip that has occurred along it. Further, the same amount of slip occurs everywhere along the length of the fault. (This makes sense if you remember that the transform fault initiated at the same time as the ridge segments—it was "born" with the length that it now has. Note also that the sense of slip on the faults must be compatible with the spreading directions on the ridge segments.) The amount of slip along an oceanic transform fault can be much greater than the length of the fault. For example, if there has been 1000 km of spreading on the ridge segments, there must be 1000 km of displacement on the 10-km long transform.

Again we note that not all plate-boundary transform faults link segments of mid-ocean ridges. Some link non-coplanar segments of subduction zones, or link other transform faults (Figure 19.2a). For example, a transform fault links the east-dipping convergent plate boundary along the west coast of South America to the west-dipping convergent plate boundary of the Scotia Sea (Figure 19.7c). Note that in cases where a transform fault links a ridge and a trench, the fault length may decrease with time (Figure 19.4b); this happens if the subduction rate is faster than the spreading rate.

---

[2]Wilson wrote his paper about transform faults in 1965. His prediction of the sense of slip for transform faults was confirmed a couple of years later by Lynn Syke's study of fault-plane solutions for earthquakes on oceanic transforms and stands as one of the proofs of plate tectonics.

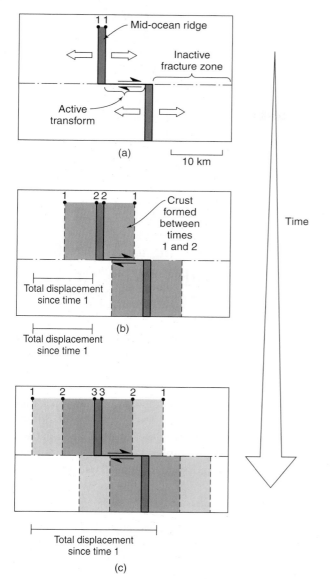

FIGURE 19.6 The amount of slip on a transform can exceed the length of the transform. (a) At time 1, the transform is 10 km long. Its length does not change over time, even though spreading at the ridges occurs continuously. (b) At time 2, the amount of displacement on the fault is already 1.5× the length of the fault. (c) At time 3, the displacement on the fault is 3.0× the length of the fault.

19.2.2 Transcurrent Faults

Transcurrent faults differ kinematically from transform faults in a number of ways:

• Transcurrent faults die out along their length. This means that at the endpoint of a transcurrent fault, the fault does not terminate abruptly at another fault, but either splays into an array of smaller faults (sometimes called a **horsetail**), or simply disappears into a zone of plastic strain. Typically, fault

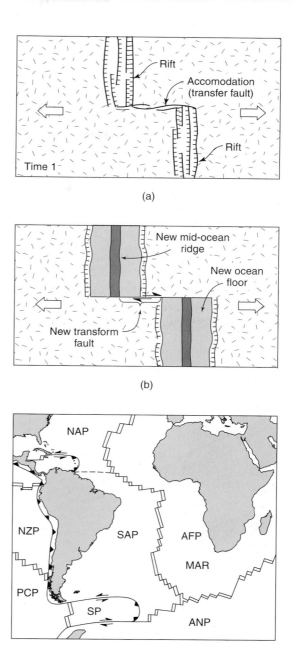

FIGURE 19.7 (a) Sketch map showing a transfer fault linking two rift segments. (b) Later, the rift segments have evolved into new mid-ocean ridges, and the transfer fault has evolved into an oceanic transform fault. Note that this fault still has the same length as when it originated. (c) Simplified map showing the transform faults along the Mid-Atlantic Ridge in the South Atlantic Ocean. SAP = South American Plate, NAP = North American Plate, AFP = African Plate, ANP = Antarctic Plate, NZP = Nasca Plate, CS = Caribbean Sea, SS = Scotia Sea, MAR = Mid-Atlantic Ridge.

splays comprising a horsetail are curved. Depending on the direction of curvature with respect to the sense of displacement, thrust or normal components of displacement occur in faults of the horsetail, and these movements will be accompanied either by folding and uplift where there is a thrust component, or by tilting and subsidence where there is a normal component. (Figure 19.8).

- Transcurrent faults initiate at a point and grow along their length as displacement increases (Figure 19.9a). As a consequence, short faults have a small amount of displacement, while long faults have a large amount of displacement. Thus, fault displacement (as measured in map view) is proportional to fault length.

- Displacement across a transcurrent fault is greatest near the center of its trace and decreases to zero at the endpoints of the fault (Figure 19.9b).

- The displacement on a transcurrent fault must always be less than the length of the fault (Figure 19.9b).

Transcurrent faults typically develop in continental crust as a means of accommodating development of regional strain. For example, slip on a conjugate system of mesoscopic transcurrent faults results in shortening of a block of crust in the direction parallel to the line bisecting the acute angle between the faults. As is the case with transform faults, transcurrent faults can form at any scale, from mesoscopic to regional.

19.3 STRUCTURAL FEATURES OF MAJOR CONTINENTAL STRIKE-SLIP FAULTS

Major continental strike-slip faults, meaning ones that have trace lengths ranging from tens to thousands of kilometers, are not simple planar surfaces. Typically, such faults have locally curved traces, divide into anastomosing (braided) splays, include several parallel branches, and/or occur in association with subsidiary faults and folds (Figure 19.10). In this section, we describe the structural complexities of large continental strike-slip faults and suggest why these complexities occur.

19.3.1 Description of Distributed Deformation in Strike-Slip Zones

Regional-scale strike-slip deformation in continental crust does not produce a single, simple fault plane. Rather, such shear produces a broad zone containing numerous individual strike-slip faults of varying lengths, as well as other structures such as normal and reverse faults, and folds.

Let's first look at the array of individual strike-slip faults that occurs in a large continental strike-slip fault zone. As an example, consider the San Andreas fault zone of California. The San Andreas Fault "proper" is one of about 10 major strike-slip faults, and literally thousands of minor strike-slip faults, that slice up western California (Figure 19.1). In some localities, faults have sinuous traces, so neighboring faults merge and bifurcate to define, overall, an anastomosing array (Figure 19.10). Elsewhere, the zone includes several subparallel faults. Locally, one fault dies out where another, parallel, but non-coplanar one initiates. The region between the endpoints of two parallel but non-coplanar

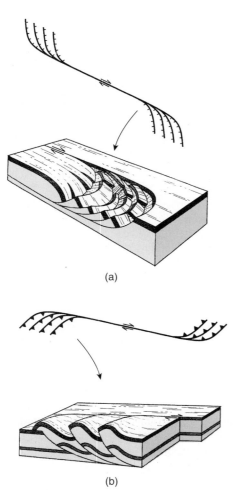

FIGURE 19.8 Terminations along transcurrent faults. (a) Here, the fault terminates in a horsetail composed of a fan of normal faults. (b) Here, the fault terminates in a horsetail composed of a fan of thrust faults.

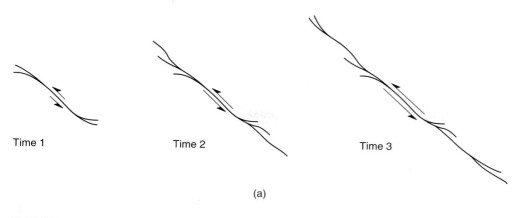

Time 1   Time 2   Time 3

(a)

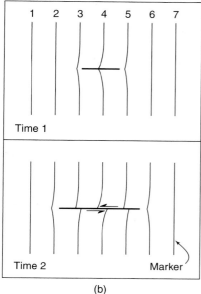

Time 1

Time 2   Marker

(b)

FIGURE 19.9 (a) Growth of a transcurrent fault. As time passes, the fault lengthens, and the displacement on the fault increases. In the process, some horsetail splays are abandoned. (b) Map sketch illustrating how the displacement varies along a transcurrent fault (horsetails are not shown). The heavier line is the fault, and the thin lines are marker lines. At time 1, the fault is short and has offset only marker 4. At time 2, the fault has grown, and now offsets markers 3, 4, and 5. Note that the markers just beyond the tips of the fault are starting to bend, prior to rupturing. Note also that the displacement on the fault is less than the length of the fault.

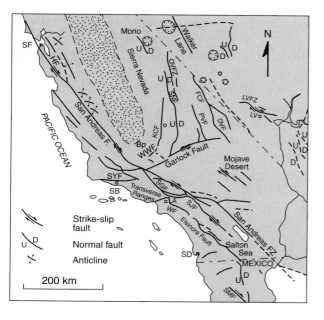

FIGURE 19.10 Faults of southern California. Note that faults anastomose and that, locally, folds form adjacent to the faults. Also note the large bend in the San Andreas Fault at its intersection with the Garlock Fault. B = Bakersfield, LA = Los Angeles, LV=Las Vegas, SB = Santa Barbara, SD = San Diego, SF = San Francisco, DVF = Death Valley Fault, FCF = Furnace Creek Fault, HF = Hayward Fault, KCF = Kern Canyon Fault, LVFZ = Las Vegas fault zone, OVFZ = Owens Valley fault zone, PVF = Panamint Valley Fault, SGF = San Gabriel Fault, SJF = San Jacinto Fault, SNF = Sierra Nevada Fault, SYF = Santa Ynez Fault, WF = Whittier Fault, WWF = White Wolf Fault.

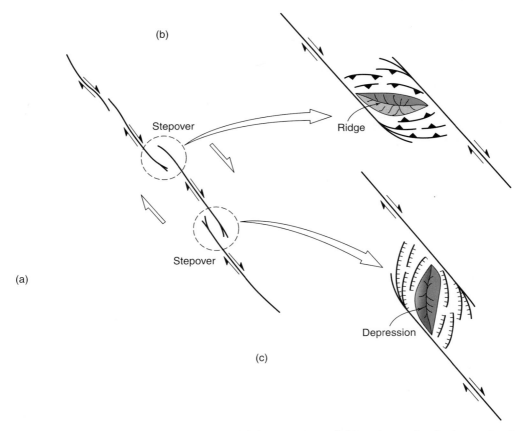

FIGURE 19.11 A stepover along a strike-slip fault. (a) Sketch map showing how regional dextral shear can be distributed along fault segments that are not coplanar. Slip is relayed from one segment to another at a stepover. (b) At a restraining stepover, compression and thrusting occur. (c) At a releasing stepover, extension and subsidence occur.

faults is called a **stepover** (Figure 19.11). Localized faulting occurs in the stepover region to accommodate the transfer of slip.

Continental strike-slip fault zones also may contain *en echelon* arrays of thrust faults, folds, and normal faults, as well as subsidiary strike-slip faults (Figure 19.12). Typically, the thrust faults and folds (Figure 19.13) trend at an angle of 45° or less to the main fault, and the acute angle between subsidiary thrust faults and the main fault (or between the fold hinges and the main fault) opens in the direction of shear. The normal faults also trend at an angle of 45° or more with respect to the main fault, but the acute angle between subsidiary normal faults and the main fault opens opposite to the direction of shear. Most subsidiary strike-slip faults trend at a shallow angle to the main fault.

19.3.2 The Causes of Structural Complexity in Strike-Slip Zones

Why do continental strike-slip fault zones contain so many subsidiary faults? To understand this complexity, we'll review an experiment that we introduced in Chapter 8. Take a homogeneous clay slab and place it over two adjacent wooden blocks (Figure 19.14). The clay cake represents the weaker uppermost crust, and the wooden blocks represent stronger crust at depth. Now, push one of the blocks horizontally so that it shears past its neighbor. As the blocks move relative to one another, the clay cake begins to deform, partly by plastic mechanisms and partly by brittle failure. Initially, this brittle deformation yields arrays of small strike-slip faults called **Riedel shears.** Two sets of Riedel shears, labeled R and R′, develop. We can picture these shears as forming a conjugate system relative to the far-field maximum compressive stress driving development of the overall fault zone. Eventually, P shears develop (see Chapter 8), which finally link with Riedel shears to form a throughgoing strike-slip fault. With this model, we can speculate that some of the subsidiary faults in a strike-slip zone initiated as Riedel shears or as P shears.

The model just described suggests that, even if the upper crust were homogeneous, one might expect strike-slip zones to contain subsidiary strike-slip faults. In reality, the upper crust is heterogeneous.

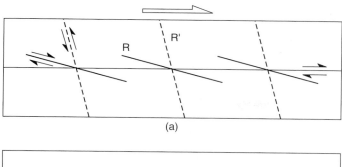

(a)

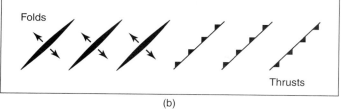

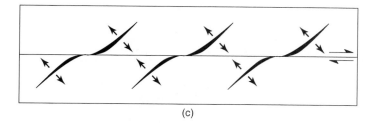

Folds

Thrusts

(b)

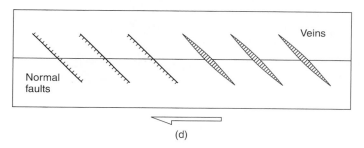

(c)

Veins

Normal
faults

(d)

FIGURE 19.12 Arrays of subsidiary structures associated with dextral shear. (a) Subsidiary strike-slip faults (Riedel shears: R and R′); (b) *en echelon* folds, and *en echelon* thrusts; (c) *en echelon* folds which formed and then were later offset by shear on a strike-slip fault; (d) *en echelon* normal faults and veins.

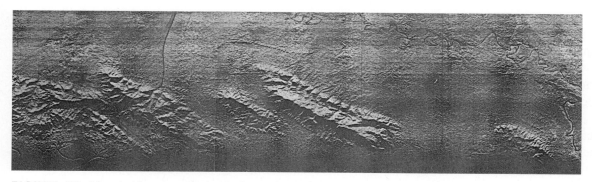

FIGURE 19.13 A side-scan radar image from the Darien Basin in eastern Panama showing an array of *en echelon* anticlines whose formation has arched up the land surface, creating a set of ridges. The field of view is about 50 km. Note that the geometry of these structures indicates left-lateral shear.

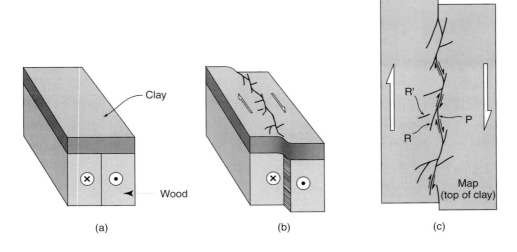

(a)   (b)   (c)

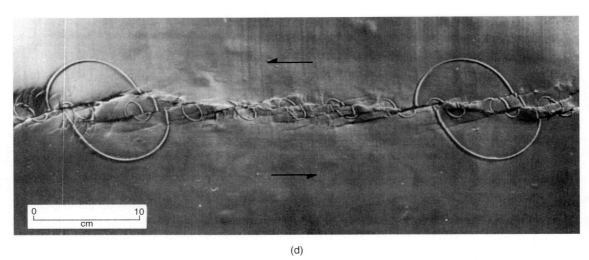

(d)

FIGURE 19.14 A laboratory model of strike-slip fault development. (a) Before deformation, a clay cake rests on two wooden blocks that were pressed together. The clay represents the weak uppermost crust, and the wood blocks represent the stronger lower crust. The vertical boundary between the two blocks represents the strike-slip fault. (b) As deformation begins, Riedel shears develop in the clay cake. (c) A map view of the top surface of the clay cake, showing a later stage of deformation, in which Riedel shears have been linked by P fractures. A throughgoing fault has just developed. (d) An example of a clay-cake experiment, this one for left-lateral shear.

Crust may contain a variety of different rock types with different strengths, and the contacts between rock units may occur in a variety of orientations. In addition, the crust contains preexisting planar weaknesses such as joints, old faults, and foliations. All these heterogeneities cause stress concentrations and local changes in stress trajectories. As a result, faults may locally bend, and they may split to form two strands on either side of a stronger block. In sum, the process by which the fault is formed, as well as the crust's heterogeneity, ensures that strike-slip fault zones include a variety of subsidiary fault splays.

To picture why *en echelon* arrays of thrusts, folds, and normal faults have the orientations that they do, picture a block of crust that is undergoing simple shear in map view. An imaginary square superimposed on the zone transforms into a rhomb, and an imaginary circle transforms into a strain ellipse (Figure 19.15a and b). In the direction parallel to the short axis of the ellipse, the crust shortens, so thrusts and folds develop perpendicular to this axis. In the direction parallel to the long axis of the ellipse, the crust stretches, so normal faults and veins develop perpendicular to this axis. As deformation pro-

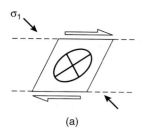

(a)

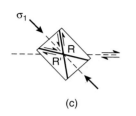

(c)

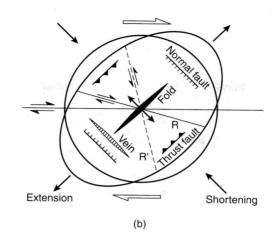

(b)

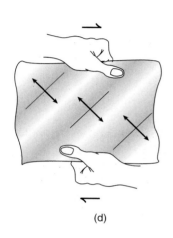

(d)

FIGURE 19.15 A strain model explaining the origin of subsidiary structures along a strike-slip fault. (a) A map view of dextral simple shear. A square becomes a parallelogram, and a circle in the square becomes an ellipse. (b) A detail of the strain ellipse showing that folds and thrusts form perpendicular to the shortening direction, while normal faults and veins form perpendicular to the extension direction. R and R' shears form at an acute angle to the shortening direction. (c) Note that R and R' are similar to conjugate shear fractures formed in rock cylinder subjected to an axial stress. (d) You can simulate formation of *en echelon* folds with a sheet of paper.

gresses, the main strike-slip fault eventually slices the block in two, and the two halves of the subsidiary faults and folds are displaced with respect to one another. Note that R and R' shears can be pictured as conjugate shear fractures whose acute bisector between the faults is parallel to σ_1, the regional maximum principle stress (Figure 19.15c). Also note that you can simulate deformation in a strike-slip fault zone by shearing a piece of paper between your hands (Figure 19.15d); the ridges that rise in the center of the paper represent folds.

19.3.3 Map-View Block Rotation in Strike-Slip Zones

Progressive slip on strike-slip faults may result in map-view rotation of crustal blocks (i.e., rotation of the blocks about a vertical axis). Such rotation can happen is one of three ways. (1) As illustrated by contrasting Figure 19.16a with Figure 19.16b, progressive simple shear causes material lines in the fault zone to rotate; (2) In places where several parallel strike-slip faults dice up a crustal block, the block as a whole can rotate around a vertical axis in the same way that books on a bookshelf tilt around a horizontal axis when they stay in contact with the shelf but shear past one another (Figure 19.16b). This process is the map-view equivalent of the process leading to rotation by slip on planar normal faults (see Chapter 9); (3) Small crustal blocks caught in a strike-slip zone may rotate about a vertical axis like ball bearings between two sheets of wood (Figure 19.16c).

19.3.4 Transpression and Transtension

The trace of the San Andreas Fault in southern California is not just a featureless line on the surface of the Earth. At some localities, the trace lies within a train of marshy depressions (called sag ponds); elsewhere, the trace is marked by 50-m high ridges

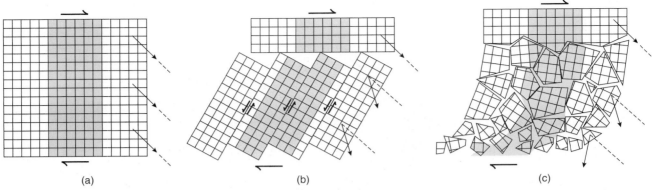

FIGURE 19.16 Mechanisms of block rotation in a right-lateral strike-slip zone. (a) A map view of a grid being subjected to dextral simple shear. (b) The grid lines rotate. Further, fault-bounded blocks may rotate intact as slip on bounding faults increases (bookshelf model). (c) Alternatively, the fault zone may break into smaller blocks that rotate by different amounts. In this case, large rotations may occur locally. The amount of rotation is tracked by the paleomagnetic declination in each block (solid arrows) relative to a reference direction (dashed line).

(called pressure ridges) that contain tight folds (Figure 19.17). The presence of such topographic and structural features tells us that motion along the fault is not perfectly strike-slip. Ridges form in response to **transpression,** a combination of strike-slip displacement and compression that yields a component of shortening across the fault. This shortening causes thrusting and uplift within or adjacent to the fault zone. Notably, where the fault zone contains a broad band of weak fault rocks (gouge and breccia), transpression squeezes the fault rocks up into a fault-parallel ridge, much like a layer of sand squeezes up between two wood blocks that have been pushed together (Figure 19.18a). Topographic depressions reflect **transtension,** a combination of strike slip and extension. The extensional component causes normal faulting and subsidence (Figure 19.18b). Transpression or transtension can occur along the entire length of a fault zone, if the zone trends oblique to the vectors describing the relative movement of blocks juxtaposed by the fault. Such a situation develops where global patterns of plate motion change subsequent to the formation of the fault, for a fault is a material plane in the Earth and cannot change attitude relative to adjacent rock once it has formed.

The dimensions of transpressive or transtensile structures forming along a strike-slip fault depend on the amount of cross-fault shortening or extension, respectively. Where relatively little transpressive or transtensile deformation has occurred, cross-fault displacement results in relatively small pressure ridges or sags, with relief that is less than a couple of hundred meters. If, however, transpression or transtension has taken place over millions of years, significant mountain ranges develop adjacent to the fault. For example, transpression along the Alpine Fault of New Zealand has resulted in uplift of the Southern Alps, a range of mountains that reaches an elevation of up to 3.7 km above sea level.

Seismic-reflection studies of large continental strike-slip faults indicate that subsidiary faults in transpressional or transtensional zones within strike-slip systems are concave downwards, and merge at depth into the main vertical fault plane (Figure 19.19a). Thus, in cross section, large continental strike-slip faults resemble flowers in profile, with the petals splaying outwards from the top of a stalk. This configuration of faults is, not surprisingly, referred to as a **flower structure.**[3] In transpressive zones, a **positive flower structure** develops, in which the slip on subsidiary faults has a thrust-sense component (Figure 19.19b); whereas in transtensile zones, a **negative flower structure** develops, in which the slip on subsidiary faults has a normal-sense component (Figure 19.19c).

---

[3]Some authors prefer the term "palm structure" for this arrangement of faults, because the shape of the faults more closely resembles fronds on a palm tree than petals on most flowers. But the term "flower structure" tends to be used more commonly. For discussion of these structures, see Sylvester (1988).

(a)

(b)

(c)

FIGURE 19.17 (a) Air photo showing the trace of the San Andreas Fault, north of San Francisco (Tomales Bay). Note that the faulting has locally caused a water-filled depression to form. (b) Photograph of pressure ridges along the San Andreas Fault, San Luis Obisbo County, California. (c) A cross section of a pressure ridge in a road cut across the San Andreas Fault near Palmdale, California.

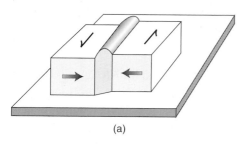

(a)

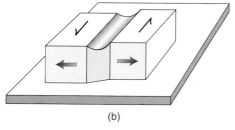

(b)

FIGURE 19.18 Simple wood block model illustrating the concept of transpression and transtension. (a) When blocks shear and squeeze together (transpression), sand is pushed up; (b) When blocks shear and pull apart (transtension), sand sags.

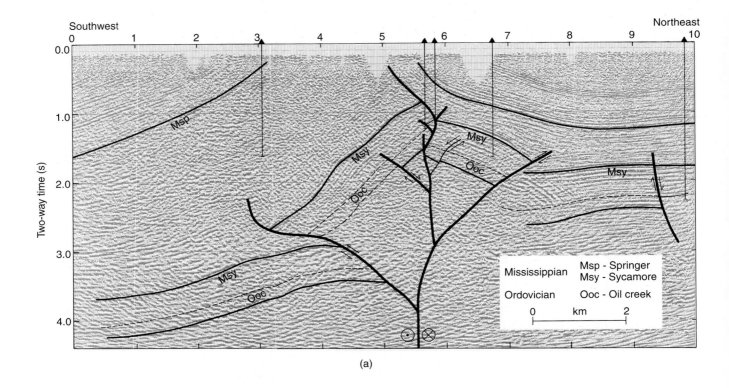

(a)

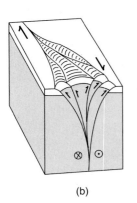

(b)

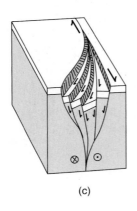

(c)

FIGURE 19.19 (a) Seismic-reflection profile across a strike-slip fault in the Ardmore Basin, Oklahoma, showing a positive flower structure. Msp and Msy are Mississippian formations, and Ooc is Ordovician. (b) Block diagram of a positive flower structure. (c) Block diagram of a negative flower structure.

19.3.5 Restraining and Releasing Bends

In many locations, transpression and transtension take place at distinct bends in the trace of a strike-slip fault.[4] To see why, picture a fault that, overall, strikes east-west, parallel to the trend of regional displace-

ment vectors (Figure 19.20a and b). Along segments of the fault that strike exactly east-west, motion on the fault can be accommodated by strike-slip motion alone, for the fault plane parallels the regional displacement vectors. But at fault bends, where the strike of the fault deviates from east-west, the fault plane does not parallel regional displacement vectors, and there must be either transpression or transtension, depending on the orientation of the bend. Let's look at the geometry of these bends in more detail.

A bend at which transpression takes place is called a **restraining bend,** because the fault segment in the

---

[4]Recall that a fault bend, in the context of discussing strike-slip faults, is a portion of a fault where the strike of the fault changes. Elsewhere in this book, we have also used the term "fault bend" when discussing dip-slip faults; a fault bend along a dip-slip fault is a location where the fault's dip changes.

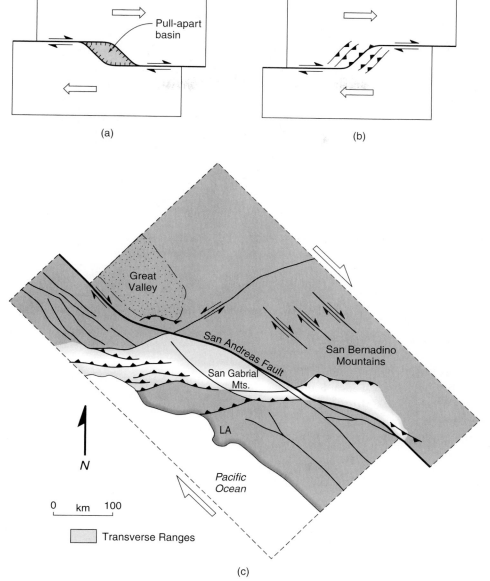

FIGURE 19.20 Map-view models of fault bends along strike-slip faults. The "edges" of the crustal blocks are provided for reference. (a) Releasing bend at which normal faults and a pull-apart basin have formed. (b) Restraining bend at which thrust faults have formed. (c) Application of this model to the San Andreas Fault north of Los Angeles (LA). The dashed lines outline imaginary reference blocks. The San Andreas Fault bends along the margin of the Mojave Desert.

bend inhibits motion. In other words, at restraining bends, pieces of crust on opposite side of the fault push together, causing crustal shortening. Along regional-scale restraining bends, a fold-thrust belt can form, causing the uplift of a **transverse mountain range** (i.e., a mountain range trending at an angle to the regional trace of the fault). For example, the Transverse Ranges just north of Los Angeles in southern California developed because of shortening across a large restraining bend along the San Andreas Fault (Figure 19.20c).

A bend at which transtension occurs is called a **releasing bend** because, at such bends, opposing walls of the fault pull away from each other (Figure 19.20b). As a consequence of this motion, normal faults develop, and the block of crust adjacent to the bend subsides. Displacement at a releasing bend yields a negative flower structure or, in cases where the bend is

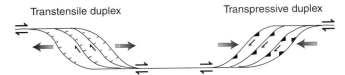

FIGURE 19.21 Map-view sketch of strike-slip duplexes formed along a dextral strike-slip fault.

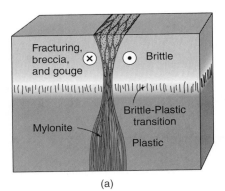

(a)

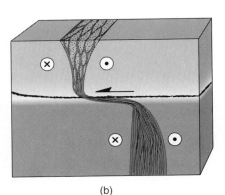

(b)

FIGURE 19.22 [a] A regional-scale strike-slip fault may consist of a broad zone of breccia and gouge at shallow crustal levels. At deeper layers, the zone may narrow into a zone of cataclasite and, at great depth, it broadens into a zone of mylonite, grading down into schist. [b] The shallower portion of some strike-slip faults may be offset, relative to the deeper portion, by a detachment at depth.

large, and large amounts of extension have taken place, a pull-apart basin. A **pull-apart basin** is a rhomboid-shaped depression, formed along a releasing bend and filled with sediment eroded from its margin. The dimension and the amount of subsidence in a pull-apart basin depends on the size of the bend and on the amount of extension. Notably, formation of small pull-apart basins involves brittle faulting only in the upper crust, but formation of large pull-apart basins involves thinning of the lithospheric mantle, so that after extension ceases, the floor of the basin thermally subsides (see Chapter 16). Examples of present-day pull-apart basins include the Dead Sea (Figure 19.2b) at the border between Israel and Jordan, and Death Valley in eastern California. In both of these basins, the land surface lies below sea-level, creating an environment in which summer temperatures become deadly hot.

Both restraining bends and releasing bends can exist simultaneously at different locations along the same fault. As a consequence, a region of crust moving along the fault may at one time be subjected to transtension and then, at a later time, be subjected to transpression. When this happens, a negative flower structure, or pull-apart basin formed at a releasing bend, becomes inverted and changes into a positive flower structure, and normal faults bordering a pull-apart basin become reverse faults. Inversion causes sediment that had been deposited in negative flower structures or pull-apart basins to thrust up and over the margins of a strike-slip fault zone.

19.3.6 Strike-Slip Duplexes

A **strike-slip duplex** consists of an array of several faults that parallels a bend in a strike-slip fault (Figure 19.21). The map-view geometry of a strike-slip duplex resembles the cross-sectional geometry of thrust-fault duplex or a normal-fault duplex. Strike-slip duplexes formed at restraining bends can also be called **transpressive duplexes,** while those formed at releasing bends can also be called **transtensile duplexes.**

19.3.7 Deep-Crustal Strike-Slip Fault Geometry

What do regional-scale strike-slip fault zones look like at progressively greater depths in the crust? Based on studies of exposed crustal sections of ancient faults, geologists surmise that the fault zone is very weak near the surface, and thus is very broad. With increasing depth, the crust becomes stronger and the fault zone becomes narrower. Then, at the brittle–plastic transition, the crust becomes weaker, and the fault zone widens (Figure 19.22a).

Recent studies suggest that the concept of a strike-slip fault zone cutting down through the entire crust as a continuous vertical zone of deformation may be an oversimplification in some locations. There is growing evidence that regional-scale strike-slip faults are locally offset at subhorizontal detachments deep in the crust, so that the strike-slip displacement in the upper crust does not lie directly over displacements in the lower crust and underlying mantle lithosphere (Figure 19.22b).

19.4 TECTONIC SETTING OF CONTINENTAL STRIKE-SLIP FAULTS

Until now, we have focused our discussion on characteristic structural features that occur in strike-slip fault zones. Now, we broaden our perspective and turn our attention to the tectonics of strike-slip faulting. Specifically, we describe the various plate settings at which strike-slip fault zones develop. Several of these occur in southern Asia (Figure 19.23).

19.4.1 Oblique Convergence and Collision

Strike-slip faults occur along convergent plate boundaries where the vector describing the relative motion between the subducting and overriding plates is not perpendicular to the trend of the convergent margin (Figure 19.24a). At such **oblique-convergent plate margins,** the relative motion between the two plates can be partitioned into a component of dip-slip motion (thrusting) perpendicular to the margin, and a component of horizontal shear (strike-slip faulting) parallel to the margin. Present-day examples illustrate that the strike-slip faults of oblique-convergent plate boundaries develop in a variety of locations across the margin, including the accretionary wedge, the volcanic arc, and the backarc region.

Partitioning of relative movement into dip-slip and strike-slip components accompanies the oblique collision of two buoyant lithospheric masses (Figure 19.24b). The strike-slip component of motion displaces fragments of crust laterally along the orogen. When the colliding mass is a small exotic terrane, the terrane may be sliced up by strike-slip faults after it docks (Figure 19.24c). For example, Wrangelia, an exotic crustal block that was incorporated into the western margin of North America during Mesozoic oblique convergence, was sliced by strike-slip faults into fragments that were then transported along the margin. As a result, bits and pieces of Wrangelia occur in a discontinuous chain that can be traced from Idaho to Alaska (see Figure 17.29).

Strike-slip faulting also develops at collisional margins where one continent indents the other. The vise created when two continents converge may cause blocks of crust caught between the colliding masses to be squeezed laterally out of the zone of collision, a process called **lateral escape.** The boundaries of these "escaping" blocks are strike-slip faults. For example, during the Cenozoic collision of India with Asia, India

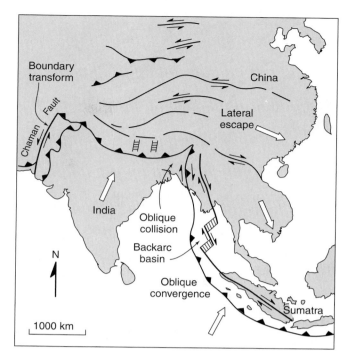

FIGURE 19.23 Sketch map of southern Asia, showing the collision of India. Strike-slip faults have developed in several settings here. A boundary transform (the Chaman Fault) delimits the northwestern edge of the Indian subcontinent. Strike-slip faults also form due to oblique collision, oblique convergence, and lateral escape. Small rifts have developed just north of the Himalayas.

pushed northward into Asia. Stress from this collision resulted in a fan-shaped pattern of strike-slip faults accommodating the escape of blocks of Asia eastward (Figure 19.23). A similar phenomenon is currently happening in the eastern Mediterranean, where the northward movement of the Arabian Peninsula along the Dead Sea transform has resulted in the westward escape of Turkey, squeezed like a watermelon seed between the North Anatolian Fault and the East Anatolian Fault (Figure 19.25).

19.4.2 Strike-Slip Faulting in Fold-Thrust Belts

The traces of thrust faults trending nearly perpendicular to the regional transport direction dominate the map pattern of fold-thrust belts. Locally, however, these belts contain strike-slip faults whose traces trend nearly parallel to the regional transport direction (Chapter 18). Some of these faults, known as **lateral ramps,** cut a thrust sheet into pieces that move relative to one another. In places where lateral

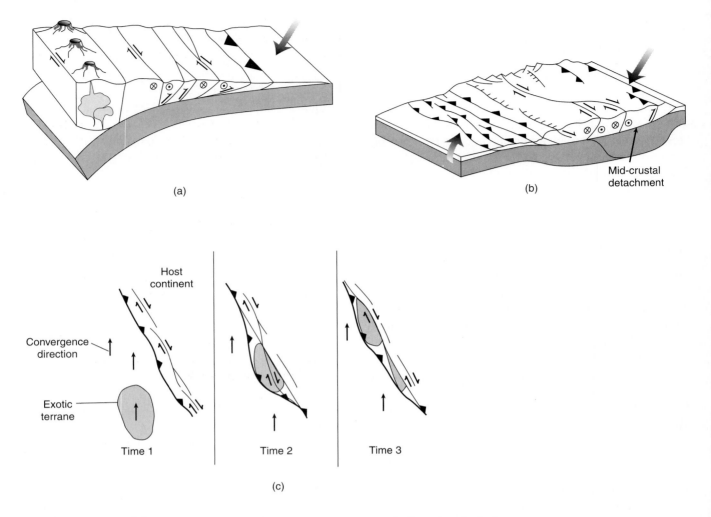

(a)

(b)

Mid-crustal
detachment

Host
continent

Convergence
direction

Exotic
terrane

Time 1

Time 2

Time 3

(c)

FIGURE 19.24 (a) Strike-slip faulting at an oblique convergent margin. Note that the faults occur at various locations across the width of the margin. The large arrow indicates the relative movement of the downgoing plate. (b) Strike-slip faulting at an oblique collisional margin. Note that in this portrayal, the faults terminate at depth at a mid-crustal detachment horizon. (c) Map-view sketches showing progressive stages during oblique docking of an exotic terrane. Note how the terrane is sliced by faults subsequent to docking, and slivers slip along the length of the orogen.

ramps have a near-vertical dip, they are also called **tear faults.** Lateral ramps or tear faults can accommodate along-strike changes in the position of a frontal ramp with respect to the foreland. Examples of lateral ramps bound the two ends of the Pine Mountain thrust sheet in the southern Appalachians (Figure 19.26).

Locally, thrust sheets contain conjugate mesoscopic strike-slip fault systems, which accommodate shortening of thrust sheets across strike and simultaneous stretching of thrust sheets along strike. Examples of conjugate strike-slip systems stand out in the Makran fold-thrust belt of southern Pakistan, and in the Jura fold-thrust belt of Switzerland and France (Figure 19.27).

19.4.3 Strike-Slip Faulting in Rifts

As we noted earlier, rifts typically consist of a chain of distinct segments. Adjacent segments in the chain may differ in amount of extension, and in the dominant dip direction of normal faults. Also, the axis of one segment may not align with the axis of the adjacent segment in the chain. Each segment in a rift is linked to its neighbor by an **accommodation zone.** In places where an accommodation zone consists of a strike-slip fault, it can also be called a **transfer fault** (see Chapter 16; also Figure 19.7). Where transtension or transpression occurs, flower structures may develop along transfer faults.

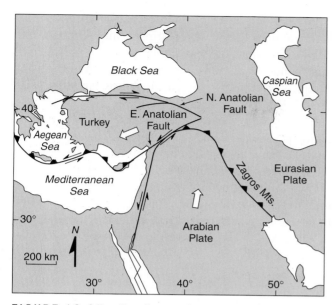

FIGURE 19.25 Sketch map of the region from the eastern Mediterranean to the Caspian Sea. Lateral escape in response to the northward movement of the Arabian Plate is squeezing Turkey out to the west. Escape is accommodated by slip on the North Anatolian Fault and the East Anatolian Fault.

The Garlock Fault in southern California is one of the largest examples of a strike-slip within a rift environment (Figure 19.22). This fault, a ~250-km long left-lateral strike-slip fault, forms the northern border of the Mojave Desert and intersects the San Andreas Fault at its western end. It defines the boundary between two portions of the Basin and Range Rift: the "Central Basin and Range" lying to the north, and the Mojave block lying to the south. The Garlock Fault exists because these two portions have not developed the same amount of extensional strain. Specifically, a greater amount of extension has occurred in the Central Basin and Range than in the Mojave block (Figure 19.28). To accommodate this difference, the crust to the north of the fault has slid westward, relative to the crust to the south.[5] Notably, this motion pushed the portion of the San

---

[5]This model was first proposed by B. C. Burchfiel and G. A. Davis (1973).

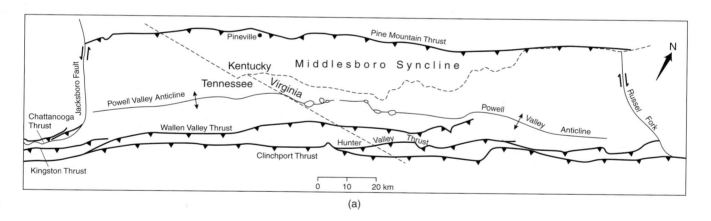

(a)

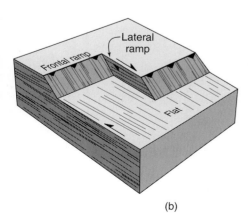

(b)

FIGURE 19.26 (a) Map of the Pine Mountain thrust system in the southern Appalachians (eastern USA), showing lateral ramps (Jacksboro Fault and Russell Fork Fault). (b) Block diagram indicating the concept of a lateral ramp or tear fault.

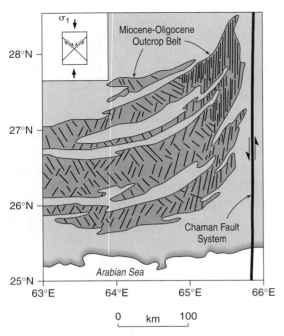

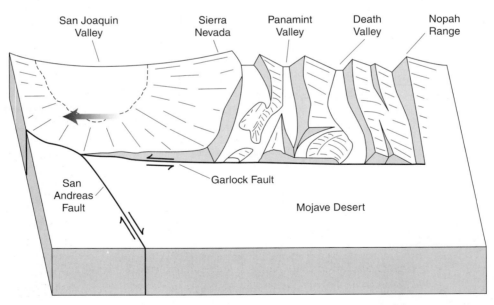

FIGURE 19.27 Simplified map of the Makran fold-thrust belt in Pakistan, illustrating the conjugate system of strike-slip faults contained within thrust slices.

Andreas Fault north of its intersection with the Garlock Fault to the west, creating a major restraining bend in the San Andreas fault zone. The presence of this bend is responsible for development of the Transverse Ranges just north of Los Angeles (Figure 19.20).

Motion between the Pacific and North American Plates across the Transverse Ranges bend in the San Andreas Fault also generates compressive stress within the Mojave block. This stress has caused some of the normal faults formed by rifting in the block in the past to be reactivated today as strike-slip faults. Motion on these faults results in rotation, around a vertical axis, of crust within the Mojave block (Figure 19.20; also Figure 19.16b).

19.4.4 Continental Transform Faults

Where transform plate boundaries cross continental lithosphere, strike-slip deformation disrupts the continental crust. Several examples of continental transforms crop out on the surface of the Earth.

FIGURE 19.28 Block model of the Garlock Fault, southern California (USA), accommodating the offset from the extended Basin and Range Province to the north and the Mojave Desert to the south.

North Americans are probably familiar with the San Andreas fault zone of California (Figure 19.1), which spans the distance between the Gulf of California rift on the south and the Mendocino oceanic transform on the north. As we noted earlier, this right-lateral fault zone is structurally complex, with numerous splays, many of which are seismically active. Subsidiary structures, such as pull-aparts, flower structures, and transverse ranges, mark locations along its trace where the trace of the fault zone is not exactly a small circle around the Euler pole describing the motion between the North American and Pacific Plates. Also, *en echelon* folds and faults form where strain has not been accommodated by strike-slip displacement. Other important examples of continental transforms include the Alpine Fault of New Zealand, which accommodates motion between the Australian and the Pacific Plates (Figure 19.2), and the Chaman Fault of Pakistan, which accommodates motion between India and Asia on the west side of India (Figure 19.23).

19.5 OCEANIC TRANSFORMS AND FRACTURE ZONES

All mid-ocean ridges (i.e., plate boundaries along which seafloor spreading occurs) consist of non-coplanar segments that range in length from tens to hundreds of kilometers. Each segment is linked to its neighbor by an oceanic transform fault. These **oceanic transform faults** are plate boundaries; strike-slip deformation within them accommodates the motion of one oceanic plate laterally past another as seafloor spreading progresses (Figure 19.7). The length of oceanic transform faults varies from 10 to 1000 km. As measured across strike, they are relatively narrow (less than a few kilometers wide); but even considering this narrow width, they are so abundant that between 1% and 10% of the oceanic lithosphere has been affected by deformation related to strike-slip faulting.

It is important to remember that transform faults occur only between ridge segments; they do not extend beyond them. There are, however, pronounced topographic lineaments, known as **fracture zones,** that extend beyond the tips of the ridge segments. Seismicity in the ocean floor shows that earthquakes occur along the ridge segments and along the transform faults between them, but that they are rare along fracture zones. Thus, fracture zones are not active fault zones.[6] Note also that when you cross a transform boundary, you pass from one plate to another. However, when you cross a fracture zone, you stay on the same plate.

Satellite gravity measurements, side-scan sonar images, dredge hauls, cores, and submarine photographs lead to the conclusion that oceanic transform zones and fracture zones are not featureless lines on the surface of the sea floor. Rather, they contain escarpments, ridges, and narrow troughs (Figure 19.29). The bathymetric complexity of oceanic transform zones and fracture zones develops in response to a variety of phenomena. First, since the depth of ocean floor depends on the age of the underlying lithosphere (see Chapter 16) and the ocean floor on one side of a zone is not the same age as the ocean floor on the other side (except at the point on a transform fault halfway between two ridge tips), there must be a change in ocean-floor depth across a zone. Second, in places where transforms are not precisely small circles around an Euler pole (see Chapter 14), transpression and transtension generate a flower structure; the flower structure becomes inactive once a transform zone becomes a fracture zone, but the faults comprising it do not disappear. Third, slip along a transform zone pervasively fractures the crust. Seawater circulates through the fracture network and reacts with crustal basalt, altering olivine to form a hydrated mineral called serpentine. This process increases the volume of the crust, because serpentine is less dense than olivine, and thus causes the sea floor to rise. Fourth, heat radiating from the tip of a ridge segment can cause isostatic uplift of the lithosphere beyond the tip. This uplifted lithosphere gradually drifts away from the ridge termination, as sea-floor spreading progresses, forming an elongate ridge (an intersection high) bordering the fracture zone (Figure 19.29). At a distance from the ridge, the intersection high cools and subsides.

Notably, the development of escarpments in oceanic transform zones and fracture zones, coupled with the weakening of rock by pervasive fracturing and serpentinization, sets the stage for submarine slope failure. The debris that tumbles down the escarpments collects

---

[6]Some authors, therefore, use the term "inactive transform," instead of fracture zone, for these lineaments.

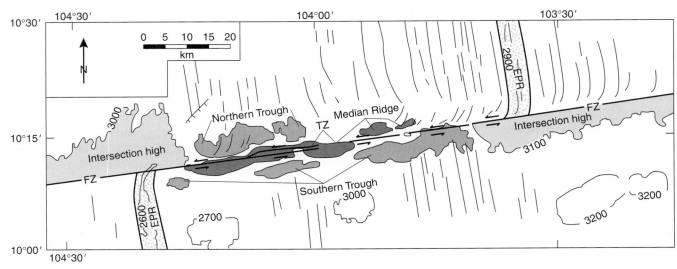

FIGURE 19.29 Topography of the Clipperton fracture zone (FZ) and transform zone (TZ) of the East Pacific Rise. Note intersection highs at ridge tips, and trough and ridges along the transform zone. Contours in meters below sea level. EPR = East Pacific Rise.

in thick piles consisting of sedimentary breccia. Thus, in contrast to normal ocean crust, the oceanic crust of transform zones and fracture zones typically has a coating of sedimentary breccia.

Examination of a map showing global plate boundaries (Figure 14.14a) reveals that not all oceanic transform faults offset ridge segments. Some are plate boundaries that link subduction zones, or link subduction zones to ridges. Examples include the north and south border of the Scotia Plate between Antarctica and South America, and the northern border of the Caribbean Plate. The Alpine Fault of New Zealand, which we discussed earlier, connects along strike to oceanic transforms.

19.6 CLOSING REMARKS

With this chapter on strike-slip tectonics, we close our discussion of tectonic settings. We have focused on describing processes and terminology, with the goal of developing a framework of concepts that we can use to describe tectonic settings. You will have noticed that we mention several features in more than one of these chapters. This necessary and appropriate redundancy emphasizes that classes of structures are not limited to

one tectonic setting. In the remainder of this book, we use our structural and tectonic framework to examine specific orogens.

ADDITIONAL READING

Aydin, A., and Nur, A., 1982. Evolution of pull-apart basins and their scale independence. *Tectonics,* 1, 91–105.

Biddle, K. T., and Christie-Blick, N., 1985. Strike-slip deformation, basin formation, and sedimentation: Society of Economic Paleontologists and Mineralogists Special Publication 37, SEPM, Tulsa, 386 p.

Burchfiel, B. C., and Davis, G. A., 1973. Garlock fault: an intracontinental transform fault. *Geological Society of America Bulletin,* 84, 1407–1422.

Freund, R., 1974. Kinematics of transform and transcurrent faults. *Tectonophysics,* 21, 93–134.

Garfunkel, Z., 1986. Review of oceanic transform activity and development. *Geological Society of London Journal,* 143, 775–784.

Naylor, M. A., Mandl, G., and Kaars-Sijpestein, C. H., 1986. Fault geometries in basement-induced wrench faulting under different initial stress states. *Journal of Structural Geology,* 7, 737–752.

Nelson, M. R., and Jones, C. H., 1987. Paleomagnetism and crustal rotations along a shear zone, Las Vegas Range, southern Nevada. *Tectonics,* 6, 13–33.

Şengör, A. M. C., 1979. The North Anatolian transform fault: its age, offset, and tectonic significance. *Geological Society of London Journal* 136, 269–282.

Sylvester, A. G., 1988. Strike-slip faults. *Geological Society of America Bulletin,* 100, 1666–1703.

Tchalenko, J. S., 1970. Similarities between shear zones of different magnitudes. *Geological Society of America Bulletin,* 81, 1625–1640.

Wilcox, R. E., Harding, T. P., and Seely, D. R., 1973. Basic wrench tectonics. *American Association of Petroleum Geologists Bulletin,* 57, 74–96.

Woodcock, N. H., and Schubert, C., 1994. Continental strike-slip tectonics. In Hancock, P. L., ed., *Continental Deformation.* Pergamon Press: Oxford, pp. 251–263.

Woodcock, N. H., and Fischer, M., 1986. Strike-Slip duplexes. *Journal of Structural Geology,* 8, 725–735.

PART E

REGIONAL PERSPECTIVES

20.1   INTRODUCTION

In Chapters 16 through 19, we introduced you to a variety of tectonic settings and we described geologic structures that are commonly associated with these environments. To put rock deformation in context, we did not limit our descriptions to structural features of these environments, but also included information concerning petrologic and sedimentologic features that accompany deformation. Our discussion focused on response to plate motion at the three basic types of plate margins, and ranged from what happens during the inception of a divergent margin (rifting) to the death of a convergent margin (collision ± strike-slip). The expression of tectonism in these various settings can be dramatic, as it results in regional belts of deformation, metamorphism, and igneous activity (Figure 20.1). These belts are known as *orogens,* and the set of processes that create them is called *orogeny.* Much of what is known about plate tectonic processes comes from the study of recent plate margins, rifts, and collision zones. Plate interactions, however, leave a permanent scar in the lithosphere, even after the associated physiography has long since eroded away. Through field study we are increasingly able to get a good understanding of ancient plate tectonics in our planet's history, as described in the subsequent chapters that contain regional perspectives.

We believe that it is best to focus on natural examples to understand the nature and consequences of plate tectonic processes, including orogeny. Thus, we devote the final chapters of this book to case studies of major deformation belts around the world. Each study is written by one or more experts, in his or her own style, so that you will get a flavor of how different geologists think and how they approach tectonics. When you talk with these seasoned geologists, chances are that you quickly share their excitement about the area in which they work. The excitement may stem from discovering key outcrops for understanding regional deformation or new insights into an aspect of fundamental significance for crustal evolution. The experts have tried to capture some of their excitement in these essays, rather than offering comprehensive review papers.

When you collect additional reading material on the areas described here or on other areas, you will rapidly learn that views vary and, sometimes, that they are contradictory. The observations, of course, stand, but alternative interpretations are often possible, especially in the case of regional tectonics. You will find that the essays which follow emphasize large-scale processes, but also that the scenarios are often based on small-scale observations. This relationship between observations on varying scales mirrors the approach we have taken throughout the book: processes on different scales are not separate and unrelated entities; rather, they form part of an integrated framework for studying the deformation of rocks and regions.

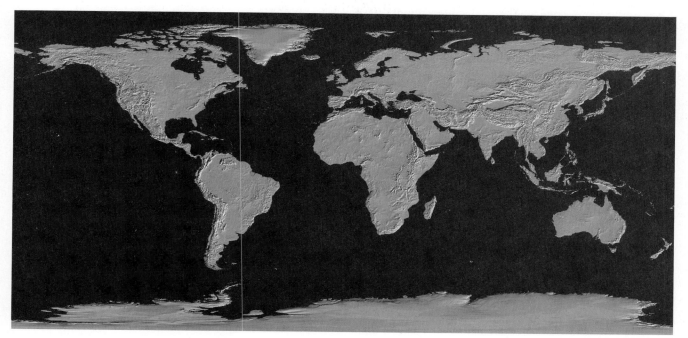

FIGURE 20.1 Shaded-relief map of the world.

20.2 GLOBAL DEFORMATION PATTERNS

The most impressive deformation features that are exposed at the Earth's surface are concentrated along relatively narrow belts at active plate margins (Figure 20.2). Today's active mountain belts, such as the European Alps, the Himalayas, the North American Cordillera, and the Andes of South America, mark convergent plate boundaries. From a human perspective the term "active" mainly means earthquakes and volcanoes, but from a geologic perspective this term implies continuing relative displacements at these margins on the timescale of millions of years. Ancient mountain belts, such as the Caledonides, the Appalachians, the Altaids, and the Tasmanides, were formed at plate boundaries and oceans that have all but disappeared; only remnants are preserved in these orogens as ophiolites, volcanic arcs, oceanic plateaus, and so on. These remnants are often able to provide us with much of the area's geologic history. Even farther back, into the Precambrian, we no longer see the mountainous physiography of orogens preserved, that is, there is little or no related topography. Yet these flat, inactive regions (called **cratons**) include deeply eroded levels of once vast mountain belts that formed from tectonic processes perhaps similar to those active today. Later in this chap-

ter we speculate on some contrasts between modern and ancient orogens, which is a topic of great interest.

Not all deformation, however, takes place at plate margins. Some of the great historical earthquakes occurred within continental interiors; for example, the 1811–1812 New Madrid earthquakes in central North America. Evidence for tectonic activity is also preserved by large intracratonic basins, such as the (mostly) Paleozoic Michigan Basin that contains as much as 5 km of sediment, by arches, and by massive rift zones, such as today's East African Rift and the Proterozoic Midcontinent Rift of North America. Thus, we conclude this set of chapters on regional perspectives by discussing the fascinating deformational features of plate interiors, focusing on the Midcontinent area of the United States.

The continental regions of Earth preserve a long history, going back as far as 4 Ga. This is in marked contrast to today's ocean basins, where the oldest rocks are on the order of 200 Ma. It is perhaps ironic that, although many of the fundamentals of plate tectonics were formulated from the study of today's ocean basins, we uncover Earth's ancient history by focusing our attention on the continents. When reading the essays you will find that we have gotten a remarkably detailed understanding of this ancient history, especially for the Phanerozoic, and increasingly for Proterozoic times also. The Archean, however, remains

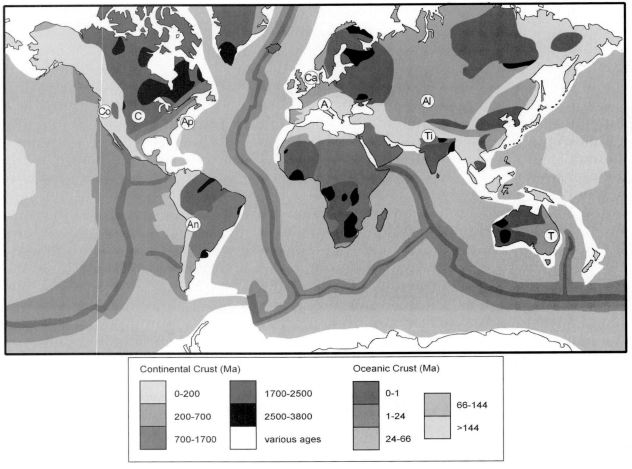

FIGURE 20.2 Areas of the world that were deformed during various geologic periods. In most continents the oldest deformed rocks (basement) are covered by younger sedimentary rocks (cover). Also shown are continental rocks on margins as in oceanic plateaus. Areas that are described in the essays are European Alps (A), Himalaya/Tibet (Ti), Altaids, North American Cordillera (Co), Andes (An), Caledonides (Ca), Appalachians (Ap), Tasmanides (T), Precambrian North America, and North American cratonic interior (c).

much less well understood, and its tectonic history is quite speculative. This temporal pattern of knowledge merely reflects the situation that, as rocks become sparser and have more complex histories with age, our ability to study ancient tectonics diminishes. Yet, the pursuit of this understanding poses new challenges to field and laboratory geologists alike.

20.3 WHAT CAN WE LEARN FROM REGIONAL PERSPECTIVES?

The next several pages contain a lot of information, as you will see. Each essay in turn condenses even more

information in only a few pages. So the answer to the question posed in the header of this section would be: *a lot!* But it makes no sense to just read all the essays and memorize the respective histories of these areas. The essays should be used as a first introduction to the continental geology of the world, but they may serve several other purposes. We identify just a few.

• The essays get you rapidly acquainted with fundamental geologic aspects of some area of the world. This probably means that you will concentrate on one or two essays as a basis for a more in-depth study of a region.

• The essays show the various approaches that may be taken in the study of (ancient) mountain belts. When you look beyond the details of individual areas, you will find that stratigraphy, geochronology, geo-

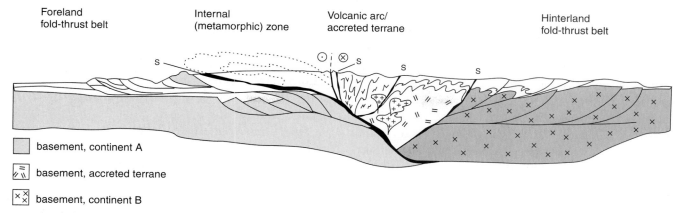

FIGURE 20.3 Idealized section through a collisional orogen, showing foreland fold-thrust belt, metamorphic hinterland (with nappes), inverted passive margin, strike-slip plate boundary, accreted volcanic arc, accreted microcontinent, and sutures (S). Most if not all mountain belts contain several of the features shown in this ideal section, but none probably contains all of them. The diagram is based on observation in Phanerozoic mountain belts; most Precambrian belts preserve only the deeper crustal levels.

chemistry, geophysics, and other earth science disciplines need to be integrated to obtain an understanding of a region's tectonic history.

• The essays allow you to recognize fundamental features that are common to most areas. We will next look at orogenic architecture as an example of one of these features.

Orogenic architecture describes the broad geometry of a mountain belt (Figure 20.3). Whereas the details of each individual mountain belt differ, they have many features in common. Generally you will find a deformed, originally wedge-shaped sedimentary sequence that was deposited at the stable continental margin. This sequence may contain marine carbonates if the area was located in the equatorial realm. Slivers of **ophiolite,** a rock assemblage containing ultramafic (mantle) rocks, gabbros, dikes, and pillow basalts, are remnants of ancient ocean floor[1] that are also preserved in an orogen. In fact, ophiolites are critical evidence for the activity of modern-day plate tectonics in ancient mountain belts. Granites, associated with volcanic arc formation or the melting of overthickened crust, are variably present. As the orogen evolves, marine clastics (sometimes called **flysch**) that are derived from the eroding mountain belt are deposited in foreland basins and at the waning stages of orogenic

activity coarse continental clastics (sometimes called **molasse**) are laid down. In young orogenic belts we find that isolated slivers of basement rocks (also called **crystalline basement**) have become exposed by faulting. In some cases, mantle rocks are similarly exposed. In ancient orogens this basement component significantly increases, and the sedimentary sequence is mainly preserved in metamorphosed and highly deformed rocks (called paragneiss). The oldest mountain belts consist nearly entirely of deformed mid-crustal to lower-crustal rocks of magmatic origin (called orthogneiss). In a way, these ancient orogens expose the roots of deeply eroded mountain belts and, in combination with modern regions, they provide a fairly complete section through orogenic crust.

Deformation is usually polyphase and each phase can contain several fold generations. Within a single orogenic phase, the deformation sequence may look something like the following: Early structures are thrusts that repeat stratigraphy, or large recumbent folds that repeat and locally invert stratigraphy (called **nappes**). These thrusts often root in a **detachment zone** (or **décollement**) at depth. In metamorphic regions this stage has produced widespread transposition. These early structures are overprinted by upright folds that may contain an axial plane foliation, and later fold generations are commonly present as kinks and crenulations. These contractional features locally overprint evidence of an initial rifting stage (normal faulting) that formed at the passive plate margin. In addition to early rifting, extensional structures often

---

[1]Geochemical evidence suggests that most ophiolites in mountain belts are obducted backarc basin oceanic lithosphere rather than main ocean basin.

form during the later stages of mountain building and during unroofing (**synorogenic** and **postorogenic extension,** respectively; not shown in Figure 20.3).

Orogens are often curved in map view, which may reflect the shape of the original plate margin, may be a result of rotation by indentation (**oroclinal bending**), or may represent differential displacements along trend. Blocks with distinct lithologies and deformation history (lithostratigraphic blocks, nowadays called **terranes**) may be incorporated in the mountain belt, reflecting the accretion of oceanic plateaus, ocean islands, or fragments of disrupted continents to the active plate margin. The boundaries of these blocks are called **sutures** and they may be marked by ophiolites, indicating that ocean floor originally separated the blocks. Other deformation characteristics, such as progressive outboard-younging of deformation, may be present and you are encouraged to search for them in the essays that follow. While you may at first be interested in the geology of only one area, a knowledge of other areas often leads to understanding your own particular region and offers alternative views; that is why we need to study the literature and that is why we offer these regional perspectives.

20.4 SOME SPECULATION ON CONTRASTING OROGENIC STYLES

Most geoscientists accept the notion that earlier in Earth's history the mantle was hotter overall than it is today, because the young Earth held more of its primordial heat and had a greater concentration of radioactive elements than it does today. For example, decay of radioactive elements produced three times as much heat at the beginning of the Archean and about 1.8 times as much heat at the beginning of the Proterozoic as it does today. Thus, **mantle convection** in the younger Earth was probably more vigorous than it is today; but whether a hotter, more vigorously convecting mantle caused young continents to be warmer than those of today remains a point of debate. If excess heat of the Earth's interior was lost, in part, by conduction through the continents, then the continents must have been warmer. However, if the additional heat of the young Earth was lost through convection involving oceanic lithosphere (because spreading rates were faster, or spreading occurred at a greater number of ridges, or there were a greater number of hot spots), then the continental crust may not have

been substantially hotter. Researchers who argue in favor of the idea that the crust was not substantially hotter in Precambrian times, point out that young continents lay above a thick lithospheric root and thus would have been insulated from the convecting asthenosphere. Uncertainty over the stability of the mantle beneath continents complicates interpretation of the thermal conditions of continents. The deep root of thickened, cooler mantle that formed beneath collisional orogens may **delaminate,** at which time hot asthenosphere flows against the base of the continent, causing an increase in heat flow into the continent. Delamination would be more likely in the Archean if mantle convection was more vigorous, which may explain the prevalence of high-temperature metamorphism in Archean terranes.

The height of a mountain range on Earth depends largely on the strength of the crust, because crust collapses and spreads laterally under its own weight if the gravitational load of the elevated region exceeds the strength of rock at depth, and exceeds the magnitude of the horizontal tectonic forces that hold up the range. Thus, a decrease in crustal strength would mean that a mountain range could not grow as high during contractional orogeny, because a weak crust would allow the range to collapse and undergo lateral spreading before it built up as high mountains. Since the strength of the crust decreases as temperature increases due to the temperature dependence of deformation mechanisms, then crust with a higher geotherm will be weaker than crust with a lower geotherm. This contrast would imply that the width and height of Precambrian orogenic belts would be less than those of Phanerozoic orogenic belts for a given amount of horizontal convergence (Figure 20.4). Thus, the cross-strike geometry of mountain ranges might have been different in the past—Archean and Early Proterozoic orogens may have contained wider belts of plastically deformed rock.

In addition to geometric observations, geologists increasingly recognize interconnections between the atmosphere, climate, erosion, and tectonics. Can changes in environmental conditions affect regional geology over time? If **atmospheric circulation** and **climate** were significantly different in the past, then ancient orogenic belts may have been different, both morphologically and structurally, from modern orogens. Earth's early atmosphere may have been more corrosive than the modern atmosphere, because of the greater concentration of volcanic gases. If so, rainfall might have caused chemical weathering at faster rates than today. If the Archean and early Proterozoic atmosphere was richer in CO_2 than the Phanerozoic atmo-

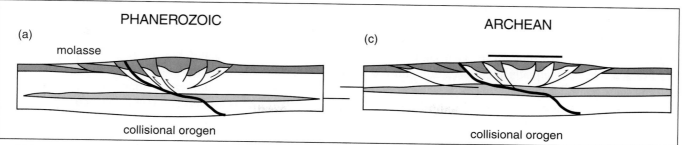

(a)

molasse

collisional orogen

(c)

collisional orogen

FIGURE 20.4 Schematic cross-sections that contrast collisional orogens of Archean/Paleoproterozoic time with those of Phanerozoic time. The shaded layer represents supracrustal rocks, the white layer represents basement, and the patterned lens represents a mid-crustal weak zone. (a) Phanerozoic collisional orogen with adjacent foreland basin. (b) Archean/Paleoproterozoic collisional orogen. The thin horizontal line above the orogen defines the comparative height of the Phanerozoic orogen.

FIGURE 20.5 View from space of the eastern Himalayas and the Tibetan Plateau.

sphere, the Earth was probably warmer most of the time, so atmospheric circulation and oceanic evaporation might have been faster, leading to greater rainfall. Because continents were smaller in Earth's early history, storms would not be calmed by movement over broad areas of land. Thus, weathering and erosion may have been faster during Earth's earlier history than they are today. So exhumation rates would be faster and isotherms in the crust would rise significantly. To replace the mass deficit resulting from erosion, rocks metamorphosed at great depth would be rapidly brought to the surface, producing wide metamorphic belts as relics of orogens. When tectonism eventually ceased, the next succession of supracrustal rocks would be deposited directly on high-grade gneiss. Further, foreland fold-thrust belts would be smaller, basement structures would be reactivated in the foreland (because uplift of isotherms would bring hot rocks to

the surface in the foreland), and deep foreland basins would not develop. Although far from proven, these speculations offer an interesting framework for exploring ancient orogens and past tectonic activity. Try to keep them in mind when reading the essays.

20.5 CLOSING REMARKS AND OUTLINE

The splendor of mountain belts has long attracted the interest of geologists and the general public both as objects of scientific investigation and for their natural beauty (Figure 20.5). You can imagine that a vast body of literature exists on regional deformation after some 150 years of regional mapping and associated laboratory work. The advent of the unifying concept of plate tectonics in the 1960s also coincides with a publication explosion in the Earth sciences (all sciences, in fact). In the Preface, we have already mentioned the enormous volume of current literature. Mercifully, each of the following essays lists only some of the more informative references and makes no attempt to offer a comprehensive reading list. To these references we only add general textbooks in this chapter, which complement the information in the essays and include many areas and topics not covered here. With all this information in hand you should not find it too difficult to explore the literature on your particular area or topic of interest.

Of course, there are many other regions of interest in the world beyond those described in the essays that follow; our choices merely represent a sampling of some reasonably well understood areas. New insights continue to be discovered everyday in these already well-studied regions, and many are waiting to be discovered in lesser known areas. As every scientist will tell you, our increasing knowledge (in our case, of deformation and tectonics) is usually accompanied by an increase in the number of unanswered questions. This ensures a continued challenge for future generations of geologists. Happy reading!

ADDITIONAL READING

Condie, K. C., 1989. *Plate tectonics and crustal evolution* (3rd edition). Pergamon Press: Oxford.

Moores, E. M., and Twiss, R. Y., 1995. *Tectonics.* W. H. Freeman and Co.: New York.

Windley, B. F., 1995. *The evolving continents* (3rd edition). J. Wiley & Sons: Chichester.

CHAPTER TWENTY-ONE

Eastern Hemisphere

21.1 THE TECTONIC EVOLUTION OF THE EUROPEAN ALPS AND FORELANDS—An essay by Stefan M. Schmid[1]

21.1.1 Introduction

The **European Alps,** located in south-central Europe, record the closure of several ocean basins in the Mediterranean region during convergence between the African and European Plates. This contribution gives a short overview of the overall architecture of the western and central Alps of Europe and their forelands (Po plain and northern foreland), based mostly on three recent geophysical-geological transects, the locations of which are given in Figure 21.1.1. The evolution of the Alpine system is discussed in time slices, starting with Cretaceous orogeny and ending with evidence for very recent movements in the area of the Rhine Graben. Some aspects of neotectonics and earthquake hazard will be addressed as well.

21.1.2 The Major Tectonic Units of the European Alps

The simplified sketch map of the Alps in Figure 21.1.1 highlights the transition from the central to the western Alps, which will be discussed along three major transects. The Insubric line marks the northern and western boundary of the southern Alps. The southern Alps are characterized by a dominantly south-verging fold-and-thrust belt whose southern tip is stratigraphically sealed by the Messinian (7 Ma) unconformity below the Po plain. At the base of this very young (Miocene) foreland prism we find the Adriatic middle and lower crust, including the Adriatic mantle, from which 10- to 15-km thick slices, consisting of basement and its Mesozoic sedimentary cover have been detached (see Figure 21.1.3c). This style of deformation points to the availability of a potential décollement (detachment) horizon within the granitic upper crust at a depth interval of 10–15 km, corresponding to temperatures in the range of 250°–375°C (assuming a gradient of 25°C/km). This depth probably corresponds to the brittle–plastic transition within granitic crust. The transition is due to the onset of crystal plasticity in quartz (at about 270°C), and/or reaction-enhanced ductility due to breakdown reactions of feldspar at about 250°C. Before this postcollisional Miocene shortening, during Paleogene plate convergence and collision, the lithosphere of the present-day southern Alps (the Adriatic Plate) formed the upper plate, under which the Penninic Valais and Piemont-Liguria Oceans, an intervening microcontinent (Briançonnais), and the European continental margin were subducted (see Figure 21.1.5a–c). Figure 21.1.1 also depicts the outlines of the Ivrea geophysical body, which represents the western edge of the Adriatic Plate. The Ivrea zone, a belt of south Alpine lower-crustal rocks, is the surface expression of the Ivrea geophysical body. Because this lower crust has been exhumed to moderate depth (corresponding to less than 300°C) during Mesozoic rifting, it represents a particularly rigid part of the south Alpine basement at the west-northwest front of the Adriatic indenter.

---

[1]Department of Earth Sciences, Basel, Switzerland.

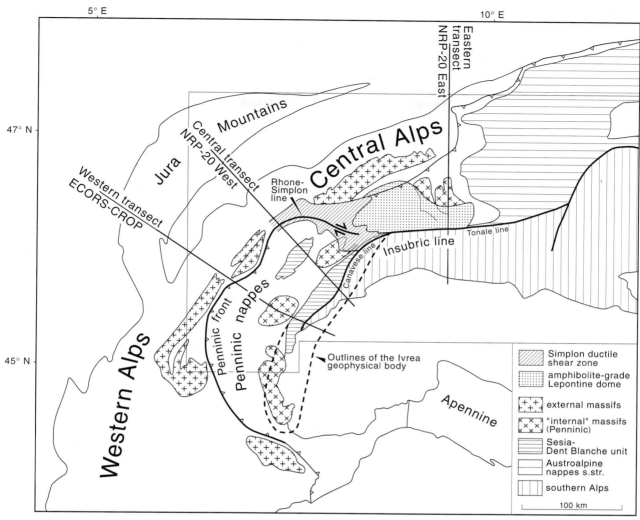

FIGURE 21.1.1 Sketch map of the Alps, indicating locations of the three geophysical-geological transects depicted in Figure 21.1.3.

Most of the roughly 100-km Oligo-Miocene dextral strike-slip along the east-west striking eastern branch of the Insubric line (the Tonale line) has been taken up by dextral strike-slip movements along the Simplon ductile shear zone and the Rhone-Simplon line. Hence, from Oligocene to probably recent times, the western Alps are kinematically part of the west-northwest moving Adriatic indenter, causing west-northwest directed thrusting along the Penninic front of the western Alps and within the Dauphinois foreland. The Rhone-Simplon line continues to act as a major discontinuity up to the present day, both in terms of seismic activity and in the character of the stress regime. During the latest stages of orogeny, this west-

northwest directed indenting by the Adriatic Plate possibly migrated further into the foreland, also affecting the western Molasse Basin and causing arcuate folding in the Jura Mountains.

The central Alps are characterized by ongoing north-south shortening that occurred during the Oligocene-Miocene, coeval with west-northwest to east-southeast shortening in the western Alps. These diverging transport directions necessitate an orogen-parallel extension, the effect of which is best documented by the Simplon normal fault and the exhumed amphibolite-grade Lepontine dome in its footwall. Oligo-Miocene exhumation of the Lepontine dome is the result of the combined effect of orogen-parallel extension, backthrusting along the Insubric line, and fast erosion.

The units north of the Insubric line consist of the Austroalpine nappes outcropping in eastern Switzerland and extending into Austria. Although paleogeographic provenance of these units is similar to that of the southern Alps, they consist of completely rootless slivers of basement and cover that have been detached (or delaminated) from their lithosphere during Cretaceous orogeny. These nappes have been stacked towards the west-northwest and their former (Cretaceous) tectonic front runs almost perpendicular to the present-day Alps in eastern Switzerland (Grisons). The Sesia-Dent Blanche unit of the western Alps underwent an Alpine tectono-metamorphic history that is different from that of the Austroalpine nappes and the southern Alps (subducted near the Cretaceous-Tertiary boundary). However, its pre-Alpine basement exhibits close similarities to that of the southern Alps.

The Penninic units are of extremely heterogeneous paleogeographic provenance, containing remnants of oceanic lithosphere, a continental fragment referred to as Briançonnais, as well as basement of the European margin. Deformation is penetrative and polyphase, and most of the Penninic units are overprinted by metamorphism, except for the Préalpes Romandes that have been detached and transported towards the northern foreland during the Eocene.

The Helvetic nappes have been detached from their former crystalline basement, which must be looked for in the lowermost Penninic nappes. The units still attached to the European lithosphere consist of the external massifs and their cover, slightly detached from the lower crust during the Miocene, when deformation started to migrate into the foreland, eventually displacing the western Molasse Basin and the Jura Mountains by up to 30 km from the Serravallian (12 Ma) onwards. The southern Rhine Graben represents an Eocene-Oligocene continental rift, kinematically linked to the Bresse Graben situated west of the Jura Mountains and ultimately to the opening of the western Mediterranean Basin (but not to the Alps). The geometry of Oligocene extensional faulting exerts a profound influence on Miocene to recent movements in the Jura Mountains and their northern margin in the southern Rhine Graben.

21.1.3 The Major Paleogeographic Units of the Alps

The major paleogeographic units of the Alps are shown in Figure 21.1.2. Many of the units are only preserved as extremely thin slivers that were detached from the subducting lithosphere and that accreted as slices (so-called nappes) to the upper plate (Austroalpine and south Alpine units). A description of the main units follows.

European margin: This consists of external massifs and their cover (extending northward underneath the Molasse Basin) and Helvetic cover nappes, whose crystalline substratum lies within the deepest part of the Lepontine dome (lowermost "Penninic" nappes). Note that the European-derived basement can be traced southward almost to the Insubric line. This demonstrates very substantial exhumation of formerly subducted and newly accreted European lithosphere during the formation of the Lepontine dome. Its later exhumation is due to a combined effect of retroflow (backfolding and backthrusting along the Insubric line) and unroofing by orogen-parallel extension and erosion during the postcollisional stages of orogeny. Some of these units also underwent Tertiary eclogitization.

Valais Ocean: The remnants of this ocean predominantly consist of Cretaceous Bündnerschiefer, grading into Tertiary flysch and at least partly deposited onto oceanic lithosphere. Eclogitic mafic rocks are preserved in the Versoyen of the western Alps, while blueschists and other low temperature-high pressure rocks are preserved in the Engadine window. The Valais Ocean opened near the Jurassic-Cretaceous boundary; its remnants presently define a northern Alpine suture zone that closed during the Late Eocene.

Briançonnais microcontinent: This microcontinent was attached to Iberia (Spain) and formed the northern passive continental margin of the Jurassic Piemont-Liguria Ocean, before it broke off the European margin in conjunction with the opening of the Valais Ocean. The Mesozoic cover of the Briançonnais microcontinent largely consists of platform sediments with frequent stratigraphic gaps ("mid-Penninic swell"). Its basement is preserved in the form of the Tambo-Suretta, Maggia, and Bernhard–M. Rosa nappes in the eastern, central, and western Penninic realm, respectively. Detached cover nappes form a substantial part of the Préalpes Romandes.

Piemont-Liguria Basin: This consists of oceanic lithosphere formed during the Middle Jurassic to Early Cretaceous and is characterized by a classic Alpine ophiolite suite. Seafloor spreading was followed by the deposition of radiolarites and aptycha limestones. During the Cretaceous, the deposition

Paleogeographic Units in the Alps

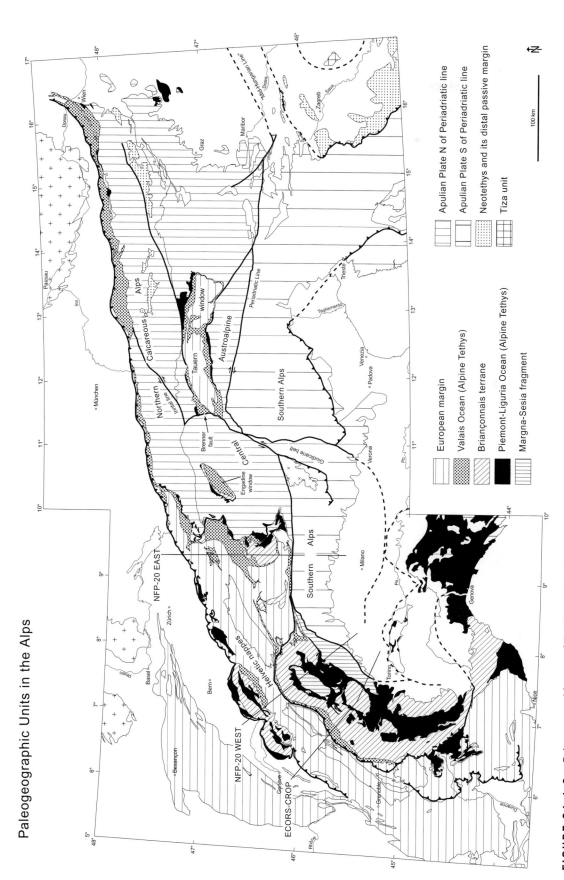

FIGURE 21.1.2 Paleogeographic map of the Alps, indicating the present-day position of the major paleogeographic units of the Alps.

of trench deposits (Avers Bündnerschiefer of Eastern Switzerland and schistes lustrées of western Switzerland) indicates that the southern (Apulian) margin of this basin had been converted into an active margin. In eastern Switzerland, the Piemont-Liguria units (Arosa and Platta unit) were involved in Cretaceous orogenic activity. However, the Piemont-Liguria Ocean did not completely close before the onset of orogeny in the Tertiary.

Margna-Sesia fragment: A small fragment of the Apulian margin, rifted off Apulia during the opening of the Piemont-Liguria Ocean and was later incorporated into the accretionary wedge along the active northern margin of Apulia.

Apulian margin: North of the Insubric line, this southern margin is only preserved in the form of rootless basement and cover slices (Austroalpine nappes). South of the Insubric line it corresponds to the southern Alps and their lithospheric substratum, the Adriatic Plate (which is part of the larger "Apulian" Plate).

Neotethys Ocean: This is a third oceanic domain, associated with the so-called Meliata-Hallstatt ocean, formed during the Triassic. In the Alps, only the distal passive margin is preserved (Hallstatt), while ophiolitic remnants (Meliata) are found in the Dinarides and the western Carpathians. This ocean is of significance mostly for the role it plays in understanding Cretaceous (or Eoalpine) orogeny.

21.1.4 Three Alpine Transects and Their Deep Structure

The major features common to all three transects, schematically sketched in Figure 21.1.3, are (1) ESE–S-directed subduction of the European lithosphere, (2) a gap between European and Adriatic Moho, and (3) the presence of wedge-shaped bodies of lower crust, largely decoupled from the piling up and refolding of thin flakes of upper-crustal material (the Alpine nappes). However, there are substantial differences in the geometry and kinematic evolution of the eastern transect (Figure 21.1.3c) as compared to the western and central transects (Figure 21.1.3a and b, respectively). Based on a study of these sections, the following observations are made:

- In the eastern (NRP-20 East) transect (Figure 21.1.3c) the Adriatic Moho descends northward and toward its contact with the lower (European) crust, while in the central and western transects (Figure 21.1.3a and b) this same Adriatic Moho rises toward the surface when approaching the contact zone with the European lithosphere. This contrast is also expressed in the surface geology. In the eastern transect, the southern Alps form an impressive south-vergent foreland fold-and-thrust belt ("retrowedge") riding above the Adriatic lower crust, while this same Adriatic lower crust is exposed in the Ivrea zone, situated at the Southeast end of the central and western transects. The Ivrea zone and Ivrea geophysical body wedge out eastward and do not extend into the area covered by the eastern transect.

- In the eastern (NRP-20 East) transect, a wedge of Adriatic lower crust is found above European lower crust and below European upper crust at its northern end. This slice of Adriatic lower crust was wedged into the European lithosphere during the Miocene, splitting apart along the boundary between the upper and lower crust. For the western (ECORS-CROP) and central (NRP-20 West) transects a geometrically different process of wedging is inferred. In these latter two cases, however, the lower crustal wedge is interpreted to be derived from the European lithosphere. Hence the wedges of lower crust seen in the eastern (Figure 21.1.3c), and the western and central transects (Figure 21.1.3a and b), respectively, are of different origin and, thus, cannot be laterally connected. This implies that the Adriatic lower crust descends below the Penninic nappe stack in the eastern profile, while it rises to the surface in the western transect, independently supports the conclusion that the two lower crustal wedges are not laterally connected.

- In the area immediately north of the Insubric line, the eastern transect (NRP-20 East) exhibits substantial backthrusting and backfolding of all the Penninic nappes, including the Valais suture zone. This was associated with exhumation of the amphibolite-grade Lepontine dome and the deep-seated Bergell Massif. In the western transect (ECORS-CROP), however, backthrusting does not affect the Valais suture zone and appears to be restricted to the units above this suture (within the Briançonnais upper crust). Note that Barrovian-type amphibolite-grade

ECORS-CROP

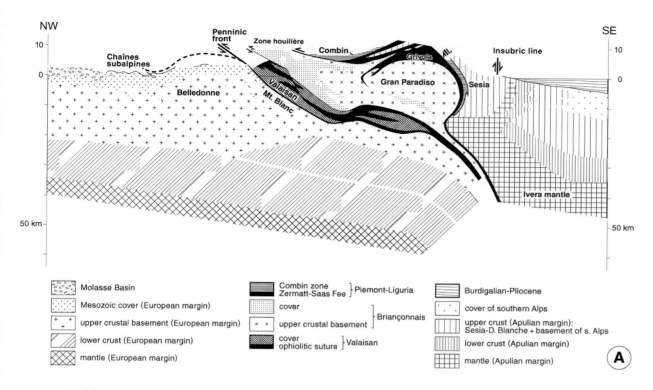

Legend for A:

- Molasse Basin
- Mesozoic cover (European margin)
- upper crustal basement (European margin)
- lower crust (European margin)
- mantle (European margin)
- Combin zone / Zermatt-Saas Fee } Piemont-Liguria
- cover } Briançonnais
- upper crustal basement } Briançonnais
- cover / ophiolitic suture } Valaisan
- Burdigalian-Pliocene
- cover of southern Alps
- upper crust (Apulian margin): Sesia-D. Blanche + basement of s. Alps
- lower crust (Apulian margin)
- mantle (Apulian margin)

(A)

NFP-20 WEST

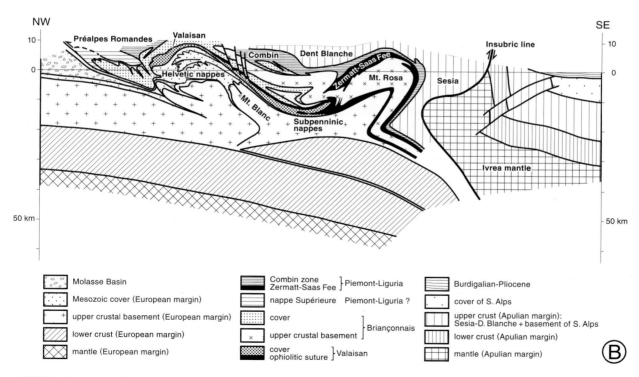

Legend for B:

- Molasse Basin
- Mesozoic cover (European margin)
- upper crustal basement (European margin)
- lower crust (European margin)
- mantle (European margin)
- Combin zone / Zermatt-Saas Fee } Piemont-Liguria
- nappe Supérieure Piemont-Liguria ?
- cover } Briançonnais
- upper crustal basement } Briançonnais
- cover / ophiolitic suture } Valaisan
- Burdigalian-Pliocene
- cover of S. Alps
- upper crust (Apulian margin): Sesia-D. Blanche + basement of S. Alps
- lower crust (Apulian margin)
- mantle (Apulian margin)

(B)

FIGURE 21.1.3   Three schematic geophysical-geological cross sections through the western and central Alps (profile traces indicated in Figure 21.1.1). Superimposed circles mark earthquake foci for the 1980–1995 time period. These foci are projected into the sections from within a 30-km traverse.

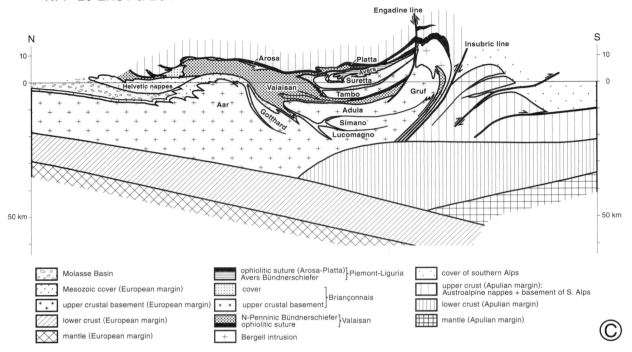

Legend:

- Molasse Basin
- Mesozoic cover (European margin)
- upper crustal basement (European margin)
- lower crust (European margin)
- mantle (European margin)
- ophiolitic suture (Arosa-Platta) Avers Bündnerschiefer } Piemont-Liguria
- cover } Briançonnais
- upper crustal basement } Briançonnais
- N-Penninic Bündnerschiefer ophiolitic suture } Valaisan
- Bergell intrusion
- cover of southern Alps
- upper crust (Apulian margin): Austroalpine nappes + basement of S. Alps
- lower crust (Apulian margin)
- mantle (Apulian margin)

FIGURE 21.1.3 (Continued)

rocks have not been exhumed to the surface in the western transect, where the Insubric line only exhibits minor vertical offset.

- The orogenic lid of the Austroalpine nappes, under which Penninic and Helvetic nappes were accreted in the eastern transect, is absent in the western and central transects.

- Accurately located earthquakes, when orthogonally projected over a maximum distance of 30-km onto the transects of Figure 21.1.3 (see Schmid and Kissling, 2000), show significant differences in the depth of seismogenic regions. The maximum depth of earthquakes is situated near the Moho in the northern and southern forelands along the eastern traverse (Figure 21.1.3c). Beneath the Penninic units (i.e., within the Lepontine metamorphic dome), they are restricted to the thickened upper crust. Also note that the Adriatic lower crustal wedge is quiescent. Coincidence of the lower limit of seismicity with predicted isotherms based on thermal modeling suggests that the 500°C isotherm controls the cataclastic–plastic transition. Quiescence within the Adriatic lower crustal wedge further suggests that, in the case of the central Alps, stress transmission between the European and Adriatic lithosphere is largely restricted to upper crustal levels. In contrast, the earthquake distribution along the western and, to a lesser degree, along the central transects exhibits a wide, east-dipping corridor of foci affecting the entire transect, including the allochthonous European lower crust (Figure 21.1.3a and b). Thus, mechanical coupling and stress transmission between the Adriatic microplate and the European lithosphere occur along a deep-reaching seismogenic zone. This indicates a contrasting (with respect to the eastern transect) thermal regime, primarily because of the following substantial differences in kinematic evolution. First, oblique convergence and collision in the western Alps before 35 Ma must have led to a significantly smaller volume of accreted radiogenic upper crustal rocks, as compared to the central Alps, the latter being characterized by head-on convergence and collision. Secondly, double-verging displacements of the central and western Alps after 35 Ma allowed for orogen-parallel extension in the central Alps (Lepontine dome), associated with updoming of the isotherms. The Penninic realms of the central and western Alps differ significantly not only in deeper crustal architecture but also in the thickness of the seismogenic zone. Some of the earthquake foci (down to a depth of about 10–15 km are known to be associated with normal faulting within the axial zone of the Alps, while strike-slip and/or thrusting modes prevail in the northern and southern forelands of the Alps. The cause of this normal faulting within the central parts of the Alps remains uncertain (gravitational collapse and/or buoyant rise of the lithospheric root). However, compression in both forelands suggests ongoing compressional coupling between the Adriatic and European Plates, although no focal solutions are available for the deep (>15 km) foci.

In summary, the three transects reveal major differences between the central Alps (Figure 21.1.3c) and western Alps (Figure 21.1.3a and b). As shown in Figure 21.1.1, the boundary between these two different segments of the Alpine chain coincides with the Rhone-Simplon line, which to this day continues to be seismically active and separates different modern stress domains.

21.1.5 Inferences Concerning Rheologic Behavior

The maximum depth of the seismogenic zone, which is assumed to coincide with the maximum depth of cataclasis (i.e., friction-controlled and dilatant deformation) is a widely disputed topic. Some partly speculative inferences can be made based on field observations and deductions from the geometry of the present-day deep structure of the Alps. Observations made by structural geologists focusing on the study of deformation microstructures indicate the onset of crystal-plasticity at different temperatures for different minerals under natural strain rates. Anhydrite (associated with the décollement horizon of Jura-type folding) may deform crystal-plastically above about 70°C; calcite exhibits significant non-cataclastic deformation above about 180°C; quartz starts to deform by crystal-plasticity above 270°C. Feldspar does not start to deform by crystal-plasticity below 450°C–500°C, but breakdown reactions in feldspar may promote reaction-enhanced ductility at lower temperatures, provided that water is available. Minerals such as hornblende and pyroxenes are more flow resistant than feldspar, and olivine does not start to deform crystal-plastically below 700°C.

At first, these data suggest a fairly shallow base for the seismogenic zone in the quartz-rich upper crust. However, elevated pore pressures are able to displace the brittle–plastic transition to greater depth (i.e., higher temperatures). On the other hand, deformation of more mafic lower crustal rocks is predicted to be controlled by cataclastic deformation at temperatures less than 450°C–500°C (i.e., down to Mohodepth for

an undisturbed geotherm within the foreland), assuming that the strength of these rocks is due to feldspar in the absence of significant amounts of quartz. Contrary to predictions based on the extrapolation of experimentally determined flow laws, therefore, lower crustal rocks may be flow resistant and deform by cataclastic processes. Hence, it is not surprising to find deep foci within lower crustal rocks, as shown in Figure 21.1.3 for the case of the Alpine forelands. However, lower crustal rocks may become weak within overthickened crustal roots of mountain belts, and/or if heat flow is elevated.

The geometry of the deep structure along the transects given in Figure 21.1.3 independently suggests that lower crustal rocks are flow resistant. This contrasts with a common assumption among geologists that the lower crust is generally "weak." Lower crustal wedging demands the lower crust to remain little deformed (or undeformed) and calls for décollement horizons at the top, as well as at the base of the lower crust. Since quartz already starts to deform by crystal-plasticity above 270°C, the upper crust may easily detach from the lower crust. Given the high strength contrast between feldspar and olivine, detachment at the base of the lower crust may also occur, provided that temperatures of around 450°–500°C (onset of crystal-plasticity in feldspar) are reached within the lowermost continental crust.

It is interesting to note that the Adriatic lower crustal wedge in the NRP-20 East profile (Figure 21.1.3c) is presently aseismic, in contrast to the European lower crustal wedges in Figure 21.1.3a–b. This points to differences in the thermal regime between the central and western Alps. In case of the NRP-20 East profile the lower limit of seismicity roughly coincides with the 500°C isotherm, as predicted by thermal modeling along this transect. This independently supports the inference that the base of the seismogenic zone in the lower crust coincides with the cataclastic–plastic transition for feldspar near the 500°C isotherm.

21.1.6 Evolution of the Alpine System and Its Forelands in Time Slices

Our brief discussion of the history of the Alpine system in central and western Europe focuses on the evolution along the eastern (NFP-20 East) transect of Figure 21.1.3c, where the timing of events is best constrained. Figure 21.1.4 gives a timetable of orogenic activity, while Figure 21.1.5 depicts cross sections along this eastern transect for different time slices.

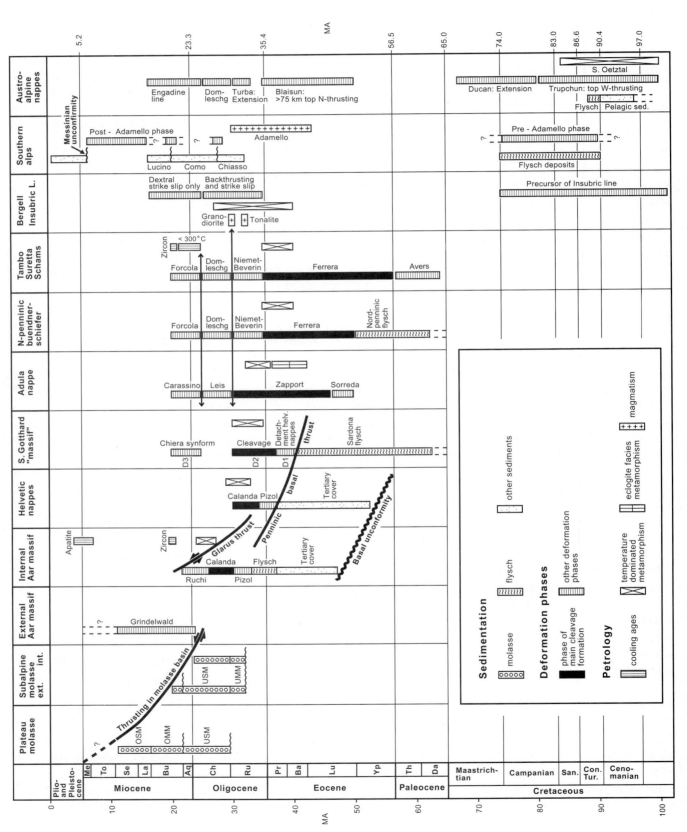

FIGURE 21.1.4 Correlation table showing an attempt to date deformation phases and metamorphism along the eastern transect.

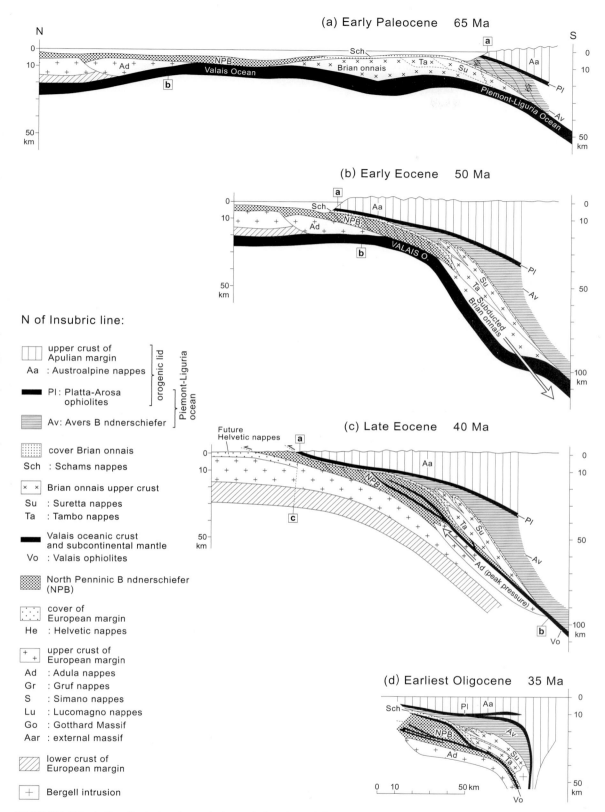

(a) Early Paleocene   65 Ma

(b) Early Eocene   50 Ma

N of Insubric line:

upper crust of Apulian margin

Aa : Austroalpine nappes

Pl : Platta-Arosa ophiolites

Av: Avers B ndnerschiefer

cover Brian onnais

Sch : Schams nappes

Brian onnais upper crust

Su : Suretta nappes
Ta : Tambo nappes

Valais oceanic crust and subcontinental mantle

Vo : Valais ophiolites

North Penninic B ndnerschiefer (NPB)

cover of European margin

He : Helvetic nappes

upper crust of European margin

Ad : Adula nappes
Gr : Gruf nappes
S : Simano nappes
Lu : Lucomagno nappes
Go : Gotthard Massif
Aar : external massif

lower crust of European margin

Bergell intrusion

(c) Late Eocene   40 Ma

(d) Earliest Oligocene   35 Ma

FIGURE 21.1.5 Scaled and area-balanced sketches of the kinematic evolution of the eastern central Alps from early Tertiary convergence (a and b), to collision (c), and postcollisional shortening (d–g).

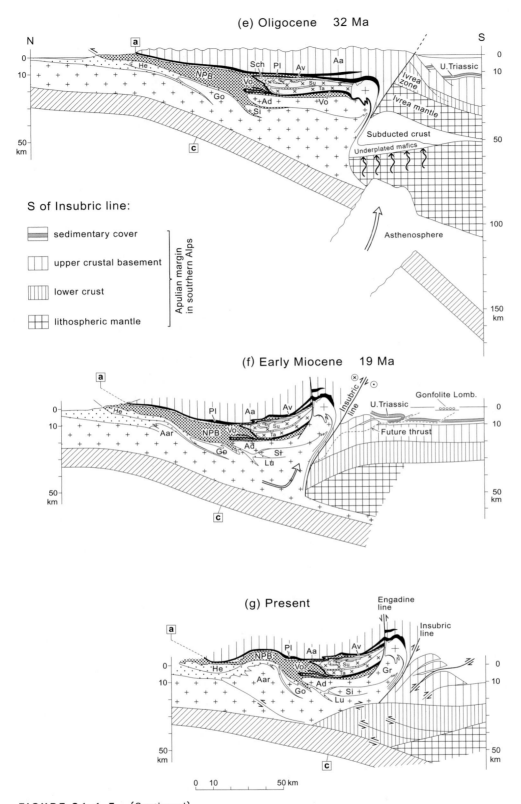

FIGURE 21.1.5 (Continued)

CRETACEOUS OROGENY Cretaceous (or Eo-Alpine) orogeny in the eastern Alps is regarded as independent and unrelated to Tertiary orogeny because of its different kinematic scenario (top to WNW, hence almost orogen-parallel thrusting) and because it is separated from Tertiary convergence by an extensional event during the Late Cretaceous ("Ducan extension" in Figure 21.1.4). Apart from the Austro-Alpine nappes, it only affects the Piemont-Liguria units of eastern Switzerland (Arosa-Platta), while the rest of the Penninic units remain largely unaffected by this orogeny, which did not propagate further to the west beyond eastern Switzerland, nor downsection into the Briançonnais units.

The attribution of a pre-Adamello phase in the southern Alps (the main deformation is of Miocene age) to Cretaceous orogenic activity is uncertain, but a precursor of the Insubric line must have been active because of the separation between the detached crustal flakes of the Austro-Alpine nappe system and the Adriatic lithosphere, which remained intact. However, the southern margin of the Piemont-Liguria margin represented an active margin, as documented by the accretionary wedge of the schistes lustrées and by the eclogitization of the Sesia unit at around the Cretaceous-Tertiary boundary.

During the various stages of Tertiary orogeny, the prestructured Austro-Alpine nappe system, together with the Arosa-Platta ophiolites, formed a rigid upper plate (referred to as "orogenic lid" in Figure 21.1.5a–d), of which the southern Alps formed part (not depicted in Figure 21.1.5a–c, but present at the southern margin of these figures).

EARLY TERTIARY CONVERGENCE AND SUBDUCTION (65–50 MA) During the Paleocene, the Briançonnais terrane enters the subduction zone, thereby closing the last remnants of the Piemont-Liguria Ocean in eastern Switzerland, the youngest sedimentary cover of which now forms an accretionary wedge consisting of the Avers Bündnerschiefer (Figure 21.1.5a). This southern ocean likely remained open for a longer period of time in the western Alps. After about 200 km of north-south convergence (13 mm/y) the distal margin of Europe (the future Adula nappe) enters the subduction zone at around 50 Ma, then closing the Valais Ocean (Figure 21.1.5b). Penetrative deformation during this time interval is largely restricted to the southernmost Penninic units, that is, the Briançonnais terrane (Tambo-, Suretta- and Schams nappes, see Figure 21.1.4) and the Avers Bündnerschiefer of the Piemont-Liguria Ocean (Figure 21.1.5b).

TERTIARY COLLISION (50–35 MA) During the middle and late Eocene (between Figure 21.1.5b and d), an additional 200-km N–S plate convergence (corresponding to 15 mm/y) was taken up by the incorporation of the Valais Ocean and the distal European margin into a growing accretionary wedge below the orogenic lid formed by the Austro-Alpine nappes. Figure 21.1.4 illustrates the migration of deformation and metamorphic events towards the northern foreland, reaching the area of the future Helvetic nappes by the end of the Eocene. Note that a total of some 400 km of north-south convergence across the central Alps involves substantial sinistral strike-slip movement across the future western Alps. Hence, the western Alps formed under a sinistrally transpressive scenario during Early Tertiary convergence and collision, with west-directed movements postdating Tertiary collision (see postcollisional stage 1).

Since the Alpine nappes in Figure 21.1.5 exclusively consist of thin slices of upper crustal basement and/or its cover, detached from their lower crustal and mantle substratum, all European (and Valaisan) lower crust (including parts of the upper crust) must have been subducted together with the mantle lithosphere (Figure 21.1.5c). Hence, north-vergent nappe stacking during this collisional stage took place within an accretionary wedge that starts to grow as more non-subductable upper crustal granitic material of the European margin enters the subduction zone. Radiogenic heat production within this granitic basement, perhaps in combination with slab break-off (depicted in Figure 21.1.5e) leads to a change in the thermal regime and to Barrovian-type (i.e., Lepontine) metamorphism.

POSTCOLLISIONAL STAGE 1 (35–20 MA) Further growth of the accretionary wedge leads to retrothrusting of part of the material entering the subduction zone above the steeply north-dipping Insubric line (Figure 21.1.5e and f). A "singularity point" develops within the lower part of the upper crust, separating the subducting part of the European crust from that part of the wedge which is backthrusted, sheared, and exhumed by erosion (this singularity point is near the bent arrow depicted in Figure 21.1.5f).

As can be seen from Figure 21.1.4, forward thrusting in the Helvetic nappes (i.e., the Glarus Thrust) is contemporaneous with retro- or backward thrusting along the Insubric line. The Alps evolve into a bivergent orogen, with a southern and northern foreland. Interestingly, the transition into bivergent thrusting coincides with increased rates of erosion due to the pop-up of the central Alps between fore- and retrothrusts, resulting in the transition from flysch-type to molasse-type sedimentation in the northern foreland.

North-south-directed plate convergence during this first postcollisional period amounts to about 60 km, slowing down to about 4.5 mm/y. In map view, this time interval coincides with the west-northwest-directed movement of the Adriatic Plate, now decoupled from the central Alps via dextral strike-slip movement along the Tonale line (about 100 km). Kinematically, the western Alps are now part of the west-northwest-moving Adriatic Plate and are separated from the central Alps along the Simplon ductile shear zone and later on by the Rhone-Simplon line (see Figure 21.1.1). Note that continental rifting in the Rhine and Bresse Grabens falls into this same time interval. However, this rifting is kinematically unrelated to shortening across the Alpine system, which remains in compression throughout.

POSTCOLLISIONAL STAGE 2 (20–7? MA) Continued crustal overthickening within the central part of the Alpine Orogen by bivergent (retro- and prowedge) thrusting eventually led to rapid propagation of the deformation front from the Insubric line towards the Po plain (southern Alps), as well as towards the northern foreland (thrusting at the base of the Aar Massif and within the southern Molasse Basin) at around 20 Ma. This is depicted in Figure 21.1.5e; the timing constraints are given in Figure 21.1.4. Regarding the southern Alps, deformation stopped at around 7 Ma (Messinian unconformity).

In the northern foreland, however, the situation is more complex. During the late Serravallian (12 Ma), deformation suddenly stepped further into the foreland, incorporating the western part of the Molasse Basin and the Jura Mountains into the orogenic wedge. While décollement along Triassic evaporites is recognized by most authors as being responsible for this forward stepping of the deformation front onto the northernmost Jura Mountains up to the southern Rhine and the Bresse Grabens, two questions remain unanswered:

• Did thin-skinned deformation stop at around 7 Ma in the Jura Mountains, that is, contemporaneous with foreland deformation in the southern Alps?
• How exactly did the arc of the Jura Mountains form? By clockwise rotation of the western part of the Molasse Basin and the northern Alps? Or by west to northwest directed indentation of the western part of the central Alps?

In regard to the first question we can argue that present-day deformation is thick-skinned, hence it is likely that Jura-folding was a short-lived event (12–7 Ma).

Regarding the second question we favor an indentation model, as there is evidence for counterclockwise rather than clockwise rotation of the Adriatic Plate during the Miocene. Assuming that relatively fast plate convergence across the Alpine system of Switzerland stopped at around 7 Ma, the 60-km plate convergence over the duration of this second postcollisional episode amounts to about 0.5 cm/y. Thus, plate convergence remained practically unchanged between 35 and 7 Ma. It is interesting to compare this figure of 5 mm/y to present-day shortening estimates across the Alpine system, as we do in the following section.

21.1.7 Recent Movements in the Upper Rhine Graben

Recent results from work in progress in the framework of EUCOR-UGENT program concerning the area of the Upper Rhine Graben in the Sundgau area west of Basel are shown in Figure 21.1.6. The Sundgauschotter have been deposited during a very short time interval from 3.2 to 2.6 Ma. Presently they outcrop within a 20-km wide corridor between the Vosges and the frontal Jura Mountains (see outlines of the base of the Sundgauschotter in Figure 21.1.6). The base of the Sundgauschotter forms an excellent reference horizon for inferring relative vertical movements during the last 3 m.y. or so (their basal part does not need to have been deposited everywhere 3.2 Ma, but they certainly must have been deposited before 2.6 Ma), provided that this basal contact is assumed to be near-horizontal at the time of deposition. The fact that these gravels were deposited in a braided river environment indicates that bases were nearly planar, with very minor slope from east to west.

The contour map of the bases of the Sundgauschotter (Figure 21.1.6) suggests substantial relative vertical uplift of the southernmost part of the depositional corridor with respect to the northernmost occurrences (in the order of 250 m). Moreover, two very pronounced en-echelon anticlines, gently folding the bases of the Sundgauschotter are inferred north of the Vendlincourt fold in the Rechésy area, these gentle folds being directly observable within Upper Jurassic limestones and Oligocene deposits below the bases of the Sundgauschotter. Note also that the bases of the Sundgauschotter are affected by at least part of the folding to be observed in the Ferrette fold.

The geometry of these folds, particularly in the area immediately east of Montbéliard, suggests thick-skinned reactivation of basement faults formed during Oligocene rifting. Hence, we propose a thick-skinned origin for the approximately north-northwest

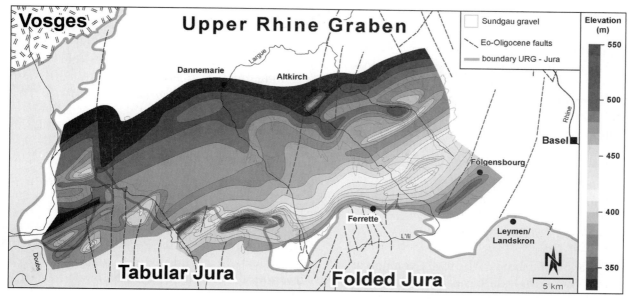

FIGURE 21.1.6 Contour map of the base of the Pliocene (3.2–2.6 Ma) Sundgau gravel deposits west of Basel. Assuming an originally planar base of these braided river deposits, the contours indicate folding with an amplitude of up to 200 m. This indicates substantial tectonic movements during the last 3 Ma, postdating thin-skinned Jura-type tectonics.

to south-southeast directed ongoing shortening, as indicated by the northernmost (post 3 Ma) folds in the Basel area affecting the Sundgauschotter, and suggest that this shortening is very probably going on at present. This pattern suggests that thin-skinned Jura folding may indeed have stopped around 7 Ma. Such a postulate is compatible with (1) the present-day stress field in the Jura Mountains, as determined by in situ stress measurements, indicating that the Jura belt is no longer an active thin-skinned fold-and-thrust belt, (2) the historical Basel earthquake, which reactivated a deep-seated basement fault, and (3) the occurrence of intracrustal earthquakes within the Molasse Basin (Figure 21.1.3c).

21.1.8 Closing Remarks

To some extent, the considerable complexities in structure and evolution of the European Alps, outlined in this essay, reflect the fact that our knowledge of their surface geology and deep structure is exceptionally detailed when compared with that of other orogens. Also, the Alps are particularly well exposed. Furthermore, the existence of along-strike culminations, such as the Lepontine dome (Figure 21.1.1), allows for along-strike projections of deep-seated structures to laterally adjacent areas. Such lateral (or axial) projections may cover a depth interval of up to 30 or 40 km (e.g., Figure 21.1.3c). Such projections have been a classical

tool of Alpine geology since the pioneering work of Argand in the beginning of the twentieth century.

Our knowledge concerning the deep structure is relatively recent and results from a series of geophysical transects carried out cooperatively by an international group of French, Italian, Swiss, German, and Austrian scientists (ECORS-CROP, NFP-20, and very recently TRANSALP in the eastern Alps). The combination of geologic and geophysical data offers the unique opportunity to unravel the crustal structure of this orogen from its former (now eroded) surface 10 km above sea level, down to the crustal roots at more than 50 km depth.

However, in spite of the "model" character the Alps have always had, there are also good reasons to believe that they are special and that many of their features should not be extrapolated to other parts of the world. Features such as, for example, the very considerable along-strike variations in the overall architecture of the orogen, the superposition of two orogenies (Cretaceous and Tertiary) in one mountain belt, and, the former existence of three oceanic domains (Valais, Piemont-Liguria, and Neotethys Oceans) may be rather special.

We will not know for certain how special the Alps are before we have data in comparable detail from other orogens. In the meantime, the Alps will continue to serve as a case study concerning the integration and interpretation of a very large set of geologic and geophysical data.

ADDITIONAL READING

Beaumont, C., Fullsack, P. H., and Hamilton, J., 1994. Styles of crustal deformation in compressional orogens caused by subduction of the underlying lithosphere. *Tectonophysics, 232,* 119–132.

Becker, A., 1999. In situ stress data from the Jura mountains—new results and interpretation. *Terra Nova,* 11, 9–15.

Burkhard, M., 1990. Ductile deformation mechanisms in micritic limestones naturally deformed at low temperatures (150–350°C). In Knipe, R. J., et al., eds., *Deformation Mechanisms, Rheology, and Tectonics, Geological Society Special Publication,* 54, 241–257.

Burkhard, M., and Sommaruga, A., 1998. Evolution of the western Swiss Molasse basin: Structural relations with the Alps and the Jura belt. In Mascle, A., et al., eds., *Cenozoic Foreland Basins of Western Europe, Geological Society Special Publication,* 134, 279–298.

Froitzheim, N., Schmid, S. M., and Frey, M., 1996. Mesozoic paleogeography and the timing of eclogite-facies metamorphism in the Alps: A working hypothesis. *Eclogae Geologicae Helveticae,* 89, 81–110.

Giamboni, M., Ustaszewski, K., Schmid, S. M., Schumacher, M., and Wetzel, A. (submitted). Plio-Pleistocene transpressional reactivation of Paleozoic and Paleogene structures in the Rhine-Bresse transform zone (Northern Switzerland and Eastern France). *International Journal of Earth Sciences.*

Meyer, B., Lacassin, R., Brulhet, J., and Mouroux, B., 1994. The Basel 1356 earthquake: which fault produced it? *Terra Nova,* 6, 54–63.

Schmid, S. M., Pfiffner, O. A., Froitzheim, N., Schönborn, G., and Kissling, E., 1996. Geophysical-geological transect and tectonic evolution of the Swiss-Italian Alps. *Tectonics,* 15, 1036–1064.

Schmid, S. M., Pfiffner, O. A., and Schreurs, G., 1997. Rifting and collision in the Penninic zone of eastern Switzerland. In Pfiffner, O. A., et al., eds., *Deep Structure of the Alps, Results from NFP 20,* Birkhauser Verlag.

Schmid, S. M., and Kissling, E., 2000. The arc of the western Alps in the light of geophysical data on deep crustal structure. *Tectonics,* 19, 62–85.

21.2 THE TIBETAN PLATEAU AND SURROUNDING REGIONS—An essay by Leigh H. Royden and B. Clark Burchfiel[1]

21.2.1 Introduction

The collision of India with Asia about 50 Ma and their subsequent convergence (Figure 21.2.1) has produced a spectacular example of active continent-continent collision. This immense region of continental deformation, which includes the Tibetan Plateau and flanking mountain ranges, contains the highest mountain peaks in the world, with many peaks rising about 8000 m (the **Himalaya**). Indeed so many mountain peaks rise above 7000 m that many of these remain unnamed and uncounted. The **Tibetan Plateau** proper stands between about 4000 and 5000 m in elevation and covers a region about 1000 by 2000 km^2, and most of it has remarkably little internal relief. It is sobering to compare the size of this region to other mountain ranges; for example, the crustal mass of the Tibetan Plateau above sea level is about 100 times greater than that of the western Alps.

The first synthesis of the tectonic evolution of southeast Asia, including the Himalaya and the Tibetan Plateau, was constructed by the Swiss geologist Emile Argand in 1924. Many of our current ideas about this region are contained in his book "La tectonique de l'Asie," which remains today one of the truly creative and imaginative works in the earth sciences. Unfortunately, from about World War I until about 1980, much of this region (in China and the former Soviet Union) was closed to foreign scientists and only since the 1980s has the area been open to international research. Thus much of our knowledge of the geology of the Tibetan plateau and surrounding regions must be considered as largely reconnaissance, particularly when compared to regions like the relatively small western Alps, where geologic research has been conducted by generations of earth scientists for more than 150 years.

In spite of our relatively sketchy knowledge of the Tibetan region, this exotic, remote, and inaccessible area has long excited the interest of scientists and nonscientists alike. For earth scientists, the rise of the Tibetan Plateau and the creation of the flanking mountain ranges and basins is a dramatic example of continental convergence and collision (Figure 21.2.1). It is exciting because of the youthful nature of the structures that accommodate deformation and the exceedingly rapid rates at which deformation is occurring. One of the great attractions of geologic study in this region is the great promise that it holds for enhancing our knowledge of continental deformation processes, as exemplified by an enormous list of still-unanswered questions (including crust–mantle interactions, driving mechanisms for local and regional deformation, and the interdependence of mountain building and global climate). In this essay we will try to summarize briefly what is known about the deformation history of the Tibetan Plateau and surrounding regions, outline some of the hypotheses and controversies that are unresolved today, and discuss some of the fundamental questions that will direct the future of geologic and geophysical studies in this region.

21.2.2 Precollisional History

At the beginning of Mesozoic time, all of the continental land masses were assembled into a giant continent called Pangea. At this time a Mesozoic ocean, called Tethys, formed a huge embayment into Pangea from the east and separated the part of Pangea that is now Eurasia (called Laurasia) from the part of Pangea that is now the continents of Africa, India, Australia,

---

[1]Massachusetts Institute of Technology, Cambridge, MA.

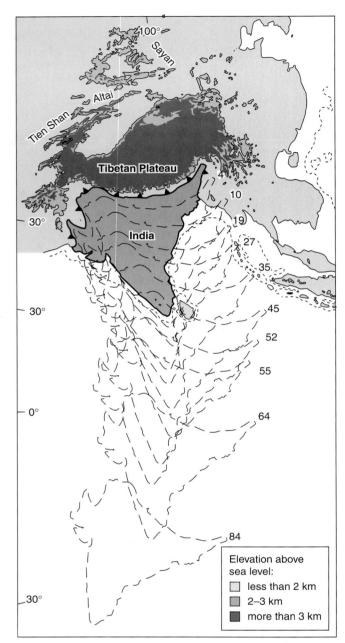

FIGURE 21.2.1   Generalized topographic map of Southeast Asia showing elevation above sealevel (from Burchfiel and Royden, 1991) and positions of India with respect to Asia from Late Cretaceous time until the present. Numbers refer to millions of years before the present. Note the decrease in convergence rate from about 100 mm/y to about 50 mm/y at about 50 Ma.

Elevation above sea level:
- less than 2 km
- 2–3 km
- more than 3 km

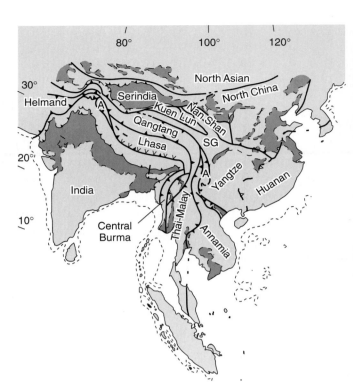

FIGURE 21.2.2   Major fragments in Southeast Asia during Late Paleozoic, Mesozoic, and Early Cenozoic time. The sutures between fragments are shown by barbed lines with the barbs indicating the upper plate of the subduction system. The suture between the Qangtang and Songpan Ganze (SG) fragments is of Mesozoic age and is thought to represent the closure of the Paleotethys Ocean. All sutures north of this are Paleozoic. The suture south of the Lhasa block is the Indus-Tsangpo suture, and represents the closure of Neotethys. A, island arc fragments, V, location of Cretaceous-Early Tertiary arc volcanism. Location of cross section in Figure 21.2.3 is also shown.

and Antarctica (called Gondwana). Tethys was perhaps a few hundred kilometers wide in the westernmost Mediterranean region, but widened toward the east, so that, at the longitude of Tibet, Tethys was approximately 6000 km wide.

At the longitude of Tibet, several sizable continental fragments collided with Asia and became accreted to the Asian continental margin before the main collision between India and Asia (Figure 21.2.2). These continental fragments were rifted from Gondwana and moved northward. As each continental fragment was rifted from Gondwana, it opened a new ocean region between itself and Gondwana. At the same time, the old ocean north of each fragment was closed by subduction. For example, the first fragment to collide with Asia closed the early Mesozoic Tethys (more properly called Paleotethys). The Indian subcontinent is the last fragment to have rifted off of Gondwana and as it moved northward, it opened the Indian Ocean to the south of India and closed the Neotethyan ocean to the north. It is the closure of this Neotethyan ocean and the collision of India with Asia that has produced the 2500-km long Alpine-Himalayan mountain chain, along which continental collision and convergence are occurring today, and many parts of it are still incom-

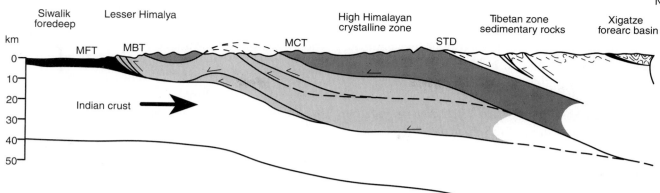

FIGURE 21.2.3 Schematic cross section through the central Himalaya (location in Figure 21.2.2). The Miocene-Quaternary molasse (heavy dots) of the modern foredeep basin are contained within the outer thrust sheets near the Main Frontal Thrust (MFT). Light shading represents low-grade metamorphic rocks north of the Main Boundary Thrust (MBT) and south of the Main Central Thrust (MCT). Dark shading represents high-grade metamorphic and crystalline rocks north of the MCT. These are separated from sedimentary rocks of the Tibetan zone (in white) by the normal South Tibetan Detachment (STD). Also shown are the Cretaceous Xigatze forearc deposits that rest on an ophiolitic basement (black sliver).

pletely closed (e.g., the Mediterranean region still contains Mesozoic seafloor of Tethys).

Prior to the main collision of India with Asia in early Tertiary time, several thousand kilometers of the Neotethys ocean floor was subducted beneath Asia along a north-dipping subduction boundary. Subduction was extremely rapid, resulting in convergence between India and Asia at about 100 mm/y. Evidence for precollisional convergence and deformation is recorded within the Himalayan mountain belt and in the southernmost part of the Tibetan Plateau. For example, within the central part of southern Tibet, Jurassic ophiolites obducted onto the Asian continental margin attest to the existence of an ancient ocean region between India and Asia. A late Cretaceous forearc sequence records subduction continuing at least into the Cretaceous (Figure 21.2.3). North-dipping subduction of an oceanic region is also recorded by a late Cretaceous to early Cenozoic volcanic arc along the southern margin of Asia (e.g., the Gandese batholith near Lhasa in south-central Tibet, which lies just north of the Xigatze forearc basin in Figure 21.2.3). Volcanism in this arc shut off in Eocene time, presumably reflecting the time of collision and the cessation of subduction of oceanic lithosphere. In the western Himalaya, precollisional subduction of Tethys must have occurred partly offshore, south of the Asian continental margin, because a Jurassic-Cretaceous volcanic arc (the Kohistan arc) was developed in a marine environment, and subsequently collided with the

southern margin of Asia in latest Cretaceous time, somewhat before the time of the main India-Eurasia collision. It is likely that events before and around the time of collision were very complicated, but at present the geologic data needed to unravel these events has not been obtained so the early evolution of the Himalayan system looks deceptively simple.

21.2.3 Postcollisional Convergent Deformation

At about 50 Ma India collided with Asia, and at about the same time the convergence rate between India and Asia slowed from about 100 mm/y to about 50 mm/y. Since that time India has continued to move northward relative to stable Eurasia at about 50 mm/y, giving a relative convergence of about 2500 km since the time of collision, and shortening and thickening the crust of Asia to produce the elevated Tibetan Plateau. Today, the continental crust beneath the high-standing plateau and flanking mountains is about 70 km, nearly double the more normal values of 30–40 km for continental crust. It is remarkable that in 1924, long before the advent of plate tectonics, Emile Argand already understood that an ocean had closed between India and Asia and that subsequent convergence between India and Eurasia has caused the intracontinental deformation of Southeast Asia. However, it is only in the last few decades that knowledge of the magnetic anomalies on the seafloor and paleomagnetism have allowed scientists to reconstruct the motions of India with respect to Asia in a quantitative manner, and to calculate the width of the subducted ocean.

The boundary where India and Asia first collided, often referred to as the Indus-Tsangpo suture, is marked by a discontinuous belt of ophiolites, mélange, and forearc sedimentary rocks of the Gandese magmatic arc. It has been impossible to reconstruct the early collisional history of this zone because postcollisional deformation involving north-directed backthrusting, strike-slip faulting, and normal faulting has completely obscured the older collisional structures. There is also very little information about postcollisional deformation within the Himalayan belt prior to about 250 Ma, although in the western Himalayan chain geologists have identified early Cenozoic metamorphosis events dated at around 35–45 Ma. This indicates that thrust faulting and crustal shortening were active in the western Himalaya at about this time, and suggests that the creation of high topography within the western Himalaya may have already been underway. However, as we shall see, the time at which high topography developed within most of the Himalaya and Tibet is extremely controversial, not least because of its implications for global climate change.

The evolution of the Himalayan Orogen is better known from about 25 Ma to the present, although there is still much that we do not know about the orogen. A schematic north-south cross section through the Himalaya and into southern Tibet (Figure 21.2.3) shows that regional shortening is thought to have been taken up on thrust faults within sedimentary rocks north of the crest of the Himalaya (called the Tibetan zone sedimentary rocks) and on three major north-dipping thrust faults south of the crest of the Himalaya, the Main Central Thrust (MCT), the Main Boundary Thrust (MBT), and the Main Frontal Thrust (MFT). All of these structures appear to be roughly continuous for more than 200 km along the entire length of the Himalayan chain. Shortening across the Himalaya was further accommodated by intense ductile strain within high-grade and crystalline rocks, as well as by folding and thrusting along less important faults.

The oldest of these thrust faults are probably within the Tibetan sedimentary sequence, but their age remains uncertain. The MCT, which may be younger, carries the high-grade crystalline and metamorphic rocks of the High Himalayan zone over the lower grade to unmetamorphosed sedimentary rocks of the Lesser Himalaya. So far, only one date has been obtained for thrusting on the MCT. This date was obtained on rocks just south of Qomolangma (Mt. Everest) in central Tibet, indicating that the MCT

was active at 21 Ma. However, we do not yet know how long before or after this time the MCT was active, although in most places it is not active today. In addition, we do not know if this data can be extrapolated to the east or west along the strike of the MCT; it is possible that although the MCT appears continuous along the length of the Himalaya, it is of different ages and different character along different parts of the belt. Metamorphic pressure and temperature data from central Tibet show that at 20–25 Ma the MCT was at a depth of about 30 km and that temperatures of High Himalayan rocks near and just above the MCT were at least 500°C–650°C. Thus the rocks of the High Himalaya must have been brought to the surface from a depth of 30 km during the last 30 m.y.

South of the MCT, the MBT carries the low-grade to unmetamorphosed rocks of the Lesser Himalaya southward over mainly Cenozoic sedimentary rocks. In some places there is evidence that the MBT may be currently active, but the main shortening and convergence across the Himalaya today is probably absorbed by motion along the structurally lower MFT and by folding within its hanging wall. The MFT marks the southern limit of deformation along most of the Himalayan chain and carries rocks and sediments of the Himalaya southward over the Ganga foredeep basin. Sediments being deposited in the Ganga Basin today are similar to those now exposed by erosion within the Himalayan foothills, suggesting that the latter were also deposited in a foredeep position in front of the frontal thrust faults, and have been subsequently incorporated into the thrust belt.

Seismic studies and examination of the rate of advance of the Ganga Basin southward over the Indian foreland indicate that about 10–25 mm/y of shortening is currently being taken up by thrust faulting and shortening within the Himalaya. Seismic studies also show that the main active thrust fault beneath the Himalaya dips gently northward by about 10°, until it becomes aseismic at a depth of about 18 km. At these depths we assume that this active, gently dipping thrust boundary is somewhat analogous to the Early Miocene MCT, but of course there are no direct observations to support this hypothesis.

Because the current rate of convergence between India and stable Eurasia averages about 50 mm/y, this means that only about one-quarter to one-half of the convergence between India and Eurasia is accommodated by shortening within the Himalaya. The rest must be taken up to the north within and around the margins of the Tibetan Plateau. The way in which the remaining convergence is absorbed remains a highly controversial topic, for which there are perhaps more models than data. However, considerable amounts of convergence are clearly absorbed by thrusting and shortening within mountain belts that lie north of the plateau, particularly in the Pamir, Tien Shan, Quilian Shan, and southern Ningxia (e.g., Madong Shan) thrust systems (Figure 21.2.4). The rates of shortening across these belts are not very well known, but they are probably at least 10 mm/y, perhaps considerably more. Although shortening across the Pamirs began in Early Cenozoic time, shortening across the other northern thrust belts probably began only in Late Miocene to Quaternary time. The onset of thrusting in these regions is primarily constrained by stratigraphic data because Cenozoic metamorphic rocks are not found in any of these young thrust belts.

The onset of uplift and shortening along the northern and northeastern margins of Tibet is probably best regarded as the northward growth of the Tibetan Plateau. We can only surmise that the early history of rocks now contained within the central part of the plateau might have been similar to those now being incorporated into the plateau by shortening along its northern boundary. It is clear that the northward growth of the plateau has not been a smooth process. The map of present-day deformation north of the plateau shows a very irregular pattern of thrust belts, with a huge undeformed region, the Tarim Basin, sandwiched between the Tibetan Plateau proper and the very active region of shortening within the Tien Shan, and overthrust from all sides. Active shortening, albeit at much lesser rates, also occurs far to the north within the Altai Ranges in north China and Mongolia. The northward growth of the plateau has not been a smooth process in time either. For example, shortening appears to have begun sometime in late Pliocene to early Quaternary time in much of the northeastern portion of the plateau. However, examination of individual ranges and thrust belts in this region does not reveal a steady northward progression of the deformation front, but rather a more haphazard onset of shortening, in part caused by the reactivation of pre-Cenozoic structures. On a scale of hundreds of kilometers, the deformation along the northern margin of Tibet appears to be controlled largely by strength heterogeneities and preexisting structural trends within the Asian lithosphere.

The structure of the thick crust beneath the Tibetal Plateau is not known, although a number of different hypotheses have been presented, each linked to a different model for plateau growth. On one hand, a number of authors, beginning with Argand, believe that the crust beneath the Tibetan Plateau has been doubled by underthrusting of the Indian crust and lithosphere beneath the Asian crust (Figure 21.2.5). In some of

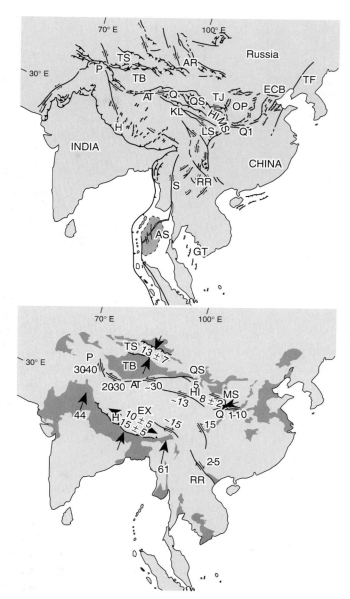

FIGURE 21.2.4 (a) Generalized Cenozoic tectonic map of Southeast Asia. Horizontal arrows indicate sense of slope on strike-slip faults; barbed lines indicate thrust faults in the continental areas and subduction zones in the oceanic areas, with barbs on the upper plates; and ticked lines indicate normal faults. Lunate shapes are folds. AR = Altai Range, AS = Andaman Sea, AT = Altyn Tagh Fault, ECB = East China Basins, GT = Gulf of Thailand, H = Himalaya, HI = Haiyuan Fault, KL = Kun Lun Fault, LS = Longmen Shan, MS = Madong Shan, OP = Ordos Plateau, P = Pamirs, Q = Qaidam Basin, QS = Qilian Shan, QI = Qinling Mountains, RR = Red River Fault, S = Sagaing fault zone, TB = Tarim Basin, TF = Lan Lu fault zone; TJ = Tianjin Shan, TS = Tien Shan, X = Xianshuhe Fault, Y = western Yunnan. (b) Slip rates (in mm/y) on active faults within Southeast Asia. Reversed arrows identify slip direction on strike-slip faults. Medium black arrows that diverge indicate areas and directions of extension; medium black arrows that converge indicate areas of shortening across thrust faults (and folds) with barbed lines. Large north-pointing arrows south of the Himalaya indicate direction and magnitude of present convergence between India and northern Asia.

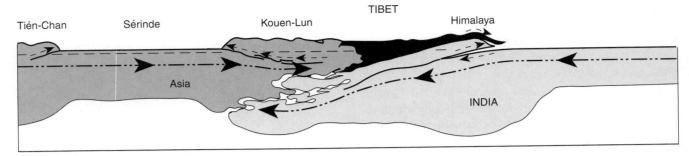

FIGURE 21.2.5 North-south cross section through Tibet and the Himalaya published by Emile Argand in 1924. Modern interpretations have not added much to our understanding of the deeper structure beneath Tibet. Light gray is Indian crust, dark gray is Asian crust, mantle material is white.

these models it is suggested that essentially no deformation of the upper (Asian) crust has occurred, while others include moderate amounts of upper crustal deformation or deformation only within the northern part of the plateau. In contrast to the first model is the suggestion that the Indian lithosphere only extends beneath the Himalaya and southernmost Tibet, and that the doubling in thickness of the Tibetan crust was mainly the result of shortening and thickening of the Tibetan crust due to the India-Eurasia collision. To date there are little conclusive data that bear directly on this problem.

21.2.4 Crustal Shortening and Strike-Slip Faulting

A heated controversy surrounds two extreme, and basically incompatible, views of how the India-Eurasia collision and crustal shortening in Tibet are related to Cenozoic deformation throughout much of Southeast Asia. One group of workers, in a ground-breaking set of papers, suggested that the postcollisional convergence of India and Asia has been responsible for most of the Cenozoic deformation of Southeast Asia. They argued that although the Tibetan Plateau did indeed form by crustal shortening, the total crustal mass beneath the plateau is too little (by about 30%) to account for all of the convergence between India and Eurasia since the time of the collision. Using earthquake seismology and newly available satellite imagery, they suggested that the remaining convergence was accommodated by eastward ejection of continental crust away from the plateau and out of the way of India as it moved northward toward Siberia (Figure 21.2.6a). Later, experiments on plasticine were used to suggest that eastward extrusion of continental fragments was also responsible for the Early Cenozoic extension within the South China Sea and the Gulf of

Thailand. This requires displacements of hundreds to a thousand kilometers on the large strike-slip faults within the Tibetan Plateau.

In a very different model, a numerical computer simulation was used by other workers to model Asia as a viscous sheet deformed by a rigid indenter (India). These numerical models predict a zone of shortening and thickening crust that grows northward and slightly to the side of the northward-moving indenter (Figure 21.2.6b). Because no deformation occurs beyond the boundaries of this zone of crustal shortening and thickening in front of the indenter, they have argued that only the Tibetan Plateau and flanking mountain belts have resulted from the collision of India and Asia. In this interpretation, the other regions of Cenozoic deformation in Southeast Asia are mainly unrelated to the India-Eurasia collision.

Both of these interpretations are based mainly on models, although the early work incorporated the geologic and geophysical data available at that time. What do current data say about this debate? The study of slip-rates on active faults shows that continental blocks within Tibet are indeed moving eastward at rates between 10 and 30 mm/y, as predicted by one set of models. However, these motions record only a snapshot of present day motions and do not answer the question of how much eastward ejection of material has occurred within Tibet. So far, geologic mapping has determined the total offset on the large strike-slip faults in only a few places. These data show about 15 km of left slip on the Haiyuan Fault, 200 km of left slip on the Altyn Tagh Fault, and 50 km of left slip on the Xianshuhe Fault (see Figure 21.2.4 for locations). Thus, while all of these strike-slip faults are moving very fast, they are also very young, and the total offset on these faults, while large, is not of the magnitude predicted by some models. While a definitive test of the extrusion model is not yet possible, the data sug-

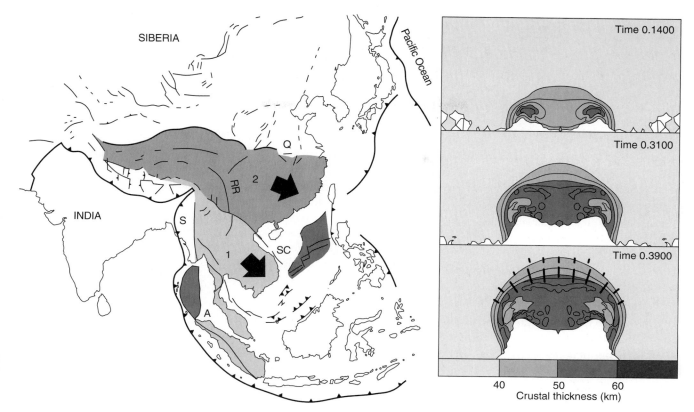

FIGURE 21.2.6 Two interpretations of the tectonic framework of Southeast Asia.
(a) The interpretation of Tapponnier and coworkers, which emphasizes the eastward extrusion of two large crustal fragments bounded by major strike-slip faults as a result of the India-Eurasia convergence. In this interpretation, eastward movement of these fragments results in the extension on the Southeast Asian continental shelf and creation of oceanic crust in the South China Sea. The first crustal fragment to move (1) is indicated by a large arrow and shown in light gray and the second to move (2) is indicated by another large arrow. RR = Red River Fault, S = Sagaing Fault, SC = South China Sea, A = Andaman Sea. (b) The interpretation of England and Houseman shows the progressive development of topography in Asia by computer modeling of a rigid indenter (India) deforming a viscous sheet (Asia). This model suggests that India-Eurasia may have little if any effect east or west of the Tibetan Plateau, and so is very different from that of Tapponnier. Contours are of increasing crustal thickness at different times (times are dimensionless, but the bottom panel would approximately correspond to the present topography of Tibet).

gest that while extrusion does occur, it is of much smaller magnitude than that predicted by this model.

Strike-slip faults within the Tibetan Plateau also have a very interesting relationship to thrust faulting and crustal shortening occurring around the margins of the plateau. At the eastern end of some of the strike-slip faults, they merge with, or end against, the active thrust faults that rim the margins of the plateau. This is best illustrated for the Haiyuan strike-slip fault in the northeastern corner of the plateau, where geologic mapping shows that 15 km of left slip on this east-west trending thrust fault is absorbed by 15 km of shortening on a north-south trending thrust belt in the Liupan Shan (Figures 21.2.4 and 21.2.7). In addition, the onset of strike-slip faulting on the Haiyuan Fault is of approximately the same age as the onset of shortening in the Madong Shan (early Quaternary). Thus the left-slip motion on the strike-slip faults is absorbed by thrusting at the plateau margins, suggesting that lateral extrusion of crust within the plateau does not extend beyond the topographically high region.

Probably one of the reasons that so much emphasis has been placed on strike-slip faulting within the Tibetan plateau and surrounding regions is that many of the studies of active faulting have used satellite photos to identify important faults. Faults that dip steeply and are straight show up very well on these photos, while gently dipping faults with a complicated outcrop pattern can be difficult to recognize. Therefore the strike-slip faults within the plateau were recognized

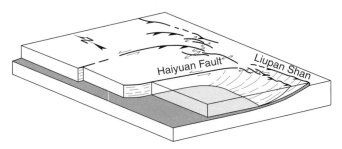

FIGURE 21.2.7 Schematic block diagram illustrating that strike-slip faults (such as the Haiyuan Fault shown here) end against thrust belts that bound the margins of the Tibetan Plateau (such as the Liupan Shan), and indicating how both thrust faults and strike-slip faults may be detached from the deeper crust by a zone of detachment within the mid-crust. See Figure 21.2.4 for locations.

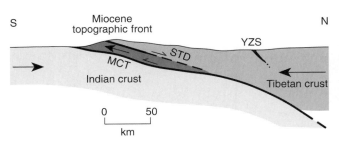

FIGURE 21.2.8 Diagrammatic cross section through the Himalayas and southern Tibet showing the southward ejection of a crustal wedge bounded below by the MCT and above by the South Tibetan Detachment (STD). Near Qomolangma (Mt. Everest) both faults were active at about 20 Ma, during the convergence of India. Faulting is thought to be due to gravitational collapse of the Miocene topographic front. The geometry shown at depth is speculative. YZS is the Indus-Tsangpo suture zone.

very quickly, while many of the important thrust faults have probably not yet been identified. Another problem is that large amounts of shortening around the margins of the plateau occur not only by faulting, but by folding, which is aseismic and not recorded by earthquakes. Thus it is easy to underestimate the rates of crustal shortening relative to the rates of strike-slip faulting from looking at first-motions from earthquakes on and around the plateau.

21.2.5 Extension of the Tibetan Plateau

Not all of the deformation of the Tibetan Plateau is compressional. A series of north-south striking grabens in southern Tibet is accommodating active east-west extension of the plateau at about 10 mm/y (Figure 21.2.4). These grabens are very young, with extension beginning at about 2 Ma, giving a total amount of east-west extension of about 20 km. The significance of this extension is unclear and is also somewhat controversial. Some workers relate the east-west extension to the eastward extrusion of material from the Tibetan Plateau and argue that the presence of east-west extension supports their interpretation. In contrast, others have suggested that east-west extension of the plateau records the delamination (falling off) of the mantle lithosphere beneath the plateau sometime in the Late Miocene or Pliocene. They argue that the removal of the dense lithosphere from beneath the plateau caused uplift of the plateau surface (as cold, dense mantle lithosphere was replaced by hot, buoyant asthenosphere), and that the surface elevation of the plateau is now decreasing by east-west extension within the plateau. At present, the significance of these north-south trending grabens remains highly uncertain

and represents a crucial point in developing new models for the deformation of the plateau. It is perhaps noteworthy that the grabens occur mainly in the western two-thirds of the plateau and that some of the grabens extend southward through the Himalaya and nearly to the Himalayan thrust front.

Within the High Himalaya there is also evidence for large-scale north-south extension that preceded the younger east-west extension and occurred on gently north-dipping normal faults (the South Tibetan Detachment zone) that parallel the MCT (Figure 21.2.3) and form a major structural break between the crystalline rocks of the High Himalaya and the rocks of the Tibetan Sedimentary Zone (Figure 21.2.3). Extension occurred simultaneously with thrusting on the Main Central Thrust zone, as indicated by dates of 20 Ma on both fault systems near Qomolangma (Mt. Everest). The normal fault zone juxtaposes rocks of mid-crustal levels in the footwall against shallow-level sedimentary rocks in the hanging wall, and has a minimum displacement of 40 km (although the total displacement could be much greater). A north-south cross section through the fault shows that it bounds the top of a mid-crustal wedge that was ejected southward at about 20 Ma (Figure 21.2.8). This wedge was bounded at its base by the Main Central Thrust zone. Southward ejection of the wedge was from an area of high topography to an area of low topography and probably reflects gravitational collapse of the steep topographic slope along the southern margin of the Tibetan Plateau. Collapse may have occurred due to weakening of the crust by melting of granites at mid-crustal depths, as suggested by the presence of ductiley deformed synkinematic granites along and just below the South Tibetan

Detachment Fault. Several other areas within Tibet contain low-angle faults with normal displacement, but at present they are only recognized on short fault segments and their regional significance remains uncertain.

The occurrence of north-south extension of large magnitude along the southern margin of the plateau is somewhat surprising, since extension was not only contemporaneous with north-south shortening along the MCT but also occurred during continued north-south convergence of India and Eurasia at about 50 mm/y. This indicates that the extensional processes that controlled deformation within the mid- to upper crust in this region were essentially decoupled from the convergent motions occurring deeper in the crust and within the mantle lithosphere. It is probable that the mid- to upper crustal deformation under many, if not all, parts of the plateau is decoupled from the lower crust and mantle. For example, the general style of deformation within the thrust belts that rim the northern and northeastern margins of the plateau, and in some cases balanced cross sections constructed across the thrust belts, show that the thrust faults sole out at depths of about 15 km. Moreover, the relationship of the large strike-slip faults (such as the Haiyuan Fault) within the plateau to adjacent thrust-fault systems also indicates that many of the strike-slip faults are probably restricted to upper and mid-crustal depths (Figure 21.2.7). Thus the motions of crustal fragments within the plateau and along the margins of the plateau may not reflect the motion or deformation of the underlying lower crust and uppermost mantle.

21.2.6 Closing Remarks

Many fundamental questions remain to be answered within the Himalayan-Tibetan region. Some of these appear straightforward, but addressing them will require many years of intensive field work. For instance, how is active crustal deformation partitioned within the plateau? How can we unravel the active and young deformation of the plateau to learn about the pre-Pliocene history of the plateau and its growth through time? Finding the answers to these and other questions will require the use of sophisticated geophysical techniques in combination with field geologic data; for example, how is the deformation partitioned vertically within the crust of the Tibetan plateau? How is crustal deformation related to motions within the underlying mantle? On what length scale is crustal deformation related to motions within the mantle? On the scale of the plate boundary system as a whole (a region 1000 by 2000 km^2) it is likely that the averaged motions of the

upper crust and the mantle lithosphere are reasonably similar. Is this also true at smaller scales, such as over regions a few hundreds of kilometers in length and width? How do we go about determining the degree of coupling between the crust and the mantle?

Lastly, one of the most controversial issues surrounding the geologic evolution of the Tibetan Plateau is the history of uplift of the plateau surface to its present elevation of about 5000 m. This is important not only to geologists, because it bears on the mechanisms by which the plateau has been deformed, but also to a wide range of scientists in fields from marine geology to climate modeling. Because the Tibetan Plateau is so large and stands so far above sea level, its uplift is thought to have influenced global circulation patterns within the atmosphere and to have been responsible for the onset of the Indian monsoons, which did not exist prior to Late Miocene time. Uplift of the plateau can also be correlated with changes in faunal distributions within the oceans and with huge changes in the isotopic ratios of elements such as strontium within the worldwide oceans. Indeed, the record of plateau uplift and erosion, as preserved within the composition of marine sediments found worldwide, as well as the geologic record from the plateau itself, suggests that collisional events of the magnitude found in Tibet and the Himalaya are probably rare events, even on a geologic timescale, and that collisional events of comparable magnitude may not have occurred within the last 600 my.

ADDITIONAL READING

Argand, E., 1977. *Tectonics of Asia* (translation by A. V. Carozzi). Hafner Press: New York.

Burchfiel, B. C., and Royden, L. H., 1991. Tectonics of Asia 50 years after the death of Emile Argane. *Eclogae Geologicae Helvetica*, 84, 599–629.

England, P. C., and Houseman, G. A., 1988. The mechanics of the Tibetan Plateau. In Shackleton, R. M., Dewey, J. F., and Windley, B. F., eds., Tectonic evolution of the Himalayas and Tibet. *Philosophical Transactions of the Royal Society of London* (A), 327, 379–420.

Lyon-Caen, H., and Molnar, P., 1985. Gravity anomalies, flexure of the Indian plate, and the structure, support, and evolution of the Himalaya and Ganga Basin. *Tectonics*, 4, 513–538.

Molnar, P., and Tapponnier, P., 1975. Cenozoic tectonics of Asia; effects of a continental collision. *Science*, 189, 419–426.

Molnar, P., 1988. A review of geophysical constraints on the deep structure of the Tibetan Plateau, the Himalaya, and the Karakorum, and their tectonic implications. In Shackleton, R. M., Dewey, J. F., and Windley, B. F., eds., *Tectonic Evolution of the Himalayas and Tibet. Philosophical Transactions of the Royal Society of London* (A), 326, 33–88.

Tapponnier, P., Peltzer, G., Le Bain, A. Y., Armijo, R., and Cobbold, P., 1982. Propagating extrusion tectonics in Asia: new insight from simple experiments with plasticine. *Geology,* 10, 611–616.

21.3 TECTONICS OF THE ALTAIDS: AN EXAMPLE OF A TURKIC-TYPE OROGEN—An essay by A. M. Cêlal Şengör[1] and Boris A. Natal'in[2]

21.3.1 Introduction

What would happen, if one were to double the entire North American Cordillera back onto itself? Or if one were to collide it with the eastern margin of Asia? Or if the North American Cordillera collided with an orogenic belt similar to itself? In all three cases, one would get a wide orogenic belt, with an extremely complicated interior consisting dominantly of rock types and structures similar to those encountered today along the convergent continental margins of North America and Asia and also along the continental strike-slip margins such as those in California, British Columbia, and the Komandorsky (Bering) Islands. Such an interior would thus be dominated by turbidites deposited in abyssal basins, trenches, and marginal basins; less abundant pelagic deposits (ribbon cherts, pelagic mudstones, some pelagic limestones); local, unconformable shallow water deposits, possibly including coral reefs, lagoonal deposits, local coarse clastic sedimentary rocks; and much volcanogenic material. All of these would appear multiply and highly deformed, tectonically intercalated with diverse rock types making up the oceanic crust and upper mantle (pillow basalts, massive diabases, gabbros, cumulate and/or massive tectonized ultramafic rocks forming the members of the ophiolite suite), and intruded by calc-alkalic plutons ranging from gabbros to granodiorites. Why would this be so? Much of the ensemble described above would be covered by the volcanic equivalents of the calc-alkalic plutonic rocks, from tholeiitic basalts to andesites, and even to local rhyolites as well as wide blankets of welded tuffs. The structural picture of the ensemble would be bewilderingly complex and would include many episodes of folding, thrusting with inconsistent (but mainly oceanward) vergence, orogen-parallel to subparallel strike-slip faulting, and even some normal faulting, alternating in time and space. Both the rock types and the structures within the internal part of such an orogen would display longitudinal discontinuity over tens to hundreds of kilometers. By contrast, the outer margins of the orogen would be marked by regular, laterally persistent (on scales of hundreds to thousands of kilometers) fold-and-thrust belts made up dominantly of shelf and epicontinental strata with dominant vergence away from the orogen.

How does such an orogen differ from other collisional orogens such as the Alps and the Himalaya? Simply in the collective width of the scrunched-up oceanic remnants between the colliding continental jaws, the amount of subduction-related magmatic rocks, and the magnitude and frequency of incidence of orogen parallel to subparallel strike-slip motion during orogeny. Both in the Alps and in the Himalaya, the width of oceanic offscrapings never exceeds a few kilometers at most; generally they are much narrower (a few hundreds of meters), in places reducing to nothing, where the original opposing continental jaws come into contact. Associated magmatic arcs usually display a single magmatic axis whose wandering across the strike in time hardly exceeds a few tens of kilometers. In some small collisional orogens, such as the Alps, the magmatic arcs are so very poorly developed as to invite suspicion of whether they ever existed.

The **Altaids** were named by the Austrian geologist Eduard Suess, after the Altay Mountains in Central Asia, shared by Russia, Mongolia, and China. The large, mainly Paleozoic orogenic complex dominates much of central Asia and extends to the Arctic in the west and to the Pacific in the east. Its tectonic units are homologous to those occupying across-strike widths of a few hundred meters in the Himalaya and

[1]Istanbul Technical University, Istanbul, Turkey.
[2]Russian Academy of Sciences, Khabarovsk, Russia.

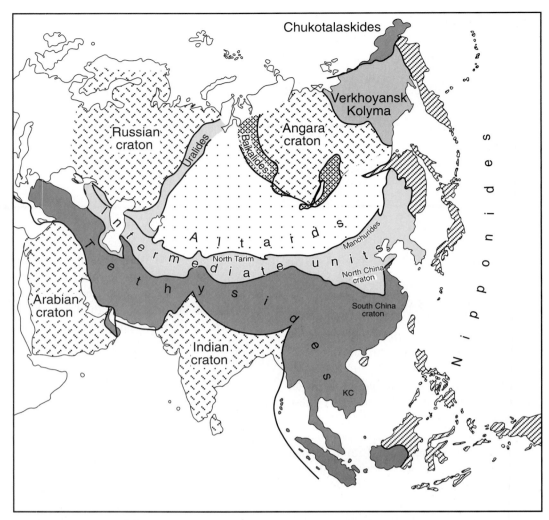

FIGURE 21.3.1 (a) Major tectonic subdivisions of Asia, showing the position of the Altaids within the structure of the continent. (b) Tectonic map of the Altaids and surrounding large-scale tectonic entities, showing their first-order tectonic units. The Baykalide and the pre-Uralide orogenic systems are not further subdivided. Numbers on the Altaid units correspond with the numbers cited for those units in the text.

the Alps but occupy widths of 1500 to 2000 kilometers (Figure 21.3.1). In the North American Cordillera, the edge of the Precambrian crystalline basement is taken to be roughly along the $Sr_i = 0.704$ line. This line delimits a band, about a 500 km wide and parallel to the coast, north of central California. If one juxtaposes the North American Cordillera against its mirror image, one obtains a band of "off-scraped oceanic material" about 1000 kilometers wide. This is comparable to the width of the internal parts of the Altaids, when one considers that the Altaid evolution lasted some 50 to 100 my longer than the Cordilleran evolution. Similarly, both in the Altaids and in the North American Cordillera, one finds that magmatic axes wandered over distances on

the order of 1000 km. In fact, if we bend the latter collisional orogen approximately 90° about a vertical axis, the picture we obtain is remarkably similar to that of the Altaids.

The term Turkic-type has been applied to collisional orogens resulting from the collision of continents that began with very wide subduction-accretion material, long-lasting subduction and rich sedimentary material fed into trenches. These orogens are named after the dominant ethnic group that has populated, in much of known history, the area of development of their best example, namely the Altaids. The purpose of this chapter is to present a synopsis of both the present structure and the history of evolution of the Altaids, which are an example of a Turkic-type development.

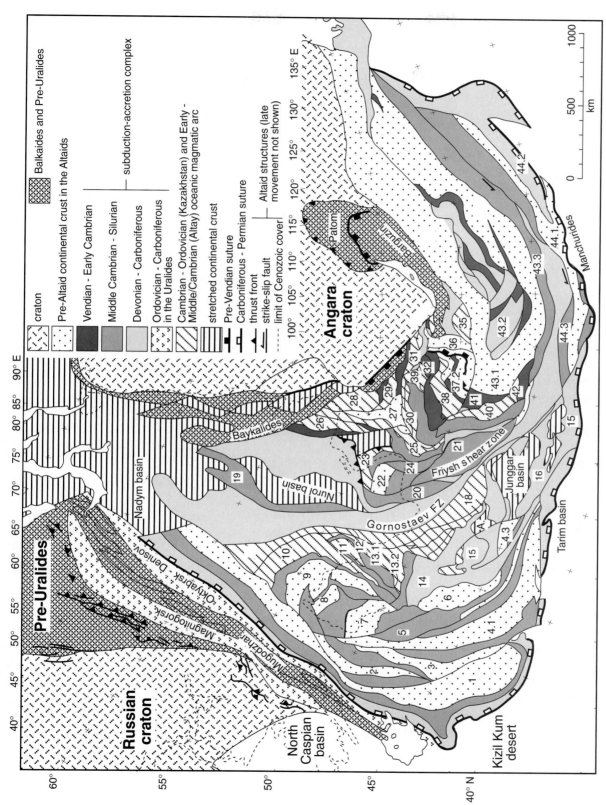

FIGURE 21.3.1 [Continued]

21.3.2 The Present Structure of the Altaids

In Figure 21.3.1a we see the Altaids within the context of the structure of Asia, while Figure 21.3.1b is a tectonic map of the Altaids. Only four main types of genetic entities are displayed in Figure 21.3.1. They are (1) units made up of continental crust that had already formed before the Altaid evolution commenced, (2) subduction-accretion complexes formed during the Altaid evolution, (3) ensimatic magmatic arc massifs formed during the Altaid evolution, and (4) continental crust (of any age) stretched and thinned as a consequence of Altaid evolution. These genetic entities are grouped into 44 main tectonic units defined on the basis of their function during the Altaid evolution, and include units such as arcs, cratons, passive continental margins, and the like. In the following list we briefly characterize their function; more detailed characterization and a summary of the rock content can be found in the readings at the end of this essay. The numbers of the units in the list correspond with the numbers shown in Figure 21.3.1.

1. *Valerianov-Chatkal unit:* Pre-Altaid continental crust, Paleozoic accretionary complex and magmatic arc.
2. *Turgay unit:* Pre-Altaid continental crust, Altaid accretionary complex and magmatic arc; all buried under later sedimentary cover.
3. *Baykonur-Talas unit:* Pre-Altaid continental crust, Early Paleozoic magmatic and arc accretionary complex.
4.1. *Djezkazgan-Kirgiz unit:* Pre-Altaid continental crust, Paleozoic accretionary complex and magmatic arc.
4.2. *Jalair-Nayman unit:* Pre-Altaid continental crust, Early Paleozoic marginal basin remnants, Early Paleozoic magmatic arc and accretionary complex.
4.3. *Borotala unit:* Pre-Altaid continental crust, Early Paleozoic magmatic arc and accretionary complex.
5. Sarysu unit: Paleozoic accretionary complex and magmatic arc.
6. *Atasu-Mointy unit:* Pre-Altaid continental crust, Early Paleozoic (including the Silurian) magmatic arc and accretionary complex.
7. *Tengiz unit:* Pre-Altaid continental crust, Vendian-Early Paleozoic magmatic arc and accretionary complex.
8. *Kalmyk Köl-Kökchetav unit:* Pre-Altaid continental crust, Vendian-Early Paleozoic magmatic arc and accretionary complex.
9. *Ishim-Stepnyak unit:* Pre-Altaid continental crust, Vendian-Early Paleozoic magmatic arc and accretionary complex.
10. *Ishkeolmes unit:* Early Paleozoic ensimatic magmatic arc and accretionary complex.
11. *Selety unit:* Early Paleozoic ensimatic magmatic arc and accretionary complex, with questionable pre-Altaid continental basement in fault contact.
12. *Akdym unit:* Vendian(?) and Early Paleozoic ensimatic magmatic arc and accretionary complex.
13. *Boshchekul-Tarbagatay unit:* Early Paleozoic (including the Silurian) ensimatic magmatic arc and accretionary complex.
14. *Tekturmas unit:* Ordovician-Middle Paleozoic accretionary complex, Middle Devonian-Early Carboniferous magmatic arc.
15. and 16. *Junggar-Balkhash unit:* Early (ensimatic) through Late (ensialic) Paleozoic magmatic arc, Middle through Late Paleozoic accretionary complex.
17. *Tar-Muromtsev unit:* Early Paleozoic ensimatic magmatic arc and accretionary complex.
18. *Zharma-Saur unit:* Early to Late Paleozoic ensimatic magmatic arc, Early Palaeozoic accretionary complex.
19. *Ob-Zaisan-Surgut unit:* Late Devonian-Early Carboniferous accretionary complex, strike-slip fault-bounded fragments of the Late Devonian-Early Carboniferous arc, Late Paleozoic magmatic arc.
20. *Kolyvan-Rudny Altay unit:* Early and Middle-Late Paleozoic magmatic arc.
21. *Gorny Altay unit:* Early Paleozoic magmatic arc and accretionary wedge, superimposed by Middle Palaeozoic magmatic arc; farther west in the "South Altay," Middle Paleozoic accretionary complex with forearc basin.
22. *Charysh-Chuya-Barnaul unit:* Pre-Altaid continental crust, Early Paleozoic magmatic arc and accretionary complex, and Middle Paleozoic magmatic arc and forearc basin.
23. *Salair-Kuzbas unit:* Pre-Altaid continental crust, Vendian-Early Paleozoic magmatic arc and accretionary complex, Ordovician-Silurian forearc basin, Devonian pull-apart basin, Late Palaeozoic foredeep.
24. *Anuy-Chuya unit:* Early Paleozoic magmatic arc and accretionary complex.

25. *Eastern Altay unit:* Pre-Altaid continental crust, Early Paleozoic magmatic arc and accretionary complex.

26. *Kozykhov unit:* Early Paleozoic magmatic arc and accretionary complex.

27. *Kuznetskii Alatau unit:* Pre-Altaid continental crust, Early Paleozoic magmatic arc and accretionary complex.

28. *Belyk unit:* Vendian-Middle Cambrian magmatic arc and accretionary complex.

29. *Kizir-Kazyr unit:* Vendian-Middle Cambrian magmatic arc and accretionary complex.

30. *North Sayan unit:* Vendian-Early Paleozoic magmatic arc and accretionary complex.

31. *Utkhum-Ota unit:* Pre-Altaid continental crust, Early Paleozoic magmatic arc and accretionary complex.

32. *Ulugoi unit:* Vendian-Early Cambrian magmatic arc and accretionary complex.

33. *Gargan unit:* Pre-Altaid continental crust, Early Paleozoic magmatic arc, Vendian-Early Paleozoic accretionary complex.

34. *Kitoy unit:* Early Paleozoic magmatic arc.

35. *Dzida unit:* Early Paleozoic magmatic arc and accretionary complex.

36. *Darkhat unit:* Pre-Baikalide continental crust, Riphean magmatic arc and accretionary complex.

37. *Sangilen unit:* Baikalide microcontinent that collided in the Riphean with the Darkhat unit (unit 36) and the Tuva-Mongol Massif (see unit 43.1) and experienced dextral strike-slip displacement during the early Altaid evolution.

38. *Eastern Tannuola unit:* Early Paleozoic magmatic arc and accretionary complex.

39. *Western Sayan unit:* Early Paleozoic magmatic arc and accretionary complex.

40. *Kobdin unit:* Early and Middle Paleozoic magmatic arc and accretionary complex.

41. *Ozernaya unit:* Vendian-Early Cambrian magmatic arc and accretionary complex.

42. *Han-Taishir unit:* Pre-Altaid continental crust, Vendian-Early Cambrian magmatic arc and accretionary complex.

43. *Tuva-Mongol unit:*
   43.1. *Tuva-Mongol arc massif:* Pre-Altaid continental crust and Vendian-Permian magmatic arc.
   43.2. *Khangay-Khantey unit:* Vendian-Triassic accretionary complex.
   43.3. *South Mongolian unit:* Ordovician to Early Carboniferous accretionary complex.

44. *South Gobi unit:* Pre-Altaid continental crust, Paleozoic magmatic arc, Early and Late Paleozoic accretionary complex.

21.3.3 Evolution of the Altaids

METHOD OF RECONSTRUCTION The problem with the Altaids has long been how to establish the trend-line of the orogen amidst the abundance of the variably orientated tectonic units listed in the preceding section. Since the late 1800s, the trend-lines of orogenic belts have been depicted as the direction, in any given cross section, of the average strike, particularly in the laterally persistent external fold-and-thrust belts. That approach worked well for narrow, linear, and/or arcuate orogens such as the Alpine system and even the Cordillera, but the strange map shape of the Altaids makes such an approach suspect. The median line of an orogen along its internal parts, consisting of median masses (Zwischengebirge) and the so-called scar-lines (Narbe), has also been used to follow the orogenic trend-line, but if one tried that method on the Altaids, the trend-line obtained from the externides and the trend of the median line would give two different, contradictory results. Try to confirm this for yourselves by using Figure 21.3.1b. So, that approach will not be satisfactory either.

The difficulty of identifying the trend-line of the orogenic edifice in the Altaids has led to proposals that they might consist of more than one independent orogenic belt. However, the great similarity of the rock material involved in their architecture, the significant uniformity of style of their internal structure, the broad correspondences between their disparate sectors in timing of tectonic evolution, and the difficulty of finding "junctures" where one orogen would join another one, make it unlikely that there are a number of independent orogenic belts tucked away within the Altaid realm.

It has been proposed that *magmatic fronts of arcs* constitute convenient markers to follow the orogenic trend owing to their easy identification, lateral persistence, and indication of facing; they are very sharp on the ocean side (i.e., the side toward which they are said to "face"), but more diffuse on the backarc side. This idea was applied to the entire Altaids to trace the first-order trend-lines of the orogen, and it was possible to show that the chaotic internal structure could be interpreted in terms of the deformation of formerly simpler arc geometries.

In Figure 21.3.2 we see how this idea is applied to the western and central sector of the Altaids (the

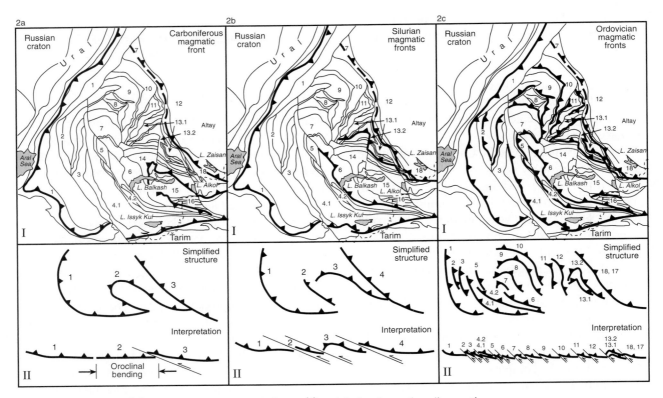

FIGURE 21.3.2   (a) Carboniferous magmatic fronts (I) and their schematic palinspastic interpretation (II). Magmatic fronts are shown with thrust symbols, and face in the opposite direction from the apices of the triangles. The numbers in I correspond to the unit identifications in Figure 21.3.1b and in the reconstructions displayed in Figure 21.3.4. In II, the numbers 1, 2, and 3 correspond to the Valerianov/Tien Shan, the central Kazakhstan, and the Zharma-Saur fronts, respectively. In II, the figure on top is a highly schematized version of the present geometry of the Carboniferous magmatic fronts, as shown in I; the bottom figure is an interpretation of the top configuration in terms of a former single arc. (b) Silurian magmatic fronts (I) and their schematic palinspastic interpretation (II). The numbers in I correspond to the unit identifications in Figure 21.3.1b and in the reconstructions displayed in Figure 21.3.4. In II, the numbers 1, 2, 3, and 4 correspond to the Valerianov/Tien Shan, the Atasu-Mointy, the central Kazakhstan, and the Zharma-Saur fronts, respectively. In II, the figure on top is a highly schematized version of the present geometry of the Silurian magmatic fronts, as shown in I; the bottom figure is an interpretation of the top configuration in terms of a former single arc. (c) Ordovician magmatic fronts (I) and their schematic palinspastic interpretation (II). The numbers in I correspond to the unit identifications in Figure 21.3.1 and in the reconstructions displayed in Figure 21.3.4. In II, the magmatic fronts can be grouped into three domains: fronts 1–6, the Tien Shan–southwest Kazakhstan domain; fronts 7–13, north-central Kazakhstan; and fronts 17 and 18, the Zharma-Saur domain. Here the enumeration of the magmatic fronts corresponds with the unit numbers in Figure 21.3.1b. In II, the figure on top is a highly schematized version of the present geometry of the Ordovician magmatic fronts, as shown in I. The bottom figure is an interpretation of the top configuration in terms of a former single arc.

Kazakhstan-Tien Shan sector). In Figure 21.3.2a, the locations of the Carboniferous magmatic fronts are shown (I). Note that, with the exception of the central Kazakhstan front, they exist only on the outer periphery of the ensemble of tectonic units and pass from one unit to another. This indicates that, by the Carboniferous, the units making up the western and central sector of the Altaids had already come together (because the magmatic front "stitches" the outer units together). Panel II shows how this picture may be interpreted in terms of a single arc. At the top of Panel II a simplified trend-line pattern is shown, which corresponds with the actual geometry in Panel I. At the bottom of Panel II, this is interpreted in terms of strike-slip disruption of a formerly continuous front. Figure 21.3.2b shows the magmatic fronts of the Silurian. In general they are not dissimilar to those in the Carboniferous, but they show a marked difference in detail. The fronts have migrated inwards along the outer periphery, whereas in central Kazakhstan they moved outward, and the front along units 5 and 6 has changed its polarity. Note how these changes affect the interpretation. Figure 21.3.2c I exhibits a much greater complication and we no longer see the neat picture of the earlier frames. As seen in Panel II, the interpretation is correspondingly more complicated. (Can you tell wherein lies the complication?)

The magmatic fronts in the Altaids may thus be interpreted in terms of a single arc. But was it really so? In other words, does this agree with the rest of their geology? In order to check that we must find tie-points (i.e., points that are now far away from one another, but that used to be adjacent to one another) on adjacent units that would allow us to bring them back to their prefaulting positions. This is not a straightforward exercise, because many geologic features can be used as tie-points. This analysis uses the arc massif–accretionary complex contacts as one set of tie-points. Faulted and displaced forearc basin parts, segments of backarc basins, and metamorphic complexes are among other features that have been used as features providing tie-point sets for reconstructions in the Altaids.

In Figure 21.3.3 we see the disassembled units in the Kazakhstan-Tien Shan sector of the Altaids, with an emphasis on the arc massif–accretionary complex contacts. The reconstructed single arc at the bottom of the figure is generated by bringing into juxtaposition the tie-points on adjacent units. Thus, much of the Altaid edifice could be interpreted as the deformed remnants of a single magmatic arc, called the Kipchak arc (after the dominant ethnic group inhabiting the area where its fragments are now found), now disrupted and squeezed between the two cratons of Russia and Angara. Altaids extending eastwards into Mongolia evolved from a second arc (Tuva-Mongol) as shown in the reconstructions in Figure 21.3.4. (Can you find this second arc in Figure 21.3.1?)

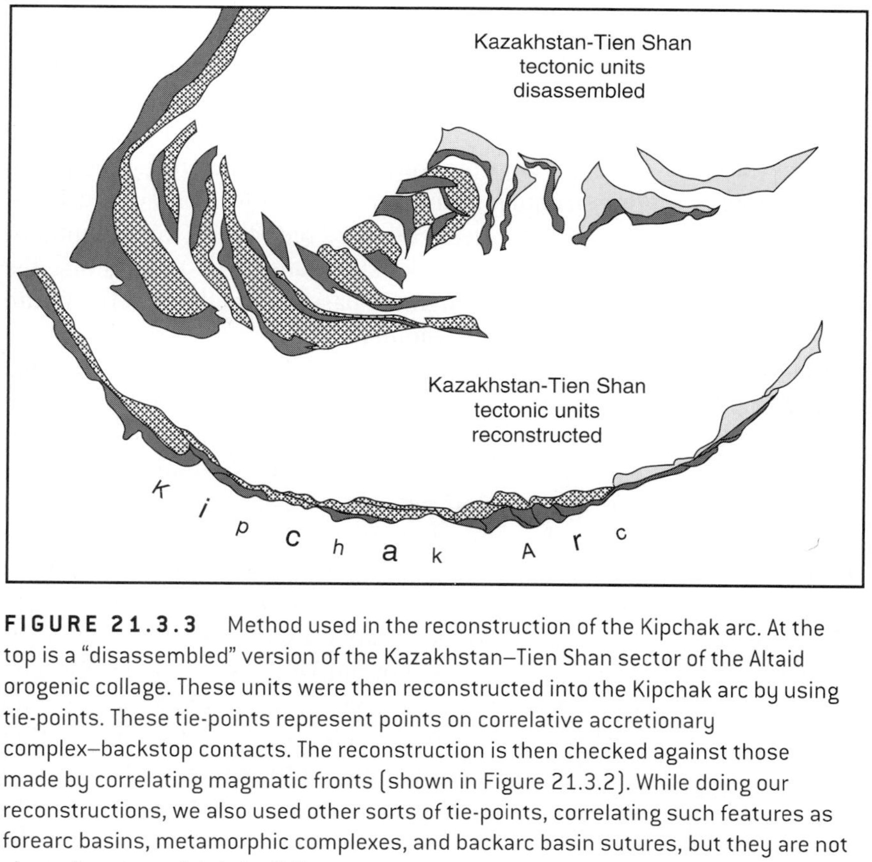

The location of the two arcs with respect to the two continental nuclei of Russia and Angara in the geologic past is established by finding a part of the arc that has remained attached to Angara (in the case of the Kipchak arc, in the vicinity of the southern tip of Lake Baykal; in the case of the Tuva-Mongol Arc, in the Stanovoy Mountains along the Stanovoy Fault; see Figure 21.3.1), and then comparing the geology to that of the margins of the Angara and the Russian cratons. Ideally one would like to support this procedure with paleomagnetic data to check paleolatitudes and paleo-orientations of the individual Altaid units, but, in the case of the Altaids,

FIGURE 21.3.3 Method used in the reconstruction of the Kipchak arc. At the top is a "disassembled" version of the Kazakhstan–Tien Shan sector of the Altaid orogenic collage. These units were then reconstructed into the Kipchak arc by using tie-points. These tie-points represent points on correlative accretionary complex–backstop contacts. The reconstruction is then checked against those made by correlating magmatic fronts (shown in Figure 21.3.2). While doing our reconstructions, we also used other sorts of tie-points, correlating such features as forearc basins, metamorphic complexes, and backarc basin sutures, but they are not shown here to maintain legibility.

reliable paleomagnetic data are very scarce. Only the positions of the two major cratons are known with any degree of confidence (Russia being better known than Angara). Figure 21.3.4 shows the result of such geologic comparisons.

First-Order Tectonic Evolution of the Altaids

In Figure 21.3.4a we show the picture obtained by reconstructing the positions of the two major cratons of Angara and Russia in the Vendian (~630–530 Ma), and placing the Kipchak and the Tuva-Mongol arcs onto the resulting large continent. Rifting in the Vendian left fields of dykes and normal-fault-bounded basins both to the north of Angara and Russia. As the two began to separate from each other, the Kipchak arc also detached itself (as shown in Figure 21.3.4b; Early Cambrian, ~547–530 Ma) in front of an opening marginal basin (the future Khanty-Mansy Ocean[3]) that may have been as large as the present Philippine Sea or the Tasman Sea, both of which also opened as backarc basins behind migratory arcs.

Units 10–18 have been formed by **ensimatic** arcs that may have nucleated along a long transform fault that connected the Kipchak subduction zone with the Tuva-Mongol subduction zone. That transform fault must have met the Kipchak arc at a triple junction as shown, otherwise the kinematics does not make sense. Can you guess why? By Late Cambrian time (~514 Ma) the triple junction migrated along the Kipchak arc, and the former transform fault turned into a subduction zone nucleating magmatic arcs along itself, thus lengthening the Kipchak arc. West-dipping subduction also started below the future Urals along the eastern margin of the present-day Mugodzhar unit, an **ensialic** magmatic arc remnant in the southern and central Urals.

Transpression along the outer Tuva-Mongol subduction zone had begun slicing up the active margin and transporting arc and accretionary complex fragments towards the Angara craton. By contrast, along the inner margin, facing the Khangai-Khantey Ocean, the geometry of subduction was much simpler. It is now impossible to reconstruct palinspastically the Tuva-Mongol fragment itself.

In the Middle Ordovician (~458 Ma) the same picture as in the Late Cambrian seems to persist (Figure 21.3.4c), except that in two places along the Kipchak arc, marginal basins (Boschekul-Tarbagatay and Djezkazgan-Kirgiz-Jalair-Nayman) resembling the present-day West Mariana Basin in position, but the Japan Sea in tectonic character, began to open. (Can you guess why they were similar to the West Mariana Basin in position, but to the Japan Sea in tectonic character?) Also rifting began tearing away a strip of land from Russia that eventually formed the Mugodzhar microcontinental arc in front of the opening marginal basin of Sakmara-Magnitogorsk in the southern and central Urals.

By Late Ordovician time, the two major cratons of Angara and Russia had rotated sufficiently towards each other and began squeezing the oceanic space between them, spanned by the Kipchak arc. This resulted in the collision of the tip of the arc with the Mugodzhar arc, similar to the collision of the Izu-Bonin arc with the Japan arc today. Further shortening led to the internal deformation of the Kipchak arc that was expressed by cutting up the arc by strike-slip faults (perhaps similar to the Philippines Fault cutting the Philippine island arc system today) and stacking its pieces beside each other along the strike-slip faults. The resulting geometry resembles thrust stacks tipped on their side.

In the Middle Silurian (about 433 Ma), the Kipchak arc broke along a left-lateral transform fault system, bounding units 5, 6 and 7, 8. In the Early Devonian (~390 Ma; Figure 21.3.4d), the transform fault became lengthened and its southern parts turned compressional. A substantial microcontinent had thus become assembled in the middle parts of the arc by the previous stacking. In the north, the eastern part of the Tuva-Mongol fragment collapsed onto the Sayan Mountains. From now on, strike-slip transfer of units along the Tuva-Mongol fragment passed directly onto the future site of the Altay Mountains.

The Late Devonian (~363 Ma) witnessed much the same sort of evolution as in the Early Devonian, except that at this time, the Angara craton began a right-lateral shear motion with respect to the Russian craton. Various backarc and pull-apart basins opened in the continuously deforming stacked Kipchak arc ensemble and this deformation began tightening the Kazakhstan orocline. Newer units were continuously being fed into the Altay from the coastwise transport along the Tuva-Mongol fragment (not unlike the Cordilleran System, especially in Canada and Alaska in the Mesozoic and Cenozoic). This further narrowed the deep gulf within the Kazakhstan orocline into which rich turbidite deposits were being fed from the surrounding mountainous frame. This was also the time in which the Sakmara-Magnitogorsk marginal basin reached its

---

[3]The name Khanty-Mansy is another name derived from the aboriginal local populations living in the northern part of the Western Siberian Lowlands.

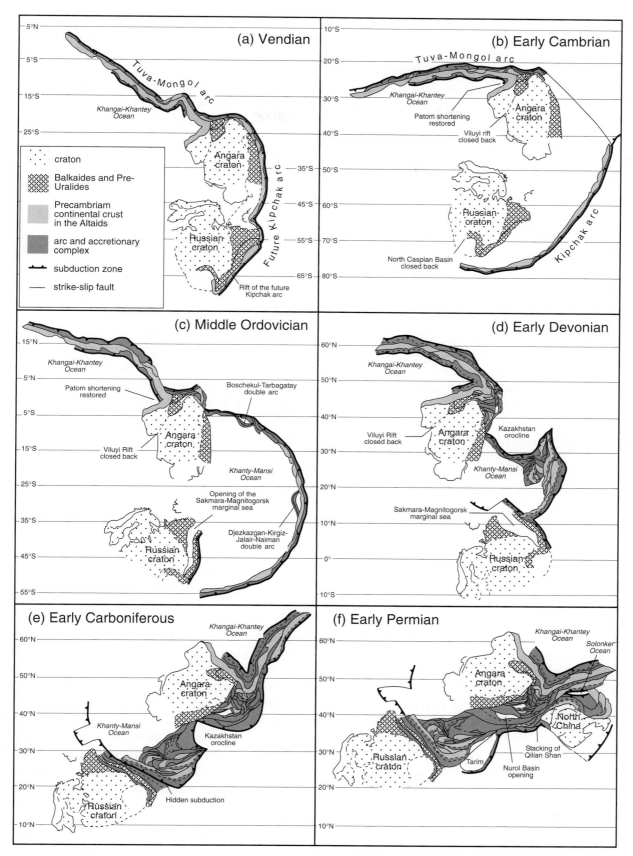

FIGURE 21.3.4 Reconstructions of the Altaids. (a) Vendian reconstruction (~630–530 Ma). The legend shown here applies to all reconstructions shown in Figure 21.3.4. (b) Early Cambrian reconstruction (~547–530 Ma). (c) Middle Ordovician reconstruction (~458 Ma). (d) Early Devonian reconstruction (~390 Ma). (e) Early Carboniferous reconstruction (~342 Ma). (f) Early Permian reconstruction (~270 Ma). The text describes the evolution during intervening times; additional maps and more detailed descriptions are found in the reading list.

apogee and began closing by northward-directed subduction (Late Devonian geographic orientation) under the Mugodzhar arc. It was during the Late Devonian that the North Caspian Basin (pre-Caspian depression in some publications) began opening as a rifted embayment very similar to the Jurassic opening of the Gulf of Mexico. In fact, the present-day Gulf of Mexico is the best analog for the North Caspian Basin.

In the Early Carboniferous (~342 Ma; Figure 21.3.4e), the Sakmara-Magnitogorsk marginal basin was almost completely closed, but because of the continuation of the arc-related magmatism on the Mugodzhar arc, some subduction probably lasted till the end of the Mississippian. To the south of the still-growing Altay Mountains (Early Carboniferous geographic orientation), units 19 (Ob-Zaisan-Surgut) and 20 (Kolyvan-Rudny Altay) were emplaced along the giant Irtysh shear zone into their present locations with respect to the Altay-Sayan Mountain complex. This further narrowed the large embayment within the Kazakhstan orocline, which was rapidly being filled with clastics eroded from the surrounding Altaid units. The right-lateral shear of Angara with respect to the Russian craton continued throughout the Early Carboniferous. In the Late Carboniferous (~306 Ma), the gulf within the Kazakhstan orocline was obliterated by the continued convergence of the Angara craton and the Russian craton. Any further accretion to the remnants of the Kipchak arc was made impossible by the collision, along the present Tien Shan Mountains, of the Tarim block with the assembled Altaid collage between the Angara and the Russian cratons. The nature of the crust underlying the Tarim block is unknown owing to lack of outcrop and paucity of geophysical data. What little is known indicates that it is a strong, old (Late Proterozoic) crust, which may be a trapped oceanic plateau similar to the present-day Ontong-Java now colliding with the Solomon Islands in the southwest Pacific Ocean.

The right-lateral motion of the Angaran craton with respect to the Russian craton and the Kazakh part of the Altaid collage continued in the Early Permian (~270 Ma; Figure 21.3.4f). This motion was being accommodated mostly along two major shear zones, namely the Irtysh and the Gornostaev. The deep Nurol Basin within the basement of the West Siberian Lowlands, just north of the Kolyvan Ranges north of the Altay Mountains began forming in the Early Permian as a pull-apart structure along the Irtysh shear zone. This was the earliest harbinger of the beginning extension in the West Siberian Lowlands that lasted until the Middle Jurassic in different places and under different strain regimes. The earlier episodes of this extension especially are directly related to the Altaid evolution and constitute one spectacular example of extensional basins forming atop former orogenic belts, such as the Eocambrian Hormuz salt basins in Arabia following the latest Precambrian Pan-African Orogeny, or the Middle to Late Cenozoic Western Mediterranean and the Late Cenozoic Aegean Basins following the Alpine Orogeny. (Can you think of similar post-orogenic basins within the United States? How do you think they formed?)

In the Late Permian (~250–255 Ma), the right-lateral motion of the Angara craton with respect to the Russian craton and the Kazakh sector of the Altaid collage reversed. The resulting left-lateral motion was largely accommodated along a broad swath of shear zones south of (Late Permian geographic orientation), and including, the Gornostaev shear zone, creating a broad **keirogen** (a deformed belt dominated by strike-slip motion). The late Permian extensional basins of Nadym, Alakol, Junggar, and Turfan opened along this keirogen as pull-apart structures, involving internal counterclockwise rotations exceeding 90° about vertical axes.

In the Kazakhstan-Tien Shan sector, the Late Permian saw the end of the Altaid orogenic evolution, although extension in the West Siberian Lowlands continued until the Early Jurassic, possibly resulting from continued limited jostling of the Angara and the Russian cratons along the large Gornostaev keirogen. Farther east (present geographic orientation), Altaid orogenic evolution continued in the Mongolian-Far Eastern sector by the ongoing closure of the Khangai-Khantey Ocean as a consequence of the collision of the North China block with the South Gobi units (unit 44). The progressive narrowing of the Khangai-Khantey Ocean lasted until the Jurassic. During this narrowing, a part of the accretionary fill, formed by the Kangai-Khantey accretionary complex (unit 43.2), was extruded northwestwards. (Can you guess why?) By Jurassic time, the Mongolia/east Russia/northeast China regions acquired more-or-less their present-day architecture; there was some shortening in the extreme far eastern Russian Altaids continuing into the early Cretaceous, but by then it was under the influence of the Nipponide evolution (i.e., orogeny resulting from the interaction of oceanic plates in the Pacific Ocean with the eastern margin of Asia) and not an integral part of the Altaid development.

21.3.4 Implications for Continental Growth

The time slices in Figure 21.3.4 show how significant volumes of new continental crust were added to the Altaid edifice as it developed from Vendian times onward. It is estimated that some 10^{12} km^3 of material may have been added to the bulk of Asia between the Vendian and the Late Permian (not taking into account the Khangei-Khantey accretionary complex), which accounts for some 40% of the total Paleozoic crustal growth rate. Clearly, the Altaids represent a major tectonic system during Earth's history in the Paleozoic. This accretionary aspect of Turkic-type orogeny may offer insights into the formation of Earth's earliest continental crust, as discussed in the following section.

21.3.5 Closing Remarks

The Altaids are the most spectacular example of a Turkic-type collisional orogen in the Phanerozoic. Turkic-type orogens form by collision of very large, subcontinent-size subduction-accretion complexes fringing two (or more) converging continents and/or island arc systems. Three main processes contribute to the consolidation of such large subduction-accretion complexes into respectable continental basement:

1. Invasion of former forearc regions by arc plutons through the trenchward migration of the magmatic axis as the trench recedes and the subduction-accretion complex becomes wider.
2. Continuing bulk shortening of the subduction-accretion complex.
3. Metamorphism of the accretionary complex up to high-amphibolite grade, either by ridge subduction or by the exposure of its bottom to hot asthenosphere by steepening of the subduction angle.
4. Further thickening of subduction-accretion complexes and melting of their bottoms as a consequence of the convective removal of the lithospheric mantle.

Recognition of Turkic-style orogeny has long been hampered by emphasis on Alpine- and Himalayan-type collisional orogens, which has been conditioned by the familiarity of the world geologic community with the Alps, the Himalaya, the European Hercynides, and the Caledonian/Appalachian mountain ranges. The study of the Altaids and their comparison with the North American Cordillera have begun to uncover the main features of their architecture and the rules of their development. Many of the familiar features guiding the geologist in Alpine- and Himalayan-type collisional orogens lose their significance in the Turkic-type orogenic systems. Although abundant ophiolitic slivers, nappes, and flakes exist in vast areas occupied by Turkic-type orogens, these do not necessarily mark sites of sutures. Intra-accretionary wedge-thrust faults and, especially, large strike-slip faults, commonly juxtapose assemblages formed in distant regions, and deformed and metamorphosed at different structural levels. Such faults are likely to mislead geologists into thinking that they are sutures, bounding different, originally independent microcontinental "terranes." The recognition of the Turkic-type orogeny has thus made necessary not only detailed and careful field mapping and description, in terms of **genetic labels,** but also detailed geochemical sampling to see how much of the accreted material is juvenile, and how much is recycled.

It is proposed that Turkic-type collisional orogeny was very widespread in the Proterozoic, possibly largely dominated the Archean development, and contributed significantly to the growth of the continental crust through time.

ADDITIONAL READING

Allen, M. B., Şengör, A. M. C., and Natal'in, B. A. 1995. Junggar, Turfan, and Alakol basins as Late Permian to ?Early Triassic sinistral shear structures in the Altaid orogenic collage, Central Asia. *Journal of the Geological Society of London,* 152, 327–338.

Burke, K., 1977. Aulacogens and continental breakup. *Annual Review of Earth and Planetary Sciences,* 5, 371–396.

Kober, L. 1921. *Der Bau der Erde.* Gebrüder Borntraeger: Berlin.

Şengör, A. M. C., 1990. Plate tectonics and orogenic research after 25 years: a Tethyan perspective. *Earth Science Reviews,* 27, 1–201.

Şengör, A. M. C., 1991. Plate tectonics and orogenic research after 25 years: synopsis of a Tethyan perspective. *Tectonophysics,* 187, 315–344.

Şengör, A. M. C., and Natal'in, B. A., 1996a. Palaeotectonics of Asia: fragments of a synthesis. In Yin, A., and Harrison, M., eds., *The tectonic evolution of Asia,* Rubey Colloquium. Cambridge University Press: Cambridge.

Şengör, A. M. C., and Natal'in, B. A., 1996b. Turkic-type orogeny and its role in the making of the continental crust. *Annual Review of the Earth and Planetary Sciences,* 24, 263–337.

Şengör, A. M. C., and Okurogullari, A. H., 1991. The role of accretionary wedges in the growth of continents: Asiatic examples from Argand to plate tectonics. *Eclogae Geologicae Helvetiae,* 84, 535–597.

Şengör, A. M. C., Natal'in, B. A., and Burtman, V. S., 1993. Evolution of the Altaid tectonic collage and Palaeozoic crustal growth in Eurasia. *Nature,* 364, 299–307.

Stille, H., 1928. Der Stammbaum der Gebirge und Vorlaender: *XIVe Congrès Géologique International (1926, Espagne),* 4. Fscl., 6. Partie, Sujet XI (Divers), Graficas Reunida S. A., Madrid, 1749–1770.

Suess, E., 1901[1908], *The face of the Earth (Das Antlitz der Erde),* v. 3, translated by H. B. C. Sollas under the direction of W. J. Sollas. Clarendon Press: Oxford.

21.4 THE TASMAN OROGENIC BELT, EASTERN AUSTRALIA: AN EXAMPLE OF PALEOZOIC TECTONIC ACCRETION—An essay by David R. Gray[1] and David A. Foster[2]

21.4.1 Introduction

The **Tasman orogenic belt** is a turbidite-dominated, composite Paleozoic "accretionary" orogenic system along the eastern margin of Australia (Figure 21.4.1). This Paleozoic orogenic system illustrates how deformational and metamorphic processes, combined with magmatism, convert deep-marine sedimentary and volcanic rocks, including large turbidite fan systems, into normal-thickness (~35 km) continental crust. Shortening and accretion occurred by stepwise deformation and metamorphism away from the cratonic core over a period of 400 m.y. from Cambrian through Triassic times. During that time eastern Australia underwent an important period of continental accretion, which added approximately 30% to the size of the ancient Australian cratonic core. The addition of recycled continental detritus in turbidite fans and of juvenile material to Australia represents a continental crustal growth mechanism that was important throughout Earth history. Originally part of a Paleozoic convergent plate margin that stretched around the supercontinent of Gondwana from South America to Australia, the present distribution of orogens and the observed structural patterns are a response to the chang-

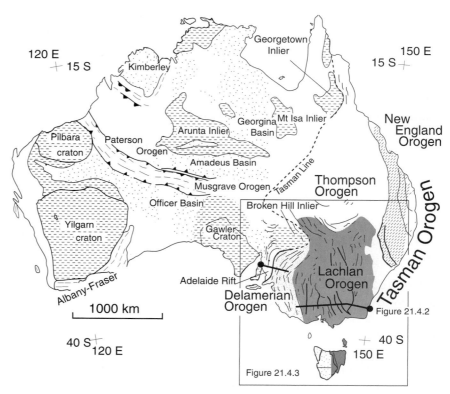

FIGURE 21.4.1 Map of major geologic elements of Australia showing the Tasman Orogen along the eastern margin. Positions of section line (Figure 21.4.2) and the more detailed map area of Figure 21.4.3 are shown.

ing character of the Gondwana margin/plate boundary from the Cambrian to the Triassic.

The features that set the Tasman Orogen apart include:

- The absence of classic sutures that are typical of continental collisional orogenic systems.
- The absence of simple craton-directed thrusting, but juxtaposition of craton-verging and oceanward-verging, marginal thrust systems (Figure 21.4.2).

[1]University of Melbourne, Australia.
[2]University of Florida, Gainesville, Florida, USA.

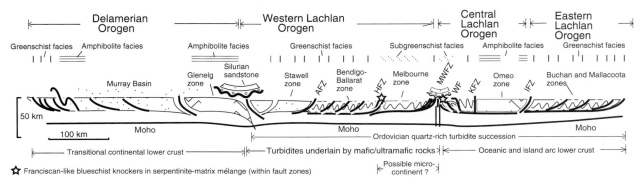

FIGURE 21.4.2 Schematic west-east crustal structural profiles showing the main geologic and metamorphic aspects of the Lachlan Orogen at 37°S latitude. Abbreviations: AFZ-Avoca fault zone, HFZ-Heathcote fault zone, MWFZ-Mt Wellington fault zone, WF-Wonnangatta Fault, KFZ-Kiewa fault zone, IFZ-Indi fault zone.

- The absence of metamorphic hinterland, but localized high-tempe/low-pressure metamorphic regions.
- The absence of thrust slices or windows exposing Proterozoic basement.
- Large volumes of granite (up to 30% exposed area) where the distribution and ages of the granites do not fit simple orogenic models involving either A or B type subduction.
- Major fault zones in the turbidite-dominated part containing slices of oceanic crust associated with blueschist blocks in serpentinite- and mud-matrix mélange.

21.4.2 Crustal Structure and Main Tectonic Elements

Eastern Australia (The Tasman Orogen) is made up of three north-south trending deformed belts (Delamerian, Lachlan/Thomson, and New England Orogens, Figure 21.4.1), which are distinguished by their lithofacies, tectonic settings, timing of orogenesis, and eventual consolidation to the Australian craton (Table 21.4.1). These represent three distinct tectonic settings: (1) deformed intracratonic rift (Delamerian Orogen), (2) deformed marginal turbidite fan system(s) in a backarc setting (Lachlan Orogen), (3) deformed arc–subduction complex belt (New England Orogen). The changing settings are part of a progressive eastwards younging of accretion along the evolving Australian continental margin, defined by their respective peak deformations of Late Cambrian–Early Ordovician, Late Ordovician–Silurian, and Permian–Triassic age (Table 21.4.1). Boundaries between the three belts are not exposed and generally they are covered by younger sequences. The inner belt (the Delamerian and the western part of the Lachlan) shows pronounced curvature and structural conformity with the promontories and recesses in the old cratonic margin (the Tasman line; Figure 21.4.1). Outboard, the central and eastern belts of the Lachlan, and the New England belt, have more continuous trends that truncate the inner belt trends and thus show no relationship to the old cratonic margin.

Delamerian Orogen

The Delamerian fold belt (Orogen) is an arcuate, craton-verging thrust belt (Figure 21.4.2 and 21.4.3), with foreland-style folds and detachment-style thrusts (external zone) to the west, and a metamorphic hinterland (internal zone) to the east, characterized by polyphase deformation, amphibolite-grade metamorphism with local development of kyanite-sillimanite assemblages, and intrusion of syn- and posttectonic granites. During the Late Cambrian–Early Ordovician (locally called the Delamerian Orogeny) allochthonous sheets consisting of northwest-verging duplexes (deformed Cambrian Kanmantoo Group) were emplaced over the less deformed and metamorphosed shelf sequence (Adelaidean) of the external zone. High-temperature metamorphism formed at 300–500 MPa (~10–17 km depth) and is spatially and temporally confined to aureoles of synkinematic granites that are conformably aligned with the structural grain. Rapid unroofing (~10 km in approximately 20 m.y.) of the belt is suggested by juxtaposition of these high-grade rocks and their syntectonic granites (520–490 Ma) with undeformed, high-level silicic granites and volcanics intruded at 486 Ma. This provides a source for the extensive Ordovician turbidite sequences of the Lachlan Orogen to the east.

TABLE 21.4.1 | OROGENS AND SUBPROVINCES OF THE TASMAN OROGENIC BELT

Characteristic	Delamerian	Lachlan			New England
		Western	Central	Eastern	
Main plutonism	Late Cambrian	Late Devonian	Late Silurian	Late Carboniferous	Late Permian-Early Triassic
Tectonic vergence	W-directed thrusting	E-directed thrusting	Overall strike-slip with SE- directed thrusting	E-directed thrusting	W-directed thrusting
Terminal folding	Mid- Cambrian	Middle Devonian	Middle Silurian	Early Carboniferous	Mid-Permian
Main facies	Platform to deep water passive margin sequence	Quartz-rich turbidite sequence on oceanic crust	Quartz-rich turbidite Sequence on oceanic crust	Platform carbonates and clastics with rhyolites and dacitic tuffs	Volcanogenic clastic
Initial record	Basic volcanics	Cambrian basic volcanics	Tremadocian chert	Orodovician andesitic volcanics	Ordovician basic volcanics

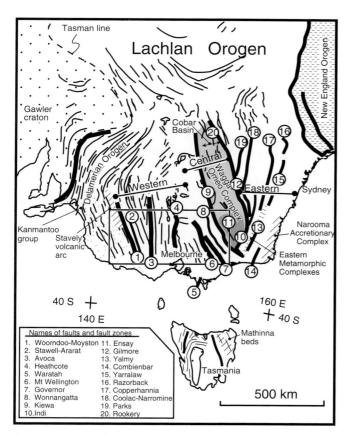

FIGURE 21.4.3 Structural trend map of the Lachlan Orogen and Tasman Orogen, combining aeromagnetic trend-lines and outcrop traces from regional maps and satellite images. The western, central, and eastern subprovinces are identified along with major faults (1–20).

The Delamerian Orogen was tectonically active from the Late Precambrian (~650–600 Ma) to the Early Ordovician (~500 Ma) and is part of the world-wide Pan-African orogenesis. It consists of a deformed Upper Proterozoic Adelaidean intracratonic rift sequence of marine to deltaic sandstones and shales, lagoonal evaporites, dolomites, and limestone, transgressed by lower Cambrian shelf sediments transitional into deep-water sandstones and mudstones. These units are progressively exposed in thrust slices as part of the craton-verging thrust system.

Lachlan Orogen

The Lachlan Orogen is a Middle Paleozoic fold belt with a 200-my history that occupies ~50% of the present outcrop of the Tasman Orogen (Figures 21.4.1 and 21.4.3). The major feature of the Lachlan fold belt is the similarity of sedimentary facies and overall structural style across this segment of the Tasman orogenic zone (Figure 21.4.2). Lower Paleozoic deep-water, quartz-rich turbidites, calc-alkaline volcanic rocks, and voluminous granitic plutons dominate the Lachlan Orogen. The turbidites overlie a mafic lower crust of oceanic affinity. They are laterally extensive over a present width of 800 km and have a current thickness upwards of 10 km. Large parts of the Lachlan Orogen represent accreted parts of a very large submarine sediment dispersal system associated with the Gondwana margin during the Early Paleozoic. This sediment

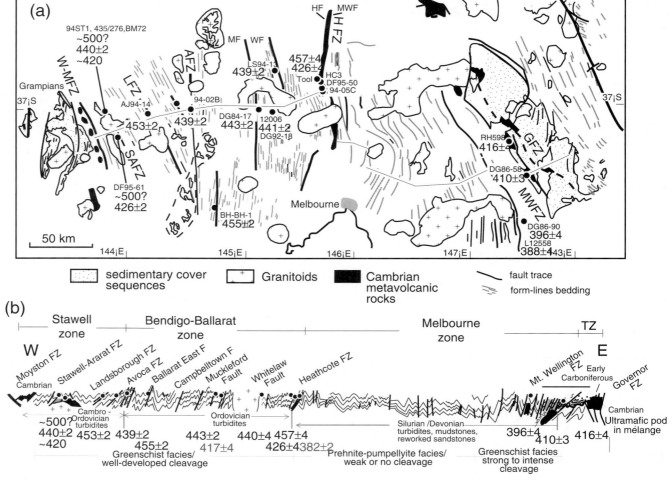

FIGURE 21.4.4 Structural trend map (a) and structural profile (b), with degree of cleavage development and grade of metamorphism of the western sub-province of the Lachlan Orogen incorporating the Stawell, Bendigo-Ballarat, and Melbourne structural zones. Faults are shown as heavy lines; abbreviations in (a) correspond to faults in (b). Dots show positions for $^{40}Ar/^{39}Ar$ ages that are given in Ma.

dispersal system must have been dimensionally comparable to the present-day Bengal fan in the Bay of Bengal. Within the turbidite succession, linear north-south trending, fault-bounded Cambrian metavolcanic belts, composed of boninite (high-Mg andesites), low-Ti andesite, and tholeiite of oceanic affinities, define the boundaries between different structural zones (Figures 21.4.2 and 21.4.3). The eastern part consists of shoshonitic (high-K) volcanics, mafic volcaniclastic rocks, and limestone, as well as quartz-rich turbidites and extensive black shale in the easternmost part.

Tight to open chevron folds (accommodating between 50% and 70% shortening), cut by predominantly west-dipping, high-angle reverse faults, are part of different thrust systems within the Lachlan Orogen (Figure 21.4.4). Chevron folds are upright and gently plunging but become inclined and poly-deformed

approaching major faults. Faults in the western part of the Lachlan fold belt are brittle faults, but they have high strain zones of varying widths that show the intense development of crenulation cleavages associated with variably but generally steeply plunging mesofolds and microfolds. Overprinting cleavages within these zones indicate complex fault movements in which early thrusting is followed by minor wrench movements. Detachments at the mid-crustal level occur at the base of the Ordovician and within the Cambrian successions in the western belt; deep crustal seismic profiling indicates a depth to detachment of approximately 15 km.

Metamorphism is greenschist facies or lower across the Lachlan Orogen, except in the shear zone-bounded Wagga-Omeo and several smaller metamorphic complexes that are part of the Eastern Metamorphic belt

(Figure 21.4.3), where high temperature–low pressure metamorphism is characterized by andalusite-sillimanite assemblages. Such assemblages are typical of thermal metamorphism, but here they occur on a regional scale. Peak metamorphic conditions in the Wagga-Omeo zone are T ≈ 700°C and P ≈ 300–400 MPa. Erosional unroofing of the metamorphic complex in the Middle to Late Silurian necessitates shallow overburden and high geothermal gradients on the order of 65°C/km. It is apparent that regional metamorphism and felsic magmatism throughout the orogen took place under very little cover, a scenario that suggests a shallow to mid-crustal heat input for melting.

Granites cover up to 36% of the exposed Lachlan Orogen. Regional aureole, contact aureole, and subvolcanic field associations, as well as sedimentary and igneous types based on geochemistry and/or mineralogy have been recognized. Most granites are posttectonic and are undeformed and have narrow (1–2-km wide) contact aureoles. Some of these are composed of subvolcanic granites associated with rhyolites, and ash flows of similar composition. The regional aureole types are less common and are associated with the high-temperature/low-pressure metamorphism, migmatites, and K feldspar-cordierite-andalusite-sillimanite gneisses. The shape distribution of the granites suggests three major granite provinces, which presumably reflect the mode and timing of emplacement and state of stress in the mid to lower crust. Elongate north-northwest to north-trending granites define the Wagga-Omeo metamorphic belt and the major part of the eastern Lachlan Orogen in New South Wales. Many of these granites are syntectonic and show the internal deformation and emplacement associated with deep-crustal shear zones. In the western Lachlan fold belt posttectonic granites of the central Victorian magmatic province, the largest granitic bodies, are east-west trending and have elongated form. The remaining granites are smaller and are more equant in shape. Spacing of the granitic bodies in Victoria fits a diapiric emplacement model with a source depth at 12–24 km, and possibly deeper, in a crust that was at least 35 km thick.

The Lachlan Orogen has undergone a complex history of amalgamation and deformation, in which there is an interplay of compressional and extensional events. Long-lived subduction in the Middle Paleozoic is envisaged along the Gondwana margin, but there is no evidence for major collision. Surface structures have been used to infer convergent margin deformation, and Ordovician volcanic rocks in New South Wales and relicts of blueschist metamorphism in fault zones in Victoria have been used to define the plate margin setting. As a consequence, the tectonic setting and evolution of the Lachlan Orogen have been contentious. Significant advances, however, have been made in the last five to ten years due to (1) better resolution on the absolute timing and patterns of deformation (see next section), (2) different approaches to understanding the origin of the granitic plutons, (3) better understanding of the Ordovician arc in the east, (4) greater knowledge nature of the lower crust, and (5) a realization of the significance of the fault zones.

New England Orogen

The New England Orogen is the youngest and most easterly part of the Tasman belt (Figure 21.4.1). A collage of deformed and imbricated terranes (Figure 21.4.6a), it consists of largely Middle to Upper Paleozoic and Lower Mesozoic marine to terrestrial sedimentary and volcanic rocks, as well as strongly deformed flysch, argillite, chert, pillow basalts, ultramafic rocks, and serpentinites. The New England Orogen was tectonically active from the Late Devonian to the Middle Cretaceous (~95 Ma) and activity of this convergent margin involves arc, forearc, and accretionary complexes.

Widespread climactic Permian-Triassic deformation, involving west-directed thrusting, interleaving and imbrication of the arc magmatic belt (Connors-Auburn belt), forearc (Yarrol-Tamworth belt) and oceanic assemblages, including subduction complexes (Wandilla-Gwydir belt) and ophiolite (Gympie belt), consolidated the terranes into Australia and caused the development of a Permo-Triassic foreland basin (Sydney-Bowen Basin). Complex outboard subduction assemblages show a strong thrust-related fabric, polyphase deformation, and greenschist to amphibolite facies metamorphism.

21.4.3 Timing of Deformation and Regional Events

The timing of orogenic events in the Lachlan Orogen has been broadly defined by the ages of strata over regional unconformities and by the age of stitching plutons. More recently $^{40}Ar/^{39}Ar$ data from metamorphic white mica in the low-grade metasedimentary rocks and associated quartz veins give precise estimates on the timing of cleavage formation and regional deformation events. This is because, in these fine-grained strongly cleaved rocks, the only significant K-bearing phase after metamorphism is metamorphic white mica that typically retains argon at temperatures greater than those under which it grew. Mica growth below the closure temperature provides a definitive

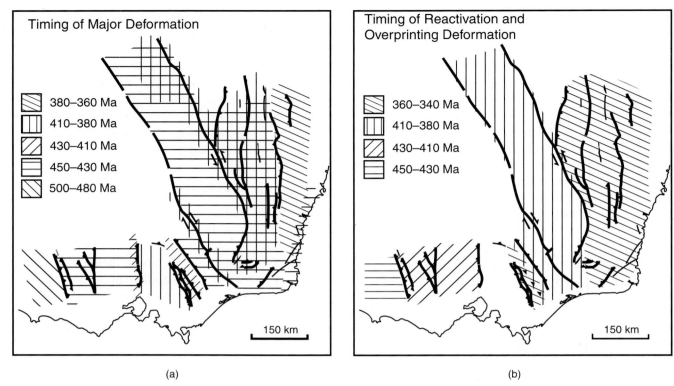

FIGURE 21.4.5 Maps of the Lachlan Orogen showing timing of deformation events and fault reactivation, based on $^{40}Ar/^{39}Ar$ geochronology and other geologic constraints.

method for dating the cleavage-forming part of the deformation, and the timing of regional metamorphism. Fine-grained metamorphic phengites close to argon loss at about 350°C, so that at very low greenschist and prehnite-pumpellyite facies metamorphic conditions, they grow at or below their closure temperatures. $^{40}A/^{39}Ar$ ratios for rocks in the high-grade metamorphic complexes give the timing of exhumation, so, wherever possible, deformation in these areas is dated by U-Pb zircon data.

The $^{40}Ar/^{39}Ar$ geochronologic and thermochronologic data, interpreted along with other geologic data, allow us to define when specific regions within the Lachlan Orogen first underwent significant deformation, metamorphism, faulting, and reactivation (Figure 21.4.5). Deformation in the Lachlan Orogen was initiated between ~455 and 430 Ma in three areas: the western subprovince, where it migrated eastward; the central subprovince, where it migrated to the southwest; and the eastern subprovince in the Narooma accretionary complex (Figure 21.4.3). The western subprovince is characterized by major deformation in the Stawell and Bendigo-Ballarat zones between ~455 and 440 Ma and fault reactivation at ~430–410 Ma. The eastern bounding fault of the western Lachlan, the Mount Wellington fault zone (Figure 21.4.3), was

active between 410 and 385 Ma and reactivated during Carboniferous time.

The central subprovince (Figure 21.4.3) underwent major deformation in the high-grade metamorphic complex between 440 and 430 Ma, and was exhumed between ~410 and 400 Ma, as shown by mica dating from the shear zones bounding the complex. To the west, in low-grade metasedimentary turbidites (Figure 21.4.2), deformation began at ~440–430 Ma and migrated southwestward until ~410–416 Ma. Later deformation took place when the eastern subprovince collided with the western subprovince at ~400–380 Ma, and again during the Carboniferous. The oldest recorded deformation in the eastern subprovince (Figure 21.4.3) took place in the Narooma complex ~455–445 Ma, when it was outboard of and separated from the rest of present eastern subprovince. The inland parts of the eastern subprovince are dominated by contractional deformation that occurred 400–380 Ma and there was additional deformation 380–360 Ma in the central and northeastern region. These younger episodes overprint the earlier fabrics such as those in the Narooma complex. The Silurian extensional event predated this contractional deformation phase. Although very widespread, intense Carboniferous deformation is almost unique to the north-

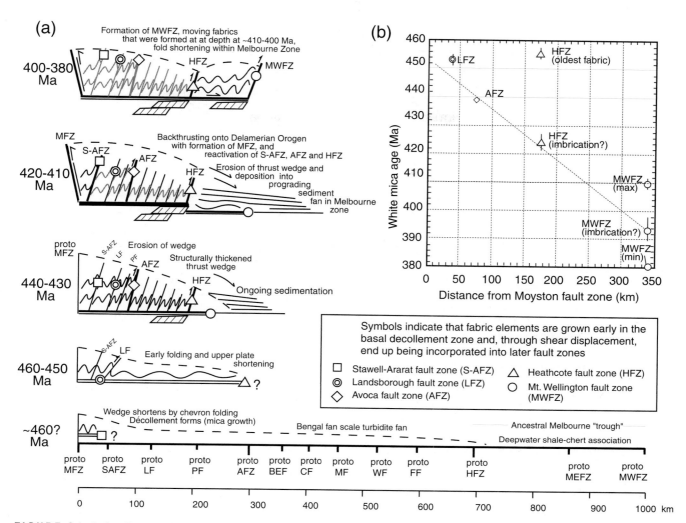

FIGURE 21.4.6 Progressive deformation sequence for the western Lachlan Orogen based on the observed thrust-belt geometry and the requirement of underthrusting from the east. There is an implied diachronous progression of chevron folding, imbrication, and unroofing from a deep level décollement. A consequence of the model is that shortening has to be achieved above the basal fault to attain critical taper prior to movement along it. This means that some mica ages within thrust sheets may be older or at least coeval with the mica ages from the basal fault zone splays. The Ar$^{40/39}$Ar data suggest that folding, quartz veining, cleavage development, and faulting progressed eastwards from Middle Ordovician (~455 Ma) through the Middle Devonian (~390–380 Ma). Fault zones record multiple ages due to multiple episodes of reactivation, crenulation cleavage, and fabric transposition. (b) Fault propagation shown as a graph of white mica age (Ma) versus distance from the Moyston fault zone (i.e., the westernmost part of the Lachlan Orogen). This is based on the youngest significant Ar$^{40/39}$Ar ages from the major fault zones (listed on the figure) of the western Lachlan. These data require diachronous but not necessarily continuous deformation. If it is assumed that deformation was continuous over this interval, then the time-averaged fault propagation rate is ~6 mm/y. Abbreviations:

ern part of the Lachlan Orogen in the eastern subprovince. Carboniferous deformation partly reflects the progressive eastward accretion of the Tasmanides and is probably related to amalgamation of the New England Orogen. The Carboniferous could even include the rapid movement of Australia toward the south pole, a process that may have also caused the Alice Springs Orogeny in central Australia.

In the past, a framework of six orogenic events was proposed for the Lachlan Orogen during the Paleozoic. It now appears that some of these events are localized and strongly time-transgressive. Because of the complex and localized pattern of deformation during the Ordovician, Silurian, and Devonian we have preferred to refer to the whole interval as the Lachlan Orogeny. Based on the present data it has been argued that two major events still dominate the Middle Paleozoic—one at ~440–430 Ma and one at ~390–380 Ma. However, because significant tectonic activity in terms of strike-slip faulting and plutonism continued between these "events" we prefer to use the broader framework.

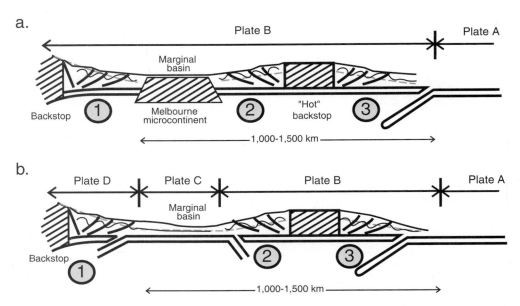

FIGURE 21.4.7 Tectonic interpretation of simultaneous, eastward and westward propagating and migrating deformation fronts in the western, central, and eastern parts of the Lachlan Orogen. Deformation is due to underthrusting of a sediment wedge between three proposed subduction systems that were from 1500 to 2000 km apart in the Late Ordovician–Early Silurian.

21.4.4 Mechanics of Deformation in Accretionary Orogens

Turbidite-dominated, accretionary orogens, such as the Lachlan Orogen of eastern Australia and those in central Asia and some Pan African belts, differ from classical orogenic belts such as the Alps, Appalachians, and North American Cordillera. The latter were constructed partly of shelf sequences developed along margins of moderately thick (~30–40 km) continents, whereas the former are characterized by monotonous, commonly chevron-folded turbidites that overlie oceanic crust and, in places, thin, attenuated continental crust.

Faults in the low-grade turbidite sequences of turbidite-dominated accretionary orogens record the kinematic evolution of this style of orogen. These fault zones are characterized by higher than average strain and intense mica fabrics, transposition foliation and isoclinal folds, poly-deformation with overprinting crenulation cleavages, and steeply to moderately plunging meso- and microfolds. They have a different character compared to the brittle-ductile fault zones of classic foreland fold-and-thrust belts such as are found in the Appalachians and the Canadian Rocky Moun-

tains. Multiple cleavages and transposition layering record a progressive shear-related deformation history. An intense mica fabric evolves initially during shortening of the overlying sedimentary wedge, but this fabric is progressively modified during rotation and emplacement to higher structural levels along the steep parts of inferred listric faults. The deformed wedge outside the fault zones generally undergoes one phase of deformation, shown by a weak to moderately developed slaty cleavage, which is parallel to the axial surface of upright, subhorizontally plunging chevron-folds. In the Lachlan Orogen we have interpreted the structural evolution constrained by ^{40}Ar/^{39}Ar dating to consist of progressive deformation associated with simultaneous, eastward propagating and migrating deformation fronts in both the western and the eastern parts of the fold belt (Figure 21.4.6). These deformation fronts are related to accretionary style deformation at the leading edges of overriding plates, in an inferred southwest Pacific-type subduction setting from the Late Ordovician to the Middle Devonian, along the former Gondwana margin (Figure 21.4.7). The fault zones effectively accommodate and preserve movements within structurally thickening, migrating, and prograding sediment wedges.

ADDITIONAL READING

Collins, W. J., 1996. Lachlan fold belt granitoids—products of three-component mixing. *Transactions of the Royal Society of Edinburgh, Earth Sciences,* 87:171–181.

Collins, W. J., 1998. Evaluation of petrogenetic models for Lachlan fold belt granitoids: implications for crustal architecture and tectonic models. *Australian Journal of Earth Sciences,* 45, 483–500.

Coney, P. J., Edwards, A., Hine, R., Morrison F., Windrum, D., 1990. The regional tectonics of the Tasman orogenic system, eastern Australia. *Journal of Structural Geology,* 125, 19–43.

Fergusson, C. L., Coney, P. J., 1992. Convergence and intraplate deformation in the Lachlan fold belt of southeastern Australia. *Tectonophysics,* 214, 417–39.

Foster, D. A., Gray D. R., 2000. The structure and evolution of the Lachlan fold belt (Orogen) of eastern Australia. *Annual Reviews of Earth and Planetary Sciences,* 28, 47–80.

Foster, D. A., Gray, D. R., Bucher, M., 1999. Chronology of deformation within the turbidite-dominated Lachlan orogen: implications for the tectonic evolution of eastern Australia and Gondwana. *Tectonics,* 18, 452–485.

Gray, D. R., 1997. Tectonics of the southeastern Australian Lachlan fold belt: structural and thermal aspects. In Burg, J. P., Ford, M., eds., *Orogeny through time, Geological Society Special Publication,* 121, 149–177.

Gray, D. R., Foster, D. A., 1997. Orogenic concepts—application and definition: Lachlan fold belt, eastern Australia. *American Journal of Science* 297, 859–891.

Gray, D. R., Foster, D. A., 1998. Character and kinematics of faults within the turbidite-dominated Lachlan orogen: Implications for the tectonic evolution of eastern Australia. *Journal Structural Geology* 20, 1691–1720

Gray, D. R., Foster, D. A., Gray C., Cull J., Gibson, G., 1998. Lithospheric structure of the southeast Australian Lachlan fold belt along the Victorian Global Geoscience Transect. *International Geology Review* 40, 1088–1117.

Spaggiari, C. V., Gray, D. R., and Foster, D. A., 2002. Blueschist metamorphism during accretion in the Lachlan Orogen, southeastern Australia. *Journal of Metamorphic Geology,* 20, 711–726.

Spaggiari, C. V., Gray, D. R., Foster, D. A., and Fanning, C. M., 2002. Occurrence and significance of blueschist in the southern Lachlan Orogen. *Australian Journal of Earth Sciences,* 49, 255–269.

CHAPTER TWENTY-TWO

Western Hemisphere

22.1 THE NORTH AMERICAN CORDILLERA—An essay by Elizabeth L. Miller[1]

22.1.1 Introduction

The broad mountainous region of western North America is known as the **"Cordillera,"**[2] an orogenic belt that extends from South America (the Andean Cordillera) into Canada (Canadian Cordillera) and Alaska (Figure 22.1.1). The youthful topography of this impressive mountain belt is closely related to ongoing crustal deformation, as indicated by the distribution of seismicity across the width of the orogenic belt (Figure 22.1). The present plate tectonic setting and the dominant style of deformation vary along strike of the orogen: folding and faulting above an active subduction zone in the Pacific Northwest of the United States and Alaska; strike-slip or transform motion along the Queen Charlotte Fault (Canada) and San Andreas Fault (California); and extension and rifting in the Basin and Range Province of the western United States and Mexico's Gulf of California. Variations in structural style are also apparent across the orogen; for example, crustal shortening and strike-slip faulting in coastal California are concurrent with crustal extension and basaltic volcanism in the adjacent, and inboard, Basin and Range Province. This diversity in plate tectonic setting and structural style of the Cordillera along and across strike, likely characterized the past history of the orogenic belt, which has been shaped primarily by Pacific–North America plate interactions. The continuity of such interactions since the Late Precambrian makes it the longest-lived orogenic belt known on Earth.

The Cordillera provides an excellent natural laboratory for studying the evolution of a long-lived active margin and the effects of subduction and plate boundary processes on the evolution of continental crust. However, the exact nature of the relationship between plate motions and continental deformation, whether mountain building or crustal extension, remains a complex question for the following reasons. The theory of plate tectonics treats the Earth's lithosphere as a series of rigid plates that move with respect to one another along relatively discrete boundaries. This simplification applies well to oceanic lithosphere, which is dense and strong and thus capable of transmitting stresses across great distances without undergoing significant internal deformation. However, it does not apply well to continents, whose more quartzo-feldspathic composition and greater radiogenic heat flow make them inherently weaker. Displacements or strain within continental crust can accumulate at plate tectonic rates (1–15 cm/y) within narrow zones of deformation, or can take place more slowly (millimeters to centimeters per year) across broad zones of distributed deformation. Thus continents can accumulate large strains, thickening over broad distances during crustal shortening, and thinning during extension. Evidence for these strain histories are at least partially preserved in the geologic record because the inherent buoyancy of continental material prevents it from being subducted into the mantle. Mantle-derived magmatism can lead to greater strain accumulation within continental crust by increasing temperatures and thus rheologically weakening the crust, allowing it to deform in a semicontinuous fashion. Because of these considerations regarding the thermal structure, composition, thickness, and rheology of continental crust, the response of the overriding continental plate to changes in subducting plate motion or to changes in the nature of plate interactions along a margin may be sluggish and may vary with time and depth in a complex fashion.

Thus, orogenic belts like the Cordillera may be at best imperfect recorders of past plate motions. Our understanding of the link between plate tectonics and continental deformation is evolving as more detailed geologic and geochronologic studies are carried out,

---

[1]Stanford University, Stanford, CA.
[2]Spanish for mountain or mountain chain.

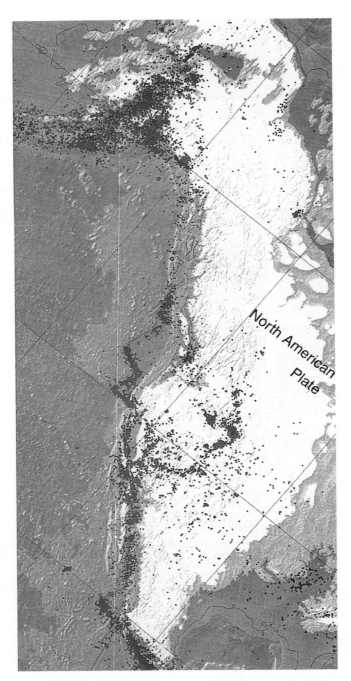

FIGURE 22.1.1 Digital topography, plate tectonic setting and seismicity map of the North American Cordillera.

22.1.2 Precambrian and Paleozoic History

Studies of modern active plate margins have played an important role in interpreting the more fragmentary geologic record of analogous events in the North American Cordillera (Figure 22.1.2). Based on these comparisons, it appears that all plate tectonic styles and structural regimes known on earth, with the exception of continent-continent collision (like the Himalaya), played a part in the creation and evolution of the North American Cordillera. The initial formation of the Cordilleran margin dates back to the latest Precambrian. The Windemere Supergroup is a thick succession of shelf-facies clastic rocks deposited between about 730 Ma and 550 Ma, its facies and isopachs define the newly rifted margin of western North America after the breakup of the Rodinia supercontinent. The Windemere Supergroup forms the lower part of a 15-km thick, dominantly carbonate shelf succession whose isopachs and facies boundaries closely parallel the trend of the Cordillera and are remarkably similar along the length of the Cordillera from Mexico to Alaska. This shelf sequence (including the older and more localized Belt-Purcell Supergroup) is now spectacularly exposed in the eastern Cordilleran fold-and-thrust belt, whose overall geometry and structure are controlled by the facies and thickness of this succession.

The Paleozoic history of the Cordillera has generally been described as one of continued passive margin sedimentation and little active tectonism. However, embedded in the western Cordillera are a multitude of tectonic fragments of island arcs and backarc basin sequences that range in age from Cambrian to Triassic (Figure 22.1.2). The common conception that these represent a collage of far-traveled terranes exotic to the Cordillera ("suspect terranes") is being reevaluated as a result of numerous studies that indicate many of these sequences developed adjacent to, but offshore of, the western edge of the continental margin. Although these "exotic" fragments are likely displaced from their site of origin by rifting, thrusting, and strike-slip faulting, their presence nonetheless argues convincingly for a long history of subduction of paleo-Pacific crust beneath the western edge of the North American plate. Study of these accreted fragments suggests that during the Paleozoic, western North America looked much like the Southwest Pacific today, with its fringing arcs separated from the main Australasian continental shelves by backarc basins. During the Paleozoic, the North American shelf itself experienced episodes of regional subsidence and uplift, but no sig-

providing quantitative information on the timescale of events and the rates of geologic processes, and increasing our ability to compare the timing of events from one part of the belt to the next. Geophysical and petrologic studies remain key tools that help us to understand how physical processes in the deeper crust and mantle are coupled to more easily studied deformation at shallower levels of the crust.

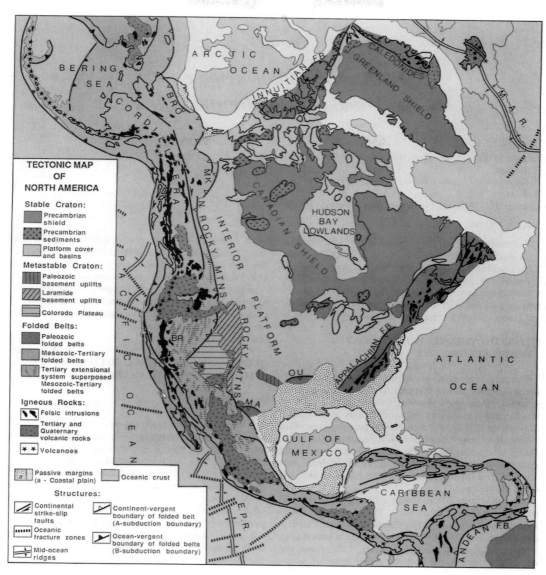

FIGURE 22.1.2 Simplified geologic/tectonic features map of North America. BR = Basin and Range Province; BRO = Brooks Range, MK = Mackenzie Mountains, MA = Marathon Uplift, OU = Ouachita Mountains, M.A.R. = Mid-Atlantic Ridge, E.P.R. = East Pacific Rise; F.B. = fold belt.

nificant deformation. Exceptions include deformation and intrusion of latest Proterozoic to Late Devonian granites along parts of the margin in Alaska and southern British Columbia, and the closure of deep-water, backarc basins by thrusting onto the shelf during the earliest Mississippian Antler Orogeny (Roberts Mountains Thrust) and during the Permo-Triassic Sonoma Orogeny (Golconda Thrust) in the western U.S. part of the Cordillera.

22.1.3 Mesozoic History

Magmatic belts related to eastward subduction beneath western North America are much better developed beginning in the Mesozoic (Figure 22.1.2). Arc mag-matism of Triassic and Early Jurassic age (230–180Ma) is recorded by thick sequences of mafic to intermediate volcanic rock erupted in an island arc (Alaska, Canada, and the northern part of the U.S.) or continental arc (southwestern U.S.) setting. Tectonism accompanying subduction during this time-span was generally extensional in nature, leading to rifting and subsidence of parts of the arc and continental margin. The Middle to Late Jurassic brought a dramatic change in the nature of active tectonism along the entire length of the Cordilleran margin. This time-span is characterized by increased plutonism during the interval 180–150 Ma and many hundreds of kilometers of crustal shortening. This shortening closed intra-arc and backarc basins, accreted arc systems to the North

American continent, and fundamentally changed the paleogeography of the Cordillera from a southwest Pacific–style margin to an Andean-style margin, a tectonic framework that persisted throughout most of the latest Mesozoic and Cenozoic. The preferred explanation for this orogeny is that it is linked to rapid westward motion of North America with respect to a fixed hot-spot reference frame (Figure 22.1.3). This westward motion occurred as the North Atlantic began to open and, in effect, caused the western margin of the continent to collide with its own arc(s) and subduction zone(s) and then to deform internally. Deformation associated with Middle Jurassic orogenesis (sometimes referred to as the Columbian Orogeny) began first in the region of elevated heat flow within and behind the arc, and migrated eastward with time towards the continental interior.

In the Cretaceous, we have a better record of the nature of Pacific Plate motions with respect to North America (in large part from magnetic anomalies on the ocean floor) and it is possible to draw some inferences about the link between orogenic and magmatic events and the history of subduction beneath the active margin. There is a general lull in deformation and a lack of evidence for significant magmatism during the time interval 150–120 Ma, which possibly corresponds to a time of dominantly strike-slip motion along the Cordilleran margin (Figure 22.1.3). Particularly rapid rates of orthogonal subduction of the Farallon and Kula Plates occurred in the later part of the Cretaceous and Early Tertiary, resulting in the emplacement of major batholiths inboard of continental margin subduction zone complexes. Depending on the configuration of subducting oceanic plates, large components of margin-parallel strike-slip faulting are also implied. How much, where, and along what faults this motion was accomplished are still controversial questions. In the western United States, the emplacement of the composite Cretaceous Sierra Nevada batholith occurred (Figure 22.1.2). On the western (forearc) side of the batholith, depressed geotherms caused by the rapidly subducting slab led to high pressure–low temperature (blueschist) metamorphism within rocks now represented by the Franciscan Complex. The intervening Great Valley Basin underwent a similar history of "refrigeration" during rapid subduction. Sediments deposited in this basin were buried as deep as 10 km, but the section reached temperatures of only about 100°C, suggesting thermal gradients of 10°C/km or less. In contrast, heat flow in the arc and backarc region to the east was high and accompanied by major shortening; the latter migrated eastward with time and resulted in the well-known Sevier foreland fold-and-

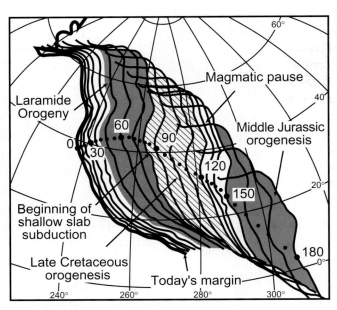

FIGURE 22.1.3 Correlation of deformational events and motion of the west coast of North America with respect to hot spots from the Late Jurassic to the Cenozoic.

thrust belt (Figure 22.1.4). At any given latitude, there are typically a series of major thrusts that displace stratified Paleozoic-Mesozoic shelf sediments eastward, with a minimum total displacement of 100–200 km. Along most of its length, the eastern front of the thrust belt closely follows the transition from thin cratonic to thicker shelf sequences, indicating important stratigraphic influence on the structures produced. Deeper parts of the crust between the thrust belt and the magmatic arc were hot and mobile and underwent thickening by folding and ductile flow (Figure 22.1.4). Crustal thickening in turn precipitated crustal melting, now represented by a belt of unusual muscovite-bearing granites that lie just west of the main thrust belt (Figure 22.1.4). Because the orogen at this latitude was later reworked by Cenozoic extension and its crust thinned, the amount of crustal thickening during the Mesozoic is still controversial. Was the western United States, at the end of the Cretaceous like the Tibetan Plateau or Andean Altiplano, underlain by 70–80-km thick crust? Or was crustal thickening more modest as evidenced by the ~50 km thick crustal root beneath the unextended Canadian Cordillera today?

22.1.4 Cenozoic History

After the latest Cretaceous, the history of the western U.S. segment of the Cordillera differs substantially from its neighboring segments to the north and south, where subduction driven arc magmatism and crustal

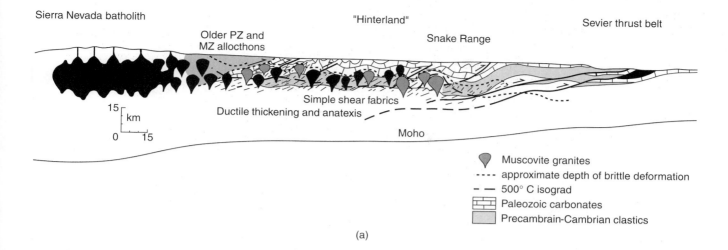

Sierra Nevada batholith   "Hinterland"   Sevier thrust belt
                                              Snake Range
     Older PZ and
     MZ allocthons

                                        Simple shear fabrics
                       Ductile thickening and anatexis

15   Moho
   km

0   15

                                                            ▽ Muscovite granites
                                                            ---- approximate depth of brittle deformation
                                                            – – 500° C isograd
                                                            Paleozoic carbonates
                                                            Precambrain-Cambrian clastics

(a)

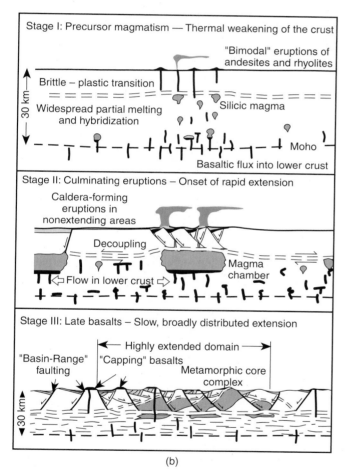

(b)

FIGURE 22.1.4 (a) Schematic crustal cross section of the western United States at the end of Mesozoic subduction-related arc magmatism and backarc crustal thickening. (b) Schematic sequence of superimposed Cenozoic Basin and Range extension-related events leading to the present (mostly young) crustal structure of the orogenic belt.

shortening continued uninterrupted into the Early to Middle Tertiary (Figure 22.1.5b). Magmatism in the western U.S. portion of the belt shut off abruptly at about 80 Ma, although subduction continued and, in fact, accelerated, achieving convergence rates of ~15cm/y. These north-to-south differences have been attributed to segmentation of the subducting slab, in which there was an extremely shallow angle of subduction beneath the U.S. portion of the belt. This hypothesis is supported by evidence for rapid cooling of the Sierra Nevada batholith as it moved into a forearc position. As the crust of the arc and backarc was "refrigerated," it regained its rheologic strength and was thus able to transmit stresses for greater distances. During this time, deformation stepped far inboard to Utah, Colorado, and Wyoming, where crustal-penetrating reverse faults caused uplift of the Rocky Mountains during the latest Cretaceous to Eocene Laramide Orogeny (Figure 22.1.5b). Their uplift was contemporaneous with continued shortening in the foreland thrust belt in Arizona and Mexico to the south, and in Montana and British Columbia to the north (Figure 21.1.5b).

Plate motions between the oceanic Kula and Farallon Plates and North America changed again at the end of the Paleocene, and the component of orthogonal convergence diminished rapidly (Figure 22.1.3). In the western United States, it is hypothesized that the shallowly dipping slab either fell away into the mantle or gradually "decomposed" due to conductive heating. Decompression melting of upwelling asthenospheric mantle into the previous region of the slab generated basalts that heated the base of the thickened continental crust, a process that caused extensive assimilation and melting of crustal rocks. This magma mixing ultimately led to eruption of large volumes of intermediate to silicic volcanic rocks (Figure 22.1.5c). Volcanism migrated progressively into the area of previously

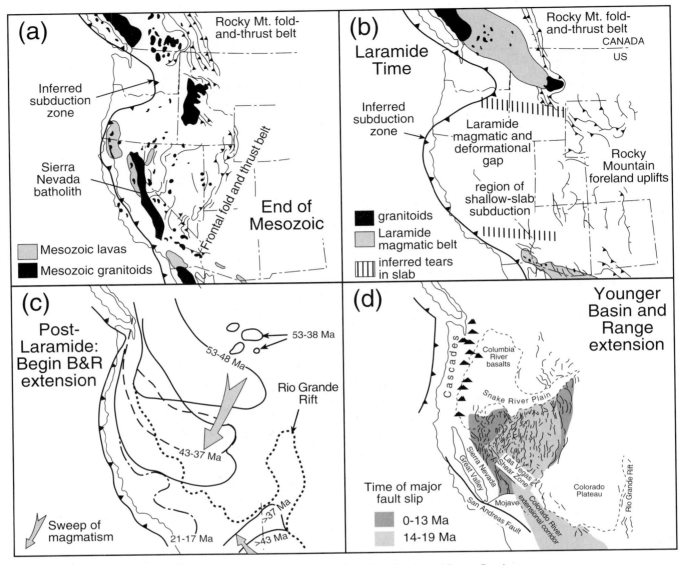

FIGURE 22.1.5 Summary of events leading up to the formation of the Basin and Range Province of the western United States.

flat slab subduction, both southeastwards from the Pacific northwest and northwards from Mexico. The large input of heat into the thick crust rheologically weakened it, and by Miocene time (about 21 Ma), when the slab finally fell away, this heat input triggered wholesale extensional collapse of much of the western United States, resulting in the formation of the present Basin and Range Province (Figure 22.1.5d). This broad zone of continental extension wraps around the southern end of the unextended but (thermally) elevated Colorado Plateau and projects as a finger northwards along the Rio Grande Rift on the eastern side of the plateau. To the west of the Basin and Range Province lies the unextended Sierra Nevada crustal block, with its thicker crustal root, and the virtually undeformed Great Valley sequence, underlain in part by oceanic crust refrigerated during the Mesozoic (Figure 22.1.5d). Volcanism and seismicity are diffuse across this broad zone of continental extension, and thermal springs abound. One of the most impressive physiographic features related to young volcanism is the depression of the Snake River plain, which is believed to represent the Miocene to Recent track of a mantle hot spot that now resides beneath Yellowstone National Park. The present Basin and Range Province, together with associated extension in the Rio Grande Rift and north of the Snake River plain, reflects ~200–300 km of east-west extension that began in the Early to Middle Tertiary and continues today. The modern, regularly spaced basin-and-range physiogeography that lends

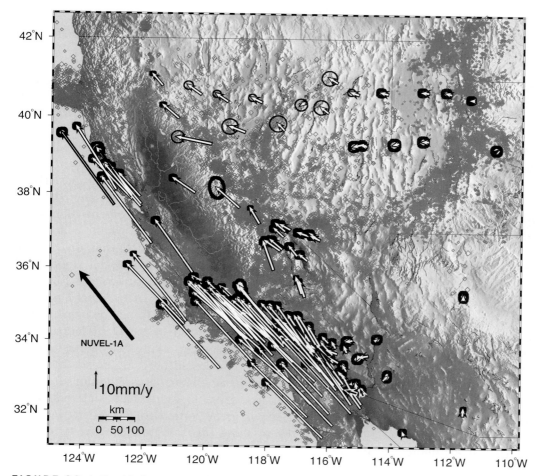

FIGURE 22.1.6 Motion vectors of various sites in the western United States relative to stable North America and NUVEL reference displacement. Site motions show strike-slip along the San Andreas fault system. Extension occurs in the Basin and Range Province north of about 36°E and changes smoothly into the strike-slip motion across a well-defined transition zone. South of 36°E, the San Andreas system accommodates most of the plate motion, and little deformation occurs in the Basin and Range.

the province its name is the surficial manifestation of the youngest system of major normal faults bounding large, tilted, upper crustal blocks (Figure 22.1.5d). Global positioning system (GPS) studies show the partitioning of strain across the western part of the U.S. Cordillera, with some strike-slip motion related to Pacific–North American plate motions taking place in the western part of the Basin and Range Province, between the relatively cold and thick crust of the Sierra Nevada and the hot and thin crust of the Basin and Range (Figure 22.1.6).

Given the long history of ocean-continent plate interaction along the western margin of North America, it may seem surprising that the actual limits and present topography of parts of the Cordillera are dictated mostly by the youngest events to affect the belt (Figure 22.1.1). For example, the Basin and Range

Province includes all or parts of the Mesozoic magmatic arc, backarc, and thrust belt, as well as older Paleozoic allochthons and sutures; it is also underlain by the Precambrian rifted western margin of North America. Despite the diversity of tectonic elements across the Basin and Range, the crust is uniformly 25–30 km thick and much of it stands >1 km above sea level, reflecting an anomalously thin and hot mantle lithosphere. The flatness of the Moho across this broad (600 km) extensional province implies that the lower crust was capable of undergoing large-scale flow during extensional deformation (Figure 22.1.4). Thus, it seems clear that the present-day structure of most of the crust and perhaps the entire lithosphere across this region reflects only the youngest event to affect this long-lived orogenic belt. This would imply that if the upper 5–10 km of the crust were removed by erosion, we would probably see very little evidence for the previous 600-m.y. history of this orogenic belt. Convergence presently occurs beneath the Alaskan-Aleutian portion of the margin and beneath the Pacific Northwest, and transform boundaries separate the North American and the Pacific Plates along most of the rest of the margin. In Alaska, shortening and associated diffuse seismicity occur in the overriding North American continent across a broad distance (1000 km). Large-magnitude subduction-zone earthquakes have occurred as recently as 1964 (beneath Anchorage) and uplift by reverse faulting has generated some of the most spectacular and rapidly rising mountains of the Cordillera, including Denali (~6,000 m) in the Alaska Range. Active shortening-related deformation extends northward to the Arctic margin of Alaska. Shortening dies out westward and is replaced by north–south extension in the Bering Strait region, where the mighty Cordillera finally ends. In the Pacific Northwest, folding and thrusting are active in the surficial part of the crust above the Cascadia subduction zone and, as predicted, a subduction-zone earthquake occurred beneath Seattle in early 2001. Detailed studies of contemporary deformation, paleoseismicity studies, and Native oral tradition suggest recurrence intervals of 300 years for such earthquakes, and raise the specter of very large (M>8.0), devastating earthquakes in the future beneath cities such as Portland and Seattle.

In California, the relative motion between the Pacific and North American Plates is partitioned into strike-slip displacement along the San Andreas Fault, and into folding and thrusting related to shortening perpendicular to the San Andreas transform plate boundary (reflected by the recent Loma Prieta and Northridge earthquakes). The exact physical explanation for the observed strain partitioning and how deformation at the surface is coupled with strain at depth in the earth's crust in such zones of continental deformation remain exciting and challenging problems for structural geologists and geophysicists.

22.1.5 Closing Remarks

This brief essay on the geologic and tectonic evolution of the North American Cordillera permit us to make several generalizations about the evolution of such orogenic belts.

- Mountain building (i.e., thickening of continental crust) is not necessarily the result of subduction and collision of allochthonous crustal fragments (terranes) along an active continental margin. Subduction occurred for long spans of time during the

history of the Cordilleran margin, but, as in the southwestern Pacific, led mostly to rifting and backarc basin development. True mountain building in the Cordillera appears to have occurred during finite time intervals of rapid convergence and increased absolute westward motion of North America, and was accompanied by magmatic activity.

- The North American Cordillera has long been cited as a classic example of continental growth by the lateral accretion of allochthonous terranes. However, this mechanism is probably not the most fundamental or important process of crustal growth, unless it involves the addition of mature island arcs. Rifting, with formation of rift basins on existing continental shelves, along continental slopes, and within island arcs, and the subsequent filling of these basins by thick prisms of sediment have contributed significantly to the formation of many terranes now incorporated in the Cordillera. Extensional thinning and reworking of continental crust or previously thickened orogenic crust, especially when accompanied by magmatic additions from the mantle, can serve to redistribute and remobilize crust across great portions of an orogen, and the results of these processes often equal or exceed estimates of crustal shortening within the same belts. The best example of this is the reworking and shape-changing of the continent during Cenozoic extension in the western United States.

- Magmatism is a process that is closely linked to deformation in mountain belts. The Cordillera provides excellent examples to illustrate that magmatism is intricately tied to deformation, in that heating causes rheologic weakening of the crust. The rise of magmas transports heat to higher levels of the crust, permitting continents to undergo large-scale deformation, whether by shortening or stretching. This is evidenced by the increasingly better documented eastward migration of magmatism and deformation of the Cordillera in the Mesozoic, as well as by the space-time relation between magmatism and extensional tectonism of the western United States in the Cenozoic.

- Many intriguing questions remain about the evolution of mountain belts such as the Cordillera. Because it is actively deforming, the Cordillera also presents us with a wonderful opportunity to study some of these questions. One of these is how strain-partitioning occurs, that is, where a particular motion vector between two plates or two parts of a continent is partitioned into different styles of deformation in different parts of the orogen (e.g., folding and thrusting in the Coast Ranges and strike-slip faulting along the San Andreas Fault in California). Other questions are understanding how seemingly incompatible strains take place within an orogen (e.g., east-west shortening in the Coast Ranges of California and east-west extension in the Basin and Range Province) and what are the driving forces for such strains. The need to answer such questions is a good reason to study contemporary deformations at the scale of the entire orogen. GPS studies measuring contemporary motion across the Cordillera (Figure 22.1.6) are an excellent way to characterize deformations at this scale.

ADDITIONAL READING

The information and ideas in this essay have been distilled from the author's own works and views, and those of many others, as represented in the various chapters of the books, *The Cordilleran Orogen: Conterminous U.S.* and *The Geology of the Cordilleran Orogen in Canada,* two of the volumes of the Geological Society of America's *Decade of North American Geology Series.* A particularly helpful review of the evolution of the western United States is given in Burchfiel et al. (see the following bibliography).

Bennett, R. A., Davis, J. L., and Wernicke, B. P., 1999. Present-day pattern of Cordilleran deformation in the western United States. *Geology* 27, 371–374.

Burchfiel, B. C., Cowan, D. S., and Davis, G. A., 1992. Tectonic overview of the Cordilleran orogen in the western United States. In Burchfiel, B. C., Lipman, P. W., and Zoback, M. L., eds., The Cordilleran Orogen: conterminous U.S., *The geology of North America,* v. G-3, Boulder, CO: Geological Society of America, 407–480.

Clarke, S. H., Jr, and Carver, G. A., 1992. Late Holocene tectonics and paleoseismicity, southern Cascadia subduction zone. *Science,* 255, 188–192.

Coney, P. J., Jones, D. L., and Monger, J. W. H., 1980. Cordilleran suspect terranes. *Nature* 288, 329–333.

Cowan, D. S., 1994. Alternative hypotheses for the mid-Cretaceous paleogeography of the western Cordillera. *GSA Today* (July), 183–186.

Dumitru, T. A., Gans, P. B., Foster, Da. A., and Miller, E. L., 1991. Refrigeration of the western Cordilleran lithosphere during Laramide shallow-angle subduction. *Geology* 19, 1145–1148.

Engebretson, D. C., Cox, A., and Gordon, R. G., 1985. Relative motions between oceanic and continental plates in the Pacific Basin. *Geological Society of America Special Paper* 206.

Gans, P. B., Mahood, G. A., and Schermer, E., 1989. Synextensional magmatism in the Basin and Range Province: a case study from the eastern Great Basin. *Geological Society of America Special Paper 233.*

Hoffman, P. F., 1991. Did the breakout of Laurentia turn Gondwanaland inside out? *Science,* 252, 1409–1419.

Humphreys, E. D., 1995. Post-Laramide removal of the Farallon slab, western United States. *Geology,* 23, 987–990.

Miller, E. L. and Gans, P. B., 1989. Cretaceous crustal structure and metamorphism in the hinterland of the Sevier thrust belt, western U.S. Cordillera. *Geology,* 17, 59–62.

Monger, J. W. H., et al. 1982. Tectonic accretion and the origin of the two major metamorphic and plutonic welts in the Canadian Cordillera. *Geology,* 10, 70–75.

Page, B. M., and Engebretson, D. C., 1984. Correlation between the geologic record and computed plate motions for central California. *Tectonics,* 3, 133–156.

Severinghaus, J., and Atwater, T. 1990. Cenozoic geometry and thermal state of the subducting slabs beneath western North America. In Wernicke, B. P., ed., Basin and Range extensional tectonics near the latitude of Las Vegas, Nevada. *Geological Society of America Memoir 176,* Boulder, CO: Geological Society of America, 1–22.

Simpson, D. W., and Anders, M. H., 1992. Tectonics and topography of the western United States—An application of digital mapping. *GSA Today* 2, 117–121.

22.2 THE CASCADIA SUBDUCTION WEDGE: THE ROLE OF ACCRETION, UPLIFT, AND EROSION—An essay by Mark T. Brandon[1]

22.2.1 Introduction

Given the constant surface area of Earth, there must be a balance between the amount of plate created at spreading centers and the amount consumed at subduction zones and other sites of plate convergence. Today, 80% of this return flow occurs at subduction zones, and the other 20% at continent-continent collision zones and diffuse oceanic convergent zones. Modern subduction zones have a total length of 51,000 km, and consume plates at an average rate of 62 km/my (or 62 mm/y), with the highest rates (210 km/my, or 210 mm/y) being found along the Tonga Trench in the southwest Pacific. Subduction zones are marked by Benioff-zone earthquakes and active-arc volcanism, which are indications of shearing and bending of the cold subducting plate and melting caused by dehydration of the plate. Seismologists have generated tomographic images that show subducting plates penetrating deep into the interior of Earth, locally reaching the core-mantle boundary (see Chapter 14).

Subduction zones are not totally efficient in removing the subducting plate. Some fraction of the plate gets left behind as **accretionary complexes** that accumulate at the leading edge of the overriding plate (Figure 22.2.1). In some cases, this accretion might be episodic, involving the collision of large lithospheric blocks, called **tectonostratigraphic terranes.** More commonly, only the sedimentary cover of the downgoing plate is accreted, while the underlying crust and mantle lithosphere are fully subducted. The thickness of this sedimentary cover varies considerably, from hundreds of meters at oceanic subduction zones, like the Mariana system, to as much as 7 km at ocean-continent subduction zones, such as the Makran margin of southwest Pakistan. There is evidence from modern subduction zones that not all of the incoming sedimentary section is accreted. The global rate of sediment subduction has been estimated in one study to be equivalent to an average thickness of 300–500 km of compacted sedimentary rock along all subduction zones combined. This analysis, however, is complicated by deeply subducted sediment that may be accreted at depth beneath the overriding plate, rather than fully subducted into the mantle.

22.2.2 Accretionary Flux

An important theme of this essay is that accretion of largely sedimentary materials has a strong influence on deformation of the overriding plate at subduction zones. We are concerned here with the **accretionary flux** into the subduction zone, which is defined as the thickness of accreted materials times the rate of plate subduction. Sedimentary rocks are quickly compacted during the subduction process, so we use the equivalent thickness of fully compacted sedimentary rock when calculating the accretionary flux. This correction for compaction ranges from ~50% for a thin sedimentary section, which would have a high average porosity, to ~20% for thick sedimentary sections, where the base of the section is already fully compacted.

The Makran subduction zone provides an upper limit for accretionary fluxes at subduction zones. There, the flux could be as high as 210 km^3/my per kilometer of subduction zone (equal to 7 km sedimentary section × 0.8 compaction factor × 37 km/my convergence velocity). It is more common to consider the

[1]Yale University, New Haven, CT.

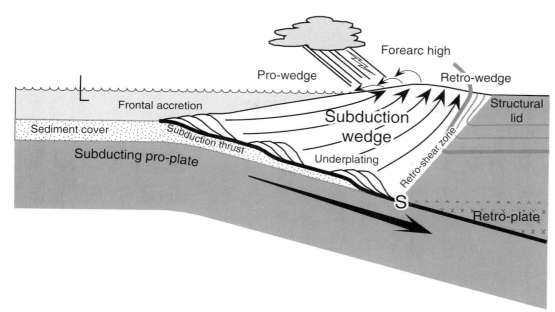

FIGURE 22.2.1 Schematic cross section of a subduction wedge. "S" refers to the S point, the subduction point, where the pro-plate is subducted beneath the retro-plate.

accretionary flux from a cross-sectional view. Thus, the flux is given as km²/my, which represents the cross-sectional area of material added to the upper plate per unit time.

22.2.3 Wedges, Taper, and Stability

The accreted material tends to accumulate in front of and beneath the leading edge of the overriding plate, forming a wedge-shaped body that grows with time (Figure 22.2.1). A useful analogy is the way that snow piles up in front of a moving plow, forming a tapered wedge that moves with the plow (representing the overriding plate). The wedge entrains snow from the roadbed, which causes the wedge to grow (see Chapter 18).

An important discovery of the 1980s was that accretionary wedges tend to maintain a self-similar form as they grow. This behavior is expected for a wedge made of frictional materials, in that the wedge taper must be blunt enough to overcome the frictional resistance to slip on the subduction thrust that underlies the wedge. A wedge that is able to slip along its base is called a stable wedge, in that the wedge does not deform or change shape as it rides above the subduction thrust. The taper geometry of the wedge introduces an important feedback. Frontal accretion tends to drive the wedge into a substable taper, where the wedge taper is now too slender to overcome friction on the subduction thrust. The wedge becomes locked to the subducting plate, and thus must deform internally to account for

the convergence velocity. The resulting horizontal shortening allows the wedge to return to its stable taper, at which point horizontal shortening stops because the wedge is now free to slip on it base. Thus, the accretionary flux into the front of the wedge controls deformation within the wedge.

Erosion will cause similar feedback, as it tends to reduce wedge taper, which causes horizontal shortening and a return to a stable taper. Erosion has a strong influence on the evolution of subaerial convergent wedges, such as the Himalaya or Alps, but subduction wedges are commonly submerged below sea level where erosion is not as active. Nonetheless, erosion can be locally important when a subduction wedge becomes emergent, which is common during the later evolution of wedges at continental subduction zones (e.g., Figure 22.2.1). Erosion acts to drive deformation and also serves to limit the growth of the wedge. This is illustrated by the examples that follow, with particular emphasis on the Cascadia margin of western North America.

22.2.4 Double-Sided Wedges

An important development in wedge theory occurred during the early 1990s, with the recognition that most convergent orogens consist of two wedges, arranged back-to-back. This idea was first proposed for collisional orogens, like the Alps, but it was also recognized as applicable for wedges that form at subduction zones.

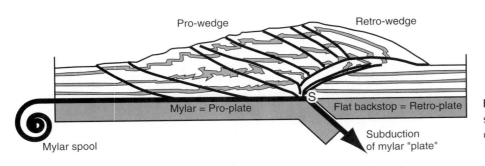

Pro-wedge
Retro-wedge

Mylar = Pro-plate
S
Flat backstop = Retro-plate

Mylar spool

Subduction
of mylar "plate"

FIGURE 22.2.2 Sketch of sandbox experiment producing a double-sided wedge.

The basic problem with the snowplow analogy was that it was hard to identify where the strong plow-blade, or backstop as it is commonly called, was located within the overriding plate. Many authors used relative strength to define backstops within the crust. However, in many cases, the backstop region could be shown to be deforming along with the accreted material in front of it. This contradicted the basic idea of a backstop, which is supposed to be a rigid part of the overriding plate.

The solution to the backstop problem was inspired by sandbox experiments (Figure 22.2.2) in which a mylar sheet serves as the subducting plate and a "fixed" flat-lying plate serves as the overriding plate. This arrangement of "plates" was covered with continuous layers of sand, representing the deformable crust above the plates. A motor was then used to draw the mylar downward through a slot in the base of the sandbox, thus simulating the motion of the subducting plate. This subduction caused the overlying sand layer to deform into a double-sided wedge centered over the subduction point, or S point. The overriding plate served as a backstop, but in this case the backstop was flat-lying. Notably, there is no visible backstop at the surface. Instead, the backstop is a deep-seated structural element, with a flat-lying geometry that is hidden from view.

In a geodynamic perspective, deformation within a convergent wedge is driven by the motion of a subducting pro-plate and an overriding retro-plate. These two plates are rigid, and correspond to relatively strong lithospheric mantle, whereas the overlying crust is deformable, and thus must accommodate the velocity discontinuity at the S point. In simple terms, the model postulates that convergent wedges overlie a deep-seated mantle subduction zone. It is the cold strong mantle in the plates that controls their plate-like behavior. The crust is draped over the zone of mantle subduction, and thus deforms in a more distributed matter to accommodate subduction.

Wedges associated with subduction zones are called **accretionary wedges,** but the term has led to the think-ing that the size of an actively deforming wedge is defined by the volume of accreted sediment. The concept of a double-sided wedge provides a different view, in that the active wedge involves both accreted sediments and upper-plate rocks. This result stems from the fact that the width of the wedge is determined by the motion of the underlying mantle plates, and not by any specific distribution of strength in the crust. Using terms like **subduction wedge** or **convergent wedge** when describing double-sided wedges helps to distinguish this model from that of a small wedge of accreted sediment bounded by a strong crustal backstop.

22.2.5 Subduction Polarity and Pro-Side Accretion

The term facing is used to describe the polarity of subduction. For instance, a west-facing subduction zone indicates that the overriding plate is moving westward relative to the subducting plate. This asymmetry can be ignored when measuring convergence velocities across a subduction zone, but it does have an important influence on how the wedge grows with time. Consider the situation where the pro-plate is subducted, while the retro-plate remains largely unconsumed. This asymmetry means that most of the accretionary flux is carried into the pro-side of the convergent wedge. When accretion occurs on the retro-side as well, it is always at a much slower rate than on the pro-side.

The dominance of pro-side accretion creates a strong tendency for all material in the wedge to move horizontally from the pro-side of the wedge to the retro-side. This situation is further influenced by the pattern of surficial erosion across the wedge. Consider accretion on the pro-side and strong erosion on the retro-side. This pattern of accretion and erosion will produce the greatest horizontal velocities within the wedge. An example of this situation is the Southern Alps of New Zealand, a beautiful active mountain range formed by oblique subduction of the Pacific Plate beneath the Australian Plate. The retro-wedge

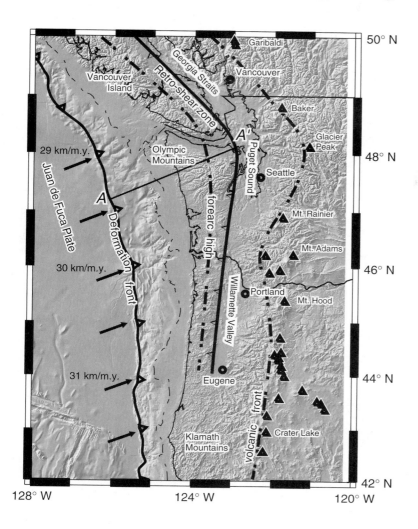

FIGURE 22.2.3 Modern tectonic setting of the Cascadia margin, emphasizing the plate boundary, volcanic arc, and physiography of the modern margin. *A-A'* shows the location of the cross section in Figure 22.2.4. Velocity vectors show the modern velocity of the Juan de Fuca Plate relative to a fixed overriding plate. Recent geodetic work has shown that the Cascade arc and forearc move as a separate plate relative to North America, due in part to deformation across the Basin and Range Province, which lies behind the southern half of the Cascade arc. The velocities shown here at the subduction zone account for this additional plate.

lies on the west side of the South Island, and is subjected to big storms coming from the Tasman Sea. Average precipitation there is more than 7 m per year, whereas the pro-wedge, located on the drier east side of the range, has an average precipitation of ~1 m per year. As a result, erosion associated with those Tasman Sea storms tends to move material from the pro-side to the retro-side of the wedge, over a horizontal distance of ~90 km.

At this point it is useful to distinguish two important geologic features within the wedge system: the structural lid and the accretionary complex. The structural lid refers to the tapered leading edge of the overriding plate, which tends to get involved in wedge deformation. The accretionary complex refers to those materials that were accreted into the wedge. The important distinction is that the structural lid originated as part of the retro-plate, whereas most, if not all, of the accretionary complex is derived by accretion from the pro-plate. From the examples that follow, we will see that, as the accretionary complex grows, the structural lid tends to get shouldered aside to the rear of the wedge.

A common result is that the structural lid is only preserved within a large backfold within the retro-wedge.

22.2.6 The Cascadia Subduction Zone

The Cascadia subduction zone has a length of 1300 km and is located along the continental margin west of northern California, Oregon, Washington, and Vancouver Island (Figures 22.2.3 and 22.2.4a). In this area, the offshore Juan de Fuca Plate is actively subducting beneath an overriding North American Plate at rates of ~30 km/m.y. (arrows in Figure 22.2.3; see caption for details on estimation of convergence velocities). The direction of convergence is approximately orthogonal to the subduction zone. The surface trace of the subduction thrust is entirely submarine (deformation front in Figures 22.2.3 and 22.2.4a), lying at ~2500 m below sea level. Most of the plate convergence is accommodated by slip on the subduction thrust, but there is also a subordinate amount of shortening in the overriding subduction wedge. Slip on the subduction thrust is thought to occur episodically during great thrust

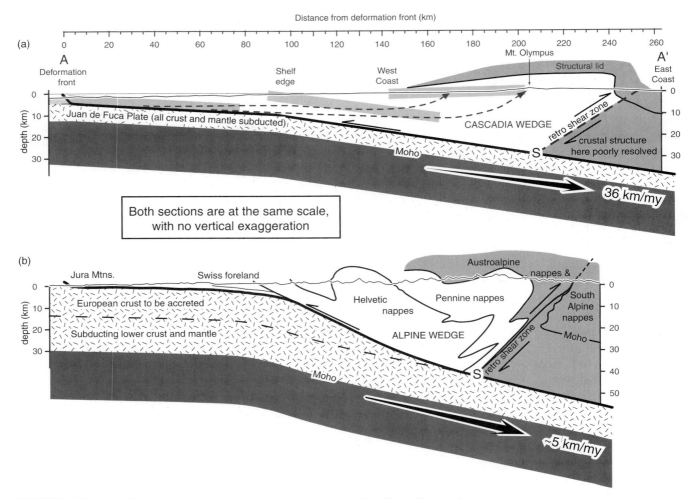

Distance from deformation front (km)

FIGURE 22.2.4 Comparison in cross section of the structure of the Cascadia margin across northwest Washington State and the structure of the European Alps across central Switzerland. The gray bands in the Cascadia section illustrate the displacement history of sediments presently exposed in the western Olympics, which were accreted at the front of the wedge at ~22 Ma and then moved rearward through the wedge, reaching the west side of the Olympics in the present.

earthquakes, with recurrence times of ~500 years. The last subduction-zone earthquake occurred on AD January 26, 1700. The Juan de Fuca Plate is covered by ~2500 m of sediment. Geophysical and geochemical evidence indicates that all of the incoming sedimentary section is accreted into the Cascadia wedge, whereas the underlying crust and mantle are fully subducted. These observations allow an estimate of the present accretionary flux of ~52 km²/my.

The subduction zone parallels the Cascade volcanic arc, which includes active volcanoes such as Mt. Rainer. Studies at modern subduction zones indicate that the subducting slab has to reach a depth of ~100 km for melting to start in the overlying mantle. Thus, the Cascade volcanic front can be viewed as an approximate indicator of the 100-km depth contour of the down-

going slab. With this in mind, note that the distance between the subduction zone and the arc is largest across the Olympic sector of the Cascadia margin. The reason is that the dip of the subducting slab varies along the subduction zone and that the shallowest dip is found beneath the Olympic Mountains. We return to this point subsequently because the shallower dip of the slab beneath the Olympics has had a strong influence on the evolution of the Cascadia wedge in that area.

The forearc, which is the region between the arc and the subduction zone, is marked by a low and a high, both of which parallel the general trend of the subduction zone. The forearc low includes the Georgia Straits adjacent to Vancouver Island, Puget Sound of western Washington State, and the Willamette Valley of western Oregon. The forearc high corresponds to the Insular

Range, which defines Vancouver Island, and the Coast Range of western Washington and western Oregon.

The forearc high represents the crest of the Cascadia subduction wedge, with a pro-wedge to the west and a smaller retro-wedge to the east. The retro-wedge shear zone, shown in Figures 22.2.3 and 22.2.4a, marks the eastern limit of permanent deformation and uplift associated with the subduction wedge. The structure exposed there is not a fault zone, but rather a large eastward vergent "backfold," which is actively accommodating slow, top-to-east shear between the retro-wedge and the forearc low. The forearc low is commonly viewed as a basin, but there has been little subsidence or sediment accumulated in this low over the last 10 to 15 my. Instead, the low is defined by the volcanic arc to the east and the actively growing forearc high to the west.

Geologic evidence indicates that the Olympic Mountains mark the first part of the Cascadia forearc high to emerge above sea level, at ~15 Ma. The forearc high apparently grew more slowly elsewhere along the margin. In fact, Vancouver Island and the southern part of the Coast Ranges may have emerged above sea level only within the last 5 to 10 my. The early emergence of the forearc high in the Olympics may reflect a slab that is shallower there. To understand this situation, consider that the growth of a subduction wedge is controlled by the accretionary flux into the wedge and erosion of the forearc high. Erosion cannot start until the forearc high becomes subaerially exposed, but once the wedge does become emergent, erosion will act to slow the growth of the wedge, until it reaches a flux steady state, when the erosional outflux becomes equal to the accretionary influx.

It was already noted that the subducting slab has a shallower dip beneath the Olympics relative to adjacent parts of the subduction zone. This archlike form of the subducting slab is apparent when one considers the depth of the slab beneath the forearc high, which is equal to ~40 km beneath Vancouver Island, ~30 km beneath the Olympics, and ~40 km beneath southwest Washington State. This variation in slab depth cannot be attributed to variations in topography, because the mean elevation of the forearc high varies by only a small amount, between ~500 to 1000 m along its length. Thus, this configuration may be a fundamental feature of the subduction geometry of the slab.

The important point here is that the wedge beneath the Olympics is much smaller than elsewhere along the margin. Thus, given a similar accretionary flux, less time was needed for the forearc high to become emergent in the Olympics relative to other parts of the subduction wedge.

Erosion of the Forearc High

The effect that erosion has on deformation and growth of the wedge was noted earlier. Several methods were used to measure the long-term erosional flux coming out of the Olympics part of the forearc high. One method is **apatite fission-track dating,** which dates when a sample cooled below the apatite fission-track closure temperature (~110°C). A thermal model is used to convert the closure temperature to a depth, with results that typically lie in the range of 3–5 km. Dividing depth by cooling age gives the average erosion rate for the time interval represented by the cooling age.

Another method involves measuring the **incision rate** of rivers, which is the rate at which a river cuts downward into bedrock. The rate of channel incision can be determined from old river deposits preserved on the hillslopes adjacent to the river. These deposits commonly overlie old remnants of the river channel where it was running on bedrock. The height of these paleochannel remnants above the modern channel divided by the age of the paleochannel gives the incision rate of the river.

In the Olympics, paleochannel remnants formed at ~65 ka and 140 ka produced 60 incision rates along the Clearwater River, which drains the west side of the forearc high. Apatite fission-track ages provide another 43 estimates of erosion rates, with average cooling ages of ~7 my. These data are shown in Figure 22.2.5, where they have been projected into the section line A-A' (Figure 22.2.4) across the subaerially exposed part of the forearc high. Particularly interesting is the similarity between incision rates and fission-track-based erosion rates, despite the fact that they represent local versus regional processes and different intervals of time as well; that is, 100 ky versus 7 my. Recently (U-Th)/He apatite dating found younger cooling ages of ~2 my, which reflect the lower closure temperature (~65°C) for this thermochronometer. The erosion rates indicated by these ages (not shown) also match closely the rates shown in Figure 22.2.5. The conclusion is that, when viewed on timescales longer than ~50 ky, the pattern and rates of erosion across the forearc high appear to have been fairly steady.

Let's compare these rates with the accretionary flux that we estimated previously. We use the data in Figure 22.2.5 to estimate the integrated erosional flux from the forearc high. The curve provides a smoothed version of the data and is used to integrate the flux. This gives an erosional flux of 51 km²/my. Our thermal-kinematic modeling gives a more precise estimate of 57 km²/my. Both estimates are similar to the accretionary flux, 52 km²/my, estimated

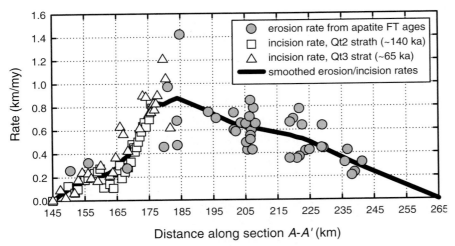

FIGURE 22.2.5 Fluvial incision rates and long-term erosion rates determined from apatite fission-track ages. The fluvial incision rates represent downcutting of the Clearwater River over a timeframe of ~100 ka, whereas the fission-track ages indicate erosion rates over a time frame of ~7 my. The similarity in rates suggests that the pattern and rates of erosion have been steady across this part of the forearc high. The black curve represents a smoothed version of the erosion rate distribution across the high. Integration of this curve indicates that the long-term erosional flux from the forearc high is ~51 km²/my.

previously, showing that there is a close balance between accretionary flux into the wedge and the erosional flux out of the wedge. Thus, the wedge cannot get any larger with time. Even so, the material within the wedge will still continue to move and deform, given that an eroding wedge must thicken to maintain a stable taper.

(U-Th)/He apatite dating in other parts of the Cascadia forearc arc high shows that everywhere else the forearc high has been only slightly eroded, by less than 3 km. Estimates of erosion rates suggest that these regions have only recently become emergent and have yet to reach the steady-state configuration observed in the Olympic Mountains. It would be interesting to predict how much time is needed for a growing subduction wedge to reach a steady-state configuration. To do so requires a better understanding of how local-scale erosional processes scale up in time and space to account for the large-scale evolution of orogenic topography. This is an area of active research, with much promise for new discoveries.

DEFORMATION AND EROSION OF THE STRUCTURAL LID

How have accretion and erosion influenced the longer-term evolution of the Cascadia wedge? Prior to initiation at ~35 Ma, the continental margin included a large oceanic terrane, Eocene in age, called the Crescent terrane. This terrane was already part of the outboard edge of the North American Plate, having been accreted to North America prior to 35 Ma. Thus, when the Cascadia subduction zone was initiated, the Crescent terrane became the structural lid for the subduction zone (Figure 22.2.4a).

Uplift and erosion in the Olympics provide a good sectional view of the Crescent terrane (Figures 22.2.4a and 22.2.6). It consists of relatively coherent Eocene oceanic crust, mostly pillowed flows of basalt, which are typical of submarine volcanism. Exposures along the north side of the Olympic Mountains show the tapered form of the structural lid (Figure 22.2.6). Across the 120-km width of the forearc high, the basaltic sequence changes from 4 km thickness in the west to more than 15 km in the east. This tapered geometry marks the original eastward dip of the Cascadia subduction zone when it was first formed. At that time, the lower part of the Crescent terrane was apparently subducted, leaving behind the tapered lid, which then became the highest structural unit within the Cascadia wedge.

In the Olympic Mountains, the structural lid is underlain by an accretionary complex, called the Olympic subduction complex. In cross section, this unit is 240 km across and extends to a depth of at least 30 km, giving it a cross-sectional area of 3600 km². Seismic studies indicate that this accretionary complex is made up entirely of sedimentary rocks; there is no evidence of basaltic crust or mantle from the subducting plate. Seismic reflection profiles across southern Vancouver Island indicate that the Crescent terrane there is also underlain by a large volume of low-velocity layered rocks, similar to the Olympic subduction complex. Given the present accretionary flux, it would take ~70 my to grow an accretionary complex this size, which is much longer than the 35 my age of the Cascadia subduction zone. Evidence in the Olympics indicates that the more internal parts of the accretionary complex are made up of sedimentary sequences that probably predated the subduction zone. This makes sense, given that the newly formed subduction zone cut across a preexisting continental margin. Thus the sedimentary units that flanked that margin were probably the first to be accreted.

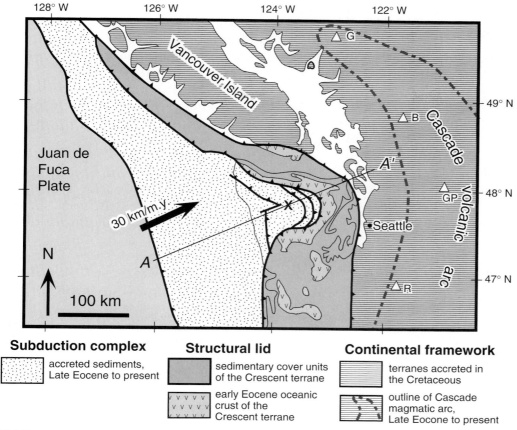

Subduction complex

	accreted sediments, Late Eocene to present

Structural lid

	sedimentary cover units of the Crescent terrane
v v v v v v v v v v v v v	early Eocene oceanic crust of the Crescent terrane

Continental framework

	terranes accreted in the Cretaceous
	outline of Cascade magmatic arc, Late Eocone to present

FIGURE 22.2.6 Geology of the Cascadia margin, emphasizing the present configuration of the structural lid. "X" marks Mount Olympus, which is the highest summit (2,428 m) in the Washington-Oregon Coast Range. A-A' indicates the location of the cross section shown in Figure 22.2.3. Major volcanoes of the Cascadia arc are: G = Mt. Garibaldi, B = Mt. Rainer, GP = Glacier Peak, R = Mt. Rainier.

In the Olympics, the structural lid has slowly been driven into the back of the wedge system as the accretionary part of the wedge has grown in size. Initially, the lid was uplifted into a broad arch, as observed in the Oregon Coast Range. When this arch became emergent, it started to erode, which allowed the lid to rise faster. As the lid was removed, the accretionary complex moved rearward to take its place. This motion was required for the wedge to maintain a stable taper, as discussed earlier.

The end result is that the lid was driven rearward, creating a large fold. This fold is actually the manifestation of a broad, west-dipping retro-shear zone, which accommodates the top-east motion of material within the retro-wedge. This structure is quite impressive in the Olympics. One can drive into the mountain range from the east, where the Crescent terrane is still flat-lying and undeformed. Moving west across the range front, one notices that the Crescent starts to dip to the east, which marks the fact that we have crossed into the

upper limb of the back fold. The dip continues to increase until the section becomes vertical, and locally overturned. This increase in limb dip is also accompanied by a dramatic topographic gradient, which marks the east-dipping surface of the retro-wedge. Within 20 km, the topography rises from sea level to over 2300 m.

The Olympics represents the most evolved part of the Cascadia forearc high, as indicated by uplift, erosion, and backfolding of the structural lid. Elsewhere along the Cascadia margin, the structural lid still covers the forearc high. This difference in evolution of the margin accounts for the large reentrant in the Crescent terrane centered on the Olympic Mountains. Other parts of the forearc high continue to grow, but that growth has only resulted so far in a broad arching of the structural lid. The prediction is that with further deformation and erosion, those other parts of the forearc high will evolve towards the structural configuration observed in the Olympics.

22.2.7 Comparison between the Cascadia and Alpine Wedges

The cross sections in Figure 22.2.4 offer a comparison between the Cascadia wedge and Europe's Alpine wedge (see Subchapter 21.1). The Swiss Alps are different in that they formed by collision of two continental plates, starting ~50 Ma. Convergence there has slowed down, and perhaps stopped within the last 5 my The average Alpine convergence rate is ~5 km/my, which is slower than the Cascadia convergence by a factor of six. However, in the European Alps, the pro-thrust cuts much more deeply into the pro-plate, resulting in accretion of the upper 15 km of crust from the pro-plate (Figure 22.2.4b). The associated accretionary flux for the Alps is therefore estimated ~75 km^2/my, which is slightly larger than for the Cascadia wedge. Nevertheless, the Alpine wedge is smaller than the Cascadia wedge, but because the Alpine wedge is entirely subaerial, erosion was able to more effectively limit the size of the wedge.

This comparison highlights the similarities in backfolding of the structural lid. In the European Alps, the structural lid is made up of crustal rocks belonging to the Southern Alpine and Austroalpine nappes (Section 21.1). These nappes are thrust sheets associated with the tapered leading edge of the overriding continental plate, which is called the Adriatic Plate. As in the Olympics, the tapered geometry of the structural lid reflects the cut made through the upper plate by the subduction thrust when subduction was initiated. Accretion from the pro-side is responsible for driving the structural lid rearward in the wedge. The Alpine geologists call this retrocharriage, meaning "to carry back." The resulting backfold underlies the southern flank of the Alps.

ADDITIONAL READING

Batt, G. E., Brandon, M. T., Farley, K. A., and Roden-Tice, M., 2001. Tectonic synthesis of the Olympic Mountains segment of the Cascadia wedge, using two-dimensional thermal and kinematic modeling of thermochronological ages. *Journal of Geophysical Research*, 106, 26731–26746.

Beaumont, C., Ellis, S., and Pfiffner, A., 1999. Dynamics of sediment subduction-accretion at convergent margins: short-term modes, long-term deformation, and tectonic implications. *Journal of Geophysical Research*, 104, 17573–17602.

Brandon, M. T., Roden-Tice, M. K., and Garver, J. I., 1998. Late Cenozoic exhumation of the Cascadia accretionary wedge in the Olympic Mountains, Northwest Washington State. *Geological Society of America Bulletin*, 110, 985–1009.

Clowes, R. M., Brandon, M. T., Green, A. G., Yorath, C. J., Sutherland Brown, A., Kanasewich, E. R., and Spencer, C., 1987. LITHOPROBE-southern Vancouver Island: Cenozoic subduction complex imaged by deep seismic reflections. *Canadian Journal of Earth Sciences*, 24, 31–51.

Dahlen, F. A., 1990. Critical taper model of fold-and-thrust belts and accretionary wedges. *Annual Review of Earth and Planetary Sciences*, 18, 55–99.

Escher, A., and Beaumont, C., 1997. Formation, burial, and exhumation of basement nappes at crustal scale; a geometric model based on the western Swiss-Italian Alps. *Journal of Structural Geology*, 19, 955–974.

Jarrard, R. D., 1986. Relations among subduction parameters. *Reviews of Geophysics*, 24, 217–284.

Malavieille, J., 1984. Modelisation experimentale des chevauchements imbriques; application aux chaines de montagnes. *Bulletin de la Societé Géologique de France*, 26, 129–138.

Pazzaglia, F. J., and Brandon, M. T., 2001. A fluvial record of long-term steady-state uplift and erosion across the Cascadia forearc high, western Washington State. *American Journal of Science*, 301, 385–431.

Stewart, R. J., and Brandon, M. T., 2003. Detrital zircon fission-track ages for the "Hoh Formation": implications for late Cenozoic evolution of the Cascadia subduction wedge. *Geological Society of American Bulletin*, 115, in press.

von Huene, R., and Scholl, D. W., 1991. Observations at convergent margins concerning sediment subduction, subduction erosion, and the growth of continental crust. *Reviews of Geophysics*, 29, 279–316.

Willett, S. D., and Brandon, M. T., 2002. On steady states in mountain belts. *Geology*, 30, 175–178.

Willett, S., Beaumont, C., and Fullsack, P., 1993. Mechanical models for the tectonics of doubly vergent compressional orogens. *Geology*, 21, 371–374.

22.3 THE CENTRAL ANDES: A NATURAL LABORATORY FOR NONCOLLISIONAL MOUNTAIN BUILDING—An essay by Richard W. Allmendinger and Teresa E. Jordan[1]

22.3.1 Introduction

One of the great early advances of plate tectonics was the realization that mountain building is associated with activity at the margins of tectonic plates. The **Andes** represent the case of mountain building produced by the convergence of an oceanic and a continental plate, a relatively simple and elegant end member in the spectrum of orogenesis (Figure 22.3.1). The other extreme is represented by continent-continent collision, for example the Himalayan-Tibetan system. Because of the association of the Andes with andesites and the chain of volcanoes that ring the Pacific, a widespread misconception among geologists outside of South America has been that the Andes Mountains are primarily a volcanic edifice. In fact, it is now clear that the crustal thickening that produced the Andes is mostly structural in origin and that the volcanoes for which the mountain belt is best known sit on top of that structural welt.

During the 1980s, many investigators sought evidence of accretionary events in the Andes that might have been responsible for Andean mountain building. Underlying this search was the concept that a major mountain system produced by horizontal contraction must be produced by collision of two buoyant crustal masses. However, nearly a decade of paleomagnetic study in Chile and Peru has turned up no evidence that accretion played even a minor role in building the modern Andes. Instead, the evidence suggests that material was removed from the continental margin during the Andean orogeny.

Thus, in the central Andes mountain building occurs by dominantly structural processes in a noncollisional

setting, in which the oceanic Nazca Plate is subducted beneath the continental South American plate (Figure 22.3.2). The purpose of this essay is to describe these processes, as well as the general tectonic setting of the Andes. Several factors make the central Andes, located between 5° and 35°S latitude, a unique laboratory of orogenesis. Because the deformation is active today, the governing boundary conditions can be identified and, in some cases, quantified. These include (1) the plate convergence rate and obliquity, (2) the geometry of the subduction zone, and (3) the dynamic topography, which reflects the interaction of tectonic and climatic processes. Furthermore, crustal seismicity gives us an idea of short-term strain rates, as well as the distribution of modes of failure in the continental crust.

The modern Andes are commonly regarded as a recent analog for ancient mountain belts, such as in the Mesozoic-Early Tertiary Cordillera of the western United States (Figure 22.3.3). Although the modern setting is quite simple, western South America has a complex pre-Andean history. Thus, the starting materials for Andean deformation were extremely heterogeneous and their responses to the stresses that produced Andean deformation are equally varied. This factor becomes particularly important when one tries to decide whether a particular structural geometry owes its existence to the modern plate setting, or to ancient anisotropies of the continental crust.

22.3.2 The Andean Orogeny

Subduction of oceanic crust beneath western South America has occurred more-or-less continuously since the Middle Jurassic. The term "Andean Orogeny" refers collectively to all tectonism that occurred since the Jurassic. Although subduction has been continuous, the

---

[1]Both at Cornell University, Ithaca, NY.

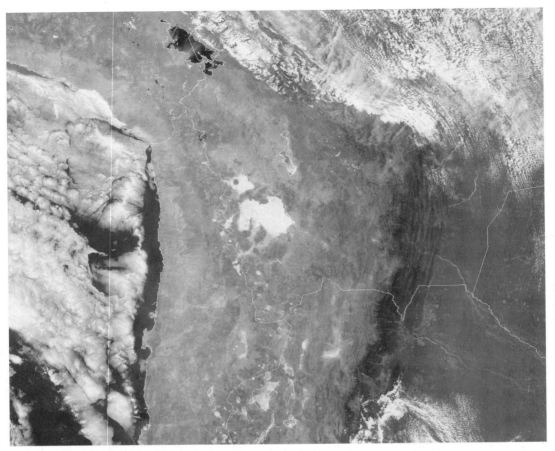

FIGURE 22.3.1 Landsat image of the central Andes in South America.

obliquity and rate of convergence, as well as the dip of the subducted plate, have varied considerably. Thus, the style and distribution of mountain building have not been continuous or uniform. Here we focus mostly on structures developed during the last 30 m.y. of the Andean Orogeny, but we begin by briefly reviewing the older events.

Arc-related igneous rocks of Jurassic and Cretaceous age occur along the present coastline in Chile and Peru and, as one moves progressively farther inland (i.e., eastward), the arc becomes progressively younger. The Mesozoic magmatic arc is anomalously close to the present-day trench (within 75 km in some cases), leading to the conclusion that the Mesozoic forearc region has been stripped off the margin and has either been subducted or perhaps moved laterally.

In many areas, the first significant deformation of continental crust associated with Andean subduction was horizontal extension. Based primarily on interpretation of lithologic associations, the Jurassic and Cretaceous has long been suspected as a time of intra-arc or backarc rifting. Both in the northern Andes of Colombia and Ecuador and in the southernmost Andes of Tierra del Fuego, this rifting culminated in the pro-

duction of new oceanic crust. In the central Andes, the clues are more subtle, but work in northern Chile at 27°S has yielded unique data on the geometry and kinematics of the extensional structures of this event. Those structures, low-angle normal faults, extensional chaos, domino blocks, and so on, are geometrically comparable to those seen in Cenozoic detachment terranes of the western United States.

Early Tertiary Andean orogenesis is commonly referred to as the "Incaic Orogeny." This phase of deformation is largely restricted to the present forearc in Chile and Peru. However, deep erosion on the eastern side of the central Andes northeast of La Paz, Bolivia, has revealed rocks that were metamorphosed (or cooled through about 300°C) at about 40 Ma, the same time as mountain building occurred farther west.

The morphologic edifice that we associate with today's Andes is a product of mountain building during just the last 30 m.y. Most of the surface-breaking structures associated with this phase of deformation are concentrated within the high topography and on the eastern side of the mountain range. Many workers refer to this young phase of deformation as the "Quechua Orogeny," although different workers use

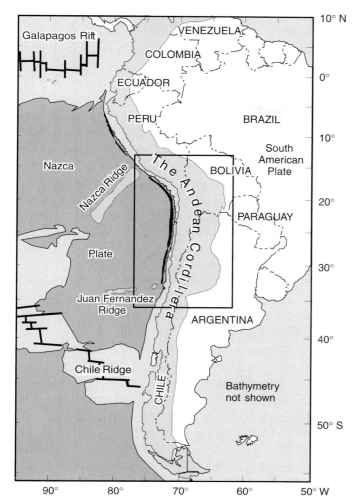

FIGURE 22.3.2 Generalized map of the Nazca-South American plate boundary showing the principal bathymetric features of the Nazca Plate and the extent of the Andean Cordillera. Bathymetric contours shown at −3500 m and −5500 m; the latter, with depths greater than −5500 m shown in black, marks the position of the Peru-Chile Trench. The box shows the position of Figure 22.3.3.

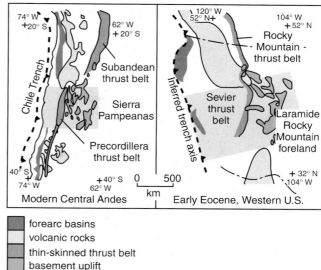

forearc basins
volcanic rocks
thin-skinned thrust belt
basement uplift
flat subduction

FIGURE 22.3.3 Comparison of the major tectonic elements of the modern central Andes (left) and the early Eocene of the western United States (right) both at the same scale.

this term to refer to events of different ages. The topography of the Andes is the result of structural and thermal processes and is not simply due to piling up of volcanic rocks; in the central Andes, deformed Paleozoic to Cenozoic rocks are commonly found as high as 4 km or more, where they form the great bulk of the high plateau known as the Altiplano. Miocene to recent volcanoes are perched on top of this plateau.

22.3.3 Late Cenozoic Tectonics of the Andes

The Nazca Plate is currently being subducted beneath western South America in a direction of 077±12° at about 10 cm/y, a rate that varies little along the strike

of the plate boundary from the triple junction at 49° S to at least 0° of latitude in Ecuador (Figure 22.3.4). Nonetheless, the geometry of the subducting slab is highly variable for reasons that are not totally understood. Between the triple junction and ~33° S, the slab dips ~30°E, but can only be tracked to a position directly beneath the modern arc; subduction zone earthquakes beneath the backarc region are virtually unknown. From 33° to 28°S, the subducted plate dips at ~30° down to depths of ~100 km, but it is nearly flat farther east; nearly 600 km east of the plate boundary beneath the city of Córdoba, Argentina, the subducted plate is only about 200 km deep. This segment of the Andes has had no volcanism since the late Miocene. To the north of 28°S, the plate gradually steepens and, by ~24 S, has resumed its uniform 30°E dip. This central steep-dipping segment correlates with the Central Volcanic Zone of the Andes as well as the high plateaus known as the Altiplano and Puna. The slab in this segment can be traced to a depth of nearly 600 km beneath the foreland of the Andes. At 15°S beneath southern Peru, the subducted slab again flattens to a near horizontal attitude below 100 km depth. This northern "flat slab" underlies most of Peru. Like their southern counterparts, the high Andes here are quite narrow and recent volcanism is absent.

It has been proposed that subduction of the Nazca and Juan Fernandez oceanic ridges (Figure 22.3.2) are responsible for flat subduction. The latter does

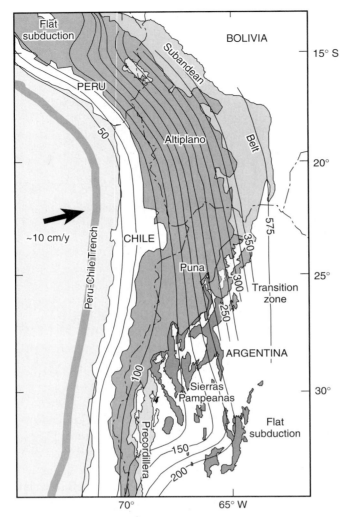

FIGURE 22.3.4 Tectonic provinces of the central Andes. Thin smooth lines are contours on the Wadati-Benioff zone (interval = 25 km). The medium gray region marks the area above 3 km elevation. Foreland thin-skinned thrust belts are shown in light gray; thick-skinned provinces in dark gray. Note that the thrust belt is not continuous but is intersected by the thick-skinned Sierras Pampeanas.

coincide closely with the boundary between flat and steep subduction at ~33°S, but the former is located 1° to 2° north of the flat-steep boundary beneath southern Peru. The plate kinematics clearly shows that the ridges have swept progressively southward along the plate boundary, but the continental geology shows no evidence of this effect.

Tectonic Segmentation of the Central Andes

The geology of the Late Cenozoic Andes to a first order reflects the lateral segmentation of the subducting plate (Figure 22.3.5). In our discussion, we will concentrate on two swaths across the Andes, one over

the nearly flat segment of the subducted Nazca Plate between 28° and 33°S and the other over the 30°E-dipping segment between 15°S and 24°S, with brief references to other parts of the Andes.

The Andes from 15°S to 24°S. This segment of the central Andes overlying a 30°E-dipping subducted slab most closely approximates the average geologist's image of the Andean orogen "type" (Figures 22.3.4 and 22.3.5a). A cross section from west to east across this segment would show the plate boundary, a forearc region dominated by the longitudinal valley of northern Chile, the active volcanic arc, a high continental plateau system bounded on the east by the Eastern Cordillera, and the low-lying Subandean belt of thrusts and folds. The segment is dominated by the Altiplano-Puna Plateau, a region of more than 500,000 km^2 elevated to an average height of 3.7 km. There has been active volcanism in this segment for about the last 25 m.y. and, although this activity is focused in the Western Cordillera, locally it reaches the eastern edge of the plateau system. The distribution of volcanic rocks, modern morphology, ancient geomorphic surfaces, and structural geometry have led to the proposal of a two-stage uplift model for the plateau. During the Miocene, the crust beneath the current plateau was thermally weakened, resulting in horizontal shortening of the entire crust. At about 10 Ma, shortening mostly ceased in the plateau and in the Eastern Cordillera and migrated eastward into the Subandean belt, a thin-skinned foreland fold-thrust belt. At that point, the cold lithosphere of the Brazilian Shield began to be thrust beneath the mountain belt.

The Andes from 28°S to 33°S. This segment overlies the southern flat subduction zone (Figures 22.3.4 and 22.3.5b). The high Andes (above 3 km elevation) are narrow, but include Mt. Aconcagua, at 7 km the highest peak in the Western Hemisphere. Although they are largely composed of volcanic rock, they are not volcanic edifices but structural uplifts; magmatism has been lacking in this segment for the last 5–10 m.y. The magmatic history of the segment indicates that the subducted plate beneath Argentina shallowed between 16 and 6 Ma. There is some evidence from the flat subduction segment in Peru to indicate that shallowing of the slab began as recently as 2 Ma. The cessation of volcanism where subduction is flat is thought to be due to the virtual lack of asthenosphere between the subducted and overriding plates.

In the Argentine flat segment, the Sierras Pampeanas are thick-skinned basement uplifts with structural geometries very reminiscent of the Laramide Rocky Mountain foreland of the western United States (Figure 22.3.3). Like the Rocky Mountain foreland,

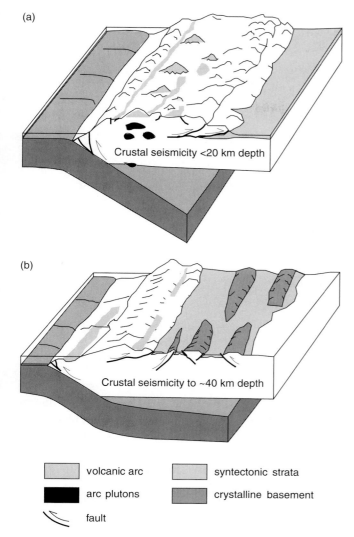

(a)

Crustal seismicity <20 km depth

(b)

Crustal seismicity to ~40 km depth

▨	volcanic arc	▨	syntectonic strata
■	arc plutons	▨	crystalline basement
⌒	fault		

FIGURE 22.3.5 Schematic block diagrams showing the tectonic segmentation of the central Andes.

many blocks of the Sierras Pampeanas display an exhumed Paleozoic-Early Mesozoic erosional surface that has been rotated, probably by movement on listric faults, during the last 5 m.y. Seismicity associated with the westernmost of these basement uplifts indicates brittle failure, at least on short timescales, to depths of 40 km, nearly the entire thickness of the continental crust. This is unusually deep for continental deformation and is well below the predicted depth for the brittle–plastic transition, using even the most conservative parameters for the frictional and power law rheologies and heat flow on which those models are based. Because of extensive jungle cover, much less is known of the style of foreland deformation located over the Peruvian flat segment north of 15°S.

Although the thick-skinned basement uplifts are the most obvious features of the foreland overlying the Argentine flat slab, thin-skinned deformation also

occurred between the Sierras Pampeanas and the High Andes in the Precordillera (Figure 22.3.4). Thrusting there began at about 20 Ma and has continued to present where it has not been buttressed by the Sierras Pampeanas. Thus, as in the western United States, thin- and thick-skinned thrusting overlapped in time, and both were active during flat subduction and during a period without significant magmatism. Shortening is about 5% in the Sierras Pampeanas and >60% in the Precordillera.

HORIZONTAL EXTENSION AND STRIKE-SLIP FAULTING Horizontal extension within convergent mountain belts has been of considerable interest since the 1980s. Young extensional deformation in the central Andes is concentrated at the northern and southern ends of the Altiplano-Puna Plateau. In the Puna, horizontal extension is mostly related to strike-slip and oblique-slip faults, whereas normal faults dominate at the northern end of the Altiplano. In both areas, however, horizontal extension is oriented approximately north-south, that is, subparallel to the strike of the orogen. There is little firm evidence for significant extension perpendicular to the belt, and the main part of the plateau system is neotectonically and seismically quiescent. There is probably significant strike-slip faulting in the Bolivian Eastern Cordillera north of the bend at ~18°S, which may accommodate the more oblique angle of convergence in this part of the belt. Even though extension is most notable around the ends of the high plateau, it is not restricted to the high topography. In northwest Argentina, horizontal extension occurs at elevations as low as 900 m in the foreland, and in southern Peru normal faults are found on the Pacific coast.

South of 28°S, several long fault zones, including the Liquiñe-Ofqui Fault in southern Chile and the El Tigre Fault in western Argentina, have undergone significant strike-slip movement during the Quaternary. These faults are thought to be due to the slight nonorthogonal convergence between the Nazca and South American Plates. The Atacama fault system of northern Chile is probably best known for its strike-slip history, but recent work has shown that it has a much more complicated and protracted history, including dip-slip and strike-slip displacements during the Mesozoic to Middle Tertiary and mostly dip-slip motion since then.

OROCLINAL BENDING The marked curvature of the Andes at about 18°S raises the question of whether this curvature reflects the initial shape of the continental margin, or whether it is a product of Andean

TABLE 22.3.1	THRUST BELT CHARACTERISTICS, BOLIVIA AND ARGENTINA				
Location	Width (km)	Shortening (km)	Topographic Slope	Wedge Taper	Annual Precipitation (mm)
N. Bolivia 13°–17°S	70	115 ± 20	3.5°	7° ± 1°	1000–2800
S. Bolivia 19°–23°S	90–110	75 ± 10	0.5–1.0°	2.5° ± 1°	400–100
Precordillera Argentina 29°–33°S	40–60	105 ± 20	~2.5°	3.5° ± 1.5°	100–200

deformation. The answer is probably both. Paleomagnetic data derived from Mesozoic rocks along the west coast of South America show clockwise rotations south of the bend and counterclockwise rotations to the north. The main debate over these results concerns whether the rotations reflect in situ block rotations or regional oroclinal bending. To date, several carefully mapped sites in the foreland provide evidence only for local, rather than regional, rotations. These preliminary results, however, do not preclude the model we have presented, in which the curvature of the central Andes has been accentuated during the last 25 m.y. by laterally variable shortening that is greatest within, and on the margins of, the Altiplano-Puna Plateau, and decreases to both north and south.

PALEOTECTONIC CONTROL Although structural style within the central Andes shows a broad correlation with the geometry of the subducted plate, as we have described, preexisting heterogeneities within the continental crust also play an important role. Thin-skinned thrust belts are restricted to thick, wedge-shaped Paleozoic basins. The Subandean belt is largely located within a previously undeformed Paleozoic passive margin and foreland basin sequence, east of a zone of preexisting deformation now occupied by the Eastern Cordillera. The Precordillera thrust belt deforms a lower Paleozoic passive margin sequence of what is called the Precordillera terrane, a narrow slice within western Argentina that bears marked similarities to the Lower Paleozoic of the southern Appalachians. The transition from the Subandean belt to the northern Sierras Pampeanas thick-skinned deformation coincides with southward thinning of Paleozoic strata, and the western boundary of the Sierras Pampeanas is the boundary of the Precordillera Terrane. Finally, many complex local structures within the Bolivian Altiplano

and in northwestern Argentina owe their geometries to reactivation of Late Cretaceous rift basins.

22.3.4 Crustal Thickening and Lithospheric Thinning

Modern estimates suggest that magmatism has contributed less than 10% to the total crustal thickening in the Andes during the last 25 m.y. Thus, the rest of the topography in the Andes must be accounted for by two mechanisms: (1) thickening of the crust by deformation, and (2) thermally controlled thinning of the lithosphere giving rise to uplift. Most of the crustal shortening responsible for the present topography is manifested at the surface as the thin-skinned thrust belts that provide the interface between the Andes and the Brazilian Shield. The thrust belts of the central Andes, including the Subandean belt and the Precordillera, are of considerable interest because they are among the few active examples of an antithetic (or foreland-verging) foreland thrust belt, in which the overall sense of shear is opposite to that in the associated subduction zone. The along-strike variations in these thrust belts allow one to identify the key first-order associations of geometry, topography, shortening, and paleotectonic setting (Table 22.3.1). There is a general correlation among high critical wedge taper, width, and a high-degree of shortening. In contrast, the order-of-magnitude variation in precipitation (and, presumably, erosion rate) shows no clear effect on the shortening.

Crustal thickening alone does not appear to be sufficient to explain the high plateau (the Altiplano) of the central Andes. About 1 to 1.5 km is probably accounted for by thinning of the lithosphere beneath the plateau. When this thinning occurred and how it relates to crustal shortening remain unresolved problems. As has been pointed out for the Alpine system, horizontal shortening should thicken not only the continental crust, but also the entire lithosphere. Yet beneath the plateau, the continental lithosphere must have thinned even as the crust was thickening. Furthermore, in the regions of flat subduction, the subducted plate is only 100–120 km beneath the surface, even though "normal" continental lithosphere is thought to be on the order of 150 km thick. Because it takes many millions of years to thin the lithosphere by conduction alone, recent proposals have invoked delamination of the base of the lithosphere to produce the necessary thinning on the timescale implied by Late Cenozoic mountain building in the Andes.

22.3.5 Closing Remarks

The Andes present a unique natural laboratory for studying mountain building that has occurred without the aid of a collision between two continental masses. The most important processes that have produced the modern topography of the Andes are structural shortening and lithospheric thinning. Volcanism, in contrast, is responsible mostly for the volumetrically minor topography above 4 to 4.5 km. Magmatism, nonetheless, probably plays a very significant role in determining the rheology of the crust, producing weak zones subject to faulting.

ADDITIONAL READING

Beck, M. E., Jr., 1988. Analysis of Late Jurassic–Recent paleomagnetic data from active plate margins of South America. *Journal of South American Earth Sciences,* 1, 39–52.

Dalziel, I. W. D., 1986. Collision and Cordilleran orogenesis: an Andean perspective. In Coward, M. P., and Ries, A. C., eds., *Collision tectonics. Geological Society of London, Special Publication* 19, 389–404.

Isacks, B. L., 1988. Uplift of the central Andean plateau and bending of the Bolivian orocline. *Journal of Geophysical Research,* 93, 3211–3231.

Jordan, T. E., Isacks, B. L., Allmendinger, R. W., Brewer, J. A., Ramos, V. A., and Ando, C. J., 1983. Andean tectonics related to geometry of subducted Nazca plate. *Geological Society of America Bulletin,* 94, 341–361.

Mpodozis, C., and Ramos, V. A., 1990. The Andes of Chile and Argentina. In Ericksen, G. E., Ca-as Pinochet, M. T., and Reinemund, J. A., eds., *Geology of the Andes and its relation to hydrocarbon and mineral resources, Earth Science Series* 11. Houston, Texas: Cricum-Pacific Council for Energy and Mineral Resources, 59–90.

Pardo-Casas, F., and Molnar, P., 1987. Relative motion of the Nazca (Farallon) and South American plates since Late Cretaceous time. *Tectonics,* 6, 233–248.

22.4 THE APPALACHIAN OROGEN—An essay by James P. Hibbard[1]

22.4.1 Introduction

The **Appalachian Orogen** (Figure 22.4.1) is a northeast-trending belt of Late Precambrian to Paleozoic rocks in eastern North America that were deformed during the Paleozoic. Its strike length is more than 3000 km, extending from Alabama to Newfoundland, and it forms a segment of a much larger Paleozoic orogenic system that encompasses the Caledonide Orogen of the British Isles, Greenland, and Scandinavia to the northeast, and the Ouachita Orogen to the southwest. The northwest limit of the Appalachians is the deformation front between rocks of the orogen and older orogens and platform rocks of North America. On its southern and southeastern flanks, the orogen is onlapped by Cenozoic sedimentary rocks of the Atlantic Coastal Plain.

The orogen is important from a historical standpoint, as many significant tectonic ideas are rooted in Appalachian bedrock. The concept of geosynclines, which dominated thought on mountain building for more than a century preceding the advent of plate tectonics, was conceived in the Appalachians. Appropriately, the geosynclinal theory was supplanted on its native Appalachian turf by plate tectonics in the 1960s and 1970s. Initial questioning of whether the Atlantic closed and then reopened was followed by the first detailed application of plate tectonic theory to an ancient orogen. In addition, the idea of "thin-skinned tectonics," meaning the deformation of cover strata above a master décollement that is independent of underlying basement, was first developed in the classic Valley and Ridge fold-and-thrust province of the southern Appalachians. Closely related to the thin-skinned concept was the realization that there is a mid-crustal detachment within the orogen that places a large portion of the deformed southern Appalachians onto the relatively undeformed North American platform. This realization lead to the general acceptance that substantial portions of orogenic belts form relatively thin, highly allochthonous sheets (or tectonic flakes) emplaced onto cratons.

Having established the prominent role of the Appalachians at the forefront of tectonic research, we turn, in the remainder of this essay, to a sketch of current thought on the orogen. Following an overview of the mountain belt, there is a brief description of the first-order crustal components, an explanation of how and when they were assembled, and finally a highlighting of potential directions for future thought and development of Appalachian tectonics.

22.4.2 Overview

The structural grain of the Appalachians is remarkably consistent, defining a series of broad, harmonically curved promontories and reentrants (Figure 22.4.1); their grace and regularity lead the renowned North American tectonicist, P. B. King, to proclaim the Appalachian Orogen as the most elegant mountain chain on earth. As we will see, this structural architecture reflects a fundamental feature of the orogen that was important in its evolution. For our purposes, the New York promontory will serve as the divide between segments referred to as the northern and southern Appalachians.

In contrast to structural divisions, lithotectonic divisions of an orogen distinguish rock associations that were either formed or deposited in a common tectonic setting during a finite time-span. These divisions are scale dependent, that is, contingent on the scale of the tectonic process considered. In this essay, the hierarchy of lithotectonic divisions consists of the realm at orogen scale and the zone at the scale of two or less reen-

[1]North Carolina State University, Raleigh, NC.

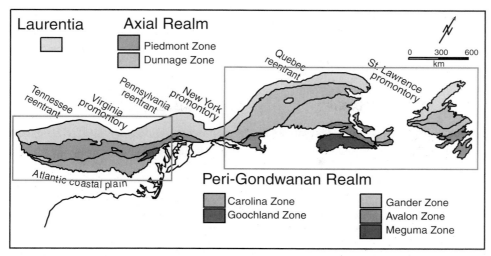

FIGURE 22.4.1 Realms and zones of the Appalachian Orogen, defined on the basis of Middle Ordovician and older geologic history. The boxes outline areas shown in Figures 22.4.4 and 22.4.5.

trants. At yet smaller scale, terranes are recognized as regional subdivisions of a zone; however, in this essay we will mainly be concerned with realms and zones.

The Appalachians are composed of three realms, the Laurentian realm,[2] the Axial realm, and the peri-Gondwanan realm (Figure 22.4.1), all of which acquired their defining geologic character before the Middle Ordovician. The Laurentian realm encompasses essentially all of the rocks deposited either on, or adjacent to, ancient North America and forms the western flank of the orogen; however, windows of Laurentian rocks occur locally among the more easterly accreted terranes (Figure 22.4.1). In contrast, peri-Gondwanan realm rocks are interpreted to have formed proximal to Gondwana[3] and thus are considered to be exotic with respect to Laurentian elements; they are distributed along the eastern flank of the Appalachians. The Axial realm is a collection of zones and terranes of mainly oceanic and volcanic arc affinity that has been caught between the Laurentian and peri-Gondwanan realms during Appalachian orogenesis. Unlike the uniformity of the Laurentian realm, both the peri-Gondwana and the Axial realms change character along strike of the orogen.

The orogen was assembled during the approximately 300 m.y. time-span between the existence of two supercontinents, the Middle Proterozoic Rodinia and the Late Paleozoic Pangea. The Appalachians formed as a result of the progressive accretion of Axial and peri-Gondwanan elements to the Laurentian realm. Classically, it has been accepted that three major events, the Taconic, Acadian, and Alleghanian Orogenies record the accretion of these elements. However, as more data are acquired, we are finding that accretionary events along the Laurentian margin were continuous for protracted periods of time and were less punctuated than is implied by the simple mantra of "Taconic, Acadian, and Alleghanian."

22.4.3 Tectonic Components

THE LAURENTIAN REALM The template for Appalachian accretionary events, the eastern Laurentian continental margin, was initiated by Late Precambrian rifting along the axis of the ~1-Ga Grenville Orogen within Rodinia; thus Grenville rocks form basement to the continental margin. The west flank of the Amazonian craton likely formed the conjugate margin to eastern Laurentia during this extensional event. The main pulse of rifting affected the entire length of the Appalachians at roughly 600–550 Ma. Sedimentation was synchronous with rifting, leading to thick deposits confined to elongate basins and characterized by abrupt changes in the thickness of strata, with most of the sediment supplied from the Laurentian craton. Bimodal magmatism accompanied rifting; however, there are two pulses of rift magmatism, an early pulse (~750–700 Ma) (Figure 22.4.2) confined to the southern Appalachians, and a later, main pulse (~600–550 Ma) along the length of the orogen. The

---

[2]Laurentia refers to early Paleozoic North America, including portions of Greenland, Scotland, and Ireland that rifted away in the Mesozoic.
[3]Gondwana is the ancient continent approximately equivalent to an amalgamation of modern southern hemisphere cratons.

early pulse appears to be coeval with the early breakup of Rodinia along the Pacific margin of Laurentia and may well be a far-field effect of this event.

Rifting led to continental breakup, the onset of seafloor spreading, thermal subsidence of the newly formed passive margin, and deposition of a drift sequence atop the rift deposits (Figure 22.4.2 and 22.4.3). The drift sequence consists of basal clastic rocks overlain by a shallow marine carbonate sequence up to 10 km thick. In contrast to the heterogeneity of the rift deposits, the drift sequence displays an orderly stratigraphy with little thickness variation and remarkable lateral continuity. Paleomagnetic studies indicate that the margin was at near equatorial latitudes during establishment of the passive margin. The seaward edge of this extensive carbonate shelf is marked by a facies change into deep water shaley rocks that locally contain carbonate blocks and boulders that spalled off the precarious edge of the shelf. The Paleozoic ocean that formed outboard of eastern Laurentia, preceding the modern Atlantic, is called Iapetus, after the mythical Greek father of Atlantis.

The geometry of the continental margin was controlled by the zigzag pattern formed by spreading and transform segments of the rift system. This shape influenced the distribution of rift and drift sequences; former ridge-transform junctions along the margin tended to form steep-sided terminations for rift basins, whereas the distribution of the drift sequence facies change from shelf to slope-and-rise was controlled by the jagged shape of the margin. Where the trace of the drift sequence facies change is preserved, it presently follows the curves of the structural promontories and reentrants in the orogen. This relationship indicates that the promontories and reentrants are inherited from the original shape of the margin at breakup.

The Axial Realm

Elements of the Axial realm record the evolution of Iapetus and its component volcanic arcs, backarc basins, and accretionary complexes. There appears to be a major change in the realm at the New York promontory; in the northern Appalachians, the Dunnage Zone records a complex history of multiple volcanic arcs and backarc basins, whereas in the southern Appalachians, the Piedmont Zone appears to record a simpler history of a single composite arc system.

The Dunnage zone is in tectonic contact with Laurentian rocks along the Baie Verte–Brompton line (Figure 22.4.4), which is a steep, relatively narrow fault system that has experienced multiple episodes of movement; in many places it is marked by narrow,

elongate ultramafic bodies. The zone is best exposed along the north coast of Newfoundland, where its entire width is at low metamorphic grade; here, the zone records the evolution of at least two distinct oceanic tracts. Contrasts in stratigraphy, fossil faunas, and paleomagnetic and isotopic data indicate that the western tract of the zone was associated with the Laurentian side of Iapetus, whereas the eastern tract developed on the Gondwanan side of the ocean realm. They are tectonically juxtaposed along the Red Indian Line. Both tracts record an early arc phase that starts in the Early to Middle Cambrian and a younger arc phase that ranges into the Late Ordovician (Figure 22.4.3b).

Elements of the Newfoundland Dunnage zone can be correlated with units in New Brunswick, Quebec, and northern Maine; however, most Dunnage zone elements from central Maine to New York are multiply deformed and have been subjected to high-grade metamorphism, thus obscuring original relationships between units. Consequently, the early evolution of the zone is not as well understood in New England, although there are strong hints that it conforms to that of the Canadian Dunnage zone.

The Piedmont zone is tectonically severed from Laurentian rocks to the west along a series of faults, most of which have multiple movement histories. Much of the zone has been subjected to intense, polyphase deformation and medium- to high-grade metamorphism and thus unraveling the depositional-magmatic history is somewhat tenuous. The zone is split into two components by the Brevard zone, a polygenetic ductile shear zone, and other faults northward along strike of the Brevard zone (Figure 22.4.5).

The western portion of the zone is dominated by metamorphosed clastic rocks and associated mélanges disposed in imbricate thrust stacks. In northern Virginia, the thrust stacks were assembled by the Early Ordovician, but locally, in southwest Virginia, radiometric ages suggest that the clastic rocks have been subjected to Early Cambrian shortening and medium-grade metamorphism (Figure 22.4.2). In North Carolina, metamorphosed clastic rocks at the western edge of the zone contain pods of eclogite. Across the Brevard zone, the eastern portion of the zone contains substantially more metamagmatic rocks than the western area. Magmatism appears to have been active from the Early Ordovician to the Early Silurian, with a Late Ordovician lull during which black slates were deposited in central Virginia (Figure 22.4.2). Where studied, the magmatic rocks are geochemically consistent with a suprasubduction zone, volcanic arc setting.

Despite the strong tectonothermal overprint, all of the characteristics just outlined are consistent with the

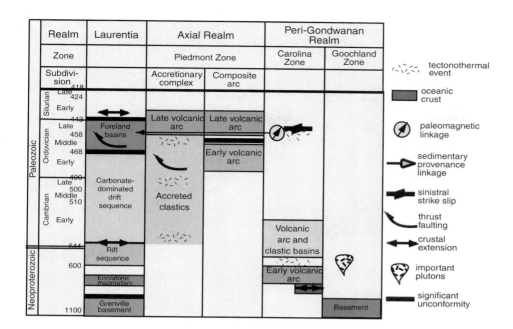

FIGURE 22.4.2 Major elements and events of the southern Appalachians from the Precambrian to the Silurian.

Legend:
- tectonothermal event
- oceanic crust
- paleomagnetic linkage
- sedimentary provenance linkage
- sinistral strike slip
- thrust faulting
- crustal extension
- important plutons
- significant unconformity

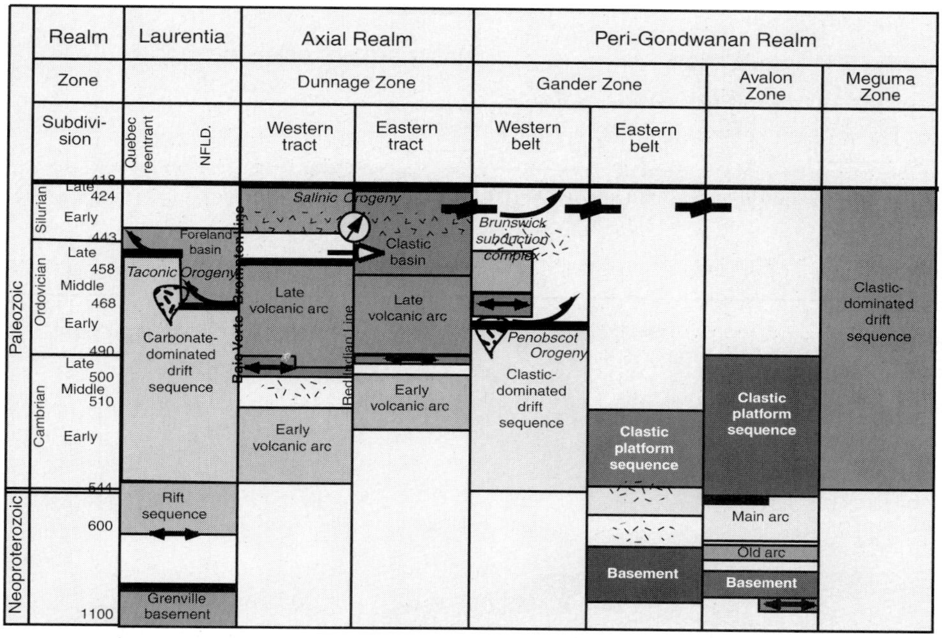

FIGURE 22.4.3 Major elements and events of the northern Appalachians from the Precambrian to the Silurian. See Figure 22.4.2 for explanation of symbols.

interpretation that the Piedmont zone encompasses a long-lived, west-facing accretionary complex in front of a more easterly suprasubduction-zone magmatic arc. If the interpretation of an Early Cambrian tectonothermal event is valid, it suggests that the Piedmont Zone formed in an ocean older than Iapetus, which was just in its rift-to-drift stage at this time.

The Peri-Gondwanan Realm

A collection of diverse crustal remnants of Proterozoic to Early Paleozoic rocks that lay across the Iapetus Ocean from the eastern Laurentian margin is grouped here as peri-Gondwanan elements. In the northern Appalachians, the realm is represented by the Gander, Avalon, and Meguma zones, whereas the Carolina and Goochland zones occupy the east flank of the exposed southern Appalachians (Figures 22.4.1, 22.4.4, and 22.4.5). The nature of the Goochland zone is controversial and it is tentatively grouped here with the peri-Gondwanan Zones.

The Gander zone is in both fault and stratigraphic contact with the Dunnage zone. For example, in central Newfoundland, the contact is marked by a thrust fault that emplaces Dunnage ophiolite on top of the Gander zone, whereas in northern Maine and New

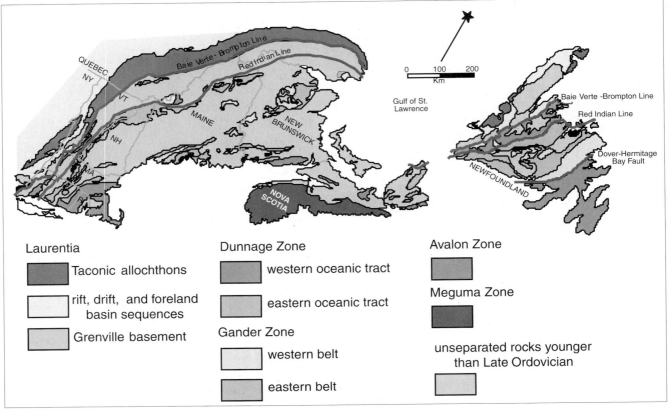

FIGURE 22.4.4 Distribution of Middle Ordovician and older elements in the northern Appalachians.

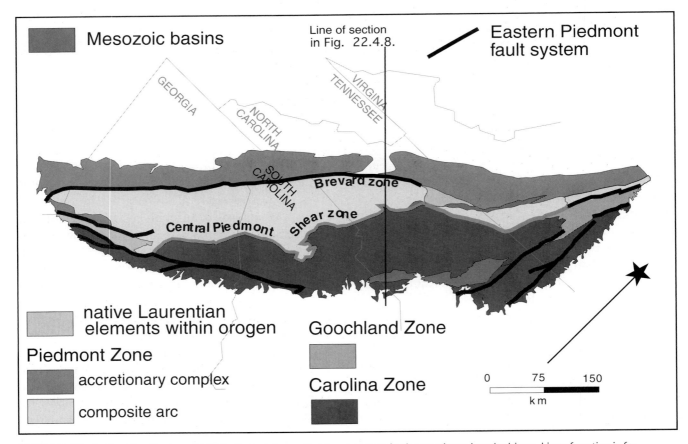

FIGURE 22.4.5 Distribution of Middle Ordovician and older elements in the southern Appalachians. Line of section is for Figure 22.4.8.

Brunswick, Dunnage volcanic rocks unconformably overlie the Gander zone (Figure 22.4.4). Elements grouped here as the Gander zone define two distinct belts on either side of the zone (Figures 22.4.3, and 22.4.4). The eastern belt (called the Avalon zone by some workers) is composed of older crystalline basement that is faulted against a younger magmatic sequence. Basement rocks consist of marble, quartzite, greenstone, and pelite that appear to have protolith ages greater than 800 Ma and that have been involved in metamorphic events prior to 600 Ma. The younger magmatic sequence includes mafic and felsic volcanic and plutonic rocks with an approximate age range of 600–545 Ma. The western belt is characterized by Cambrian-Early Ordovician continentally derived quartz arenite and pelite with minor mafic magmatic rocks that, collectively, have been interpreted as representing a passive margin sequence. The eastern belt may form the basement on which the western belt passive margin was deposited, but the contact between the two is unknown.

The Avalon zone is in fault contact with the Gander zone; in Newfoundland the contact is represented by the Dover-Hermitage Bay Fault (Figure 22.4.4), which is vertical; it is documented on seismic reflection profiles as reaching the base of the crust. The zone is dominated by Neoproterozoic magmatic rocks that exhibit diverse compositions and have mainly suprasubduction zone signatures. Geochemical studies indicate that the volcanic pile likely formed on thin continental crust, although this basement does not appear to be exposed. Magmatism extended over the broad time period of 685–540 Ma (Figure 22.4.3), with a preponderance of activity in the range of 630–580 Ma. Locally, deformation was synchronous with deposition and appears to have been dominated by extension.

The magmatic rocks are overlain by a Lower Paleozoic transgressive, shallow marine platform sequence (Figure 22.4.3); the lack of substantial carbonate in this sequence as well as paleomagnetic data attests to the platform's being deposited at high paleolatitudes. Fossil faunas in the platform sequence are of "Avalonian" affinity, distinct from those of the Laurentian and Axial realms.

The Meguma zone underlies most of southern Nova Scotia and forms the southeastern most exposed crustal block in the orogen (Figures 22.4.1 and 22.4.4). It is faulted against Carboniferous cover rocks to the north, which in turn are unconformable upon the Avalon zone. The zone is dominated by Early Paleozoic turbidites that have been interpreted as being deposited in an abyssal fan setting along a passive continental margin. On the basis of sedimentology, stratigraphy, paleontology, petrology, and geophysics, the zone has been correlated with rocks in Morocco.

The Carolina zone is in tectonic contact with the Piedmont zone along the central Piedmont shear zone, a Late Paleozoic thrust fault (Figure 22.4.5). The zone is an amalgamation of Neoproterozoic to Early Paleozoic volcanic arcs and associated sedimentary rocks that have an approximate age range of 675–530 Ma (Figure 22.4.2). It appears that one or more deformational events coincided with magmatism, although the nature of these events is poorly known. The Carolina zone resembles the Avalon zone, but appears to be distinct from its northern Appalachian counterpart. Neoproterozoic magmatism in the Carolina Terrane peaked at 630–610 Ma and again at ~550 Ma, whereas in Avalon, peak magmatism is in the period 630–580 Ma. Although the Carolina zone contains an Early Paleozoic clastic sequence, it does not appear to represent a transgressive platformal sequence as found in Avalon. Also, Carolina fossil faunas have a peri-Gondwanan affinity, but are not "Avalonian."

The Goochland zone is in tectonic contact with the Piedmont zone and is likely faulted against the Carolina zone (Figure 22.4.5). The zone comprises orthogneiss and paragneiss that have been intruded by anorthosite dated at ~1 Ga. In addition, this package is intruded by alkalic granite dated at ~630 Ma. The zone may represent a structural window into Laurentian basement, with the younger granite representing rift magmatism; however, the 630-Ma age of this granitoid does not coincide with known rift magmatism on the Laurentian margin. Alternatively, the zone represents peri-Gondwanan basement with the granitoid equivalent with Neoproterozoic plutons in the Carolina zone.

22.4.4 Assembly

EARLY PALEOZOIC DESTRUCTION OF IAPETAN PASSIVE MARGINS

The eastern Laurentian passive margin came to an abrupt demise in the Early to Middle Ordovician. This event, termed the Taconic Orogeny, is marked by a regional unconformity on the continental shelf, a change in sedimentation along the margin, the development of a submarine thrust belt, and accompanying metamorphism; it is best preserved in the northern Appalachians (Figure 22.4.3).

Carbonate sedimentation along the Laurentian shelf, slope, and rise was choked off in the Early to Middle Ordovician, and a Middle Ordovician erosional to slightly angular unconformity was developed along the length of the carbonate shelf (Figure 22.4.3). The new sedimentary regime was marked by deep water, foreland deposition of easterly derived clastic sediments. The continental shelf and clastic foreland basin were overridden by thrust sheets containing Laurentian slope and rise- and rift-related sediments; the highest thrust sheets in Newfoundland and Quebec are composed of ophiolite. Where they are emplaced upon the passive margin, there is a striking contrast between autochthonous rocks of the shelf and allochthonous, deeper water, rocks of the thrust sheets; the conspicuous thrust sheets are termed the "Taconic allochthons." Perhaps one of the most inspirational geologic sights in the Appalachians is the view of the barren, flat-topped ophiolite sheet in eastern Newfoundland (now a UNESCO World Heritage Site).

Taconic events are interpreted as reflecting the introduction of the Laurentian margin into a subduction zone beneath the eastern tract of the Dunnage zone (Figure 22.4.6a). The unconformity represents the flexural bulge due to loading of the continental margin by an overriding accretionary complex, the clastic sedimentation represents foreland basin, or trench, sedimentation on top of the downgoing continental margin, and the thrust sheets represent an accretionary wedge. However, this tectonic system must have been more complex than the simple subduction of Laurentia beneath the Dunnage zone, for some of the obducted ophiolites were just forming while obduction was in progress elsewhere along the margin. Additionally, Early Ordovician plagiogranites intrude obducted ophiolites; in contrast to the obduction process, these plutons and other geologic evidence require a rapid change in subduction polarity involving west-directed subduction beneath the margin and volcanic arc (Figure 22.4.6b).

The Taconic Orogeny in the southern Appalachians appears to represent the attempted subduction of the Laurentian margin beneath the Piedmont zone accretionary prism and arc. However, the foreland clastic wedge in the southern Appalachians is overlain by that of the northern Appalachians, indicating that the Taconic event was slightly older in the south. Also, the hallmark Taconic thrusting and metamorphism have been severely overprinted and obscured by younger tectonothermal events.

Nearly synchronous with the Taconic Orogeny along the Laurentian margin, the Gander passive margin of Iapetus was also tectonically terminated. This event, the Penobscot Orogeny, is recognized by an unconformity of Early Ordovician volcaniclastic rocks affiliated with the Dunnage Zone atop Cambrian-Early Ordovician Gander quartzose clastic rocks. This unconformity persists along strike from northern Maine to Newfoundland. In Newfoundland, the Penobscot Orogeny also involved the eastward obduc-

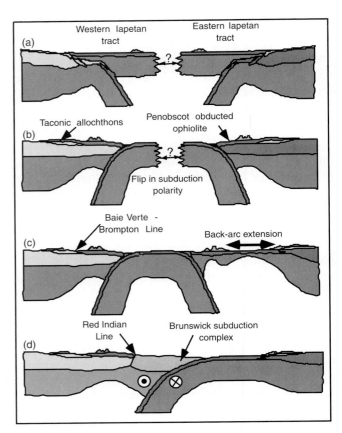

FIGURE 22.4.6 Cartoon depicting possible tectonic evolution of the northern Appalachians: (a, b) Early to Middle Ordovician; (c, d) Middle to Late Ordovician.

tion of Dunnage zone ophiolite onto the Gander passive margin (Figure 22.4.6b). The timing of Penobscot obduction is tightly constrained by Early Ordovician fossils in the overthrust oceanic rocks and by an Early Ordovician granitoid that intrudes and "stitches" both Gander and Dunnage zone rocks. Thus, Iapetus commenced closure from both margins in the Early Ordovician.

MID-PALEOZOIC CLOSURE OF IAPETUS Late Ordovician to Late Devonian events that contributed towards the closure of Iapetus and construction of the Appalachians are best recorded in the northern Appalachians, where Silurian and younger strata blanket the orogen. In the southern Appalachians, Middle Paleozoic strata are largely confined to covering the Laurentian realm. However, the earliest interaction of Laurentia with the peri-Gondwanan realm is apparently recorded in the southern Appalachians, where circumstantial evidence from across the orogen indicates that the Carolina zone commenced docking in the Middle to Late Ordovician, immediately on the "coattail" of the Taconic accretion of the Piedmont zone (Figure 22.4.2).

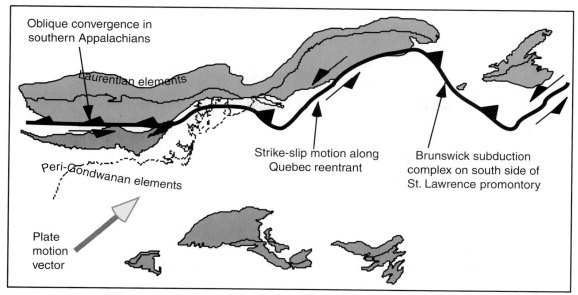

FIGURE 22.4.7 Cartoon of possible tectonic setting during the Late Ordovician to Silurian Salinic Orogeny. The interaction between Laurentia and peri-Gondwana elements involved a strong component of sinistral shear, which may have produced oblique subduction in the southern Appalachians, and sinistral shear and convergence along reentrants and promontories, respectively, of the northern Appalachians. Bold arrow shows approximate plate motion vector for peri-Gondwana elements relative to Laurentia.

In native Laurentian rocks of the southern Appalachians, an unconformity at the Ordovician-Silurian boundary has been attributed to tectonic loading of the post-Taconic margin (Figure 22.4.2). Furthermore, the post-Taconic margin of Laurentia, the Piedmont zone, was intruded by a pulse of Late Ordovician granodioritic to tonalitic plutons that likely reflects subduction beneath the Laurentian margin (Figure 22.4.2). Finally, in the Carolina zone, Late Ordovician upright folding, greenschist facies metamorphism, and uplift probably mark the initiation of the collision of Carolina with Laurentia, as paleomagnetic data indicate that it was at Laurentian paleolatitudes by this time (Figure 22.4.2). Folds in the Carolina zone define an *en echelon* array that is consistent with a component of sinistral shear during collision.

In the northern Appalachians, closure of Iapetus continued after the Late Ordovician flip in subduction polarity (Figure 22.4.6b), but it was counteracted by the generation of a backarc basin on the peri-Gondwanan side (Figure 22.4.6c). However, the western and eastern oceanic tracts of the Dunnage zone, which had distinct Early Paleozoic faunas, shared a mixed Late Ordovician fauna, indicating that they were proximal to one another. Additionally, similar Silurian paleomagnetic data from each tract as well as an Early Silurian stitching pluton along the trace of the Red

Indian Line further support a Late Ordovician juxtapositioning of the two tracts (Figures 22.4.3 and 22.4.6c).

Evidence for the convergent closure of the Middle to Late Ordovician peri-Gondwanan backarc basin is preserved only along the east-west trending portions of the orogen in the vicinity of the St. Lawrence promontory; there, the Brunswick subduction complex, a southeast vergent stack of thrust sheets that includes Late Ordovician to Silurian blueschist and ophiolitic mélange, records this closure. Elsewhere, along more northeast-trending segments of the northern Appalachians, Silurian sinistral shear is recorded, from the Avalon zone across to the Laurentian margin (Figure 22.4.7). Thus, it appears that by the end of the Silurian, most components along the length of the orogen had been assembled along the Laurentian margin through the closure of Iapetus with a strong component of sinistral shear displacement (Figure 22.4.7); this kinematic regime, with regional shortening oriented approximately north-south, has been termed the Salinic Orogeny in the northern Appalachians.

At the end of the Silurian, the Laurentian margin underwent an abrupt change in kinematic character. Following a Late Silurian unconformity found in many places in the northern Appalachians, regional shortening reoriented to a position more at right angles to the trend of the orogen, with a component of dextral strike slip; this new kinematic regime is responsible for the

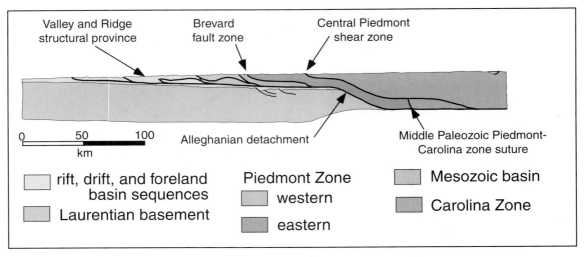

FIGURE 22.4.8 Cross section of the southern Appalachians showing major features of the Alleghanian thrust system. Line of section shown in Figure 22.4.6; no vertical exaggeration.

Acadian Orogeny. The change in kinematics is heralded by an Early Devonian westward transgressive clastic wedge in northern New England. However, the most intense manifestation of the Acadian event is recorded on the New York promontory, in southern New England, where Middle Paleozoic and older rocks were deformed into regional-scale recumbent fold nappes at high metamorphic grade. Subsequent rapid uplift resulted in the removal of up to 20 km of crust in the southern New England area and the deposition of the thick Devonian Catskill clastic wedge to the west. The plate-scale process responsible for the Acadian Orogeny may have been the docking of the Meguma zone to the orogen, for the Acadian is the first deformation shared by the Meguma rocks and the remainder of the orogen. Regardless, it is noteworthy that at the scale of the northern orogen, intense Acadian tectonism and uplift appear to be limited to the region of the New York promontory.

Late Paleozoic Formation of Pangea

The final phases of Appalachian orogenesis took place from the Mississippian to the Permian; events within this time frame are ascribed to the Alleghanian Orogeny. The orogeny is penetratively developed in the southern Appalachians as well as in southern New England; in both areas Alleghanian events strongly overprint earlier deformation and metamorphism; in contrast, it is more limited in development in most of the northern Appalachians.

In the southern Appalachians, the Valley and Ridge Province along the western flank of the orogen records the Late Paleozoic westward-directed thrusting of the

Cambrian to Permian strata onto the Laurentian platform. This thrust belt was studied in detail for more than a century before it was discovered that it represented merely the toe of what is now recognized as an orogen-scale thrust wedge (Figure 22.4.8). The full magnitude of this thrust wedge was only realized in a seismic reflection profile study across the southern orogen. It revealed that the Laurentian platform sequence extends in the subsurface to at least as far east as the central Piedmont shear zone. The crystalline thrust sheet encompasses the Piedmont zone and is separated from the underlying Laurentian platform by a major detachment fault; this geometry resolves into at least 175 km of shortening along the Alleghanian detachment.

The oldest documented west-directed thrusting is within the crystalline sheet along the central Piedmont shear zone in northern North Carolina; here Middle Mississippian granitoids are syntectonic with respect to ductile thrusting. This early thrusting is roughly coeval with the initiation of Carboniferous clastic wedges shed out over the Laurentian platform. Although thrusting continued into the Permian in the Valley and Ridge, there was a major change in kinematics in the Early Pennsylvanian in the eastern Piedmont and Carolina zones. There, thrusting was replaced by dextral strike-slip motion along a network of large faults, termed the eastern Piedmont fault system. This kinematic change appears to be reflected in the clastic wedges by a Morrowan (Early Pennsylvanian) unconformity. Partitioning of deformation between shortening in the Valley and Ridge and strike-slip motion in the eastern portion of the orogen is likely related to dextral transpression between Laurentia and Gondwana. In the eastern portion of the orogen, both thrusting and younger strike-slip motion were accompanied by medium-grade metamorphism and plutonism that are spatially related to the major Alleghanian fault zones.

In most of the northern Appalachians, Alleghanian deformation is manifested mainly by dextral strike-slip faults. Sedimentation was generally localized in narrow elongate basins associated with these faults, and multiple unconformities in the basins attest to synkinematic deposition. The Alleghanian Orogeny is attributed to the oblique collision of Laurentia with Gondwana, associated with the assembly of Pangea, the Late Paleozoic supercontinent. Clearly, in the northern Appalachians the event was more of a "grazing" of Laurentia by Gondwana, whereas in the southern Appalachians, the two crustal blocks collided more head-on, although the partitioning of strain in the southern Appalachians attests to the transpressive nature of the collision there.

Just as the Grenville Orogen served as one of the seams along which Rodinia broke up, the Appalachian Orogen formed the locus of Mesozoic rifting that led to the breakup of Pangea and the formation of the modern Atlantic. Elongate basins containing rift facies clastic sedimentary rocks and mafic magmatic rocks are preserved along the length of the orogen, much as in their ancestor Iapetan rift basins.

22.4.5   Closing Remarks

The orogen has provided fodder for many tectonic concepts and continues to lure us with the many stones still unturned in its mountains, hollows, and coves. Some of the first-order observations and questions that arose as I composed this essay are:

- Does the New York promontory mark the end of a major transform in the Iapetus Ocean—perhaps one that split the ocean into two major domains, as reflected in the difference in accretionary history between the northern and southern Appalachians?
- The Taconic Orogeny, one of the oldest events in the orogen, is well preserved in the north but strongly overprinted by the Alleghanian Orogeny in south. These relations suggest that during the Middle and Late Paleozoic, accretion in the northern Appalachians mainly involved a strong strike-slip component and that areas of intense Salinian and Acadian deformation and metamorphism were localized collisions (at the scale of the orogen) where strike-slip motion was impeded by promontories.

- The traditional "mantra" of "Taconic, Acadian, Alleghanian" is giving way to the realization that tectonic activity was ongoing along the Laurentian margin and that it is somewhat naive to view the development of the orogen exclusively in the time frames of three named events.
- The nature of Late Mesozoic and Cenozoic erosional and epeirogenic events that are responsible for the modern form of the mountain range are not well known and await the attention of future researchers.

ADDITIONAL READING

Colman-Sadd, S. P., Dunning, G. R., and Dec, T., 1992. Dunnage-Gander relationships and Ordovician orogeny in central Newfoundland: a sediment provenance and U-Pb age study. *American Journal of Science*, 292, 317–355.

Hatcher, R. D., Jr., Thomas, W., Geiser, P., Snoke, A., Mosher, S., and Wiltschko, D., 1989. Alleghanian orogen. In Hatcher, R. D., Jr., Thomas, W. A., and Viele, G. W., eds., *The Appalachian-Ouachita orogen in the United States, The geology of North America, v. F-2.* Boulder, CO: Geological Society of America, 233–318.

Hibbard, J. P., 1994. Kinematics of Acadian deformation in the northern and Newfoundland Appalachians. *Journal of Geology,* 102, 215–228.

MacNiocaill, C., van der Pluijm, B. A., Van der Voo, R., 1997. Ordovician paleogeography and the evolution of the Iapetus Ocean. *Geology,* 25, 159–162.

Murphy, J. B., Keppie, D., Dostal, J., and Nance, R. D., 1999. Neoproterozoic—early Paleozoic evolution of Avalonia. In Ramos, V., and Keppie, D., eds., *Laurentia-Gondwanan connections before Pangea, Geological Society of America Special Paper 336,* 253–266.

Rankin, D. W., 1994. Continental margin of the eastern United States: past and present. In Speed, R. C., ed., *Phanerozoic evolution of North American continent-ocean transitions, DNAG continent-ocean transect volume.* Boulder, CO: Geological Society of America, 129–218.

Rodgers, J., 1968. The eastern edge of the North American continent during the Cambrian and Early Ordovician. In Zen, E., White, W., Hadley, J., and Thompson, J., eds., *Studies of Appalachian geology, northern and maritime.* New York: Wiley & Sons.

Thomas, W. A., 1977. Evolution of Appalachian-Ouachita salients and recesses from reentrants and promontories in the continental margin. *American Journal of Science,* 277, 1233–1278.

van Staal, C. R., Dewey, J. F., MacNiocaill, C., and McKerrow, W. S., 1998. The Cambrian-Silurian tectonic evolution of the northern Appalachians and British Caledonides: history of a complex, west and southwest Pacific-type segment of Iapetus. In Blundell, D. J., and Scott, A. C., eds. *Lyell: the past is the key to the present, Geological Society of London, Special Publication 143,* 199–242.

Williams, H., 1995. Taconic allochthons. In Williams, H., ed., *Geology of the Appalachian-Caledonian Orogen in Canada and Greenland, Geological Survey of Canada, Geology of Canada,* no. 6, 99–114.

22.5 THE CALEDONIDES—An essay by Kevin T. Pickering[1] and Alan G. Smith[2]

22.5.1 Introduction

In this essay we review the early Paleozoic history of the **European Caledonides** from Svalbard (Spitsbergen) through the northwest European and British Caledonian belts. These orogenic belts mark the edges of lithospheric plates and originated from crustal stresses associated with the subduction of former oceans, and the collision of the adjacent continents. In the Early Paleozoic, or Cambrian to Silurian (~ 570–408 Ma) the present-day Caledonides region consisted of essentially three large continental blocks separated by one or more oceans (Figure 22.5.1): Gondwana (South America and Africa being the most important continents in relation to the Caledonides), Laurentia (North America, Greenland, and northwestern Scotland), and Baltica (northwestern Europe to the Ural Mountains in the east, and south to the poorly defined southern edge in the region of the Tornquist Teisseyre lineament, which stretches from the North Sea—via the Polish Caledonides and its concealed eastward extension—to the Urals).

Three collisional belts formed between these three continents. The first, the Caledonides of Norway, western Sweden, and eastern Greenland, lies between western Baltica and eastern Laurentia; the second, a poorly exposed branch of the Caledonides under the North Sea, continues into eastern Europe situated between northern Gondwana and southern Baltica; the third consists of the Caledonides of the British Isles between northwestern Gondwana and southern Baltica. Each collisional belt represents the site of a former ocean. The name "Iapetus Ocean" has been given to the ocean area in general. For convenience, the ocean between Baltica and Laurentia will be referred to as the Eastern Iapetus, that between Gondwana and Baltica as the Tornquist Ocean, and the ocean between Gondwana and Laurentia as the Western Iapetus.

In detail, the histories of the collisions are complex. In particular, continental slivers rifted away from the northern margin of Gondwana, moved away from it across the Iapetus Ocean, and collided with the margins of Laurentia and Baltica before Gondwana itself collided with these continents. In Europe, the best defined of these slivers are Eastern Avalonia (southern Britain and much of France); in North America, they are western Avalonia (the Avalon Peninsula and Gander Zone of Newfoundland, New Brunswick, and Nova Scotia), and the Piedmont Terrane and Carolina Slate Belt in the southern Appalachians. Thus there was a stage in which short-lived new oceans lay between the Gondwanan fragments and Gondwana itself. These new oceans can be regarded as parts of the Iapetus in the broad sense, but one of them, the Rheic Ocean, is highlighted in this account. It lay between the northern margin of Gondwana and parts of central and western Europe to the south of eastern Avalonia, and is discussed in the following account.

[1]University College London, London, UK.
[2]University of Cambridge, Cambridge, UK.

FIGURE 22.5.1 Computer-generated plate reconstructions, based on a synthesis of paleomagnetic data, for the following time intervals: Late Precambrian (570–560 Ma); Cambrian (530 Ma); Early Ordovician (490–480 Ma); Middle Ordovician (460 Ma); latest Ordovician (440 Ma); Middle Silurian (420 Ma), and early Middle Devonian (390 Ma). Positions of major magmatic arcs are shown schematically. The initial Late Precambrian reassembly joins western South America to eastern Laurentia (present coordinates). Baltica's position is uncertain and is not shown on this map. The positions have been obtained by interpolating between this initial reassembly and positions suggested by Ordovician paleomagnetic data. The paleomagnetic evidence suggests that during Cambrian time Gondwana approached Laurentia, and that northwestern South America possibly collided with an oceanic arc or arcs that fringed southeastern Laurentia in Early Ordovician time. Gondwana then rotated anticlockwise in later Ordovician to Early Devonian time, bringing opposite one another those parts of Gondwana and Laurentia (e.g., Florida) that were to collide in the later Paleozoic. Uncertainties in Baltica's position relative to other continents in Cambrian to Early Ordovician time mean that it has been omitted from maps of 490 Ma and older periods. It is first shown on the 460-Ma map (d) separated from Laurentia by a relatively narrow branch of the Eastern Iapetus Ocean. For compatibility with the events in Svalbard, the distance between the two continents is shown as decreasing between 460 and 440 Ma.

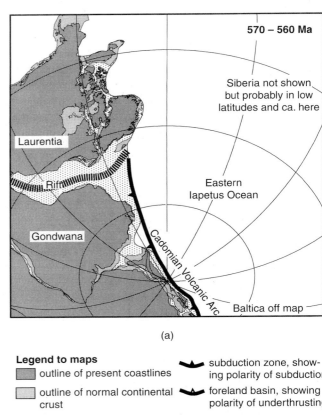

(a)

Legend to maps

▨	outline of present coastlines
▨	outline of normal continental crust
▨	inferred thinning and/or anomalous thickness of continental and arc crust
◣▬	location of accreted volcanic arc terranes

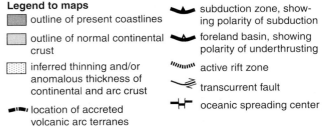

⌄⌄	subduction zone, showing polarity of subduction
⌃⌃	foreland basin, showing polarity of underthrusting
⁗⁗	active rift zone
⫽	transcurrent fault
╫	oceanic spreading center

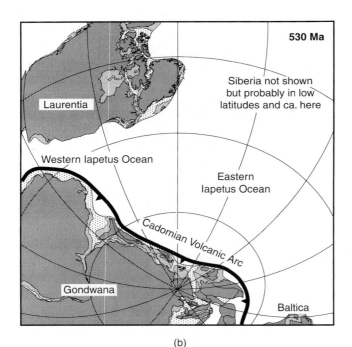

(b)

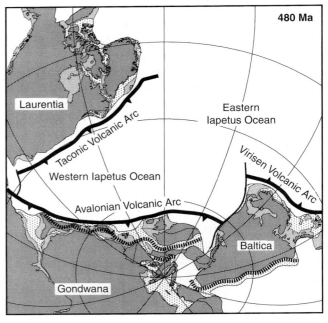

(c)

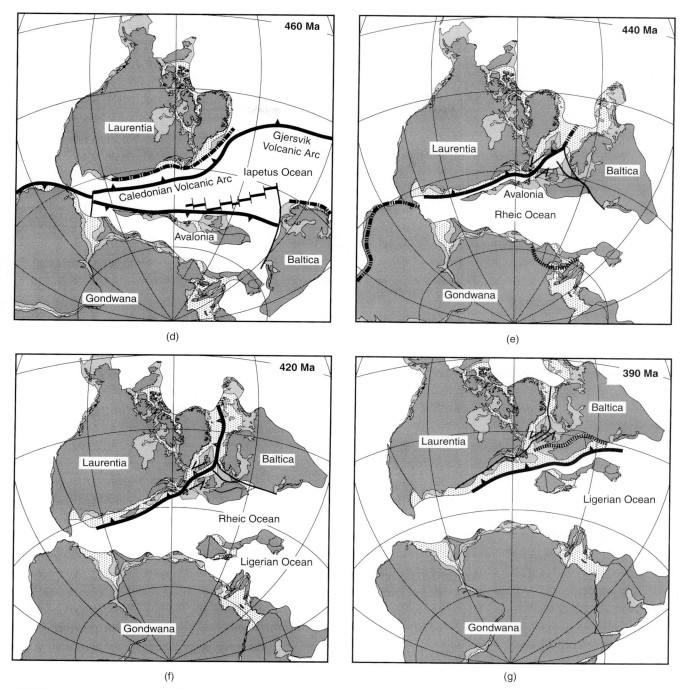

FIGURE 22.5.1 (Continued)

The orogenic belts formed by these collision events have long-established names related to their present-day geographic positions. Going back in time, the youngest phase of orogenesis (Early Devonian–Silurian) is known as Acadian in North America, Late Caledonian in East Greenland and the British Isles, and Ligerian in mainland Europe, except in Scandinavia where the term Scandian is used. In North America, the orogenic activity that peaked in the Middle Ordovician is known as Taconic, whereas in north-western Europe, including the British Isles, East Greenland, and Scandinavia, it is referred to as the Early Caledonian; however, in Svalbard (Spitsbergen), it is known as the M'Clintock Orogeny. Late Precambrian orogenic deformation is known as the Famantinian in South America, and was for several decades known as the Grampian in northwestern Scotland (now recognized as Early Ordovician), and Cadomian in Brittany (northwestern France) and southern Britain. Although the interpretation is still controversial, we

TABLE 22.5.1	MAJOR TECTONOTHERMAL EVENTS, IAPETUS OCEAN

620–570 Ma Breakup Gondwana Phase I

Andean margin of South America rifts from eastern margin of Laurentia to create the *Western Iapetus Ocean.*

Eastern Iapetus Ocean already in existence in the Late Precambrian, with landward-dipping, peri-Gondwanan, mainly Andean-like subduction zone and subduction-related tectonothermal events, as the:

(i) *Cadomian* along northern margin of Gondwana.

(ii) *Penobscotian,* 560–510 Ma in Exploits-Gander zones of Newfoundland.

(iii) *Famantinian* in South American Andes.

(iv) *Finnmarkian* in Scandinavia, 540–490 Ma.

(v) *Grampian* in northwestern Britain, 475–460 Ma (synchronous with ophiolite obduction).

490–460 Ma Ophiolite Obduction along Northern and/or Northwestern Margin of Laurentia (Precursor to Arc-Continent Collision)

Probably due to major change from transtensional to transpressional motion between Laurentia and South America, following short-lived generation of oceanic crust in peri-Laurentian and peri-Gondwanan marginal basins.

490–470 Ma Breakup Gondwana Phase II

Avalonia (including Carolina Slate Belt, Piedmont, parts of Nova Scotia and New Brunswick, Avalon Peninsula, southern Britain) and other microcontinental terranes rift away from Gondwana. Late Arenig rift event along entire Urals.

Peak ~450-Ma Orogenesis Involving Arc Collision(s)

Taconic in U.S. and Canadian Appalachians, 480–440 Ma (arc-continent collision along eastern margin of Laurentia).

Early Caledonian in northern Britain, 530–430 Ma (arc-continent collision along eastern margin of Laurentia).

From ~470 Ma, collision of magmatic arc(s) and western margin of Baltica as it rotates around to collide with Laurentia during Silurian time.

M'Clintock in Svalbard, 500–450 Ma (arc-continent collision).

435–370-Ma Destruction of Iapetus Ocean and Associated Events

Late Caledonian, 460–380 Ma (Baltica-intervening arc(s)-Laurentia collision).

Scandian, 430–400 Ma (Baltica-Laurentia collision).

Ligerian, 390–370 Ma (collision of Gondwana-derived Aquitaine-Cantabrian blocks with eastern Avalonia-Baltica).

suspect that the Late Precambrian events represent lithospheric extension rather than orogenesis. We have endeavored to show what is meant by the terms in Table 22.5.1, which summarizes the time span for the events, their location, and cause. The end result is that remnants of small continental fragments of island arcs and of backarc basins that originally bordered the continental margins have been swept up and incorporated into all these orogenic belts as "terranes" and ophiolites during the closure of these branches of the Iapetus Ocean.

The processes that led to the eventual welding of the three supercontinents to form Pangea in Permian time gave rise to a perfect match: there are no gaps in the reassembly, such as unfilled remnant ocean basins.

The continents are most unlikely to have matched perfectly before collision and one must therefore expect processes to have taken place that allow imperfectly matching continents to fit together, such as thickening of continental crust, indentation by promontories, and lateral movement of slices along "escape structures." One might also anticipate that the area around the triple junction in the North Sea, where all three continents meet, is likely to be one of the more tectonically complex areas and one of long-lived activity.

Paleomagnetic poles from the stable parts of the three large continents provide the principal quantitative data for repositioning the continents relative to one another in early Paleozoic time. The data are poor, particularly for Cambrian time (see text to follow).

Fortunately, the Late Precambrian to Early Cambrian was a time when new continental margins formed along Laurentia, Baltica, and parts of Gondwana. By joining what are believed to be the original opposing margins together it is possible to make a plausible reassembly of the circum-Iapetus continents, and this reassembly serves as a starting configuration for its evolution through the remainder of early Paleozoic time. Where paleomagnetic evidence is inconclusive, geologic evidence permits an independent inference about the nature of each continental margin from its stratigraphic record. The geologic evolution of the area is therefore discussed next and is related to the maps where possible.

22.5.2 Late Precambrian–Cambrian Extension and Passive Margins

The Late Precambrian–Cambrian history of Laurentia, Baltica, and western South America appears to reflect Late Precambrian breakup and the development of passive continental margins. Continental breakup of the Late Precambrian supercontinent of Gondwana was diachronous. In northwestern Britain, the mainly Late Precambrian Dalradian Supergroup accumulated between Laurentia and South America (Gondwana). In West Africa, the earliest recorded tectonic event is the westward rifting of a continental fragment from the West African craton about 700 Ma, and the development of a rift-drift stratigraphy (including tholeiitic and alkaline basalts). Dyke swarms, reflecting crustal extension, are known along the Laurentian margin in Labrador dated at 615 Ma. Similar dyke swarms are known from the Baltica margin in Scandinavia, dated at 665 Ma and ~640 Ma. In themselves, the dykes merely indicate stretching, but in all the areas studied they pass upward into, or are closely associated with, the development of extensive carbonate platforms (believed to have accumulated in low latitudes), and/or were intruded into very thick deep-water clastics, that are interpreted as passive continental margin successions. The eastern Laurentian passive margin extended from Greenland through the Durness sequence of northwestern Scotland to fringe most of the United States. Similarly, the Late Precambrian–Cambrian margin of Baltica is interpreted as a 200-km wide passive margin sequence. A comparable passive margin of the same age is known in northwestern Argentina and the western margin of the South American craton.

Margins formed by lithospheric extension have a characteristic subsidence curve determined both by the amount of stretching and the time it began. From such curves, the breakup and rifting ages of the new passive margins of Laurentia, Baltica, and northwestern Argentina occurred between 625 and 555 Ma, in agreement with all the other evidence. Also, in western Newfoundland, a rift-drift transition has been proposed for western Newfoundland at about 570–550 Ma.

Eastern North America (Laurentia) and western South America (Gondwana) separated from one another in the Late Precambrian to form opposing margins. This suggestion is supported by the otherwise puzzling distribution in northwest Argentina of early Cambrian *olenellid* trilobites that are similar to those in eastern Laurentia and not known elsewhere. Rift events recorded in dyke swarms from Baltica and Greenland suggest that rifting occurred about 100 million years earlier than in the Appalachians, so that the Eastern Iapetus Ocean was much older than the Western Iapetus Ocean.

22.5.3 Late Precambrian–Cambrian Arcs, Northern and Northwestern Gondwana

By contrast with the passive margins of Gondwana, the Late Precambrian–Cambrian history of northern and northwestern Gondwana is one of arc formation and orogenesis. The orogeny, known as the Cadomian, is a Late Precambrian–Cambrian (650–500 Ma) belt exposed on the northern and northwestern edge of Gondwana, with its type area in Brittany, northwestern France. It includes arc and arc-related rocks that record 650–500 Ma age tectonothermal events, also exposed in southern Britain, Spain, southeastern Ireland, and the "Avalonian" of the northern Appalachians. To the east of Brittany, Cadomian deformation is recognized in Czechoslovakia and the south Carpathian-Balkan region, so this was a time of extensive new crustal growth. The Cadomian events probably represent the vestiges of a major arc system formed by subduction under the northern and northwestern margin of Gondwana.

Many areas affected by the Cadomian Orogeny were later detached from the margin of Gondwana and migrated across the intervening ocean to become "exotic" terranes attached to Laurentia, before Gondwana itself collided with them and Laurentia. Thus, during the Late Precambrian, what were to become exotic microcontinental blocks, such as eastern and western Avalonia, the Piedmont Terrane, and the Carolina Slate Belt, probably formed the outboard parts of Gondwana along a margin that bordered an existing ocean.

22.5.4 Early to Middle Ordovician Arcs, Marginal Basins, and Ophiolites

Throughout northwestern Europe, regional chemical and isotopic signatures in the Ordovician-Devonian igneous suites north of the Iapetus suture, or its inferred along-strike continuation, show a subduction-related affinity associated with both southeast- and northwest-directed subduction, which was active from Early Ordovician to Middle Silurian times. By late Tremadoc to early Arenig time (~490 Ma) the Laurentian craton was fringed by marginal basins and arcs. Two subparallel arcs appear to have existed along much of the western/northwestern margin of the Iapetus Ocean (Figures 22.5.2a and b): (1) an inboard island-arc system that was developed mainly on continental crust, (2) a second island-arc system occupying a more oceanic setting, and developed above a north- to northwest-dipping subduction zone associated with the destruction of the Iapetus Ocean *sensu stricto*. This latter arc system appears to have been slightly younger in age and locally lasted into Silurian time. Destruction of the marginal basins took place throughout the Early Ordovician to Late Silurian interval along most of its length, but was restricted to Middle Ordovician time in Svalbard (M'Clintock Orogeny) and to latest Ordovician–earliest Silurian time in central Newfoundland (Taconic Orogeny).

A Taconic (~470–460 Ma) high-pressure granulite event has been identified in the Moine Supergroup in

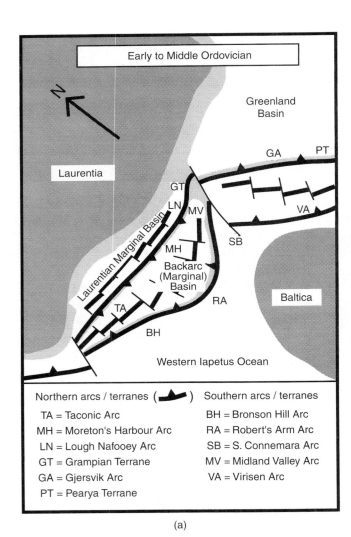

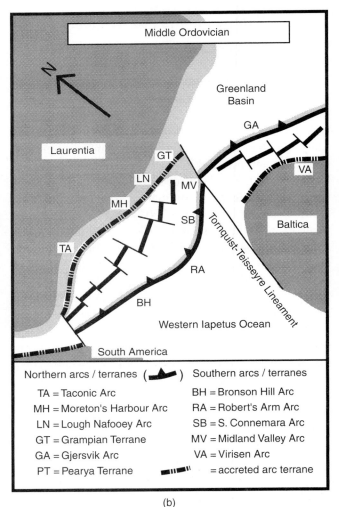

Northern arcs / terranes ( )	Southern arcs / terranes
TA = Taconic Arc	BH = Bronson Hill Arc
MH = Moreton's Harbour Arc	RA = Robert's Arm Arc
LN = Lough Nafooey Arc	SB = S. Connemara Arc
GT = Grampian Terrane	MV = Midland Valley Arc
GA = Gjersvik Arc	VA = Virisen Arc
PT = Pearya Terrane	

(a)

Northern arcs / terranes ( )	Southern arcs / terranes
TA = Taconic Arc	BH = Bronson Hill Arc
MH = Moreton's Harbour Arc	RA = Robert's Arm Arc
LN = Lough Nafooey Arc	SB = S. Connemara Arc
GT = Grampian Terrane	MV = Midland Valley Arc
GA = Gjersvik Arc	VA = Virisen Arc
PT = Pearya Terrane	= accreted arc terrane

(b)

FIGURE 22.5.2 Schematic diagram showing Early to mid-Ordovician arcs and marginal basins in the Iapetus Ocean. These were subsequently accreted, in part, to Laurentia and South America (Taconic-Grampian-Early Caledonian Orogeny) and to Baltica (final emplacement in Scandian [i.e., Late Caledonian] orogeny). Note that the main Scandinavian Caledonide arcs are shown as originally located more oceanward of the Taconic arc(s). Solid black arrows indicate the probable polarity of subduction; light gray areas indicate thinned continental crust. Not to scale.

northern Scotland. U-Pb geochronologic data from Connemara, in the western Irish Caledonides, suggest that continental arc magmatism along the southern Laurentian margin was short-lived, lasting from ~475 to ~463 Ma. A corollary of these ages is that the Grampian Orogeny in Connemara was considerably younger than is generally acknowledged. The Grampian Orogeny, therefore, was synchronous with the Taconic Orogeny in the northern Appalachians.

Like the Mesozoic Tethys, the Western Iapetus Ocean was associated with a relatively brief phase of arc development and oceanic spreading, rapidly followed by plate convergence leading to extensive, Early Ordovician ophiolite obduction along the Laurentian margin immediately preceding arc-continent collision (Taconic-Grampian–Early Caledonian). From west to east, early Ordovician ophiolites fall into three categories: (1) Laurentian marginal basin oceanic crust, as in western Newfoundland and the Shetland Islands; (2) intra-arc, backarc marginal basin oceanic crust, as in the central Newfoundland, western Ireland, and Highland Border Group ophiolites (and possibly in slivers along the Highland Boundary fault zone), and (3) forearc ophiolites, including accreted seamount material (as in eastern central Newfoundland and southwest Scotland), which may include seamount fragments. In the Scandinavian Caledonides, arc fragments appear to have been involved in collisional events in Early Ordovician times (Finnmarkian), but were finally emplaced onto Baltica during the Silurian–Early Devonian Scandian orogeny.

The marginal basins that fringed Laurentia in the Early Ordovician were closed by the end of Middle Ordovician time. In North America their closure caused extensive east-west shortening (today's coordinates), mainly accommodated by westward transport of thrust slices (Taconic Orogeny). The Taconic Orogeny of eastern North America lasted from about 480 to 430 Ma, with a peak between 470 and 450 Ma. The Taconic Orogeny overlaps in time with the Early Caledonian (M'Clintock) events in Svalbard (Spitsbergen) on the west side of Baltica.

The closure of the marginal basins that fringed Laurentia in the Early Ordovician was caused by arc-continent collision, driven by the proximal approach of the northern margin of Gondwana (probably northern South America) to the Appalachian margin of Laurentia, and/or a major reorientation of relative plate vectors because of plate-tectonic processes associated with other ocean basins. The collision events along the Laurentian margin (including the Grampian Orogeny) on the northern side of the Eastern Iapetus Ocean, may be related to the narrowing of the ocean between

Baltica and Laurentia such that, by the late Llandeilo, generic faunal links were established across a Galapagos-like island chain between both continents.

The maps for this period (Figure 22.5.2, 480–460 Ma) show western South America moving past eastern Laurentia. The motion appears to have been quite oblique, without actual collision, but caused arc accretion in Middle Ordovician time in the absence of an intense Himalayan-style continent-continent collision between Laurentia and western Gondwana. South America (or "Occidentalia"), which is bordered by a Cambrian carbonate platform similar to that of eastern North America, may have been a continental margin opposing Laurentia during Late Ordovician time. In northwest Argentina, the early Paleozoic *olenellid* trilobite faunas in the Famantinian Orogen are similar to those found in eastern Laurentia, which may suggest geographic linkage between these areas.

As noted previously, the position of relative to other continents, is uncertain, therefore it has been omitted from maps of 490 Ma and older periods. It is first shown on the 460-Ma map, where it appears separated from Laurentia by a relatively narrow branch of the Eastern Iapetus Ocean (Figure 22.5.2). For compatibility with the events in Svalbard, the distance between the two continents is shown as decreasing between 460 and 440 Ma. This presumed narrowing of the ocean between Baltica and Laurentia is supported by the establishment of faunal links between the two continents by late Llandeilo time (about 465 Ma), where none had existed previously.

Throughout northwestern Scotland and northern Ireland, north of the Iapetus suture, geologic data, including regional chemical and isotopic signatures in the Ordovician-Devonian igneous suites, are associated with northwest-directed subduction that was active from Early Ordovician to Middle Silurian times. This subduction zone appears to have been active in accommodating the subduction of most of the Iapetus oceanic crust, although southeast-directed subduction may have been important locally, and even in the final stages of closure.

22.5.5 Early Ordovician Breakup of the Northwest Margin of Gondwana

In the late Arenig, there was a second major episode of continental fragmentation in Gondwana (Phase II breakup, to distinguish it from the Late Precambrian Phase I breakup. The Avalonian, Piedmont, and Carolina Slate Belt Terranes, all of which contain Cadomian-age arc basement, broke away from the northwestern edge of the Gondwana continent. The partial separation of eastern Avalonia (including northwestern France, also called Armorica) from Gondwana during the late Arenig, has been documented using sedimentological criteria, including subsidence curves, from the Sahara, Middle East, Nova Scotia, Ibero-Armorica, Ireland, England, and Wales.

In addition to the separation of these fragments, a new ocean basin may have been created on the eastern edge of Baltica in the present region of the western Urals. A more speculative view is that the Eastern Iapetus Ocean (i.e., the Tornquist Sea) was created at about the same time by the separation of southern Baltica from northern South America, that is, Phase II breakup could have resulted from Baltica rifting away from Gondwana along with Avalonia, and the Piedmont and Carolina Slate Terranes. After rifting, Baltica may have moved toward the equator, from a more southerly position in Early Ordovician time. There is no clear evidence for the creation of new passive margins in these areas like that for those of Late Precambrian to Cambrian age discussed previously. Avalonia is not linked with Baltica at this time.

22.5.6 Middle–Late Ordovician Subduction, Continental Fragmentation, and Collisions

The late Arenig Phase II breakup of Gondwana was probably temporally linked with the extensive ophiolite obduction initiated in the Llanvirn. Much of the northern Gondwanan margin of the Iapetus Ocean was an Andean-type covergent plate margin with a subduction polarity towards the continental interior. Remnants of this arc include the calc-alkaline igneous rocks of the English Lake District, southern Welsh Basin, and southern Ireland. The Welsh Basin was initiated as a marginal basin on the southern side of the Eastern Iapetus Ocean during the Arenig, at the same time as eastern Avalonia rifted off northwestern Gondwana. Because the late Arenig Phase II breakup of Gondwana immediately preceded the extensive ophiolite obduction initiated in the Llanvirn, both processes may be causally related.

During Cambrian to Early Ordovician time, northwestern Britain was part of Laurentia and located at about 15°–20°S. By contrast, the paleolatitude of eastern Avalonia (southern Britain) in Early Ordovician time was about 60°S. The paleolatitude of eastern Avalonia had changed to about 45°S in the Middle Ordovician, and ~15°–25°S in the latest Ordovician–Early Silurian, suggesting it had a steady northward drift across the Iapetus Ocean. The latitudinal separation

across the Iapetus Ocean in the Early Ordovician, between the part of Laurentia containing northwestern Scotland (where eastern Avalonia eventually docked), and the Gondwanan margin with eastern Avalonia, changed from about 5000 km in the late Tremadoc–early Arenig to ~3300 km by the Llanvirn-Llandeilo. The underlying plate-tectonic causes for this northward motion are not immediately obvious. The paleomagnetic data independently support the faunal arguments for the northward movement of eastern Avalonia across the Iapetus Ocean during the Late Ordovician.

Subduction-related igneous activity occurred in the British Isles, south of the Iapetus suture, from the Tremadoc (earliest Ordovician) to earliest Caradoc (Middle-Ordovician) with a rapid change to a more alkaline and peralkaline signature and abrupt cessation in the Longvillian (Middle-Caradoc). Ridge subduction, and the creation of a slab window below the northern margin of Gondwana, (i.e., Eastern Avalonia; see Figure 22.5.1d), provide an elegant mechanism to explain (a) the abrupt switch-off in subduction-related igneous activity in eastern Avalonia in the early Caradoc; (b) the changed geochemical signature of the Caradoc compared with earlier igneous activity in eastern Avalonia; (c) the subduction of thermally warm ridge-flanks millions of years prior to ridge subduction as a reason for the widespread Llandeilo hiatus, or thin stratigraphies, throughout much of eastern Avalonia, and (d) a fundamental cause for the transference of eastern Avalonia to a north-moving plate and eastern Avalonia's rifting away from Gondwana, as the ridge is spreading center jumped southward of the microcontinent. An analogy can be found in the present-day Pacific where the small continental fragment of Baja California is now attached to the Pacific Plate and moving with it.

The protracted collision events contemporaneous with the Taconic Orogeny of North America culminated in high-grade metamorphism and major uplift at about 460–440 Ma in the Western Iapetus Ocean, and along other parts of the Laurentian margin and the associated marginal basins. Parts of the Scandinavian Caledonides (Figure 22.5.3) record an uplift in 450–435 Ma. Stable argon isotope studies of rocks from northern Sweden in the Upper Allochthon (Lower Koli Nappe, the Seve-Koli shear zone, the Seve Nappe) and the shear zones of the Middle Allochthon, reveal high-grade metamorphism and associated deformation of the Seve units as a Late Cambrian–Early Ordovician event in which the rocks cooled below the respective closure temperatures for hornblende at ~490 Ma and muscovite at ~455 Ma. The structurally

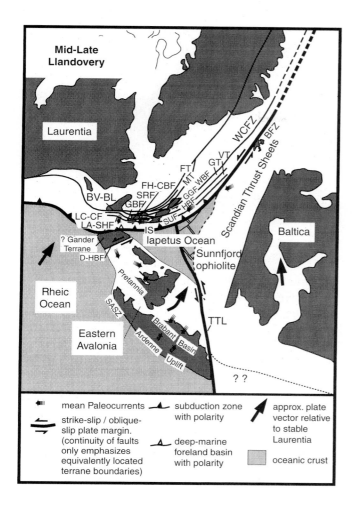

FIGURE 22.5.3 Mid-late Llandovery reconstruction. Plate motion arrows are shown for the Baltica and eastern Avalonia Plates relative to a fixed Laurentian (North American) plate. Closely spaced stipple outlines parts of Armorica, Britain, and the Gander Terrane south of the Iapetus suture, and fragments of western Newfoundland and northwestern Britain with Laurentian crustal affinities prior to ocean closure. Abbreviations for major geologic features: BV-BL = Bay Verte-Brompton Lineament, LC-CF = Lobster Cove-Chanceport Fault, LA-SHF = Lukes Arm-Sops Head Fault, GBF = Galway Bay Fault, SRF = Skerd Rocks Fault, FH-CBF = Fair Head-Clew Bay Fault, SUF = Southern Uplands Fault, HBF = Highland Boundary Fault, GGF = Great Glen Fault, WBF = Walls boundary Fault, FT = Flannan Thrust, MT = Moine Thrust, IS = Iapetus suture (i.e., Cape Ray-Reach Fault in Newfoundland); BFZ = Billefjorden fault zone (Svalbard), WCFZ = Western Central Fault Zone (Svalbard); TTL = Tornquist Teisseyre Lineament; SASZ = South Armorican Shear Zone, D-HBF = Dover-Hermitage Bay Fault.

Paired Ordovician arc systems in the northerly parts of the Iapetus Ocean correlated from Newfoundland to Svalbard and separated by ophiolites are as follows: (1) Taconic island arc, Early–Middle Ordovician developed above oceanward-dipping subduction zone, Moreton's Harbour arc (Newfoundland), Lough Nafooey arc (Ireland), Grampian Terrane (Scotland), and Gjersvik arc (Norway), Pearya/Northwestern Svalbard (including Biskayerhalvoya) Terranes; (2) Southern island arc, Ordovician-Silurian arc developed above Laurentia-ward dipping subduction zone, Bronson Hill arc, Robert's Arm arc (Newfoundland), South Connemara arc, Midland Valley terrane (Scotland) Virisen arc (Norway), and western Svalbard.

lower rocks of the Middle Allochthon, inferred to have been more proximal to Baltica prior to emplacement, show only the ~430-Ma event(s), whereas the Upper Allochthon records the older Finnmarkian event(s). In the Seve Nappes, there is evidence for Middle–Late Ordovician 450–440-Ma shear zones, showing that a pre-Scandian deformation affected rocks outboard from, or marginal to, Baltica. During this time, Seve Nappes of different P-T-t (pressure, temperature, time) histories were juxtaposed. Subsequently, during the Scandian Orogeny, the Seve and Koli Nappes were juxtaposed, and the Middle Allochthon mylonites formed as these nappes were emplaced over the Baltic Shield. All these tectonic units were assembled prior to regional cooling through the Ar closure temperature of muscovite.

22.5.7 Middle Ordovician–Silurian Closure of the Eastern Iapetus Ocean

The Ordovician-Silurian history of the Midland Valley, Scotland, records the evolution of an arc and backarc basin. Throughout the Late Ordovician and Early Sil-

urian, the Southern Uplands of Scotland and the along-strike Wexford-County Down area (Longford Down inlier) of Ireland were part of an active accretionary prism developed above a northward-dipping subduction zone on the northern margin of the Iapetus Ocean. For example, the Southern Uplands accretionary prism, which developed over at least 50 m.y. from the Llanvirn to Wenlock, was associated with the Midland Valley forearc basin further to the north. The arc massif of older metamorphic basement in the Grampian Highlands northwest of the forearc basin was capped by calc-alkaline arc volcanics and intrusive igneous suites, and supplied most of the sediments to the trench-forearc accretionary system to the south.

Closure of the Eastern Iapetus Ocean by oblique (overall sinistral) collision took place between the island arc(s) sandwiched between the converging continents of Laurentia and Baltica. The closure may have begun, during the latest Llandeilo to Caradoc in the region of northern Norway-Svalbard, to incorporate M'Clintock-Finnmarkian orogenic crustal fragments. In places the setting probably resembled that between mainland Southeast Asia and northern Australia today.

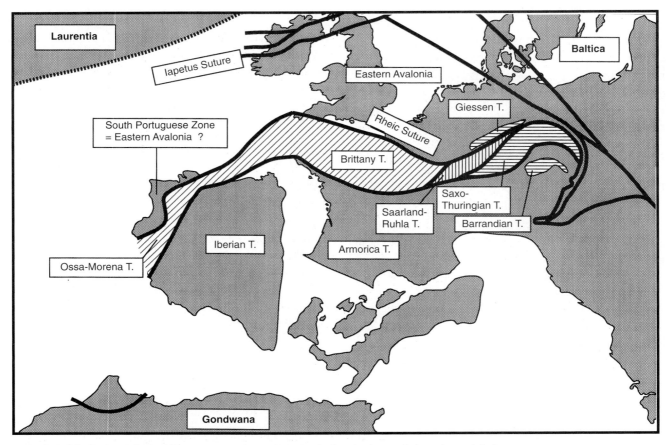

FIGURE 22.5.4   Principal European continental blocks during the Lower Paleozoic, and their suture sites (heavy lines).

Elsewhere, continent-continent collision may have created a situation like that between the present-day Himalayas and the Bengal Fan of eastern India. Voluminous flysch sediments were shed away from the collision zone to form the axial-trench wedges of sandy turbidites preserved in the Ordovician tracts of the Southern Uplands accretionary prism in northwestern Britain, akin to the present-day Bengal Fan.

The Silurian-Devonian Scandian Orogeny, caused by the collision of Laurentia and Baltica above a northwest-dipping subduction zone, resulted in the final emplacement of thrust sheets eastwards onto the Scandinavian crystalline basement with its Cambrian-Ordovician shelf successions.

Collision of eastern Avalonia with Baltica and Laurentia occurred in the latest Ashgill–earliest Llandovery, with the microcontinent behaving as a rotating rigid indentor, probably in the region of present-day central Newfoundland, above a north-dipping subduction zone (Figure 22.5.4). Collision was oblique (or "soft") with a sinistral component along the margin. The major phase of bimodal Silurian magmatism in central Newfoundland, New Brunswick, south-central

Britain, and western Ireland implies that there was a component of extension shortly after the initial collision. Extension is also suggested by the kinematic history of syn-deformation granites in northwestern Britain. Even faunal evidence from Middle Silurian ostracodes suggests a phase of extension after which oblique convergence continued. The phase of Middle Silurian oblique extension was probably caused by the rotation of eastern Avalonia against Laurentia during the final stages of suturing. The main sinistral displacement of eastern Avalonia took place throughout the Silurian and up until the peak Acadian deformation in the Emsian (Early Devonian time).

Final welding of eastern Avalonia took place in the Wenlock and was associated with a prolonged phase of deep-marine foreland-basin development. By Llandovery time, the Tornquist-Teisseyre lineament, formed a major, probably active, backarc strike-slip fault to the arc associated with closure of the Eastern Iapetus Ocean between Baltica and eastern Avalonia as a submarine (subshelf) lineament. The southern continental plate boundary of Baltica remains poorly defined, but was south of the arc complex associated with northward subduction in the northwestern European Caledonides.

Silurian-Devonian sinistral shear was associated with the amalgamation of the Western, Central, and Eastern Provinces of Svalbard. These provinces have pre-Devonian histories that involved complete separation of the terranes. Eastern Spitsbergen and Nordaustlandet, for example, may have originated along the Laurentian margin far to the south of their present position, to be juxtaposed against the Central Province along the Billesfjorden fault zone by the Late Devonian. In Svalbard, argon ages suggest an Early Devonian cleavage formation at ~400 Ma.

In north-central Newfoundland, the age of the slaty cleavage is constrained as latest Silurian (Ludlow-Pridoli), whereas in eastern Avalonia the slaty cleavage appears to range in age from Early Silurian to early Middle Devonian. In Wales and the Welsh Borderland, the Middle Devonian is commonly missing, with postorogenic sedimentation resuming in the Late Devonian (Famennian). Also, in north Devon (southwestern England), the early Middle Devonian involved the influx of clastics derived from the Welsh Basin, which is interpreted as substantial uplift (orogeny) to the north. In the Midland Valley, Scotland, the Middle Devonian is also missing, but it was the main period of chiefly fluvio-lacustrine sedimentation in the Orcadian Basin, northeastern Scotland.

Within the Caledonian slate belt, south of the Iapetus suture, there is a major arcuate trend in the orientation of the strike of cleavage, from an Appalachian (northeast) trend in Ireland and Wales to a more Tornquist (east-southeast) trend that is typical of northern Germany and Poland, with intermediate trends in northern England. Based on the clockwise cleavage transection of related fold axial surfaces, and associated sinistral displacement on strike-slip faults in northwestern England, there was an episode of Late Caledonian ("Acadian") sinistral transpression. The microgranite dyke swarm associated with the emplacement of the Shap granite intrudes folded and cleaved Silurian sediments, but the dykes themselves are weakly cleaved. In the English Lake District the formation of this cleavage was contemporaneous with the emplacement of various syn- and post-deformational igneous suites, dated at about 394–392 Ma or early Middle Devonian (Emsian). The cleavage arcuation (from north-northeast in northwestern England to a more easterly trend further east in northern England) is explained as a consequence of the anticlockwise rotation of eastern Avalonia relative to Laurentia during collision and final suturing.

22.5.8 Late Ordovician Icehouse

In Late Ordovician time (Hirnantian stage of the Ashgill ~443–444 Ma), a major short-lived glaciation affected western Gondwana (mostly northwest Africa). The icehouse may have lasted for as little as 0.5 m.y., or it may have lasted 4–6 m.y., beginning in the Caradoc, with peak glaciation during the Hirnantian stage of the Ashgill. Stable-isotope data from brachiopods show a dramatic positive isotope excursion in the $\delta^{13}C$ and $\delta^{18}O$ record (PDB scale ~2‰) for eastern North America, central Sweden, and the Baltic States. In the Baltic States, the magnitude of these isotopic excursions is up to ~4‰, which is equivalent to the combined effects of a sea-level fall of 100 m and a drop of 10°C in tropical sea surface temperatures.

Why this icehouse came into being in the middle of a greenhouse period and then lasted for such a short time has always been a puzzle. It has been proposed that its initiation and demise are attributable to the action of a Central American or similar oceanic gateway. This moved some continents into polar latitudes; while there, they opened and closed oceanic low-latitude gateways that changed global oceanic circulation from one with important circumequatorial currents (greenhouse) to one with inhibited circumequatorial deep-water currents (icehouse) and more restricted oceanic gyres. The latter scenario, favored the export of moisture to high polar latitudes, where it could accumulate as snow and ice to form the continental ice sheets on the polar parts of Gondwana.

The paleomagnetic data suggest that northwestern Gondwana and southwestern Laurentia were closer to one another at 440 Ma than they were just before or just after the icehouse. Thus there was a potential gateway in the correct period; it could have closed in the Late Ordovician and could have re-opened shortly afterwards. Plate-tectonic motions have the appropriate scales of time and length to account for the movements required: about 100 km/m.y. Also, zircon and monazite U-Pb data, together with tectonic mapping and petrologic studies in the Acatlan Complex, southern Mexico, have led workers to invoke a Late Ordovician–Early Silurian continental collision orogeny, which they ascribe to the collision of eastern Laurentia with western South America along the entire Andean margin. Thus, there appears to have been more than one important site at which oceanic gateways were active at the time of the Late Ordovician icehouse Earth.

22.5.9 Ordovician-Silurian Magmatic Arcs Elsewhere in Europe

In contrast to the Lower Paleozoic rocks of Norway and Sweden, rocks of this age elsewhere in continental Europe are largely concealed beneath younger strata or have been overprinted by intense Variscan (Hercynian or Armorican) and Alpine deformation. Exposure is poor and tectonic boundaries are difficult to define. Nevertheless, the Ordovician-Silurian outcrops of the European Caledonides can be assigned to three continental plates (Figure 22.5.4): (1) eastern Avalonia (considered above); (2) the southern parts of Baltica; and (3) parts of the northern margin of Gondwana (including, from west to east, the Brittany, Saarland-Ruhr, Tepla-Barrandian, Saxo-Thuringian, Bavarian, Gory Sowie, and Moravo-Silesian Terranes, and, farther south, the Ibero-Armorica-Moldanubian Terranes).

By Late Ordovician time, the Eastern Iapetus between eastern Avalonia and Baltica (also known as the Tornquist Sea) had probably closed by eastward and/or northeastward subduction. The southern margin of eastern Avalonia is marked by a major Early Paleozoic suture zone with ophiolites in Ibero-Armorica and Alpine Europe. The zone also includes arc-related igneous-volcanic suites and thick marine successions that can be traced through Iberia, into Armorica (along the South Armorican Shear Zone, or SASZ), and across to southern Austria. The SASZ includes thrust-bound slices of metasedimentary and igneous rocks metamorphosed to eclogite facies, 420–375-Ma blueschists, and a high-temperature low-pressure migmatite belt, interpreted as remnants of an accretionary complex formed above a subduction zone, which was almost certainly active during the Silurian. There was also a major Late Ordovician–Silurian (450–415 Ma) high-pressure event with little deformation, in France, Iberia and Morocco.

In the Ossa Morena zone, central Iberia, the Middle Devonian emplacement of northeast-vergent nappes was followed by major sinistral transpression. This created the central Iberian fold belt separating the Aquitaine-Cantabrian microcontinent to the east from the South Portuguese block to the west; the latter can be regarded as belonging to the southern part of eastern Avalonia.

Further east, from Early Silurian to Middle Devonian time, the mid-European and Tepla-Barremian "terranes" appear to have been subject to considerable crustal extension, and the extrusion of voluminous, within-plate, alkali basalts. In the eastern Alps of southern Austria, there are subduction-related volcanics and sediments that formed in an island arc and active continental margin setting. These are probably vestiges of an arc on the edge of Gondwana, but their precise relation to Gondwana is unclear.

The Ligerian Orogeny involved the collision of these Gondwana-derived fragments with the southern margin of eastern Avalonia, which was itself a fragment broken off Gondwana at an earlier period. It involved the development and destruction of a volcanic arc complex, with a backarc marginal basin to the north, by continent-continent collision in the younger Variscan (Hercynian) Orogeny. A Middle Devonian intermediate-pressure metamorphic event associated with major tectonism is well known, not only in France, but also in Morocco and, more speculatively, in Iberia. This event is interpreted as reflecting the amalgamation of other continental fragments with eastern Avalonia, such as Saxo-Thuringia and Moldanubia, in addition to those in Aquitaine and Cantabria.

The Rheic Ocean lay to the south of the Ligerian orogenic belt and north of Gondwana, that is, south of the Tepla-Barremian Plate, where the Ibero-Armorica-Moldanubian Plate was converging northwards throughout the Devonian. Essentially, the Rheic Ocean is synonymous with the vestiges of the Eastern Iapetus Ocean. The ocean appears to have been closed by Middle Devonian (Givetian) time, though its closure may have been preceded by the creation of a small backarc basin to the north. All oceanic areas between Gondwana and Laurussia (Laurentia and Baltica) had been eliminated by Late Carboniferous (Namurian) time. Subduction of the Rheic Ocean crust probably was initiated in the Early Devonian, and ocean closure occurred mainly by the northward subduction of oceanic crust below the southern arc-related margin of the Baltica Plate (which includes cratonic Russia west of the Urals, a Permian collisional orogen).

The southeastern margins of some of the microcontinental terranes that were accreted to Laurentia-Baltica during the Ordovician-Devonian are marked by Late Devonian ophiolites; for example, the obduction of the ~400–375-Ma Lizard ophiolite around 370 Ma. After unfolding the major Variscan flexure from Armorica to Iberia, and allowing for significant dextral offset along faults, such as the fault that displaces the Haig Fras from the rest of the Cornubian Batholith, it appears that the approximately 390-Ma Morais ophiolite, one of a series of crystalline complexes exposed in northern Iberia, represents the along-strike equivalent of the Lizard complex. These ophiolites were obducted northwards as fragments of the Rheic or Ligurian Ocean crust.

22.5.10 Postorogenic Continental Sedimentation and Igneous Activity

The collision of Baltica with Laurentia united them into a single continent known as Laurussia and created a high mountain chain. The collision of Gondwana and Laurussia to form Pangea was not completed until Late Carboniferous time. Molasse accumulated in fault-controlled intermontane basins, which were preserved along the major tectonic lineaments and suture zones.

By the Early to Middle Devonian, most of the major strike-slip between terranes appears to have occurred along many lineaments, for example, the Great Glen Fault of northwestern Scotland is a major strike-slip fault but does not appear to have been significantly active during Old Red Sandstone (ORS) deposition. Furthermore, the Emsian-Eifelian Orcadian ORS next to the fault shows a net offset today of 25–29 km, which has a dextral sense, rather than the sinistral sense implied by the geometry of docking and collision. Post-ORS dextral offsets are also known in the Shetland Isles and are much larger—on the order of 120 km.

The high topography was supported by thickened continental crust. Temperatures in the lower part of the crust rose sufficiently to partially melt it and produce late-stage granites and granitoid bodies. These are the late-orogenic and postorogenic intrusions that characterize the final stages of continent-continent collisions. By earliest Carboniferous time, the crust on the southern margin of Laurentia was extending and spreading laterally by gravitational collapse, which induced faulting at the surface. Such faults should be at right angles to the greatest stress, that is, parallel to the mountain chain.

The tectonic vergence divide between the major northward (Late Paleozoic coordinates) obduction and thrusting events associated with closure of the Western Iapetus Ocean, and the southward tectonic transport onto the Baltic Shield due to closure of the Eastern Iapetus Ocean, was situated in the region of central Newfoundland to the British Caledonides. This latter region probably was the site of a triple junction associated with transforms and spreading centers during the opening of the Western and Eastern Iapetus Oceans, evolving into a triple junction involving subduction zones and transforms. The obliquity of the collision events may have been a major contributory factor in the preservation of the low-grade slate belts, in contrast, a Himalayan-style collision, which occurred farther north, resulted in high-grade metamorphic rocks being common at the surface.

22.5.11 Closing Remarks

The Late Precambrian–Early Cambrian was marked by the opening of an essentially east-trending Western Iapetus Ocean between southeastern Laurentia and western South America. The Eastern Iapetus Ocean, separating the Greenland area of Laurentia from Baltica opened earlier than the western ocean and was elongated in a north-south direction. Both oceans had different histories of opening and closure. Ocean closure involved overall sinistral shear between Baltica and Laurentia during the Late Ordovician to Devonian, whereas the U.S. Appalachians were influenced by dextral shear (first transpressive, then transtensional) between South America and Laurentia during the Ordovician.

Like the Mesozoic breakup of Pangea, the breakup of Gondwana (Phase I, ~620–570 Ma; Phase II, ~490–470-Ma) may have been associated with plume activity though, the evidence for this is not clear. Continental breakup would probably have increased the length of the global ocean-ridge system and caused a reduction in the mean age of the ocean floor. This would have resulted in a global rise in sea level, leading to a widespread flooding of continents. This is consistent with the preponderance of wide Cambro-Ordovician shelf seas, which were commonly sites for the accumulation of organic-rich muds (now pyrite-rich black shales). The lack of glaciogenic sediments or striated pavements, at least until the latest Ordovician (Ashgill), suggests that there were no substantial, if any, polar ice caps. The Cambrian-Ordovician was probably a greenhouse period induced by enhanced atmospheric CO_2 levels, which, in turn, are attributable to increased oceanic-ridge and mantle-plume activity.

Future research needs to better constrain the timing and nature of basin-forming events, and their subsequent histories and destruction, a task that can be accomplished through improved radiometric dating techniques and careful (macro- to micro-) structural and stratigraphic and/or sedimentologic studies. Careful reevaluation and new measurements of paleomagnetic data, for example, those using stepwise thermal demagnetization techniques, are giving us an improved understanding of the movement history of continental fragments. Sophisticated geochemical and isotopic arguments are helping to define the plate-tectonic settings of igneous rocks (supra-subduction zone, extensional intraplate, etc.), and to infer past global and/or regional climates from sediments. Undoubtedly, the Lower Paleozoic remains an area of fruitful research.

ADDITIONAL READING

Bond, G. C., Nickeson, P. A., and Kominz, M. A., 1984. Breakup of a supercontinent between 625 Ma and 555 Ma: new evidence and implications for continental histories. *Earth and Planetary Science Letters,* 70, 325–345.

Franke, W., 1989. Tectonostratigraphic units in the Variscan belt of central Europe. Dallmeyer, R. D., ed., *Terranes in the Circum-Atlantic Paleozoic orogens, Geological Society of America Special Paper,* 230, 67–90.

Friedrich, A. M., Bowring, S. A., Martin, M. W., and Hodges, K. V., 1999. Short-lived continental magmatic arc at Connemara, western Irish Caledonides: implications for the age of the Grampian orogeny. *Geology,* 27, 27–30.

Friend, C. R. L., Jones, K. A., and Burns, I. M., 2000. New high-pressure granulite events in the Moine Supergroup, northern Scotland: implications for Taconic (early Caledonian) crustal evolution. *Geology,* 28, 543–546.

Frisch, W., and Neubauer, F., 1989. Pre-Alpine terranes and tectonic zoning in the Eastern Alps. In Dallmeyer, R. D., ed., *Terranes in the Circum-Atlantic Paleozoic Orogens, Geological Society of America Special Paper,* 230, 91–100.

Harland, W. B, Armstrong, R. L., Cox, A. V., Craig, L. E., Smith, A. G., and Smith, D. G., 1989. *A Geologic Time Scale 1989.* British Petroleum Company plc and Cambridge University Press: London.

Murphy, J. B., van Staal, C. R., and Keppie, J. D., 1999. Middle to late Paleozoic Acadian orogeny in the northern Appalachians: a Laramide-style plume-modified orogeny? *Geology,* 27, 653–656.

Pickering, K. T., and Smith, A. G., 1995. Arcs and back-arc basins in the Lower Palaeozoic circum-Atlantic. *The Island Arc,* 4, 1–67.

Ryan, P. D., and Dewey, J. F., 1991. A geological and tectonic cross-section of the Caledonides of western Ireland. *Journal of the Geological Society, London,* 148, 173–180.

Salda, L. H. D., Cingolani, C., and Varela, R., 1992. Early Paleozoic orogenic belt of the Andes in southwestern South America: result of Laurentia-Gondwana collision? *Geology,* 20, 617–620.

Smith, A. G., and Pickering, K. T., 2003. Oceanic gateways as a critical factor to initiate icehouse Earth. *Journal of the Geological Society, London,* 160, 337–340.

Soper, N. J., and Hutton, D. H. W., 1984. Late Caledonian sinistral displacements in Britain: implications for a three-plate model. *Tectonics,* 3, 781–794.

Stephens, M. B., and Gee, D. G., 1985. A tectonic model for the evolution of the eugeoclinal terranes in the central Scandinavian Caledonides. In Gee, D. G., and Sturt, B. A., eds., *The Caledonide Orogen—Scandinavia and related areas.* Chichester: Wiley and Sons.

Sturt, B. A., and Roberts, D., 1991. Tectonostratigraphic relationships and obduction histories of Scandinavian ophiolitic terranes. In Peters, T., et al., eds., *Ohpiolite Genesis and Evolution of the Oceanic Lithosphere.* Sultanate of Oman: Ministry of Petroleum and Minerals.

Torsvik, T. H., Olesen, O., Ryan, P. D., and Trench, A., 1990. On the palaeogeography of Baltica during the Palaeozoic: new palaeomagnetic data from the Scandinavian Caledonides. *Geophysical Journal International,* 103, 261–279.

van der Pluijm, B. A., Johnson, R. J. E., and Van der Voo, R., 1993. Paleogeoghraphy, accretionary history, and tectonic scenario: a working hypothesis for the Ordovician and Silurian evolution of the northern Appalachians. *Geological Society of America Special Paper,* 75, 27–40.

Wilson, J. T., 1966. Did the Atlantic close and then re-open? *Nature,* 211, 676–681.

Zonenshain, L. P., Kuzmin, M. I., and Natapov, L. M., 1990. *Geology of the USSR: a plate-tectonic synthesis, Geodynamics Series,* 21. American Geophysical Union: Washington, DC.

22.6 TECTONIC GENEALOGY OF NORTH AMERICA—An essay by Paul F. Hoffman[1]

22.6.1 Introduction

Continents are complex tectonic aggregates that evolved over hundreds of millions of years. Today's continents are drifted fragments of the mid-Phanerozoic **supercontinent** Pangea, more or less reshaped by accretion and ablation at subduction zones. The creation of Pangea involved the fusion of many older continents, each of which was itself a drifted fragment of some still older continental assembly. Orogenic belts, marking the sites of ocean opening and closing, are the basic elements that are used to unravel the tectonic genealogy of continents.

Today's giant continent, Eurasia, was assembled in the Phanerozoic eon (545–0 Ma) through the piecemeal convergence of many pre-Phanerozoic continental fragments, roped together by Phanerozoic subduction complexes. The assembly of Eurasia is ongoing today. Other continents are fragments of former giant continents assembled at various times in the pre-Phanerozoic. The southern continents, for example, are derived from Gondwanaland, which was assembled in the Neoproterozoic era (1000–545 Ma). North America (Figure 22.6.1) is the largest fragment of a continent assembled in the Paleoproterozoic (2500–1600 Ma); other fragments exist in Eurasia and probably elsewhere. Since the Paleoproterozoic, North America (Laurentia, exclusive of fragments lost to Europe when the Atlantic opened) has twice collided to form supercontinents (all continents gathered together). The older collision is represented by the Mesoproterozoic

(1600–1000 Ma) Grenville Orogen and the resulting supercontinent, named Rodinia, had an approximate age span of 1050–750 Ma. The younger supercontinent is Wegener's Pangea, which had an age span of 300–150 Ma. It was conjoined with North America along the Appalachian orogen and its connecting orogenic belts around the Gulf of Mexico (Ouachitas), East Greenland (Caledonides), and Arctic Canada (Franklin). Ancestral North America, therefore, participated in most of the salient tectonic events of the past three billion years.

The Phanerozoic evolution of the continents is reasonably well understood and key Phanerozoic orogens are the subjects of other essays in this book. Earth's pre-Phanerozoic history is less well known, despite having produced over 80% of existing continental crust, but it is a subject in healthy ferment. Before discussing the role of ancestral North America in pre-Phanerozoic continental evolution, I should clarify some terminology. A collisional orogen implies a fusion of mature (>200 million years old) continental blocks. An accretionary orogen implies the addition of juvenile (<200 million years old) oceanic material—accretionary prisms, magmatic arcs, and volcanic plateaus, for the most part. For brevity, I will use the numerical geon time scale, where geon 0 equals 0–99 Ma, geon 1 equals 100–199 Ma, geon 10 equals 1000–1099 Ma, and so on. The divisions of the Proterozoic eon, defined above, are those recognized by the International Union of Geological Sciences and differ slightly from the parallel divisions in the Geological Society of America DNAG time scale.

[1]Harvard University, Cambridge, MA, USA.

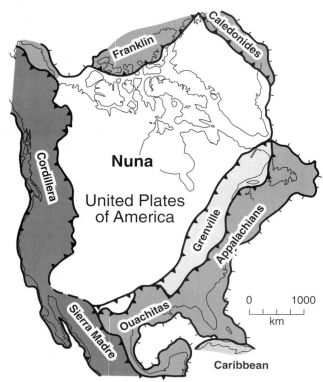

FIGURE 22.6.1 Simplified orogenic structure of North America: a Paleoproterozoic nucleus, Nuna, is discontinuously bordered by the Meosproterozoic Grenville and Racklan (far northwest) Orogens; the Paleozoic Ouachita, Appalachian, Caldonide, and Franklin Orogens, and the Mesozoic-Cenozoic Cordilleran and Caribbean Orogens. Greenland is restored to pre-rift (>90 Ma) position.

22.6.2 Phanerozoic (545–0 Ma) Orogens and Pangea

The stable interior of North America is framed by two great Phanerozoic orogenic systems. The Cordilleran system (Figure 22.6.1) borders the Pacific Ocean basin and is still active. The Pacific continental margin first opened in geon 7, but the main phase of tectonic accretion occurred in geon 1, coeval with rapid northwesterly drift of the continent (relative to hot spots) following the breakup of Pangea and opening of the North Atlantic Basin. Cordilleran crust that was thickened during Mesozoic accretion collapsed in extension in the Paleogene, when convergence between North America and the Pacific Basin slowed. Dextral strike-slip deformation became increasingly important in the Neogene, when North America began to override the East Pacific spreading ridge. The Cordillera will continue to be active until the Pacific Basin closes. If the Atlantic continues to open, North and South America

will eventually collide with eastern Asia, which by then will have incorporated Australasia.[2]

The other Phanerozoic orogenic system formed during the Paleozoic assembly of Pangea. It includes the Appalachian, Ouachita, Caledonide, and Franklin orogenic belts (Figure 22.6.1). They evolved from a continuous continental margin that opened diachronously in geons 6 and 5. Parts of the margin collided with island arcs and became active margins in geon 4 and, by the end of geon 3, the northern Appalachian and Caledonide segments had collided with Baltica, and the southern Appalachian and Ouachita sectors had done the same with northwest Gondwanaland. The resulting orogenic system was dismembered when the North Atlantic Basin opened.

22.6.3 Neoproterozoic (1000–545 Ma) Orogens and Gondwanaland

Gondwanaland (Figure 22.6.2), the former giant continent that broke up in the Mesozoic to form Africa, South America, Antarctica, Australasia, and southern Eurasia, was assembled in the Neoproterozoic era. Gondwanaland was an aggregate containing at least five older continents—West Africa, Amazonia, Congo, Kalahari, and East Gondwanaland (Australia, East Antarctica, India). They were welded together by a network of Neoproterozoic collisional orogens and bordered by Neoproterozoic-Paleozoic accretionary orogens. Pre-Phanerozoic North America lacks Neoproterozoic orogens, but displaced continental slivers that originated on the northwest margin of Gondwanaland were incorporated into the Appalachians in Middle and Late Paleozoic time. A two-way land trade apparently occurred in the Middle Paleozoic, suggesting a glancing encounter between eastern North America and western South America. Part of the southern Appalachians ended up in northwest Argentina and a strip of northern South America (the Avalon Terrane) was added to the coast of New England and eastern Canada. Later, a piece of northwest Africa (the Florida Peninsula and panhandle) was transferred to North America during the climactic Late Paleozoic collision with Gondwanaland.

---

[2]The new supercontinent is dubbed Amasia.

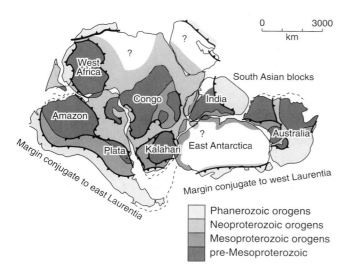

FIGURE 22.6.2 Aggregate structure of Gondwanaland, cemented by Neoproterozoic collisional and accretionary orogens. The reconstruction is well constrained by dated Mesozoic-Cenozoic seafloor magnetic anomalies. Dotted lines show edges of extensive modern continental shelves. Margins conjugate to Laurentia based on the Rodinia restoration in Figure 22.6.3. Note the disaggregated nature of Mesoproterozoic orogenic segments and the Neoproterozoic orogenic belt entering East Antarctica opposite Sri Lanka.

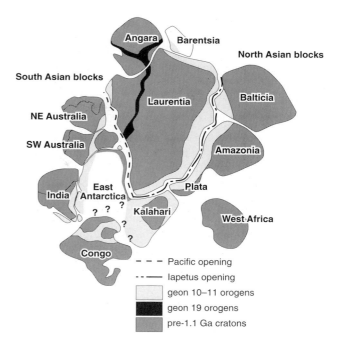

FIGURE 22.6.3 Hypothetical reconstruction of Rodinia, showing lines of Neoproterozoic opening of the Pacific and Iapetus Ocean Basins. As shown, Rodinia was an aggregate of cratons cemented by geon 10–11 (Grenvillian) orogens. Present-day north arrows for Laurentia (North America plus Rockall Bank and northwest Britain) and Kalahari cratons point to top and bottom of page, respectively.

22.6.4 Mesoproterozoic (1600–1000 Ma) Orogens and Rodinia

The Mesoproterozoic Grenville Orogen lies inboard of the Appalachians and extends for 5000 km from Mexico to Labrador (Figure 22.6.1). It truncates Archean and Paleoproterozoic structural fabrics and tectonic boundaries to the northwest, implying that the orogen evolved as a rifted or sheared continental margin. The orogen comprises an outer (northwestern) zone consisting of reactivated Archean and Paleoproterozoic basement, and an inner (southeastern) zone of juvenile Mesoproterozoic crust. The outer zone is characterized by northwest-directed crustal-scale thrust shears and exposes metamorphic rocks that underwent >30 km of post-1050-Ma exhumation. Thrusting occurred in geons 11–10 and is presumably related to accretion of the inner zone and terminal collision of ancestral North America with an outboard continent(s). The hypothetical Grenvillian hinterland must have broken away when the Iapetus paleocean basin opened, initiating the Paleozoic Appalachian orogenic cycle. Likely candidates for the Grenvillian hinterland are the Amazonia-Plata craton of South America (Figure 22.6.2) and the Baltica craton of northern Europe. Both are flanked by Grenville-age orogenic and plutonic belts, consistent with those cratons belonging to the overriding plate of the terminal Grenvillian collision.

Rifted segments of orogenic belts of Grenville age occur throughout Gondwanaland, except for the West African craton (Figure 22.6.2). In addition to the belts in South America mentioned previously, deeply eroded geon 11–10 orogenic belts occur in central and southern Africa, including Madagascar; in Sri Lanka and the Eastern Ghats of India; and in eastern Antarctica and Australia (Figure 22.6.2). Under current investigation is the hypothesis that all or most of these segments originally belonged to a continuous, 10,000-km-long system analogous to the Neogene Alpine-Himalayan system. The product of these collisions was the supercontinent **Rodinia** (Figure 22.6.3), the exact configuration of which is still conjectural, but which should make a structurally compatable restoration of the Grenvillian orogenic segments. The Rodinia reconstruction (Figure 22.6.3) implies that Gondwanaland was turned inside-out following the breakup of Rodinia at about 750 Ma. The requisite anticlockwise rotation of East Gondwanaland, relative to Laurentia, and convergence with West Gondwanaland before about 500 Ma is consistent with paleomagnetic data.

22.6.5 Paleoproterozoic (2500–1600 Ma) Collisional Orogens and Nuna

The vast region between the Grenville and Cordilleran Orogens (Figure 22.6.1), including the marginal parts of the orogens, was assembled in Paleoproterozoic time. It incorporates at least four Archean microcontinents—Churchill, Superior, Nain, and Slave (Figure 22.6.4). Paleomagnetic data indicate >4000 km of Late Paleoproterozoic convergence between the Churchill and Superior cratons, broadly contemporaneous with significant relative motion between the Churchill and Slave cratons. The Wyoming craton may be an extension of the Churchill or a fifth independent microcontinent; the mutual boundary is buried by thick Phanerozoic platform cover. The Churchill has two major divisions—Rae and Heane (Figure 22.6.4)—previously thought to have fused in the Paleoproterozoic. However, recent studies indicate an earlier, Late Archean time of assembly. The overall Paleoproterozoic assembly has been called the United Plates of America, and is commonly believed to be continuous with Baltica (Figure 22.6.4). An appropriate name for the entire continent assembled by the end of the Paleoproterozoic is **Nuna,** an eskimo name for the lands bordering the northern oceans and seas.

The subduction zones that accommodated the converging Archean microcontinents dipped predominantly beneath the Churchill continent, with far-reaching structural and magmatic consequences. The Churchill margins have well-developed Paleoproterozoic plutonic belts, representing the eroded roots of continental magmatic arcs (Figure 22.6.5). These are lacking on the Churchill-facing margins of the Superior, Nain, and Slave cratons. The Churchill continent was far more severely and extensively deformed by the collisions than were the cratons. Large-scale strike-slip and oblique-slip shear zone systems developed as weak Churchill crust was extruded laterally in response to indentation by the three more-rigid cratons (Figure 22.6.4). The Churchill continent also experienced unique intraplate magmatic events during and after the collisions. Ultrapotassic alkaline volcanism occurred in

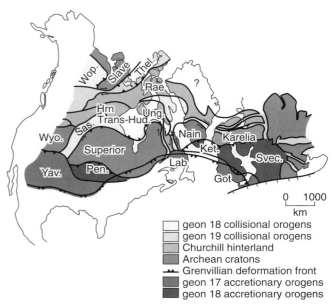

geon 18 collisional orogens
geon 19 collisional orogens
Churchill hinterland
Archean cratons
Grenvillian deformation front
geon 17 accretionary orogens
geon 18 accretionary orogens

FIGURE 22.6.4 Existing aggregate structure of Nuna: the United Plates of America and its extension in Baltica after Gorbatschev and Bogdanova (1993). Nuna was cemented by geon-18 orogens, which are truncated at the present margins of Nuna, and by the peripheral geon-17 accretionary orogens. Other extensions of Nuna exist on other continents or have been destroyed. Got = Gothian Orogen; Hrn = Hearne craton; Ket = Ketilidian orogen; Lab = Labrador Orogen; Pen = Penokean Orogen; Sas = Saskatoba syntaxis; Svec = Svecofennian Orogen; Thel = Thelon Orogen; Trans-Hud = Trans-Hudson Orogen; Ung = Ungava syntaxis; Wop = Wopmay Orogen; Wyo = Wyoming craton; Yav = Yavapai Orogen.

FIGURE 22.6.5 The charnokitic Cumberland batholith in central Baffin Island is a product of Paleoproterozoic arc magmatism on the margin of the Churchill hinterland. Mount Asgard, the nunatak in the center of the photo, rises over 800 meters above the surrounding glacier.

a 100,000-km^2 area west of Hudson Bay close to the time of the Churchill-Superior collision. Almost 80 million years later, after most of the collision-related deformation and metamorphic unroofing had occurred, the same region underwent high-silica rhyolite volcanism and associated rapakivi-type granite emplacement over an area of 250,000 km^2. Analogies have been drawn between the Churchill hinterland and the Neogene Tibetan Plateau, hinterland of the Himalayan collisional orogen.

The margins of the cratons facing the Churchill are characterized by large-scale thrust and nappe structures, directed away from the Churchill hinterland. Paleoproterozoic sedimentary and volcanic rocks deposited on the rifted margins of the cratons are discontinuously preserved and exposed. Volcanism and dike swarms related to initial continental breakup occurred mainly in geons 21–20. Subsequent passive-margin sediments (platformal carbonates and mature fine clastics) are overlain disconformably by foredeep sediments (in complete ascending sequence: ironstones, black shales, greywacke turbidites, and redbeds). The change occurred as the leading edge of each craton entered a peri-Churchill subduction zone. The stratigraphic transition and hence the onset of collision can be precisely dated if suitable material (e.g., air-borne volcanic ash layers) is present.

The structurally higher levels of the overthrust belts bordering the Superior craton are composed of quasi-oceanic material. At the Ungava syntaxis (Figure 22.6.4) in northern Quebec, parts of an oceanic plateau (1.92 Ga), an imbricated ophiolite (2.00 Ga), and an immature island arc (1.87–1.83 Ga) were thrust southwards across the rifted margin onto the Superior craton. The ophiolite, one of the world's oldest, includes a sheeted dike complex, ultramafic cumulates, and volcanic suites chemically and isotopically correlated with mid-ocean ridge and ocean island basalts. At the Saskatoba syntaxis in northern Saskatchewan and Manitoba (Figure 22.6.4), the craton is tectonically juxtaposed by juvenile island-arc-type volcanic and plutonic rocks (1.93–1.85 Ga) and derived metasediments. The craton at first formed the structural footwall, but was later thrust toward the hinterland over Paleoproterozoic juvenile rocks of the intervening Trans-Hudson Orogen (Figure 22.6.4). The northeast-facing lateral margin of the craton and the arcuate embayment between the Ungava and Saskatoba syntaxes (Figure 22.6.4) contain allochthonous mafic sill–sediment complexes formed in syncollisional pull-apart basins (also called rhombochasms). The thrust-fold belts on the lateral margins of all three indented cratons are flanked by crustal-scale strike-shear zones bordering the hinterland (Figure 22.6.4). Structurally, the belts have much in common with Phanerozoic collisional orogens, but lithologically they are richer in volcanic rocks, consistent with higher mantle temperatures in the Paleoproterozoic.

The timing of the Paleoproterozoic collisions is best constrained by U-Pb geochronology of (1) the passive-margin to foredeep stratigraphic transition described earlier, (2) the cessation of arc magmatism or change from arc-type to collision-type magmatism on the Churchill margins, and (3) the exhumation of metamorphic rocks having pressure-temperature-time trajectories characteristic of collisional origin. The geochronological data show that the Slave-Churchill collision occurred first, beginning about 1.97 Ga. The Nain-Churchill collision occurred about 100 m.y. later, and the Superior craton joined the assembly at about 1.84 Ga.

22.6.6 Paleoproterozoic Accretionary Orogens Add to Nuna

Around the time the Superior, Nain, and Slave cratons collided to nucleate Nuna, juvenile Paleoproterozoic crust began to be accreted onto their trailing margins (i.e., those facing away from the Churchill hinterland). The respective accretionary orogens are the Penokean of the Great Lakes region, the Ketilidian of South Greenland and adjacent Labrador (where it is called Makkovik), and the Wopmay around Great Bear Lake, Northwest Territories (Figure 22.6.4). All three orogens evolved from passive margins that collided with island arcs and were converted to Andean-type margins as a result of subduction-polarity reversal (meaning that subduction zones first dipped away from the cratons and later beneath them). Arc-continent collision in the Wopmay Orogen occurred at 1.88 Ga, almost 90 m.y. after the Slave-Churchill collision. In the Ketilidian Orogen, chronometrically less well defined, the arc-continent collision is placed at about 1.84 Ga, about 30 m.y. after the Nain-Churchill collision. Arc-continent collision in the Penokean orogen occurred about 1.85 Ga, close to the time of the Superior-Churchill collision. The confluence of cratons and accretion at their trailing margins suggests a large-scale pattern of lithospheric flow converging on the Churchill hinterland. A possible explanation is that the hinterland was situated over a vigorous downwelling region of the sublithospheric mantle self-sustained by the descent of cold oceanic slabs.

Accretion on the (present) southern and southeastern margins of Nuna was renewed in geon 17, following apparent truncation of geon 18 structures on those margins. The accretionary orogens of geon 17 are principally exposed in Labrador and the southwestern United States, but, based on studies of drill core, they also make up the buried basement across most of the southern midcontinent (Figure 22.6.4). Juvenile crust, at least 1300 km wide, was accreted in geon 17 between the Wyoming craton and the Grenville Orogen of West Texas. The accreted material proved to be a fertile source for Mesoproterozoic crustal-melt granites and rhyolites, which are extensively encountered in the midcontinent subsurface.

The widely held belief that Nuna originally included Baltica (the Baltic Shield and East European platform) is based on proposed continuity between the Archean Nain and Karelia cratons, the Ketilidian and Svecofennian Orogens of geon 18, and the Labrador and Gothian Orogens of geon 17 (Figure 22.6.4). A more tentative connection is proposed between Nuna and the Angara craton of Siberia, based on extensions of the Slave-Churchill collision zone (Thelon Orogen) across the Arctic. Even more tenuous links exist between Nuna and major geon-18 orogenic belts in northern and western Australia. In addition, there are tantalizing similarities in Late Paleoproterozoic to Early Mesoproterozoic platform cover sequences worldwide that have contributed to the notion, as yet undemonstrated, that Nuna was a giant continent predating Rodinia.

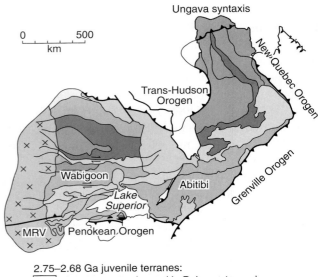

2.75–2.68 Ga juvenile terranes:
□ deeper crust exhumed in Paleoproterozoic
▨ sediment-dominated accretionary prisms
▨ volcanic-dominated island arcs and plateaus
▨ 2.73–2.69 Ga magmatic arc within older proto-arc
▨ 3.0–2.8 Ga volcanic-rich composite proto-arc
■ ca. 3.6 Ga continental foreland
⊠ post-Archean cover on Superior craton

FIGURE 22.6.6 Accretionary structure of the Superior craton, the largest Archean craton in Nuna (Figure 22.6.4). Note change in tectonic strike from east-west to north-south and the truncation of tectonic boundaries at the edge of the craton. MRV = Minnesota River Valley Terrane.

22.6.7 Archean Cratons and Kenorland

Archean cratons have dimensions below which their plate tectonic settings are difficult to determine. The largest one continuously exposed is the Superior craton of the Canadian Shield (Figure 22.6.6), measuring 2500 km east to west and 1500 km north to south. It was constructed in the latter half of geon 27. It exposes, in zonally varying proportions, deformed plutonic and volcanic rocks of island-arc and oceanic-plateau affinities and derived sediments. The regional tectonic strike swings from east-west in the western and southeastern parts of the craton, to north-south in the northeastern part (Figure 22.6.6). In the northwest of the craton, a long-lived composite protoarc had evolved since geon 30. Bilateral accretion of juvenile material onto the protoarc began about 2.75 Ga. A progressively southward docking sequence of south-facing volcanic arc and accretionary prism couplets occurred until 2.69 Ga. Accretion was terminated by the collision of the entire assemblage with an old (~3.6 Ga) continent, the Minnesota River Valley (MRV) Terrane (Figure 22.6.6). The southward docking sequence and the structural evidence of persistent

dextral strike-slip displacement accompanying the docking events imply overall oblique northwest-directed subduction. In the northeast of the craton, where the tectonic strike is north-south, a relatively high proportion of plutonic rocks, deep level of exhumation, and absence of strike-slip displacements indicate a less oblique subduction regime. The tectonic zonation and associated structural grain of the craton is clearly truncated at its western, southeastern, and northeastern margins (Figure 22.6.6), showing that the entire craton is a rifted fragment of a Late Archean continent, referred to as **Kenorland.**

There is a good possibility that the Wyoming craton (Figure 22.6.4) was originally connected to the south margin of the Superior craton in Kenorland. The Wyoming craton is old, like the MRV, and it preserves erosional remnants of a highly distinctive Early Paleoproterozoic sequence of uraniferous conglomerates, tropically weathered quartz arenites, and glacial diamictites that are remarkably similar to the Huronian sequence on the north shore of Lake Huron in the southern Superior craton. Separation of the Superior and Wyoming cratons occurred in the latter half of geon 21, about 300 m.y. before the two cratons were reunited in Nuna.

A remarkable number of cratons worldwide underwent major accretion or collision in geon 27. It was a time of oblique plate convergence and rapid continental accretion in the Yilgarn craton, the largest in Australia, and the time of collision between the Kaapvaal and Zimbabwe cratons of southern Africa, to name just two of the more important overseas examples. In North America, the Hearne Province of the Churchill hinterland was accreted in geon 27, but the Slave and Nain cratons were assembled in geon 26. Precise U-Pb dating of large-scale mafic dike swarms, emplaced during breakup events associated with mantle plumes, holds promise as a means of identifying formerly contiguous cratons. This should yield important insights into the extent and configuration of Kenorland, possibly the first giant continent.

22.6.8   Closing Remarks

North America consists of an aggregate nucleus that assembled in the Paleoproterozoic, discontinuously bordered by Mesoproterozoic and Phanerozoic orogens. Parts of North America participated in perhaps five giant continents: Kenorland, assembled in geons 27–26; Nuna, assembled in geons 18–17; Rodinia, assembled in geons 11–10; Gondwanaland, assembled in geons 6–5; and Pangea, assembled in geons 3–2.

The recurrence interval for giant assemblies seems to have become shorter with time—900, 700, 500, and 300 m.y. This may merely reflect our inferior knowledge of the earlier part of the record. If these intervals are real, on the other hand, how can this pattern be reconciled with secular cooling of the earth, which would result in mantle convection of diminishing vigor. It seems unlikely that continental collisions have become more frequent. However, collision zones tend to reopen, particularly when young. This tendency could have been even stronger in the distant past. To make a giant continent piecemeal, the disassembly processes must be checked. Therefore, the growing incidence of giant continents may reflect increased stability, rather than increased frequency, of continental collisions.

ADDITIONAL READING

Card, K. D., 1990. A review of the Superior province of the Canadian shield, a product of Archean accretion. *Precambrian Research,* 48, 99–156.

Condie, K. C., and Rosen, O. M., 1994. Laurentia-Siberia connection revisited. *Geology,* 22, 168–170.

Dalziel, I. W. D., 1992. On the organization of American plates in the Neoproterozoic and the breakout of Laurentia. *GSA Today,* 2, 237–241.

Dalziel, I. W. D., Dalla Salda, L. H., and Gahagan, L. M., 1994. Paleozoic Laurentia-Gondwana interaction and the origin of the Appalachian-Andean mountain system. *Geological Society of America Bulletin,* 106, 243–252.

Dewey, J. F., and Burke, K. C. A., 1973. Tibetan, Variscan, and Precambrian basement reactivation: products of continental collision. *Journal of Geology,* 81, 683–692.

Gorbatschev, R., and Bogdanova, S., 1993. Frontiers in the Baltic shield. *Precambrian Research,* 64, 3–21.

Gower, C. F., Rivers, T., and Ryan, A. B., eds., 1990. *Mid-Proterozoic Laurentia-Baltica, Special Paper 38.* Geological Association of Canada: St. John's.

Hoffman, P. F., 1988. United Plates of America, the birth of a craton: Early Proterozoic assembly and growth of Laurentia. *Annual Reviews of Earth and Planetary Sciences,* 16, 543–603.

Hoffman, P. F., 1989. Speculations on Laurentia's first gigayear (2.0–1.0 Ga). *Geology,* 17, 135–138.

Hoffman, P. F., 1991. Did the breakout of Laurentia turn Gondwanaland inside-out? *Science,* 252, 1409–1412.

Lewry, J. F., and Stauffer, M. R., eds., 1990. The *Early Proterozoic Trans-Hudson orogen of North America, Special Paper 37,* Geological Association of Canada: St. John's.

Lucas, S. B., Green, A., Hajnal, Z., White, D., Lewry, J., Ashton, K., Weber, W., and Clowes, R., 1993. Deep seismic profile across a Proterozoic collision zone: surprises at depth. *Nature,* 363, 339–342.

Moores, E. M., 1991. Southwest U.S.—East Antarctic (SWEAT) connection: a hypothesis. *Geology,* 19, 425–428.

Powell, C. McA., Li, Z. X., McElhinny, M. W., Meert, J. G., and Park, J. K., 1993. Paleomagnetic constraints on timing of the Neoproterozoic breakup of Rodinia and the Cambrian formation of Gondwana. *Geology,* 21, 889–892.

Reed, J. C. Jr., Bickford, M. E., Houston, R. S., Link, P. K., Rankin, D. W., Sims, P. K., and Van Schmus, W. R., eds., 1993. Precambrian: conterminous U.S. *The Geology of North America,* v. C-2. Geological Society of America: Boulder, CO.

Rivers, T., Martignole, J., Gower, C. F., and Davidson, A., 1989. New tectonic divisions of the Grenville province, southeastern Canadian shield. *Tectonics,* 8, 63–84.

Sengör, A. M. C., Natal'in, B. A., and Burtman, V. S., 1993. Evolution of the Altaid tectonic collage and Paleozoic crustal growth in Eurasia. *Nature,* 364, 299–307.

Thurston, P. C., Williams, H. R., Sutcliffe, R. H., and Stott, G. M., eds., 1991. *Geology of Ontario: Parts 1 and 2, Ontario Geological Survey Special Volume 4.*

Van der Voo, R., 1988. Paleozoic paleogeography of North America, Gondwana, and intervening displaced terranes: comparisons of paleomagnetism with paleoclimatology and biogeographical patterns. *Geological Society of America Bulletin,* 100, 311–324.

Weil, A. B., Van der Voo, R., Mac Niocaill, C., and Meert, J. G., 1998. The Proterozoic supercontinent Rodinia: paleomagnetically derived reconstructions for 1100 to 800 Ma. *Earth and Planetary Science Letters,* 154, 13–24.

Williams, H., Hoffman, P. F., Lewry, J. F., Monger, J. W. H., and Rivers, T., 1991. Anatomy of North America: thematic geologic portrayals of the continent. *Tectonophysics,* 187, 117–134.

22.7 PHANEROZOIC TECTONICS OF THE UNITED STATES MIDCONTINENT

22.7.1 Introduction

If you've ever driven across the United States, or have looked at a relief map of the country, you can't help but notice that the nature of topography varies radically with location. Along the East Coast, the land rises to form the long ridges of the Appalachian Mountains, whereas in the west, numerous chains of rugged peaks together compose the broad North American Cordillera. In between these two mountain belts, a region known geographically as the **Midcontinent,** the land surface is relatively flat and low-lying (Figure 22.7.1). Roads wind among dramatic cliffs and deep valleys in the mountains, but shoot like arrows across the checkerboard of farmland and rangeland in the Midcontinent. These contrasts in topography reflect contrasts in the character and geologic history of the continent's crust.

Following the definitions of Table 22.7.1, we classify the Midcontinent region as a **continental-interior platform.** Here, a cover of Phanerozoic strata was deposited in wide but shallow seaways over a basement of Precambrian (crystalline) rock. A continental-interior platform is one of two kinds of crustal provinces that can occur in a craton. We define a **craton** as continental crust that has not developed penetrative fabrics and tight folds, and has not been subjected to regional metamorphism since the beginning of the late Neoproterozoic (i.e., since about 800 Ma). In addition to continental-interior platforms, a craton can include a **shield,** a region where Precambrian basement rocks crop out extensively at the ground surface, either because cover strata were never deposited or because they were eroded away after deposition. In North America, much of the interior of Canada is a shield—not surprisingly, the region is known as the Canadian Shield.

The mountain ranges of the United States consist of two kinds of crust. Specifically, the Appalachians and the North American Cordillera can be considered Phanerozoic orogens, for during the Phanerozoic these regions developed penetrative deformation and, in places, regional metamorphism. They were also the site of widespread igneous activity and significant uplift. As a consequence of Phanerozoic tectonism, these orogens became warm and relatively weak. What is often overlooked is that the North American Cordillera also includes a large area of crust that is identical in character to that of the continental-interior platform, even though it has been uplifted significantly and has been locally subjected to deformation and igneous activity during the Phanerozoic. This crust underlies the Rocky Mountains and Colorado Plateau of the United States. In this essay, you will see that the tectonic behavior of the Midcontinent resembles that of the Rocky Mountains and Colorado Plateau, except on a more subdued scale.[1]

Cratons differ from younger orogens not just in terms of topography, but also in terms of physical characteristics such as strength, seismicity, heat flow, and crustal thickness. Specifically, orogens have less strength,[2] more seismic activity, higher heat flow, and thicker crustal roots, than do cratons. The contrast in strength reflects the contrast in heat flow, for warmer rock tends to be more plastic, and therefore weaker,

---

[1]Note that the Canadian Rockies, in contrast, comprise a fold-thrust belt and, as such, are part of a younger orogen.

[2]The collision between India and Asia illustrates the strength contrast between a craton and a younger orogen. India consists of old, cold Precambrian crust, while the southern margin of Asia consists of warm, relatively weak crust. During their collision, India pushed deeply into Asia.

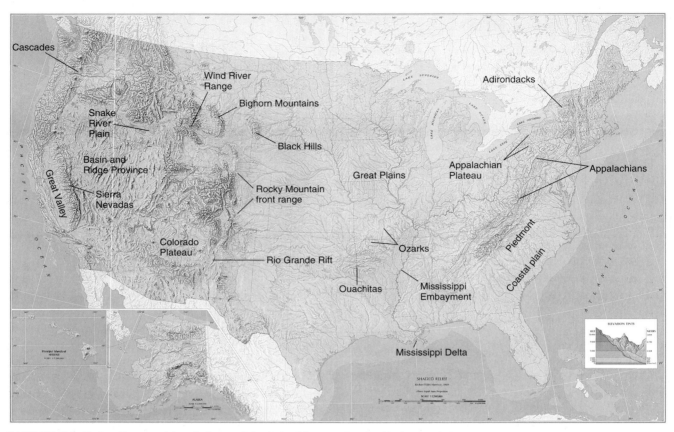

FIGURE 22.7.1 Topographic relief map of the United States, showing the contrast between Phanerozoic marginal orogens and the Great Plains.

than cooler rock of the same composition. Rock that makes up cratons initially formed either at volcanic arcs bordering convergent plate boundaries, or at hotspot volcanoes above mantle plumes. Thus, the continental crust that eventually became the craton grew by successive collision of these buoyant blocks. This fact is well illustrated by studies in the Canadian Shield, where outcrops contain spectacular tectonite fabrics, folds, and faults that tell of a long and complex history of accretionary orogeny. If crust of the cratons began its life in orogenic belts, then how did it become strengthened and stabilized or, in other words, cratonized? Researchers suggest that cratonization comes in part from aging, because continental lithosphere loses heat and becomes stiffer as it ages. In fact, the lithospheric mantle beneath cratonic crust is thicker than that of other kinds of continental crust. Researchers also suggest that relatively buoyant asthenosphere has stayed attached to the base of the lithosphere and insulates the lithosphere from warmer parts of the asthenosphere. This special asthenosphere has been called the root of the craton.

In this essay, we discuss the Phanerozoic tectonics of the Midcontinent region of the United States. At

first glance, the title may seem like an oxymoron, for the lack of topography and of penetrative structures gives the impression that this region has been tectonically stable during the Phanerozoic. However, while the Midcontinent has been *relatively* stable, it has not been *completely* stable. In fact, during the Phanerozoic, faults in the region have been reactivated, and broad regions of the surface have subsided to form sedimentary basins or have been uplifted to form domes or arches. Manifestations of Phanerozoic tectonism in the Midcontinent might not be as visually spectacular as that of North America's mountain belts, but tectonism has, nevertheless, occurred there.

22.7.2 Classes of Structures in the Midcontinent

To structural geologists raised on a diet of spectacular folds, faults, and deformation fabrics exposed in Phanerozoic orogens, the Midcontinent may seem, at first glance, to be structureless. But, in fact, the region contains four classes of tectonic structures: (1) epeirogenic structures, (2) Midcontinent fault-and-fold

TABLE 22.7.1	CATEGORIES OF CONTINENTAL CRUST
Active rift	A region where crust is currently undergoing horizontal stretching, so that crustal thicknesses are less than average crustal thicknesses. In active rifts, continental crust has a thickness of only 20–25 km, the crust has been diced up by normal faults, and volcanism occurs. Examples include the Basin and Range Province of the western USA, and the East African Rift.
Inactive rift	A belt of continental crust that underwent stretching, and became a narrow trough that filled with sediment, but never succeeded in breaking a continent in two. The Midcontinent rift of the central USA, and the Rhine Graben of Europe are examples.
Active orogen	A portion of the continental crust in which tectonism (faulting ± igneous activity ± uplift) currently takes place or has taken place in the recent past (Cenozoic). Such orogens tend to be linear belts, in that they are substantially longer than they are wide. Examples include the North American Cordillera and the European Alps.
Continental shelf	A belt fringing continents in which a portion of the continent has been submerged by the sea. Water depths over shelves are generally less than a few hundred meters. Continental shelves are underlain by passive-margin basins, which form subsequent to rifting, as a consequence of the subsidence of the stretched continental crust that bordered the rift. Sediment washed off the adjacent land buried the sinking crust. The stretching occurred during the rifting event and predated formation of a new mid-ocean ridge. Examples include the East Coast and Gulf coast of the USA.
Craton	A portion of a continent that has been relatively stable since the late Neoproterozoic (since at least about 800 Ma). This means that penetrative fabrics, regional metamorphism, and widespread igneous activity have not occurred in the craton during the Phanerozoic. The crust of a craton includes the eroded roots of Precambrian mountain belts.
Shield	A broad region, typically of low relief (though some shields have been uplifted and incised since the Mesozoic), where Precambrian crystalline rocks are exposed. In North America, a shield area (the Canadian Shield) is part of the craton. In South America, shield regions include cratons and Neoproterozoic orogens.
Continental platform	A broad region where Precambrian rocks (basement) have been covered by a veneer of unmetamorphosed Phanerozoic strata. Examples include the Midcontinent region of the USA, and large portions of northern Europe.
Inactive Phanerozoic orogen	Orogenic belts that were tectonically active in the Phanerozoic, but are not active today. Some inactive orogens, however, have been uplifted during the Cenozoic, so they are topographically high regions. Examples include the Appalachians of the USA and the Tasmanides of Australia.
Phanerozoic orogen	A general name for active Phanerozoic orogens and inactive Phanerozoic orogens, taken together.
Neoproterozoic orogen	Orogen active at the end of the Precambrian. Examples include the Pan-African/Brasiliano Orogens of Gondwana.

zones, (3) intragranular strain, and (4) regional joint systems. We'll describe these in succession.

EPEIROGENIC STRUCTURES When geologists first mapped the Midcontinent, they noted that Paleozoic strata of the region are nearly, but *not exactly* flat-lying. As a consequence, outcrop patterns of Paleozoic strata on a regional geologic map of the Midcontinent United States display distinctive bull's-eye patterns (Figure 22.7.2). In some examples, the youngest strata occupy the center of the bull's-eye, while in others, they occur in the outer ring. Bull's-eyes with the youngest strata in the center define intracratonic basins, in which strata are warped downward to form a bowl shape. Those with the youngest strata in the outer ring define intracratonic domes or, if elongate, intracratonic arches (by **"intracratonic"** we mean "within the craton"). As new data defining subsurface

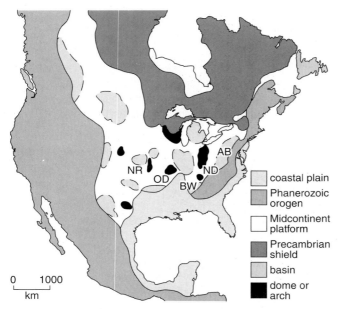

FIGURE 22.7.2 Regional tectonic map of North America showing the distribution of Phanerozoic orogenic belts, shield, continental-interior platform, and basins and domes.

coastal plain
Phanerozoic orogen
Midcontinent platform
Precambrian shield
basin
dome or arch

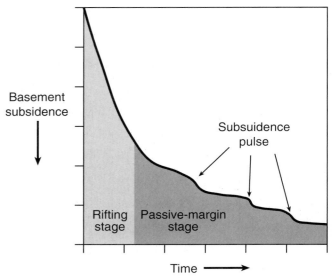

FIGURE 22.7.3 A graph illustrating how the rate of subsidence changes with time for an intracratonic basin. Rapid subsidence occurs at first, probably in association with rifting. Then, as the basin subsides due to cooling, the rate of subsidence gradually decreases. Pulses of rapid subsidence may occur during this time, perhaps due to the effects of orogeny along the continental margin. For example, an increase in stress may cause the lithosphere to weaken, so uncompensated loads beneath the basin may sink.

stratigraphy became available, primarily from correlation of well logs[3] and seismic-reflection profiles, researchers realized that stratigraphic formations thicken toward the center of a basin, and thin toward the center of a dome. Specifically, the Phanerozoic sedimentary cover of continental interior platforms thickens to about 5–7 km in basins, because that is where subsidence occurred most rapidly during deposition. In domes or arches, cover decreases in thickness to zero because these locations were emergent or were shallow shoals during deposition. A drop in sea level of the shallow oceans that covered the interior could expose the crest of a dome or arch, while the center of a basin could remain submerged. Thus, many more unconformities occur in the Phanerozoic section of domes and arches than in the interior of basins.

In effect, the lateral variations in sediment thickness that we observe in the Midcontinent indicate that there has been differential uplift and subsidence of regions of the Midcontinent during deposition. In other words, some regions went up while others went down as sediments were accumulating. Vertical movement affecting a broad region of crust is called **epeirogeny,** and the basins, domes, and arches that form as a result are **epeirogenic structures.** Significantly, the general position of basins and domes in the Midcontinent has remained fixed through the Phanerozoic. For example, the Illinois Basin and Michigan Basin have always been basins while the Ozark Dome and the Cincinnati Arch have been relatively high since their initial formation in the Late Proterozoic or Early Cambrian. Thus, these epeirogenic structures represent permanent features of continental-interior lithosphere.

Lack of stratigraphic evidence, either due to nondeposition or erosion, makes it impossible to constrain the rate of uplift of arches and domes precisely, but we can constrain rates of epeirogenic movement in basins by studying subsidence curves, which are graphs that define the rate at which the basement-cover contact at the floor of the basin moved down through time. Geologists generate subsidence curves by plotting sedimentary thickness as a function of time, after taking into account the affect of compaction (Figure 22.7.3). Subsidence curves for intracratonic basins demonstrate

---

[3]When drilling for oil, exploration companies use a rotating drill bit, which penetrates into the earth by grinding the rock into a mixture of powder and small chips called "cuttings." Circulating drilling "mud," pumped down into the hole, flushes the cuttings out of the hole. Since this process does not yield an intact drill core, the only way to determine the precise depth at which specific rock units lie in the subsurface is to lower instruments down the hole to record parameters such as electrical resistively and gamma-ray production, parameters that correlate with rock type. A record of resistivity versus depth, or gamma-radiation versus depth, is called a well log.

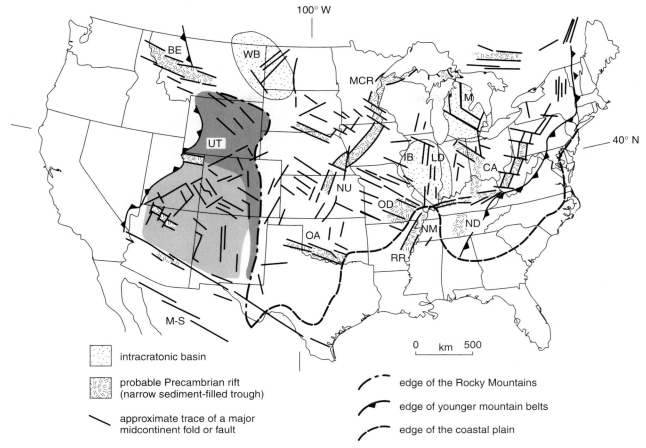

FIGURE 22.7.4 Map of the United States showing the distribution of Midcontinent fault-and-fold zones and of documented intracratonic rifts. BE = Beltian embayment, UT = Uinta trough, MCR = Midcontinent Rift, OA = Southern Oklahoma aulacogen, RR = Reelfoot Rift, LD = La Salle deformation belt, WB = Williston Basin, IB = Illinois Basin, MB = Michigan Basin, ND = Nashville dome, CA = Cincinnati arch, M-S = Mojave-Sonora megashear.

that basins have subsided overall through most of the Phanerozoic, probably due to slow cooling of lithosphere beneath the basins, but emphasize that the rate of subsidence for a basin varies with time. For example, pulses of rapid subsidence occurred in the Michigan and Illinois Basins during Ordovician, Late Devonian–Mississippian, and Pennsylvanian–Permian time. Notably, the timing of some, but not all, epeirogenic movements in the Midcontinent roughly corresponds with the timing of major orogenic events along the continental margin. For example, the time of Late Paleozoic subsidence in the Midcontinent U.S. basins roughly corresponds with the time of the Alleghanian Orogeny (the collision of Africa with the eastern margin of North America).

MIDCONTINENT FAULT-AND-FOLD ZONES Bedrock geology of much of the Midcontinent has been obscured by Pleistocene glacial deposits and/or thick

soils. In the isolated exposures that do occur, geologists have found occurrences of folds and faults. The regional distribution of such structures has slowly emerged from studies of well logs (compiled on structure-contour and isopach maps), potential-field data (compiled on gravity- and magnetic-anomaly maps), and seismic-reflection profiles. Abrupt steps and ridges on structure-contour maps or isopach maps, linear anomalies on potential field maps, abrupt bends or breaks in reflectors on seismic-reflection profiles—along with outcrop data—reveal that the Phanerozoic strata of the Midcontinent United States has been disrupted locally by distinct belts of deformation that we refer to here as Midcontinent fault-and-fold zones (Figure 22.7.4).

Individual fault-and-fold zones range in size from only a few hundred meters wide and several kilometers long, to 100 km wide and 500 km long. The north-northwest trending La Salle fault-and-fold zone of

Illinois serves as an example of a larger zone—it effectively bisects the Illinois Basin (Figure 22.7.5). Larger zones include numerous non-coplanar faults that range in length from <5 km to as much as 50 km. At their tips, these faults overlap with one another in a relay fashion. Locally, a band of *en echelon* subsidiary fault segments borders the trace of principal faults. In the upper few kilometers of the crust, major faults of a Midcontinent fault-and-fold zone dip steeply and divide upwards into numerous splays. The resulting array of faults resembles that of a flower structure. At depth, major faults decrease in dip (i.e., some faults appear to be listric) and penetrate basement (Figure 22.7.6). Some, but not all, faults clearly border narrow rift basins that contain anomalously thick sequences of sediments and volcanics. The largest of these intracratonic rifts, the Midcontinent Rift, consists of two principal arms, one running from Lake Superior into Kansas and the other running diagonally across Michigan (Figure 22.7.4). Faults along these rifts initiated as normal faults, but later reactivated as thrust or strike-slip faults.

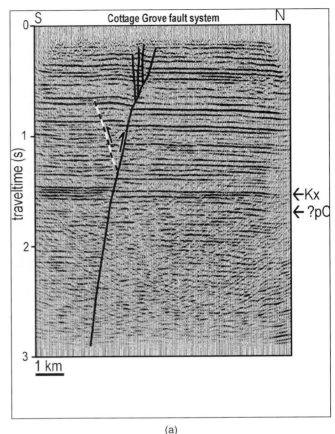

(a)

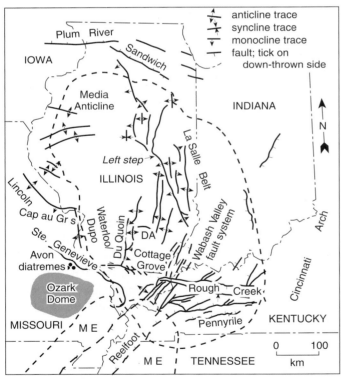

FIGURE 22.7.5 Structure map of the Illinois Basin region showing the map traces of Midcontinent fault-and-fold zones. Note that the Cottage Grove Fault is bordered by short *en echelon* faults, indicative of strike-slip displacement.

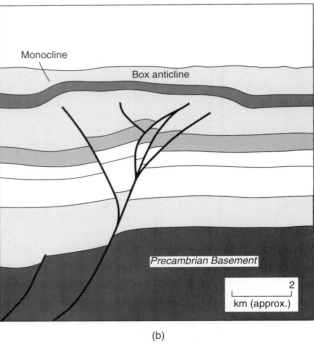

(b)

FIGURE 22.7.6 Cross-sectional structure of a Midcontinent fault zone. (a) Seismic-reflection profile of the Cottage Grove fault system; (b) Schematic cross section showing the principal features of a Midcontinent fault zone.

Cross sections indicate that major faults in the Midcontinent locally have vertical throws of as much as 2 km, but more typically throws are less—no more than tens to hundreds of meters. Strike-slip components of displacement across continental-interior faults are difficult to ascertain because of lack of exposed shear-sense indicators on fault surfaces and the lack of recognizable offset markers. But geologists have found strike-slip lineations on faults exposed in coal mines, and the *en echelon* map pattern of subsidiary faults adjacent to larger faults resembles the *en echelon* faulting adjacent to continental strike-slip faults. Such features suggest that a component of strike-slip motion has occurred on some Midcontinent faults. Where such motion has occurred, the faults can be considered to be oblique-slip faults. The occurrence of oblique slip, in turn, suggests that transpression or transtension has taken place in some Midcontinent fault-and-fold zones.

In general, folds of the continental interior are monoclinal in profile, meaning that they have one steeply dipping limb and one very shallowly dipping to subhorizontal limb. Locally, oppositely facing monoclinal folds form back-to-back, creating "box anticlines" (Figure 22.7.6b). Though geologists have not yet obtained many clear images of Midcontinent folds at depth, several studies document that folds lie above steeply dipping faults and that structural relief on folds increases with depth.

Detailed study of spatial variations in the thickness and facies of a stratigraphic unit relative to a fault-and-fold zone, documentation of the timing of unconformity formation, as well as documentation of local slump-related deformation, permits determination of when the fold-and-fault zone was tectonically active. Such timing constraints suggest that the structures, in general, became active during more than one event in the Phanerozoic. Activity appears to have been particularly intense during times of orogenic activity along the continental margin, but occurred at other times as well. The most significant reactivation occurred during the late Paleozoic, at the same time as the Alleghanian Orogeny. This event triggered reactivation of faults across the entire interior platform. Fault reactivation in the region that now lies within the Rocky Mountains yielded localized uplifts (now eroded) that have come to be known as the Ancestral Rockies. Some authors now use this term for Late Paleozoic uplifts associated with faulting across the width of North America (Figure 22.7.7).

Midcontinent fault-and-fold zones do not have random orientations, but rather display dominant trends over broad regions of the craton. As indicated by the map in Figure 22.7.4, most of the zones either trend north-south to northeast-southwest, or east-west to northwest-southeast, and thereby outline rectilinear blocks of crust. Along-strike linkage of fault-and-fold zones seems to define transcratonic belts of tectonic reactivation, which localize seismicity today.

Of note, Midcontinent folds closely resemble Laramide monoclines of the Colorado Plateau, and Laramide basement-cored uplifts of the U.S. Rocky Mountain region, though

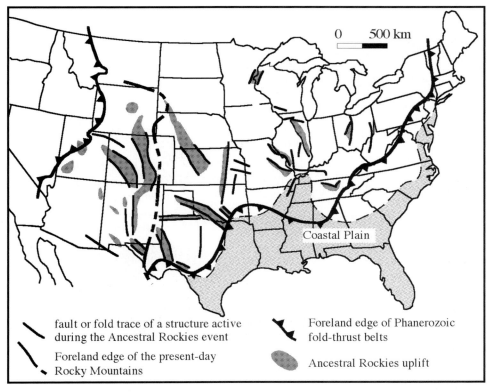

FIGURE 22.7.7   Late Paleozoic uplifts of the United States. The large ones west of the Rocky Mountain front have traditionally been referred to as the Ancestral Rockies.

at a smaller scale. This is not surprising because, as we noted earlier, the Colorado Plateau and Rocky Mountain region were once part of the craton's interior platform. These regions differ from the Midcontinent only in that they were uplifted and deformed during the Late Mesozoic–Early Cenozoic Laramide Orogeny. During this event, slip on some faults of the region exceeded 15 km, an order of magnitude more than occurred during Paleozoic reactivation of faults in the Midcontinent. Because of substantial Laramide and younger uplift, and because of the dry climate of the Plateau, fault-and-fold zones of the Plateau are brazenly displayed. But keep in mind that, if the Midcontinent region were stripped of its glacial sedimentary blanket and its prairie soils, it would look much like the Colorado Plateau.

INTRAGRANULAR STRAIN AND REGIONAL JOINTING

When you traverse a large fold in the Appalachian fold-thrust belt, you will find that outcrops of argillaceous (clay-rich) sandstones and limestones contain well-developed spaced cleavage to slaty cleavage. But, if you walk across a large fold in the Midcontinent (or

on the Colorado Plateau), you will find a distinct lack of cleavage. Layer-parallel shortening strains sufficient to form a regional cleavage apparently did not develop in the Midcontinent. Microstructural studies, however, indicate that subtle layer-parallel shortening did, in fact, develop in Midcontinent strata. This strain is manifested by the development of calcite twinning, a type of intragranular, crystal-plastic strain that can be seen only with a microscope and has developed in limestone.

Calcite twinning forms under the relatively low pressure and temperature conditions that are characteristic of the uppermost crust. Regional studies of calcite twinning in Midcontinent limestones indicate that the maximum shortening direction remains fairly constant over broad regions and trends roughly perpendicular to orogenic fronts, though more complex shortening patterns occur in the vicinity of Midcontinent fault-and-fold zones (Figure 22.7.8). Notably, strain and differential stress magnitudes in the eastern Midcontinent decrease progressively from the Appalachian-Ouachita orogenic front to the interior—strain magnitudes are 6% at the front but decrease to 0.5% in the interior,

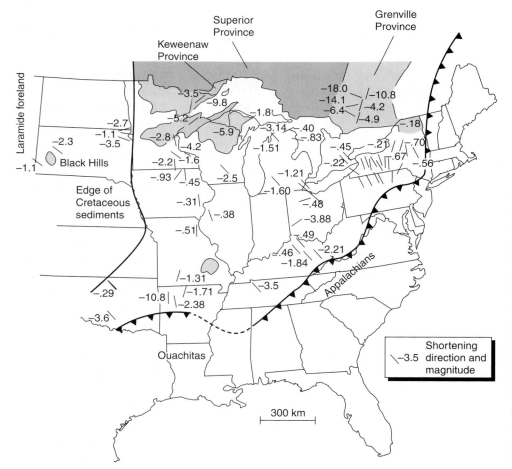

FIGURE 22.7.8 Calcite twinning strains in eastern North America. Regional tectonic provinces are labeled, except for the Paleozoic cover sequence inland from the Appalachian-Ouachita thrust front (bold, toothed line). Strains are presented by orientation (short lines) and magnitude in percent (negative is shortening); typically, maximum shortening is horizontal and perpendicular to the Appalachian-Ouachita thrust front. Twinning data from other tectonic provinces show patterns that are unrelated to Paleozoic deformation.

with associated differential stresses exponentially decreasing from ~100 MPa to less than 20 MPa. This pattern suggests that calcite twinning in strata of the eastern Midcontinent formed during the Alleghanian Orogeny, the Late Paleozoic collision of Africa with North America. In the western part of the Midcontinent, twinning strains also generally lie in the range of 0.5% and 3%, and the direction of maximum shortening trends roughly at right angles to the Rocky Mountain front. This geometry implies that layer-parallel shortening in strata of the western Midcontinent developed in association with compression accompanying the Sevier and Laramide Orogenies.

No one has yet compiled joint-trend data for the entire Midcontinent, but the literature does provide data from numerous local studies. Joint frequency diagrams suggest that there are two dominant joint sets (one trending generally northwest and one trending generally northeast) and two less prominent sets (one trending east-west and one trending north-south) in the Devonian strata of northern Michigan. Similar, but not exactly identical, trends have been documented in Ohio, Indiana, Illinois, and Wisconsin. Taken together, regional studies suggest that systematic vertical joint sets do occur in platform strata of the Midcontinent, and that in general there are east-west sets, northwest sets, north-south sets, and north-east sets, but that orientations change across regions and that different sets dominate in different locations. The origin of this jointing remains enigmatic.

22.7.3 Some Causes of Epeirogeny

Over the years, geologists and geophysicists have proposed many mechanisms to explain continental interior epeirogeny. The candidates that may explain epeirogeny will be briefly discussed and are illustrated in Figure 22.7.9.

- *Cooling of Unsuccessful Rifts.* Intracratonic basins may form because of thermal contraction due to cooling of unsuccessful rifts that opened during the Late Proterozoic. When active extension ceased, these rifts cooled and subsided, much like the rifts that underlie passive margins. Such epeirogeny continued through the Phanerozoic at ever-decreasing rates.
- *Variations in Asthenosphere Temperature.* Modern seismic tomography studies of the Earth's interior demonstrate that the Earth's mantle is thermally heterogeneous. As lithosphere drifts over this heterogeneous asthenosphere, it conceivably warms when crossing hot asthenosphere and cools when crossing cool asthenosphere. These temperature changes could cause isostatic uplift (when warmed) or subsidence (when cooled) of broad regions of the lithosphere.

- *Changes in State of Stress.* Changes in stress state in the lithosphere may cause epeirogenic movement in many ways. For example, as differential stress increases, plastic deformation occurs more rapidly, so that the viscosity of the lithosphere effectively decreases. Thus, an increase in differential stress weakens the lithosphere. If this were to happen, denser masses in the crust (e.g., a lens of mafic igneous rock below a rift), which were previously supported by the flexural strength of the lithosphere, would sink, whereas less dense masses (e.g., a granite pluton) would rise. Thus, differential epeirogenic movements may be localized by preexisting heterogeneities of the crust, which is set free to move in an attempt to attain isostatic equilibrium by the weakening of the lithosphere that accompanies an increase in differential stress. Some geologists have suggested that changes in horizontal stress magnitude may also cause epeirogeny by buckling the lithosphere, or by amplifying existing depressions (basins) or rises (arches).
- *Flexural Response to a Load.* Creation of a large load, such as a volcano or a stack of thrust sheets, results in flexural loading on the surface of the continent, and thus bending down of the continent's surface. Flexural loading due to emplacement of thrust sheets leads to the development of asymmetric sedimentary basins, called foreland basins, on the craton side of fold-thrust belts. In addition to causing a depression to form, the levering effect of the loaded lithosphere may cause an uplift, or outer swell, to form on the cratonic-interior of the depression. This effect may have caused Midcontinent uplifts, like the Cincinnati arch, to rise in response to loading the continental margin by thrust sheets of the Appalachian orogen.
- *Block Tilting.* As noted earlier, Midcontinent fault-and-fold zones divide the upper crust into fault-bounded blocks. Changes in the stress state in the continental interior may cause tilting of regional-scale, fault-bounded blocks of continental crust relative to one another. These could cause uplift or subsidence of the corners and edges of blocks.
- *Changes in Crustal-to-Lithosphere Mantle Thickness Ratio.* Continental elevation is controlled, regionally, by isostasy. Since crust and mantle do not have the same density, any phenomenon that causes a change in the proportion of crust to lithospheric mantle in a column from the surface of the Earth down to the level of isostatic compensation

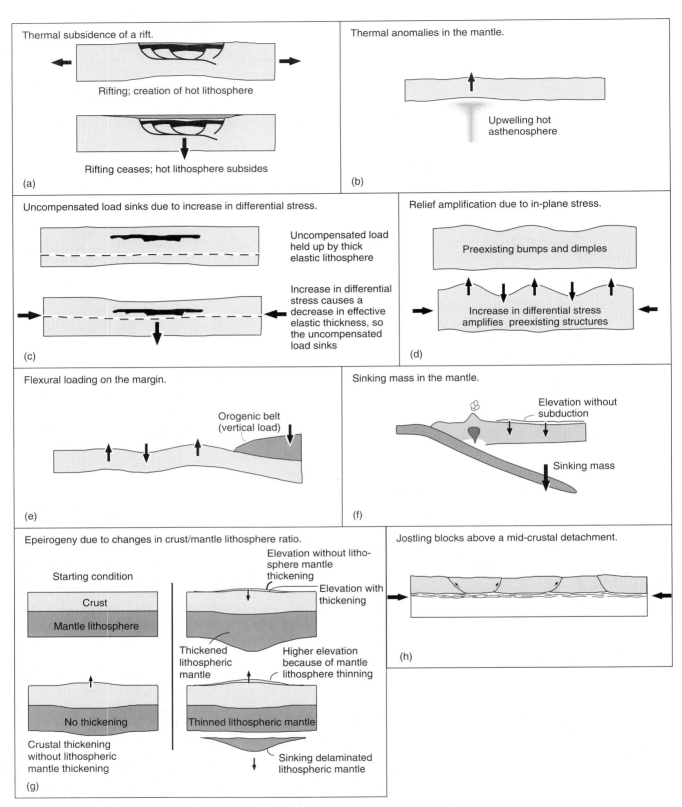

FIGURE 22.7.9 Models of epeirogeny. (a) Thermal cooling over an unsuccessful rift (before and after); (b) Uplift related to thermal anomalies in the mantle; (c) Vertical movement of an uncompensated load due to changes in the elastic thickness of the lithosphere; (d) Amplification of preexisting bumps and dimples due to in-plane stress; (e) Flexural loading of a lithospheric margin; (f) Epeirogeny related to subduction; (g) Epeirogeny due to changes in the ratio of crustal thickness to lithosphere mantle thickness; (h) Epeirogeny due to tilting of regional fault-bounded blocks.

FIGURE 22.7.10 Schematic drawing showing the sinking ball model for epeirogeny caused by subduction (or dynamic topography). As the ball sinks through the honey (shaded area), the thin plastic film on the surface of the honey is pulled down.

has the potential to cause a change in the elevation of the continent's surface. For example, thickening of the crust, perhaps due to plastic strain in response to tectonic compression relative to the lithospheric mantle, would cause a rise in elevation; decreasing the thickness of the lithospheric mantle in response to delamination (separation of lithospheric mantle from the base of the plate) could also cause a rise in crustal elevation; adding basalt to the base of the crust (a process called underplating) would thereby thicken the crust and could cause uplift.

- *Subduction.* Subduction of oceanic lithosphere can cause epeirogenic movement because, as the subducted plate sinks, it pulls down the overlying continent. To picture this phenomenon, imagine a bucket filled with honey (representing the asthenosphere), in which an iron ball (representing oceanic lithosphere) has been suspended just below the surface (Figure 22.7.10). Now, place a film of plastic wrap (representing the continental lithosphere) over the top of the honey, and then release the ball. As the ball sinks (representing subduction of the oceanic lithosphere), the plastic wrap is pulled down. This downward motion is epeirogenic subsidence.

22.7.4 Speculations on Midcontinent Fault-and-Fold Zones

As noted above, stratigraphic evidence demonstrates that Midcontinent fault-and-fold zones were reactivated multiple times during the Phanerozoic. But how did the zones originate in the first place? Did the faults initiate during the Phanerozoic by brittle rupturing of intact crust in response to compression caused by orogeny along the continental margin? If so, then the faults started out as reverse or transpressional faults. Or did the structures form earlier in Earth history? We

favor the second proposal, and suggest that Midcontinent fault-and-fold zones initiated as normal faults during episodes of Proterozoic extension. Thus, displacements in the zones during the Phanerozoic represent fault reactivation. We base this statement on the observation that Midcontinent fault-and-fold zones have the same trends as Proterozoic rift basins and dikes in the United States. They are not systematically oriented perpendicular to the shortening directions of marginal orogens.

If the above hypothesis is correct, then the fault-and-fold zones of the Midcontinent, as well as of the Rocky Mountains and Colorado Plateau, are relicts of unsuccessful Proterozoic rifting. Once formed, they remained as long-lived weaknesses in the crust, available for reactivation during the Phanerozoic, when the interior underwent slight regional strain. A reverse component of displacement occurred on the faults if reactivation was caused by regional shortening, whereas a normal sense of displacement occurred if reactivation was caused by regional extension. On faults that were not perpendicular to shortening or extension directions, transpression or transtension led to a component of strike-slip motion on faults. Regarldless of slip direction, displacement on the faults generated fault-propagation folds in overlying strata. Note that reverse or transpressional motion represents inversion of Proterozoic extensional faults, in that reactivation represents reversal of slip on the faults. In cases where the faults bounded preserved rift basins, the inversion led to thrusting of the rift's contents up and over the rift's margin. But rift basins do not occur in all fault-and-fold zones, either because Proterozoic displacements were too small to create a basin, or because the basin was eroded away during latest Proterozoic rifting, before Phanerozoic strata were deposited.

The hypothesis that Midcontinent fault-and-fold zones are reactivated Proterozoic normal faults is appealing because it explains how these structures could have formed with the orientations that they have, and how they formed without the development of regional cleavage. Zone orientation simply reflects the trends of preexisting Proterozoic normal faults, not the orientation of a regional stress field during the Phanerozoic. The lack of regional cleavage reflects the fact that development of these structures is not associated with significant shortening above detachments within the Phanerozoic section. This hypothesis also explains the timing of movement—faults were reactivated primarily during marginal orogenies or rifting events, when displacement of the continental margin caused a slight strain in the interior, and this strain was accommodated by movement on faults. Perhaps a way to envision Phanerozoic Midcontinent faulting and folding is to think of the upper crust in the craton as a mosaic of rigid fault-bounded blocks that jostle relative to one another in response to changes in the stress state of the continental interior (Figure 22.7.11). Depending on the geometry of the stress during a given time period, blocks may move slightly apart, move slightly together, or move laterally relative to one another. Movements tend to be transpressional or transtensional, for the belts are not oriented appropriately for thrust, reverse, or strike-slip faulting alone to occur.

The greatest amount of movement in Midcontinent fault-and-fold zones occurred in Late Paleozoic time, when Africa collided on the east, South America collided on the south, and a subduction zone had formed along the southwest. This pulse resulted in the formation of the Ancestral Rockies of the Rocky Mountains Province as well in the kilometer-scale displacements in fault-and-fold zones across the Midcontinent (Figure 22.7.7). The intracratonic strain that formed at this time is significantly less than that in Asia during the Cenozoic collision of India with Asia. The contrast in continental-interior response to collision probably reflects the respective strengths of these two continents. The interior of Asia consists of weak continental crust of the Altaids Orogen, while the interior of the United States is a strong craton.

While major movements appear to have accompanied major marginal orogenies, movement on these faults can occur during nonorogenic times as well. For example, historic intraplate earthquakes (earthquakes occurring in a plate interior, away from plate boundaries) at New Madrid, Missouri, result from movements at the intersection of two Midcontinent fault-and-fold zones. This movement may be a response to the ambient stress in the continental lithosphere, caused by ridge-push force and/or basal traction, or to stress resulting from epeirogenic movements.

22.7.5 Closing Remarks

The speculative tone used in this essay emphasizes that geologists need to obtain more data on structures in cratonic interiors before we can confidently explain them and assess their significance. However, it has become increasingly clear that cratonic interiors were not tectonically dead during the Phanerozoic. Rather, they were sensitive recorders of plate interactions, which may have caused jostling of upper-crustal blocks. Although we have focused on examples of structures in the Midcontinent of the United States,

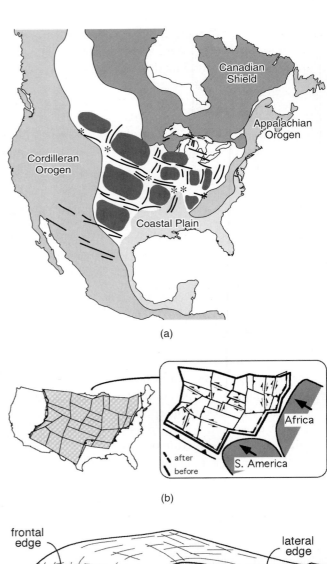

(a)

(b)

(c)

FIGURE 22.7.11 (a) Map showing block model of intracratonic tectonism in the United States. Stars indicate seismically active regions. The diagram in (b) illustrates how thrust and strike-slip motions can be reactivated in response to regional compression, while the diagram in (c) illustrates how uplift along a reverse fault leads to oblique slip along the side of the block.

keep in mind that these structures can be viewed as type examples for continental-interior platforms throughout the world. Finally, it is important to emphasize once more that the dramatic basement-cored uplifts that developed in the Rocky Mountains, as well as in the Sierra Pampeanas of Argentina, and the Tien Shan of southern Asia, are similar in style to those of the U.S. Midcontinent; all may have been formed by reactivation of preexisting faults.

ADDITIONAL READING

Bally, A. W., 1989. Phanerozoic basins of North America. In Bally, A. W., and Palmer, A. R., eds., *The Geology of North America—An Overview, the Geology of North America, v. A,* Boulder, CO: Geological Society of America.

Bond, G. C., 1979. Evidence for some uplifts of large magnitude in continental platforms. *Tectonophysics,* 61, 285–305.

Cathles, L. M., and Hallam, A., 1991. Stress-induced changes in plate density, Vail sequences, epeirogeny, and short-lived global sea level fluctuations. *Tectonics,* 10, 659–671.

Craddock, J., Jackson, M., van der Pluijm, B. A., and Versical, R. T., 1993. Regional shortening fabrics in eastern North America: far-field stress transmission from the Appalachian-Ouachita orogenic belt. *Tectonics,* 12, 257–264.

Gurnis, M., 1992. Rapid continental subsidence following the initiation and evolution of subduction. *Science,* 255, 1556–1558.

Howell, P. D., and van der Pluijm, B. A., 1990. Early history of the Michigan basin: subsidence and Appalachian tectonics. *Geology,* 18, 1195–1198.

Karner, G. D., 1986. Effects of lithospheric in-plane stress on sedimentary basin stratigraphy. *Tectonics,* 5, 573–588.

Lambeck, K., 1983. The role of compressive forces in intracratonic basin formation and mid-plate orogenies. *Geophysical Research Letters,* 10, 845–848.

Marshak, S., Nelson, W. J., and McBride, J. H., 2003. Phanerozoic strike-slip faulting in the continental interior platform of the United States: examples from the Laramide orogen, Midcontinent, and Ancestral Rocky Mountains. *Geological Society of London Special Publication* (in press).

Marshak, S., and Paulsen, T., 1996. Midcontinent U.S. fault and fold zones: a legacy of Proterozoic intracratonic extensional tectonism? *Geology,* 24, 151–154.

Park, R. G., and Jaroszewski, W., 1994. Craton tectonics, stress, and seismicity. In Hancock, P. L., ed., *Continental Deformation.* Oxford: Pergamon Press.

Paulsen, T., and Marshak, S., 1995. Cratonic weak zone in the U.S. continental interior: the Dakota-Carolina corridor. *Geology,* 22, 15–18.

Quinlan, G. M., and Beaumont, C., 1984. Appalachian thrusting, lithospheric flexure, and the Paleozoic stratigraphy of the eastern interior of North America. *Canadian Journal of Earth Sciences,* 21, 973–996.

Sloss, L. L., 1963. Sequences in the cratonic interior of North America. *Geological Society of America Bulletin,* 74, 93–114.

Van der Pluijm, B. M., Craddock, J. P., Graham, B. R., and Harris, J. H., 1997. Paleostress in cratonic North America: implications for deformation of continental interiors. *Science,* 277, 792–796.

Spherical Projections

Spherical projections are used in geology to present three-dimensional orientation data in two-dimensional space for geometric elements (such as bedding planes, foliations, and hinge lines) and crystallographic orientation data (such as c-axis orientations, glide planes). Generally, we use the lower hemisphere for these data, which can be imagined as slicing the Earth in half along a plane containing the poles (i.e., a meridian). The Earth's lines of latitude and longitude are projected in this sectional plane, which produces a gridded net.

In a spherical projection, a plane appears as an arc. To picture this, imagine that the half-sphere represented by the spherical projection is a bowl. Now pass your hand through a point in space that is the center of the full sphere, while it intersects the lower surface of the bowl. You will find that the trace of the intersection between the plane and the bowl is a curved line. Analogously, a line is represented as a point in spherical projection. To picture this, imagine passing your finger through a point in space through the center of the full sphere, where it intersects the surface of the bowl.

In the equal-area net (or Schmidt net; A1a), the projection of lines of latitude and longitude are elliptical arcs. The main advantage of the equal-area projection is that the area of a $1° \times 1°$ grid segment does not change with position on the net; a $1° \times 1°$ grid segment occupies the same area at the center of the net as it does at the edge (hence the name equal-area net). The equal-area net is, therefore, particularly useful for analyzing the distribution of spatial data. In contrast, the equal-angle net (or Wulff net; A1b) projects lines of latitude and longitude as segments of circular arcs. As a consequence, the area size varies with position on the net. The main advantage of the equal-angle net lies in the fact that angular relationships are preserved (hence its name), which means that circular elements are not distorted in this projection (as opposed to the equal-area projection, where circles become ellipses). For example, the projection of a cylinder remains circular regardless of the position of the equal-angle plot.

The procedures for plotting planar and linear elements are essentially the same for both the equal-area and the equal-angle nets, and you should consult a laboratory manual for step-by-step instructions. Nowadays several very good projection programs are available for personal computers, which also allow increasingly sophisticated data analysis methods (such as contouring, clustering analysis, and rotations). While these computer programs are quick and powerful, it is useful to have your first experience with projection techniques through manual plotting.

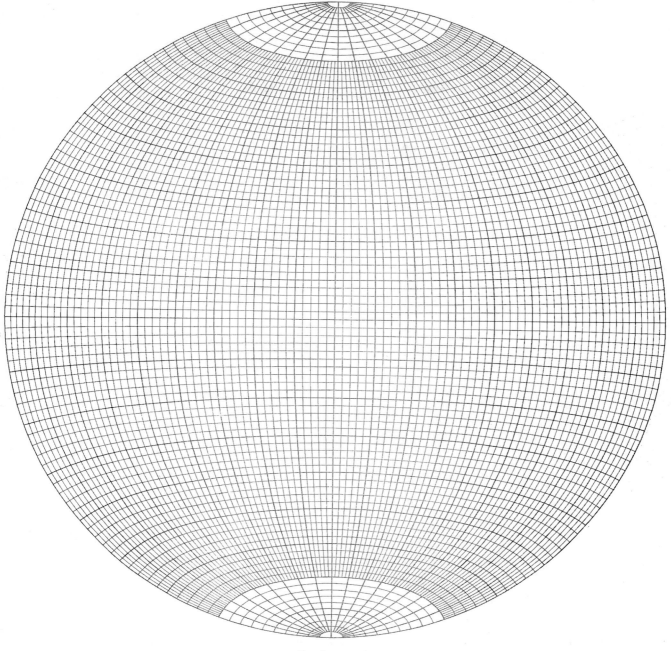

Equal-area net
(a)

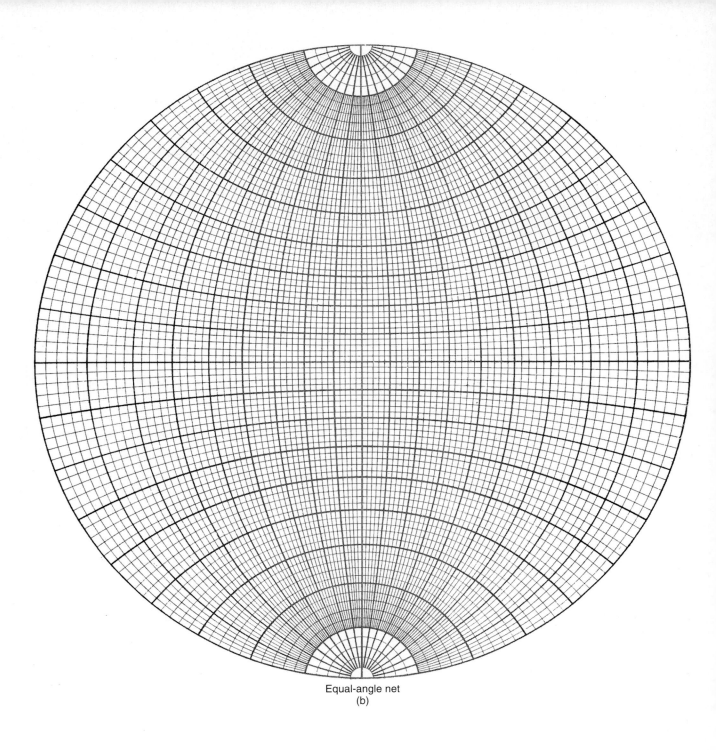

Equal-angle net
(b)

Geologic Timescale

GEOLOGIC TIME SCALE

CENOZOIC

PERIOD	EPOCH		AGE	PICKS (Ma)
QUATER-NARY	PLEISTOCENE	L	CALABRIAN	0.01 / 1.8
NEOGENE (TERTIARY)	PLIOCENE	L	PIACENZIAN	
		E	ZANCLEAN	3.6
				5.3
	MIOCENE	L	MESSINIAN	7.1
			TORTONIAN	11.2
		M	SERRAVALLIAN	14.8
			LANGHIAN	16.4
		E	BURDIGALIAN	20.5
			AQUITANIAN	23.8
PALEOGENE (TERTIARY)	OLIGOCENE	L	CHATTIAN	28.5
		E	RUPELIAN	33.7
	EOCENE	L	PRIABONIAN	37.0
			BARTONIAN	41.3
		M	LUTETIAN	49.0
		E	YPRESIAN	54.8
	PALEOCENE	L	THANETIAN	57.9
			SELANDIAN	61.0
		E	DANIAN	65.0

MESOZOIC

PERIOD	EPOCH		AGE	PICKS (Ma)	PICKS UNCERT. (m.y.)
CRETACEOUS	LATE		MAASTRICHTIAN	65	±2
			CAMPANIAN	71.3	1
			SANTONIAN	83.5	1
			CONIACIAN	85.8	1
			TURONIAN	89.0	1
			CENOMANIAN	93.5	4
	EARLY (NEOCOMIAN)		ALBIAN	99.0	1
			APTIAN	112	2
			BARREMIAN	121	3
			HAUTERIVIAN	127	3
			VALANGINIAN	132	4
			BERRIASIAN	137	4
				144	5
JURASSIC	LATE		TITHONIAN	151	6
			KIMMERIDGIAN	154	7
			OXFORDIAN	159	7
			CALLOVIAN	164	8
	MIDDLE		BATHONIAN	169	8
			BAJOCIAN	176	8
			AALENIAN	180	8
	EARLY		TOARCIAN	190	8
			PLIENSBACHIAN	195	8
			SINEMURIAN	202	8
			HETTANGIAN	206	8
TRIASSIC	LATE		RHAETIAN	210	8
			NORIAN	221	9
			CARNIAN	227	9
	MIDDLE		LADINIAN	234	9
			ANISIAN	242	9
	EARLY		OLENEKIAN	245	9
			INDUAN	248	±10

PALEOZOIC

PERIOD	EPOCH		AGE	PICKS (Ma)
PERMIAN	L		TATARIAN	248
			UFIMIAN-KAZANIAN	252
			KUNGURIAN	256 / 260
	E		ARTINSKIAN	269
			SAKMARIAN	282
			ASSELIAN	290
CARBONIFEROUS (PENNSYLVANIAN)	L		GZELIAN	296 (S)
			KASIMOVIAN	303
			MOSCOVIAN	311 (W)
			BASHKIRIAN	323 (N)
CARBONIFEROUS (MISSISSIPPIAN)	E		SERPUKHOVIAN	327
			VISEAN	342
			TOURNAISIAN	354
DEVONIAN	L		FAMENNIAN	364
			FRASNIAN	370
	M		GIVETIAN	380
			EIFELIAN	391
	E		EMSIAN	400
			PRAGHIAN	412
			LOCKHOVIAN	417
SILURIAN	L		PRIDOLIAN	419
			LUDLOVIAN	423
			WENLOCKIAN	428
	E		LLANDOVERIAN	443
ORDOVICIAN	L		ASHGILLIAN	449
			CARADOCIAN	458
	M		LLANDEILIAN	464
			LLANVIRNIAN	470
	E		ARENIGIAN	485
			TREMADOCIAN	490
CAMBRIAN*	D		SUNWAPTAN*	495
			STEPTOEAN*	500
	C		MARJUMAN*	506
	B		DELAMARAN*	512
			DYERAN*	516
	A		MONTEZUMAN*	520
				543

PRECAMBRIAN

EON	ERA	BDY. AGES (Ma)
PROTEROZOIC	LATE	543 / 900
	MIDDLE	1600
	EARLY	2500
ARCHEAN	LATE	3000
	MIDDLE	3400
	EARLY	3800?

*International ages have not been established. These are regional (Laurentian) only. Boundary Picks were based on dating techniques and fossil records as of 1999. Paleomagnetic attributions have errors, Please ignore the paleomagnetic scale.

Sources for nomenclature and ages: Primarily from Gradstein, F., and Ogg, J., 1996, *Episodes*, v. 19, nos. 1 & 2; Gradstein, F., et al., 1995, SEPM Special Pub. 54, p. 95–128; Berggren, W. A., et al., 1995, SEPM Special Pub. 54, p. 129–212; Cambrian and basal Ordovician ages adapted from Landing, E., 1998, *Canadian Journal of Earth Sciences*, v. 35, p. 329–338; and Davidek, K., et al., 1998, *Geological Magazine*, v. 135, p. 305–309. Cambrian age names from Palmer, A. R., 1998, *Canadian Journal of Earth Sciences*, v. 35, p. 323–328.

PHOTOGRAPHS

Chapter 1

Figure 1.1: From The Royal Collection © Her Majesty Queen Elizabeth II; **Figure 1.2:** © Jonathan Blair/Corbis; **Figure 1.9:** Stephen Marshak

Chapter 2

Figure 2.1: John G. Dennis; **Figure 2.3 b:** Stephen Marshak; **Figure 2.4:** The McGraw-Hill Companies, Inc./Doug Sherman, Photographer; **Figure 2.5:** Stephen Marshak; **Figure 2.6:** Ben van der Pluijm; **Figure 2.7:** Stephen Marshak; **Figure 2.8:** Ben van der Pluijm; **Figure 2.10:** Ben van der Pluijm; **Figure 2.11:** Stephen Marshak; **Figure 2.13:** Ben van der Pluijm; **Figure 2.14:** Ben van der Pluijm; **Figure 2.15:** Stephen Marshak; **Figure 2.17:** Henry McQuillan; **Figure 2.22:** Stephen Marshak; **Figure 2.23:** Stephen Marshak; **Figure 2.24:** Ben van der Pluijm; **Figure 2.25 a:** © Charles O'Rear/Corbis; **Figure 2.26:** Stephen Marshak

Chapter 3

Figure 3.1: © Galen Rowell/Corbis; **Figure 3.7 a:** From Daubree, G. A., *Etudes Synthetiques de Geologie Experimentale*, 1879, Paris; **Figure 3.10:** From *Structural Analysis and Synthesis* (2nd ed.) by S. M. Rowland and E. M. Duebendorfer (Figure 13.5)

Chapter 4

Figure 4.1: U.K. Crown copyright, reproduced by permission of the director, British Geological Survey, © NERC. All rights reserved; **Figure 4.10:** Ben van der Pluijm; **Figure 4.22:** Ben van der Pluijm; **Figure 4.28:** Ben van der Pluijm

Chapter 5

Figure 5.1: From Sharp, 1958, Photograph by United States Coast Guard

Chapter 6

Figure 6.1: Stephen Marshak; **Figure 6.13:** Stephen Marshak; **Figure 6.29:** © Roger Ressmeyer/Corbis

Chapter 7

Figure 7.1 a: U.S. Geological Survey; **Figure 7.1 b:** Stephen Marshak; **Figure 7.2 a:** Stephen Marshak; **Figure 7.2 b:** Stephen Marshak; **Figure 7.5 a:** Stephen Marshak; **Figure 7.7:** Martin Miller; **Figure 7.20:** John G. Dennis; **Figure 7.23 a:** Stephen Marshak; **Figure 7.25:** Martin Miller; **Figure 7.28:** Royal Canadian Air Force, trimmed to reproduce

Chapter 8

Figure 8.2 a: Carol Simpson; **Figure 8.2 b:** A. Keith, U.S. Geological Survey; **Figure 8.2 c:** © C. C. Plummer; **Figure 8.9:** Glacier National Park; **Figure 8.16 a:** Stephen Marshak; **Figure 8.16 b:** Ben van der Pluijm; **Figure 8.17:** Jerry Magloughlin; **Figure 8.18:** Stephen Marshak; **Figure 8.19 b:** F. Arthaud and M. Maltauer, *Bulletin Societe Geologique de France*, Series 7, 1969;11:738-744. Reprinted by permission of Societe Geologique de France, 77 rue Claude Bernard, Paris; **Figure 8.23:** Stephen Marshak; **Figure 8.24:** Stephen Marshak; **Figure 8.35:** © Michael S. Yamashita/Corbis

Chapter 9

Figure 9.1: Switzerland Geological Survey; **Figure 9.3:** From D. T. Griggs, and J. Handin, Observations of fracture, a hypothesis of earthquakes. Rock Deformation, a Symposium, *Geological Society of America memoir,* 1960;79:347-364; **Figure 9.5:** J. M. Christie, © American Geophysical Union Geophys. Monogr. Photomicrograph, 1972;16:117-138. Courtesy J. M. Christie; **Figure 9.9:** Ben van der Pluijm; **Figure 9.15:** Ben van der Pluijm; **Figure 9.21:** Stephen Marshak; **Figure 9.23:** Ben van der Pluijm; **Figure 9.26:** Ben van der Pluijm; **Figure 9.27:** Ben van der Pluijm; **Figure 9.28:** Ben van der Pluijm; **Figure 9.38:** W. T. Lee, U.S. Geological Survey

Chapter 10

Figure 10.1: From J. Haller, Die Strukturelemente Ostgroenlands Zwischen 74 und 78 N. 1956 Medd Groenland, 1956;154: 3:153; **Figure 10.5:** R. Y. Anderson; **Figure 10.7:** John S. Shelton; **Figure 10.8:** Mary Hill; **Figure 10.18:** Ben van der Pluijm; **Figure 10.19:** Ben van der Pluijm; **Figure 10.20:** Ronadh Cox; **Figure 10.21:** Ben van der Pluijm; **Figure 10.22:** John G. Dennis; **Figure 10.26:** Ben van der Pluijm; **Figure 10.39:** Topographical Survey of Switzerland

Chapter 11

Figure 11.4: J. B. Woodworth, U.S. Geological Survey; **Fig-ure 11.5 a:** John G. Dennis; **Figure 11.5 b:** John G. Dennis; **Figure 11.8:** J. M. Dares; **Figure 11.10:** A. Keith, U. S. Geological Survey; **Figure 11.11 a:** Ben van der Pluijm; **Figure 11.11 b:** Reprinted from K. Weber, *Tectonophysics*, 78. with kind permission of Elsevier Science – NL; **Figure 11.12:** John G. Dennis; **Figure 11.14:** John G. Dennis; **Figure 11.16:** Stephen Marshak and Ben van der Pluijm; **Figure 11.18:** Ben van der Pluijm; **Figure 11.22:** John G. Dennis; **Figure 11.26:** Stephen Marshak; **Figure 11.28:** John G. Dennis; **Figure 11.30:** Nei-Che Ho

Chapter 12

Figure 12.1: Ben van der Pluijm; **Figure 12.3:** Ben van der Pluijm; **Figure 12.5 a:** Ben van der Pluijm; **Figure 12.5 b:** Carol Simpson; **Figure 12.7 a:** Carol Simpson; **Figure 12.7 b:** Carol Simpson; **Figure 12.10:** Carol Simpson; **Figure 12.13:** Mikrotektoniek Collection, University of Utrecht; **Figure 12.16:** Ben van der Pluijm; **Figure 12.25:** Ben van der Pluijm; **Figure 12.27 d:** Ben van der Pluijm; **Figure 12.28:** Jerry Magloughlin

Chapter 13

Figure 13.1: Ben van der Pluijm; **Figure 13.4 a:** Mary Ellen Tuccillo; **Figure 13.6 a-c:** Cees Passchier

Chapter 14

Figure 14.1: Land Processes Distributed Active Archive Center (LP DAAC), U. S. Geological Survey's EROS Data Center; **Figure 14.2 a-b:** JOIDES Resolution (Ocean Drilling Program)

Chapter 16

Figure 16.2: SPL; **Figure 16.13 e:** Stephen Marshak; **Figure 16.20:** Stephen Marshak; **Figure 16.22 a:** Stephen Marshak; **Figure 16.22 b:** Stephen Marshak

Chapter 17

Figure 17.1 a-b: Stephen Marshak; **Figure 17.15:** Ben van der Pluijm; **Figure 17.23:** Ben van der Pluijm; **Figure 17.27:** NASA

Chapter 18

Figure 18.1: M. S. Wilkerson; **Figure 18.10 b:** M. S. Wilkerson; **Figure 18.13 b:** M. S. Wilkerson; **Figure 18.16 a:** NASA; **Figure 18.18 b:** M. S. Wilkerson; **Figure 18.20 b:** Chris Hedlund; **Figure 18.26:** M. P. Fischer

Chapter 19

Figure 19.13: From R. E. Wilcox, T. P. Harding, and D. R. Seely, 1973, American Association of Petroleum Geologists Bulletin, v. 57/1, Basic wrench tectonics, p. 74-96, Fig. 11 (p. 88); imagery acquired by Westinghouse Electric Corp., p. 74-96, Fig. 11. AAPG©1973, reprinted by permission of the AAPG whose permission is required for further use; **Figure 19.14 d:** From R. E. Wilcox et al., 1973, American Association of Petroleum Geologists Bulletin, v. 57/1, Basic wrench tecton-ics, p. 74-96, Fig. 8, p. 84. AAPG©1973, reprinted by permis-sion of the AAPG whose permission is required for further use; **Figure 19.17 a:** R. E. Wallace and Parke D. Snavely, U.S. Geological Survey; **Figure 19.17 b:** R. E. Wallace, U.S. Geological Survey; **Figure 19.17 c:** Ben van der Pluijm

Chapter 20

Figure 20.1: NOAA; **Figure 20.5:** NASA

Chapter 22

Figure 22.3.1: NASA; **Figure 22.6.5.:** Paul F. Hoffman

LINE ART

New and modified line art was prepared by Stan Maddock and Dale Austin

Chapter 1

Figure 1.3: From G. P. Scrope, *Considerations on Volcanoes*, W. Philips, 1825, p. 270; **Figure 1.4:** After B. Issacks et al., Seismology and the New Global Tectonics, *Journal of Geophysical Research*, 1968;73:5855-5899, American Geophysical Union

Chapter 2

Figure 2.2: From A. H. Bouma, *Sedimentology of some Flysch Deposits*, 1962, Elsevier Science Publishers, B. V., Amsterdam. Used by permission of the author; **Figure 2.18:** From M. P. A. Jackson and C. J. Talbot, Advances in salt tectonics. In *Continental deformation*, P. L. Hancock (ed.), 1994, Fig. 8.2; **Figure 2.19:** Redrawn From W. E. Galloway et al., *Atlas of Major Texas Reservoirs*, 1983, Bureau of Economic Geology, University of Texas at Austin, used by permission; **Figure 2.20:** From M. P. A. Jackson and C. J. Talbot, 1994, Advances in salt tectonics. In *Continental deformation*, P. L. Hancock (ed.), Fig. 8.2; **Figure 2.21:** From Cloos, *Einfürung in die Geologie*, 1936, Gebr, Borntraeger Publishers Berlin, Stuttgart, Germany.

Chapter 3

Figure 3.4: Modified From Hobbs, Means, and Williams, *An Outline of Structural Geology*, p. 7, copyright © John Wiley & Sons, Inc., New York; **Figure 3.13:** Modified from Hafner, in *Geological Society of America Bulletin*, 1951;62:373-398; **Figure 3.14:** Based on Townend and Zoback, *Geology*, 2000;28, Fig. 1; **Figure 3.15 a-b:** M. L. Zoback, First and second order patterns of stress in the lithosphere: the world stress map project, *Journal of Geophysical Research*, 1992:11703-11728, copyright 1992 by the American Geophysical Union. Courtesy of M. L. Zoback; **Figure 3.16 a-b:** G. Ranalli, *Rheology of the Earth*, copyright 1986 Allen and Unwin, Boston, Figs. 12.1 and 12.2

Chapter 4

Figure 4.15: From J. G. Ramsay and M. I. Huber, *The Techniques of Modern Structural Geology*, vol. 1, Strain Analysis, Figs. 11.9 and 11.0, copyright © 1983 Academic Press, London. Reprinted by permission; **Figure 4.17 a-b:** C. Richter and B. van der Pluijm; **Figure 4.19:** From E. Cloos, Oolite Deformation in the South Maintain fold, Maryland, *Geological Society of America Bulletin,* 1947;58:843-918; **Figure 4.25:** Modified from J. Ramsay, *Folding and Fracturing of Rocks*, 1967, Figs. 5.60 and 5.62, modified with permission, copyright 1967 McGraw Hill, Inc. All Rights Reserved; **Figure 4.26:** John G. Dennis; **Figure 4.27:** Modified from J. G. Ramsay and M. I. Huber, *The Techniques of Modern Structural Geology*, vol. 1: Strain Analysis, 1983, Fig. 6.6, Academic Press, London, Reprinted and modified by permission; **Figure 4.29:** From C. Richter, B. A. van der Pluijm, and B. A. Housen, The quantification of crystallographic pre-ferred orientation using magnetic anisotrophy, *Journal of Structural Geology*, 1993;15:113-116, Fig. 3, copyright 1993, with permission from Elsevier Science, Ltd., The Boulevard, Langford Lane, Killington OX5 16B, UK; **Figure 4.30:** From Pfiffner and Ramsay, 1982, Fig. A1

Chapter 5

Figure 5.4: After Jeanloz and Morris, *Annual Review of Earth and Planetary Sciences*, 1986;14, Fig. 2; **Figure 5.7:** From G Ranalli, *Rheology of the Earth,* copyright 1987 Allen and Unwin, Boston; **Figure 5.8:** From Handin J. W., Strength and Ductility, *Geological Society of America Memoir,* 1966; 97:223-290, copyright 1966 John W. Handin; **Figure 5.9:** From C. T. Walker and John G. Dennis, Explosive phase transitions in the mantle, *Nature,* 1966;209:182-183, Macmillan Magazines, Ltd.; **Figure 5.10:** After H. C. Heard, Transition from brittle fracture to ductile flow in Solenhogen limestone as a function of temperature, confining pressure, and interstitial fluid pressure, *Geological Society of America,* 1960;79:193-226; **Figure 5.11:** From F. A. Donath, Some information squeezed out of rock, *American Scientist,* 1970;58:54-72, Used by permission; **Figure 5.12:** After H. C. Heard, Transition From brittle fracture to ductile flow in Solenhogen limestone as a function of temperature, confining pressure, and interstitial fluid pressure, *Geological Society of America,* 1960;79:193-226; **Figure 5.13:** After D. T. Griggs and J. Hardin, Rock deformation—A Symposium, *Geological Society of America memoir* 1960;79:282; **Figure 5.14:** From H. C. Heard, Effect of large changes in strain rate in the experimental deformation of Yule marble, *Journal of Geology,* 1963;71:162-195, University of Chicago; **Figure 5.15:** From H. C. Heard and C. B. Raleigh, Steady state flow in marble at 500 to 800 C, *Geological Society of America Bulletin,* 1972;83:935-956, Fig. 17; **Figure 5.16 a:** From H. H. Robinson, The effect of pore and confining pressure on the failure process in sedimentary rock, *Colorado School Mines Quarterly,* 1959;50:177; **Figure 5.16b:** F. Donath, Some information squeezed out of rock, *American Scientist* 1970;58:54-72; **Figure 5.17:** From D. T. Griggs, Hydrolitic weakening of quartz and other silicates, *Geophysical Journal* 1967;14:19-31, Royal Astronomy Society, Oxford, England; **Figure 5.18:** After John W. Handin, Strength and Ductility, *Geological Society of America memoir,* 1966; 97:223-290; **Figure 5.20:** From Rutter, *Geology Today,* 1993;9:61-65, Fig. 7; **Figure 5.21:** Reprinted from J. G. Ramsay, Shear zone geometry: a review, *Journal of Structural Geology,* 1980;2: 83-99, Fig. 26.1, copyright 1980 with permission From Elsevier Science, Ltd. The Boulevard; Langford Lane, Kidlington OX5 16B, UK

Chapter 6

Figure 6.4: From T. Engelder, *Stress Regimes in the Lithosphere,* Princeton University Press, 1993; **Figure 6.23:** From L. H. Scholz, 1990, *The Mechanics of Earthquakes and Faulting.* Copyright © 1990 Cambridge University Press, reprinted with the permission of Cambridge University Press

Chapter 7

Figure 7.4: Adapted From D. Bahat and T. Engelder, Surface Morphology on Cross-Fold Joints of the Appalachian Plateau, New York, and Pennsylvania, *Tectonophysics,* 1984;104:299-313, Copyright 1984, with kind permission of Elsevier Science—Amsterdam, the Netherlands; **Figure 7.6 c:** After Pollard and Aydin, Progress in understanding jointing over the last century, *Geological Society of America Bulletin,* 1988:100;1181-1204; **Figure 7.13:** T. Engelder and P. Geiser, *Journal of Geophysical Research,* 1980;85:6319-6341, copyright by the American Geophysical Union; **Figure 7.13:** From John G. Dennis, 1987, Structural Geology: An Introduction, W. C. Brown, Dubuque. All Rights Reserved. Reprinted by permission

Chapter 8

Figure 8.10 c: Based on J. E. Gill, Fault Nomenclature, *Royal Society of Canada Transactions,* 1991:35:71-85; **Figure 8.14:** From J. G. Ramsay and M. I. Huber, *The Techniques of Modern Structural Geology,* vol. 2, Folds and Fractures, copyright 1987 Academic Press Ltd., London, p. 507; **Figure 8.15 c:** From P. A. Lowie and C. H. Scholz, Displacement—length scaling relationships for faults: data synthesis and conclusion, *Journal of Structural Geology,* 1992:14:1149-1156, Pergamon Press; **Figure 8.26:** Modified From L. H. Scholz, 1990, *The Mechanics of Earthquakes and Faulting,* p. 29, Cambridge University Press, Cambridge; **Figure 8.28:** After M. K. Hubbert, Mechanical basis for certain familiar geologic structures, *Geological Society of America Bulletin,* 1951:355-372; **Figure 8.30:** Modified from L. H. Scholz, *The Mechanics of Earthquakes and Faulting,* 1990, Fig. 3.18, Cambridge University Press. After Sisbon, 1974; **Figure 8.32:** L. H. Scholz, Microfracturing and the inelastic deformation of rock in compression, *Journal of Geophysical Research,* 1968;73:1417-1432, American Geophysical Union; **Figure 8.36:** From Scholz, *Journal of Geophysical Research,* 1968, Fig. 7

Chapter 9

Figure 9.6: From Hayden et al, 1965, p. 65; **Figure 9.7:** From Hull, 1975, p. 20; **Figure 9.8:** From J. Suppe, 1985, Figs. 4.8 and 4.12; **Figure 9.10:** From Hayden et al, 1965, p. 69; **Figure 9.11:** From Hayden et al, 1965, p. 75; Modified from Nicholas and Poirier, 1976, pp. 75-76; **Figure 9.13:** From Hayden et al., 1965, p. 72; **Figure 9.16:** From Schedl and van der Pluijm, *Journal of Geological Education,* 1988;36:111-121, copyright 1988 National Association of Geology Teachers, Inc.; **Figure 9.18:** After R. H. Groshong, Jr., Strain calculated from twinning in calle, *Geological Society of America Bulletin,* 1972;83:2025-2038. Used by permission; **Figure 9.20 a:** From H. J. Frost and M. F. Ashby, *Deformation-mechanism maps. The Plasticity and creep of metals and ceramics,* Pergamon Press, 1982. Used by permission of the author; **Figure 9.20 b:** From H. J. Frost and M. F. Ashby, 1982, *Deformation-mechanism maps. The Plasticity and creep of metals and ceramics,* Pergamon Press, 1982. Used by permission of the author; **Figure 9.24:** Modified from J. Suppe, 1985, Fig. 4.17; **Figure 9.25:** From Moffatt et al, 1965, p. 94; **Figure 9.29 a:** From J. P. Poirier, *Creep in Crystals: High-temperature Deformation Processes in metals, Ceramics and Minerals,* p. 66, copyright 1985 Cambridge University Press, Cambridge; **Figure 9.29 b:** J. L. Urai et al., Dynamic recrystallization of minerals in Mineral and Rock Deformation: Laboratory Studies (The Paterson Volume), *Geophysical Monograph,* Hobbs and Heard (eds.), 1986;36:161-200; **Figure 9.30:** After S. White, The Effects of Strain on the microstructures, fabrics, and deformation mechanisms in quartzites, *Philosophical Transactions of the Royal Society of London,* Series A, 1976; 283:69-86, Fig. 2; **Figure 9.32:** Modified from M. F. Ashby and R. A. Verrall, Micromechanisms of flow and fracture, and their relevance to the rheology of the upper mantle, *Philosophical Transactions of the Royal Society of London,* Series A, 1973;288:59-95; **Figure 9.33:** From T. G. Langdon, p. 219-232, Fig. 5, in *Preferred Orientation in deformed metals and rocks: An introduction to modern texture analysis,* H. R. Wenk (ed.), Academic Press, Orlando, 1985; **Figure 9.34:** From E. H. Rutter, The Kinetics of rock deformation by pressure solution, *Philosophical Transactions of the Royal Society of London,* 283:203-219; **Figure 9.35:** From E. H. Rutter, The Kenetics of rock deformation by pressure solution, *Philosophical Transactions of the Royal Society of London,* 283:203-219; **Figure 9.36:** After M. F. Ashby and R. A. Verrall, Micromechanisms of flow and fracture, and their relevance to the rheology of the upper mantle, *Philosophical Transactions of the Royal Society of London,* Series A, 1973;288:59-95

Chapter 10

Figure 10.3: From F. J. Turner and L. E. Weiss, *Structural Analysis of Metamorphic Tectonites,* Fig. 4.17a-b. Copyright © 1963 The McGraw-Hill Companies, Inc., All Rights Reserved. Reprinted by permission; **Figure 10.6 e-f:** From G. J. Borradaile, Structural facing (Shackelton's rule) and the Paleozoic rocks of the malagmide complex near Velez Rubio,

SE Spain, *Proc. Kon. Nederl. Akademie Wetenschnappen,* 1976;7-9:330-336; **Figure 10.10:** From Richard, M. J., A classification diagram for fold orientations, *Geological Magazine,* 1971;108:23-26. Copyright 1971 Cambridge University Press. Reprinted by permission of Cambridge University Press; **Figure 10.11:** From J. G. Ramsay, The Geometry and Mechanics of Formation of similar type folds, *Journal of Geology,* 1962;70:309-327. Copyright © 1962 by the University of Chicago. Used by permission; **Figure 10.12:** Modified with permission from J. G. Ramsay, *Folding and Fracturing Rocks.* Copyright © 1967 The McGraw Hill Companies, Inc. All Rights Reserved; **Figure 10.24:** From J. G. Ramsay and M. I. Huber, *The Techniques of Modern Structural Geology,* vol. 2: Folds and Fractures, Fig. 22.15, Academic Press, London. Reprinted by permission; **Figure 10.25:** Reprinted From R. L. Thiessen and W. D. Means, Classification of fold interference patterns: a reexamination, *Journal of Structural Geology,* 1980;2:311-316, Figs. 1 and 5. Copyright 1980, with permission From Elsevier Science Ltd., The Boulevard, Langford Lane, Kidlington OX5; **Figure 10.29:** Topographical Survey of Switzerland; **Figure 10.30:** From J. B. Curie et al., Development of folds in sedimentary strata, *Geological Society of America Bulletin,* 73:655-674; **Figure 10.31:** From J. H. Dietrich, *Canadian Journal of Earth Sciences,* 1970;7, Fig. 5.7; copyright © 1970 National Research Council of Canada. Reprinted by permission of NRC Research Press; **Figure 10.32:** From John G. Ramsay, *Folding and Fracturing Rocks.* Copyright © 1967 The McGraw-Hill Companies, Inc. All Rights Reserved. Reprinted by permission; **Figure 10.33:** From John G. Ramsay, *Folding and Fracturing Rocks.* Copyright © 1967 The McGraw-Hill Companies, Inc. All Rights Reserved. Reprinted by permission; **Figure 10.34:** From Hobbs et al, 1976; **Figure 10.36:** From John G. Ramsay, *Folding and Fracturing Rocks.* Copyright © 1967 The McGraw-Hill Companies, Inc. All Rights Reserved. Reprinted by permission; From P. Y. Hudleston and T. B. Holst, Strain Analysis and fold shape in a limestone layer and implications for later rheology, *Tectonophysics,* 1984;106:321-347, with kind permission of Elsevier Science—Amsterdam, The Netherlands

Chapter 11

Figure 11.20: From Stephens et al., *American Journal of Science,* 1979; 279. **Figure 11.23:** Adapted From G. J. Borradaile, Transected folds; a study illustrated with examples from Canada and Scotland, *Geological Society of America Bulletin,* 1978;89:481-493; **Figure 11.24:** Adapted with permission from J. G. Ramsay and M. I. Huber, *The Techniques of Modern Structural Geology,* vol. 1: Strain Analysis, pp. 236-280. Copyright © 1987 Academic Press; **Figure 11.29:** From Pares et al., *Tectonophysics,* 1999;37:9, Fig. 6

Chapter 12

Figure 12.2: Based on various sources, including R. H. Sibson, Fault rocks and fault mechanisms, *Journal of the Geological Society of London,* 133:190-213; and C. H. Scholz, *The Mechanics of Earthquakes and Faulting,* 1990, Cambridge University Press, Cambridge; **Figure 12.4 a:** From Hammer S. and Passchier C. W., Shear-sense indicators: a review paper, *Natural Resources Canada,* 1991:90-117. Courtesy of Geological Survey of Canada. Reproduced with permission of the Minister of Public Works and Government Services of Canada, 1996; **Figure 12.4 b:** Modified from S. Marshak and G. Mitra, *Basic Methods of Structural Geology,* 1988, Fig. 11.31d, Prentice-Hall, Upper Saddle River, NJ; **Figure 12.9:** Reprinted From G. Lister and A. W. Snoke, S-C mylonites, *Journal of Structural Geology,* 1984;6:617-638, Copyright 1984, with permission from Elsevier Science, Ltd., The Boulevard, Langford Lane, Kidlington OX5 1BG, UK. From Lister and Snoke (1984); **Figure 12.17:** From C. Simpson and D. G. De Paor, Strain and Kinematic Analysis in General Shear Zones, *Journal of Structural Geology,* 1993;15:1-20; **Figure 12.22:** From R. D. Law, in Deformation Mechanisms, Rheology and Tectonics, R. J. Knipe and E. H. Rutter (eds.), *Geological Society Special Publication,* 1990:54:335-352. Used by permission; **Figure 12.23:** From R. D. Law, in Deformation Mechanisms, Rheology and Tectonics, R. J. Knipe and E. H. Rutter (eds.), *Geological Society Special Publication,* 1990:54:335-352. Used by permission; **Figure 12.24:** From Williams, Geologische Rundschau, 1982;72:602; **Figure 12.26:** From Hudleston and Lan, *Journal of Structural Geology,* 1993;15, figs 7b and 7c; **Figure 12.27 a-c:** After Hobbs, et. at. 1976; **Figure 12.29:** Modified with permission from Hudleston, *Journal of Geological Education,* 1986;34:24. Copyright © 1986 National Association of Geology Teachers.

Chapter 13

Figure 13.2: Modified From Yadley, 1989, Fig. 2.8; **Figure 13.4 b-d:** From M. E. Tuccillo et al., Thermobarometry, geosynchronology, and the interpretation of P-T-t data, *Journal of Petrology,* 1992;33:1225-1259, Oxford University Press, Oxford, UK. By permission of Oxford University Press; **Figure 13.5:** From H. Zwart, On the determination of polymetamorphic mineral associations, *Geologische Rundschau;* 1962;52:38-65. Used by permission of Springer-Verlag GmbH & Co. KG, Germany; **Figure 13.7:** After A. Spry, *Journal of Petrology,* 1963;4:211-222. Used by permission of Oxford University Press, Oxford; **Figure 13.8:** Adapted From Gunter Faure, *Principles of Isotope Geology,* 2nd ed., Fig. 8.2. Copyright © 1986 John Wiley & Sons, New York; **Figure 13.9:** Adapted from Gunter Faure, *Principles of Isotope Geology,* 2nd ed., Fig. 8.5. Copyright © 1986 John Wiley & Sons, New York; **Figure 13.10:** From M. A. Cosca

Chapter 14

Figure 14.3: From S. Marshak, *Earth: Portrait of a Planet,* 2001, W. W. Norton & Company, New York; **Figure 14.4 a:** From S. Marshak, *Earth: Portrait of a Planet,* 2001, W. W. Norton & Company, New York, Fig C-13; **Figure 14.4 b:** From S. Marshak, *Earth: Portrait of a Planet,* 2001, W. W. Norton & Company, New York, Fig 2.12a; **Figure 14.5 a:** Data from J. G. Cogley, 1984; *Reviews of Geophysics and Space Physics,* v. 22, p. 101-122. Copyright 1984 American Geophysical Union. Modified by permission of American Geophysical Union; **Figure 14.6:** Modified from M. J. DeWit and C. Stern, 1978, *Journal of Volcanology and Geothermal Research,* v. 4, p. 55-80, Fig. 4. Reprinted with permission from Elsevier; **Figure 14.7 a:** From R. O. Mooney and W. D. Meissner, 1991, EOS, *Transactions American Geophysical Union,* 72, 537-541, Fig. 1. Copyright 1991 American Geophysical Union. Reproduced by permission of American Geophysical Union; **Figure 14.7 b:** Based on a concept by N. Christensen; **Figure 14.8 a:** Modified from A. M. Goodwin, 1996, *Principles of Precambrian Geology,* Academic Press, London, p. 4. Reprinted with permission from Elsevier; **Figure 14.8 b:** From S. Marshak, *Earth: Portrait of a Planet,* 2001, W. W. Norton & Company, New York; **Figure 14.9:** From A. M. Dziewonski, 1984, *Journal of Geophysical Research,* v. 89, p. 5929-5952. Copyright 1984 American Geophysical Union. Reproduced by permission of American Geophysical Union; **Figure 14.10 a:** From P. Keary and F. J. Vine, 1990, *Global Tectonics,* Blackwell Scientific Publishers, Oxford, Fig. 2.39; **Figure 14.13:** From S. Marshak, *Earth: Portrait of a Planet,* 2001, W. W. Norton & Company, New York; **Figure 14.14 a:** Modified from C. M. R. Fowler, *The Solid Earth,* 1990, Cambridge University Press, Fig. 2.2. Reprinted with permission of Cambridge University Press; **Figure 14.14 d:** Modified from C. M. R. Fowler, 1990, *The Solid Earth,* Cambridge University Press Fig. 2.2, p. 6; **Figure 14.16:** Adapted from C. R. Scotese, 1997, *Continental Drift, The PALEOMAP Project, 7th ed.:* Dept. of Geology, Univ. of Texas, Arlington, maps for 250 Ma, 150 Ma, 70 Ma, 9 Ma; **Figure 14.19:** In S. Marshak, *Earth: Portrait of a Planet,* 2001, W. W. Norton & Company, New York. Modified from A. Cox and R. B. Hart, 1986, *Plate Tectonics: How It Works,* Blackwell, Oxford, Fig. 10.18; **Figure 14.22:** Modified from C. M. R. Fowler, 1990, *The Solid Earth: An Introduction to Global Geophysics,* Cambridge University Press, Cambridge, Fig. 2.14. Reprinted with the permission of Cambridge University Press

Chapter 15

Figure 15.2 a-c: Data recorded by COCORP (Consortium for Continental Reflection Profiling) in 1977 and by the Canadian LITHOPROBE program in 1984; **Figure 15.3 a-d:** Data record-ed by the Canadian LITHOPROBE program in 1994; **Figure 15.5:** Data recorded by the Canadian LITHOPROBE program; **Figure 15.6:** Data recorded by the Canadian LITHO-PROBE program; **Figure 15.7:** Data recorded by the Canadian LITHOPROBE program

Chapter 16

Figure 16.4 a-b: Based on a concept from D. P. McKenzie, 1978; **Figure 16.4 c:** From P.A. Ziegler, 1982, Philosophical Transactions of the Royal Society, London, pp. 13–143: **Figure 16.6:** Adapted from B. Wernicke and B. C. Burchfiel, 1982, *Journal of Structural Geology,* v. 4, p. 105-115, Fig. 7, p. 107. Reprinted with permission from Elsevier; **Figure 16.8:** Modified from A. D. Gibbs, 1984, Structural evolution of exten-sional basin margins, *Journal of the Geological Society of London,* v. 141, p. 609-620, Fig. 5, p. 611; Fig. 9, p. 614; **Figure 16.10:** From G. S. Lister, M. A. Etheridge, and P. A. Symmonds, 1986, *Geology,* v. 14, p. 246-250, Fig. 3; **Figure 16.11 a:** From P. A. Ziegler, 1982; *Philosophical Transactions of the Royal Society,* London, v. A305, p. 113-143, Fig. 11, p. 131; **Figure 16.11 b:** From W. Bosworth, 1994, *Geologische Rundschau,* v. 83, p. 671-688, Fig. 10, p. 684. Reprinted with permission from Springer-Verlog; **Figure 16.12:** J. H. Stewart, 1978, *Geological Society of America Memoir* 152, p. 1-31, Fig. 1.1, with data added from P.J. Coney, 1980, *Geological Society of America Memoir* 153, p. 7-31; **Figure 16.13 a-d:** After G. S. Lister and G. A. Davis, 1989, *Journal of Structural Geology,* v. 11, p. 65-94, Fig. 20. Reprinted with permission from Elsevier; **Figure 16.14:** Adapted from B. R. Rosendahl, 1987, *Annual Review of Earth and Planetary Science Letters,* v. 15, p. 445-503, Fig. 3. Used with permission by Annual Reviews; **Figure 16.15 b:** Modified from C. J. Ebinger, 1989, *Tectonics,* v. 8, p. 117-133, Fig. 4, p. 121; **Figure 16.15 c:** From A.W. Bally and S. Snelson, 1980, *Canadian Society of Petroleum Geologists Memoir* 6, p. 9-75, Fig. 34, p. 66; **Figure 16.16:** After W. Bosworth, 1994, *Geologische Rundschau,* 83, 671-688, Fig. 10, p. 684. Reprinted with permission from Springer-Verlog; **Figure 16.17 b:** Modified from D. R. McClay and M. J. White, 1995, *Marine and Petroleum Geology,* 12, 137-151, Fig. 1a; **Figure 16.21 a:** Based on data From W. W. Arwood, *The Physiographic Provinces of North America,* 1940, Ginn and Co.; **Figure 16.21 b:** Based on data From T. H. Dixon et al., 1989; **Figure 16.24:** Data from A. Cox and R. B. Hart, 1986, *Plate Tectonics. How It Works.* Blackwell Scientific Publications, Oxford; **Figure 16.26 a, c:** Modified from G. Davis and S. J. Reynolds, *Structural Geology,* John Wiley and Sons, New York; **Figure 16.26 b:** Modified from K. C. MacDonald, 1982, *Annual Review of Earth and Planetary Science Letters,* v. 10, p. 155-190. Used with permission by Annual Reviews; **Figure 16.26 d:** Open University Course

Team, 1989, *The Ocean Basins: Their Structure and Evolution*. Pergamon Press, Oxford, Fig. 4.15. Reprinted with permission from Elsevier; **Figure 16.27:** Modified from R. Twiss and E. Moores, 1992, *Tectonics,* W. H. Freeman, New York, Fig. 5.14. © 1992 by W. H. Freeman and Company. Used with permission; **Figure 16.28:** Inset from A. W. Erxleben and G. Carnahan, 1983, *American Association of Petroleum Geologists Studies in Geology Series,* no. 15, v. II, Seismic Expression of Structural Styles: A Picture and Work Atlas, ed. A. W. Bally (1983). Slick ranch area, Starr County, Texas by A. W. Erxleben and G. Carnahan, p. 2.3.1-22— 2.3.2-27, Fig. Detached Sediments in Extensional Provinces/ Growth Faults (p. 2.3.1-23). AAPG©1983, reprinted by permission of the AAPG whose permission is required for further use

Chapter 17

Figure 17.5 a: From D. L. Turcotte et al., 1978, *Tectonophysics,* v. 47, p. 193-205, Fig. 5. Reprinted with permission from Elsevier; **Figure 17.6 a:** Modified from E. R. Oxburgh and D. L. Turcotte, 1970, *Geological Society of America Bulletin,* 81, 1665-1688, Fig. 3; **Figure 17.6 b-c:** Modified from K. C. Condie, 1989, *Plate Tectonics and Crustal Evolution, 3d ed.,* Pergamon Press, Oxford, Fig. 6.9, p. 108. Reprinted with permission from Elsevier; **Figure 17.7:** Modified from Schubert et al., 1975, *Geophysical Journal of the Royal Astronomical Society,* 42, p. 705-735, Fig. 11; **Figure 17.9:** Modified from L. Kellogg, 1999, *Science,* v. 283, Fig. 1, p. 1882; **Figure 17.12 b:** From G. F. Moore et al., 1991, *Proceedings of the Ocean Drilling Program, Initial Reports,* v. 131, p. 15-23, Fig. 5; **Figure 17.19:** Adapted from A. Nur and Z. Ben-Avrahem, 1982, *Journal of Geophysical Research,* v. 87, 3644-3661, Fig. 1, p. 3649. Copyright 1982 American Geophysical Union. Modified by permission of American Geophysical Union; **Figure 17.23 a:** Modified from J. F. Dewey et al., 1986. In M. P. Coward and A. C. Ries (eds.), Collision tectonics, *Geological Society of London Special Publication* 19, 3-36, Fig. 4A, p. 8; **Figure 17.26 a-b:** From P. Tapponnier et al., 1982, *Geology,* 10, 611-616, Fig. 1, p. 612; Fig. 3, p. 615; **Figure 17.26 c:** From P. Tapponnier and P. Molnar, 1976; *Nature,* 264:319-324, Fig. 2a, p. 321. Reprinted with permission from Nature Publishing Group; **Figure 17.29 a:** Based on a concept from R. D. Hatcher and R. T. Williams, 1986, *Geological Society of America Bulletin,* 97, 975-985, Fig. 1, p. 976-977; **Figure 17.29 b:** Based on P. Coney, D. L. Jones, and J. W. H. Monger, 1980, *Nature,* 288, 329-333, Fig. 1, p. 330. Reprinted with permission from Nature Publishing Group; **Figure 17.30 a-b:** Adapted from Willett, Beaumont, and Fullsack, 1992, *Geology,* v. 21, no. 4, p. 371-374, Fig. 4

Chapter 18

Figure 18.5 b: Seismic data courtesy of the PGS-IDSL partnership and the Nigerian Department of Petroleum Resources; **Figure 18.7 a-c:** Adapted from J. Suppe, 1985, *Principles of Structural Geology,* Prentice-Hall, Englewood Cliffs, NJ, Fig. 9-42; **Figure 18.10:** Adapted from R. Twiss & E. Moores, 1992, *Tectonics,* W. H. Freeman, New York, Fig. 6.4. © 1992 by W.H. Freeman and Company. Used with permission; **Figure 18.11 c:** Adapted from F. Royse, Jr., 1993, in A.W. Snoke et al. (eds.), *Geology of Wyoming: Geological Survey of Wyoming memoir* 5, p. 272-311, Map sheet 1; **Figure 18.12 a:** Modified from S. Mitra, 1986, Duplex structures and imbricate thrust systems: Geometry, structural position, and hydrocarbon potential, *American Association of Petroleum Geologists Bulletin,* v. 70/9, p. 1087-1112; Fig. 7 (p. 1095). AAPG©1986, modified by permission of the AAPG whose permission is required for further use; **Figure 18.13a:** Modified from S. Boyer and D. Elliott, 1982, Thrust Systems, *American Association of Petroleum Geologists Bulletin,* v. 66, p. 1196–1230, Fig. 19. AAPG © 1982, modified by permission of the AAPG whose permission is required for further use; **Figure 18.15:** Simplified from J.G. Ramsay & M.I. Huber, 1983, *The Techniques of Modern Structural Geology, v. 1: Strain Analysis,* Academic Press, London, Fig. 11.10, p. 205. Reprinted with permission from Elsevier; **Figure 18.16 b:** S. Mitra, 1986, Duplex structures and imbricate thrust systems: Geometry, structural position, and hydrocarbon potential, *American Association of Petroleum Geologists Bulletin,* v. 70/9, p. 1087-1112; Fig. 10 (foldout). AAPG©1986, reprinted by permission of the AAPG whose permission is required for further use; **Figure 18.17 b:** Modified from B. Willis, 1923, *Geologic Structures,* McGraw-Hill Book Company, New York, Fig. 60; **Figure 18.18 a:** Modified from J. Suppe, 1983, Geometry and kinematics of fault-bend folding, *American Journal of Science,* v. 283, p. 684-721, Fig. 3; **Figure 18.19:** M. S. Wilkerson, D. A. Medwedeff, S. Marshak, 1991, Geometrical modeling of fault-related folds: a pseudo-three-dimensional approach, *Journal of Structural Geology,* v. 13, p. 801-812, Fig. 1. Reprinted with permission from Elsevier; **Figure 18.20 a:** J. Suppe and D. Medwedeff, 1990, Geometry and kinematics of fault-propagation folding, *Eclogae Geologicae Helvetiae,* v. 83, p. 409-454, Fig. 6; **Figure 18.21:** Adapted From Zehnder and Allrendinger, *Journal of Structural Geology,* 2000;22:1099-1014, based on a concept by Erslev, 1991; **Figure 18.22:** Adapted from H. P. Laubscher, 1961, *Eclogae Geologicae Helvetiae,* v. 54, Fig. 3, p. 233; **Figure 18.24:** Modified from Dahlstrom, 1970, *Bulletin of Canadian Petroleum Geology,* v. 18, p. 332-406l, Fig. 26; **Figure 18.25 b:** D. Elliott, 1976, The energy balance and deformation mechanisms of thrust sheets, *Philosophical Transactions of the Royal Society,* London, v. A283, p. 289-312, Fig. 4; **Figure 18.28 a-b:** S. Boyer and D. Elliott, 1982, Thrust systems, *American Association of Petroleum Geologists Bulletin,* v. 66/9, p. 1196-1230, Fig. 17 (p.1206), Fig. 19 (p.1208). AAPG©1982, reprinted by permission of the AAPG whose permission is required for further use; **Figure 18.31 a-c:** Adapted from K. Dejong and R. Scholten, 1973, *Gravity and Tectonics,* John Wiley & Sons, New York, Fig. 1, p. xii

Chapter 19

Figure 19.1 a: Modified from J. C. Crowell, 1987, in Episodes v. 10, p. 278-282, Fig. 1, p. 279; **Figure 19.1 b:** Modified from J. C. Crowell, 1979, *Journal of the Geological Society of London,* v. 136, p. 293-302, Fig. 2, p. 295; **Figure 19.2 a:** After G. W. Grindley, 1974. New Zealand; in A.M. Spencer (ed.) *Geological Society of London Special Publication,* v. 4, p. 387-416, Fig. 1, p. 388; **Figure 19.2 b:** Modified from Z. Garfunkel, 1981, *Tectonophysics,* v. 80, p. 81-108, Fig. 1, p. 82. Reprinted with permission from Elsevier; **Figure 19.8 a-b:** Modified from R. Twiss & E. Moores, 1992, *Tectonics,* W. H. Freeman, New York, Fig. 7.10, p. 121. © 1992 by W. H. Freeman and Company. Used with permission; **Figure 19.10:** Modified From Goddard Space Flight Center, 1976; **Figure 19.14 d:** From R. E. Wilcox, T. P. Harding, and D. R. Seely, 1973, *American Association of Petroleum Geologists Bulletin,* v. 57/1, Basic wrench tectonics, pp. 74-96, Fig. 8, p. 84 AAPG©1973; **Figure 19.16 a-c:** Modified from M.R. Nelson and C. H. Jones, 1987, *Tectonics,* v. 6, p. 13-33, Fig. 6. Copyright 1987 American Geophysical Union. Modified by permission of American Geophysical Union; **Figure 19.19 a:** From T.P. Harding and J.D.Lowell, 1979, *American Association of Petroleum Geologists Bulletin,* v. 63/7, Structural styles, their plate-tectonic habitats . . . , p. 1016-1058, Fig. 6 (p. 1025). AAPG©1979, reprinted by permission of the AAPG whose permission is required for further use; **Figure 19.21:** From Woodcock and Fischer, 1986; **Figure 19.23:** Modified from P. Tapponnier et al., 1982, *Geology,* v. 10, p. 611-616, Fig. 1, p. 612; **Figure 19.24 b:** See Dewey J. F. et al, in *Geological Society of London Special Publication 9,* 1974:11; **Figure 19.26 a:** Modified from S. Mitra, 1988, *Geological Society of America Bulletin,* v. 100, p. 72-95, Fig. 1, p. 72; **Figure 19.27:** Modified from S. Marshak, 1988, *Tectonics,* v. 7, p. 73-86, Fig. 15, p. 83. Copyright 1988 American Geophysical Union. Modified by permission of American Geophysical Union; **Figure 19.28:** From B. C. Burchfiel and G. A. Davis, 1973, *Geological Society of America Bulletin,* v. 84, p. 1407-1422, Fig. 4, p. 1417; **Figure 19.29:** from I. Barany and J. A. Karson, 1989, *Geological Society of America Bulletin,* v. 101, p. 204-220, Fig. 2, p. 205

Chapter 20

Figure 20.1: NGDC; **Figure 20.2:** Modified from Burchfiel, *Scientific American,* September, 1983, Fig. 2; **Figure 20.3:** Modified from Hatcher and Williams, *Geological Society of America Bulletin,* 1986

Chapter 21

Figure 21.1.2: Modified after Froitzheim et al., 1996; **Figure 21.1.3:** Profile modified after Schmid and Kissling, 2000; **Figure 21.1.4 b:** Profile modified after Schmid and Kissling, 2000; **Figure 21.1.5 c:** Profile modified after Schmid and Kissling, 2000

Chapter 22

Figure 22.1.1: From a map compiled by T. Simkin et al., *This Dynamic Planet, World Map of Volcanoes, Earthquakes, and Plate Tectonics,* U.S.G.S., 1989, Map Distribution Box 25286 Federal Center Denver CO 80225; **Figure 22.1.2:** From Balley et al., North America; Plate-tectonic setting and tectonic elements. *Dec N. Am. Geol,* vol. A., Fig. 1; **Figure 22.1.3:** Modified from Engebretson et al., 1985; **Figure 22.1.4 a:** After Miller and Gans, 1989; **Figure 22.1.4 b:** After Gans et al., 1989; **Figure 22.1.5:** Modified and simplified after Burchfiel et al., 1992; **Figure 22.1.6:** Based on Bennet et al., from Unavco Poster (NSF-NASA), 1999; **Figure 22.2.1:** After Willett et al., 1993; **Figure 22.2.2:** Malavieille, 1984; **Figure 22.2.4 a:** After Brandon, 1998 and Stewart and Brandon, 2003; **Figure 22.2.4 b:** After Escher and Beaumont, 1997; **Figure 22.2.5:** From Pazzaglia and Brandon, 2001; **Figure 22.2.6:** After Clowes et al., 1987; **Figure 22.5.3:** Modified From Pickering et al., 1988; **Figure 22.7.1:** From S. Marshak, *Earth: Portrait of a Planet,* 2001, W. W. Norton & Company, New York, Fig. 11.34; **Figure 22.7.4:** Adapted from S. Marshak and T. Paulsen, 1996, *Geology,* v. 24, p. 151-154; **Figure 22.7.7:** Adapted from J. H. McBride and W. J. Nelson, 1999, *Tectonophysics,* v. 305, p. 275-286, Fig. 11, p. 268. Reprinted with permission from Elsevier; and S. Marshak et al., 2003, *Geological Society of London Special Publication* 210, p. 159-184, Fig. 4, p. 164; **Figure 22.7.8:** From J. P. Craddock et al., *Tectonics,* 1993, v. 12, p. 257-264, Fig. 2, p. 260. Copyright 1993 American Geophysical Union. Reproduced by permission of American Geophysical Union

Page numbers in *italics* refer to figures.

relative, *8*
rigid-body (RBR), 63
rotational normal faults, 388, 391–92
rotation recrystallization, 226–27
R' shears, 184, *184,* 484, *485,* 486, *486*
Rubey, William, 193*n,* 470*n*
rubidium, 325
Rundle Thrust, *445*
Russell Fork Fault, *495*
rutile, 319

sag ponds, 188, 487
salient, 466
Salinic Orogeny, 589, *589*
salt intrusions, 26
salt structures, 26–30, *27–30*
    faulting and folding of, 29–30, *29–30*
    fault movements and, 407–8
    geometry of, 27–29
    halokinesis and, 26–27, 28
    importance of, 30
    rifting and, 397
    stages in formation of, *28*
    terminology of, *28*
San Andreas Fault, 90, 91, *137, 168,* 177,
    194, 452, *491,* 557
    displacement across, 188, *189,* 564
    earthquakes along, 564
    fault bends in, 491, 496
    magnitude of velocity on, 363–64
    net slip on, 176
    strike-slip along, 564–65
    subsidiary faults in, 184
    subsidiary structures and, 482, *483*
    trace of, 487, *491,* 497
Sander, Bruno, 311*n*
sandstone, 103, 105, 148, 154, 162, 186,
    219, 260, *260,* 262, 265, *266,* 274,
    275, 278, 281, 283, 286, 289, 337,
    374, 397, 424, 452, 453, 549, 622
    Cambrian, *144*
    cleavage refraction in, *287*
    Entrada, 158, *158*
    fabric of, 271
sand volcanoes, 19
Saudi Arabian Plate, 438
sawtooth model, 332–333
Sayan Mountains, 542
Scandian Orogeny, 601, 602
schist, 272, 279–80, 289, 433, 492
schistosity, 273, 278, 279–80
Scotia Plate, 358, 428, 498
screw dislocation, 207–10, *208, 212,* 221
Scrope, G. P., *4*
seafloor spreading, 4, 355, 356, 358, *359,*
    366, 384, *395,* 396, 401–5, 410, 425,
    497, 512
seal, clay, 181
seamounts, 342, 428, *428,* 440
secondary (steady-state) creep, 93, *93*
secondary grain growth, 226*n*
Second Law of Motion, 43
second-order folds, 464
second-rank tensor, 48, 53, 66
sectional strain ellipses, 76, *76*
sedimentary basin, 26, 433, *433,* 505
sedimentary structures, 14–30
    stratification in, *see* bedding
    see also structural analysis
seismic discontinuities, 340, 342
seismic fault, 199
seismic imaging, 368–70
    interpreting data of, 370, *371*
    method of, 369–70, *369*

seismicity, 199–200
seismic profiles, 374
seismic pumping (fault valving), 192
seismic tomography, 339–50, *350,* 418
seismic velocity, 292, 337, 369–70
    depth profile vs., 337, 369–70
seismic waves, 93, 98, 337, 349
    P—(compression), *340*
    S—(shear), *340*
seismites, 201
seismology, 93
selection method of field study, 150, *151*
selvage, 276
sense of displacement, 298
sense of slip, 169
serpentine, 181, 182, 405, 497–98, 551
Serra do Mar, 401, *401*
Seve Nappe, 600–1
Sevier/Laramide fold-thrust belt, 449, 560
Sevier Orogeny, 623
S-foliation, 302–3, *303,* 305, 310
shale, 16, 23, 81, 103, 105, 140, *158,* 186,
    199, 250, 262, 265, *266, 280,* 283,
    289, 397, 452, 453, 549, 605, 611
    cleavage in, 277–78, 281, 286–87
shallow fold, 244
shatter cones, 37, *37*
shear, 307
    angular, 71
    failure strength for, 125–26
    general, 68
    intermediate principle stress and, 136
    pure, 68
    R and R', 184, *184,* 484, *485,* 486, *486*
    simple, 68, *68*
    sinistral, 287
shear bands, 303
shear-direction lineations, 290
shear-fracture criteria, 127–32, *128–31*
shear fractures, 116, 124–26, *125*
shear heating, 194
shear modulus (rigidity), 95, *95,* 229
shear reaction, 123
shear rupturing, 123
shear sense, 169–70
shear-sense indicators, 176, *187,* 298–304
    domino model for, 301–2, *301*
    foliations as, 302–3, *302, 303*
    fractured grains as, 299–302, *301*
    grain-tail complexes as, 299, *301*
    mica fish as, 299, *301–2,* 302
    plane of observation and, 298–99
    textures as, 310–11, *310*
shear strain, 71
shear strain rate, 91, 96, 98, 99
shear stress, 44–45, 48, *48,* 54, 133, *133,*
    192–93
shear zones, 111, 116, 166, *167*
    defined, 167
    ductile, *see* ductile shear zones
    strain in, 304–7, *305*
sheath folds, 251–52, *254,* 264, *264,*
    265, 266
    in ductile shear zones, 313, *314*
Sheep Mountain Anticline, *244*
sheeted-dike layer, 402
sheeted dikes, 342
sheeting (exfoliation) joints, 146, *146*
    formation of, 153–54, *153*
sheet intrusion, 31–32, *32*
Sherwin-Chapple equation, 261*n*
shield, defined, 615
shock metamorphism, 37
Sibson-Scholz fault model, *296*

Siccar Point, *23*
Sierra Nevada Mountains, 146, *146*
Sierras Pampeanas, 578–79, 627
sign convention, 47
Sigsbee Escarpment, *406*
sills, 32, 370, 399
    columnar jointing in, 146
siltstone, 140, *142,* 397
Siluro-Devonian "Acadian" phase, 252
similar fold, 246
simple shear, 68, *68*
simple-shear model, 389, *389*
sinistral (left-lateral) fault, 170, *170*
sinistral shear, 287
SI units, 42
"slab graveyards," 418, *419*
slab-pull force, 365–66, *365,* 409, 414, 417
slate, 103, 219, 250, 272, 280, *293,* 424, 584
    pencil cleavage in, 277–78
    reduction spots in, 284, *284*
slaty cleavage, 274, 278, *278, 279,* 281,
    286–87, 465, 554, 603, 622
Slave craton, 610, 611–13
Slave Province, 375, *376, 379*
slickenlines, 183, *183*
slickensides, 182–84, *182,* 189
    anisotropy on, 183
slickolites, 183–84, *183*
slip (shear displacement), 166
    sense of, 168
slip bands, 210
slip fibers, 183–84, *183*
slip lineations, 150, 176, 182, *182,* 189,
    290, *290*
slip-line field, 436–38, *437*
slip (glide) planes, 210
slow ridges, 403, *404,* 405
slump blocks, 422
slumping, 24
    intraformational intervals and, 25
smectite, 278
"Snake outcrop," *312*
snowball garnet, 304–5, *304,* 323–24,
    324, 325
solid-state diffusion, *see* Coble creep;
    Nabarro-Herring creep
solution cleavage, *see* disjunctive
    cleavage
Sonoma Orogeny, 559
Sorby, Henry C., 292
source layer, 27
South American Plate, 358, 360, 448,
    575, *577*
South American Shear Zone (SASZ), 604
South Atlantic Ridge, 408
Southern Alps, 488, 568
spaced cleavage, 274, *275–76,* 284
spaced fabric, 271, *272*
spherical geometry, 359
spherical projections, 628–630, *629–30*
spinel, 200
stable backarc, 425–27, *427*
stable stress state, 130
stable triple junction, 364
stable wedge, 567
stacking, 12
stair-step thrust, 462
Stanovoy Fault, 541
Stanovoy Mountains, 541
static friction, 132
static recrystallization (annealing), 226
staurolite, *324*
steady-state (secondary) creep, 93, *93*
steady-state flow, 220, 233

S-tectonite, 271–72, *272*
steerhead basins, 397, *399*
Steno, Nicholas, 2, 241
stepover, 482–84, *484*
stepped fault array, 184
stick-slip behavior, 199
stishovite, 37
stockwork vein arrays, 160, *160*
stored strain energy, 224
straight dislocations, 221
strain, 6–8, *7*, 62–89
  angular, 70-71
  cleavage and, 284–85
  coaxial accumulation and, 67–69, *68*
  defined, 7, 12, 90
  deformation and, 63–65, *64*
  in ductile shear zone, 304–7
  elastic behavior and, 95–96
  finite, 66
  Flinn diagram for, 76–77, *77*, 78
  in folding, 265, *265*, 266, 268
  frame of reference and, 64
  grain-scale, 465
  heterogeneous, 7, 65, *65*
  homogeneous, 7, 65–66, *65*, *82*
  incremental, 66
  infinitesimal, 72
  intragranular, 617, 622–23
  joint spacing and magnitude of, 149
  longitudinal, 70–71
  measurement of, *see* strain
    measurement
  mesoscopic, 465
  Mohr circle for, 73–75, *74*
  natural, 72
  non-coaxial, 67–69, *68*, 305–8, *307*, 310
  non-recoverable (permanent), 96, 117
  path of, 66–67, *66*
  quantities of, 70–72, *71*
  Ramsay diagram for, 77–78
  recoverable, 95, 117
  regional, 123
  representation of, 75–78
  shape and intensity of, 76–78
  in shear zones, 304–7, *305*
  states of, *74–75*, 75
  superimposed, 69–70, *69*, 265, *265*
  in superplastic creep, 228–29
  tangential longitudinal, 264
  tectonic, 266*n*
  in thrust sheets, 465
  types of, *69*
  volumetric, 70–71
  see also rheolitic relationships;
    rheology
strain curves, *103–5, 107*
strain-dependent viscosity, 99
strain ellipsoid, 65–66, *66*, 79, 81
  orientation of, 75–76
  in pencil cleavage, 277
strain measurement, 78–89
  active markers and, 80–81
  of angular changes, 84–85, *85*, 86
  center-to-center method of, 83
  information learned from, 87–89, *88*
  of initially non-spherical objects,
    82–84, *83–84*
  of initially spherical objects, 81–82, *82,*
    83–84
  of length changes, 85–86, *86*
  passive markers and, 80
  questions to consider, 79–80
  Rf/Φ method of, 83–84
  rock textures and other gauges for,
    86–87
strain-producing mechanism, 216

strain rate, 6, 89, 91–92, 104–5, 108, 185
  in buckling experiment, 260
  pressure solution and, 219
stratification, *see* bedding
stratigraphic facing (younging
    direction), 17
strength, strength curves, 59, *60*, 110
  tensile, 130–31, 136
strength paradox, 118–19
stress, 6–8, 40–60
  anisotropic, 47, 53, 55
  at a point, 46, *46*, 54
  biaxial, 52, *52*
  in collisional tectonics, 438–40
  components of, 46–47, *47*
  compressive, 47
  defined, 31
  deviatoric, 52–53, *53*, 54
  differential, 50, 53, 56, 59, 60
  in Earth, 57–58
  effective, 136, 471
  elastic behavior and, 95–96
  epeirogeny and, 623
  experiment for, 47–48, *47*
  failure, 93
  far-field, 439
  faulting in relation to, 191–94
  in fold-thrust belts, 470–72, *471*
  isotropic, 47, 52–53, *52–53*, 55, 102,
    134, *135*, 470–71, *471*
  local, 118
  mean, 52–53, *53*, 54
  measuring, 56–60, *56, 57*
  membrane, 409
  Mohr diagram for, 49–51, *49–51, 52,*
    102–3
  nonrecoverable, 117–18
  normal, 44–45, 46, 48, *48*, 54, 133,
    133, 471
  present-day, 56–57
  principal, 45–46, 47, 55
  principal axes of, 47, 54, 55
  principal planes of, 45–46, 47, 55
  remote, 118, 119–20, 122, 127
  residual, 154
  shear, 44–45, 48, *48*, 54, 133, *133*,
    192–93
  states of, 47–48, 51–52, *52*
  strength and, 110
  summarized, 54–55
  tensile, 47, 127
  terminology and symbols of, *42*
  three-dimensional, 45–46
  triaxial, 52, *52*
  two-dimensional, 44–45, *44–45*
  units and quantities of, 42–43, *43*
  yield, 99
  see also rheolitic relationships;
    rheology
stress concentration, 118–21, *119*
stress curves, *103–5, 107*
stress-dependent viscosity, 99
stress drop, 194
stress ellipse, 46, 53–55, 73
stress fields, 55–56, 146
  around dislocation, 209, *209*
stress intensity factor, 127
stress provinces, 58
stress shadow, 158
stress tensor, 44, 53–55
  reduced, 57
stress trajectories, 55–56, *55*, 191–92,
    197, 198
stretching lineation, 290, *295*
strike-parallel joints, 146, 465
strike-slip component, 169, *169*

strike-slip faults, 169, *170*, 172, *173*, 188,
    191, 197, 394, 410, 4126, 427–28,
    440, 603, 605
  in Alps, 511, 516
  in Altaids, 535, 544
  in Andes, 579
  block rotation in, 487, *488*
  causes of complexity in, 484–87
  collisional margins and, 492, *493*
  in collisional tectonics, 436–38,
    437, 492
  continental tectonic setting of, 493–97
  continental transform faults and, 496
  in convergent plate margins, *426*, 492
  deep-crustal geometry of, 492
  defined, 476
  distributed deformation in, 482–84
  distribution of, 476
  duplexes, 492, *492*
  earthquakes along, 476, *477*
  fault bends in, 490–92, *491*
  in fold-thrust belts, 493, *493*
  fracture zones and, 497–98
  in Midcontinent U.S., 620–21, *620*
  in North American Cordillera, 560,
    560–64
  oblique-convergent plates and, 493
  oceanic transform faults and, 497–98
  restraining bends along, 452, *452,*
    490–92, *491*
  in rifts, 495–96, *496*
  structural features of, 482–92
  subsidiary structures and, 482, *483,*
    484–86, *485, 486*
  terminology of, *478*
  in Tibetan Plateau, 528, 530–33, *531, 532*
  transcurrent faults and, 481–82
  transform faults and, 478–81, *479–81*
  transpression in, 487–88, *489*
  transtension in, 487–88, *489*
strike-slip fault systems, 197, *197*
strong cleavage, 277
strontium, 533
structural analysis, 8–12
  categories of, *9*
  compaction and, 23–24
  of diagenetic structures, 23–24
  guidelines for, 10–12, *11*
  of penecontemporaneous structures,
    24–25, *25*
  of salt structures, 26–30, *27–30*
  scale of observation and, 8–9
  stratification in, *see* bedding
  surface markings and, *18*, 19, *19*
  terminology of, *10*
structural grain, 465*n*
stylolites, 24, *24*, 219, *219*, 276, 480
  tectonic, 276–77
stylolitic cleavage, *see* disjunctive
    cleavage
stylolitic pitting, 150
subcritical crack growth, 135
subduction, 356, 414, 429, 431, 449, 554
  abortive, 433–34
  in Alps, 510, 514, 521
  in Altaids, 536, 542, 544
  in Andean Orogeny, 575–78, 579, 581
  in Caledonides, 600–601
  in Cascadia subduction wedge,
    569–71, *570*
  epeirogenic movement and, 625
  global rate of, 566
  in India-Asia collision, 526–27
  in North American Cordillera,
    558–61, 564
subduction wedge, 568